Biostatistics

Basic and Advanced

Biostatistics

Basic and Advanced

Manju Pandey

MV Learning

London • New Delhi

MV Learning

3, Henrietta Street
London WC2E 8LU
UK

4737/23, Ansari Road,
Daryaganj, New Delhi 110 002
India

A Viva Books imprint

ISBN: 978-81-309-2753-4

Printed and bound in India.

SUB-0030, SSUB-0072

Dedicated to My Loving Father

Dadaji, author of many still read books, I could ultimately fulfill your repeatedly expressed desire of my writing books instead of research papers.

Preface

It is well-known that the foundation and development of the statistical theory came up with a search for interpretation of results of research experiments and, hence, decision making in biological, agricultural and medical fields. This endeavour gave birth to a new stream of mathematical sciences named as "Statistics". The statistical concepts are independent of various subject areas, since the procedures developed for agricultural or biological experiments are equally applicable in industrial, technical, psychological, medical, educational, social and commercial fields.

This led to recognition of need for formal courses on statistical methods in biological, medical and agricultural fields and they were started in UK and USA around the end of 19th and beginning of 20th century. In India, the foundation of development in agriculture, food and nutrition, population control and human health was laid soon after. The importance of biostatistics is now deeply felt all over India. A few decades ago this course was introduced in various life science, earth science, education and technology degree syllabi at Masters level also, in most of the Universities and Institutes. Some of these institutions are offering short term and degree courses in Biostatistics.

Explaining the simplest concepts of Statistics, which stands on the foundation of mathematics, to students belonging to non-mathematics fields, is an arduous task. The present book aims to provide basic knowledge and applications of Biostatistics to Masters level students pursuing courses in different disciplines and also to help young researchers in solving their problems. This book effectively summarizes the author's many years of teaching and research experience. While herself possessing a pure mathematics background, about three-fourth of the author's professional career has been devoted to teaching biostatistics to Masters level students of Zoology with non-mathematics background. The author has also drawn ideas from her interaction with researchers of a variety of disciplines such as education, social science, life sciences, agricultural sciences, medical sciences, earth sciences and engineering.

The book covers almost all parts of the recent UGC Model Syllabus in biological sciences relating to Quantitative Zoology. Each chapter contains relevant figures and numerical examples to help clarify the concepts. The book explains how to choose an appropriate statistical analysis for various biological experimental data. The computational steps and interpretation of the findings are also given in detail. Some of the statistical techniques, often used by researchers of all the life sciences disciplines as well as those belonging to social sciences, psychological sciences, earth sciences and education are also included, briefly, if not in detail.

This book may be used as a textbook by the undergraduate and postgraduate students of Biostatistics in biological, agricultural and health sciences. It will help the researchers in these and other aforesaid disciplines, after acquiring sound knowledge of basic concepts of statistical techniques, to choose advanced topics for a proper analysis of their research data and draw valid conclusions.

This book is divided into two parts: Part I on "Basic" consists of chapters 1 to 11 and Part II on "Advanced" consists of chapters 12 to 23. The book contains solved examples to illustrate the computational steps of various statistical procedures. Tables and Figures supplement the text for clear illustration of the theory behind the procedures and their applications.

The material in the book assumes the readers to have the knowledge of High School algebra and arithmetic in order to understand the logic behind the statistical procedures. Applications of these on experimental data require simple calculators for computations. However, if a researcher is handling huge project data, easy access to computer systems enables large or repeated computations. Therefore, two chapters (Chapters 11 and 23) are included, which provide a fundamental idea of computer systems, methods of using these for common simple problems and also give some idea of using statistical softwares.

Abbreviations used in the book are listed in the beginning. For performing various tests of significance and discussing some advanced topics, some Statistical and Mathematical Tables are required. These are covered under Appendix A. Many students and researchers of life sciences may be unfamiliar with various mathematical symbols and expressions, use of which could not be avoided in the book. Therefore, Appendix B is included to define and explain these. It becomes very easy to understand concepts of probability theory using Venn diagram and rules of Set Theory. Appendix C on Matrix Algebra contains the results (without proof) which enable the logic behind complex multivariate techniques and a description of these procedures in an easy and concisely presentable form. Appendix D provides meaning of the rules and principles of Set Theory used in the chapter on Probability.

Now a brief description of the various chapters is provided for the reader. Chapter 1 is introductory. It defines Statistics and Biostatistics, types of data that arise for analysis of experiments and a brief history of growth of statistical theory and practice in India and abroad. Chapter 2 gives the method of summarising and presenting data in tables and graphs. Chapter 3 defines and determines the descriptive statistics viz. measures of location, dispersion, skewness and kurtosis. Chapter 4 discusses the concepts of probability and Chapter 5 defines the random variables and associated functions giving their nature and characteristics. The concepts of probability distributions, expected values, raw and central moments of continuous and discrete random variables are also given in this chapter.

Chapter 6 discusses the distributions of various continuous and discrete random variables which are extensively used in data analysis in almost all the fields. Chapter 7 deals with the fundamental theory behind problems of estimation of parameters and testing of hypotheses. The criteria of good estimators, concepts of two types of error and choice of rejection region are also explained. Chapter 8 gives the one and two sample procedures of testing means and variances. Equality of more than two means and variances are also considered in this chapter.

Chapter 9 deals with measuring and testing relationships in bivariate and multivariate data in terms of correlation and regression. Chapter 10 deals with analysis of categorical data by establishing independence and measuring association in 2×2 and $r \times c$ contingency tables. Chapter 11 considers data handling electronically, indicating components and types of computers, machine and high level computer languages. Basic ideas of working on DOS, Windows, MS Office and computer network are also given. Chapter 12 illustrates the methods of planning medical and biological studies in lab and field, while Chapter 13 considers complete enumeration or census (of human and animal populations) and commonly used sampling methods in brief.

Chapter 14 gives methods of comparison of two means when the commonly assumed equality of two variances does not hold. Subsequently analysis of variance in various designs, such as randomized block, Latin square, nested design, BIBD, PBIBD and factorial experiments have been described. Lastly, multiple comparison methods, if hypothesis of equality of means is rejected, have been given.

Chapters 15, 16 and 17 consider one sample, two sample and k-sample non-parametric tests when the distribution, from which the samples have come, may not be known. Chapter 18 defines components of time series and provides procedure for analysis of these under conditions of being stationary or autocorrelated. Chapter 19 considers the problem and analysis of bioassays.

Chapters 20 and 21 deal with analysis of multivariate data. The Hotelling's T^2, Mahalonobis D^2, Discriminant Analysis, Principal Component Analysis and Cluster Analysis have been explained with examples of real data and use of statistical packages.

Chapter 22 deals with concepts of bioinformatics and some procedures of sequence and microarray data analysis. Chapter 23 on Computer Techniques gives some ideas of programming in FORTRAN, C and C^{++} for use of enthusiastic readers willing to develop their own program for simple but repeatedly required computations. For other problems a brief description of three standard statistical software packages (SPSS, BMDP and SAS) with illustration of their use is included.

In the course of preparation of the manuscript some of the standard books on Statistics and Biostatistics were found to be very helpful, especially the works by Steel and Torrie; Milton and Tsokos; Daniel; Zar; Sokal and Rohlf; Kramer; Finney; Mood and Graybill; Meyer; Yule and Kendall; Croxton, Cowden and Klein; Cochran and Cox; Anderson; Press; Dillon and Goldstein; Hohn; Sukhatme and Sukhatme; Goon, Gupta and Dasgupta; Gupta and Kapoor, to name a few.

The author gratefully acknowledges some of these authors and their publishers for granting permission to use the data and figures in their books for illustrating the application of statistical tools. The statistical and mathematical tables required for completing various statistical testing procedures have been included from some of the said books and from the Formula and Tables for Statistical Work by Rao, Mitra, Mathai and Ramamurthy.

The infrastructural and administrative support extended by the respective Head of the Departments of Zoology and Statistics, Banaras Hindu University in this endeavour, is highly appreciated.

The author very respectfully acknowledges her highly esteemed Ph.D. supervisor Prof. J. Singh, former Vice Chancellor, Vikram University, Ujjain, India for his painstaking efforts in reading many of the book's chapters and for giving his expert suggestions to improve the original draft.

The author is also grateful to Late Prof. M.S. Kanungo and Prof. B.N. Singh for devoting their valuable time to read some of the initial chapters and for their valuable suggestions to make the text more comprehensible.

Critical comments by Dr. G.K. Shukla, former Professor, Department of Mathematics, Indian Institute of Technology Kanpur, and former President, International Biometric Society (Indian Region), on Chapters 12 and 14 were very useful in giving final form to the same. The author deeply expresses her regards to him.

Mr. P.K. Chakravorty, Maintenace Engineer, Computer Center; Prof. K.K. Shukla; Prof. A.K. Agrawal (both from Department of Computer Engineering, Institute of Technology, BHU); Dr. Rajnikant Mishra and Dr. A.K. Maurya, my colleagues in the department, helped by extending their critical advice on technical points, respectively, pertaining to Chapters 11, 22, 23 and a part of Chapter 13. The author expresses her sincere thanks to all for their expert suggestions.

Last but not the least, the elegant typing and preparation of the figures by Dr. Vinod Kumar Upadhyay is sincerely and thankfully acknowledged. Without his patient and meticulous work, this book may not have taken this form. High appreciation is accorded to Mr. P.K. Sinha for his willing cooperation in speedy mathematical typing of Part II of the book.

The author hopes that the good presentation of the book and its well-arranged contents will make for useful reading not only to researchers in biological sciences, but also to researchers in various other application areas like education, social sciences, life sciences, agricultural sciences, medical sciences, earth sciences and engineering etc.

The basic bio-statistical concepts in Part I, as well as the new and advanced topics contained in Part II of this book will remain useful for all times to come. It is, therefore, hoped that the book will prove useful as a textbook for Graduate and Postgraduate students as well as a reference book for researchers.

MANJU PANDEY

Acknowledgements

The author thankfully acknowledges the Department of Science and Technology (DST), Ministry of Science and Technology, Government of India, New Delhi, India for extending financial support towards preparation of this book under Utilisation of the Scientific Expertise of Retired Scientists (USERS) Scheme vide Grant No. HR/UR/15/2006 dated 26 October 2006. The financial support extended by DST has made it possible to complete the book well in time.

Abbreviations

a	Adenine
ACF	Autocorrelation Function
ADF	Augmented Dickey-Fuller
AIC	Akaike Information Criterion
ALU	Arithmetic/Logical Unit
ANCOVA	Analysis of Covariance
ANOVA	Analysis of Variance
ANSI	American National Standard Institute
AR	Attributable Risk
AR(q)	Autoregressive Model of Order q
ARIMA	Autoregressive Integrated Moving Average
ARMA	Auto Regressive Moving Average
ASCII	American Standard Code for Information Interchange
BASIC	Beginners' All purpose Symbolic Instruction Code
BIBD	Balanced Incomplete Block Design
BJ	Box-Jenkins
BLAST	Basic Local Alignment Search Tool
BLOSUM	BLOcks SUbstitution Matrix
BLUE	Best Linear Unbiased Estimator
BMDP	BioMeDical Program
c	Cytosine
CATH	Class Architecture, Topology, Homologous Super-family
CBR	Crude Birth Rates
CD	Compact Disks
cdf	Cumulative Distribution Function, $F(x) = P[X \le x]$
CF	Characteristic Function
C.F.	Correction Factor
CI	Confidence Interval
C.I.	Class Interval
COBOL	Common Business Oriented Language
CPU	Central Processing Unit
c.q.d.	Coefficient of Quartile Deviation
CRD	Completely Randomized Design
CRT	Cathode Ray Tube

C.V. Coefficient of Variation
DDBJ DNA Data Bank of Japan
d.f. Degrees of Freedom
DF Dickey Fuller
DNA Deoxyribo Nucleic Acid
DoE Design of Experiment
DS Difference Stationary
EBCDIC Extended Binary Coded Decimal Interchange Code
ED Effective Dose
EDSAC Electronic Delay Storage Automatic Calculator
EDVAC Electronic Discrete Variable Automatic Computer
EF Etiologic Function
EMBL European Molecular Biology Laboratory
ENIAC Electronic Numerical Integrator and Calculator
ExPASy Expert Protein Analysis System
FA Factor Analysis
FORTRAN FORmula TRANslation
fpc Finite Population Correction
ftp File Transfer Protocol
g Guanine
GB Giga Byte
Gbp Giga Base Pairs
GHz Giga Hertz
GUI Graphics User Interface
Hb Hemoglobin
Huges Human Genome Equivalents
IBD Incomplete Block Designs
IBM International Business Machines
ID Incidence Density
I/O Input Output
ISP Internet Service Provider
IT Information Technology
IUGR IntraUterine Growth Retarded
JIPSD Japanese International Protein Sequence Database
KB Kilo Byte
LAN Local Area Network
LB Ljung-Box
LCD Liquid Crystal Display
LD Lethal Dose
L.H.S. Left Hand Side
LISP List Processing
LSD Least Significant Difference
LS_qD Latin Square Design
LSI Large Scale Integrated
MA Moving Average

MANOVA	Multivariate Analysis of Variance
MB	Mega Byte
MCSE	Microsoft Certified Systems Engineer
m.d.	Mean Deviation
m.e.	Mutually Exclusive
MGF	Moment Generating Function
MIPS	Munich Information Centre for Protein Sequences
mips	Mega Instructions Per Second
m.l.	Maximum Likelihood
m.l.e.	Maximum Likelihood Estimator
MNIC	Multipurpose National Identity Card
MRT	Multiple Range Test
MS	Mean Squares
MSA	Multiple Sequence Alignment
MSE	Mean Square Error
MSGr	Mean Sum of Square due to Groups
MSS	Mean Sum of Square
MSTr	Mean Sum of Square due to Treatments
MTBF	Mean Time Between Failure
MVN	Multivariate Normal
NBRF	National Biomedical Research Foundation
NCBI	National Centre for Biotechnology Information
NIH	National Institute of Health
NMR	Nuclear Magnetic Resonance
NTD	Non Tea Drinkers
NW	Needleman-Wunsch
OOPS	Object Oriented Programming Languages and Systems
OR	Odds Ratio
OR_o	Outcome Odds Ratio
OR_E	Exposure Odds Ratio
OS	Operating System
PACF	Partial Autocorrelation Function
PAM	Percent Accepted Mutation
PBIBD	Partially Balanced Incomplete Block Design
PCA	Principal Component Analysis
PDB	Protein Data Bank
pdf	Probability Density Function, $f(x)$ so that $f(x)dx = P[x \leq X \leq x + dx]$
PE	Prevalence Exposure
PGF	Probability Generating Function
PIAR	Percentage of Identically Aligned Residues
PIR	Protein Information Resource
pmf	Probability Mass Function, $p(x) = P[X = x]$
Q.d.	Quartile Deviations
QS	Quadrant Sum
RBD	Randomized Block Design

RC	Renal Cancer
RCT	Randomized Clinical Trials
RHS	Right Hand Side
RNA	Ribo Nucleic Acid
RR	Relative Risk
SAS	Statistical Analysis System
SBC	Schwartz Bayesian Criterion
SCOP	Structural Classification of Proteins
s.d.	Standard Deviation
SE	Standard Error
SI	Seasonal Index
SIB	Swiss Institute of Bioinformatics
SLP	Serum Lipid Peroxidase
SM	Scoring Matrix
SNOBOL	StriNg Oriented SymBOlic Language
SPSS	Statistical Package for the Social Sciences
SRS	Sequence Retrieval System
SRSWOR	Simple Random Sampling Without Replacement
SRSWR	Simple Random Sampling With Replacement
SS	Sum of Squares
SSB	Sum of Square due to Blocks
SSE	Sum of Square due to Error
SSGr	Sum of Square due to Groups
SSR	Sum of Square due to Regression
SST	Sum of square due to Treatments
SSTr	Sum of Square due to Treatments
SW	Smith-Waterman
t	Thymine
TD	Tea Drinkers
TMC	Tabulating Machine Company
TS	Trend Stationary
TSS	Total Sum of Squares
URL	Uniform Resource Locator
UT	Union Territory
VAR	Vector Auto Regression
VLSI	Very Large Scale Integrated
WAN	Wide Area Network
WBC	White Blood Cell
WS	Weighted Sum
www	World Wide Web

Contents

Part II: ADVANCED

APPENDICES

Part I
Basic

1. Introduction

The present chapter provides the definition of statistics and biostatistics, as well as a brief history of how this subject came into existence, got its name and, subsequently, evolved into a full-fledged science. The foundation of this young science and art is mathematics. It has a systematic logic to find application in various real-world problems.

1.1 Statistics and Biostatistics

The term *statistics* is commonly used with two meanings. One of these meanings is the data or information collected, or gathered from a controlled experiment, or as prevalent due to some phenomena, or due to some interaction of various forces operating, apparently, beyond human intervention. In this sense, statistics is plural and represents a set of information. However, the singular word *statistic* refers to a single piece of information. The second meaning is the theory or techniques of (i) collection, (ii) classification or summarisation, (iii) analysis, and (iv) interpretation of data.

Each of the above mentioned techniques in itself is a broad area of statistics. The part dealing with collection of data is called *sampling theory* and the common basic and easy methods form a full semester course for undergraduates. Advanced and complex sampling methods used under special situations form another semester course for masters level students.

The data as collected or noted down by an investigator is called *raw data* and it neither conveys much nor leads to a clear picture of the nature of data. Hence, some way of summarising or classification is necessary and an appropriate method is to be used for a specific type of data to give a first hand picture of its inherent nature. Broadly, *data* is of two types: *quantitative* and *qualitative*. Slightly different methods of handling such data are needed and these are discussed in subsequent chapters. A brief description of the types and nature of data is given in this chapter. The next step of data handling, viz. graphical representation of data and its summarisation by descriptive statistics, are dealt with in Chapters 2 and 3. The third step, viz. analysis of data, is also a broad topic called *theory of estimation and tests of hypotheses*. The third and fourth steps, viz. analysis and interpretation of data may be termed as *statistical inference*. The nature of available data and purpose of study are factors that govern the choice of a proper method of analysis (or a statistical technique or tool) for reliable and valid interpretation and for taking correct decision for further course of action.

For any scientific study, its purpose or objective must be specified first and the relevant information or data must be collected in a proper way so that the study objective is achieved. Before actually starting the study, one must carefully think of some points – How to collect the data? How much data is to be collected?

More care is needed if the study is about living organisms, e.g. humans, animals, plants, crops, microorganisms etc., because living organisms have inherent differences even though they may be of same species, age, sex and geographical location. There should be a method to separate out this inherent variability from the differences due to the factor or phenomenon being studied.

Thus, special statistical techniques are needed in biological studies, whether controlled ones in the laboratory or those involving field experiments. Even in controlled laboratory experiments, almost always, measurement errors creep in. Eliminating the errors by averaging the replicated observations and providing an estimate of such errors is now a universal convention, even if the facts, as observed in the experiment, are narrated as the first step. Further, deeper insight may be attempted in a subsequent study, which should be planned very carefully on the basis of the nature of the experimental objects, evident from the initial (or pilot) study.

Let us now understand the possible nature of data.

1.1.1 Types of Data

In any type of experiment, whether in a laboratory or in the field, the experimenter observes or measures a characteristic of the experimental objects. These objects may either be living (e.g. children, rats, mice, birds, insects, hospital patients etc.) or non-living (e.g. light bulbs, metal wires or strings, tables, chairs, transistors, talcum powder, washing detergent, television sets, jotter pen etc.). Data on objects, whether living or non-living, can either be quantitative and qualitative. Observations or measurements on some characteristics of the objects under study result in raw material on which an experimenter bases his/her study.

Data resulting from *observations alone* are known as *qualitative data*, e.g. eye colour is blue or black or brown; colour of a pigeon is white, or grey, or brown, or mixed; a building is of white, or yellow, or ash or some other colour; a boy is tall, or average, or short; a girl is fair, or wheatish, or dark; the porridge is very tasty, or good, or not bad; a man is very fat, or average, or slim; a walking stick is very long, or short; etc.

However, if the characteristics are *measured* then the data is called *quantitative*, e.g. the height of one of the undergraduate boys is 152 cm, that of another is 162 cm, and that of third is 142 cm. Other examples may be the recovery times of patients from a disease; or head circumferences and lengths of human palms and feet. Also, the number of full teeth of an eight-year-old child, the number of flowers in a rose plant, number of defects in a 1 m piece of cloth, number of offsprings of a given species of dog or goat during last one year, etc., are all *quantitative* data.

1.1.2 Variables and their Types

When an experimenter observes a characteristic then it is not the same for its different possessors and it takes on different values on different people, or places, or things. Such varying characteristics are called *variables*, and they are broadly of two types, viz. quantitative and qualitative. *Quantitative variables* may be further classified as *discrete* and *continuous*.

Discrete variables take only some specified values, e.g. number of full teeth in an 8-year-old child, number of defects in a piece of cloth, number of outgoing phone calls from a phone booth in 5 min, number of α-particles emitted in 15 min from a radioactive source.

Continuous variables take any value in a range, e.g. the height of human being is a continuous variable as it changes continuously in the range say 1 to 8 feet. Similarly, weight of an adult mice may be any value between 50 and 500 g; or life hour of a tubelight may be any value between 0 hr and 10 yrs.

However, data on variables need to be treated differently, depending on their types. Sometimes it also depends upon the question to be answered on the basis of the data on variable under study. There arises a need to identify a proper scale of the data on the considered variables. There are four scales – ratio, interval, ordinal, and nominal. Only one of these scales is associated with a variable.

Ratio Scale

Suppose that while studying a group of plants, our interest lies in the height of the plants and number of leaves per plant. Using height measuring devices, it is possible to assign a numerical value to the height of each plant and the counting of leaves allows a numerical value to be assigned to the number of leaves on each plant.

Whatever be the units of height measurements and whatever be the number system (decimal or any other), there are two important *fundamental properties* of these data. The first property is that there is *a constant size interval between two heights with equal differences anywhere on the scale*, i.e. difference in the heights of 10 and 11 cm plants is same as that between plants of heights 34 and 35 cm. Similarly, the difference between 4 and 6 leaves is same as that between 5 and 7 leaves. The second property is that *there exists a zero point on the measurement scale with a physical significance*. This enables us to talk about the ratio of the measurements in a meaningful way. Thus, a plant of 20 cm height can be said to be four times as tall as one of 5 cm height. A plant with 36 leaves has twice as many leaves as one with 18 leaves.

Such measurement scales are said to be *ratio scales* if they have a *true zero point* and *constant interval size*. Besides lengths and count of items, weights (mg, kg, lb etc.), volumes (cc, cu. ft etc.), capacities (ml, l etc.), rates (ft/s, mph, mg/min, ppb etc.), and time duration (s, min, h, yrs etc.) have ratio scales.

Interval Scale

Measurement scales that possess constant interval size but do not have a true zero are called *interval scales*. Very common examples of such scale are the two temperature scales, viz. Celsius (°C) and Fahrenheit (°F). The difference in temperature between 20°C (68°F) and 25°C (77°F) is the same as that between 5°C (41°F) and 10°C (50°F). However, it cannot be said that an object with temperature 40°C (104°F) is twice as hot as one with 20°C (68°F). This is because the zero point is arbitrary*.

Absolute† or Kelvin (K) scale has a physically meaningful zero and, thus, constitutes a ratio scale, e.g. clock time of the day and the time of year. The time interval between 2.30 pm (14.30 hr) and 4 pm (16.00 hr) is same as that between 9 am (09.00 hr) and 10.30 am (10.30 hr). But we cannot talk about ratios of the time of day because the zero of clock time (midnight) is arbitrary, as it could be set at 12.00 noon or 3 pm or any other time.

Interval scales of biological data may be said to have circular scale, and are like compass points that may be set in a direction in which an animal or plant under study is oriented. Biological records on such a circular scale (circular data) require special statistical techniques that are different from those applicable to other data.

---

*The Fahrenheit Scale devised by German Dutch Physicist G.D. Fahrenheit in 1714 was between freezing point of pure water 32°F and its boiling point 212°F. Centigrade scale devised by Swedish astronomer in 1742 was of 100°C between the boiling point of water as zero and its freezing point as 100°. A year later in 1743 it was revised i.e. zero at the freezing point and 100 at the boiling point of water.

†Absolute scale of temperature devised by Lord Kelvin in 1800 has zero at – 273°C, the lowest possible temperature in the universe. The freezing and boiling points of water are 273°K and 373°K.

Ordinal Data

Sometimes the data are not measurements on ratio or interval scale that give an idea of the numerical differences between two or more data, but they indicate only a relative difference. For example, exact weight of A, B and C are not given but only that A weighs more than B and B weighs more than C. Thus, the data are only an ordering or ranking of the measurements of a characteristic. Such data are said to be on an *ordinal scale* (the term *ordinal* is a Latin word for 'order').

We may describe a biological entity as being shorter, faster, darker, more active than another; the sizes of cell types may be labelled as 1, 2, 3, 4, 5 and 6; the success in learning a technique may be recorded as A, B, C or D. Biological data expressed on an ordinal scale can sometimes be expressed on ratio or interval scale, provided the exact measurements are available or can be obtained. Sometimes data originally on interval or ratio scale may be changed to ranks (i.e. ordinal data), e.g. examination percentages 99, 85, 73 and 66% may be recorded as A, B, C and D (on ordinal scale), respectively.

Ordinal data contains and conveys less information than the ratio or interval scale data, because only relative magnitudes are known. Hence, quantitative comparisons are not possible. It cannot be said that grade C is half as good as grade A, or that sizes of cell types 1 and 2 differ as much as those of types 4 and 5. However, many useful statistical techniques exist for applications on ordinal data to draw valid conclusions.

Nominal Scale

Sometimes the subjects under study are classified by some quality that they possess and not by some numerical measurements. In such cases the generated variable may be called an *attribute* and the data is on a *nominal scale* (the term *nominal* is a Latin word for 'name').

Genetic phenotypes are common biological attributes. An animal's eye colour (blue or brown), human hair colour (black, brown, blonde or red), sex of animals (male or female), left or right handedness in human beings, classification of plants with or without thorns are all on nominal scale. Taxonomic categories also form a nominal classification scheme.

Sometimes data on ordinal, interval or ratio scales may be transformed into nominal scale categories, e.g. plant height data may be changed into tall and short; class test scores of students may be changed into either passed and failed (or 1st division, 2nd division, 3rd division, and fail).

Statistical methods applicable on ratio, interval, and ordinal scale data cannot be used for nominal data. Different techniques are applied on nominal or categorical data. It is now evident that ratio, interval and ordinal scale data may be either continuous or discrete. However, nominal scale data are only discrete.

A scientific investigator must identify the nature of the available data, choose an appropriate statistical technique to analyse and interpret the data with a view towards objective evaluation of the conclusions from the data. During data collection, statistical considerations aid to design the experiments and to set up the hypotheses to be tested. Sometimes an experimenter finds, after all the experimentation, that too few data are available, which are insufficient for valid reliable conclusions to be drawn. Or, the experimenter may detect that much extra effort was put in collection of data that cannot be of ready use for a proper analysis. Thus, knowledge of basic statistical principles and procedures is important even before starting an experiment.

Procedures to organise and summarise the collected data, leading to an orderly and informative presentation, are called *descriptive statistics*. One may wish to make some generalisation from the collected data. The ability to make generalised conclusions about the characteristics of the whole from the characteristics of the part, belongs to the realm of *statistical inference*.

Chapters 2 and 3 and some parts of Chapter 9 deal with the first of the said procedures. Chapters 7 to 10 give elementary and commonly used inferential procedures. Chapters 4 to 6 give the basic foundation of the discussed procedures. Part II of the book deals with the advanced topics of inferential procedures needed in specific situations.

1.2 History and Applications

In this section, a brief history of the origin of the theory and practice of statistics is given and, subsequently, various application areas are enumerated, though nowadays there is no discipline which does not use statistics in some or the other form.

1.2.1 History

Statistics – the subject of utility in almost all the disciplines of theory and application – is quite young and started taking shape in early 20th century. For scientists, especially biological scientists, formal beginning of statistics may be attributed to the publication of the book "Statistical Methods for Research Workers" by R.A. Fisher in 1925.

Statistics and *biostatistics* are rapidly growing with the need to find answers to more and more problems posed by the research workers. Many of the earliest contributors or founders of statistics are still productive and current researchers find manifold opportunities for their talents.

Initially, statistics appeared to have been started for assisting a ruler who needed to know about the revenue, number of soldiers, ammunitions, food stock, and subjects under him, in order to impose taxes or wage a war. Historical records also contain statistics of that period. It is known that Caesar Augustus gave an order which required all his subjects to report to the nearest tax collector – the then statistician. Lands of England were also surveyed for taxation and military service.

The word *statistik* was originally introduced in 1749 by Gottfried Achenwall of Germany designating the *analysis* of the data about the state. The same was signified as "science of state" or "political arithmetic" (in English). The collection of data to be used by governmental and administrative bodies still continues and censuses are held regularly to take stock of populations.

Statistical methods emerged from probability theory, which originated in 1654 from discussions between Pierre de Fermat and Blaise Pascal to explain the gambling experiences of the Chevalier de Mere'. Jacob Bernoulli's *Ars Conjectandi* (1713) and Abraham de Moivre's *Doctrine of Chances* (1718) treated the subject as a branch of mathematics. Normal curve of errors has been very important in the development of statistics. It was first published in 1733 by de Moivre but he had no idea of its applications until Karl Pearson noticed it in 1924. Two mathematical astronomers P.S. Laplace (1749-1827) and Carl Friedrich Gauss (1777-1855) derived the same result independently of one another. The method of least squares was published independently by Adrien-Marie Legendre (1805), Robert Adrain (1808) and C.F. Gauss (1809). Gauss used this method for predicting the location of a dwarf planet *Ceres*.

In the 19th century, Karl Pearson (1857-1936) (initially, a mathematical physicist), Laplace, Morgan and Boole improved the exposition of statistical theory. Karl Pearson applied his mathematics to the theory of evolution put forward by the famous biologist Charles Darwin (1809-1882). He devoted about half a century to serious statistical research and also founded the journal *Biometrika*.

Pearson's large sample theory was proving somewhat inadequate for experiments with necessarily small samples. One of his students W.S. Gossett (1876-1937) took up the task of finding the exact distribution of statistics from small samples and published "Student's t-distribution" (*Biometrika*, (1908)) under the pseudonym 'Student', which is widely used in all branches of science.

R.A. Fisher (1890-1962) was influenced by Pearsons' work and that of the 'Student'. Fisher has made numerous and important contributions to statistics, agriculture, biology and genetics. His book "Statistical Methods for Research Workers" is still considered a guidebook in these areas. J. Neymann (1894-1981) and E.S. Pearson (1895-1980) laid down the theory of the test of the statistical hypotheses in 1936 and 1938, which has practical application. Abraham Wald (1902-1952) laid the foundation of the sequential analysis and decision functions.

In the 20th century statistics was no more a sub-field of mathematics but emerged as a distinct, though allied field. This period also saw the creation of precise statistical instruments that are used to understand data and make informal decisions in the natural and social sciences, medicine, business, industry, public health, pharmaceutics and space programmes.

The above description is based on the writings by Box J.F. (1978), Mather (1971), Mc Mullen (1947), Menger (1952), Tintner (1952) and Walker (1958) etc.

Now let us look at the development of statistics in India. The country has a long historical tradition of both collection and use of various kinds of statistics. Kautilya's (321-296 BC) *Arthashastra*, a great treatise of economics, indicates a system of data collection relating to agriculture, population and other economic activities, covering villages and towns. The concept of cross-checking and validation by independent agents was part of data collection. Thus, there exists pre-historic evidence of the co-existence of economics and statistics for welfare of states in India.

Subsequently, Huen Tsang in his historic writings during late 7th to early 8th century, has given accounts of city plans, houses, common products of India, data on area of kingdoms and distances between them. During the regime of Mughal Emperor Akbar, the application of statistical knowledge is well documented in *Ain-i-Akbari* by Abul Fazal. Todar Mal systematised the records of land tenure, land use and the land revenue to be collected.

During the British rule the statistical system was strengthened and geared up for the administration and tax collection, to determine the status of the revenue, trade, commerce and related activities. In this process, the East India Company emphasised the need of a statistical survey of India in 1807. In 1847, a small department of statistics was founded. In 1848, the first census relating to area and revenue of North West Province was undertaken. The Director General of Statistics in India was appointed in 1869 and operation of a decennial census for the whole country was started in 1881 and this process is still continuing.

Modern statistics came to India with Prof. P.C. Mahalanobis who returned to India from England. In 1933, he founded the Indian Statistical Institute (ISI) and dazzled the world with his associates S.N. Roy, R.C. Bose, C.R. Rao, S.S. Bose, K.R. Nair, and B.D. Lahiri, to name a few. Mahalanobis launched a series of statistical researches concerning critical social issues like management of floods, assessment of yields of jute and other crops, and anthropometrical analysis of ethnic elements of Indian society. By 1960, India secured a place in the top League of Nations along with UK, USA, USSR etc. in the field of statistics. Perhaps the first Indian to get formal education from a team of founders of statistical theory was Late Prof. P.V. Sukhatme who completed his Ph.D from London in 1936. He not only founded and led statistical research in food and nutrition, eradication of poverty and growth of agriculture in India, but also in the United Nations.

Presently, India has a central government ministry of Statistics and Program Implementation, a separate arm of bureaucracy (Indian Statistical Service), world class set up for data collection called National Sample Survey Organisation, specialised research institutes (ISIs, Indian Agricultural Research Institute, Institute of Research in Medical Statistics, etc.) and many Universities running postgraduate and research programmes in statistics.

In the 21st century of outsourcing, many pharmaceutical industries have established centres of statistical analysis of their data and centres for business intelligence that use the services of statisticians for analytical study of stock market, retail sales and insurance.

1.2.2 Applications

After giving a short history of the theory of Statistics and Biostatistics, this subsection lists some of the applications of the theory. Application of statistical theory for problem solving in various disciplines have been given special names to the techniques for a specific purpose. Table1.1 below gives the list of some of the techniques, their purpose and the disciplines where the techniques are often applied.

Table 1.1 Purpose of statistical techniques and major disciplines of their use

S. No.	Statistical techniques	Purpose	Discipline(s)
1	Descriptive statistics	Summarization of data	Field or lab. Experiments in all the disciplines
2	Statistical inference	Estimating population characteristics or parameters of distribution of r.v. and testing statistical hypotheses	...do... with large and small samples
3	Correlation and regression	Study of relationship between two or more variables and prediction of dependent variable(s)	All disciplines, especially economics, commerce and meteorology
4	Bayes procedures	Inferences using prior knowledge, even for small samples	Biology, Medical Science, Sensitive equipments in industry
5	Time series	To determine trend, cyclic and seasonal components and hence prediction	Economics,industry, meteorology commerce and agriculture
6	Reliability theory	Determination of average life and reliability of components and systems (classical and Bayes methods)	Industry, defence and space programs
7	Actuarial statistics	Expected life of humans and risk of life at various ages and geographical areas, framing insurance policies	In life insurance companies
8	Educational and psychological statistics	Scaling of intelligence, education and knowledge status	In education and psychology
9	Multivariate analysis	Comparing mean vectors, classification, clustering, exploring factors, principal components etc. and reducing dimension of multivariate data for decision making	For multivariate data in any disciplines
10	Design of experiments and ANOVA	Designing experiment and choosing proper analysis to take a valid decision	Agriculture and Medical sciences
11	Sampling theory	To choose proper strategy for sample selection	All disciplines
12	Bioassay and microarray analysis	For potency determination and mapping gene sequence	Pharmaceutics, Biological and Medical Research

2. Data Handling I: Graphical Methods

Once the data have been collected the huge mass of data as recorded (raw data) must be summarised or presented in such a way that it is comprehensible to the presenter and to the reader. The presentation should enable the reader to draw some definite conclusion from the data. As a first step the data must be classified or tabulated. The next step of presentation is graphical. The following section explains classification and tabulation of data.

2.1 Classification and Tabulation

This section deals with first step of data summarisation. Qualitative data needs to be counted to know how many belong to various categories. This process of presentation is called classification and it is discussed in Section 2.1.1. Preparation of frequency table for quantitative data has been described in Section 2.1.2.

2.1.1 Classification

If the data is qualitative, counting of subjects with different qualities is a common way of summarising the data. The following examples clarify this statement.

Example 2.1 Suppose a plant ecologist observes 100 trees of a rare species in 900 sq. km area. It is recorded if the tree is with pubescent or smooth leaves. The collected record of 100 letters (P or S) or words pubescent or smooth can be summarised by presenting the findings as given in Table 2.1. Suppose the ecologist also records whether the tree is rooted in serpentine soil or not, then the resulting data can be presented as given in Table 2.2.

Table 2.1 Distribution of trees according to leaf type (source: Sokal and Rohlf (1973), p. 300)

Trees with pubescent leaves	Trees with smooth leaves	Total
28	72	100

Table 2.2 Contingency table of trees according to leaf and soil type (source: Sokal and Rohlf (1973), p. 300)

Soil type	Trees with		Total
	Pubescent leaves	Smooth leaves	
Serpentine	12	22	34
Not serpentine	16	50	66
Total	28	72	100

Example 2.2 A species of tiger beetle was observed for presence of bright red colour patterns during early and late spring and summer. The recorded data are summarised in Table 2.3.

Table 2.3 Colour pattern of tiger beetle in various seasons (source: Sokal and Rohlf (1973), p. 302)

Season	Colour pattern		Total
	Bright red	Not bright red	
Early spring	29	11	40
Late spring	273	191	464
Early summer	8	31	39
Late summer	64	64	128
Total	374	297	671

Example 2.3 1,000 nursing school applicants are observed for the type of high school from which they completed their schooling (state board, *Kendriya Vidyalaya* (KV) and convent) and their entrance test scores (low, medium and high). The data can be summarised as given in Table 2.4.

Table 2.4 Types of passout schools and scores of applicants

Score	Type of high school			Total
	Board	KV	Convent	
Low	100	65	50	215
Medium	80	180	150	410
High	20	55	300	375
Total	200	300	500	1000

Thus, the qualitative data is classified into various categories. The resulting tables are called *contingency tables*. Such tables are called $r \times c$ contingency tables, where r represents the number of rows and c the number of columns.

2.1.2 Frequency Tables

If the data is quantitative the first step of summarisation is preparation of a frequency table. In general, in a frequency table there are three columns. The first contains the *class number*, the second contains the *classes* and the third contains *frequencies* of various classes. The *frequency* of a class is an integer showing the number of observations, the values of which lie in that class.

As mentioned earlier, quantitative data may be discrete or continuous. A frequency table can be prepared for both types and the relevant methods are given as follows:

Discrete Data

The values taken by a discrete variable are some specified points only and not just any value in an interval. For example, consider that 50 flowers are observed for the number of petals in them. The raw data consisting of 50 numbers from 6 to 10 indicating the number of petals do not convey anything definite but if the data are arranged in a frequency table (Table 2.5), the 50 numbers are reduced to only 10 and give a better idea.

The first column lists the possible values of the variable and the second column contains the frequency, i.e. the number of flowers with said number of petals. Thus, it is clear from Table 2.5 that the flower under study has 6 to 10 petals and 8 petals are in the highest number of flowers, while 7 to 9 petals are contained in all the flowers except in 13.

Table 2.5 Distribution of number of petals in flowers (source: Chandel (1964), p. 16)

No. of petals in a flower	No. of flowers
6	8
7	12
8	15
9	10
10	5

However, this type of situation does not prevail always. The number of possible values of a variable may not be as few as only 5 to 6. For example, consider the following:

Example 2.4: Monthly rental values of 30 houses in Indian Rupees are as follows (source: Chandel (1964) p. 17):

5, 7, 6, 9, 8, 9, 7, 8, 12, 8, 9, 10, 10, 9, 9, 10, 12, 16, 14, 13, 10, 10, 15, 16, 12, 14, 10, 12, 14, 15

A frequency table may be prepared in which various possible values of rent are listed in Column 1 giving 12 possible rents.

Such a frequency table given in Table 2.6 has 12 rows (or classes). The frequencies are fluctuating, i.e. increasing, decreasing and, again, increasing. This nature of frequency has emerged because there are too many classes compared to the total number of houses (objects). One class has no frequency. In a frequency table, in general, one should have 7 to 12 classes. If observations are few (≤ 30), 5 to 7 classes are enough so that proper summarisation is achieved, as larger number of classes may give 0 frequency and also irregular fluctuation.

Table 2.6 Monthly rent of houses

Rent (Rs.)	No. of houses
5	1
6	1
7	2
8	3
9	5
10	6
11	0
12	4
13	1
14	3
15	2
16	2
Total	30

The number of classes (say K) may be decided on the basis of total number of observations or total frequency N. Struges suggested using the following formula (cf. Gupta and Kapoor (1999)) to determine approximate number of classes:

$$K = 1 + 3.322 \log_{10} N$$

Another point that should be kept in mind is that once the classes are decided the frequencies may be obtained by putting tally marks (a tally mark is one vertical line for one data value in the class) in appropriate class, proceeding from 1st observation to the next, one by one. It may be noted that the fifth tally mark is put diagonally or horizontally over the last four tally marks, for ease of counting. Counting how many times a value occurs in a class in all the observations may result in wrong value of frequency due to some human error and will require a re-look at the observations as many times as the number of classes (i.e. 12 times in the current problem). In the procedure of putting tally mark one has to look over all the data only once and the tally mark is put as soon as the value is read one by one in the class where the value is contained in the classes in the 1st column.

Let us consider this method of finding frequency. Let the number of classes be chosen as 5. As there are 12 values and there are only 5 classes there will be more than one value in a class. Hence, there will be an interval of values in a class. To determine the class width h we need to follow the steps given below after deciding the number of classes K:

1. Find the minimum and the maximum of the observations (5 and 16 in Example 2.4).
2. Find the range, R = Maximum – Minimum value ($R = 16 - 5 = 11$).
3. To find the class width h determine R/K (= 11/5 = 2.2, for Example 2.4). Choose h = integer or fraction nearest to the value of R/K.
4. Determine the class intervals starting from the minimum value and rounding h to 2. If we start from minimum value (= 5), the classes will be as follows:

   (A) 5-7, 7-9, 9-11, 11-13, 13-15, 15-17 (one more than K = 5) to include the maximum value in the last class.

   However, if we choose h = 3, by rounding to nearest upper integer, the classes will be as follows:

   (B) 5-8, 8-11, 11-14, 14-17 (one less than K = 5).

If the classes are chosen as given by (A) or (B), they are called *class boundaries*.

A formal *definition of class boundaries* may be given as, "Lower values and upper values of a class are class boundaries of the class, if *upper value* of the *lower class* is same as *lower value* of the *upper class*".

In this type of classes a problem arises: where do we include a boundary value? For example, should value 7 be included in the lower class 5-7 or in the upper class 7-9?

Therefore, only one convention should be followed. One boundary value in a lower class and another boundary value in an upper class should not be included. As the lowest boundary is 5, value 5 of the data should be included in the lowest class 5-7. Hence, upper boundary value of the class 7 should be included in the next upper class 7-9. Thus, the correct convention, in this case, will be to include the boundary value in the upper class.

However, if the lower boundary of the lowest or first class is chosen less than the minimum value (i.e. taken equal to 4), then the classes will be 4-6, 6-8, 8-10, 10-12, 12-14, 14-16 and the boundary value has to be included in the lower class, so that the maximum value (= 16) can be included in the table in the last class. Another convention may be followed, that if maximum of the observations is same as the upper boundary of the highest class, one more class may be formed. Once the class intervals and the convention for boundary values are decided, one may look at the data once only and put the tally marks in the appropriate class. For Example 2.4 we get a frequency table given by Table 2.7.

Table 2.7 Frequency table of monthly rents

Class no.	Class interval	Tally marks	Frequency
1	05-07	||	2
2	07-09	𝍸	5
3	09-11	𝍸 𝍸 |	11
4	11-13	||||	4
5	13-15	||||	4
6	15-17	||||	4
	Total		30

Note: Boundary values are included in the upper class.

One way to obviate the difficulty of including a boundary value is to use class limits instead of class boundaries. In such classes, observation values equal to both the upper and lower limits are included in the class. In this example h is 2 and, thus, class limits may be chosen as 5-6, 7-8, 9-10, 11-12, 13-14, 15-16. If the upper limit of lower class is not equal to the lower limit of upper class the frequency table is said to have *class limits* and not class boundaries.

Continuous Data

When a characteristic can assume any value in an interval, its measured values constitute continuous data.

Example 2.5 In a study designed to correlate seasonal change in plasma testosterone with the reproductive cycle in lizards, the following data on testosterone levels (in nano gram per milliliter) were obtained from a sample of 40 lizards in October (source: Milton and Tsokos (1983), p. 19):

43.7	37.2	29.0	31.6	47.5	48.3	38.3	29.7
32.5	45.2	36.1	30.5	37.2	50.5	36.9	44.5
35.9	28.7	37.5	30.2	36.9	43.2	27.0	26.2
41.8	26.4	34.3	28.6	35.9	22.0	45.4	30.3
29.8	46.1	42.7	31.5	37.4	25.1	27.2	45.0

It is required to construct a frequency table for this data. As there are 40 observations, let the frequency table have 6 classes. Thus, $K = 6$. Looking at the data, we find that maximum value = 50.5 and minimum value = 22.0. So, $h = (50.5 - 22.0)/6 = 28.5/6 = 4.75 \approx 4.8$ (rounded off to one decimal).

If we start from a value lower than the minimum (= 22.00), say 21.8, the upper boundary of the highest class will be $6 \times 4.8 + 21.8 = 50.6$ (greater than the maximum). Hence, we can safely take the lowest boundary in the frequency table as 21.8 (less than the minimum) and proceed as mentioned for Table 2.7 to get the following frequency table (Table 2.8)

Table 2.8 Distribution of plasma testosterone level in lizards

Class no.	Class interval (nano gram per milliliter)	Tally marks	Frequency
1	21.8-26.6	||||	4
2	26.6-31.4	~~||||~~ ~~||||~~	10
3	31.4-36.2	~~||||~~ ||	7
4	36.2-41.0	~~||||~~ ||	7
5	41.0-45.8	~~||||~~ |||	8
6	45.8-50.6	|||||	4
	Total		40

2.2 Graphical Methods

One of the aims of statistics is to reduce the huge mass of data so that it can be comprehended and some definite conclusion can be drawn about the objects which have been observed. The first step towards this aim is classification or tabulation of data to give contingency tables or frequency tables.

These do not make the whole thing clear at a single glance. If as the second step, the data are presented as some kind of a diagram, a pictorial representation makes a clearer impression about the nature of the data, than the columns of figures. Thus, we now consider different methods of graphical or diagrammatic representation of data. Various types of graphs are described and their use is given. Different types of graphs are appropriate to answer a different type of question arising in the mind of an investigator. The type of available data also influences the choice of an appropriate graphical presentation. The following qualities of graphical presentation make it useful in a scientific study:

(i) *Attractiveness* and *effectiveness* of a diagram is greater than a bunch of numeric figures in a table.
(ii) *Readily intelligibleness* of a graph arrests the attention of readers, immediately.
(iii) *Comparability* of two or more groups is readily evident if presented in a graph.
(iv) *Clarity at a glance* of a graphical presentation saves a lot of space, time and energy to grasp the significance of data.

However, despite these outstanding effectiveness, the following are some limitations of charts or graphs:

(i) A chart or graph cannot show many facts that a table can show by including sub-rows and sub-columns. A graph can show only limited information.
(ii) A table can show exact values but a chart or graph is plotted by taking some approximation.
(iii) A chart or graph requires some time to construct after the table is prepared.

It is, therefore, worth discussing the methods of various types of graphical presentations. Let us first consider graphs, charts or diagrams for qualitative data or contingency tables and then those from frequency tables.

2.2.1 Graphical Methods for Qualitative Data

In this section construction of bar diagrams and pie charts, which are more suitable for qualitative data, is considered.

Bar Diagrams

These are all one-dimensional graphs and provide more accurate impression of magnitudes than if two- or three-dimensional graphs are prepared. A bar diagram can be classified as:

(a) Simple
(b) Multiple or compound
(c) Divided

(a) Simple Bar Diagram

If a characteristic is observed on several groups or places or times, a bar chart can be used to present the data. For proper comparison the bar diagram should be scaled. Thus, Y-axis, which indicates the count of the characteristics is scaled. On the X-axis the bars are based for various groups. If the grouping of the observations is according to some quantitative variable such as distance, age, time or years, the X-axis may also be scaled and bars are put on the X-axis at proper intervals on this scale. Consider the following examples for simple bar diagrams.

Example 2.6 The following table (Table 2.9) gives the populations of three states of India in 1951 in millions (source: Chandel (1964), p. 23).

Table 2.9 Population of the three states in 1951

State	Population (millions)
Uttar Pradesh	63.22
Andhra Pradesh	31.25
Madras	29.98

This information can be represented in the following bar diagram (Fig. 2.1):

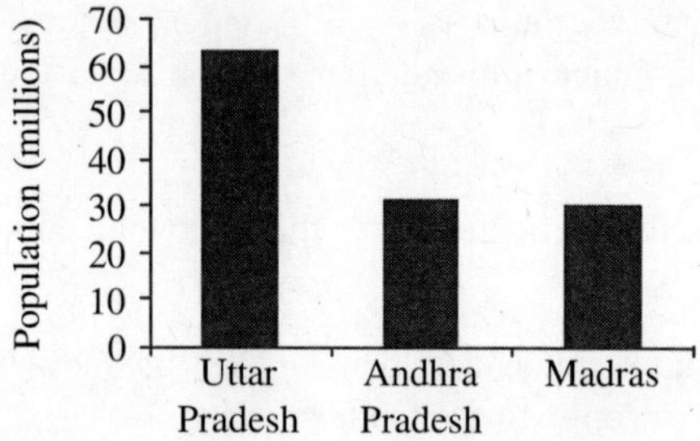

Fig. 2.1. Bar diagram for population of the three states.

Figure 2.1 clearly indicates that the population of Uttar Pradesh is almost double that of Andhra Pradesh and Madras and also that the latter two are very close to each other.

Example 2.7 Table 2.10 gives the total expenditure on the science and technology sector in India during 5th, 6th and 7th five-year plan periods (source: India (1993), p. 144).

Table 2.10 Expenditure on India's science and technology sector

	5th plan period (1974-79)	6th plan period (1980-85)	7th plan period (1985-90)
Expenses (Rs. crores)	693.16	2016.21	5086.53

A break on the X-axis indicates that zero on this scale is not to scale in plotting the 5th Five Year Plan period: 1974-79. Both the bar diagrams are vertical. We may also obtain a horizontal bar diagram by interchanging X and Y axes, i.e. groups or categories shown on Y-axis and magnitude of characteristics on the X-axis. Figure 2.2 shows the bar diagram corresponding to Example 2.7 (see Table 2.10).

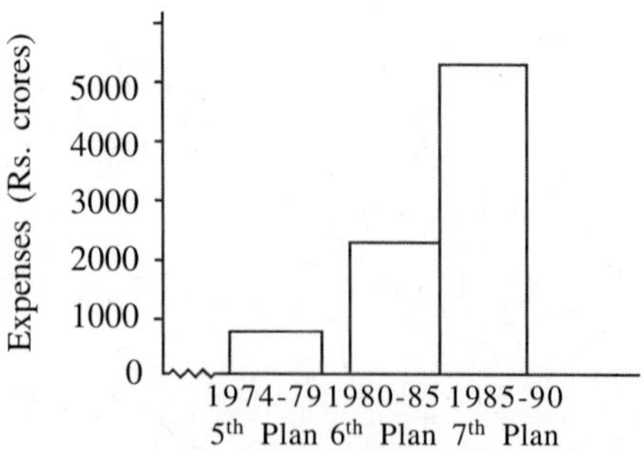

Fig. 2.2. Bar diagram of expenditure on India's science and technology sector.

Example 2.8 Table 2.11 shows the population of workers by category in India according to the 1991 census. The corresponding horizontal bar diagram is shown in Fig. 2.3 (source: India (1993), p. 291).

Table 2.11 Population of workers in 1991 under various categories

Category	No. of workers (crores)
Marginal workers	2.95
Cultivators	11.06
Agricultural labourers	7.46
Household industry workers	1.04
Other workers	8.98

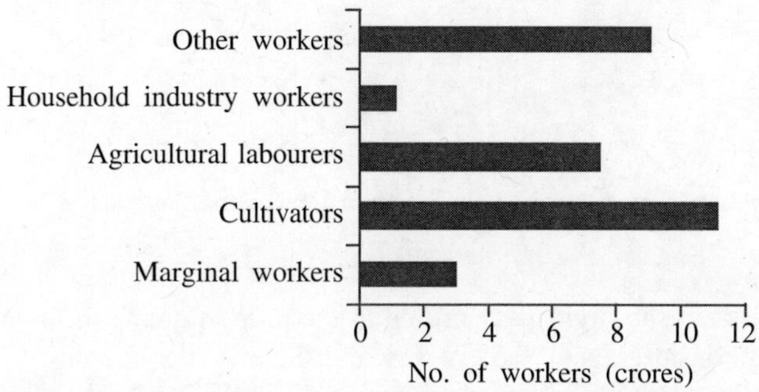

Fig. 2.3. Horizontal bar diagram for populations of various types of workers in 1991.

Multiple or Compound Bar Diagram

For comparison of a characteristic relating to items from different locations or at different times, this type of bar diagram is useful. To illustrate this, let us consider the following example:

Example 2.9 Population (millions) of Uttar Pradesh, Andhra Pradesh and Madras State for 1951 and 1961 is shown in tabulated form (Table 2.12) and also as a multiple bar diagram (Fig. 2.4) (source: Chandel (1964), p. 23).

Table 2.12 Population (millions) of the three States in 1951 and 1961

Year ↓ States →	Uttar Pradesh	Andhra Pradesh	Madras State
1951	63.22	31.25	29.98
1961	73.75	35.98	33.65

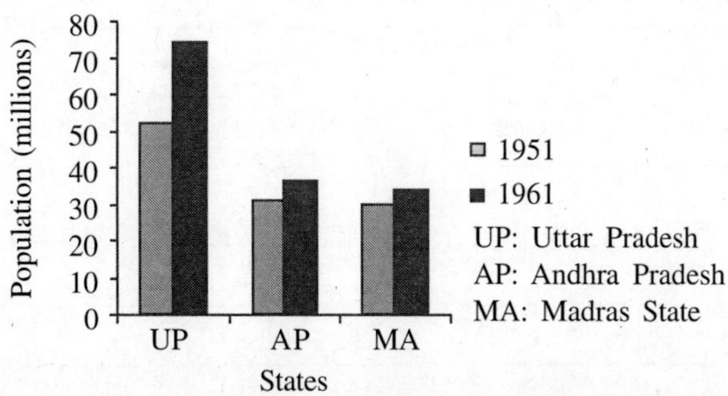

Fig. 2.4. Multiple bar diagram for population of the three states in 1951 and 1961.

This graph can also be represented by horizontally drawn bars. Both of these clearly show growth of population in the three states from 1951 to 1961. If a comparison between the states is desired during the said two years, the compound bar diagram should be drawn as in Fig. 2.5.

Different colours for different bars may also be used instead of different symbols.

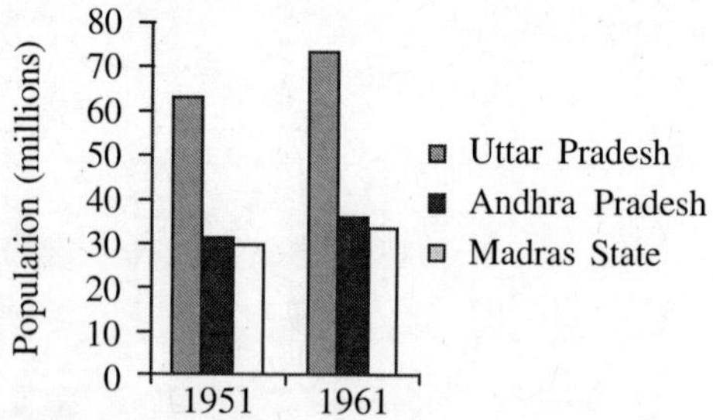

Fig. 2.5. Compound bar diagram showing population of the three states in 1951 and 1961.

Divided Bar Diagram

In this diagram the bars are sub-divided according to the proportions in different categories of objects. As an example, let us consider again Table 2.4 giving the 3 × 3 contingency table of nursing school entrants according to their scores and type of high school they passed out from (Example 2.3).

In ordinary sub-divided bar diagram the bars are of unequal length and divisions are according to count with the different scores (see Fig. 2.6). A variant of the sub-divided bar diagram is percent sub-divided bar diagram. In a per cent sub-divided bar diagram the bars are of equal height of 100 units. Sub-divisions are according to the percentages of the counts with different scores. For preparing this type of sub-divided bar diagram the sub-divisions (counts) are converted into percentages of the total. Thus, the resulting percentages for Example 2.3 are given in Table 2.13.

All the three bars will be of equal length of 100 units and the sub-divisions will be percentages calculated in the three bars (Table 2.13 and Fig. 2.7).

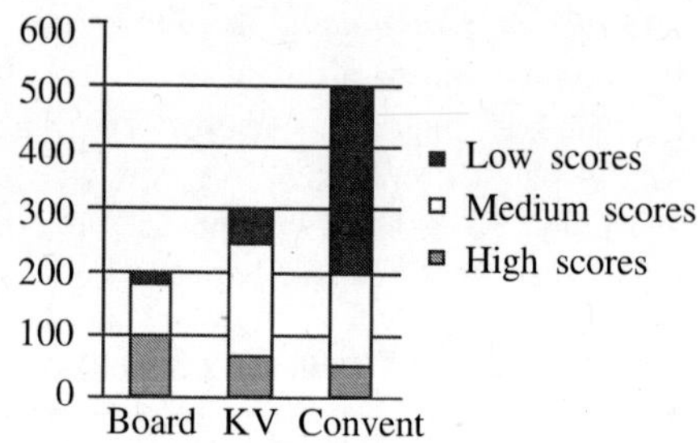

Fig. 2.6. Divided bar diagram for scores of various high school pass outs.

Table 2.13 Percentage scores of various high school pass outs

	Type of high school		
Scores	Board (%)	KV (%)	Convent (%)
Low	50	21.7	10
Medium	40	60	30
High	10	18.3	60
Total	100	100	100

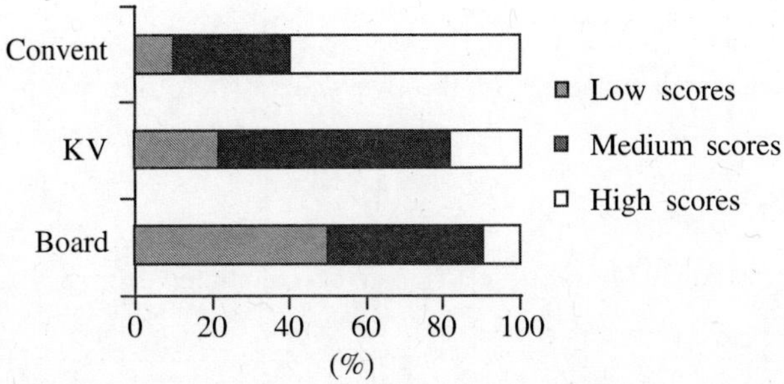

Fig. 2.7. Per cent sub-divided bar diagram for scores of various high school pass outs.

Angular Diagram or Pie Chart

This may be considered as a two-dimensional graph. If more than one groups of measures or count of items are sub-divided into categories according to certain characteristics (qualitative or quantitative) it is possible to present the information in a pie chart. The circles are drawn with areas proportional to the magnitude of the first characteristic. Each circle is sub-divided into angular sectors according to the magnitude of the second characteristic. Area of the sector is proportional to the angle between the two lines joining the center of the circle to the two points on its circumference. The sector angles (°) are obtained by using the formula

$$\text{Sector angle} = \frac{(\text{magnitude of the } 2^{\text{nd}} \text{ characteristic in sub-division}) \times 360}{\text{Total of characteristic magnitude}}$$

This can be better understood by considering the following example:

Example 2.10 Table 2.14 gives the agricultural production of Uttar Pradesh in 1952-53 and 1953-54 (source: Chandel (1964), p. 28).

Table 2.14 Agricultural production in Uttar Pradesh

Commodities	1952-53		1953-54	
	Yield (lac maunds)	Sector angle (°)	Yield (lac maunds)	Sector angle (°)
Cereals	2837	63	2865	75
Pulses	389	9	378	10
Fodder	4657	107	4490	117
Oilseeds	246	5	238	5
Fibers	19	0.5	14	0.3
Others	7666	175.5	5900	152.7
Total	15814	360	13885	360

Thus, the first angle in Column $3 = \frac{2837}{15814} \times 360$ is proportional to the quantity in the first row of Column 2. Similarly all other angles are also computed as shown above.

A pie chart is drawn having circle with area proportional to the total quantity. The area of a circle is πr^2, where r is radius of the circle. Hence radii of the two circles r_1 and r_2 are chosen such that

$$\pi r_1^2 : \pi r_2^2 = Q_1 : Q_2$$

where Q_1 and Q_2 are the total quantities of the two years. Therefore, we get

$$r_1/r_2 = \sqrt{\left(\frac{Q_1}{Q_2}\right)}$$

Now choosing some convenient value of r_2, r_1 can be calculated from

$$r_1 = r_2 \sqrt{\left(\frac{Q_1}{Q_2}\right)}$$

After drawing the circles, using a protractor, sectors are drawn starting from right hand horizontal radius. The order in both circles should be same for easy comparison. The sectors may be of different colours or with different symbols. An index has to be given for each item of produce.

If percentages are calculated or available for sub-divisions, a protractor calibrated from 0 to 100 may be used very easily to draw the pie chart (Fig. 2.8).

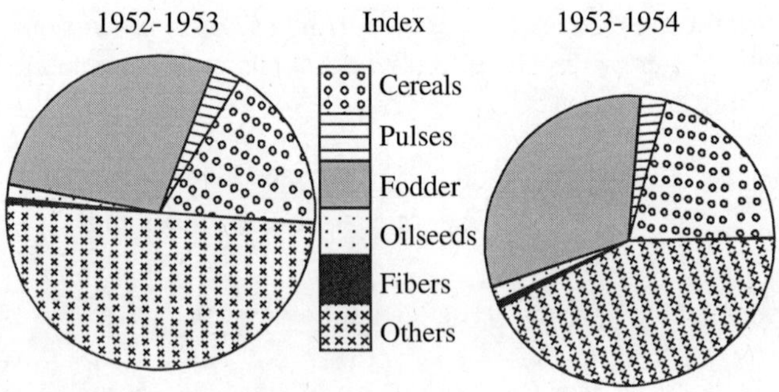

Fig. 2.8. Yields of various commodities in Uttar Pradesh.

2.2.2 Graphical Methods for Quantitative Data

As already mentioned, if quantitative data is available, the first step of summarising the raw data is preparation of a frequency table. A graphical representation is the second step in the direction of better comprehension of the nature of data, to understand how the observations are distributed over the range of its possible values. It also leads to an ease of comparison. Now, the following methods of graphical presentations from frequency table are described:

(a) Histogram
(b) Frequency polygon
(c) Frequency curve
(d) Ogives or cumulative frequency polygon and curve

Histogram

Consider the following frequency table (Table 2.15) obtained from Example 2.4:

To draw a histogram for this frequency table with *equal class widths* follow the steps given below:

(i) Put the class boundaries on the X-axis after deciding a scale.

(ii) On the Y-axis put frequencies after choosing a proper scale so that maximum of the class frequencies (11 in this case) can be drawn on Y-axis.

(iii) Draw rectangular bars on the class boundaries taking height of the bars proportional to the class frequencies.

(iv) Give graph heading, figure number, scale for X- and Y-axis and a label for X- and Y-axis.

After following the above steps we get Fig. 2.9.

Table 2.15 Frequency table of monthly rent

Classes	House rent (Rs.)	No. of households
1	5-7	2
2	7-9	5
3	9-11	11
4	11-13	4
5	13-15	4
6	15-17	4
	Total	30

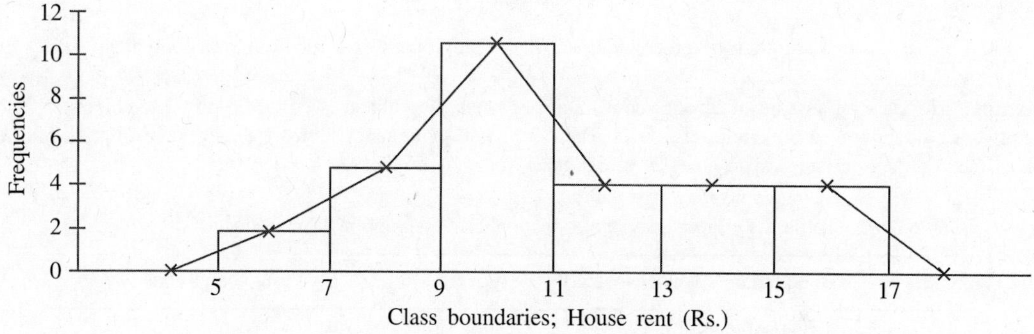

Fig. 2.9. Histogram and frequency polygon for Table 2.15.

If a frequency table has *unequal class widths* the histogram has to be drawn after calculation of relative frequencies per unit of variable. Let us consider the following example:

Example 2.11 The gross incomes (in whole dollars only) of 100 employees, including part-time, of a medium size company is given in Table 2.16.

Table 2.16 Frequency table of the income of employees of a medium size company

Income level	Frequency	Frequency per US$ 5000
1-4,999	2	2.00
5,000-9,999	7	7.00
10,000-14,999	12	12.00
15,000-19,999	16	16.00
20,000-24,999	18	18.00
25,000-34,999	24	12.00
35,000-49,999	11	3.67
50,000-99,999	8	0.80
100,000-149,999	2	0.20

Class intervals are of unequal widths in Table 2.16. So frequencies per \$ 5000 have been obtained in column 3. Now, X-axis, class boundaries are plotted and rectangular bars of heights equal to frequency per US \$ 5000 are drwan in Fig. 2.10.

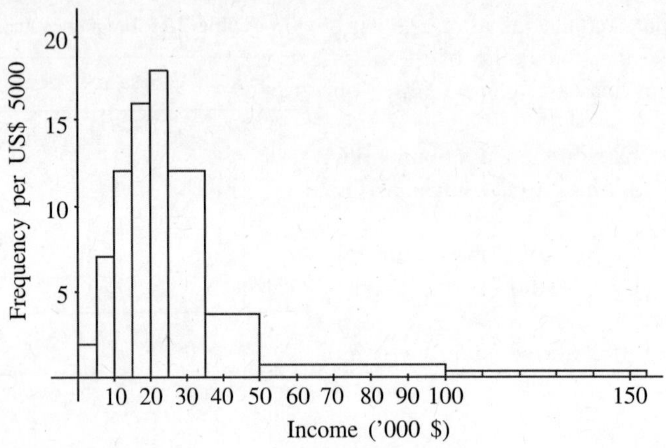

Fig. 2.10. Histogram of gross incomes of the employees of a medium size company.

Example 2.12 Data on amount of alcohol consumed per week per subject is collected on 923 subjects to assess the effect of alcohol on serum cholesterol level. Table 2.17 is the frequency table of alcohol consumption (ounces/week) (source: Milton and Tsokos (1983), p. 34 data).

Table 2.17 Frequency table of alcohol consumption (ounces/week)

Class	Class boundaries (ounces/week)	Frequency f_i	h_i	Frequency density f_i/h_i
1	0.0-0.5	201	0.5	402.00
2	0.5-3.5	372	3.0	124.00
3	3.5-9.5	260	6.0	43.30
4	9.5-19.5	80	10.0	8.00
5	19.5-49.5	10	40.0	0.25
	Total	923		

Drawing of histogram with height of rectangles over class intervals proportional to frequencies is misleading because the area under the histogram should be proportional to the total frequency (= N). If the classes are of equal width, the area under the histogram is $N \times h$. For a table with unequal class width the steps given below are followed:

1. Frequencies are first divided by the corresponding class width giving frequencies per unit of variables (= ounce/weeks in Example 2.12) or frequency density.
2. On X-axis mark class boundaries.
3. On Y-axis plot rectangles of height = f_i/h_i; $i = 1, 2, 3, ..., K$ (= 5 in Example 2.12) on i^{th} class of width h_i; where K is the number of classes, f_i the frequency in i^{th} class and h_i the width of ith class.

Hence, the total area below the resulting histogram is N (= 923 in Example 2.12).

A third situation may arise when the frequency table does not have class boundaries as class interval but has class limits as class intervals as given in the following example:

Example 2.13 Table 2.18 shows the age distribution of cases of a certain disease reported during a year in a particular state (source: Daniel (1987), p. 35, Q. 25).

Table 2.18 Frequency table of ages of cases of a disease

Age (yrs)	Number of cases
5-14	5
15-24	10
25-34	20
35-44	22
45-54	13
55-64	5
Total	75

To draw histogram class boundaries are required to be plotted on X-axis. These are obtained as follows:

(i) Find the difference between upper limit of lower class and the lower limit of the next higher class. Table 2.18 has equal class widths, so we can consider any two consecutive classes giving difference = 15 – 14 = 1 (from 1st and 2nd classes) or = 45 – 44 (from 4th and 5th class) = 1.

Now ½ of this difference is added to the upper class limits and the same is subtracted from all the lower class limits. Thus, ½ × 1 = 0.5 is added to the upper class limits and subtracted from the lower class limits to give Table 2.19.

Table 2.19 Class boundaries from class intervals of Table 2.18

Age (class limits)	Number of cases f	Age (class boundaries)
5-14	5	4.5-14.5
15-24	10	14.5-24.5
25-34	20	24.5-34.5
35-44	22	34.5-44.5
45-54	13	44.5-54.5
55-64	5	54.5-64.5
Total	75	

Hence, the histogram is drawn as shown in Fig. 2.11.

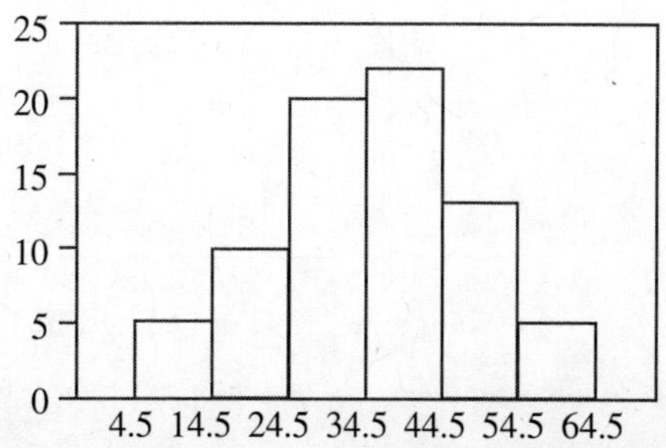

Fig. 2.11. Histogram for frequency table of ages of cases.

Frequency Polygon

It is a polygon obtained by joining the points by a scale on the graph. These points are plotted at X = mid-points of the classes and Y = frequencies of the classes. The polygon is joined down to the base at one class interval below the first mid-point and at one class interval above the last mid-point.

Example 2.14 Consider the frequency table of marks in English obtained by 200 students (Table 2.20)

Table 2.20 Frequency table of marks of students in English

Class	Marks in English (class boundaries)	Number of students	Mid-points
1	10-18	22	14
2	18-26	44	22
3	26-34	80	30
4	34-42	32	38
5	42-50	14	46
6	50-58	5	54
	Total	200	

The mid-points of the classes are calculated from the following formula:

Mid-point or class mark of ith class

$$X_i = (\text{lower class boundary of the } i^{th} \text{ class + upper class boundary of the same class})/2$$

An alternative formula is

$$X_i = (\text{lower class limit of the } i^{th} \text{ class + upper class limit of the } i^{th} \text{ class})/2$$

The points where the polygon joins the base are

$$1^{st} \text{ mid-point} - \text{class width} = 14 - 8 = 6$$

and

$$\text{Last mid-point} + \text{class width} = 54 + 8 = 62$$

Thus, the frequency polygon is obtained as shown in Fig. 2.12.

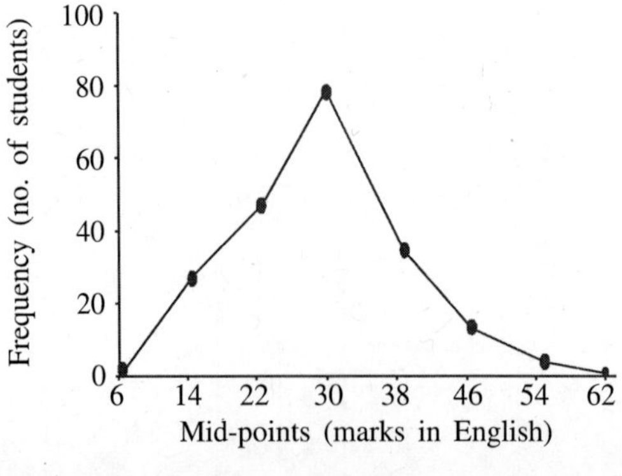

Fig. 2.12. Frequency polygon for distribution of marks in English.

It can be easily seen that for Example 2.13, if the frequency polygon is to be drawn using the second formula, the mid-points will be 9.5, 19.5, 29.5, 39.5, 49.5 and 59.5.

Frequency Curve

If the points plotted for drawing the frequency polygon are joined free hand as a curve, then the resulting figure gives the frequency curve. Now let us consider the frequency table (Table 2.18) for Example 2.13. Calculating mid points from class limits of Table 2.18 gives Table 2.21.

Table 2.21 Mid-points from class limits of Table 2.18

Age (yrs) (class limits)	Number of cases f_i	Mid-points
5-14	5	9.5
15-24	10	19.5
25-34	20	29.5
35-44	22	39.5
45-54	13	49.5
55-64	5	59.5
Total	75	

Figure 2.13 shows the frequency curve for Example 2.13.

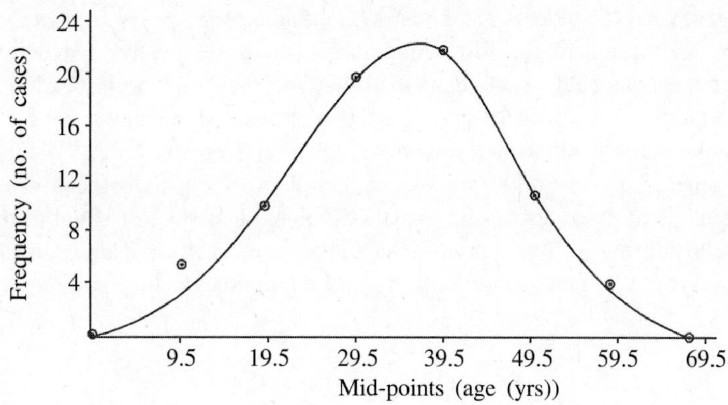

Fig. 2.13. Frequency curve for age distribution of cases.

Note: A frequency polygon or frequency curve can also be drawn on the histogram by taking the points on the middle of upper edge of rectangles in the histogram. Joining these points by scale gives frequency polygon and joining free hand by a smooth curve gives frequency curve. Joining the ends at the base line is done as explained earlier. See histogram of Example 2.4 from Table 2.15 in Fig. 2.9.

Ogives or Cumulative Frequency Curve

There are two types of ogives – *less than* and *more than*. These are also called *cumulative frequency curves* if the points are joined free hand by a smooth curve. If the points are joined by scale by straight lines then the figure is called *cumulative frequency polygon*. To draw these the cumulative frequencies have to be found. Let $F_<$ and $F_>$ denote the cumulative frequency less than and cumulative frequency greater than, respectively. Let $F_<$ denotes the number of observations (or frequency) with values less than a specified value of X while $F_>$ denotes the number of observations with values greater than X. Before plotting the ogives both the cumulative frequencies have to be calculated. These are called cumulative frequencies because they are obtained by successive addition (cumulation) of frequencies in the table. For $F_<$ successive addition is done from the lowest (or first) class and for $F_>$ successive addition is done from the last class of the table. As an illustration take Table 2.20 for Example 2.14.

Table 2.22 Distribution of marks in English

Classes	Marks (class boundaries)	No. of students f	Cumulative frequency less than ($F_<$)	Cumulative frequency greater than ($F_>$)
1	10-18	22	22	200
2	18-26	44	66	178
3	26-34	80	146	134
4	34-42	32	178	54
5	42-50	14	192	22
6	50-58	8	200	8

Ogives are drawn by plotting class boundaries on the X-axis and the cumulative frequencies on Y-axis. $F_<$ are plotted on the class boundaries and are joined either by scale in a straight line to give less than type cumulative frequency polygon or by free hand curve to give less than cumulative frequency curve or less than ogive. In Table 2.22, $F_<$ values correspond to the upper boundaries of the respective classes, i.e. 146 students have marks less than or equal to 34 (marks $\leq$ 34). The $F_>$ values in the last column correspond to the lower boundaries of the respective classes. Thus, 54 students have marks greater than or equal to 34 (marks $\geq$ 34). One may use the last value in $F_<$ column and first value in $F_>$ column as a check to verify the correctness of calculations of $F_<$ and $F_>$.

These two values should always be equal to the total frequency (= N) since all the students (observations, in general) have marks (values) greater than or equal to the lowest boundary in the table, and all the students (observations) have marks (values) less than or equal to the upper boundary of the last class. Considering all these points, the ogives are drawn as shown in Fig. 2.14. To avoid confusion different symbols may be used for $F_<$ and $F_>$ points on the graph.

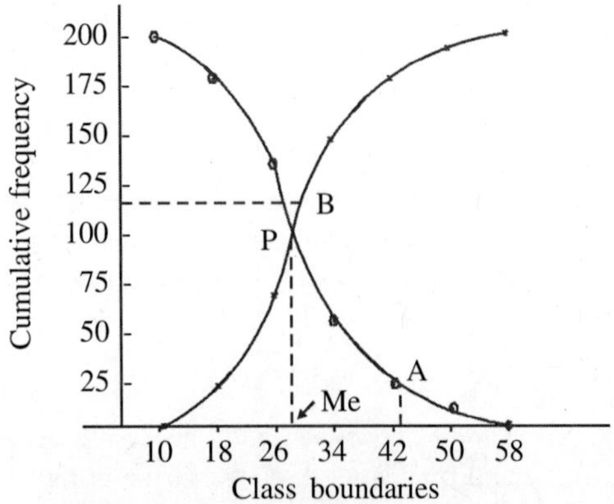

Fig. 2.14. Ogives for marks in English.

The $F_<$ and $F_>$ values will always be zero for lower boundary of the lowest class (= 10 in Table 2.22 for Example 2.14) and upper boundary of the last class (= 58), respectively. Thus, four points of the ogives are always known:

(i) (Lowest class boundary value, 0) on less than type ogive (10, 0) for Example 2.14.
(ii) (Lowest class boundary value, N) on greater than type ogive (10, 200) for Example 2.14.
(iii) (Highest class boundary value, N) on less than type ogive (58, 200) for Example 2.14.
(iv) (Highest class boundary value, 0) on greater than type ogive (58, 0) for Example 2.14.

From the less than type ogive one can read from the graph the number of observations (frequencies) less than any value of X. Similarly, from the greater than type ogive one can read number of observations (frequency) greater than any value of X. One may draw a vertical line from X upto the appropriate curve and read the Y-coordinate of the point where the vertical line meets the ogive.

One can also determine the value of X below which a specified number of observations lie or above which a specified number of observations lie, by drawing a line parallel to X-axis from the specified value of cumulative frequency on Y-axis.

Suppose it is desired to find:

(a) How many students have secured marks less than 45?
(b) What is that value of marks which is secured by more than 120 students?

For (a) , a vertical line will be drawn from X = 45 up to less than type ogive at A and Y-coordinate of A will give the answer. For (b), a horizontal line will be drawn from Y = 120 up to greater than type ogive at B and the value of X for B will be the marks secured by more than 120 students. It may also be noted that the two ogives intersect at a point P with coordinates (Me, $N/2$).

3. Data Handling II: Descriptive Statistics

Nature of long series of collected data becomes comprehensible, when as a first step it is summarized into a frequency table and presented graphically. It is evident that the observations seem to center round a value. High concentration of observations is around a value inside the range and the frequency tapers off near the two ends. An experimenter wishes to define quantitatively the characteristics of a frequency distribution in as few numbers as possible. Moreover, he wishes to compare two or more sets of data.

Distributions of similar characteristics on two groups of objects are of similar form and one is interested to compare such distributions. For instance height of college students in two states say Punjab and Tamil Nadu, weights of employees between ages 45 to 60 years in two or more Educational Institutes and Administrative Offices, yield of wheat crop per acre in two states from various villages, daily milk produced by cows of two species in dairy farms of state, number of wealthy people in two countries in the last decade etc. may be needed to be compared.

By comparing two distributions one wants to bring out the differences and the similarities between the two in a precise quantitative way. This requires some measures computed from the data, that can be used to characterize the data set. Such measures are definite numbers so are comparable and are called descriptive statistics, since they describe the data set almost completely.

The two fundamental characteristics in which frequency distributions may differ are (a) their locations and (b) extent of their dispersion.

Figures below clarify this idea.

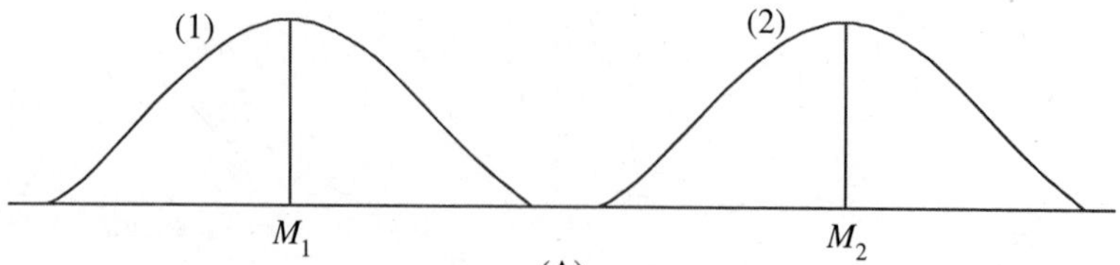

(A)

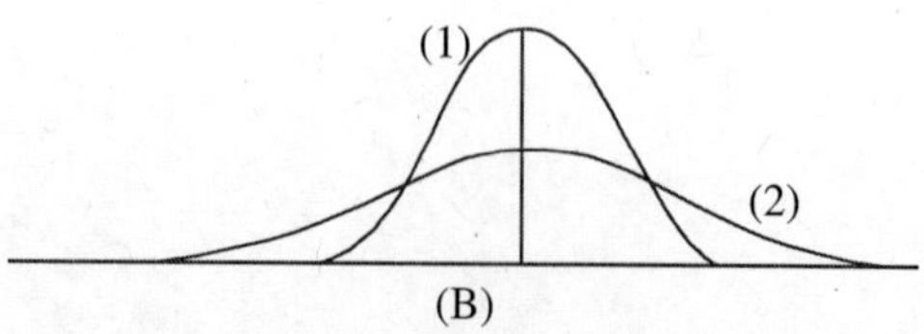

(B)

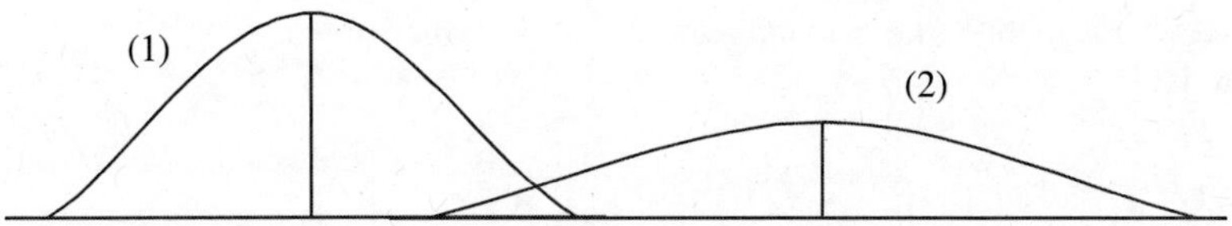

(C)

Fig. 3.1. Distributions with different locations and dispersions (A) gives two distributions which have markedly different locations but almost same dispersion, (B) both the distributions have the same location *M*, but the distribution (2) is much more dispersed than (1) and (C) the two distributions differ in locations as well as dispersions (source: A, B, C Yule and Kendall (1958), p. 103.

The measures of these two qualities of distribution are called (a) Measures of location or measures of central tendency and (b) Measures of dispersion.

Apart from these two, the distributions many differ in two more characteristics the departure from symmetry (called skewness) and steepness (or peakedness).

Frequency curve (3) is symmetric, whereas curve (1) is skewed to the right and curve (2) is skewed to the left.

In (E) frequency curve (1) is normal or mesokurtic, the curve (2) is steep or leptokurtic and the curve (3) is flat or platykurtic.

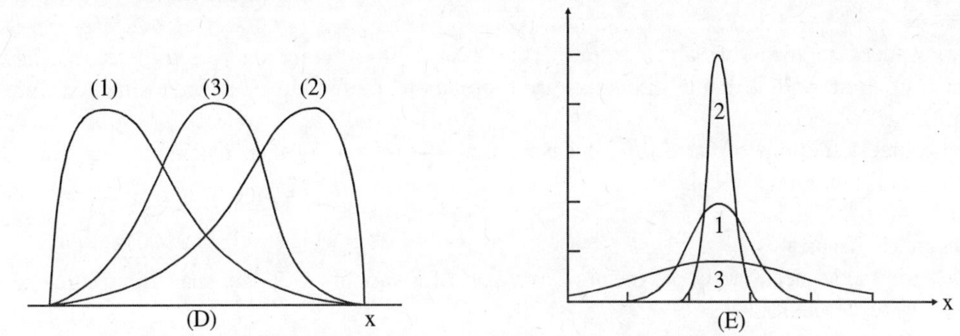

Fig. 3.2. Distributions with different skewness and kurtosis.

The quantitative measures for these two characteristics are also defined. In the next two sections of this chapter, the measures of location and dispersion are defined. The last two of the four measures, viz. the coefficients of skewness and kurtosis are discussed in the subsequent section of this chapter. These can be calculated from the frequency table (or grouped data) as well as for the raw or ungrouped data. Methods to compute these are given for both grouped and ungrouped data.

For more details of the ideas and numerical examples discussed in this and the previous Chapter 2, the interested reader may also refer the books by Sokal and Rohlf (1973), Daniel (1987), Milton and Tsokos (1983), Yule and Kendall (1958), Croxton, Cowden and Klein (1975), Goon et al. (1983) and Gupta and Kapoor (1999) etc.

3.1 Measures of Location or Measures of Central Tendency

There are three measures of central tendency used commonly:

(a) The arithmetic mean or simple average, commonly called mean
(b) Median and
(c) Mode

A measure of central tendency is merely a value of the variable, so each of these three has the same unit as the unit of the original variable.

Each measure has a physical significance and is useful under specific situations. Out of these mean is most commonly used because of its convenient and useful properties.

It is expected that a measure of central tendency should have some criteria. The *desired properties* of a satisfactory measure of central tendency are:

(i) It should be uniquely defined
(ii) It should be based on all the observations
(iii) It should be readily comprehensible and should not be an abstract mathematical quantity.
(iv) It should be easily and rapidly calculated.
(v) It should not be affected by fluctuation of sampling. Thus if several samples are taken from the same population, though the measure from different samples are likely to differ but a good measure should not differ too much due to sampling.
(vi) A measure of central tendency should be amenable to algebraic treatment. Thus it is desired that if measures for two series of data from similar objects are available one should be able to calculate measures of central tendency for the two or more combined series.

As the first measure, mean has all these properties, it is the most commonly used measure. The other two are also used, and in some situations they are more preferable. But such situations are less abundant.

The actual definitions of these three measures and the method of their calculation is given below for ungrouped and grouped data.

(a) Arithmetic Mean

Mean (or arithmetic mean or simple mean or average) of a variable X is denoted by putting a bar over X, and may be defined as

$$\bar{X} = \frac{\text{Sum of all the observations on } X}{\text{Number of observations}} \tag{3.1}$$

Calculation of Mean

Ungrouped Data

Let $X_1, X_2, \ldots\ldots, X_n$ be n observations on a random variable X, then

$$\bar{X} = (X_1 + X_2 + \ldots\ldots + X_n)/n$$

$$= \frac{1}{n}\sum_{i=1}^{n} X_i \tag{3.2}$$

If n workers in all, are paid monthly Rs. P, P/n is the average payment of these workers. Physical meaning of this is that if the total payment of Rs. P is shared equally by all the n workers, each will receive the payment of P/n rupees.

Moreover, if the average payment to a group of n workers is M, the total money paid to the n workers will be nM rupees.

Grouped Data

Let N observations on a random variable X, be grouped in k classes such that the frequency of ith class is f_i (i = 1, 2, 3, …k) and mid point or class mark of ith class is X_i, then

$$\bar{X} = \frac{1}{N}\sum_{i=1}^{k} f_i X_i, \text{ where } N = \sum_{i=1}^{k} f_i \tag{3.3}$$

This calculation is based on the assumption that all the f_i observations in ith class are centered (i.e. have common value) at X_i. This assumption is not a reality so the computed mean is based on some loss of information due to grouping. It is because, the actual values in a class are not available, and it is assumed that all the observations in i^{th} class have value X_i, the mid point of the said class.

Example 3.1: Following are the number of seriously injured athletes of 25 schools, while participating in basketball match:

Injured basketball athletes: 1 3 2 4 5 2 3 4 4 6 4 2 3 3 4 4 4 5 6 6 7 5 3 5 5

(source: Milton and Tsokos (1983), p. 20, Example 1.3.1).

Solution: Let X denote the number basketball injured athletes, then the average or mean number of injured athletes per school is

$$\bar{X} = \frac{1}{25}\sum_{i=1}^{25} X_i, \quad = \frac{100}{25} = 4$$

Hence average number of injured athletes per school = 4

Example 3.2: Consider the frequency table of number of petals in 50 flowers. Find the average number of petals in the said flowers (source: Chandel (1964), p. 17).

Table 3.1 Calculation of mean number of petals (source: Chandel, p. 17)

Classes i	No. of petals X_i	No. of flowers f_i	$f_i X_i$
1	6	8	48
2	7	12	84
3	8	15	120
4	9	10	90
5	10	5	50
	Total	Σf_i = 50	$\Sigma f_i X_i$ = 392

$$\bar{X} = \frac{1}{N}\sum_{i=1}^{k} f_i X_i, \quad = \frac{1}{50} x \quad 392 = 7.84$$

So average number of petals per flower is 7.84.

Example 3.3: Consider the frequency table on age in years of diseased cases (Example 2.10) for calculation of average age in years of diseased cases.

$$\bar{X} = \frac{1}{N}\sum_{i=1}^{6} f_i X_i \quad = \frac{2642.5}{75} = 35.23 \text{ years}$$

Table 3.2 Average age in years of diseased cases (source: Daniel (1987), p. 35, Q. 25)

Age in years (class limits)	Number of cases (f_i)	Mid points X_i	$f_i X_i$ Col (2)*x* Col (3)	U_i	$f_i U_i$
5-14	5	9.5	47.5	– 3	– 15
15-24	10	19.5	195.0	– 2	– 20
25-34	20	29.5	590.0	– 1	– 20
35-44	22	39.5	869.0	0	0
45-54	13	49.5	643.5	1	13
55-64	5	59.5	297.5	2	10
Total	$\Sigma f_i = 75$		$\Sigma f_i X_i = 2642.5$		– 32

Short Cut Method

Calculation of $f_i.X_i$ is a bit time taking. So an *easy method* is given below, *if classes are of equal width:*

$$\text{Choose } U = (X - A)/h \tag{3.4}$$

where U is the transformed variable obtained by change of origin and scale of X.
h = common class width (= 10 in this example)
A = an assumed mean, is some value of X_i to be chosen so that value of U for that class becomes zero.
The values of U are given in the above table for $A = 39.5$.
It may be noted that U_i corresponding to $X_i = A$ is 0 and for classes lower than this i, U_i are –1, –2, –3, ... and for higher than this i, U_i are 1, 2, 3.... for any table of equal class width.
Now computations of f_iU_i instead of f_iX_i is much easy.
$\sum_i f_iU_i$ is obtained for all –ve U_i's and for all +ve U_i's, and the appropriate algebraic addition is performed to get $\sum_{i=1}^{k} f_iU_i$.

In this problem $\sum_{i=1}^{6} f_iU_i = -32$ and $\bar{U}$ can be obtained as

$$\bar{U} = \frac{1}{75}\sum_{i=1}^{6} f_iU_i = -32/75 = -0.4267$$

Now $\bar{X}$ can be obtained from $\bar{U}$ using the reverse transformation

$$\bar{X} = A + h\bar{U} \tag{3.5}$$

(can be proved by substituting U_i in terms of X_i)

In this problem $\bar{X} = 39.5 + 10\ \bar{U}$
$= 39.5 + 10\ (-.4267) = 39.5 - 4.267$
$= 35.23$ years

Hence average age of diseased cases is 35.23 years

Properties of Mean

(i) It is uniquely defined
(ii) It is based on all the observations
(iii) It is easily comprehensible
(iv) It is least affected by fluctuations of sampling
(v) Algebraic treatment is possible.

Illustration of Algebraic Treatment

(a) *Combined mean of the two series* Let means of two series of N_1 and N_2 observations on similar variables are $\bar{X}_1$ and $\bar{X}_2$. Then, combined mean of the two series of $N_1 + N_2$ observations, M can be obtained as

$$M = (N_1.\ \bar{X}_1 + N_2.\ \bar{X}_2)/(N_1 + N_2) \tag{3.6}$$

(b) *Combined mean of several series* In general let there are k series of N_1, N_2 .... N_k observations on similar variables with means $\bar{X}_1, \bar{X}_2 \ldots \bar{X}_k$. Then mean M of the combined series of $N_1 + N_2 + \ldots + N_k = N$ observations is given by

$$M = \left(N_1.\bar{X}_1 + N_2.\bar{X}_2 + \ldots\ldots + N_k\bar{X}_k\right)/\left(N_1 + N_2 + \ldots\ldots + N_k\right)$$

$$= \frac{1}{N}\sum_{i=1}^{k} N_i\bar{X}_i \tag{3.7}$$

(c) If there are two or more *series of equal number of observations* and Y is sum or differences of these, i.e. if

$$Y = X_1 \pm X_2 \pm \ldots \pm X_k \tag{3.8}$$

then M, the mean of Y is given by

$$M = M_1 \pm M_2 \pm \ldots \pm M_k \tag{3.9}$$

where, M_1, M_2 ...M_k are means of the series X_1, X_2 ...X_k, respectively.

(d) Let a constant number C *is added or subtracted* from each value of a random variable X with mean $\bar{X}$. Then mean M of the resulting series is also increased or decreased by C from $\bar{X}$. In symbols

$$\text{If } Y = X \pm C, \text{ Then } M = \bar{Y} = \bar{X} \pm C \tag{3.10}$$

(e) If $Y = CX$ then

$$\bar{Y} = C\bar{X} \tag{3.11}$$

for any constant C.

(vi) The *sum of deviations* of the variate values *from mean is zero* i.e.

$$\sum_{i=1}^{n}\left(X_i - \bar{X}\right) = \sum_{i=1}^{n} X_i - n\bar{X} = n\bar{X} - n\bar{X} = 0 \tag{3.12}$$

For grouped data

$$\sum_{i=1}^{n} f_i\left(X_i - \bar{X}\right) = \sum_{i=1}^{n} f_iX_i - \bar{X}\sum_{i=6}^{k} f_i$$

$$= N\bar{X} - \bar{X}N = 0 \tag{3.13}$$

(vii) Sum of squares of deviations of observations from mean is minimum, i.e.

$$S^2 = \sum_{i=1}^{n}(X_i - A)^2 \textit{ is minimum if } A = \bar{X}$$

Demerits of Arithmetic Means

(i) It is very much affected by one or two extremes values.

(ii) It can not be obtained graphically, as the two other measures median and mode can be obtained by proper inspection from graph.

(iii) Some times the first and last classes are open i.e. are of the form *less than 1 year* or *more than 50 years*. The mean of such frequency tables with one or both ends open, can not be determined, since, mid points of the open classes can not be found. However, if the values of observations in these classes are known, mean can be found out from the table, since total of observations in the open classes can be calculated.

(b) Median

The median is defined as the value of the middle most observation when they are arranged in the order of their magnitude. It is the value of the variable such that ordinate over it divides the area of the frequency curve in two equal parts. Thus it is a value such that half of the observed values are greater than median and half are less than median.

Median is a measures of central tendency which has the property of being based on all the observations and has comprehensible physical meaning.

Calculation of the Median

Ungrouped Data

Let $X_1, X_2 \ldots X_n$ be a set of n observations on a random variable X.

Arrange the observations in ascending or descending order of magnitude. Let the resulting values be $X_{(1)} \le X_{(2)} \le \ldots \le X_{(n)}$ in ascending order.

Thus $X_{(1)}$ = Minimum $(X_1, X_2, \ldots X_n)$ and

$X_{(n)}$ = Maximum $(X_1, X_2, \ldots X_n)$

(i) *Let n be odd* If n is odd = $(2k + 1)$ say. The median

$$\text{Me} = \text{Value of } \left(\frac{n+1}{2}\right)^{\text{th}} \text{ observation} = X_{(k+1)} \tag{3.12}$$

It will be the value of the middle most observation.

(ii) *Let n be even* If n is even = $2k$, say, the median is taken as the average of two middle most observations. Thus

$$\text{Me} = ½\left[\text{Value of the } \left(\frac{n}{2}\right)^{\text{th}} \text{ observation + value of } \left(\frac{n}{2}+1\right)^{\text{th}} \text{ observation}\right]$$

$$= \frac{1}{2}\left(X_{\left(\frac{n}{2}\right)} + X_{\left(\frac{n}{2}+1\right)}\right) = ½\left[X_{(k)} + X_{(k+1)}\right] \tag{3.13}$$

Note: Any value between $X_{(k)}$ and $X_{(k+1)}$ may be median. But by convention, average of these two, is taken to define median.

Example 3.4: The following are the serum lipid peroxide (SLP) levels of 9 patients of *Diabetes mellitus*: 5.85, 6.17, 6.09, 7.70, 3.17, 3.83, 5.17, 4.31, 3.09. Find median of the data (source: Daniel (1987), p. 35, Q. 27).
Solution: First the values are arranged in the ascending order as:

3.09, 3.17, 3.83, 4.31, 5.17, 5.85, 6.09, 6.17, 7.70

$n = 9$ in this problem, which is odd. So

$$\text{Me} = \text{Value of} \left(\frac{9+1}{2}\right)^{\text{th}} \text{observation}$$

$$= \text{Value of 5th observation} = 5.17.$$

Example 3.5: Following are the hemoglobin values (in gm/100 ml) of 10 children under treatment for anemia. Find the median of the hemoglobin values (source: Daniel (1987), p. 34, Q. 23).

9.1, 10.0, 11.4, 12.4, 9.8, 8.3, 9.9, 9.1, 7.5, 6.7

Solution: The values in ascending order are:

6.7, 7.5, 8.3, 9.1, 9.1, 9.8, 9.9, 10.0, 11.4, 12.4

Here $n = 10$, so

$$\text{Me} = \text{Average of} \left(\frac{10}{2}\right)^{\text{th}} \text{and} \left(\frac{10}{2}+1\right)^{\text{th}} \text{observations}$$

$$= \frac{1}{2}\left(\text{Value of } 5^{\text{th}} \text{ observation} + \text{Value of } 6^{\text{th}} \text{ observation}\right)$$

$$= \frac{1}{2}\left(X_{(5)} + X_{(6)}\right) = \frac{1}{2}(9.1 + 9.8)$$

$$= \frac{1}{2}(18.9) = 9.45 \text{ gm/100 ml}$$

Grouped Data

For grouped data value of $\left(\frac{N}{2}\right)^{\text{th}}$ observation is taken as median. Grouped data may be on a discrete series or a continuous series. Let us consider as an example the following discrete series again, given in Example 3.2.

Table 3.3 Number of petals in flowers

Classes	No. of petals in flowers X_i	No. of flowers f_i	Cumulative frequency less than ($F_<$)
1	6	8	8
2	7	12	20
3	8	15	35
4	9	10	45
5	10	5	50
	Total	50	

In this example N = 50 and $N/2$ = 25. The 25th observation is between cumulative frequencies 20 to 35. All the 21st to 35th observations have value 8 petals. This is also the value of 25th observation. Hence Me = 8 petals.

Continuous Series

Consider again the frequency table, given in Example 3.3.

Table 3.4 Average age in years of diseased cases (source: Daniel (1987), p. 35, Q. 25)

Class	Class limits	Number of cases (f_i)	Class boundaries	Cumulative frequency ($F_<$)
1	5-14	5	4.5-14.5	5
2	15-24	10	14.5-24.5	15
3	25-34	20	24.5-34.5	35
4	35-44*,**	{22}	{34.5-44.5}	57
5	45-54	13	44.5-54.5	70
6	55-64	5	54.5-64.5	75
	Total	75		

Note: * shows the median class and ** shows the modal class

For this type of grouped data median is again the value of $\left(\frac{N}{2}\right)^{th}$ observation, when observations are arranged in order of magnitude. Frequency table values are already in ascending order.

As N = 75 So $N/2$ = 37.5

Next locate the class in which median lies i.e. the median class. For this see the cumulative frequencies and mark the class in which $F<$ just exceeds $N/2$. In this example this cumulative frequency is 57 and 34.5-44.5 is the median class. Since cumulative frequencies less than is related to upper class boundaries, the class boundaries are found and entered in 4th column.

It is assumed that the 22 observations in this class are uniformly distributed and the value of variable corresponding to 37.5[th] observation is determined.

The general formula to find median is

$$\text{Me} = l_o + \frac{N/2 - F_<}{f_m} \times h \tag{3.12}$$

where l_o is lower boundary of median class; f_m is frequency of median class; h is width of median class and $F_<$ is cumulative frequency less than l_o.

In this example l_o = 34.5, f_m = 22, $F_<$ = 35, h = 10. Hence,

$$\text{Me} = 34.5 + \frac{37.5 - 35}{22} \times 10 \quad = 34.5 + \frac{2.5 \times 10}{22}$$

$$= 34.5 + \frac{25}{22} \quad = 34.5 + 1.136 \quad = 35.636 \text{ years}$$

Graphical Method

Median can be determined graphically for grouped data from the ogives as follows:

(i) Two ogives intersect at cumulative frequency = $N/2$. Draw a perpendicular from this point on X-axis, and the value of X where the perpendicular touches X-axis is Me. See Fig. 3.3(a).

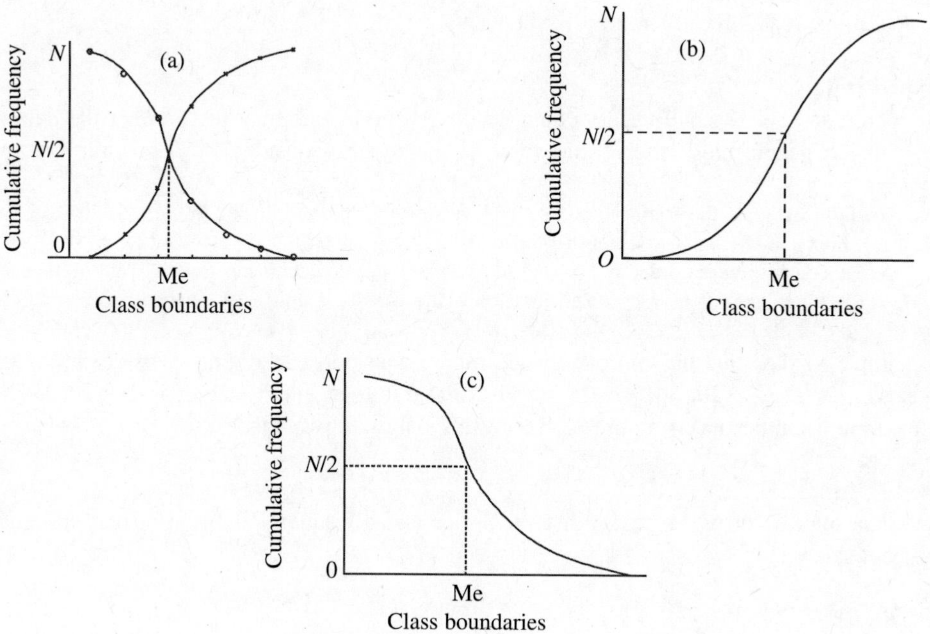

Fig. 3.3. Graphical determination of median: (a) Ogives; (b) less than type ogive and (c) greater than type ogive.

(ii) Draw a line parallel to X-axis from the value $N/2$ on Y-axis to intersect less than type ogive or more than type ogive. Drop a perpendicular on X-axis from this point on the ogives. The value where this perpendicular touches the X-axis is Me. See Figs. 3.3(b) and (c).

Merits and Demerits of Median

Merits

(i) Median is rigidly defined, based on all the observations and is readily comprehensible.
(ii) It can be computed for frequency table with open ends unless it lies in the lowest class and it is open.
(iii) Median is not affected by one or two very large or very small values. Stature of a giant does not affect value of median height of normal men more than any stature just greater than median height.

Demerits

(i) It is incapable of being given to algebraic treatment. Thus medians of two or more series can not be used to find median of combined series.
(ii) Median is more likely to be affected by fluctuations of sampling.

(c) Mode

Mode is the value of the variable which corresponds to the maximum of the frequency curve. It represents the value which is most typical or most frequent or in fashion.

Calculation of Mode

In this section we describe how to calculate mode denoted by (M_0) for grouped and ungrouped data and to determine it graphically for a grouped data.

Ungrouped Data

If ungrouped data is very small i.e. less than 20 observations, one may just look at the data and find the value which occurs most often. This value is the mode. Consider the following example.

Example 3.6: Following are the fasting blood sugar levels of 10 children: 56, 62, 63, 65, 65, 65, 65, 68, 70, 72 in mg/100 ml. Find mode (source: Daniel (1987), p. 20, Exercise 1.6.1).
Solution: Here the data is already ordered. So one can find that value 65 occurs most often or has highest frequency, Hence, M_0 = 65 mg/100 ml. It may be noted that mode has the same unit as the variable.

As an example of data that has more than one mode, consider a laboratory whose employees are of ages (years): 20, 21, 20, 20, 34, 22, 24, 27, 27, and 27. It is found that two ages 20 and 27 years occur maximum number (= 3) of times. Hence this data has two modes.

Grouped Data

Grouped data may be of two types: a discrete series and a frequency table. Calculation of mode for both are considered.

Discrete Series

Looking at the frequency column, find the maximum frequency and the value in the series with this frequency is Mode.

Consider again the Example 3.2 on distribution of number of petals in flowers. It can be seen that 15 is maximum frequency and the value 8 in the series has this frequency. So M_0 = 8 i.e. Mode number of petals in flower is 8.

Frequency Table

To calculate mode from a frequency table first look for highest frequency. The class with highest frequency is called modal class. The following formula is used now to find mode

$$M_0 = l_0 + \frac{\Delta_1}{\Delta_1 + \Delta_2} \times h \tag{3.13}$$

where l_0 *is* lower boundary of the modal class; h *is* width of the modal class; Δ_1 is $f_0 - f_{-1}$; Δ_2 is $f_0 - f_{+1}$; f_0 is highest frequency; f_{-1} is frequency of the class preceding the modal class; f_{+1} is frequency of the class following the modal class.

As an illustration of the procedure, consider again Example 3.3.

An observation of the table indicates that 22 is highest frequency and it is of the class 35-44. Hence 35-44 is the modal class. It is seen that the frequency table has class limits, so the class boundaries are found (already calculated for finding median, in column 4 of the table).

Thus $l_0 = 34.5,\ h = 10,\ f_0 = 22$

$f_{-1} = 20,\ f_{+1} = 13$

So, $\Delta_1 = f_0 - f_{-1} = 22 - 20 = 2$

$\Delta_2 = f_0 - f_{+1} = 22 - 13 = 9$

Hence $M_0 = 34.5 + \dfrac{2}{2+9} \times 10$

$= 34.5 + \dfrac{20}{11} = 34.5 + 1.81$

$= 36.31$ years.

Graphical Method to Find Mode

Graphically the mode is found as follows, by drawing histogram from the frequency table:

The logic for formula (3.13) is this graph. The rule of Geometry using properties of similar triangles ABC and ADE[1] gives the value of Mode.

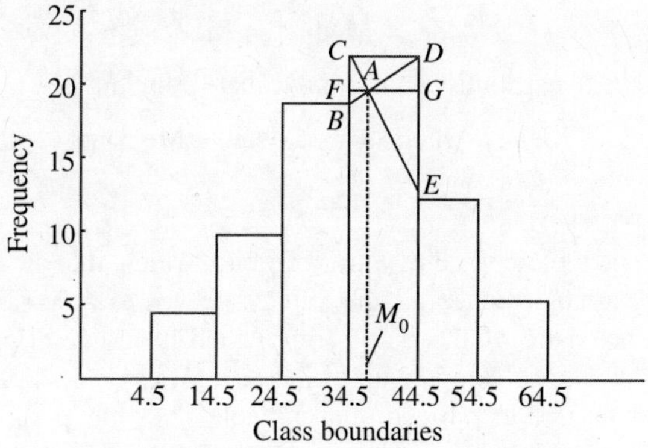

Fig. 3.4. Graphical determination of mode.

Merits and Demerits of Mode

Merits

1. It is easily comprehensible and can be computed easily.
2. It can be computed from frequency tables with open ends, if the mode does not lie in these classes.

Demerits

1. Many times there is no single mode. In such a case it is not uniquely defined.
2. No algebraic treatment is possible on modes from different groups of observations.

---

1. From the properties of similar triangles ABC and ADE

$$\frac{AF}{BC} = \frac{AG}{DE} \text{ or } \frac{AF}{\Delta_1} = \frac{AG}{\Delta_2} \text{ or } \frac{AF}{AG} = \frac{\Delta_1}{\Delta_2} \text{ or } \frac{AF}{AF + AG} = \frac{\Delta_1}{\Delta_1 + \Delta_2} \text{ or } AF = \frac{\Delta_1 x\, h}{\Delta_1 + \Delta_2}$$

Effect of the change of origin and scale of variable

Suppose X is transformed to $Y = \dfrac{X-A}{C}$ then

$$\text{Me}(Y) = \frac{\text{Me}(X) - A}{C} \tag{3.14}$$

and

$$M_0(Y) = \frac{M_0(X) - A}{C} \tag{3.15}$$

Thus mode and median are also transformed as the change of origin and scale in the variable.

Relation between Mean, Median and Mode

There is a relation between the three measures of location.

(a) If the distribution of variable is symmetric, all the three measures are equal i.e.

$$\text{Mean} = \text{Median} = \text{Mode}$$

(b) For a moderately skewed distribution, an approximate relation is

$$\text{Mean} - \text{Mode} = 3\ (\text{Mean} - \text{Median}) \tag{3.16}$$

3.2 Measures of Dispersion

Measures of central tendency indicate the values around which the observations are centered or located. The data is on a variable; hence all the values are not exactly same but vary. Variation in the values is another characteristic of the data series which needs description in a precise numeric value. These values enable a precise comparison of the data series on a single variable on two or more groups or of the data series on two or more similar variables. For this purpose measures of dispersion or scatter or variability are used.

Table 3.5 An example of the data series on four groups

Groups	Values of variable							Total	Mean
Group A	9	9	9	9	9	9	9	63	9
Group B	6	7	8	9	10	11	12	63	9
Group C	1	2	3	4	12	13	28	63	9
Group D	2	16						18	9

All the four series have the mean 9. Group A has no variation. Group B has variation but still all values are close to 9. Third group values are scattered quite much, ranging from 1 to 28. Fourth group has only two observations with mean 9. Nature of data is clearer if a measure of its dispersion or scatter are also given along with its measures of location.

In this section the following measures of dispersion are considered, which are commonly used

(a) Range (b) Standard deviation (c) Mean deviation and (d) Quartile deviation or semi-interquartile range.

(a) Range

It is the simplest measure and is used mostly in industrial work as in quality control. It is normally denoted as R and is defined as

$$R = \text{Maximum value} - \text{Minimum value}$$

It has the same unit as the unit of variable.

Though measure is simple but is highly susceptible to change in one or two extreme values. If there is even one very large or very small value, it changes the range disproportionately to rest of the observations.

For the four groups given above the ranges are:

Groups	Range
A	0
B	6
C	27
D	14

(b) Standard Deviation

If deviations of observations from their mean are calculated, it is well known that sum of these will always be zero. Hence average of deviations from mean will not give a measure of dispersion. To overcome this the deviation from mean are squared so that positive and negative deviations do not cancel, but all squares become positive. To bring the dimension of the squared variable to normal dimension the square root of the average of squared deviations is taken and is used as a measure of dispersion.

This measure is called standard deviation. It is denoted by σ in symbol. It is defined below for ungrouped or raw data and for a grouped data. It is most commonly used measure of dispersion in all scientific studies.

Ungrouped Data

Let $X_1, X_2, \ldots X_n$ be a set of observations on random variable X. Then standard deviation of X is defined as

$$\sigma = \left[\frac{1}{n} \sum_{i=1}^{n} (X_i - \bar{X})^2\right]^{\frac{1}{2}} \tag{3.17}$$

Square of Standard deviation σ^2 is called variance and can be written as

$$\sigma^2 = \left[\frac{1}{n} \sum_{i=1}^{n} (X_i - \bar{X})^2\right] \tag{3.18}$$

which can also be written as

$$\sigma^2 = \frac{1}{n}\sum_{i=1}^{n} X_i^2 - \bar{X}^2$$

$$= \left(\frac{\text{Sum of squares of } X}{\text{Number of observations}} \right) - \text{Square of mean of } X \tag{3.19}$$

Unit of σ^2 is square of unit of X. Thus if X's are in kg, σ^2 will be in kg^2. Similarly if X is in cm, σ^2 will have unit cm^2 and that of σ will be cm.

Grouped Data

Let there be a grouped data i.e. a frequency table with k classes having mid points $X_1, X_2, \ldots X_k$ and frequencies $f_1, f_2, \ldots f_k$ with $\sum_{i=1}^{n} f_i = N$.

Then σ^2 is defined as

$$\sigma^2 = \frac{1}{N} \sum_{i=1}^{k} f_i \left(X_i - \bar{X} \right)^2 \tag{3.20}$$

σ^2 can also be written as

$$\sigma^2 = \frac{1}{N} \sum_{i=1}^{k} f_i X_i^2 - \bar{X}^2 \tag{3.21}$$

The standard deviation σ is square root of variance σ^2 and is given by

$$\sigma = \left[\frac{1}{N} \sum_{i=1}^{k} f_i \left(X_i - \bar{X} \right)^2 \right]^{\frac{1}{2}}$$

$$= \left[\frac{1}{N} \sum_{i=1}^{k} f_i X_i^2 - \bar{X}^2 \right]^{\frac{1}{2}} \tag{3.22}$$

Computation of Standard Deviation

Ungrouped Data

Consider the following observations on X and it is desired to find σ.

X: 3, 6, 8, 11

To obtain σ, using formula (3.19), find $\sum_{i=1}^{4} X_i = 28$.

Since $n = 4$, $\bar{X} = 7$.

Now X^2: 9, 36, 64, 121, so $\sum_{i=1}^{4} X_i^2 = 230$

Hence $\sigma = \left[\frac{1}{4} \times 230 - 7^2 \right]^{\frac{1}{2}}$

$= [57.5 - 49]^{\frac{1}{2}}$

$= \sqrt{8.5} = 2.9$

Grouped Data

From formula (3.22) it is evident that calculations of σ requires $\sum_{i=1}^{k} f_i X_i^2$ and $\sum_{i=1}^{k} f_i X_i$.

$\sum_{i=1}^{k} f_i X_i$ is needed to compute $\bar{X}$ appearing in (3.22).

It has been seen that short cut method using a transformed variate U in place of X makes calculations of $\bar{X}$ very easy. It can be seen that the same method can be used for calculations of σ also. Using the transformed variable

$$U = \frac{X - A}{h},\ \bar{X} = A + h\bar{U}$$

Hence
$$X - \bar{X} = A + hU - (A + h\bar{U}) = h(U - \bar{U})$$

So,
$$\sum_{i=1}^{k} f_i (X_i - \bar{X})^2 = \sum_{i=1}^{k} f_i h^2 (U_i - \bar{U})^2$$

$$= h^2 \sum_{i=1}^{k} f_i (U_i - \bar{U})^2 \tag{3.23}$$

Dividing by $\sum_{i=1}^{k} f_i = N$ one obtains

$$\sigma_x^2 = h^2 \sigma_U^2 \tag{3.24}$$

where, σ_x^2 is variance of X and σ_U^2 is variance of U from the definition of variance given in equation (3.20). Now to obtain σ_U^2, again one can use:

$$\sigma_U^2 = \frac{1}{N}\sum_{i=1}^{k} f_i U_i^2 - \bar{U}^2, \text{ where } \bar{U} = \frac{1}{N}\sum_{i=1}^{k} f_i U_i \tag{3.25}$$

Hence it is seen that instead of performing computations on X_i the similar computations on the transformed variable U_i gives σ_U^2. Subsequently, σ_x^2 can be very easily obtained from (3.24).

As an illustration of this short cut method consider frequency table of Example 3.3.

Table 3.6 Short cut method of calculating σ^2

Classes	Mid points X_i	Frequency f_i	$U_i = (X_i - 39.5)/10$	$f_i U_i$	$f_i U_i^2$
1	9.5	5	– 3	– 15	45
2	19.5	10	– 2	– 20	40
3	29.5	20	– 1	– 20	20
4	39.5	22	0	0	0
5	49.5	13	1	13	13
6	59.5	5	2	10	20
	Total	$N = 75$		– 32	138

Calculation of σ_U^2 requires one more column of $f_i U_i^2$ and the table with short cut method will be as Table 3.6.

$$\sigma_U^2 = \frac{138}{75} - \left(\frac{-32}{75}\right)^2$$

$$= 1.84 - (-0.4267)^2 = 1.84 - 0.1821 = 1.6579$$

$$\sigma_x^2 = 100 \times 1.6579 = 165.79$$

$$\sigma_x = 12.876 \text{ years}$$

Properties of Standard Deviation

1. Standard deviation is independent of change of origin. In other words if a constant is added or subtracted from all the values of the variable X, the standard deviation of new variable will remain the same as that of X.

   This is because let $Y = X \pm C$

$$Y - \bar{Y} = (X \pm C) - (\bar{X} \pm C) = X - \bar{X}$$

Hence $\sigma_y^2 = \sigma_x^2$ so $\sigma_y = \sigma_x$.

2. If there is change of scale of variable X, standard deviation of resulting variable also changes proportionately. In symbols if

$$Y = CX$$

$$Y - \bar{Y} = C(X - \bar{X})$$

   Hence, $\sigma_y^2 = C^2\sigma_x^2$, so $\sigma_y = C\sigma_x$.

3. Standard deviation is root mean square deviation from mean, and is least among all the root mean square deviations from constant A. In symbols $\left[\frac{1}{n}\sum_{i=1}^{n}(X_i - A)^2\right]^{\frac{1}{2}}$ is least if A = $\bar{X}$.

4. It is uniquely defined and is based on all the observations.

5. If all the values of the variable are same, standard deviation will be zero. In symbols if

$$X_1 = X_2 = \ldots = X_n, \; \sigma_x = 0$$

6. If variances and means of *two series* are known the *variance of the combined series* can be obtained by an algebraic formula as follows.

Let σ_1^2 and σ_2^2 be the variances of two series of N_1 and N_2 observations with means M_1 and M_2, respectively. Then variance σ^2 of the combined series of $N = N_1 + N_2$ observations is given by

$$\sigma^2 = \frac{N_1\left(\sigma_1^2 + d_1^2\right) + N_2\left(\sigma_2^2 + d_2^2\right)}{N_1 + N_2} \tag{3.26}$$

where $d_1 = M_1 - M$ and $d_2 = M_2 - M$ and

$$M\text{: mean of the combined series} = \frac{N_1M_1 + N_2M_2}{N_1 + N_2}$$

This formula can be generalized for more than two (say r) series also, as follows:

Let σ_1^2, σ_2^2 ... σ_m^2 ... σ_r^2 be the variance of the r series of N_1, N_2, ... N_m ... N_r observations, respectively, having means M_1, M_2, ... M_m ... M_r.

Let d_1, d_2 ... d_m ... d_r be the deviations of the individual series' means from the combined series mean M given by

$$M = \left(\sum_{m=1}^{r} N_m M_m\right) \Big/ \left(\sum_{m=1}^{r} N_m\right)$$

i.e. $d_m = M_m - M$.

Then σ^2, the variance of the combined r series of $N = N_1 + N_2 \ldots + N_r$ observations is

$$\sigma^2 = \frac{1}{N}\left[\sum_{m=1}^{r} N_m\left(\sigma_m^2 + d_m^2\right)\right] \tag{3.27}$$

As is evident, standard deviation can be easily found by taking square root of σ^2 given in equations (3.26) and (3.27).

(c) Mean Deviation

Since sum of deviations of observations from their mean is zero, it can not be used as measure of dispersion. Standard deviation was obtained after getting rid of the negative sign of deviations by squaring them. Another way of overcoming this difficulty is to remove the negative signs by taking the magnitude of deviations without sign and using the arithmetic mean of these absolute deviations as measure of dispersion. This measure is called mean deviation (m.d.). It has the same unit as the variable and it is defined for ungrouped and grouped data as follows.

Ungrouped Data

Let X_1, X_2, ..., X_n be n observations on a random variable X, then

$$\text{m.d.} = \frac{1}{n}\sum_{i=1}^{n}\left|X_i - \bar{X}\right| \tag{3.28}$$

Grouped Data

Let there be k classes with mid points X_1, X_2, ... X_k and frequencies f_1, f_2, ... f_k. Then

$$\text{m.d.} = \frac{1}{N}\sum_{i=1}^{k} f_i\left|X_i - \bar{X}\right| \tag{3.29}$$

where $N = \sum_{i=1}^{k} f_i$.

Computation of Mean Deviation

Ungrouped Data

Computation of mean deviation can not be simplified either algebraically as by opening brackets in standard deviation, nor by short cut method using transformed variable from the original variable. For illustration of computation, the following examples are considered.

Example 3.7: Calculate mean deviation of the following observations

10, 4, 6, 11, 3, 0, 1

Solution: First we determine m.d. about mean, and next about median. So first find mean

$$\bar{X} = \left(\sum_{i=1}^{7} X_i\right)\Big/7 = 35/7 = 5$$

Absolute deviations from mean $|X_i - \bar{X}|$ = 5, 1, 1, 6, 2, 5, 4

$$\text{So m.d.} = \frac{1}{7}(5+1+1+6+2+5+4)$$

$$= \frac{24}{7} = 3.42$$

To obtain median the observations arranged in ascending order of magnitude are: 0, 1, 3, 4, 6, 10, 11.
So that median Me = 4

Absolute deviations about Me: $|X_i - \text{Me}| = |X_i - 4|$ are 6, 0, 2, 7, 1, 4, 3. Mean deviation about median

$$\text{md}_{\text{me}} = \frac{1}{7}\sum_{i=1}^{7}|X_i - M_e|$$

$$= \frac{1}{7}(6+0+2+7+1+4+3) = \frac{1}{7}(23)$$

$$= 3.28$$

It can be noted that mean deviation about median is smaller than mean deviation about mean. In fact mean deviation about median is least among mean deviations from any constant A.

i.e. $$\sum |X_i - \text{A}| \text{ is least if A} = \text{Me} \tag{3.30}$$

Grouped Data

Consider again frequency table of Example 3.3 as

Table 3.7 Calculation of m.d. from frequency table of age in years

Classes	Mid points X_i	Frequency f_i	$\lvert X_i - \bar{X}\rvert$	$f_i\,\lvert X_i - \bar{X}\rvert$
1	9.5	5	25.73	128.65
2	19.5	10	15.73	157.3
3	29.5	20	5.73	114.6
4	39.5	22	4.27	93.94
5	49.5	13	14.27	185.51
6	59.5	5	24.27	121.35
	Total	75		801.35

As already computed $\overline{X}$ = 35.23 years. From equation (3.29)

$$\text{m.d.} = \frac{1}{N}\sum_{i=1}^{k} f_i \left|X_i - \overline{X}\right|$$

$$= \frac{1}{75} \times 801.35 = 10.68 \text{ years}$$

Properties of Mean Deviation

1. It is uniquely defined and is based on all the observations.
2. Mean deviation about median is least among the mean deviations about any value.
3. It is not as easy to compute as standard deviation.
4. For a symmetrical or moderately skew distribution the mean deviation is about four fifth of the standard deviation.

   In symbols
   $$\frac{\text{m.d.}}{\text{s.d.}} \approx \frac{4}{5} \tag{3.31}$$

5. Mean deviation is not suitable for algebraic treatment. Thus mean deviation of combined series can not be determined from the mean deviations of the component series.

(d) Quartile Deviation or Semi-interquartile Range

If there is an open end or both ends are open in a frequency table none of the above three measures of dispersion can be calculated, because neither the mid points of all the classes nor minimum and/or maximum values of variable are known. In such a case only this measure of dispersion can be calculated. This measure of dispersion is not based on deviations of observations from any specific value of the variable but is based on values of quartiles and is called quartile deviations (Q.d.) or semi-interquartile range. It is defined as

$$\text{Q.d.} = \frac{Q_3 - Q_1}{2} \tag{3.32}$$

where, Q_3 is the third quartile of the distribution of variable and
Q_1 is the first quartile of the distribution

A definition of the quartile is appropriate at this place. Quartiles are the values of the variable which divide the distribution in four equal parts. One can recollect that median is the value of variable that divides the frequency distribution in two equal parts. Thus 2/4 (= ½) of the observations lie below median and an equal number of observations lie above it. Thus median can be well understood as second quartile of a frequency distribution. The three quartiles Q_1, Q_2 and Q_3 are given in Fig. 3.5.

Thus, in Fig. 3.5, the three quartiles Q_1, Q_2 and Q_3 divide the area under frequency curve in four equal parts.

First quartile Q_1, is therefore, the value of the variable such that one fourth of the total observations lie below it. Third quartile Q_3 is the value of the variable such that three fourth of the total observations lie below it.

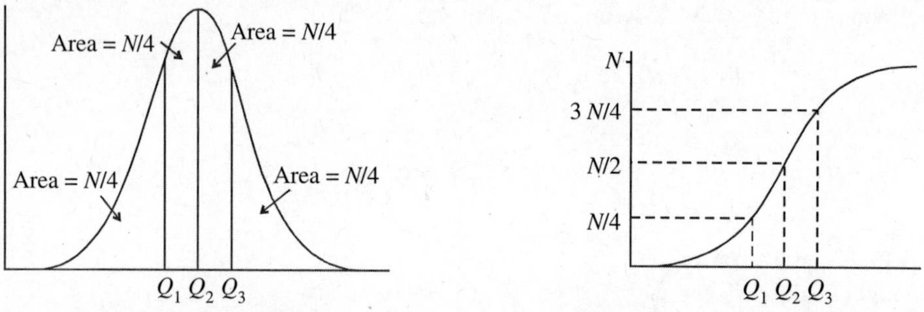

(a) Quartiles in frequency curve (b) Ogive (less than type) giving quartiles

Fig. 3.5. Quartiles from frequency curve and from less than type ogive.

Computation of Quartile Deviation

The steps for finding median, which is second quartile is followed for calculating the first and third quartiles. Thus the methods for ungrouped and grouped data are given below.

Ungrouped Data

Let $X_1, X_2, \ldots, X_n$ be a set of n ungrouped observations from X. As a first step these are arranged in ascending order of magnitude as $X_{(1)}, X_{(2)}, \ldots X_{(i)} \ldots X_{(n)}$. If n is divisible by four, the $(n/4)^{th}$ value will be Q_1 and $(3n/4)^{th}$ value will be Q_3.

Example 3.8: Following are the strengths of right grip scores (kilograms) of 12 post pubescent adolescent males: 30, 21, 62, 18, 53, 29, 65, 19, 62, 56, 63, 64. It is desired to compute quartile deviation (source: Daniel (1987), p. 26, Exercise 1.7.5).

Solution: As a first step the observations are arranged in order of magnitude:

18, 19, 21, 29, 30, 53, 56, 62, 63, 64, 65

In this example n is divisible by four.

$$\frac{n}{4} = \frac{12}{4} = 3 \quad \text{and} \quad \frac{3n}{4} = 9$$

21 is the 3rd observation and 62 the 9th observation.

Any value between 21 to 29 will have 3 observations below it.

So one may take average of 21 and 29 (as was done for median if n was even)

Thus to get Q_1

$$Q_1 = \frac{21+29}{2} = 25 \text{ kg}$$

and

$$Q_3 = \frac{62+63}{2} = 62.5 \text{ kg}$$

and the

$$\text{Q.d.} = \frac{62.5-25}{2} = \frac{37.5}{2} = 18.75 \text{ kg}$$

Grouped Data

For data grouped in frequency table, the same process is followed as to find the median or Q_2. Thus the formula for Q_3 and Q_1 are:

$$Q_1 = l_1 + \frac{\frac{N}{4} - F_1}{f_1} \times h \tag{3.33}$$

where l_1 = Lower boundary of the first quartile class
f_1 = Frequency of the first quartile class
F_1 = Cumulative frequency of the class preceding first quartile class (or cumulative frequency less than l_1)
h = Width of the first quartile class

$$Q_3 = l_3 + \frac{\frac{3N}{4} - F_3}{f_3} \times h \tag{3.34}$$

where l_3 = Lower boundary of the third quartile class
f_3 = Frequency of the third quartile class
F_3 = Cumulative frequency less than l_3
h = Width of the third quartile class

The first quartile class is one where the cumulative frequency less than ($F_<$), just exceeds $N/4$ and the third quartile class is one where this cumulative frequency just exceeds $3N/4$.

Table 3.8 Frequency table of age in years

Class boundaries	Frequency f_i	Cumulative frequency ($F_<$)
4.5-14.5	5	5
14.5-24.5	10	15
24.5-34.5*	20	35
34.5-44.5**	22	57
44.5-54.5	13	70
54.5-64.5	5	75
Total	$N = 75$	

Note: * shows Q_1 class and ** shows Q_3 class

As a numerical illustration consider again frequency table of Example 3.3.

$$\frac{N}{4} = \frac{75}{4} = 18.75 \quad \text{and} \quad \frac{3N}{4} = \frac{3 \times 75}{4} = 56.25$$

Looking at cumulative frequencies in third column and the values of $N/4$ and $3N/4$, it is found that $F_<$ in the class 24.5 – 34.5 (= 35) just exceeds $N/4$ (= 18.75) and $F_<$ in the class 34.5 – 44.5 (= 57)) just exceeds $3N/4$ (= 56.25). So 24.5 – 34.5 is Q_1 class and 34.5 – 44.5 is Q_3 class.

Hence $l_1 = 24.5$ and $l_3 = 34.5$, $f_1 = 20$, $f_3 = 22$

$$F_1 = 15,\ F_3 = 35,\ h = 10$$

From (3.33) and (3.34)

$$Q_1 = 24.5 + \frac{18.75 - 15}{20} \times 10$$

$$= 24.5 + \frac{3.75}{2} = 24.5 + 1.875 = 26.375 \text{ years}$$

$$Q_3 = l_3 + \frac{\frac{3N}{4} - F_3}{f_3} \times h$$

$$= 34.5 + \frac{56.25 - 35}{22} \times 10 = 34.5 + \frac{21.25}{22} \times 10$$

$$= 34.5 + \frac{212.5}{22} = 34.5 + 9.659 = 44.159 \text{ years}$$

So $$\text{Q.d.} = \frac{44.159 - 26.375}{2} = \frac{17.784}{2} = 8.892 \text{ years}$$

In the next example consider a frequency table with open ends.

Table 3.9 Hb level (gm %) of college students

Hb level (gm %)	No. of students	Class boundaries	$F_<$
≤ 10.0	10	≤ 10.05	10
10.1-11.0	21	10.05-11.05	31
11.1-12.0*	67	11.05-12.05	98
12.1-13.0##	170	12.05-13.05	268
13.1-14.0**	84	13.05-14.05	352
14.1-15.0	29	14.05-15.05	381
> 15.0	4	≥ 15.05	385

Note: * Q_1 class, ## median (Q_2) class and modal class, ** Q_3 class

Example 3.9: Following data gives the distribution of college students according to their Hb level in gm %. To find out suitable measure of location and dispersion.

Solution: Since the frequency table has open ends mean, standard deviation, mean deviation and range can not be calculated. Mode and median as measures of central tendency and only quartile deviation can be calculated for measuring dispersion. In this example median and quartile deviation will be calculated. Former as measure of location and the latter as measure of dispersion. Thus all the quartiles will be calculated, which require class boundaries and cumulative frequency less than (= $F_<$). Column 3 of the above table contains the class boundaries and fourth column has the values of $F_<$.

$$N = 385,\ N/4 = 385/4 = 96.25 \text{ and } 3N/4 = 3 \times 385/4 = 288.75$$

$$N/2 = 385/2 = 192.5$$

Median is given by

$$Q_2 = \text{Me} = l_0 + \frac{\frac{N}{2} - F_<}{f_m} \times h$$

$$= 12.05 + \frac{192.5 - 98}{170} \times 1$$

$$= 12.05 + \frac{94.5}{170} = 12.05 + 0.559 = 12.609 \text{ gm } \%$$

$$Q_1 = l_1 + \frac{\frac{N}{4} - F_1}{f_1} \times h$$

$$= 11.05 + \frac{96.25 - 31}{67} \times 1$$

$$= 11.05 + \frac{65.25}{67} = 11.05 + .974 = 12.024 \text{ gm } \%$$

$$Q_3 = l_3 + \frac{\frac{3N}{4} - F_3}{f_3} \times h$$

$$= 13.05 + \frac{288.75 - 268.00}{84} \times 1$$

$$= 13.05 + \frac{20.75}{84} = 13.05 + .247 = 13.297 \text{ gm } \%$$

Hence

$$\text{Q.d.} = \frac{13.297 - 12.024}{2}$$

$$= \frac{1.273}{2} = .6365 \text{ gm } \%$$

For a frequency table with open ends, if mode is not in the open class, this also can be calculated as measure of central tendency. It can be verified that modal class is same as the median class in this table. Hence

$$M_0 = l_0 + \frac{\Delta_1}{\Delta_1 + \Delta_2} \times h$$

$$l_0 = 12.05,\ \Delta_1 = 170 - 67 = 103,\ \Delta_2 = 170 - 84 = 86$$

Hence

$$M_0 = 12.05 + \frac{103}{103 + 86} \times 1 = 12.05 + \frac{103}{189} = 12.05 + .580$$

$$= 12.63 \text{ gm } \%$$

Relationship between Quartile Deviation and Standard Deviation

For a moderately skewed distribution, quartile deviation is about 2/3 of the standard deviation i.e.

$$\frac{Q.d.}{s.d.} = \frac{2}{3} \quad \text{or} \quad Q.d. = \frac{2}{3} \times \sigma$$

For the data in Example 3.3

$$\frac{Q.d.}{s.d.} = \frac{8.892}{12.876} = .6906 \cong \frac{2}{3}$$

3.3 Measures of Skewness and Kurtosis

As already mentioned departure from symmetry of the distribution of variables is called skewness. A histogram or frequency polygon or frequency curve enables to visualize that skewness is present or not. Skewness or asymmetry may be either of two types, positive or negative.

Positive Skewness

The curve has a long tail on the right side and a greater concentration of observations on the left of the frequency curve. (see Figure 3.2 (D)). The distribution is then said to be positively skewed.

Negative Skewness

If the left tail is longer and there is a high concentration of frequency on the right side of the frequency curve, distribution is said to be negatively skewed (see Figure 3.2 (D)).

Kurtosis

Even though two or more distributions may have same locations, the dispersions may also be almost same, they may be symmetric about their means but still differ in peakedness. One may be almost bell shaped or normal, the other flat like a plato and the third may have a steep peak, like the high neck of a Kangaroo. This nature of a curve is called kurtosis.

Just as a precise numeric measures of location and dispersion have been defined and discussed, precise measures of skewness and kurtosis are also needed for proper comparison. These measures are based on moments of a distribution. So before defining these measures, first a few lines about moments at this place are essential.

3.3.1 Moments

The word moment in statistics has been derived from its origin in physics which talks about moments of forces. The moment of a force about a certain point is product of the force and its distance from the point.

In statistics moments are obtained about any arbitrary value (say $X = A$) or about the origin ($X = 0$) or about the mean. Let us consider the moments about origin, called raw moments and about the mean called central moments.

r^{th} raw moment of X denoted by μ'_r is defined for ungrouped and grouped data as follows:

(a) Ungrouped Data

Let $X_1, X_2 \ldots X_n$ be the raw data on a random variable X. Then

$$\mu'_r = \frac{1}{n}\sum_{i=1}^{n} X_i^r \tag{3.40}$$

(b) Grouped Data

Let $X_1, X_2 \ldots X_k$ be the mid points of the k classes and $f_1, f_2, \ldots, f_k$ be the frequencies of these classes, respectively. Then

$$\mu'_r = \frac{1}{N}\sum_{i=1}^{k} f_i X_i^r, \text{ where } N = \sum_{i=1}^{k} f_i \tag{3.41}$$

Hence

$$\mu'_1 = \frac{1}{n}\sum_{i=1}^{n} X_i \quad \text{for ungrouped data}$$

$$= \frac{1}{N}\sum_{i=1}^{k} f_i X_i \quad \text{for grouped data} \tag{3.42}$$

which is same as the arithmetic mean. Thus arithmetic mean is first moment about origin.

r^{th} Central moment of X is denoted by μ_r and is defined as

$$\mu_r = \frac{1}{n}\sum_{i=1}^{n} (X_i - \mu_1')^r \quad \text{for ungrouped data}$$

$$= \frac{1}{N}\sum_{i=1}^{k} f_i (X_i - \mu_1')^r \quad \text{for grouped data} \tag{3.43}$$

Let the arithmetic mean of X be denoted by μ

Now putting $r = 2$

$$\mu_2 = \frac{1}{n}\sum_{i=1}^{n} (X_i - \mu)^2 \quad \text{for ungrouped data}$$

$$= \frac{1}{N}\sum_{i=1}^{k} f_i (X_i - \mu)^2 \quad \text{for grouped data} \tag{3.44}$$

is defined earlier as σ^2. Hence the second central moment is variance, the measure of dispersion.

Some results:

(i) μ_1' = arithmetic mean = μ (if deviations of X are taken from the origin)
(ii) $\mu_1 = 0$ (always) since sum of deviations from mean is always 0.
(iii) μ_2 = variance.

It may be noted that the 3rd and 4th central moments μ_3 and μ_4 are used to define measures of the skewness and kurtosis.

Relationship between Raw Moments and Central Moments

For

$r = 1$ $\mu_1 = 0$ $\quad$ $\mu_1' = \mu$ (Mean if deviations are from origin) (3.45)

$r = 2$ $\mu_2 = \mu_2' - \mu^2$ $\quad$ $\mu_2' = \mu_2 + \mu^2$ (3.46)

$r = 3$ $\mu_3 = \mu_3' - 3\mu\, \mu_2' + 2\mu^3$ $\quad$ $\mu_3' = \mu_3 + 3\mu\, \mu_2 + \mu^3$ (3.47)

$r = 4$ $\mu_4 = \mu_4' - 4\mu\, \mu_3' + 6\mu^2\, \mu_2' - 3\mu^4$ $\quad$ $\mu_4' = \mu_4 + 4\mu\, \mu_3 + 6\mu^2\mu_2 + \mu^4$ (3.48)

For calculation of 3rd and 4th central moments all the first four raw moments are calculated first. If data is frequency table with equal class intervals the variable X is transformed into $U = (X - A)/h$, where A is mid point of a class. This gives U as simple integers, positive and negative, as illustrated for calculation of standard deviation.

The central moments can be evaluated from all the four raw moments of U, using the relation

$$X - \mu = hU + A - \mu = hU - d \text{ where } d = \mu - A,\ \mu'_{ru} = r^{th} \text{ raw moment of } U$$

$$\mu_1 = h\bar{U} - d = 0 \Rightarrow \mu = A + h\bar{U} \;\therefore\; d = \mu - A = h\bar{U} \text{ and } X - \mu = h(U - \bar{U})$$

Hence $\mu_r = h^r \mu_{ru}$ and for $r = 2, 3, 4$ central moments of X can be evaluated from raw moments of U using (3.46), (3.47), and (3.48).

3.3.2 Sheppard Correction for Moments

The moments are calculated assuming that all the observations in a class have value equal to its class mark (or mid point). This is just an approximation and not a correct situation. Hence Sheppard suggested some corrections in the moments from grouped data to remove error due to grouping. The corrections suggested are as follows

$$\mu_2 \text{ (corrected)} = \mu_{2\,cal} - h^2/12$$

$$\mu_3 \text{ (corrected)} = \mu_{3\,cal}$$

$$\mu_4 \text{ (corrected)} = \mu_{4cal} - \frac{1}{2}h^2\mu_2 + \frac{7}{240}h^4 \qquad (3.49)$$

These corrections are used for variables which are continuous and for moderately skewed distributions i.e. when tails on both sides decrease to zero.

3.3.3 Measures of Skewness and Kurtosis

(a) Measures based on the Moments

$$\text{Coefficient of skewness} = \sqrt{\beta_1} = \gamma_1 \qquad (3.50)$$

Here $\beta_1 = \dfrac{\mu_3^2}{\mu_2^3}$

Hence $\sqrt{\beta_1} = \dfrac{\mu_3}{\mu_2^{3/2}} = \dfrac{\mu_3}{\sigma^3}$

Coefficient of kurtosis $= \beta_2 - 3 = \gamma_2$

where
$$\beta_2 = \frac{\mu_4}{\mu_2^2} \qquad (3.51)$$

It may be noted that both these measures are unit free numbers. $\sqrt{\beta_1}$ takes the sign of μ_3 as σ is always positive.

γ_1 has positive value if μ_3 is positive, the distribution is then a positively skewed i.e. with long right tail. γ_1 is negative if μ_3 is negative, the distribution is then negatively skewed and has long left tail.

γ_1 is zero if μ_3 is zero, the distribution is then symmetric about mean.

If $\beta_2 > 3$, the distribution is leptokurtic i.e. has a high peak.
If $\beta_2 < 3$, the distribution is platykurtic i.e.. has flat shape.
If $\beta_2 = 3$, distribution is mesokurtic i.e. has normal peak.

(b) Other Measures of Skewness

(i) Karl Pearson defined a measure of skewness based on distance of mean and mode and of mean and median. The first coefficient of skewness S_k is given by

$$S_{k1} = (\text{Mean} - \text{Mode})/\sigma \tag{3.52}$$

For a moderately skewed distribution

$$\text{Mean} - \text{Mode} = 3(\text{Mean} - \text{Median})$$

So a second coefficient of skewness is

$$S_{k2} = 3(\text{Mean} - \text{Median})/\sigma \tag{3.53}$$

These coefficients are positive and negative if distributions are positively or negatively skewed as is evident from the following figures.

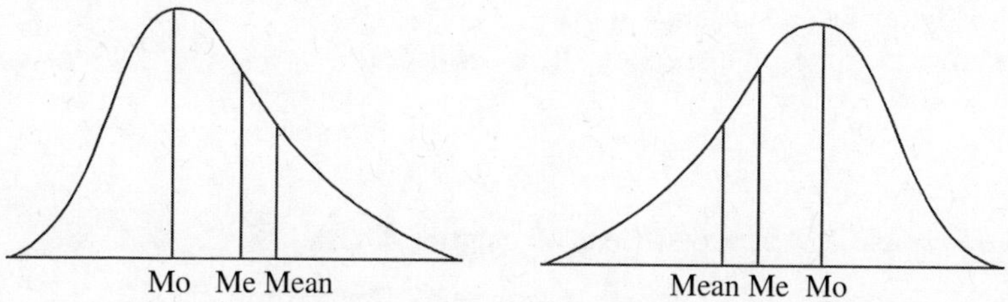

Fig. 3.6. Mean, median and mode in moderately skewed distributions (source: Yule and Kendall, p. 109).

Third measure of skewness is defined as

$$S_{k3} = \frac{(Q_3 - M_e) - (M_e - Q_1)}{(Q_3 - M_e) + (M_e - Q_1)}$$

$$= \frac{Q_3 + Q_1 - 2M_e}{Q_3 - Q_1} \tag{3.54}$$

3.4 Absolute Measures of Dispersion

All the measures of skewness and of kurtosis are unit free hence are comparable for distribution of various variables. The measures of dispersion are all with unit of the variable. Standard deviation of weight (kg) can not be compared with standard deviations of hemoglobin level in gm %. For proper comparison it is appropriate that a unit free measure of dispersion is defined. Two such measures are defined below:

(a) Coefficient of Variation (CV)

It is defined as

$$CV = \frac{\sigma}{\mu} \times 100 \tag{3.54}$$

Coefficient of variation is expressed as percentage. It is unit free and can be compared as two numeric quantities.

Example 3.10: Suppose each day a technician A completes on the average 40 analyses with standard deviation of 5. Technician *B* completes per day on the average 160 analyses with standard deviation 15.

On the first glance it appears that technicians *B* has 3 times more variability in the number of analyses per day. But instead of comparison of s.d., coefficients of variation should be computed for comparison, since technician *B* is four times faster.

$$\text{CV (A)} = \frac{5}{40} \times 100 = 12.5\%$$

$$\text{CV (B)} = \frac{15}{160} \times 100 = 9.4\%$$

Though technician *B* has more standard deviation, relative variation of *A* is more.

(b) Coefficient of Quartile Deviation (c.q.d.)

It is another relative measure of dispersion. It is defined as

$$\text{c.q.d} = \frac{Q_3 - Q_1}{Q_3 + Q_1} \tag{3.55}$$

For the data of Example 3.9 the coefficient of quartile deviation is

$$\text{c.q.d} = \frac{13.299 - 12.024}{13.299 + 12.024} = \frac{1.275}{25.323}$$

$$= 0.05$$

4. Concepts of Probability

Probability may be considered to be the basic foundation of statistical science. Especially it is an indispensable tool of statistical inference. Decisions for future steps in any discipline by the project manager or the engineer of a manufacturing process or administrative head are taken on the basis of conclusions from statistical inference with probabilistic support.

Probability is not used by statisticians alone. Even in day to day life, people talk about chances of occurrence of events of interest. For example, a school teacher may tell students that it is very likely that tomorrow will be a rainy day of school so the theorem of geometry must be finished today.

Cricket lovers may bet that chances are very high that team A will beat team B. Crop forecasters may predict that it is very likely that this year paddy crop will give about one and half times more yield than that of the last year.

A friendly advice may be given by a tourist to another that chances are high that an air ticket is available for tomorrow in an air line A among seven others for Delhi from Toronto.

In all of these a statement about chance is present, which the statisticians and mathematicians named as probability. They assign a numeric value called probability to the chance of occurrence of an event. If an event is almost impossible to occur (e.g. a candidate for election is almost sure not to win) its chance of occurrence is assigned a number close to zero and if the event is almost sure to occur it will be assigned a value close to one. In other words probability of getting 8 points in a throw of dice is zero, whereas getting less than or equal to 6 points, has probability one.

The theory of probability which forms the base of statistical inference was considered a branch of mathematics and had its formal beginning with games of chance in seventeenth century.

In this chapter relationship of uncertainty and probability is explained. Some different meanings attached to probability and formal definition with three approaches are given. Its basic properties and some rules to calculate probability of events are explained. For the theoretical concepts and the numerical examples the books by Meyer (1975), Uspensky (1937), Goon et al. (1983), and Milton and Tsokos (1983) may also be referred.

4.1 Uncertainly and Random Experiments

There are some phenomena which can be explained as a natural consequence of the existing conditions. For example an iron piece if pushed forward beyond the edge of table will fall down on earth. Every day the sun rises, darkness of night disappears and in the evening, in tropical countries the sun sets and darkness again wraps the surroundings. This is because of the natural rotation of earth on its axis in about 24 hours. For all practical purposes the events caused by natural laws are certain or definite to occur.

Thus position of the smooth ball on the plane surface moving with a constant speed of 4 ft/second can be correctly found after 10 seconds, from the formula $S = u\,t$. If a gas is compressed with a definite pressure its final volume can be determined from the formula PV^{γ} = a constant. Thus if initial volume V_1 and initial pressure P_1 is known, the final volume V_2 under a pressure P_2 can be calculated. If a battery is inserted into a simple circuit, the observable flow of current can be determined from the Ohm's Law as $I = E/R$, if E and R are given.

In all these experiments, if the initial conditions are known, the outcome or the result of the experiment is known. One can predict the outcome of the experiment even without performing the experiment. Such experiments are known as *deterministic experiment.* There are many phenomena or experimental situations the outcomes of which can be precisely predicted with certainty.

However, there are also conditions when results or outcomes of an experiment or a natural phenomenon can not be predicted with certainty. For example when throwing a dice, tossing a coin, drawing a card from a well shuffled deck, drawing a ball blindfolded from an urn containing 10 red, 40 black and 50 white balls of same size and smoothness, the outcome of the experiment can not be predicted.

In the science of genetics, it is uncertain whether a pea plant from a cross of round and wrinkled seeds will have round seeds.

Outcome of such experiments have an element of uncertainty associated with it. But it is quite obvious that there exists a long term regularity in these outcomes.

If the die is homogeneous and cubic, one sixth of the times a six will appear. If the coin is balanced, symmetrical, not bent, about one half of the tosses will give heads. About one fourth of the cards drawn (after replacement) will be of a given suit (say heart). About one tenth of the balls drawn (after replacement and mixing) will be red and about one half will be white. About three fourth of the pea plants in the above genetical experiment will have round seeds.

A life insurance company can find exactly how many of its clients will attain age 60 years in the coming calendar year. It can also predict almost accurately what proportion will die within one year of attaining 60 years, but can not predict whether a particular client (say A) of the said age will or will not die in the following one year.

These types of experiments with uncertain outcomes are called *random* or *non-deterministic experiment.*

The deterministic experiments have the idea of certainty associated with them, whereas the random or non-deterministic experiments have closely associated with them an idea of chance. The words 'probability' and 'probable' used in common language have different shades of meaning. Saying that "Probably it will rain tomorrow" means that there are more signs of rainy weather than fair for tomorrow. In the statement "There is little probability in the cause of his absence", the word probability is in the sense of credibility.

The word probability may be considered as a measure or degree of *credence* or *certainty* with which an event may materialize. This notion meets the almost instinctive desire to compare probabilities of different events. Consider the following example to confirm that such a comparison is possible.

A person living in the 3rd floor can reach the ground either by using the stairway or by jumping from the window. How do the probabilities of being injured in two cases compare? Quite obviously every one will say that probability of getting injured by jumping is much greater than the probability of getting injured by using the stairway to walk down to ground.

As another example consider an urn containing an equal number of white and black balls similar to each other except the colour. One can say without hesitation that probabilities of drawing a white or black ball from this urn are equal.

If another urn contains 1 black and 10 white balls then one will again say without hesitating that probability of drawing a white ball is much greater than that of drawing a black ball.

Following are some of the examples of random experiments denoted by E with appropriate suffix. It may be noted that while describing a random experiment, one should also specify what is being observed apart from describing the process of performing the experiment.

E_1: Throw a die and observe the number of the points on the upper face.
E_2: Toss a coin three times and count the number of heads.
E_3: Toss a coin three times and observe the sequence of heads and tails
E_4: Count the number of defective items in the lot of items produced in a production line during a 24 hour period.
E_5: Count the defective rivets in the assembly of rivets in an airplane wing.
E_6: A light bulb is tested for its life length by putting it to use till it burns out
E_7: A board (2 ft^2) has a circle of 3" diameter, second concentric one of 6" diameter with centre at intersection of diagonals of board to be used for an artillery competition. The distance where the bullet hits from the centre is recorded.
E_8: In the set up for E_7 the distances from horizontal (x) and vertical (y) lines from the center of the circles are measured.
E_9: A missile is launched and at time t, velocity components in three directions (horizontal, vertical and Z-axis) are observed.
E_{10}: In the set up of E_9 the heights of missile at times t_1, t_2, ... t_n are observed.
E_{11}: At a specified locality a thermograph records temperatures over a 24 hours period.
E_{12}: In the set up of E_{11}, the maximum and minimum temperatures x and y are read.
E_{13}: Count the number of α-particles emitted by a radioactive source in 30 minutes.

These experiments are all different but have some things in common. The following are the common features:

(i) Unless the experiments are performed, the outcome of a particular trial can not be stated. However, one can describe the set of all possible outcomes of the experiments.
(ii) Each experiment can be repeated indefinitely, with an effort to maintain similar conditions.
(iii) The outcomes of the experiment appear to occur haphazardly. But after a large number of repetitions a pattern or regularity begins to emerge. It is this regularity which makes it possible to attach a measure of certainty (called probability) with various outcomes of the experiment.

Before defining probability few more related concepts are discussed in the next section.

4.2 Sample Space and Events

Probability is always talked for one or the other event. Hence a proper definition of event is required before defining its probability. Event is related to the concept of sample space. So, first a definition and subsequently some examples of sample space are given.

For defining events and sample space some notions about set theory are essential and may be looked into the Appendix A, if the reader is not already having its knowledge.

4.2.1 Definition of Sample Space

A set of all possible outcomes of a random experiment E, is called sample space and is denoted by Ω. Each of the above random experiments are again considered and their sample spaces are given as:

Ω_i is sample space of E_i. To describe a sample space, all the elements (outcomes of the random experiment) are listed and put in braces. Thus

$\Omega_1 = \{1, 2, 3, 4, 5, 6\}$
$\Omega_2 = \{0, 1, 2, 3\}$
$\Omega_3 = \{HHH, HHT, HTH, THH, HTT, THT, TTH, TTT\}$
$\Omega_4 = \{0, 1, 2, 3, ..., N\}$, where N is the maximum number of items that could be produced in 24 hours duration.
$\Omega_5 = \{0, 1, 2, ..., K\}$, where K is the number of rivets installed on the airplane wing.
$\Omega_6 = \{t \mid t \geq 0\}$
$\Omega_7 = \{d \mid 0 \leq d \leq 12\sqrt{2}\}$
$\Omega_8 = \{(x, y) \mid -12 \leq x \leq 12, -12 \leq y \leq 12\}$
$\Omega_9 = \{v_x, v_y, v_z\}$, where all three are real numbers.
$\Omega_{10} = \{h_1, h_2, ..., h_n \mid h_i \geq 0, i = 1, 2, 3, ..., n\}$
Ω_{11} = A thermostat graph will be a continuous function f between the minimum (m) and maximum (M) possible temperatures of the locality. So $\{f \mid m \leq f(t) \leq M\}$
$\Omega_{12} = \{x, y \mid m \leq x \leq y \leq M\}$
$\Omega_{13} = \{0, 1, 2, ..., \infty\}$

Looking at the above examples of sample space the following characteristics may be noted:

(a) To describe a sample space associated with a random experiment. One must be not only clear about how the experiment is being performed, but also about what is being observed or measured, since outcomes of the random experiment depend on what is being observed.
Thus E_2 and E_3 are same experiment of tossing a coin three times, but the observations noted being different, these are designated as two experiments and the sample spaces Ω_2 and Ω_3 have different elements since the set of outcomes are different.
(b) Sample space may consist of non-numeric elements also, as in Ω_3.
(c) Sample space may consist of discrete elements as in Ω_1, Ω_2, Ω_4, Ω_5, and Ω_{13}. The number of the outcomes may be finite in number as in Ω_1, Ω_2, Ω_4 and Ω_5 or it may be countably infinite as in Ω_{13}.
(d) Sample space may be a function as the Ω_{11}.
(e) Sample space may be continuous (non-countably infinite) also e.g. Ω_6, Ω_7, Ω_8, Ω_9, Ω_{10}, Ω_{11}, Ω_{12}. A continuous sample space may be finite (as Ω_7, Ω_8, Ω_{11} and Ω_{12}) or infinite as (Ω_6, Ω_9 and Ω_{10}).
(f) A sample space may be of one dimension (all except Ω_8, Ω_9, Ω_{10}, and Ω_{12}) or may be two dimensional, as Ω_8 and Ω_{12} or may be three dimensional as Ω_9 or even multidimensional as Ω_{10}, which is n-dimensional.
(g) A sample space may be mathematically idealized one or an experimentally realizable one. The sample spaces Ω_6 and Ω_{13} are mathematically idealized ones. The measuring instrument for a continuous variable (time till the bulb burns out) has an accuracy and therefore, experimentally the times measured will be discrete values. But actually a continuous variable like time changes continuously not at intervals of two or three or more decimal places.

Similarly a bulb can not function for an infinite time, nor a radioactive material can emit infinite number of α particles in 30 minutes. These values may be very large and hence for mathematical convenience and simplicity these are taken as $t \geq 0$ and $t \leq \infty$, to accommodate all the possibilities.

The sample spaces considered henceforth will be any of the above types and mathematically simplest and most convenient.

4.2.2 Definition of Events

An event is a subset of sample space possessing some properties.

Capital letters A, B, C, ... are used for denoting events. Events are always associated with a random experiment (E) and there are several events associated to a given random experiment E.

As examples of some events consider the following events A_i associated with random experiments E_i ($i = 1, 2, 3, ..., 13$).

A_1: An even number appears on upper face, i.e. $A_1 = \{2, 4, 6\}$
A_2: Number of heads is 3 i.e. $A_2 = \{3\}$
A_3: More heads appear than tails, i.e. $A_3 = \{HHH, HHT, HTH, THH\}$
A_4: No defective item produced in 24 hours i.e. $A_4 = \{0\}$
A_5: The number of defective rivets of airplane are between 5 to 8 i.e. $A_5 = \{5, 6, 7, 8\}$
A_6: The light bulb burns out in less than 5 hours so A_6: $\{t \mid t < 5\}$
A_7: The bullet hits between the two circles i.e. A_7: $\{d \mid 3 < d < 6\}$
A_8: The bullet hits 4" on the left of center at a distance of 2" to 3" from the horizontal line from the center A_8: $\{(x, y) \mid x = -4\ 2 < y < 3\}$
A_{12}: Maximum temperature is 10° F more than minimum temperature i.e. A_{12} $\{x, y \mid y = x + 10\}$
A_{13}: α-particles emitted are more than 5 so A_{13}: $\{5, 6, 7, ...\infty\}$

After defining an event, various types of events, are defined now, which occur commonly.

4.2.3 Types of Events (Definitions)

1. Simple Events

If the event is a subset consisting of only one element of finite or countably infinite sample space (or consists of a single outcome of the random experiment), the event is called a simple event. For example:

(a) Getting a six on throw of a dice.
(b) Getting all heads in three tosses of a coin.
(c) Getting four defective rivets in airplane wing.

Simple events $A_i = \{w\}$

2. Mutually Exclusive Events

Two or more events are said to be mutually exclusive (m.e.), if no two of them can occur together.

Thus A and B are mutually exclusive if they can not occur together i.e. $A \cap B = \phi$ (the null set) indicating that there are no points common in the sets A and B.

As some examples, consider

(a) Getting head and tail in a toss of coin are mutually exclusive events.
(b) Getting 1 and 6 in a throw of dice are mutually exclusive events.
(c) Getting 1, 2, 3, 4, 5 and 6 in a throw of dice are six mutually exclusive events.

(d) Getting of a white and a black pigeon in one selection, from a cage of pigeons are two mutually exclusive events.
(e) An offspring being a male and a female are mutually exclusive events.

3. Exhaustive Events

A set of events are said to be exhaustive, if one of them must occur. For example:

(a) Getting a head and getting a tail in toss of a coin are exhaustive events.
(b) In a toss of two coins

 (i) Let A_1 = Getting no head, A_2: Getting one head, A_3: Getting two heads. Then A_1, A_2, A_3 are exhaustive, as well as mutually exclusive events.
 (ii) Let B_1: {TT}-both coins fall tail
  B_2: {HT}-1st coin has head and 2nd has tail
  B_3: {TH}-1st coin has tail and 2nd has head
  B_4: {HH}-both coins fall head

 Then B_1, B_2, B_3, B_4 are exhaustive events.

(c) In a cross with respect to single gene with two alleles with dominant recessive relationship the events
 A: Offspring has dominant phenotype.
 B: Offspring has recessive phenotype are exhaustive and mutually exclusive.

4. Equally Probable or Equally Likely Events

A set of events A_1, A_2, ..., A_n are said to be equally likely or equally probable, if none of them occurs in preference to the others. Examples of such events are:

(i) If a coin is unbiased, getting head or tail in a toss are equally likely.
(ii) In 3 (b) (ii) above, B_1, B_2, B_3, B_4 are equally likely but events A_1, A_2, A_3 in 3 (b) (i) are not equally likely.
(iii) In a test cross with respect to a single gene with two alleles having dominant recessive relationship, the events: (A) Offspring has dominant phenotype and (B) Offspring has recessive phenotype are equally probable or equally likely.
(iv) In a cage there are 10 black, 10 white, 10 grey and 10 spotted pigeons. One is selected from it, the events A_1: getting a black pigeon A_2: getting a white pigeon, A_3: getting a grey pigeon and A_4: getting a spotted pigeon are equally probable events or are equally likely events.

5. Sure Events

If all the outcomes of a random experiment favour the occurrence of the event A, it is called a sure event. It is an event which must occur if the experiment is performed. Some examples of sure events are:

(a) Getting number of points less than or equal to six in a throw of a dice is a sure event.
(b) Weight of a graduate student being less than 150 kg is a sure event.
(c) If an urn contains only white balls, then getting a white ball in 3rd draw from the urn is a sure event.

A sure event A has all the elements of sample space. Thus a sure event $A = \Omega$.

6. Impossible Events

An event which can not occur or if none of the outcomes of a random experiment favour the occurrence of an event A, it is called an impossible event. Some examples of such events are:

(a) Getting 8 points in a throw of a dice is an impossible event.
(b) Getting black ball in a draw from an urn which contains only white balls is an impossible event.
(c) If five cards are drawn from a pack of well shuffled cards, getting all aces is an impossible event.

Thus an impossible event $A = \phi$, a null set or an empty set.

7. Complementary Events

Non occurrence of an event A is called complementary event of A and is denoted by $\bar{A}$ or A^C. Thus all the elements of sample space which do not favour occurrence of A constitute the elements of A^C. The elements of A and A^C together constitute all the elements of sample space Ω. In symbols $A + A^C = \Omega$, A and A^C are mutually exclusive.

As an example consider

(a) If A: getting even numbers in throw of a dice, then A^C: getting odd numbers in throw of a dice.
(b) In a test cross of single gene with two alleles with dominant recessive relationship the events

A: Offspring is of dominant phenotype
B: Offspring is of recessive character are complementary events i.e. $A^C = B$ and $B^C = A$

Also it may be noted that $(A^C)^C = A$.

8. Compound Events

It was noted that a single element of a sample space or a single outcome of a random experiment constitutes a simple event. A compound event consists of a combination of several events. The following are some forms of compound events.

(a) Occurrence of A or B or both A and B (in other words occurrence of at least one of A and B) is a compound event called union of A and B and is denoted as $A \cup B$.
(b) Simultaneous occurrence of A and B both is also a compound event, called intersection of A and B and is denoted by $A \cap B$ or AB.

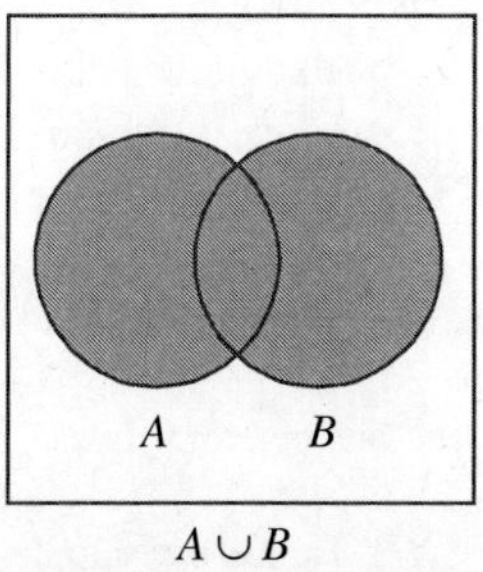

Fig. 4.1. $A \cup B$ (source: Meyer (1975), p. 5).

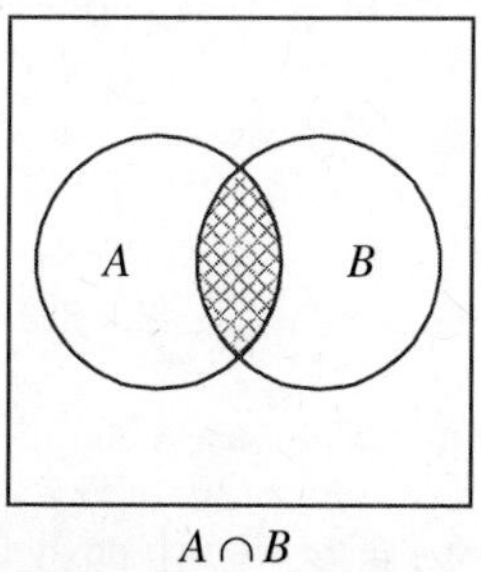

Fig. 4.2. $A \cap B$ or AB (source: Meyer (1975), p. 5).

(c) If $A_1, A_2 \ldots A_n$ are n events then $A_1 \cup A_2 \cup \ldots \cup A_n = \bigcup_{i=1}^{n} A_i$ denotes occurrence of at least one of the n events and $A_1 \cap A_2 \cap \ldots \cap A_n = \bigcap_{i=1}^{n} A_i = A_1\, A_2 \ldots A_n$ denotes occurrence of each of the n events.

Some examples of $A \cup B$ and AB are given below:

(i) Let A_1 be the event that a person in a city is overweight and A_2 that he is diabetic, then the event that a person is either overweight or diabetic or both is $A_1 \cup A_2$ and the event that the person is overweight and diabetic is

$$A_1 \cap A_2 \text{ or } A_1A_2$$

(ii) Let A_1 be the event that a college student suffers from depression and A_2 be the event that a college student considers to commit suicide. Then $A_1 \cup A_2$ is the event that a college student either suffers from depression or considers to commit suicide.

$A_1 \cap A_2$ is the event that a college student suffers from depression and considers to commit suicide.

4.3 Definitions of Probability

After defining the basic concepts required to define probability of an event, in this section three definitions are given. First one may be called a classical or a priori definition. The second one may be called a statistical definition and it is with a relative frequency approach. The third one is with an axiomatic approach.

4.3.1 Classical Definition

As already stated the origin of probability is games of chance. Through this association emerged the classical definition of probability as given below.

Classical probability of an event. If there are n mutually exclusive, equally likely and exhaustive outcomes of a random experiment E and of these, m favour the occurrence of an event A. Then probability of the event A is given by the ratio m/n which is the ratio of number of favourable outcomes and total number of outcomes. In symbols

$$P(A) = \frac{m}{n} = \frac{\text{Number of outcomes favourable to event } A}{\text{Total number of outcomes of } E} \tag{4.1}$$

Thus evaluation of probability of an event requires counting of all the possible outcomes of the random experiment E associated with the event. Thus all the elements of Ω need to be counted and care should be taken that all of these form a set which are mutually exclusive, equally likely and exhaustive. Out of these one should be able to find the number of outcomes which are favourable for occurrence of A.

Example 4.1: Find the probability that a woman with four sons, who is carrier of haemophelia will have one son with the disease, if her husband is normal.

Solution: Haemophilia is a sex linked disease and affects male even if they are heterozygous. Thus for a given son probability is ½ to receive the gene for haemophilia.

For each son there are two possibilities either diseased (D) or healthy (H). Hence the sample space for birth of 4 sons will be

Ω = {HHHH, HHHD, HHDH, HDHH, DHHH, HHDD, HDHD, HDDH, DHHD, DHDH, DDHH, HDDD, DHDD, DDHD, DDDH, DDDD}

Each of the 16 are equally likely

So $$n = 16$$

Looking at Ω $m = 4$, the 2nd 3rd, 4th and 5th outcomes favour occurrence of A: one son is diseased. Hence $P(A) = 4/16 = 1/4$.

Example 4.2: Find the probability of picking a heart from a pack of well shuffled deck of cards.

$n = 52$ since there are 52 cards in a deck
$m = 13$ since there are 13 hearts in a pack.

Any one of the card can be picked from the deck so all the 52 outcomes are equally likely. Let A be getting a heart in a draw from the deck of cards.

So $$P(A) = 13/52 = 1/4$$

Properties of Probability

1. It can be seen that smallest value of $m = 0$ i.e. none of the outcomes favour the event and the maximum value of m may be n i.e. all the outcomes favour the occurrence of an event. Hence
$$0 \leq P(A) \leq n/n \ (= 1) \tag{4.2}$$
Thus *range of probability* is 0 to 1. It can never be less than 0 i.e. negative, nor more than 1.
2. If $P(A) = 0$, A is an impossible event and *probability of an impossible event* is 0. If $P(A) = 1$, A is a sure event and that *probability of a sure event* is 1.
3. If A is an event, A^C is a complementary event of A.
So $$P(A \cup A^C) = P(\Omega) = 1$$

If there are n elements in Ω and m favour occurrence of A, $n - m$ will favour A^C.

Hence $$P(A) = \frac{m}{n} \text{ and } P(A^C) = \frac{n-m}{n}$$

$$P(A) + P(A^C) = \frac{m+(n-m)}{n} = \frac{n}{n} = 1 \tag{4.3}$$

Thus $$P(A^C) = 1 - P(A) \tag{4.4}$$

Disadvantages of Classical Definition

It is true that classical definition can not be used always to determine probability of an event. It seems that it can be used for very simple and unimportant cases like games of chance. But this is not true. This definition is very useful, can be applied in many fields but can not be applied in the following conditions.

(a) If the outcomes of the experiment can not be claimed to be equally likely, this definition can not be applied. For example if it is known that the coin is loaded on one side, it can not fall head or tail equally often. In this case, it is not possible to determine probability of getting a head by classical definition.

(b) There are situations when it is not possible to count the total number of outcomes of an experiment. In such a case also classical definition can not be applied. For example to find the probability of getting an albino mouse or probability of getting a parrot with lining in the neck in a city, classical definition can not be used. Since it is almost impossible to count the total number of mice or parrots in a city limit and to count the number of albino mice and parrots with lining in neck, classical definition of probability can not be used.

Hence an alternative approach was considered and the second definition of probability is as follows.

4.3.2 Statistical Definition

There are some random experiments for which the probability of an outcome or of an event related to it can not be determined by classical definition. However, some of these experiments are capable of being repeated under similar conditions. If all the outcomes are not equally likely or if the total cases and those favourable to event, can not be counted, one may obtain an estimate of the probability of such an event by statistical definition. This can be done by repeating the random experiment under identical conditions and finding out the relative frequency of the event and taking it as the probability of the event.

Thus formally statistical definition is as follows:

Let there be n independent trials of the random experiment. E under identical conditions, of which r result in the occurrences of the event A. Then r/n is called the relative frequency of the event A and the probability of the event A is the limiting value of this relative frequency, if the experiment is repeated very large number of times. In symbols

$$P(A) = \underset{n \to \infty}{Limit} \left(\frac{r}{n} \right) \tag{4.5}$$

It may be noted that the exact probability of an event A can not be determined by this definition, since repeating the experiment indefinitely will never end and so limiting value of relative frequency can not be found. So in practice only a limited or finite number (= n) of trials are performed and the relative frequency obtained from it is taken as the probability, even though it is only an estimate or approximation of the probability. Larger the number of repetitions, better is the estimate. With larger and larger repetitions, the relative frequency of event A normally comes closer and closer to a definite numeric value which is probability of event A.

The disadvantages of this approach are (a) the experiment must be repeatable and the determination of probability is possible only after repeating the experiment and observing the occurrences of the event. (b) Another disadvantage is that *exact probability of the event can not be determined.* Advantage of *classical method* is that wherever it *is applicable the probability is exact* and it *does not require even a single trial of the experiment to be performed,* nor it requires noting down occurrences of the event under consideration. For further discussions and illustrations on these ideas books by Uspensky (1937), Feller (1966) and Meyer (1975) may also be consulted.

Example 4.3: The following table shows 1000 nursing school applicants classified according to scores made on a college entrance examination and type of high school they passed.

Type of high school of applicants (source: Daniel (1987), p. 62, Q. 14)

Scores on entrance test	Govt. (G)	K.V. (K)	Convent (C)	Total
Low (L)	105	60	55	220
Medium (M)	70	175	145	390
High (H)	25	65	300	390
Total	200	300	500	1000

Find the probability that an applicant selected at random from the above group

(i) Made a low score in entrance exam.
(ii) Passed out from a KV

Solution

(i) Let L denote the event given in (i), then the required probability is

$$P(L) = \frac{220}{1000} = 0.22$$

(ii) Let K denote the event in (ii), then

$$P(K) = \frac{300}{1000} = 0.3$$

These probabilities are obtained using statistical or empirical definition of probability. The numerators are the frequencies of occurrence of the above two events, and the denominator, the total number of picked up applicants.

4.3.3 Axiomatic Definition of Probability

Consider a random experiment with sample space Ω. Probability of an event A in Ω denoted by $P(A)$ is a number which satisfies the following three axioms:

(i) $P(A) \geq 0$ for all A (4.6)
(ii) $P(\Omega) = 1$ (4.7)
(iii) If $A_1, A_2, \ldots A_n, \ldots$ be a sequence of mutually exclusive events i.e. $A_i \cap A_j = \phi$ for $i \neq j$ then

$$P(A_1 \cup A_2 \cup \ldots \cup A_n \cup \ldots) = [P(A_1) + P(A_2) + \ldots P(A_n) + \ldots]$$

or

$$P\left(\bigcup_{i=1}^{\infty} A_i\right) = \sum_{i=1}^{\infty} P(A_i) \tag{4.8}$$

This definition indicates what type of function is probability of an event. It does not tell how a value is to be assigned to it.

4.4 Additive Rule of Probability

Through this rule, probabilities of the events $A \cup B$ or $A \cup B \cup C$ or $A_1 \cup A_2 \cup \ldots \cup A_n$ or $A_1 \cup A_2 \cup A_3 \cup \ldots \cup A_n \ldots$ can be found.

These events are already defined in Section 4.2. The following theorems give the methods of finding probabilities of these events.

Theorem 1 *Theorem of total probability (for two events)*
Probability of occurrence of A or B or both A and B, is given by

(a) $P(A \cup B) = P(A) + P(B) - P(A \cap B)$ (4.9)
From the Venn diagram the rule is evident
(b) If A and B are mutually exclusive, A and B can not occur together so from (4.9)

$$P(A \cup B) = P(A) + P(B) \tag{4.10}$$

Since $AB = \phi$ an impossible event, A and B being mutually exclusive.

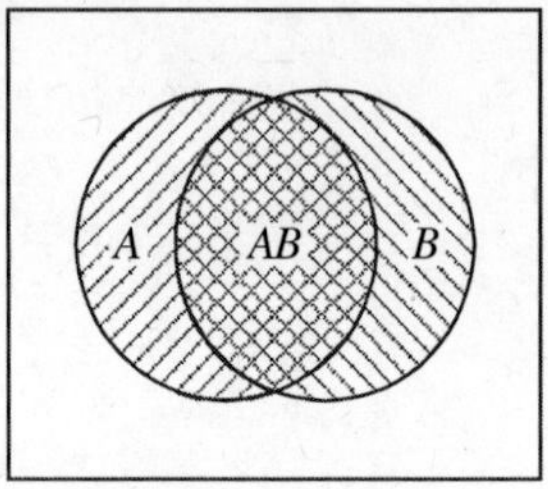

Fig. 4.3. Theorem of total probability (source: Daniel (1987), p. 43).

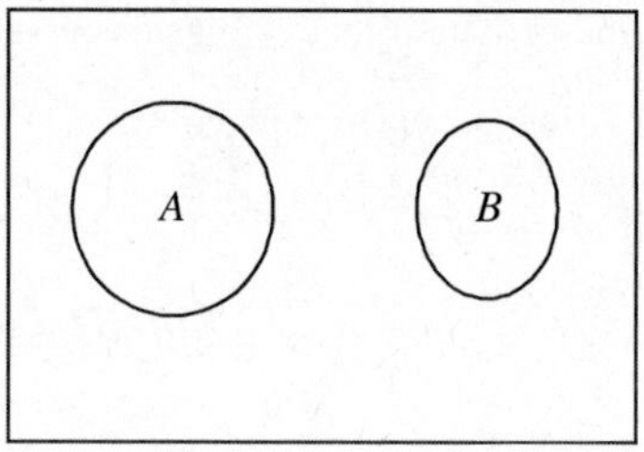

Fig. 4.4. Theorem of additive probability (source: Daniel (1987), p. 43).

In such a case, $P(A \cup B)$ is denoted by $P(A + B)$
So (4.10) can be written as

$$P(A + B) = P(A) + P(B) \tag{4.11}$$

(c) For three events A, B, C the additive rule is

$$\begin{aligned} P(A \cup B \cup C) &= P(A) + P(B) + P(C) - P(AB) - P(AC) - P(BC) + P(ABC) \\ &= P(A) + P(B) + P(C) \text{ if } A, B, C \text{ are mutually exclusive} \end{aligned} \tag{4.12}$$

or

$$P(A \cup B \cup C) = P(A + B + C) = P(A) + P(B) + P(C) \quad \text{if } A, B, C \text{ are mutually exclusive} \tag{4.13}$$

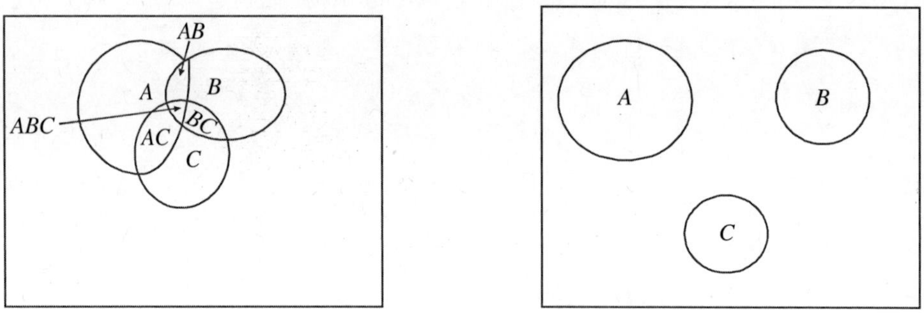

Fig. 4.5. Total probability for three events.

(d) For n events, the additive rule can be written as an extension of (4.12) as

$$P(A_1 \cup A_2 \cup \ldots \cup A_n)$$

$$= \sum_{i=1}^{n} P(A_i) - \sum_{i<j=2}^{n} P(A_i A_j) + \sum_{i<j<r=3}^{n} (A_i A_j A_r) + \ldots + (-1)^{n-1} P(A_i A_j A_r \ldots A_n) \tag{4.14}$$

Example 4.4: As some examples for use of this rule, the compound events (i) and (ii) of Section 4.2.3 are revisited (source: Milton and Tsokos (1983), p. 72, Exercise 3.2.1 and 3.2.2).

(i) A_1: a person in a city is obese,
A_2: a person in a city is diabetic and
A_1 A_2: a person is both obese and diabetic.

It is given that 30% of persons are obese, 3% suffer from diabetes and 2% are obese and diabetic both. Find the probability that a person in the city is either obese or diabetic or obese and diabetic both.
Solution: From statistical definition of probability

$$P(A_1) = .3,\ P(A_2) = .03, \text{ and } P(A_1\ A_2) = .02$$

Hence to find probability that a person in the city is either obese or diabetic or obese and diabetic both, equation (4.9) of theorem is used to give the required probability.

$$\begin{aligned} P(A_1 \cup A_2) &= P(A_1) + P(A_2) - P(A_1\ A_2) \\ &= .30 + .03 - .02 \\ &= .31 \end{aligned}$$

(ii) A_1 = a college student suffers from depression
A_2 = a college student considers suicide
$A_1\ A_2$ = a college student suffers from depression and considers to commit suicide

Given that 18% of college students suffer from depression
2% of college students consider suicide
19% suffer from depression or consider suicide

to find the probability of the event that a college student suffers from depression and considers suicide.
Solution: It is required to find $P(A_1 \cap A_2)$. By statistical definition

$$P(A_1) = 0.18\ P(A_2) = 0.02,\ P\ (A_1 \cup A_2) = 0.19$$

Using theorem 1, $$P(A_1 \cup A_2) = P(A_1) + P(A_2) - P(A_1 \cap A_2)$$

Hence $$\begin{aligned} P(A_1 \cap A_2) = P(A_1\ A_2) &= P(A_1) + P(A_2) - P(A_1 \cup A_2) \\ &= .18 + .02 - .19 \\ &= .01 \end{aligned}$$

4.5 Multiplicative Rule

This rule is used to find out the probability of simultaneous occurrence of two or more events. The notations for denoting these events have already been described in compound events (c) of section 4.2.3.

Theorem 2 *Multiplication rule for two events* states that probability of simultaneous occurrence of two events A and B is given by

$$P(A \cap B) = P(AB) = P(A).\ P(B/A) \tag{4.15}$$

$$= P(B).\ P(A/B) \tag{4.16}$$

$$= P(A).P(B) \text{ if } A \text{ and } B \text{ are independent events} \tag{4.17}$$

Conditional Probability and Independent Events

To understand the multiplicative rule the concepts of conditional probability and independent events should be understood.

4.5.1 Conditional Probability

Conditional probability is being discussed first in this section and the independent events are defined in the next.

The symbols *P(B/A)* and *P(A/B)* are conditional probabilities of the occurrence of the events *B* given *A* has occurred and that of *A* given that *B* has occurred, respectively.

The first event listed is one whose occurrence probability is to be determined when it is 'given that' (the slash is read in this way) the second event is known to have already occurred.

There are some situations when occurrence or nonoccurrence of one of the events affects occurrence or nonoccurrence of the other event.

As already mentioned if occurrence or nonoccurrence of one of the events has no effect on occurrence or nonoccurrence of the other event, the two events are said to be independent and formula (4.17) is used to calculate probability of simultaneous occurrence of the two events. But if the occurrence of one affects the occurrence of the other, one should use the conditional probability instead of unconditional probability of the second event.

To explain the concept of conditional probability, the following two examples are considered.

Example 4.5 (a): A cage contains 10 albino and 20 normal mice. Another cage has 20 albino and 10 normal mice. One mouse is selected from first cage, and put in the second cage, and a mouse is taken out from the second. Desired to find the probability that the mice from the first and the second cage are both albino.

Solution: To solve this problem let the event *A*: mouse from cage 1 is albino and the event *B*: mouse from cage two is albino.

Getting an albino mouse from cage 1 and putting it in cage 2 changes the probability of albino mouse from cage 2, since the number of albino mice in cage 2 increases by 1 (becomes 21) and number of total mice also increases from 30 to 31

Thus we find that $$P(B/A) = \frac{21}{31}$$

Hence, required probability is *P(AB)* = *P(A)*. *P(B/A)* from (4.15)

$$= \frac{10}{30} \times \frac{21}{31} = 7/31 \tag{4.18}$$

However, if mouse from cage 1 is normal the probability of albino mouse from cage 2 will be 20/31 and will be denoted by

$$P(B/A^C)$$

Example 4.5 (b): If the mouse selected from cage 1 is not put in the second cage, selection of mouse from 2nd cage is not affected by which type of mouse is selected from cage 1.

Thus under this situation *P(AB)* = *P(A)*. *P(B)*

$$= \frac{10}{30} \times \frac{20}{30} = \frac{2}{9} \tag{4.19}$$

In this case *P(B)* = 20/30 = 2/3. Also *P(B/A)* = 20/30 = 2/3.

Since there is no change in cage 2, it is seen that occurrence of *A* i.e. taking out albino mouse from cage 1 does not affect occurrence of *B*.

So $$P(B) = P(B/A) = 2/3 \tag{4.20}$$

A and *B* are, therefore, independent in this case.

Example 4.6: Suppose a tank contains 10 red and 20 golden fishes and two fishes are taken out from the tank one by one (a) after replacing the first drawn fish (b) without replacing the first fish.

Let A be the event that first fish is red and B be the event that second fish is red.

(a) If one chooses the second fish after replacing the first one

$$P(A) = \frac{10}{30} \quad P(B) = \frac{10}{30}$$

(b) If the second is chosen without returning the first into the tank

$$P(A) = \frac{10}{30}$$

But occurrence of B now depends on the composition of the fishes in the tank. The total number of fishes after first is drawn will be only 29 and the number of red fishes will depend on whether event A has occurred or not. If A has occurred there will be only 9 red fishes.

So the
$$P(B/A) = \frac{9}{29}$$

Thus it can be seen that, if 2nd draw is with replacement $P(B)$ can be determined without consideration of occurrence of A. This will be always $\frac{10}{30}$.

$P(B)$ is unconditional probability of B and it can be determined irrespective of occurrence or nonoccurrence of A. Whereas in the draw without replacement, B occurs in conjunction with A. These two conditional probabilities are

$$P(B/A) = \frac{9}{29} \text{ and } P(B/A^C) = \frac{10}{29} \tag{4.21}$$

For unconditional probability of B one should consider sample points of B in whole of Ω. But conditional probability of B given A has occurred, restricts the total outcomes from Ω to A.

One asks how probable it is for B to occur only in the part of the sample space where A can occur. Thus leaving 29 fishes out of 30 original number, before the second is drawn. Hence for $P(B/A)$ total fishes are 29 with only 9 red fishes in tank, so (4.21) is used for calculating this probability.

Formal definition of conditional probability $P(B/A)$ can be given from the equation (4.15) as

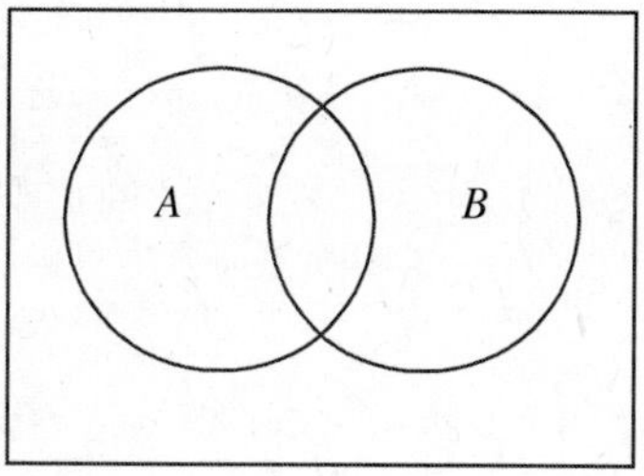

Fig. 4.6. Restricted sample space for $P(B/A)$.

$$P(B/A) = P(AB)/P(A) \tag{4.22}$$

which holds if A is not an impossible event or $P(A) > 0$.

4.5.2 Independent Events

Two events A and B are said to be independent if occurrence or nonoccurrence of one of them does not affect occurrence or nonoccurrence of the other. For example consider that A is getting a head in first toss of a coin and B is getting a tail in the second toss of the coin. It is evident that A and B are independent since getting head or tail in first toss does not have any effect on the result of second toss.

As another example, let A be the event that Mrs. S gave birth to a baby boy and B the event that Mrs. R gave birth to a baby girl. In this set up, A and B are independent since birth of the baby boy or girl to Mrs. S is irrespective of birth of baby boy or girl to Mrs. R.

Formal definition of two independent events can now be given as: Two events A and B are independent if and only if any one of the following holds

(i) $P(AB) = P(A). P(B)$
(ii) $P(A/B) = P(A)$ if $P(B) > 0$
(iii) $P(B/A) = P(B)$ if $P(A) > 0$

Example 4.7: It is estimated that among a locality inhabitants' population 55% are overweight, 20% have blood pressure and 60% are overweight or have high blood pressure. Is the event that a person is overweight independent of the event that person is having high blood pressure (source: Milton and Tsokos (1983), p. 84, Exercise 3.5.1)?
Solution: Let A: person is overweight and B: person has high blood pressure.
Given that $P(A) = .55$ $P(B) = .20$ and $P(A \cup B) = .60$.
If A and B are independent, $P(AB)$ should be equal to $= P(A) \times P(B)$

$$= .55 \times .20 = .1100$$

Given $$P(A \cup B) = .60 = P(A) + P(B) - P(AB)$$

$$= .55 + .20 - P(AB)$$

So $$P(AB) = .55 + .20 - .60$$

$$= .75 - .60 = .15$$

Since $P(AB) = .15 \neq .11 = P(A). P(B)$, A and B are not independent.

Example 4.8 (Example 4.3 (revisited)): Consider Example 4.3 again and suppose it is desired to find the probability that an applicant passed out from convent secures high score in entrance examination. The required probability is probability of simultaneous occurrence of C and H i.e. $P(CH)$ (source: Daniel (1987), p. 62, Q. 14).

(a) This can be found directly from the table using statistical definition as follows.
Since out of 1000 total applicants, 300 are from convent schools securing high score, so $P(CH) = 300/1000 = .3$.
(b) One can also apply multiplicative rule to find this probability

$$P(CH) = P(C). P(H/C) = 500/1000 \times 300/500$$

To find $P(H/C)$ the total cases are those from convent (since securing high score from convent pass outs only are to be looked into) = 500. High scores from convent are secured by 300 which is the frequency of the event H/C.

Hence $P(H/C) = 300/500$ and $P(CH) = 300/1000 = .3$ which is same as in (a).

Example 4.9: Studies in human population genetics indicate that 39% of available genes for determining Rh blood factor are negative. Based on this information, what is the probability that a randomly selected individual will have Rh negative blood (source: Milton and Tsokos (1983), p. 85)?
Solution: Since Rh negative is recessive allele, unless both the genes are negative, a person can not have Rh negative blood group. Since one gene of an individual is from mother and another from father, the gene type inherited from father is independent of that from mother. Hence

P (both genes are Rh negative) = P (mother's gene is Rh negative) $\times$ P (father's gene is Rh negative)
$= .39 \times .39$
$= 0.15$

Example 4.10: Of 50 patients on the third floor of a hospital, 35 are females and 12 are over 70 years of age. Among those over 70 years of age 8 are females. What is the probability that a patient among 50 is male and above 70 years of age (source: Daniel (1987), p. 46, Exercise 24.2)?

Solution: Number of persons above 70 years = 12
Number of females over 70 years = 8
Number of males over 70 years = 12 – 8 = 4

Let *A*: A patient is male,
B: A patient is above 70 years.

Required probability $P(AB) = 4/50$.

4.6 The Bayes Rule or Bayes Theorem

This theorem is also called the formula for the probability of 'causes' and gives the $P(B_i/A)$, where $B_1, B_2, \ldots B_k$ are a partition of Ω. Since B_i's are a partition of Ω, so one of the events B_i must occur and only one of the events can occur. Now *A* is also an event in Ω. The Bayes theorem gives the formula to give probability of a particular B_i (a 'cause') given that *A* has occurred as:

$$P(B_i/A) = P(AB_i)/P(A)$$

$$= \frac{P(A/B_i)\,P(B_i)}{\sum_{i=1}^{k} P(A/B_i)\,P(B_i)}, \quad i = 1, 2, \ldots k \tag{4.23}$$

It may be noted that if *A* is an event on Ω and B_i are partitions of Ω

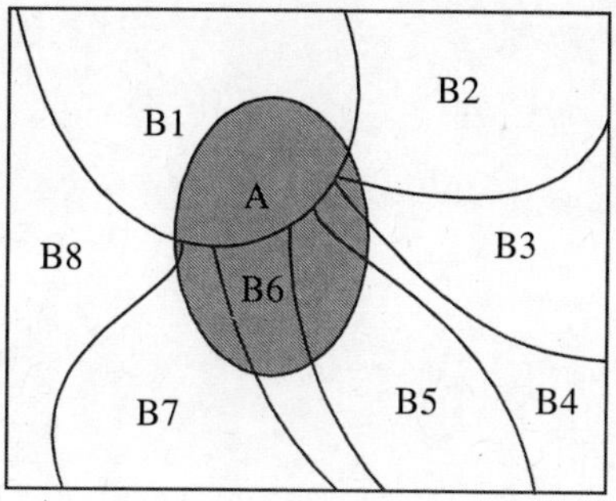

Fig. 4.7. Partitions of Ω and *A*.

$$A = AB_1 \cup AB_2 \ldots \cup AB_i \ldots \cup AB_k$$

and $P(A) = \sum_{i=1}^{k} P(AB_i)$, since B_i's are mutually exclusive

$$= \sum_{i=1}^{k} P(B_i).P(A/B_i) \text{ from multiplicative rule (4.16)}$$

Example 4.11: The blood group type distribution in US is 41% type *A*, 9% type *B*, 4% type *AB* and 46% type *O*. It is estimated during World War II that 4% of inductees with type *O* blood group were typed as having type *A*. 88% of those with type *A* were correctly typed, 4% with type *B* were typed as *A*, and 10% with type *AB* were typed as *A*. A soldier was wounded and brought to surgery. He was typed as having *A* blood group. What is the probability that this is true blood group type (source: Milton and Tsokos (1983), p. 93, Example 3.6.2)?

Solution: Let A_1: a person has type A blood group
A_2: a person has type B blood group
A_3: a person has type AB blood group
A_4: a person has type O blood group
B: person is typed as blood type A.

The desired probability is $P(A_1/B)$, when given that

$$P(A_1) = 0.41,\ P(A_2) = 0.09,\ P(A_3) = 0.04 \text{ and } P(A_4) = 0.46$$

$$P(B/A_1) = .88,\ P(B/A_2) = .04,\ P(B/A_3) = .10,\ P(B/A_4) = .04$$

By Bayes Theorem
$$P(A_1/B) = \frac{P(A_1B)}{P(B)} = \frac{P(A_1)P(B/A_1)}{\sum_{i=1}^{4} P(A_i)P(B/A_i)}$$

$$= \frac{.41\,x\,.88}{041x.88 + .09x.04 + .04x.10 + .46x.04}$$

$$= 0.93$$

5. Random Variables and their Characteristics

After defining probability of an event and understanding the concepts of sample space and events, the next step is to consider implication of probability theory in data analysis. Before seeing how probability theory is utilized to draw precise conclusions about a population on the basis of a sample drawn from it, the concept of random variable must be clear. It will be seen that almost all the scientific studies are related to properties of one or more random variables. Statistical methods employed help to understand properly the properties of random variables of interest. This chapter, therefore, gives the definition and types of random variables, various functions related to these which give all the characteristics or properties of the random variable.

Books by Meyer (1975), Milton and Tsokos (1983), Goon et al. (1983) and Steel and Torrie (1981) are recommended for further reading.

5.1 Definition and Types of Random Variables

Once the concepts of random experiment and sample space are clear, it is not difficult to understand what a random variable is. It can be recalled that the elements of a sample space were not all numeric. However, even for sample spaces with non numeric elements, a numeric value can be assigned. For example consider the sample space Ω: {HH, HT, TH, and TT}. Numeric values assigned to this sample space may be {2, 1, 1, 0}, which are the number of the heads.

Similar convention can be adopted for any sample space. This type of rule or function which assigns a numeric value to sample points is called a random variable.

5.1.1 Definition of Random Variable

Thus a definition of random variable can be given as: A numeric valued function defined on sample space is called a random variable.

Capital letters $X, Y, Z, T, U, V, \ldots$ are used to denote random variables and the corresponding small letters $x, y, z, t, u, v, \ldots$ are used to denote the values of the respective random variables.

Thus X is a random variable since to each element of Ω, X assigns a numeric value

$$X(\omega) = x$$

Thus for the sample space considered above, the random variable may be defined as X: Number of heads in two tosses of a coin.

Hence, $$X(\text{HH}) = 2,\ X(\text{HT}) = X(\text{TH}) = 1 \text{ and } X(\text{TT}) = 0 \qquad (5.1)$$

It may be noted that the random variable is a *single valued function* i.e. for every ω in Ω there is only one value of $X(\omega)$. However, different elements of Ω may lead to the same value of X.

For example for ω_1 = HT and ω_2 = TH, the value of X is same (= 1).

The set of all possible values of X, is called the range of X. If the random variable is an identity function i.e., $X(\omega) = \omega$ the sample space and the range of X are same. This is possible only if the sample space is numeric. One such sample space may be for the following random experiment.

From a pond collect 5 ml of water and count the number of bacteria present. In this case

$$\Omega = \{0, 1, 2, \ldots\} \tag{5.2}$$

For this sample space $X(\omega) = \omega$ may be the most natural random variable. Thus X: number of bacteria in 5 ml of pond water.

After defining a random variable and knowing its range, one may be interested in the probabilities associated with the values of the random variable. Thus in the experiment the investigator might be interested to find $P[X \leq 5]$ i.e. that number of bacteria in collected 5 ml of the pond water, is less than 5.

It is worth noting that $[X \leq 5]$ is also an event since $X \leq 5$ for the sample points 0, 1, 2, 3, 4, 5 in Ω of equation (5.2).

Let $[X \leq 5] = A$ and $B = \{0, 1, 2, 3, 4, 5\}$ in Ω. Then it is evident that A and B are equivalent events. Thus to find $P[X \leq 5]$ one may find $P[B]$ which is not so easy, as finding of $P[X = 1]$ for the random variable defined on Ω of (5.1). Since $[X = 1]$ = {HT, TH} and so it can be easily seen that $P\{X = 1\} = 2/4 = 1/2$, from classical definition of probability.

This illustrates why such variables are called random variables and how probability is associated with their values.

5.1.2 Types of Random Variables

It is very clear that since there is exactly single value associated to a sample point, there can not be more values of a random variable than the number of sample points in the sample space of the associated random experiment. Hence the random experiments with finite or countably infinite elements in the sample space can have only finite or countably infinite values of the random variable defined on it.

Such random variables which can assume finite or countably infinite number of values, are called *discrete random variables*. The range of such variable may be $x_1, x_2, \ldots x_n \ldots$ The list terminates if X is finite discrete random variable and continues indefinitely, otherwise.

Another type of random variables are continuous and take values on a finite interval such as $a \leq X \leq b$ or on infinite intervals such $X \geq 0$ or $-\infty \leq X \leq \infty$. All such variables can attain infinite number of values, even if their range is a finite interval.

Examples of discrete and continuous random variables are considered below:

Example 5.1: Consider a variable Y, the number of oil spills per year in Mumbai sea shore. Y is a random variable, it does not have a definite value and chance factors play to determine its value. If in a particular year there were 3 spills one may write $y = 3$. Y is a discrete random variable.

Example 5.2: Let Z denote the doze (in *cc*'s) of a drug prescribed by a doctor to a patient to control epileptic seizures. Since the doze changes from patient to patient and also on one patient from time to time, depending on metabolic conditions, Z is a random variable, and may be continuous with range $.5 \leq z \leq 3.5$ (source: Milton and Tsokos (1983), p. 95, 4.1.2).

Example 5.3: Let D denote the body diameter of an adult Octopus. Then D is a random variable since all the adult Octopus will not have same diameter but will vary due to genetic and environmental effects. Diameter of body is a continuous variable, as it can be any value in a finite range. Only accuracy of measuring instrument may restrict the diameter values to one, two or three decimal places (source: Milton and Tsokos (1983), p. 96, 4.1.3).

Example 5.4: Consider the total coliform count T in 1 ml of a river water in front of an untreated city sewage discharge at various distance and depths. Then T is a discrete random variable with range $t = 0, 1, 2, 3, \ldots$.

Example 5.5: Consider X: the time in minutes it takes a nurse to attend a patient's call. Suppose that hospital administration is strict that every call should be attended within 5 minutes. Then X is a continuous variable with range $0 < x < 5$.

5.2 Functions for Probability Distribution of a Random Variable

Now it is clear what is a random variable, what are different types of random variables, and that the values attained by it are based on chance. But only this knowledge is not enough. One is also interested to know the probabilities with which various values are attained by the random variable.

It has been seen that since the random variable is defined on a sample space, there is a correspondence (rule) between the sample space and range space of X. Hence one way of determining the probability that X takes different values is to find the probability of the sample points associated with the value of X. But this is possible only for some simple sample spaces and random variables defined on them. To enable one determining the probability of different values taken by a random variable, a function is needed which can associate with various values (x) of X, the probabilities with which these values are attained. In other words a function of x is needed which for various values of x will enable to calculate the required probability of random variable taking value equal to x or around x, if X is continuous.

In this section such functions are defined and the methods to use these to calculate the required probability are given.

There are two functions defined for this purpose, one for discrete and the other for continuous random variables. The function for the discrete random variable is called *probability mass function* (pmf) denoted by $p(x)$ and for continuous random variable, it is called *probability density function* (pdf) and is denoted by $f(x)$. In the next two subsections these two are discussed for a univariate random variable X. A random variable must attain some value in its range, so interest lies in how total probability of one is distributed over the range of values of X.

5.2.1 Probability Mass Function (pmf)

Probability mass function (pmf) $p(x)$ is used for any discrete random variable X. As the values taken by a discrete random variable are at most countably infinite, $p(x)$ is also defined for these values.
Definition: If X is a discrete random variable, a function $p(x)$ is probability mass function (pmf) of random variable X, if it satisfies the following conditions:

(i) $p(x) \geq 0$ for all x

(ii) $p(x) = P[X = x]$ for all x

(iii) $\sum_{all\ x} p(x) = 1$

(iv) $P[a \leq X \leq b] = \sum_{x=a}^{b} p(x)$ (5.3)

$p(x)$ takes positive values only at the values assumed by random variable X. Hence pmf is not a continuous function but is a line diagram. Lines are at the values assumed by X. Length of these lines are proportionate to the values of $p(x)$ at the assumed values of X.

To illustrate the nature and use of pmf, some examples are discussed below.

Example 5.6: Two coins are tossed and the number of heads is counted. Then X: the number of heads is a random variable and can take the following values:

$$x = 0, 1, 2, \text{ Applying classical definition}$$

$$p(x) = 1/4, 1/2, 1/4$$

and the graph of the pmf will be as follows

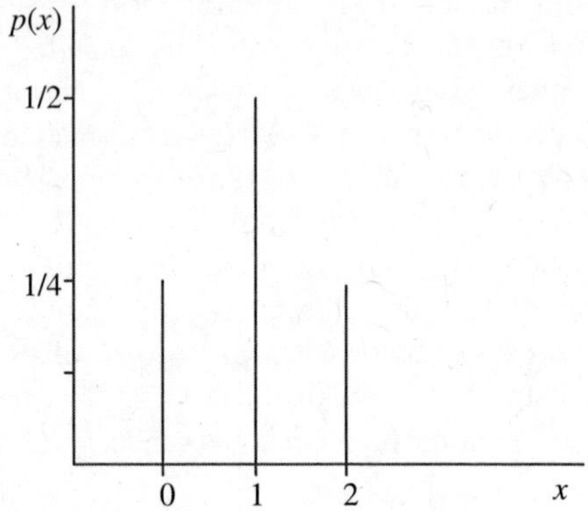

Fig. 5.1. pmf of X: number of heads in toss of two coins.

5.2.2 Probability Density Function (pdf)

If X is a continuous random variable, it takes values on a finite or infinite interval. It can not assume any specific value, i.e. $P[X = x]$ is always zero. One can talk about $P[x \le X \le x + dx]$, howsoever small dx may be. This probability is non negative and a function $f(x)$ is needed to evaluate this probability. Such a function is called probability density function (pdf) of the random variable X.

Definition: A function $f(x)$ may be probability density function (pdf) of the continuous random variable X, if it satisfies the following properties:

(i) $f(x) \ge 0$ for all x

(ii) $\int_{-\infty}^{\infty} f(x)dx = 1$

(iii) $P[a \le X \le b] = \int_a^b f(x)dx$ (5.4)

Property (i) of pmf and pdf indicate that these should always be non negative as probability of any event is always non negative.

Properties (ii) of (5.4) and (iii) of (5.3) imply that the total area under the function $f(x)$ over the whole range of X is always 1 as well as the sum of pmf over all the possible values of X is also 1.

Probabilities of the random variable lying between a to b can be found if the pmf or pdf is known, by adding pmf over all the values between a to b, if X is discrete. If X is continuous, it will be the area under the curve $f(x)$ between a to b as shown in the figure.

Two examples of pdf and method to compute probability of some values of X, are given below.

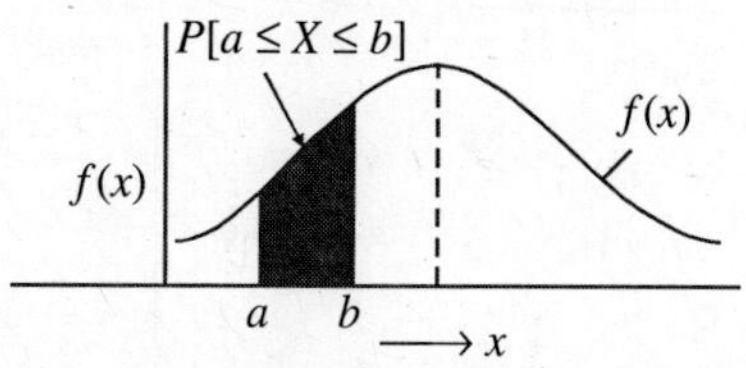

Fig. 5.2. pdf of X (total area below $f(x)$ is always 1) (source: Milton and Tsokos (1983), p. 104 (modified)).

Example 5.7: Let $f(x) = \begin{cases} 1 & \text{if } 0 < x < 1 \\ 0 & \text{otherwise} \end{cases}$

Prove that it is a pdf and find $P(X < ½)$

Solution: $f(x) \geq 0$ for all x so first condition is satisfied

For property (ii): $\int_{-\infty}^{\infty} f(x)dx = \int_{-\infty}^{0} f(x)dx + \int_{0}^{1} f(x)dx + \int_{1}^{\infty} f(x)dx$

$$= 0 + \int_0^1 1\, dx + \int_1^{\infty} 0\, dx$$

$$= x\Big/_0^1 = 1$$

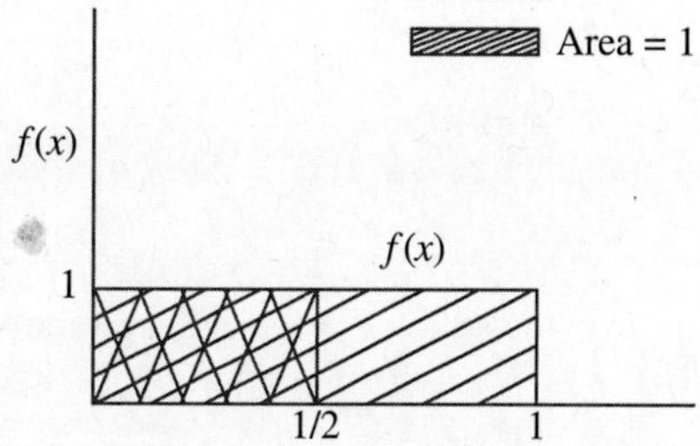

Fig. 5.3. pdf of X in Example 5.7.

From the graph of pdf, total area below $f(x)$ = 1.1 = 1

$$P[X \leq ½] = \int_0^{1/2} f(x)dx = \int_0^{1/2} 1.dx = x\Big/_0^{1/2} = 1/2$$

From the Figure 5.3, $P[X \leq ½]$ = area of rectangle between $x = 0$ to $x = ½ = ½.\ 1 = ½$

Example 5.8: Suppose pdf of a random variable is

$$f(x) = \begin{cases} 2x & 0 < x < 1 \\ 0 & \text{otherwise} \end{cases}$$

(Source: Meyer (1983), p. 69, Ex. 4.1.2)

(i) Prove that it is pdf.
(ii) Find $P[X < ½]$

Solution: (i) Since $f(x)$ is non negative for all x condition (i) is satisfied

For condition (ii), consider $\int_{-\infty}^{\infty} f(x)dx = \int_{-\infty}^{0} 0\, dx + \int_0^1 2x\, dx + \int_1^{\infty} 0\, dx$

$$= 0 + x^2\Big/_0^1 + 0 = 1$$

Hence condition (ii) is satisfied

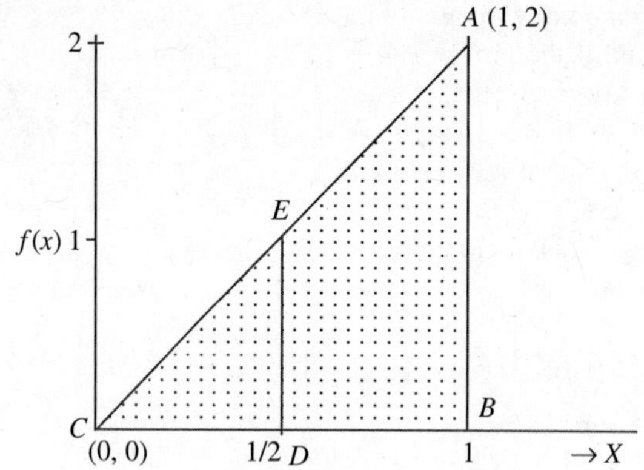

Fig. 5.4. pdf of Example 5.8.

From Graph:
Total area below $f(x)$ = Area of the triangle ABC

$$= \tfrac{1}{2} BC.\ AB = \tfrac{1}{2} \times 1 \times 2 = 1$$

Hence the given $f(x)$ is pdf of random variable X.
(ii) $P[X < \tfrac{1}{2}]$ = Area of triangle $ECD = \tfrac{1}{2}\, DC \times DE$

$$= \tfrac{1}{2} \times \tfrac{1}{2} \times 1 = \tfrac{1}{4}$$

Mathematically from condition (iii) $P[X < \tfrac{1}{2}] = \int_0^{1/2} 2x\, dx = x^2 \Big/_0^{1/2} = \tfrac{1}{4}.$

5.2.3 Probability Distribution of Random Variable

All the characteristics of the random variable can be seen or computed if its pmf or pdf is known. The first of the characteristics is its distribution i.e. how total probability 1 is distributed over the values taken by the random variable X.

For a discrete random variable, the couplet$\{x, p(x)\}$ and for a continuous random variable, $f(x)$ and range of x are called the *distribution or probability distribution* of random variable X.

Some points to be noted are

(a) pmf is also called probability function for a discrete random variable X.
(b) For a continuous random variable X, $P[X = x_0] = 0$ for all x_0. This is because $P[X = x_0]$ can be written as $\int_{x_0}^{x_0} f(x)\, dx$ and the width of the interval on which area under $f(x)$ is to be determined (which is probability under consideration) is 0. So the area will be always zero and hence a continuous random variable can not take any specified value even within the range of X on which $f(x)$ is positive.
(c) Because of (b) for a continuous random variable X

$$P[a \le X \le b] = P[a < X \le b] = P[a \le X < b] = P[a < X < b] \tag{5.5}$$

For a discrete random variable X, if values a and b are not assumed, $P[X = a]$ and $[X = b]$ will be zero, so (5.5) will be true for such discrete *X* also. But if *a* and *b* are attained values of a discrete random variable *X*, (5.5) will not hold for it.

(d) pmf may be a formula in *x* with list of possible values of *x* or a table containing values of *x* and *p*(*x*).

5.2.4 Cumulative Distribution Function (cdf)

One more function denoted by *F*(*x*) is also useful for finding probabilities of different values of random variables. This function is called cumulative distribution function (cdf) or distribution function of *X*.

Definition of cdf: A function *F*(*x*) called cdf of a random variable *X*, discrete or continuous, is defined as

$$F(x) = P[X \le x]$$

$$= \sum_{x_i \le x} p(x_i), \text{ if } X \text{ is discrete, and}$$

$$= \int_{-\infty}^{x} f(u)\, du, \text{ if } X \text{ is continuous} \qquad (5.6)$$

Some examples and properties of cumulative distribution function follow:

Example 5.9: A random variable *X* takes values 0, 1 and 2 with probabilities 1/3, 1/6 and 1/2 respectively

Table 5.1 pmf and cdf for Example 5.9
(source: Meyer (1975), p. 70, Ex. 4.14)

x	*p*(*x*)	*F*(*x*)
0	1/3	1/3
1	1/6	1/3 + 1/6 = 1/2
2	1/2	1/2 + 1/2 =1
Total	1	

If a graph is drawn, it is of the following nature

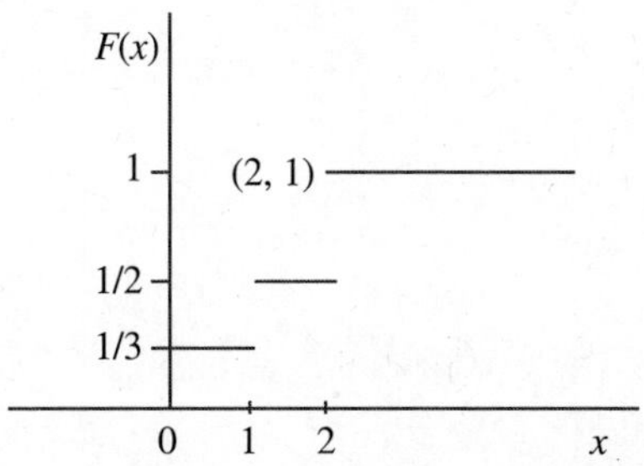

Fig. 5.5. cdf of *X* in Example 5.9 (source: Meyer (1975), p. 70).

Example 5.10: Suppose that *X* is a continuous random variable with pdf

$$f(x) = \begin{cases} 1/2 & 0 < x < 2 \\ 0 & \text{otherwise} \end{cases}$$

Then cdf of X will be

$$F(x) = \begin{cases} 0 & \text{if } x < 0 \\ \int_0^x \tfrac{1}{2}\, du = x/2 & \text{if } 0 < x \le 2 \\ 1 & \text{if } x > 2 \end{cases}$$

Graphically $F(x)$ is given by

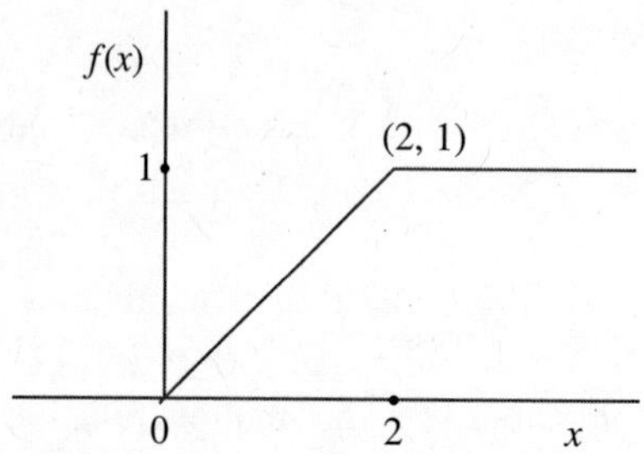

Fig. 5.6. cdf of X in Example 5.10.

Thus one can see that

(i) $F(x)$ is *defined for all the values of* x between $-\infty$ to ∞.
(ii) *For a discrete X, F(x) is a step function but for a continuous X, it is a continuous function.*

In general the cdf for a continuous X is like an elongated S as below

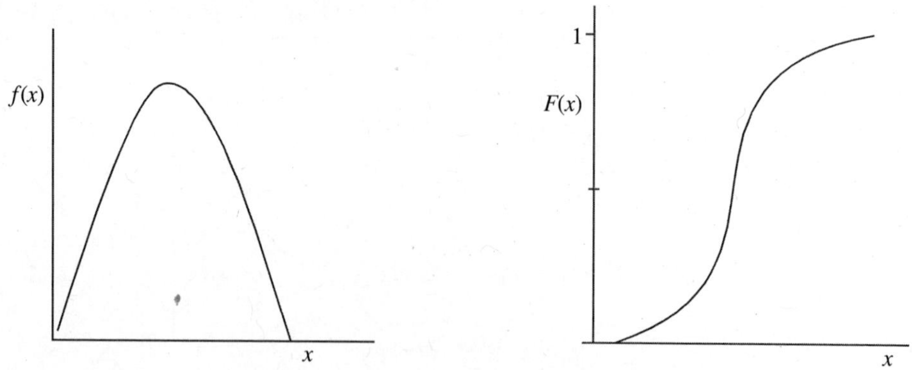

Fig. 5.7. pdf and cdf of a continuous random variable.

Properties of cdf

1. $F(-\infty) = 0$ and $F(\infty) = 1$
2. $F(x)$ is a non decreasing function of x; that is $F(a) \le F(b)$ if $a < b$.

3. $F(x)$ is continuous from right, that is $F(x + h) = F(x)$, where h is very small positive number.
4. $p(x_j) = F(x_j) - F(x_{j-1})$ if x is discrete, and

$$f(x) = dF(x)/dx = F'(x) \text{ if } X \text{ is continuous}$$

   Thus pmf or pdf can be found from cdf.
5. $F(b) - F(a) = P[a < X \leq b]$ if X is discrete

$$= P[a \leq X \leq b] \text{ if } X \text{ is continuous}$$

   Thus cdf can also be used to find the probabilities of various values of X.

5.3 Joint pmf, Joint pdf, Marginal and Conditional pdf and Independent Random Variables

pmf, pdf and cdf of a univariate random variable have been discussed in the last section. Some times instead of observing one characteristic from a random experiment two or more characteristics are observed. In other words if for various outcomes of a random experiment (i.e. to a sample point observed) more than one numeric characteristics are observed, the result will be a set of two or more numeric quantities associated to a sample point. Consider for example an experiment of tossing two coins and let the following numeric functions be defined for the outcomes.

X_1: Number of heads
X_2: Net gain from the prize given to the experimenter as Rs. 5 for each head and the penalty of Rs. 2 for each tail.
X_3: Shortest distance between the two coins
X_4: Shortest distance of the coins from the edge of the table near the experimenter.

Thus, assuming that table on which coins are tossed is of dimension 1 feet square for the four outcomes of the experiment, the values of the four random variables are:

	HH	HT	TH	TT
$X1$:	2	1	1	0
$X2$:	10	3	3	– 4
$X3$:	$0 \leq x_3 \leq 12''$	for all the four outcomes		
$X4$:	$0 \leq x_4 \leq 12''$	for all the four outcomes		

In such a case the 4 random variables are associated with each of the outcomes. Hence $X = \begin{pmatrix} X_1 \\ X_2 \\ X_3 \\ X_4 \end{pmatrix}$, the random variable defined on the above sample space is 4 dimensional. Thus the formal definitions of a two dimensional (also called bivariate) and of a multidimensional or multivariate random variables are as follows.

Bivariate Random Variable

Let X_1 (ω) and X_2 (ω) be two numeric functions (or rules) assigning two numeric values to each outcome ω in the sample space then (X_1, X_2) is called a bivariate random variable.

Multivariate Random Variable

Let X_1 (ω), X_2 (ω), …, X_p (ω) are p numeric functions (or rules) assigning p numeric values to each outcome ω in the sample space, then $(X_1, X_2,\ldots, X_p)$ is a p-dimensional random variable. All of these p variates are defined on the same sample space Ω.

Both may be called random vectors also and denoted by

$$\underline{X}_{2x1} = \begin{pmatrix} X_1 \\ X_2 \end{pmatrix} \tag{5.7}$$

and

$$\underline{X}_{px1} = \begin{pmatrix} X_1 \\ X_2 \\ \vdots \\ \vdots \\ X_p \end{pmatrix} \tag{5.8}$$

$\underline{X}_{2x1}$ and $\underline{X}_{px1}$ are bivariate and multivariate random variables, respectively and take values in 2-dimensional and p-dimensional Euclidean or real space. These variables may be discrete or continuous.

If $\underline{X}_{2x1}$ takes *finite or countably infinite* values (x_1, x_2), it is a discrete bivariate random variable, otherwise it is continuous. Similarly if $\underline{x}_{px1}$ is finite or countably infinite, $\underline{X}_{px1}$ is a discrete multivariate random variable. If it takes infinite values it is continuous.

For simplicity denote bivariate random variable by (X, Y) and its values by (x, y). If (X, Y) is discrete its values may be represented as (x_i, y_j), $i = 1, 2, \ldots m$ and $j = 1, 2,\ldots n$. If (X, Y) is continuous and suppose it can take any value in the rectangle on the interval $[a, b]$ on X-axis and that on $[c, d]$ on Y-axis, then its range can be represented as

$$\{(x, y) \,|\, a \le x \le b;\;\; c \le y \le d\}$$

Though there may be situations when one component of (X, Y) is discrete and the other continuous, such variables are not discussed in this book. Considered (X, Y) are either both discrete or both continuous.

As X and Y considered here are functions defined on the same outcomes of an experiment, it is natural to consider the two jointly. As an analogy of probability distribution of univariate random variable, joint probabilities of (X, Y) need to be considered. In this effort, the definitions and examples of the joint pmf, joint pdf and joint cdf are given below.

5.3.1 Joint pmf and joint pdf

(a) If (X, Y) is a discrete bivariate random variable a function $p(x, y)$ is called the joint pmf or joint probability function of (X, Y), if it satisfies the following conditions:

(i) $p(x, y) \ge 0$ for all (x, y)

(ii) $\displaystyle\sum_{all\ x}\sum_{all\ y} p(x, y) = 1$

(iii) $$P[a \le X \le b \text{ and } c \le Y \le d] = \sum_{x=a}^{b}\sum_{y=c}^{d} p(x, y) \tag{5.9}$$

(b) If (X, Y) is a continuous bivariate random variable, a function $f(x, y)$ is said to be the joint pdf of (X, Y) if it satisfies the following conditions:

(i) $f(x, y) \geq 0$ for all (x, y)

(ii) $\int_{-\infty}^{\infty}\int_{-\infty}^{\infty} f(x, y)dx.dy = 1$ and

(iii) $P[a \leq X \leq b \text{ and } c \leq Y \leq d] = \int_a^b\int_c^d f(x, y)dy\, dx$ (5.10)

(c) If $(X_1, X_2, \ldots X_p)'$ is p-dimensional discrete random variable, then $p(x_1, x_2, \ldots x_p)$ is joint pmf or joint probability function of $(X_1, X_2, \ldots X_p)$ if the following properties are satisfied:

(i) $p(x_1, x_2, \ldots x_p) \geq 0$ for all $x_1, x_2, \ldots x_p$
(ii) $\Sigma p(x_1, x_2, \ldots x_p) = 1$, where summation is over all the values of $(x_1, x_2, \ldots x_p)$
(iii) $p(x_1, x_2, \ldots x_p) = P[X_1 = x_1, X_2 = x_2, \ldots X_p = x_p]$ (5.11)

(d) If $(X_1, X_2, \ldots X_p)'$ is p-dimensional continuous random variable then $f(x_1, x_2, \ldots x_p)$ is joint pdf of $(x_1, x_2, \ldots x_p)$ if the following properties are satisfied.

(i) $f(x_1, x_2, \ldots x_p) \geq 0$ for all $x_1, x_2, \ldots x_p$

(ii) $\int_{-\infty}^{\infty} \ldots \int_{-\infty}^{\infty} f(x_1, x_2 \ldots x_p)\, dx_1\, dx_2 \ldots dx_p = 1$

(iii) $P[\{X_1, X_2, \ldots X_p\} \in A] = \int \ldots_A \int f(x_1, x_2 \ldots x_p) dx_1\, dx_2 \ldots dx_p$ (5.12)

where A is a subset of p-dimensional range space of the said p-dimensional random variable.

It may be mentioned that as for pmf and pdf of univariate random variable, the range space of bivariate and multivariate random variables should also be given along with their pmf and pdf.

(e) Some examples of pmf and pdf of bivariate random variable (X, Y) are given below:

Example 5.11: Two production lines manufacture certain type of item. Line I has capacity to produce 5 items per day and line II has capacity of 3 items per day. Number of items actually produced per day by line I (X) and by line II (Y) are random variables. The joint pmf of (X, Y) is given below (source: Meyer (1975), p. 96, Ex. 6.1):

Table 5.2 Joint pmf of number of items produced by line I (X) and line II (Y)

$y\downarrow$ $x\rightarrow$	0	1	2	3	4	5
0	0	0.01	0.03	0.05	0.07	0.09
1	0.01	0.02	0.04	0.05	0.06	0.08
2	0.01	0.03	0.05	0.05	0.05	0.06
3	0.01	0.02	0.04	0.06	0.06	0.05

$$p(2, 3) = 0.04 = P[X = 2 \text{ and } Y = 3]$$

Suppose it is desired to find the probability of the event that more items are produced by line I than by line II i.e.

$$\begin{aligned} P[X > Y] = \Sigma P[x_{ij} > y_{ij}] &= .01 + .03 + .05 + .07 + .09 \\ &\quad + .04 + .05 + .06 + .08 \\ &\quad + .05 + .05 + .06 \\ &\quad + .06 + .05 \\ &= .75 \end{aligned}$$

Example 5.12: Consider the bivariate function

$$f(x,y) = \begin{cases} K(x+y) & 0<x<1 \text{ and } 0<y<1 \\ 0 \text{ otherwise} \end{cases}$$

Can $f(x, y)$ may be joint pdf of (X, Y) for a proper choice of K? Find that value of K and find the probability that $X < ½$ and $Y < ½$ i.e. $P[X < ½$ and $Y < ½]$ (source: Mood et al. (1974), p. 141, Example 7).
Solution: For positive values of K, $f(x, y)$ is positive. Value of K is obtained from the result that

$$\int_{-\infty}^{\infty}\int_{-\infty}^{\infty} f(x,y)dy\, dx = 1$$

or

$$K\int_0^1\int_0^1 (x+y)dy\, dx = 1$$

On integrating LHS of above and solving for K, $K = 1$.
Hence for $K = 1$, $f(x, y)$ is joint pdf of (X, Y)

$$P[X < ½, Y < ½]$$

$$= \int_0^{1/2}\int_0^{1/2} (x+y)dx\, dy = \left(\frac{x^2}{2} + \frac{y^2}{2}\right)_{x,y=0}^{1/2} = \frac{1}{4} + \frac{1}{4} = \frac{1}{2}$$

5.3.2 Marginal and Conditional Distributions

(a) Let (X, Y) be a discrete bivariate random variable with joint pmf $p(x, y)$, then *marginal pmf's* of X and Y are denoted by $p_1(x)$ and $p_2(y)$ and are given by

$$p_1(x) = \sum_{all\ y} p(x,y) \tag{5.13}$$

and

$$p_2(y) = \sum_{all\ x} p(x,y) \tag{5.14}$$

(b) If (X, Y) is a continuous bivariate random variable with joint pdf $f(x, y)$, then *marginal pdf's* of X and Y are denoted by $f_1(x)$ and $f_2(y)$ and are given by

$$f_1(x) = \int_{-\infty}^{\infty} f(x,y)\, dy \tag{5.15}$$

$$f_2(y) = \int_{-\infty}^{\infty} f(x,y)\, dx \tag{5.16}$$

(c) If (X, Y) is a discrete bivariate random variable with joint pmf $p(x, y)$, then *conditional pmf's* of $X/_{Y=y}$ and of $Y/_{X=x}$ are given by

$$h(x/y) = [p(x, y)]/p_2(y), \text{ if } p_2(y) > 0$$
$$= P[X = x/Y = y] \tag{5.17}$$

and

$$g(y/x) = [p(x, y)]/p_1(x) \text{ if } p_1(x) > 0$$
$$= P[Y = y/X = x] \tag{5.18}$$

(d) If (X, Y) is a continuous bivariate random variable with joint pdf $f(x, y)$, then *conditional pdf's* of $X/_{Y=y}$ and $Y/_{X=x}$ are defined by

$$h\left(x/_{y}\right) = \frac{f(x, y)}{f_2(y)}, \text{ if } f_2(y) > 0 \tag{5.19}$$

$$g\left(y/_{x}\right) = \frac{f(x, y)}{f_1(x)}, \text{ if } f_1(x) > 0 \tag{5.20}$$

Some times though a bivariate random variable (X, Y) is being considered, need may arise to make probability statements about only X or about only Y. The marginal pmf's and marginal pdf's are used for those evaluations. One may also be interested to make probability statements about one of the variables X or Y for some known values of the other.

For example if $p_2(2)$ or $f_2(2)$ is not zero, one may be interested in probability distributions of X for known value of $Y = 2$. Using the conditional pmf $h\left(X/_{y=2}\right)$ from (5.17), if (X, Y) is discrete or using the conditional pdf from (5.19) if (X, Y) is continuous one can make probability statements about X when $Y = 2$.

Example 5.13: Consider again the joint pmf of discrete bivariate random variable of Example 5.11. In addition to the entries in Example 5.11, the 'marginal' totals are also computed and entered below.

Table 5.3 Joint pmf of number of items produced by lines I and II (source: Meyer (1975), p. 96, Ex. 6.1)

$y\downarrow$ $x\rightarrow$	0	1	2	3	4	5	Total $p_2(y)$
0	0	0.01	0.03	0.05	0.07	0.09	0.25
1	0.01	0.02	0.04	0.05	0.06	0.08	0.26
2	0.01	0.03	0.05	0.05	0.05	0.06	0.25
3	0.01	0.02	0.04	0.06	0.06	0.05	0.24
Total $p_1(x)$	0.03	0.08	0.16	0.21	0.24	0.28	1.00

Solution: Totals in the last column (or margin of the table), when all the respective row values are summed up, give the marginal pmf of Y, $p_2(y)$ as defined in (5.14). It can be used to find probability of $Y = 3$ irrespective of what X is.

It can be used to find $P[Y < 2]$.

From the marginal pmf $p_2(y)$, $P[Y < 2] = p_2(0) + p_2(1)$

or
$$P[Y < 2] = .25 + .26 = .51$$

Similarly the column totals of Table 5.3 give marginal pmf of $X = p_1(x)$. It can be used to find the probability that

$$P[1 \le X < 4] = p_1(1) + p_1(2) + p_1(3) = .08 + .16 + .21$$
$$= .45$$

Now suppose it is required to find the conditional pmf $h\left(x/_{y=2}\right)$. Using formula (5.17) first $h\left(x/_{y=2}\right)$ is obtained.

$$h\left(x/_{y=2}\right) = \frac{p(x,y)}{p_2(2)} \quad \text{since } p_2(2) \text{ is positive}$$

If $P[2 \le X < 4 /y = 2]$ is to be evaluated, $h\left(x/_{y=2}\right)$ will be used as

$$P[2 \le X < 4 /y = 2] = \sum_{x=2}^{3} \frac{p(x,y)}{p_2(2)} = \frac{p(2,2)+p(3,2)}{p_2(2)}$$
$$= (.05 + .04)/0.25 = .09/.25 = 9/25$$

Example 5.14: Suppose the joint pdf of continuous bivariate random variable (X, Y) of Example 5.12 is considered again. It is defined as

$$f(x,y) = \begin{cases} (x+y) & 0 < x < 1 \text{ and } 0 < y < 1 \\ 0 & \text{otherwise} \end{cases}$$

Obtain the marginal pdf's $f_1(x)$ and $f_2(y)$ and the conditional probability $P[0 < Y < ½/x = ½]$ (source: Mood and Graybill, p. 141, Example 7).
Solution: The marginal pdf's $f_1(x)$ and $f_2(y)$ are obtained first and then the conditional probability.

$$f_1(x) = \int_{-\infty}^{\infty} f(x,y)\,dy = \int_0^1 (x+y)\,dy$$

$$= \left(xy + \frac{y^2}{2}\right)\Bigg|_{y=0}^{1} = x + \frac{1}{2} \quad 0 < x < 1$$

$$f_2(y) = \int_{-\infty}^{\infty} f(x,y)\,dx = \int_0^1 (x+y)\,dx = \left(\frac{x^2}{2} + xy\right)\Bigg|_{x=0}^{1}$$

$$= \frac{1}{2} + y \quad 0 < y < 1$$

Conditional density of $Y/X = 2$ is given by

$$h(y/x) = \frac{f(x,y)}{f_1(x)} = \frac{x+y}{x+1/2} \quad 0 < y < 1$$

Hence $P[0 < Y < 1/2/x = 1/2]$

$$= \int_0^{1/2} h\left(y/_{x=1/2}\right)dy = \int_0^{1/2} \frac{(x+y)}{x+1/2}\Bigg|_{x=1/2} dy$$

$$= \int_0^{1/2} \frac{(1/2+y)}{1/2+1/2}\,dy = \int_0^{1/2} \frac{(1/2+y)}{1}\,dy = \left(1/2y + y^2/_2\right)\Bigg|_0^{\frac{1}{2}}$$

$$= \frac{1}{4} + \frac{1}{8} = \frac{3}{8}$$

5.3.3 Independent Random Variables

Conditional probability of an event and independence of events has been already defined in previous chapter. In this section independence of random variables is defined.

(a) Let (X, Y) be discrete bivariate random variable with joint pmf $p(x, y)$. X and Y are defined as stochastically independent if and only if

$$p(x, y) = p_1(x).\ p_2(y) \tag{5.21}$$

(b) Let (X, Y) be continuous bivariate random variable with joint pdf $f(x, y)$. X and Y are defined to be stochastically independent if and only if

$$f(x, y) = f_1(x).\ f_2(y) \tag{5.22}$$

(a) and (b) indicate that (X, Y) are independent random variables if their joint pmf or pdf can be factored into marginal pmf's or marginal pdf's of the random variables X and Y.

(c) If $(X_1, X_2, \ldots, X_k)$ is a k-dimensional discrete random variable with joint probability mass function $p(x_1, x_2, \ldots, x_k)$, the random variables $X_1, X_2, \ldots, X_k$ are independent if and only if

$$p(x_1, x_2, \ldots, x_k) = p_1(x_1).\ p_2(x_2), \ldots, p_k(x_k) \tag{5.23}$$

(d) If $(X_1, X_2, \ldots, X_p)$ is a p-dimensional continuous random variables with joint pdf $f(x_1, x_2, \ldots, x_p)$, $X_1, X_2, \ldots, X_p$ are independent if and only if

$$f(x_1, x_2, \ldots, x_p) = f_1(x_1).\ f_2(x_2), \ldots, f_p(x_p) \tag{5.24}$$

Thus it is clear that two or more random variables are independent if and only if their joint pmf or pdf can be factored into marginal pmf's or pdf's of the set of random variables.

(e) Some examples of independent bivariate random variables (X, Y) follow:

Example 5.15: A machine is used for one type of task in the morning and for another type of task in the evening. Let X, and Y represent the number of times the machine breaks down in the morning and in the evening, respectively. Table below gives the joint pmf of (X, Y).

Table 5.4 Joint pmf of number of break downs in the morning and evening (Source: Meyer (1975), p. 104, Example 6.10)

$Y\downarrow$ $X\rightarrow$	0	1	2	Total = $p_2(y)$
0	0.1	0.2	0.2	0.5
1	0.04	0.08	0.08	0.2
2	0.06	0.12	0.12	0.3
Total = $p_1(x)$	0.2	0.4	0.4	1.0

Are X, and Y independent?

Solution: To answer this question the marginal distribution of $X(= p_1(x))$ and $Y(= p_2(y))$ are obtained by adding the columns and rows, respectively.

The last row gives the $p_1(x)$ for various x (= 0, 1, 2) and the last column gives the $p_2(y)$ for various y(= 0, 1, 2).

For each of $p(x_i, y_j)$, i = 1, 2, 3 and j = 1, 2, 3, it is verified whether

$$p(x_i, y_j) = p_1(x_i).\ p_2(y_j)$$

It is evident that this relation is satisfied for each row and column element.

Hence since joint pmf of (X, Y) is product of marginal pmf's for all x and y, the variables X and Y are independent.

Example 5.16: Let X and Y be the life lengths of two electronic devices. Suppose that their joint pdf is given by

$$f(x, y) = e^{-(x+y)} \quad x > 0,\ y > 0$$

Are X, and Y independent?

Solution: To answer this one can see that $f(x, y) = e^{-x}.\ e^{-y}$.

It can be seen that e^{-x} is marginal density of X and e^{-y} is marginal density of Y. Hence X and Y are independent.

5.4 Expected Values of Random Variables and their Rules

The probability distribution of any random variable completely describes the behavior of the random variable. Associated with any random variable or its distribution, there are some constants called parameters which help in describing their nature. Knowledge of these gives a quick insight to experimenter about the random variable. Some of the descriptive statistics associated with a random variable are based on a general concept of *mathematical expectation* or *expected values*.

Before a formal definition of expected value of a random variable X, consider the following experiment.

Example 5.17: A single fair die is rolled over and over. Each time X, the number of points on the upper face is recorded. Let one such sequence of values of X be

$$2,\ 5,\ 4,\ 1,\ 6,\ 3,\ 2\ \ldots$$

From this sequence of values of X, a sequence of arithmetic means of X may be obtained based on 1, 2, 3, 4 ... observations as:

Table 5.5 Sequence of arithmetic mean

Number of X's	Arithmetic mean
1	2/1 = 2
2	(2 + 5)/2 = 3.5
3	(2 + 5 + 4)/3 = 3.6
4	(2 + 5 + 4 + 1)/4 = 3
5	(2 + 5 + 4 + 1 + 6)/5 = 3.6
6	(2 + 5 + 4 + 1 + 6 + 3)/6 = 3.5
7	(2 + 5 + 4 + 1 + 6 + 3 + 2)/7 = 3.28

The values of average in the sequence are not constant. However, if the experiment is repeated a large number of times, the sequence of average tends to settle around a definite value = 3.5. This definite value is called 'expected value' or the long run average value of X.

Example 5.18: A manufacturer is producing items such that 10% are defective and 90% are non defective. For each defective he loses Rs. 1 and for each non defective item, he earns the profit of Rs. 5. If X denotes the net profit per item, it is a random variable taking values – 1 Rs. with probability 0.1 and 5 Rs. with probability 0.9.

Thus the manufacturer looses Rs. 1 with probability 0.1 and earns Rs. 5 with probability 0.9. In the long run 1/10 of times he loses Rs. 1 and 9/10 of times he earns Rs. 5. So the expected or long run profit earned

$$= (-1).\ (1/10) + 5\ (9/10)$$

$$= 4.5 - .1 = 4.4 \text{ Rs. per item}$$

On the basis of these examples the expected value or the long run average value of a random variable is defined below.

Expected value of a random variable X is denoted by *E*(*X*) and is defined as

$$E(X) = \sum_{all\,x} x\ p(x), \text{ if } X \text{ is discrete with pmf } p(x) \tag{5.25}$$

$$= \int_{-\infty}^{\infty} x\ f(x)dx, \text{ if } X \text{ is a continuous with pdf } f(x) \tag{5.26}$$

It may be noted that the expected value of a random variable has the same unit as the variable. Moreover, the expected value of a random variable *E*(*X*) may not be a possible value of *X*, as the expected value of number of points on the upper face is 3.5 which can never come up in any roll of the dice.

Example 5.19: The following table shows the pmf of random variable *X*, the number of adult females in a band of howler monkeys

Table 5.6 Probability distribution of adult female monkeys (source: Milton and Tsokos (1983), p. 120, Q. 2 of Exercises 4.4)

x	1	2	3	4	5
p(*x*)	.1	.15	.5	.15	.1

Find expected number of adult female howler monkeys.
Solution:

$$E(X) = \sum_{x=1}^{5} x\ p(x) = 1\times.1 + 2\times.15 + 3\times.5 + 4\times.15 + 5\times.1$$

$$= .1 + .30 + 1.5 + .6 + .5 = 3.0$$

Example 5.20: Consider the random variable *Z*, the number of *cc*'s of a drug to be prescribed to control epileptic seizures. Its density is given as

$$f(z) = \frac{200}{9} z \quad 0 < z < .3$$

Find the expected value of the drug administered to epileptic patients (source of pdf only: Milton and Tsokos, p. 101, Example 4.2.2).
Solution:

$$E(Z) = \int_{-\infty}^{\infty} z\ f(z)dz = \int_{0}^{.3} \frac{200}{9} z^2\, dz = \frac{200}{9} \frac{z^3}{3}\Big|_0^{.3}$$

$$= \frac{200}{9} \frac{(.3)^3}{3} = .2\ cc$$

5.4.1 Rules for the Expected Values

Following are some of the rules followed by expected values of a random variable:

1. $E(k) = k$ (since $X = k$ with probability 1, so $E(k) = k \times 1 = k$) (5.27)
2. $E(X \pm k) = E(X) \pm k$ (5.28)
3. $E(k\,X) = k\,E(X)$ (5.29)
4. $E(X \pm Y) = E(X) \pm E(Y)$ (5.30)
   where X and Y are random variables
5. $E(X_1 \pm X_2 \pm \ldots \pm X_k) = E(X_1) \pm E(X_2) \pm \ldots \pm E(X_k)$ (5.31)
   where $X_1, X_2, \ldots, X_k$ are random variables
6. $E\left(\sum_{i=1}^{k} a_i X_i\right) = \sum_{i=1}^{k} a_i\, E(X_i)$ (5.32)
   where X_i, $i = 1, 2, \ldots, k$ are random variables and a_i, $i = 1, 2, \ldots, k$ are constants
7. $E(XY) = E(X).\, E(Y)$, if X and Y are independent (5.33)

5.5 Expected Values of Function of Random Variables

Let $y = H(x)$ is a real valued function of x, where x is value attained by a random variable X defined on sample space Ω. Then $Y = H(X)$ is also a random variable, since for every $\omega \in \Omega$, $H(X(\omega))$ can be determined say y. Thus for every $\omega \in \Omega$, range space of Y i.e. $\{y\}$ can be determined. So Y is also a random variable.

Thus if X is a random variable, $X + 1$, $X^2 - 2$, e^x, $\log_e (X + 5)$, $X^3 + X^2 + X + 1$ etc. are all random variables. Often one is interested in expected values of functions of a random variable. It can be determined if the probability distribution of X is known using the following result:

Expected value of $Y = H(X)$ is denoted by $E(Y) = E\,(H(X))$ and is given by

$$E(Y) = E(H(X)) = \sum_{all\, x} H(x)\, p(x), \text{ if } X \text{ is discrete} \qquad (5.34)$$

$$= \int_{-\infty}^{\infty} H(x)\, f(x)\, dx, \text{ if } X \text{ is continuous} \qquad (5.35)$$

Example 5.21: Consider the Example 5.19 on X: number of adult female monkeys. Determine $E(X^2)$ (source: Milton and Tsokos (1983), p. 120).

Solution: Since X is discrete, using (5.34)

$$E(X^2) = \sum_{all\, x} x^2 p(x)$$

$$= 1^2 \times .1 + 2^2 \times .15 + 3^2 \times .5 + 4^2 \times .15 + 5^2 \times .1$$

$$= .10 + .60 + .45 + 2.40 + 2.50 = 6.05$$

Example 5.22: Let X be a random variable with pdf $f(x) = \begin{cases} 1 & 0 < x < 1 \\ 0 & \text{otherwise} \end{cases}$

Find $E(X^2 + X)$.

Solution: Since X is a continuous random variable, from (5.35)

$$E(X^2 + X) = \int_0^1 (x^2 + x).1dx = \left(\frac{x^3}{3} + \frac{x^2}{2}\right)_0^1$$

$$= \frac{1}{3} + \frac{1}{2} = \frac{5}{6}$$

5.6 Generating Functions

Generating functions were originally used for solving problems in probability theory. These are useful for generating probabilities, moments etc. of distributions and for study of probability distributions. In this section the probability generating function, moment generating function and characteristic function are defined and their properties and uses are given.

5.6.1 Probability Generating Function

It was first introduced by Laplace and has systematic application in probability theory. This is abbreviated as PGF denoted by $P_X(t)$ and is specially meant for a discrete random variable with non negative integer values.

Suppose $p(x) = P[X = x]$ for $x = 0, 1, 2, 3\ldots$

Then PGF of X is given by

$$P_X(t) = E\left(t^X\right) = \sum_{x=0}^{\infty} t^x p(x) \tag{5.36}$$

The coefficient of the t^x in the PGF gives the pmf of X at x or $P[X = x]$ is coefficient of t^x in $P_X(t)$.

For every integer valued random variable X, the PGF exists and is unique.

Thus

1. Every PGF determines a unique pmf.
2. If X_1 and X_2 are two independent random variables then

$$P_{(X_1+X_2)}(t) = E(t^{X_1+X_2}) = E(t^{X_1} \cdot t^{X_2}) = E(t^{X_1}).E(t^{X_2})$$

$$= P_{X_1}(t).P_{X_2}(t) \tag{5.37}$$

Thus PGF of sum of two independent random variables is product of the individual PGF's

3. The result of property 2 can be generalized to the case of a number of independent random variables.

Example 5.23: Consider the *binomial distribution*

$$p(x) = \binom{n}{x} p^x (1-p)^{n-x} \quad x = 0, 1, 2, 3, \ldots, n;\ p, q > 0 \text{ and } p + q = 1$$

PGF of $X = P_X(t) = E(t^X)$

$$= \sum_{x=0}^{n} t^x p(x) = \sum_{x=0}^{n} \binom{n}{x} t^x p^x q^{n-x}$$

$$= \sum_{x=0}^{n} \binom{n}{x} (pt)^x . q^{n-x} = (q + pt)^n \tag{5.38}$$

5.6.2 Moment Generating Function

The moment generating function of a random variable is denoted by $M_X(t) = E(e^{tX})$ where X is a real valued continuous random variable. $M_X(t)$ is abbreviated as MGF. MGF may not exist for all random variables for any value of t (including t = zero).

Thus $$M_X(t) = \sum_{all\,x} e^{tx} p(x), \text{ if } X \text{ is a discrete} \tag{5.39}$$

$$= \int_{-\infty}^{\infty} e^{tx} f(x)\,dx, \text{ if } X \text{ is a continuous} \tag{5.40}$$

Properties

1. Relation between $P_X(t)$ and $M_X(t)$ as

$$M_X(t) = P_X(e^t) \tag{5.41}$$

i.e. replacing t by e^t in $P_X(t)$, $M_X(t)$ is obtained.

2. If $M_X(t)$ is MGF of a random variable X, then r^{th} raw moment of X is given by

$$\mu_r' = \frac{d^r M_x(t)}{dt^r}\Big|_{t=0} \tag{5. 42}$$

3. $M_{(X\pm a)}(t) = e^{\pm at} M_X(t)$ (5.43)
4. $M_{(a+bX)}(t) = e^{at} M_X(bt)$ (5.44)

Replacing t by bt in $M_x(t)$ and multiplying it by e^{at} gives MGF of $a + bX$.

5. Let X_1 and X_2 be two independent random variables then

$$M_{(X_1+X_2)}(t) = M_{X_1}(t).M_{X_2}(t) \tag{5.45}$$

The result can be generalized as

$$M_{\sum_{i=1}^{n} X_i}(t) = \prod_{i=1}^{n} M_{X_i}(t) \tag{5.46}$$

Example 5.24: MGF of Binomial distribution $B(n, p)$ is given by

$$M_X(t) = E\left(e^{tX}\right) = \sum_{x=0}^{n} e^{tx}\binom{n}{x} p^x q^{n-x} = \sum_{x=0}^{n}\binom{n}{x}\left(pe^t\right)^x q^{n-x}$$

$$= (q + pe^t)^n \tag{5.47}$$

5.6.3 Characteristic Function

The MGF does not exist always. A function which always exists is characteristic function. It is abbreviated as CF and is denoted by

$$\phi_X(t) = E(e^{itX}) \tag{5.48}$$

$$= \sum_{all\,x} \left(e^{itx}\right)p(x), \text{ if } X \text{ is a discrete variable} \tag{5.49}$$

$$= \int_{-\infty}^{\infty} e^{itx} f(x)\,dx, \text{ if } X \text{ is continuous} \tag{5.50}$$

where, $i = \sqrt{-1}$.

Properties

1. If MGF of X exists, then it is related to CF of X by
$$M_X(it) = \phi_X(t) \tag{5.51}$$
2. Given a random variable with distribution function $F(x)$ then $\phi_X(t)$, determined from (5.49) and (5.50), is unique.
3. A CF $\phi_X(t)$ uniquely determines a distribution function from the following
$$f(x) = F'(x) = \frac{1}{2\pi}\int_{-\infty}^{\infty} e^{-itx}\phi_x(t)\,dt \tag{5.52}$$
4. If $\phi_X(t)$ is CF of random variable X then CF of $a + bx$, where a and b are real constants is given by
$$\phi_{(a+bX)}(t) = e^{iat}\ \phi_X(bt) \tag{5.53}$$
5. X is called a symmetric random variable about $X = 0$ if distribution of $-X$ is same as that of X. Hence if $\phi_X(t)$ is CF of X, it is also CF of $-X$, and it must be real. Thus CF of a symmetric variable is real and if $\phi_X(t)$ is real corresponding X must be symmetric random variable.
6. r^{th} raw moment of X, is given by
$$\mu'_r = \frac{1}{i^r}\frac{\partial^r \phi_X(t)}{dt^r}\Big|_{t=0} \tag{5.54}$$
7. If X and Y are independent random variables CF of $X + Y$ is product of the CF's of X and Y. In symbols
$$\phi_{(X+Y)}(t) = \phi_X(t).\ \phi_Y(t) \tag{5.55}$$

5.7 Raw and Central Moments

In chapter 3 raw and central moments have been defined for ungrouped and grouped (or frequency table) data. If the data is on a part (a sample) not on all the elements in the population, the calculated moments will not be actual (or will not be true) moments but will be only estimates of the moments.

The actual or true moment of a random variable can be obtained if the probability distribution of the random variable is known. The definition of these is given below:

5.7.1 Raw Moments

If X is a random variable with pdf $f(x)$ then r^{th} raw moment of X is denoted by μ'_r and is defined as
$$\mu'_r = E(X^r)$$
Taking $(X^r) = H(X)$ of equation (5.34) and (5.35)
$$\mu'_r = \sum_{all\ x} x^r p(x), \text{ if } X \text{ is discrete} \tag{5.56}$$
$$= \int_{-\infty}^{\infty} x^r f(x)\,dx, \text{ if } X \text{ is continuous} \tag{5.57}$$
It is customary to denote $\mu'_1 = E(X)$ by μ, the mean of the random variable. In symbols
$$\mu'_1 = E(X) = \mu \tag{5.58}$$

It is worth seeing that first moment about origin is mean of the distribution and there is an analogy of the definitions of mean given for the grouped and ungrouped data and those given in this section for discrete random variable. If a discrete random variable X is taking finite (say N) values $x_1, x_2, \ldots, x_N$ then mean of these is

$$\frac{x_1 + x_2 + \ldots x_N}{N} = \sum_{i=1}^{N} \frac{1}{N} x_i$$

Since each of X_i is appearing once in N possible values of X, $P[X = x_i] = \dfrac{1}{N}$

Hence $\sum_{i=1}^{N} \frac{1}{N} x_i = \sum_{i=1}^{N} x_i \, P[X = x_i]$ is same as (5.56)

If N is large so that data is grouped in k classes with x_i as the mid point and f_i as the frequency of the ith class with $\sum_{i=1}^{k} f_i = N$ then, the mean as defined for grouped data is $\frac{1}{N}\sum_{i=1}^{k} f_i \, x_i$ which may also be written as $\sum_{i=1}^{k} x_i \left({}^{f_i}\!/_{N} \right)$.

According to statistical definition of probability f_i/N, the relative frequency of x_i is taken as the probability $P[X = x_i]$. Hence this definition is also same as that in equation (5.56).

$x_1, x_2, \ldots x_i, \ldots$ are the values of discrete random variable and unit mass is distributed over these points such that mass $p(x_i)$ is at x_i, then $\sum_{all\ x_i} x_i p(x_i)$ is the center of mass. Similarly if X is a continuous random variable $f(x)\, dx$ is the density of mass around x. So $\int_{-\infty}^{\infty} xf(x)\, dx$ is the center of mass of the random variable X i.e. center of gravity of the lamina cut along the curve $f(x)$ and X-axis below $f(x)$.

Thus it is evident that μ_1' is the actual mean of the random variable X. $E(X)$ is therefore, a measure of central tendency of the random variable X.

In a similar way μ_r', the rth raw moment of X is physically moment about origin when horizontal distances from origin (0, 0) are raised to r^{th} power.

Example 5.21 has given calculation of 2nd raw moment of X: number of all adult female monkeys which is a discrete random variable.

Example 5.25: Consider again the Example 5.22 and determine $E(X)$, for the random variable X with pdf

$$f(x) = \begin{cases} 1 & \text{if } 0 < x < 1 \\ 0 & \text{otherwise} \end{cases} \tag{5.59}$$

Solution: From (5.26), putting $r = 1$.

$$E(X) = \int_0^1 x\, f(x) dx = \int_0^1 x\, dx = x^2/2\Big|_0^1 = \frac{1}{2}$$

5.7.2 Central Moments

Already indicated that the central moments defined in Chapter 3 are estimates of actual or true central moments, unless the observations on the random variable X are available on all the elements of a finite population. However, if probability distribution of the random variable X is known the actual or true central moments can be determined from its definition given below.

$$\mu_r = E[(X - E(X)]^r \tag{5.60}$$

$$= E(X - \mu)^r \text{ from (5.58)}$$

It is evident that

$$\mu_1 = E(X - \mu) = 0$$

$\mu_2 = E(X - \mu)^2 = \sigma^2$, the variance of $X = E(X)^2 - [E(X)]^2$
$\mu_3 = E(X - \mu)^3$ and $\mu_4 = E(X - \mu)^4$ are the third and 4th central moments of X. It will be seen in the next section that these are used to measure skewness and kurtosis of the distribution of X.

5.8 Coefficients of Skewness and Kurtosis

Departure from symmetry is measured by coefficient of skewness denoted by $\sqrt{\beta_1}$ and defined by

$$\beta_1 = (\mu_3^2/\mu_2^3) \tag{5.61}$$

So that, coefficient of skewness

$$\sqrt{\beta_1} = (\mu_3/\mu_2^{3/2}) = (\mu_3/\sigma^3) \tag{5.62}$$

Kurtosis or steepness is measured by coefficient of kurtosis denoted by β_2 and defined by

$$\beta_2 = (\mu_4/\mu_2^2) = (\mu_4/\sigma^4) \tag{5.63}$$

Using these relations all the characteristics of the distribution of the random variable X can be obtained precisely, from its pdf (or pmf, if X is discrete)

Hence one can conclude that by knowing the pdf (or pmf) of a random variable all the information about the random variable including its location, dispersion, skewness and steepness becomes known.

6. Distributions

Last chapters gave idea about use of data, types of random variables and the way probability statements about possible values of a random variable can be given. In this chapter, families of random variables are considered which have been found useful in describing various phenomena in scientific studies. The variables in each group form a family in the sense that each of them possesses the probability density function of the same mathematical form, differing only in numeric values of some pertinent parameter(s). Thus the random variables of a family are said to have a scientific probability distribution. The following sections give the mathematical form and properties of some such distributions.

6.1 Discrete and Continuous Distributions

It is already clear that random variables may be discrete or continuous. Hence first some distributions are considered which are associated with discrete random variables and next those which are associated with continuous random variables.

The distributions associated with discrete random variables are called discrete distributions and those associated with continuous random variables are called continuous distributions.

The discrete distributions considered in this chapter are (i) The Binomial Distribution (ii) The Poisson Distribution (iii) The Hypergeometric Distribution (iv) The Geometric Distribution and (v) The Negative Binomial Distribution.

Continuous distributions discussed here are (i) The Normal Distribution (ii) The Uniform Distribution (iii) The Bivariate Normal Distribution (iv) The Chi-square Distribution (v) The Student's t-Distribution and (vi) The F-Distribution.

For further reading, clarification of concepts and examples, the interested reader may refer the books by Mood et al. (1974), Meyer (1975), Milton and Tsokos (1983), Hoel (1971), Daniel (1987), Goon et al. (1983), Zar (2004), Johnson and Kotz (1969, 1970, 1994, 1995).

6.2 Binomial Distribution

Random variables with binomial distribution may arise from experiments which appear quite different. But the following common conditions are essential when a random variable has a binomial distribution.

1. A finite number (= n) of trials of an experiment are performed under essentially same conditions.
2. The trials are independent so that outcome of any trial does not affect outcome of any other trial.

3. There are only two mutually exclusive outcomes of each trial: occurrences or non-occurrences of an event. Occurrences may also be designated as 'success' S and non-occurrence as 'failure' F.
4. The probability of success, (denoted by p) remains the same for all the trials. So is the probability of failure: $1 - p$ $(= q)$.p is called the parameter of the binomial distribution and n is called its index.

Let X denote the total number of successes in n trials. Then X is a random variable of interest and follows the Binomial distribution. Symbolically one can write that the said $X \sim B\ (n,\ p)$.

A precise definition of binomial distribution may be given as: A random variable X is said to have binomial distribution $B(n, p)$ if its probability mass function (pmf) is given by

$$p(x) = {}^nC_x\, p^x\, (1 - p)^{n-x};\ x = 0, 1, 2, ..., n \tag{6.1}$$

Appendix A.6 at the end of this chapter gives the definition of the mathematical symbol ${}^nc_x$, also denoted by $\binom{n}{x}$. Figure 6.1 gives the probability distribution of binomial random variables with n = 10, 5 and 20 for various values of p.

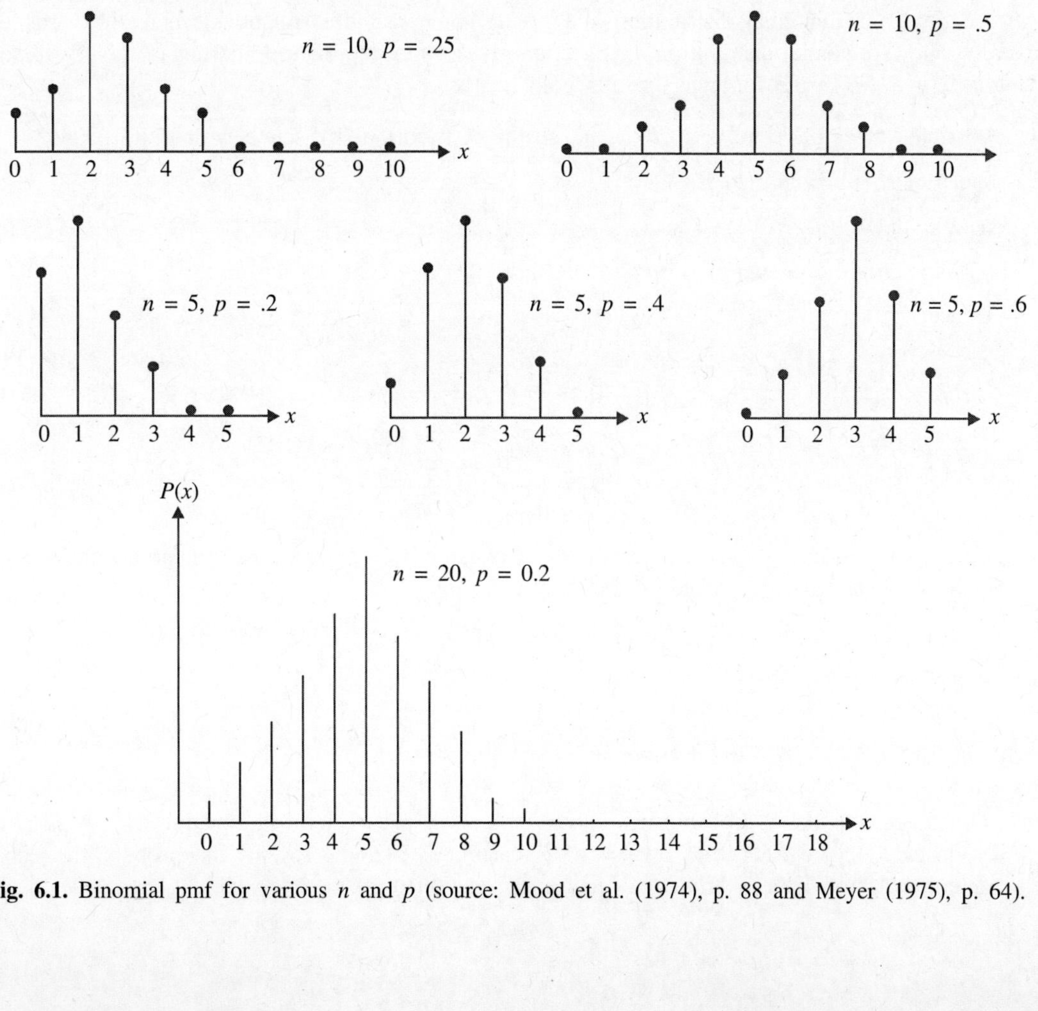

Fig. 6.1. Binomial pmf for various n and p (source: Mood et al. (1974), p. 88 and Meyer (1975), p. 64).

Some practical examples when binomial distribution is found to arise are given below.

Example 6.1: A man and woman, each with one recessive gene for blue eye colour and one dominant gene for brown eye colour parent four children. Then X: the number of children with blue eye colour follows binomial distribution $B(4, 1/4)$.

Example 6.2: A new variety of wheat is being developed and it is hoped that its germination rate is 80%. If 20 seeds are planted in soil of identical composition and given the same care, then X: the total number of germinated seeds follows binomial distribution $B(20, .8)$. It will be seen that the expected number of germinated seeds are 16.

Example 6.3: In a population of insects, 40% are infected with a given virus, V (say). Suppose samples of size n (= 5) insects each are selected and each insect is examined for infection of virus V. Count of the total number of infected insects in the sample say X, follows binomial distribution $B(5, .4)$.

It is assumed that the population is so large that the proportion of infected insects does not change for all practical purposes when sample is selected.

It may be noted that the probabilities of various values of X, $(0, 1, 2, \ldots, n)$ are the terms in the expansion of $(q + p)^n$ according to Binomial Theorem. The term containing p^x is the probability of X taking value x, and is given in equation (6.1).

6.2.1 Properties of Binomial Distribution

After defining the binomial distribution and giving some real life situations where the variable of interest follows binomial distribution, this section gives some important properties of this distribution. Thus if $X \sim B(n, p)$ the following results hold:

1. Possible values of X are $0, 1, 2, \ldots, n$, so the distribution is discrete with finite range.
2. Mean of $X = n\,p$ and (b) $V(X) = n\,p\,q$ (6.2)
3. $\mu_3(X) = n\,p\,q\,(q - p)$ (6.3)

   $\mu_4(X) = 3n^2\,p^2\,q^2 + npq\,(1 - 6pq)$ (6.4)

   Hence

$$\beta_1 = \frac{\mu_3^2}{\mu_2^3} = \frac{(p-q)^2}{npq} \tag{6.5}$$

$$\beta_2 = \frac{\mu_4}{\mu_2^2} = 3 + \frac{1-6pq}{npq} \tag{6.6}$$

   From (6.5) and (6.6) β_1 and $\beta_2 - 3$ tend to 0 for large n. In other words, for large n the binomial distribution tends to be symmetric with zero kurtosis.

   Moreover, if $p = 1/2 = q$ binomial distribution is symmetric for all n since $P[X = x] = P[X = (n - x)] = {}^n c_x\,(1/2)^n$ and also since $\beta_1 = 0$.

4. *Recurrence Relation*

$$p\,(k+1) = \frac{n-k}{k+1} \cdot \frac{p}{q} \cdot p(k), \quad k = 0, 1, 2, \ldots, n-1 \tag{6.7}$$

   Thus to find binomial probabilities for $x = 0, 1, 2, \ldots, n$ each term in equation (6.1) need not be evaluated. Only $P[X = 0] = p(0) = q^n$ is calculated and the rest of the probabilities can be determined subsequently from the recurrence relation (6.7).

5. (i) $p(x - 1) < p(x)$, if $x < (n + 1)\, p$
   (ii) $p(x - 1) > p(x)$, if $x > (n + 1)\, p$ and
   (iii) $p(x - 1) = p(x)$, if $x = (n + 1)\, p$.

6. If $n \to \infty$ and $p \to 0$ such that

$$np = \lambda \text{ is finite, then} \tag{6.8}$$

   binomial distribution of X tends to Poisson distribution with parameter λ.
   In symbols,

$$\text{if } X \sim B(n, p) \text{ then if (6.8) holds, then } X \to P(\lambda) \tag{6.9}$$

7. If p is neither close to 0 nor close to 1, then for large n the distribution of

$$Z = (X - np)/\sqrt{npq} \to N(0,1) \tag{6.10}$$

   It can be said, under above conditions, that X has approximately $N(np, npq)$ distribution.
8. *Moment Generating Function* of binomial distribution $B(n, p)$ is given by

$$M_x(t) = (q + pe^t)^n \tag{6.11}$$

6.3 Poisson Distribution

It was mentioned in the properties of binomial distribution that under certain conditions, it can be said that a binomial random variable has approximately a Poisson distribution. However, it may be noted that Poisson distribution arises not only from a binomial distribution as its limiting form, but it provides a realistic probability model for many random phenomena. Any random phenomenon for which a count of some sort is of interest may be tested, if it followed Poisson distribution. This family of distribution has been named for the French mathematician Siméon Denis Poisson (1781-1840) who was a pioneer in probability theory.
Such counts may be:

(i) Number of fatal traffic accidents at a busy highway crossing in a month.
(ii) Number of radioactive particles emitted in 5 minutes duration.
(iii) The number of telephone calls per hour duration coming to an exchange of a business concern.
(iv) The number of total coliforms in 5 ml of water collected from a pond.
(v) The number of defects per meter of cloth.

All the counts may not follow the Poisson distribution if certain conditions or assumptions are not satisfied by the phenomenon.

The events whose count are noted may happen in time, space, region or length. But without loss of generality, though the events occur in space or length, they may be treated as taking place with passage of time. Following are the common conditions that the occurrences of events must satisfy, so that their count may follow Poisson distribution. Before giving the conditions, consider that there exist a positive constant say ν and a small interval of the time say Δt.
Conditions when Poisson distribution arises are:

1. The probability that exactly one event occurs in a small time interval $(t, t + \Delta t)$ is $\nu\Delta t$. In symbols,

$$P[\text{One happening in } (t, t + \Delta t)] = \nu\Delta t \tag{6.12}$$

2. The probability of two or more happenings in a small time interval $(t, t + \Delta t)$ is negligible or

$$P[\text{two or more happenings in } (t, t + \Delta t)] = 0 \tag{6.13}$$

3. The events happening in two non overlapping time intervals are independent.

Under these Conditions

Let X be the number of events occurring in a specified time interval of length T_0 with occurrence rate ν. Then X has Poisson distribution with parameter $\lambda = \nu T_0$.

Definition: A discrete random variable X is said to have *Poisson distribution* with parameter $\lambda = \nu T_0$ if its pmf is given by

$$p(x) = e^{-\nu T_0} (\nu T_0)^x / x!; \quad x = 0, 1, 2, \ldots\ldots, \infty \tag{6.14}$$

If $\lambda = \nu T_0$, pmf of X is given by

$$p(x) = e^{-\lambda} \lambda^x / x!; \quad x = 0, 1, 2, \ldots\ldots, \infty \tag{6.15}$$

The constant ν can be interpreted as *mean rate of occurrence of events per unit time* or *mean occurrence rate* of events.

In symbols, it is denoted as $X \sim P(\lambda)$. Figure 6.2 gives nature of Poisson random variable for various λ (values of $p(x)$ on y-axis).

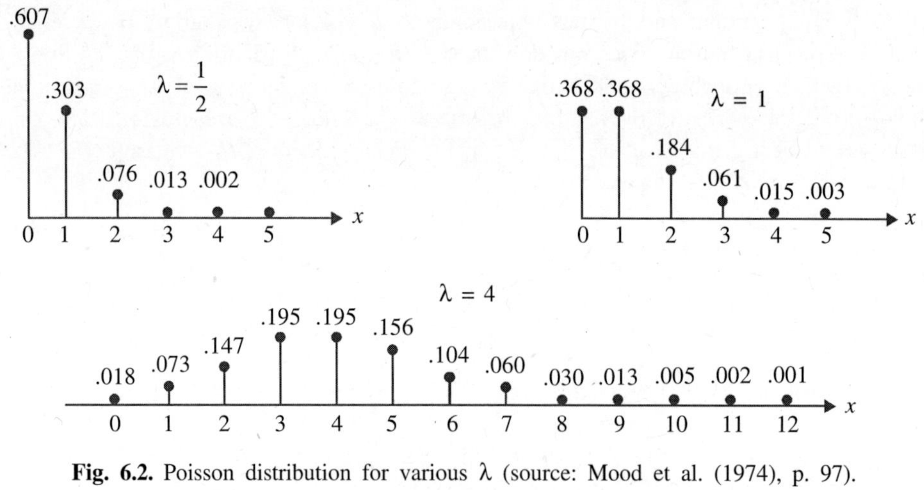

Fig. 6.2. Poisson distribution for various λ (source: Mood et al. (1974), p. 97).

Some real examples of Poisson random variable are given below.

Example 6.4: Suppose that the average number of telephone calls arriving per hour at the switch board of a small business firm in 30. The number of telephone calls in a specified duration, X has Poisson distribution and one may ask

(a) What is the probability that no call will arrive in a 3 minutes period?
(b) What is the probability that more than 5 calls arrive in a 5 minutes interval?

(source: Mood et al. (1974), p. 97)

Solution: Let X: number of telephone calls in 3 minutes interval then $X \sim P(\nu T_0)$, where $\nu = 30/60$ per minute and $T_0 = 3$ minutes.
Hence $\nu = .5$ per minute

(a) Required probability $= P$ [no calls in 3 minutes interval]

$= p(0) = e^{-.5 \times 3} (.5 \times 3)^0/0! = e^{-1.5}$

$\simeq .223$

(b) Required probability $= P$ [more than 5 calls in 5 minutes interval]

$$= \sum_{x>5} p(x) = 1 - \sum_{x=0}^{5} p(x)$$

$$= 1 - \sum_{x=0}^{5} e^{-.5\times 5} (.5 \times 5)^x / x!$$

$$= 1 - \sum_{x=0}^{5} e^{-2.5} (2.5)^x / x!$$

$= 1 - e^{-2.5} (1 + 2.5/1! + (2.5)^2/2! + (2.5)^3/3! + (2.5)^4/4! + (2.5)^5/5!)$

$\simeq .042$

Example 6.5: Suppose that flaws in a plywood occur at random with an average of one flaw per 50 sq. feet. Then the number of flaws in a sheet of specified area will follow a Poisson distribution and one may enquire that what is the probability that a 4 ft × 8 ft sheet will have (a) no flaws? (b) at most one flaw? (source: Mood et al. (1974), p. 98)
Solution: Let the number of flaws per square feet = $\nu = 1/50$ and assume that number of flaws in 32 sq feet sheet, X has Poisson distribution with parameter

$$\nu T_0 = 1/50 \times 32 = 16/25$$

i.e. $$p(x) = e^{-16/25} (16/25)^x / x!, \quad x = 0, 1, 2, ..., \infty$$

So

(a) P [no flaw in the sheet of 32 sq. feet] $= p(0) = e^{-16/25} = e^{-.64} \simeq .527$ and
(b) P [at most one flaw in the sheet of 32 sq. feet] $= p(0) + p(1) = e^{-.64} + .64\, e^{-.64} \simeq .865$.

Example 6.6: The white blood cell (WBC) count of a healthy individual can average as low as 6000 per cubic millimeter of blood. To detect a WBC deficiency, a .001 cubic millimeter drop of blood is taken and the count of WBC, X is determined.

(a) How many WBC's are expected in a healthy individual?
(b) If at most 2 are found, is there evidence of WBC deficiency?

(source: Milton and Tsokos (1983), p. 132)

Solution: The discrete random variable of interest is X: The count of WBC in .001 cu. millimeter of blood, which follows a Poisson distribution with parameter λ, determined below.
ν: The average number of WBC per cu millimeter in healthy individuals = 6000.

(a) Hence λ: the average number of WBC per .001 cu. millimeter of healthy person's blood $= \nu \times .001$ $= 6000 \times .001 = 6.00$
(b) Probability that at most 2 WBC are found

$= P[\text{in } .001 \text{ cu. millimeter of blood } X \leq 2]$
$= p(0) + p(1) + p(2)$

$$= e^{-6} + e^{-6}\frac{6}{1!} + e^{-6}\frac{6^2}{2!}$$

$= e^{-6}\,[1 + 6 + 18] = 25 \times e^{-6} = 25 \times .00248$
$= .06200$

Since probability .062 is moderate in size, there appears no clear cut answer to the question (b). However, if as will be seen later, since this probability is greater than .05 (normally chosen value to decide rarity of an event), one may consider that the observed count of 2 may not be taken as evidence of WBC deficiency.

6.3.1 Properties of Poisson Distribution

If X follows Poisson distribution with pmf (6.14) i.e. if $X \sim P(\lambda)$ then

1. Range of X is 0, 1, 2, ..., ∞. Thus X can take countably infinite values.
2. $E(X) = \lambda$, $V(X) = \lambda$.
3. *Recurrence relation*: For evaluating Poisson probability it is not necessary that each term is evaluated using equation (6.14). Instead the following recurrence relation may be used

$$p(x+1) = \frac{\lambda}{x+1}\,p(x), \quad x = 0, 1, 2, ..., \infty \tag{6.16}$$

Thus one needs to calculate

$$p(0) = e^{-\lambda}\,\lambda^0/0! = e^{-\lambda} \tag{6.17}$$

only and the subsequent terms $p(1)$, $p(2)$, ..., can be easily calculated using the above recurrence relation.

4. *Additive Property*

(a) If X_1 and X_2 are independent Poisson random variables, with parameters λ_1 and λ_2 then $X_1 + X_2$ has Poisson distribution with parameter $\lambda_1 + \lambda_2$
In Symbols:
If $X_1 \sim P(\lambda_1)$ and $X_2 \sim P(\lambda_2)$ are independent, then $X_1 + X_2 \sim P(\lambda_1 + \lambda_2)$ (6.18)

(b) This property may be generalized for k variables also as given below:
If $X_i \sim P(\lambda_i)$, $i = 1, 2, ..., k$ and are independent

Then
$$\sum_{i=1}^{k} X_i \sim P\left(\sum_{i=1}^{k} \lambda_i\right) \tag{6.19}$$

5. $p(x-1) < p(x)$ if $x < \lambda$ (6.20)
$p(x-1) > p(x)$ if $x > \lambda$
$p(x-1) = p(x)$ if $x = \lambda$ (an integer) (6.21)

This gives the nature of Poisson pmf of first increasing and then decreasing with increase in x.

6. $\mu_3(x) = \lambda$ and $\mu_4(x) = 3\lambda^2 + \lambda$ (6.22)
Hence

$$\beta_1 = \mu_3^2/\mu_2^3 = \frac{\lambda^2}{\lambda^3} = \frac{1}{\lambda} \tag{6.23}$$

and

$$\beta_2 = \frac{\mu_4}{\mu_2^2} = 3 + \frac{1}{\lambda} \tag{6.24}$$

From values of β_1 and β_2 in (6.23) and (6.24) it is evident that if λ is large, the Poisson distribution tends to be symmetric and with no steepedness (or kurtosis).

7. *Moment Generating Function* of Poisson distribution $P(\lambda)$ is given by

$$M_x(t) = e^{\lambda\left(e^t - 1\right)} \tag{6.25}$$

6.4 Hypergeometric Distribution

This distribution has some use in biological problems also. The conditions under which it arises are considered first in the context of an urn and balls again.

Let there be an urn containing N balls of the same size, shape and smoothness; but are of two colours. n are of red colour and $N - n$ are of white colour. Let m balls are drawn without replacement and without noting their colour.

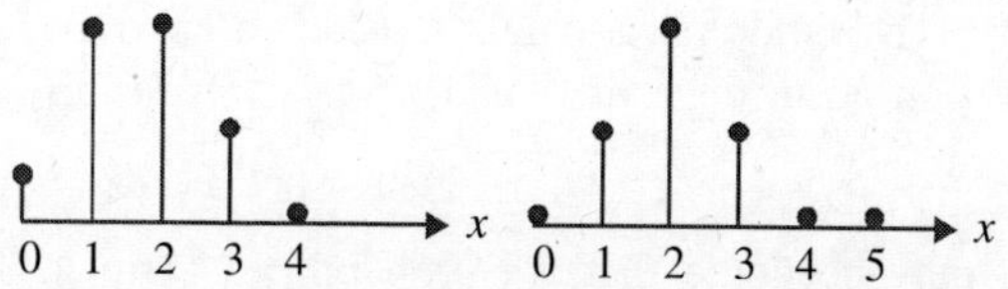

(a) $(N, n, m) = (10, 4, 4)$ and (b) $(N, n, m) = (10, 4, 5)$

Fig. 6.3. Hypergeometric pmf (source: Mood et al. (1974), p. 92).

Define a random variable X: The number of red balls drawn. Then the pmf of X is given by

$$p(x) = \frac{\binom{n}{x}\binom{N-n}{m-x}}{\binom{N}{m}} \qquad x = 0, 1, 2, 3, \ldots\ldots, B \tag{6.26}$$

where $B = \min(n, m)$, N is a positive integer, n is a non negative integer that is at most N and m is a positive integer that is at most N.

Any random variable with pmf given by (6.26) is said to have a hypergeometric distribution. Figure 6.3 gives the hypergeometric pmf's for $(N, n, m) = (10, 4, 4)$ and $(10, 4, 5)$.

This experiment also has a fixed number m, of trials of drawing a ball from the urn. There are only two outcomes, drawn ball is red or white only. These are conditions of binomial distribution. But the trials are not independent since the draw is without replacement.

Example 6.7: Suppose that 20 laboratory mice, 7 of which are males and 13 females are irradiated. It is desired to test whether a certain mutation is more likely to arise in male mice than in females. After the radiation it is found that, in all there are 8 mutant mice among the total N mice. In this case Y: the number of mutant male mice has hypergeometric distribution with $(N, n, m) = (20, 7, 8)$.

6.4.1 Properties of Hypergeometric Distribution

$$\text{Mean} = E(X) = \frac{n}{N} \tag{6.27}$$

$$\text{Variance}(X) = m \cdot \frac{n}{N} \cdot \frac{N-n}{N} \cdot \frac{N-m}{N-1} \tag{6.28}$$

6.5 Geometric Distribution

Two other families of discrete distributions that play important roles are the geometric (or Pascal) and negative binomial distribution. The geometric distribution is described in this section and the negative binomial in the next section. It will be seen in the next section that the geometric distribution may be considered a special case of negative binomial distribution.

Suppose that an experiment E is performed and the concern is only about occurrence or non occurrence of some event A as in binomial distribution. Moreover, similar to binomial trial the experiment is performed repeatedly and the repetitions are independent and in each trial $P(A) = p$ and $P(A^C) = 1 - p = q$.

Unlike binomial trials (where the number of trials are predetermined) the experiment is repeated until A occurs for the first time. Thus the number of repetitions (= X) required up to and including the first occurrence of A is a random variable assuming possible values 1, 2, ...

The random variable X is said to have geometric distribution. $X = x$ if $(x - 1)$ trials of E result in non-occurrence of A and x^{th} trial results in A. The pmf of the geometric random variable is

$$p(x) = q^{x-1}\, p; \quad x = 1, 2, 3, ... \qquad 0 < p \leq 1 \tag{6.29}$$

It may be noted that the probabilities form a geometric series hence the name of this distribution. The following Fig. 6.4 gives the pmf of geometric random variable with (a) $p = 3/4$, (b) $p = 1/2$ and (c) $p = 1/4$.

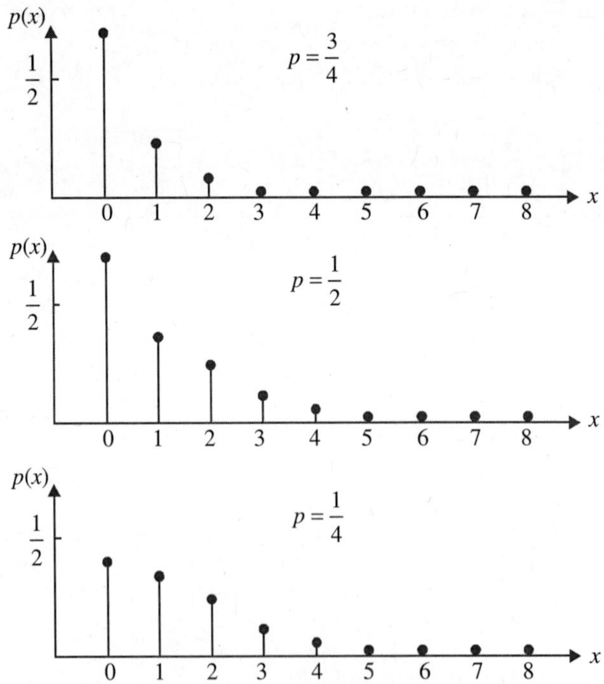

Fig. 6.4. Geometric pmf's (source: Mood et al. (1974), p. 100).

Example 6.8: In a certain locality, the probability distribution that a thunderstorm will occur on any given day during August and September equals 0.1. Assuming independence from day to day, let X be the number of days from

August 1, until the first thunderstorm occurs. Then X has geometric distribution (6.29) with $p = .1$ (source: Meyer, p. 171).

Suppose one wants to determine the probability that first thunderstorm occurs on August 20. It may be seen that the required probability is $p(20) = (.9)^{19}\ .1$

Example 6.9: Probability that a certain test yields a 'positive' reaction equals 0.4. What is the probability that fewer than 5 'negative' reactions occur before the first positive one (source: Meyer (1975), p. 171).

Solution: Let Y be the number of negative reactions before the first positive one. Then $X = Y + 1$ has the geometric distribution (6.29), so the required probability is

$$P[Y < 5] = P[X < 6] = \sum_{x=1}^{5} p(x) = \sum_{x=1}^{5} q^{x-1} p$$

$$= p \sum_{x=1}^{5} q^{x-1} = .4 \sum_{x=1}^{5} (.6)^{x-1} = .4\left(.6^0 + .6^1 + .6^2 + .6^3 + .6^4\right) = 0.92$$

6.5.1 Properties of Geometric Distribution

If X has the geometric distribution given by equation (6.29)

1. Mean of $(X) = E(X) = 1/p$ and variance of $X = q/p^2$ (6.30)
2. For any two positive integers r and s

$$P[X \geq r + s \mid x > s] = P[X \geq r] \tag{6.31}$$

   This result implies that the geometric distribution has 'no memory' in the following sense. This is because the probability that at least $r + s$ repetitions are required to observe the first success given that there have been s failures is equal to the unconditional probability that at least r repetitions are needed to observe the first success.
3. The moment generating function of X is given by

$$M_x(t) = pe^t \big/ \left(1 - qe^t\right) \tag{6.32}$$

4. The random variable X which has a geometric distribution may also be called a discrete *waiting-time* random variable. It represents the number of trials one has to wait to observe a success.

6.6 Negative Binomial Distribution

An obvious generalization of geometric distribution results if an experiment is repeated until the event A occurs r^{th} time. The conditions of the experiment are same as those of geometric distribution.

As in section 6.4 on each independent repetition

$$P(A) = p \text{ and } P(A^c) = 1 - p = q \tag{6.33}$$

Let X be the number of repetitions needed to have A occur exactly r times, then the pmf of X is given by

$$P[X = x] = p(x)$$

which is probability of the event that A occurs in x^{th} trial, and in the previous $(x - 1)$ trials, A occurred exactly $(r - 1)$ times.

Hence

$$p(x) = p\binom{x-1}{r-1}p^{r-1}q^{x-r} \qquad x = r, r+1, r+2, \ldots \qquad 0 < p \le 1$$

or
$$p(x) = \binom{x-1}{r-1}p^{r}q^{x-r} \qquad x = r, r+1, r+2, \ldots \qquad 0 < p \le 1 \tag{6.34}$$

It can be easily seen that if $r = 1$, negative binomial pmf reduces to geometric pmf (6.29). The random variable with pmf (6.34) is also called Pascal distribution Fig. 6.5 gives the pmf of the negative binomial distribution.

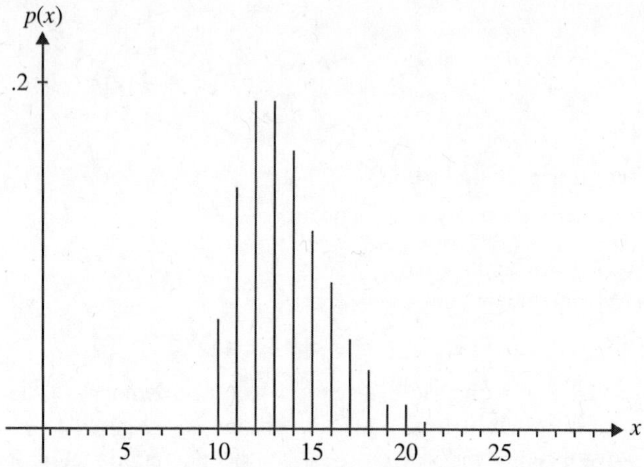

Fig. 6.5. Negative binomial distribution $r = 10$, $p = .75$ (source: Ewens and Grant (2001, 2005), p. 16).

Note: Reason for the name *negative binomial* may be seen from the requirement that

$$\sum_{x=r}^{\infty} p(x) = 1$$

$$\text{L.H.S.} = p^r \sum_{k=r}^{\infty} \binom{k-1}{r-1} q^{k-r} \tag{6.35}$$

It can be seen that
$$(1-q)^{-r} = \sum_{n=0}^{\infty} \binom{-r}{n}(-q)^n = \sum_{k=r}^{\infty} \binom{k-1}{r-1} q^{k-r}$$

after some algebraic simplification and definition of generalized binomial coefficients. Hence (6.35) gives

$$\sum_{x=r}^{\infty} p(x) = p^r (1-q)^{-r} = 1 \tag{6.36}$$

The negative exponent ($-r$) in (6.36) led to the name *negative binomial* to this distribution.

6.6.1 Properties of Negative Binomial Distribution

If X has negative binomial distribution given by (6.34) then

1. Mean $(X) = E(X) = rq/p$ and variance $(X) = \sigma_x^2 = rq/p^2$ (6.37)
2. *Moment generating function* of X is

$$M_x(t) = \left[\frac{p}{\left(1 - qe^t\right)}\right]^r \tag{6.38}$$

3. Relation between binomial and negative binomial distribution
   Let $Y \sim B(n, p)$ and let X have a negative binomial distribution (6.34) with parameters r and p (That is X = number of Bernoulli trials to obtain r successes)
   Then (a) $P(X \leq n) = P(Y \geq r)$ and (b) $P(X > n) = P(Y < r)$.

   Property 3 enables to use binomial distribution table for evaluating probabilities associated to negative binomial distribution. Thus to find probability that more than 10 repetitions are necessary to obtain 3 successes, when the P(success) = .2, it can be evaluated using

$$P(X > 10) = P(Y < 3)$$

   When $Y \sim B(10, .2)$

   Thus $$P(X > 10) = \sum_{x=0}^{2} \binom{10}{x} .2^x (.8)^{10-x} = .678$$

6.7 Normal Distribution

Normal family is a family of continuous random variables which has great many applications in the data analysis in almost every field of science, engineering and humanities. The distribution was published by Abraham De Moivre (1667-1754) in 1733 as limiting form of binomial distribution.

The distribution was rediscovered by C.F. Gauss (1777-1855) after about half a century as distribution of errors in measurements of orbit of heavenly bodies. Because of Gauss's work, the normal distribution is also called *Gaussian curve*.

In this section first a formal definition of normal distribution is given and its properties are given subsequently.

Definition: A random variable X, assuming all real values in the interval $-\infty < x < \infty$ has normal (or Gaussian) distribution if its pdf is of the form

$$f(x) = \frac{1}{\sqrt{2\pi}\,\sigma} \exp\left[-(x-\mu)^2/2\sigma^2\right] \tag{6.39}$$

$$-\infty < x < \infty, \ -\infty < \mu < \infty, \ \sigma > 0$$

The constants μ and σ are the two parameters of a normal distribution. In symbols this distribution can be denoted as $N(\mu, \sigma^2)$ and if X has this distribution it is denoted by $X \sim N(\mu, \sigma^2)$. Figure 6.6 gives the graph of a typical normal density curve given in (6.39).

As for any continuous random variable, the total area bounded by $f(x)$ and X axis is 1 and probability can be found by determining appropriate areas.

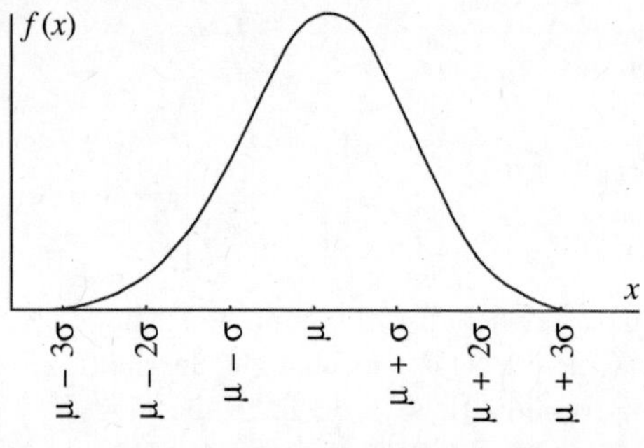

Fig. 6.6. Normal density with mean μ and variance σ^2 (source: Hoel (1971), p. 76).

Some examples:

Example 6.10: Let X denote the grams of hydrocarbons emitted by an automobile per mile. If X has normal distribution with mean 1 gm and standard deviation .25 gm per mile. Then $X \sim N(1, .0625)$ and its density is given by (source: Milton and Tsokos (1983), p. 137 (data only))

$$f(x) = \frac{1}{\sqrt{2\pi}(.25)} \exp\left[-\frac{1}{2}\left\{\frac{(x-1)}{.25}\right\}^2\right] \tag{6.40}$$

It will be seen in the properties of normal distribution that attaining negative value of this X is always almost improbable. Since values of $X \le 0$ are less than value of $X = 1 - 3 \times 0.25 = 0.25$, which is 3σ below mean.

If $(X - \mu)/\sigma = Z \sim N(0, 1)$, where X is a random variable which can take only positive values, then it can be said that $X \sim N(\mu, \sigma^2)$. The same applies on emission from automobiles, which can never be negative.

Example 6.11: A physical therapist feels that scores on a certain manual dexterity test are approximately normally distributed with a mean of 10 and a standard deviation of 2.5. If a randomly selected person takes the test, what is the probability that he will make a score of 15 or more (source: Daniel (1987), p. 89).
Solution: Let X be the score on the above test, then $X \sim N(10, 2.5^2)$ and it is required to find $P[X > 15]$. This will be the area under the density curve of X for $X > 15$ (The method will be explained after giving the properties).

6.7.1 Properties of Normal Distribution

If $X \sim N(\mu, \sigma^2)$, then the properties of this normal distribution are as follows:

1. The distribution of X is bell shaped, i.e. symmetric about $X = \mu$. Thus

$$f(\mu + x) = f(\mu - x) \tag{6.41}$$

2. The range of X is the whole real line i.e., theoretically X can take any value from $-\infty$ to $+\infty$.
3. The mean, median and mode of X are μ and standard deviation of X is σ. In symbols

$$E(X) = \mu \quad \text{and} \quad V(X) = \sigma^2$$

4. Let $Z = \dfrac{(X-\mu)}{\sigma}$ then $Z \sim N(0, 1)$ and Z is called *Standard Normal* variate. Let $\phi(z)$ denote the ordinate of density of Z (see Fig. 6.7) and $\Phi(z)$ denote the cdf of Z at z (see Fig. 6.8). In symbols $\Phi(z) = P[Z < z]$ = Area under standard normal curve from $-\infty$ to z.

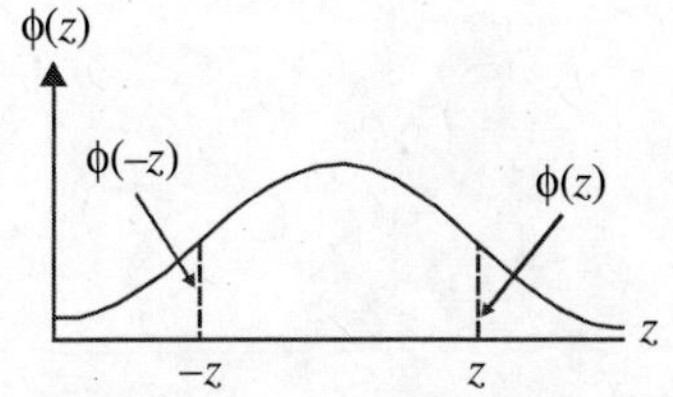

Fig. 6.7. Ordinates $\phi(z)$ and $\phi(-z)$ (source: Milton and Tsokos (1983), p. 139).

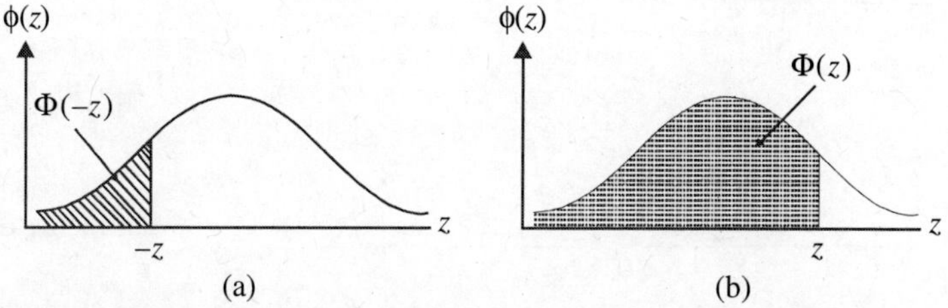

Fig. 6.8. Area or cdf of standard normal distribution (source: Milton and Tsokos (1983), p. 139).

5. Standard normal distribution is symmetrical about $Z = 0$, so

$$\phi(z) = \phi(-z) \tag{6.42}$$

and

$$\Phi(z) = 1 - \Phi(-z) \tag{6.43}$$

The above relation (6.43) can be easily seen from Fig. 6.8 in which $\Phi(z)$ the cdf at $Z = z$ and $\Phi(-z)$ the cdf at $Z = -z$ are shown.

6. To find probability: $P[a \le X \le b]$ the area tables for standard normal distribution can be easily used, since

$$\begin{aligned} P[a \le X \le b] &= P\left[\frac{a-\mu}{\sigma} \le \frac{X-\mu}{\sigma} \le \frac{b-\mu}{\sigma}\right] \\ &= P\left[\frac{a-\mu}{\sigma} \le Z \le \frac{b-\mu}{\sigma}\right] \\ &= P[a^* \le Z \le b^*] \\ &= \Phi(b^*) - \Phi(a^*) \end{aligned} \tag{6.44}$$

where, $a^* = \dfrac{a-\mu}{\sigma}$ and $b^* = \dfrac{b-\mu}{\sigma}$.

Thus if area tables from $-\infty$ to z i.e. cdf of Z are available (6.44) can be easily determined by reading cdf's at a^* and b^*.

7. If $z > 0$

$$\Phi(z) = .5 + \int_0^z \phi(x)\,dx = .5 + A(z) \tag{6.45}$$

and if $z < 0$ $\Phi(z) = .5 - A(-z)$, where $A(z) = P[0 < Z < z]$ = area under $N(0, 1)$ from 0 to z. Moreover, because of symmetry, area under standard normal curve from 0 to z is same as that from $-z$ to 0 i.e. $A(z) = A(-z)$ (see Fig. 6.9). Hence some books contain tables of $A(z)$ instead of the cdf $\Phi(z)$. The following relations for positive c and $d (d > c)$ may be used for determining various probabilities of values of X.

(i) $P[-c \le Z \le d] = A(d) + A(c)$
(ii) $P[c \le Z \le d] = A(d) - A(c)$
(iii) $P[-d \le Z \le -c] = A(d) - A(c)$

due to symmetry. (See Figs. 6.10-6.12)

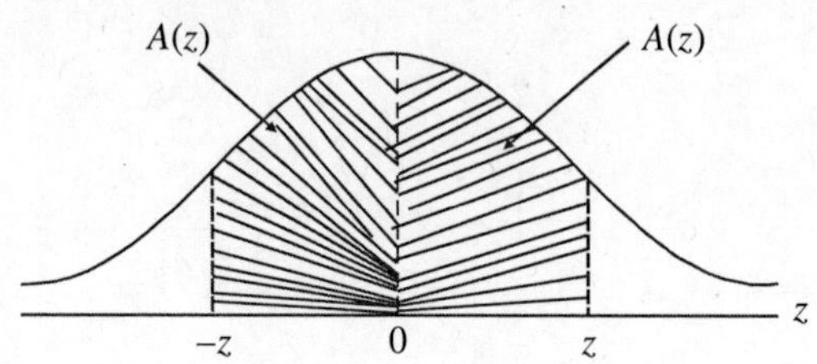

Fig. 6.9. Area $A(z)$ (source: Milton and Tsokos (1983), p. 139).

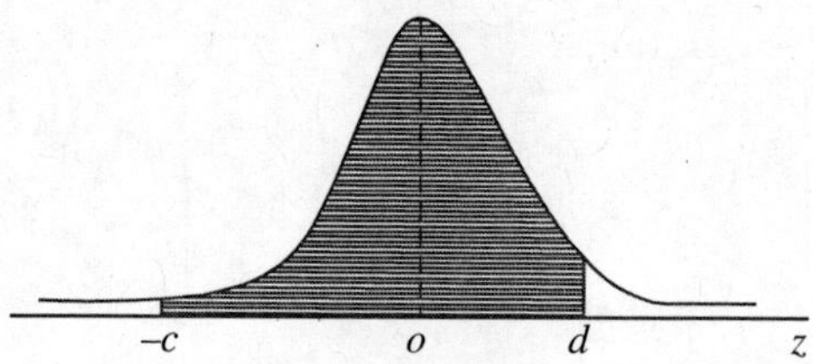

Fig. 6.10. Probability of standard normal variable from $-c$ to d (source: Daniel (1987), p. 87 (labels different)).

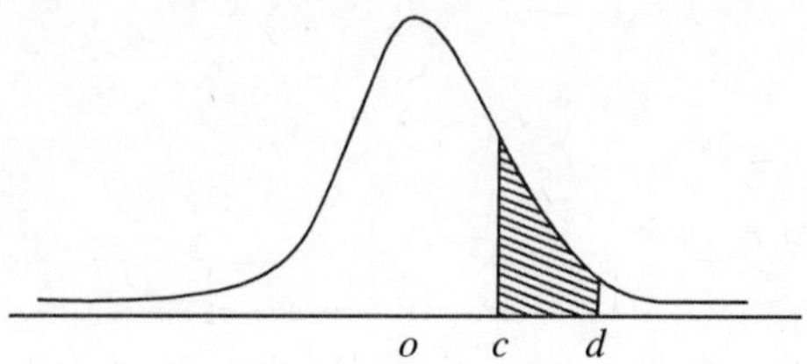

Fig. 6.11. Probability of standard normal variable from c to d (source: Daniel (1987), p. 87 (labels different).

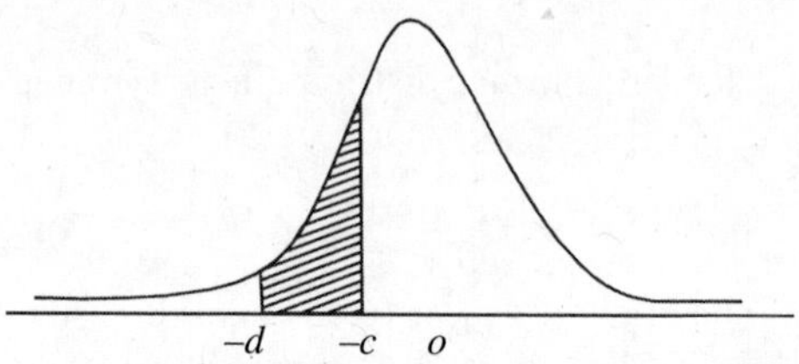

Fig. 6.12. Probability of standard normal variable from $-c$ to $-d$ (source: Daniel (1987), p. 82 (labels different)).

8. (i) $P[\mu - \sigma < X < \mu + \sigma] = .6826 = P[-1 < Z < 1]$
   (ii) $P[\mu - 2\sigma < X < \mu + 2\sigma] = .9544 = P[-2 < Z < 2]$
   (iii) $P[\mu - 3\sigma < X < \mu + 3\sigma] = .9973 = P[-3 < Z < 3]$
   (iv) $P[\mu - 1.96\sigma < X < \mu + 1.96\sigma] = .95 = P[-1.96 < Z < 1.96]$
   (v) $P[X < \mu + 1.645\sigma] = .95 = P[Z < 1.645]$

9. On the basis of 8 (iv) and (v) it can be noted that 1.645 is upper 95% significant point and ±1.96 are two sided 95% significant points of $N(0, 1)$.

10. If X_i, $i = 1, 2, \ldots\ldots, n$ are iid $N(\mu, \sigma^2)$ then $\sum_{i=1}^{n} X_i \sim N\left(n\mu, n\sigma^2\right)$ and $\bar{X}_n \sim N\left(\mu, \sigma^2/n\right)$, where

$$\bar{X}_n = \frac{1}{n}\sum_{i=1}^{n} X_i \cdot$$

11. If $X \sim N(\mu, \sigma^2)$ then $(aX + b) \sim N(a\mu + b, a^2\sigma^2)$, where a and b are constants.

12. If X_i, $i = 1, 2, \ldots\ldots, n$ are iid $N(\mu, \sigma^2)$, then $\sum_{i=1}^{n} a_i X_i \sim N\left(\mu \sum_{i=1}^{n} a_i, \sigma^2 \sum_{i=1}^{n} a_i^2\right)$, where a_i, $i = 1, 2, \ldots\ldots, n$ are constants.

Some examples given below illustrate the use of normal area tables for obtaining desired probabilities of values taken by a normal random variable.

Example 6.12: Suppose it is known that weights of a certain group of individuals are approximately normally distributed with mean of 140 pounds and a standard deviation of 25 pounds. What is the probability that a person selected at random from the group will weigh between 100 and 170 pounds (source: Daniel (1987), p. 91)?

Solution: Let X denote the weight of an individual in the group.
Then $X \sim N(140, 25^2)$
Hence $Z = (X - 140)/25 \sim N(0, 1)$

$$\text{Desired probability} = P[100 \le X \le 170]$$

$$= P\left[\frac{100-140}{25} \le \frac{X-140}{25} \le \frac{170-140}{25}\right]$$

$$= P[-1.6 \le Z \le 1.2]$$

$$= \Phi(1.2) - \Phi(-1.6)$$

From the standard normal cdf table

$$\Phi(1.2) = .8849 \text{ and } \Phi(-1.6) = .0548$$

Hence required probability = .8849 – .0548 = .8301.

Example 6.13: The cholesterol values for a certain population are approximately normally distributed with a mean of 200 mg/100 ml and a standard deviation of 20 mg/100 ml. Find the probability that an individual picked at random from this population will have a cholesterol value (source: Daniel (1987), p. 93):

(a) Between 190 and 210 mg/100 ml
(b) Greater than 225 mg/100 ml
(c) Less than 150 mg/100 ml

Solution: Let X denote the cholesterol value of an individual in the population. Then $X \sim N(200, 20^2)$

(a) It is desired to find $P[190 < X < 210]$

$$= P\left[\frac{190-200}{20} \le \frac{X-200}{20} \le \frac{210-200}{20}\right]$$

$$= P[-0.5 \le Z \le 0.5] = 2P[0 \le Z \le .5]$$

$$= 2A(.5) = 2 \times .1951 = .3902$$

(b) $P[X > 225] = P\left[\left[\frac{X-200}{20} > \frac{225-200}{20}\right] = P[Z > 25/20]\right]$

$$= P[Z > 1.25] = 1 - \Phi(1.25)$$

$$= 1 - .8944 = .1056$$

(c) $P[X < 150] = P\left[\frac{X-200}{20} < \frac{150-200}{20}\right] = P[Z < -2.5]$

$$= \Phi(-2.5) = .0062$$

6.8 Uniform and Rectangular Distributions

A very simple distribution for a continuous random variable is the Rectangular distribution. It is useful in some statistical problems and is very easy to deal mathematically.

Definition: A random variable X, is said to have a rectangular distribution in the interval $[a, b]$ if its pdf is given by

$$f(x) = \left\{\frac{1}{b-a}\right\} \quad \text{if } a \le x \le b$$

$$= 0 \quad \text{otherwise} \tag{6.46}$$

In symbols one can write that $X \sim R(a, b)$, X has a pdf which is constant over the finite range $[a, b]$. The figure below gives the pdf of X. The constant value of the pdf is the reciprocal of the length of the interval.

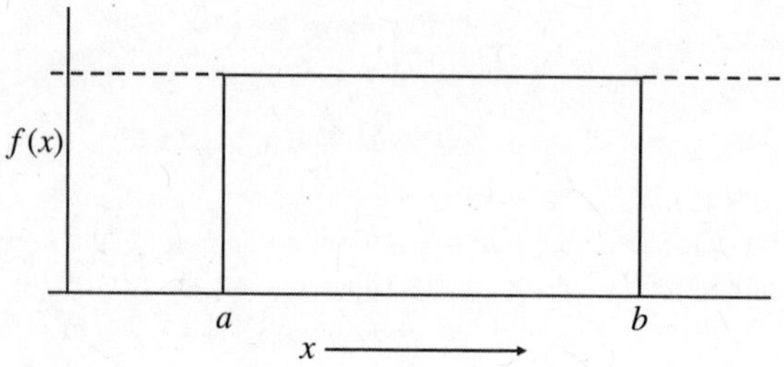

Fig. 6.13. Rectangular pdf on (a, b) (source: Goon et al. (1977), Vol. I, p. 170).

Example 6.14: The hardness, say H of a specimen of the steel measured on the Rockwell scale, is assumed to be a continuous random variable with rectangular distribution over [50, 70]. So

$$f(h) = \frac{1}{20} \quad \text{if } 50 < h < 70$$

$$= 0 \quad \text{otherwise.}$$

(Source: Meyer (1975), p. 72).

6.8.1 Properties of Rectangular Distribution

If $X \sim R\ (a, b)$

1. Mean of $X = E(X) = \dfrac{a+b}{2}$
2. $V(X) = \dfrac{(b-a)^2}{12}$
3. For any sub interval $[c, d]$ in $[a, b]$

$$P[c \le X \le d] = \int_c^d f(x)\, dx = \frac{d-c}{b-a} = \text{area of the rectangle of length } (d-c) \tag{6.47}$$

4. The cdf of the rectangular distribution (6.46) is given by

$$F(x) = \int_{-\infty}^{x} f(u)\, du$$

$$= 0 \quad \text{if } x < a$$

$$= \frac{x-a}{b-a} \quad \text{if } a \le x < b$$

$$= 1 \quad \text{if } x \ge b \tag{6.48}$$

The graph of the cdf is shown below.

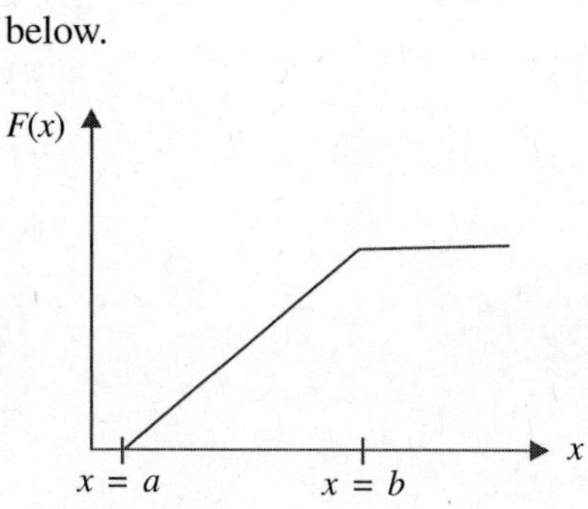

Fig. 6.14. cdf of rectangular distribution on (a, b) (source: Meyer (1975), p. 75).

5. $\mu_3 = 0$ and $\mu_4 = \dfrac{(b-a)^4}{80}$ so that $\gamma_1 = 0$ and $\gamma_2 = -1.2$. Thus the distribution is symmetrical and highly platykurtic.
6. The *moment generating function* of $X \sim R[a, b]$ is given by

$$M_x(t) = (e^{bt} - e^{at})/(b - a)t \tag{6.49}$$

7. If a random variable X has rectangular distribution on the interval [0, 1] i.e. $X \sim R(0, 1)$ it is normally denoted as $X \sim U(0, 1)$ and is called a Uniform random variable. If $U \sim U[0, 1]$, it can be seen from (6.48) that

$$\begin{aligned} F(u) &= u \qquad \text{if } 0 \le u \le 1 \\ &= 0 \qquad \text{otherwise} \end{aligned} \tag{6.50}$$

The uniform random variable $U(0, 1)$ is very much used in simulation and Monte Carlo studies.

6.9 Bivariate Normal Distribution

Just as normal distribution for a single continuous random variable is very useful, it may be very likely that a joint normal density function for two continuous random variables may also be a useful model. Especially, if relationship between such variables arising from experiment is to be explored.

In this section, therefore, as a simplest generalization to multivariate normal distribution, a bivariate normal distribution is considered.

Definition: If the two continuous random variables X and Y have the joint pdf

$$f(x, y) = \frac{1}{2\pi\sigma_x\sigma_y\sqrt{1-\rho^2}} \exp\left[-\frac{1}{2(1-\rho^2)}\left\{\left(\frac{x-\mu_x}{\sigma_x}\right)^2 + \left(\frac{y-\mu_y}{\sigma_y}\right)^2 - 2\rho\left(\frac{x-\mu_x}{\sigma_x}\right)\left(\frac{y-\mu_y}{\sigma_y}\right)\right\}\right]$$

$$-\infty < x, y < \infty, -\infty < \mu_x, \mu_y < \infty, \sigma_x > 0, \sigma_y > 0 \text{ and } -1 < \rho < 1 \tag{6.51}$$

then the random variable (X, Y) is defined to have a bivariate normal distribution.

In this case $E(X) = \mu_x$, $E(Y) = \mu_y$, $V(X) = \sigma_x^2$, $V(Y) = \sigma_y^2$ and ρ = correlation between X and Y. The density in equation (6.51) is a bell shaped surface $z = f(x, y)$ as shown in the Fig. 6.15 (a) and (b): (a) is for $\mu_x = \mu_y = 0$, $\sigma_x = \sigma_y$ and (b) for $\rho \ne 0$, $\mu_x \ne 0$, $\mu_y \ne 0$, $\sigma_x \ne \sigma_y$.

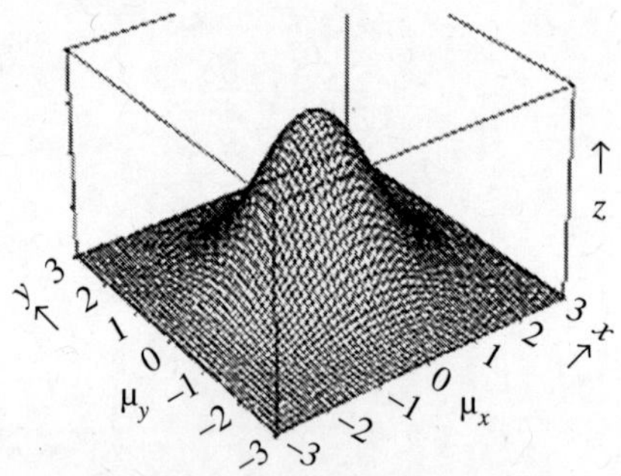

(a) (source: Zar (2004), p. 113)

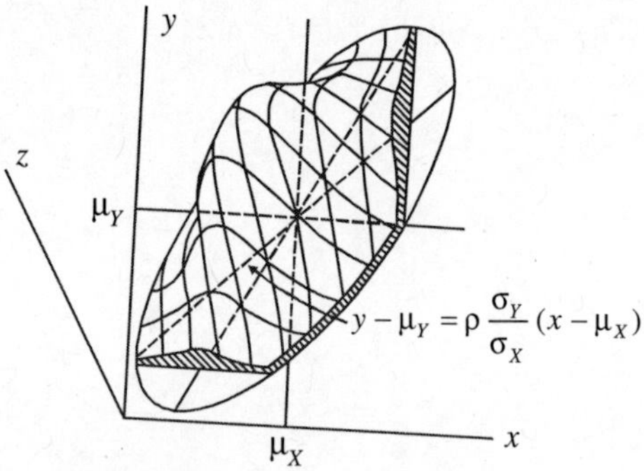

(b) (source: Hoel (1971), p. 135)

Fig. 6.15. Bivariate normal distributions.

Any plane parallel to (X, Y) plane cuts this surface in an ellipse (or a circle if $\sigma_x = \sigma_y$). Figure 6.16 (a) and (b) shows the two concentric circles due to intersection by two parallel planes. The larger one is due to intersection near the tails.

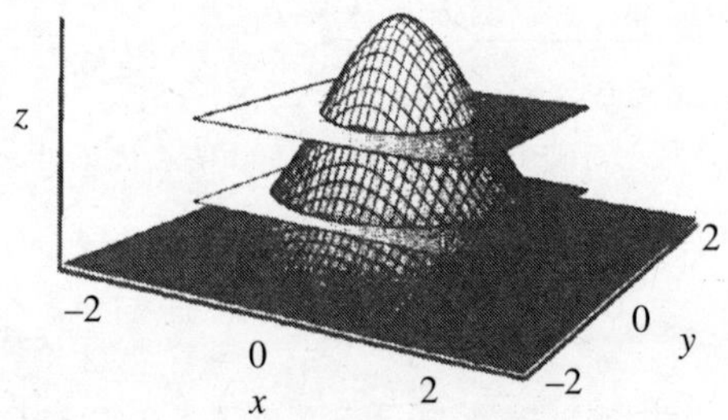

(a) Two intersecting (X, Y) planes of bivariate normal surface.

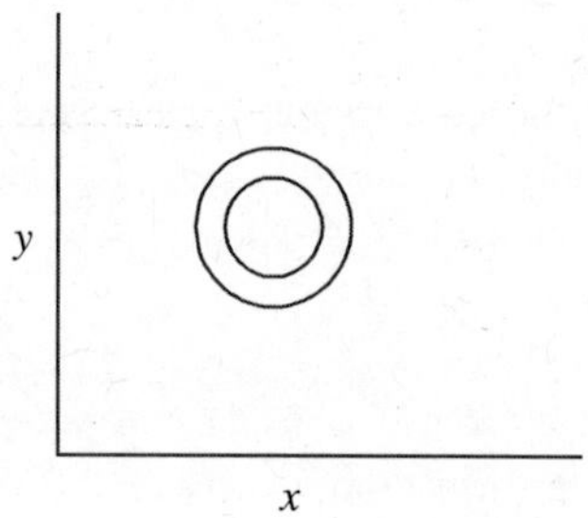

(b) Concentric circles due to intersection by parallel (X, Y) planes

Fig. 6.16. Intersection of normal surface by (X, Y) plane (source: Zar (2004), p. 314).

As shown in the Fig. 6.17 (a), (b), (c) and (d) three intersecting planes will result in three ellipses if $\sigma_x \neq \sigma_y$, (and three circles, if $\sigma_x = \sigma_y$).

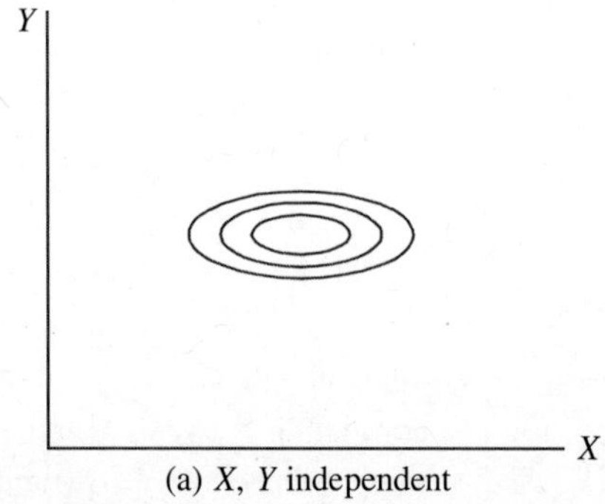

(a) X, Y independent

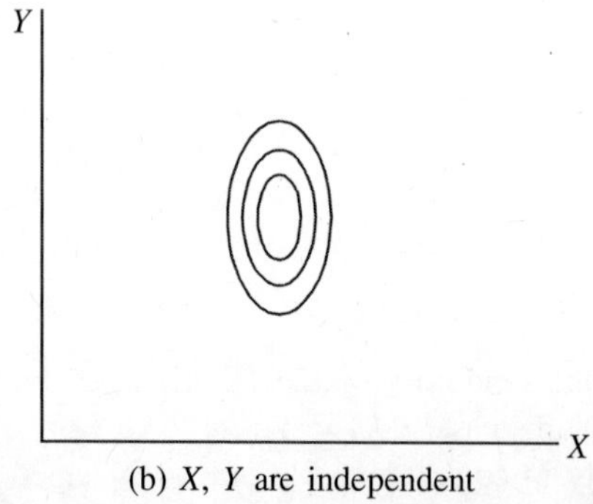

(b) X, Y are independent

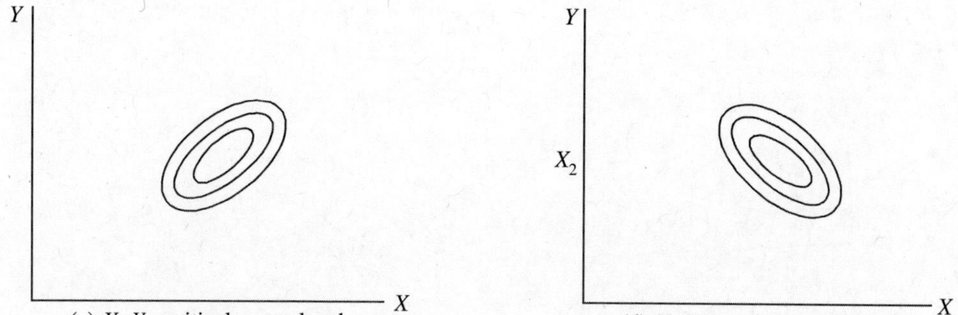

(c) X, Y positively correlated (d) X, Y negatively correlated

Fig. 6.17. Intersection ellipses of bivariate normal surface with unequal variances (source: Zar (2004), p. 315).

Properties: If (X, Y) has a bivariate normal distribution given by (6.51)

1. Probability that a point (X, Y) lies in any region R of (X, Y) plane is given by

$$P\left[(X, Y) \in R\right] = \iint_R f(x, y)\, dy\, dx \tag{6.52}$$

2. The constants appearing in the density function (6.51) have specific significance given by $E(X) = \mu_x$, $E(Y) = \mu_y$, $V(X) = \sigma_x^2$, $V(Y) = \sigma_y^2$ and correlation between X and $Y = \rho$.
3. If (X, Y) has a bivariate normal distribution, then X *and* Y *are independent* if and only *if* X *and* Y *are uncorrelated.*
4. Marginal density of X and Y are given by

$$f(x) = \int_{-\infty}^{\infty} f(x, y)\, dy = \frac{1}{\sqrt{2\pi}\,\sigma_x} \exp\left[-\frac{1}{2}\left(\frac{x-\mu_x}{\sigma_x}\right)^2\right] \quad \text{and} \tag{6.53}$$

$$f(y) = \int_{-\infty}^{\infty} f(x, y)\, dx = \frac{1}{\sqrt{2\pi}\,\sigma_y} \exp\left[-\frac{1}{2}\left(\frac{y-\mu_y}{\sigma_y}\right)^2\right] \tag{6.54}$$

which are univariate normal densities.

5. The conditional distribution of X given $Y = y$ is normal i.e.

$$\left(X/_{Y=y}\right) \sim N\left(\left\{\mu_x + \left(\rho\sigma_x/\sigma_y\right)\left(Y-\mu_y\right)\right\}, \sigma_x^2\left(1-\rho^2\right)\right) \tag{6.55}$$

and that of Y given $X = x$ is also normal i.e.

$$\left(Y/_{X=x}\right) \sim N\left(\left\{\mu_y + \left(\rho\sigma_y/\sigma_x\right)\left(X-\mu_x\right)\right\}, \sigma_y^2\left(1-\rho^2\right)\right) \tag{6.56}$$

See Fig. 6.18 below.

Figure 6.19 gives the intersection of normal surface

(a) by a plane parallel to (Y, Z) plane at $X = x$ giving the normally distributed sub population of Y for a given $X = x$ and

(b) by a plane parallel to (*X*, *Z*) plane at $Y = y$ giving the normally distributed sub population of *X* for a given $Y = y$.

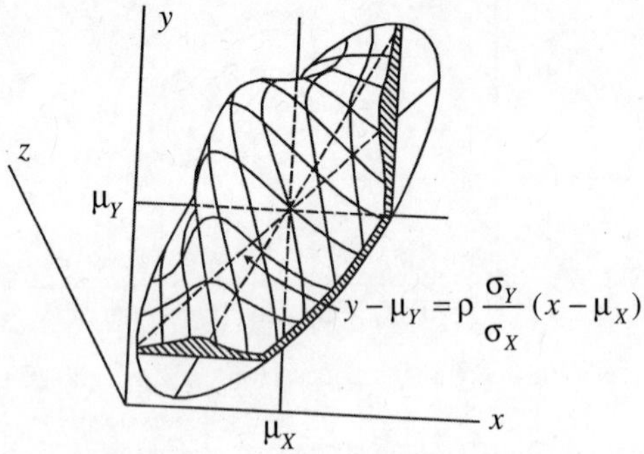

Fig. 6.18. Normal surface and conditional mean of *Y* (source: Hoel (1971), p. 155).

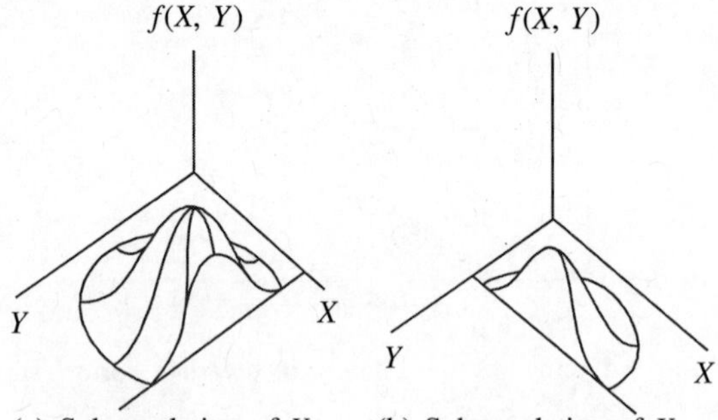

(a) Subpopulation of *Y* given *X* (b) Subpopulation of *X* given *Y*

Fig. 6.19. Normally distributed subpopulations (source: Daniel (1987), p. 402).

6. *Moment generating function* of (*X*, *Y*) is given by

$$M_{x,y}(t_1, t_2) = \exp\left[t_1\mu_x + t_2\mu_y + \frac{1}{2}\left(t_1^2\,\sigma_x^2 + t_2^2\sigma_y^2 + 2\rho t_1 t_2 \sigma_x \sigma_y\right)\right] \tag{6.57}$$

7. The *mean* of a random variable in a *conditional distribution* is *regression curve* when regarded as a function of the fixed variable in conditional distribution. Hence for the bivariate normal distribution the *regression curve of X on Y* is the straight line in (6.55) given by

$$X = \mu_x + (\rho\sigma_x/\sigma_y)(Y - \mu_y) \tag{6.58}$$

and *that of Y on X* is the straight line in (6.56) given by

$$Y = \mu_y + (\rho\sigma_y/\sigma_x)(X - \mu_x) \tag{6.59}$$

6.10 Chi-square Distribution

It will be found in the Chapters 7 and 8 that the normal distribution plays a very important role in statistical analysis of data. The central limit theorem ensures the use of the normal distribution for large samples. Apart from this there are other important reasons also.

One of the most important reason is the fact that if the sampled data is from a parent normal population, the distributions of sample statistics are easily manageable analytically. For drawing inferences about populations from samples, it is necessary to know the distributions of various functions of sample observations (called statistics). In this chapter three more distributions are described, which also arise as sampling distribution from normal population. In this section the chi-square distribution is considered which is also distribution of sample variance.

Definition: If *X* is a random variable with pdf

$$f(x) = \frac{1}{\Gamma(n/2)}\left(\frac{1}{2}\right)^{n/2} x^{\frac{n}{2}-1}\, e^{-x/2}, \quad x > 0 \quad n > 0 \tag{6.60}$$

then X is defined to have a chi square distribution with n degrees of freedom (d.f.), where n is positive integer.

In symbol $X \sim \chi_n^2$. The Figure 6.20 below gives the graph of pdf of χ_n^2 for various values of n. It may be noted that this is a particular case of gamma density with $n/2$ degrees of freedom and scale parameter equal to 2.

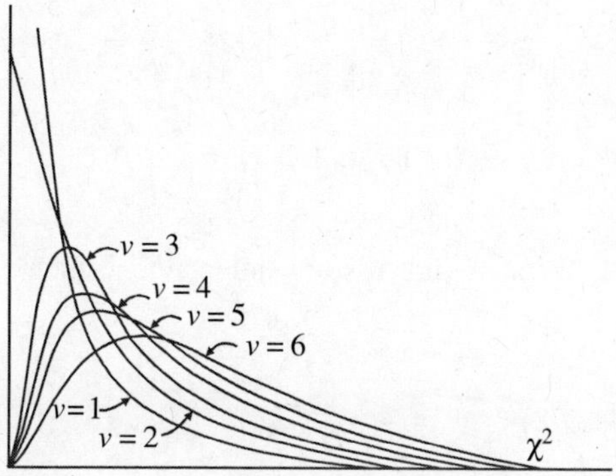

Fig. 6.20. pdf of χ^2 distribution with various degrees of freedom (source: Daniel (1987), p. 154).

6.10.1 Properties of Chi-square Distribution

If $X \sim \chi_n^2$ then

1. Range of X is 0 to ∞ i.e. X takes positive values only.
2. Distribution is positively skewed.
3. Mean of X, $E(X) = n$ and variance of X, $V(X) = 2n$.
4. *Additive property* (a) if $X \sim \chi_{n1}^2$ and $Y \sim \chi_{n2}^2$ are independent then

$$X + Y \sim \chi_{n1+n2}^2$$

In words if X and Y are independent χ^2 distribution with n_1 and n_2 d.f. respectively, then $X + Y$ has χ^2-distribution with $n_1 + n_2$ degree of freedom.

(b) In general if $X_i \sim \chi_{ni}^2$ for $i = 1, 2, ..., k$ and X_i's are independent then

$$\sum_{i=1}^{k} X_i \sim \chi_n^2, \text{ where } n = \sum_{i=1}^{k} n_i$$

5. If $X \sim N(0, 1)$ then $X^2 \sim \chi_1^2$. In words, square of a standard normal variable has Chi-square distribution with 1 degree of freedom.
6. If $X_i \sim N(0, 1)$ for $i = 1, 2, ..., k$ and X_i's are independent then $\sum_{i=1}^{k} X_i^2 \sim \chi_k^2$. In words, the sum of squares of independent standard normal random variable has a Chi-square distribution with degrees of freedom equal to the number of variables in the sum.
7. If $X_i \sim N(\mu_i, \sigma_i^2)$, $i = 1, 2, ..., k$ and X_i's are independent then

$$U = \sum_{i=1}^{k} \left(\frac{X_i - \mu_i}{\sigma_i} \right)^2 \sim \chi_k^2$$

It is evident from property 6, since $\dfrac{X_i - \mu_i}{\sigma_i}$; $i = 1, 2, ..., k$ are standard normal variables.

8. *Moment Generating Function* of X following Chi-square distribution with n degree of freedom is given by

$$M_x(t) = (1 - 2t)^{-n/2} \tag{6.61}$$

9. For sufficiently large n, (a) $\sqrt{2X}$ has approximately normal distribution with mean $\sqrt{2n-1}$ and variance 1. In symbols, if $X \sim \chi_n^2$ then

$$\text{as } n \to \infty, \sqrt{2X} \rightsquigarrow N\left(\sqrt{2n-1}, 1\right) \tag{6.62}$$

(b) Since $E(X) = n$, $V(X) = 2n$ so for large n, $\dfrac{X-n}{\sqrt{2n}} \sim N(0, 1)$.

10. If $X_i \sim N(\mu, \sigma^2)$ for $i = 1, 2, ..., n$ and X_i's are independent, then

$$\text{(i) } \sum_{i=1}^{n}\left(\frac{X_i - \mu}{\sigma}\right)^2 \sim \chi_n^2 \text{ and (ii) } \sum_{i=1}^{n}\left(\frac{X_i - \bar{X}}{\sigma}\right)^2 \sim \chi_{n-1}^2, \text{ where } \bar{X} = \frac{1}{n}\sum_{i=1}^{n} X_i. \tag{6.63}$$

6.11 Student's *t*-Distribution

Another distribution of practical importance is Student's *t*-distribution, especially when sample size is small and population may be known to be normal. This distribution came up as a result of work by W.S. Gosset (1908) who published it with a pseudo name of "Student".

Definition: If X is a random variable having probability density function given by

$$f(x) = \frac{\Gamma\left[{(k+1)}/{2}\right]}{\Gamma(k/2)\sqrt{k\pi}}\left(1 + x^2/k\right)^{-(k+1)/2}, \quad -\infty < x < \infty \tag{6.64}$$

then X is said to have Student's *t*-distribution with k degrees of freedom. In symbols $X \sim t_k$.

It may be noted that Student's *t*-distribution has range $(-\infty, \infty)$ same as that of the normal distribution. Figure 6.21 gives graph of pdf (6.64) for various degrees of freedom.

A result: If Z is a standard normal random variable, U has a Chi-square distribution with k degrees of freedom and Z and U are independent, then $t = \dfrac{Z}{\sqrt{U/k}}$ has Student's *t*-distribution with k degrees of freedom.

It is this result which is the base of inferences about mean, when sample size n is small ($n < 30$).

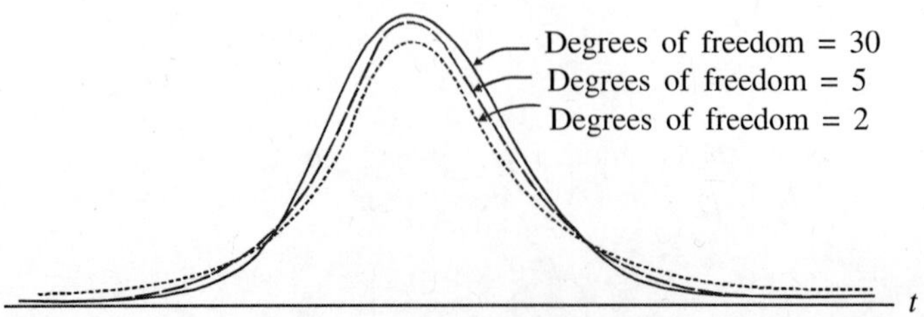

Fig. 6.21. pdf of Student's *t*-distribution for various k (source: Daniel (1987), p. 137).

6.11.1 Properties of *t*-distribution

If $X \sim t_k$ i.e. X follows Student's t-distribution with k degrees of freedom then

1. Range of X is entire real line i.e. $(-\infty, \infty)$.
2. Distribution of X is symmetric about $x = 0$.
3. The mean, median and mode of X are all equal to 0.
4. $E(X) = 0$ if $k > 1$ and $V(X) = k/(k-2)$ if $k > 2$, so $V(X) > 1$.
   Thus it is worth looking at normal and Student's t-distribution and to find that t-distribution is more dispersed than a standard normal distribution.

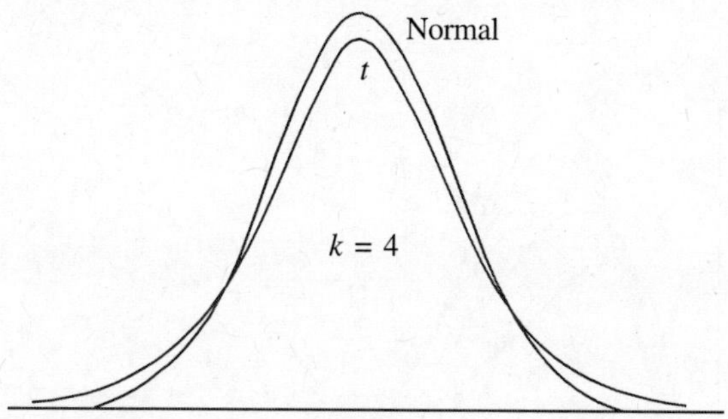

Fig. 6.22. Student's t and standard normal distribution (source: Hoel (1971), p. 261).

5. If $X_1, X_2, ..., X_n$ is a random sample from $X \sim N(\mu, \sigma^2)$, then let

$$Z = (\bar{X}_n - \mu)/(\sigma/\sqrt{n}) \sim N(0,1), \text{ and } U = \sum_{i=1}^{n} (X_i - \bar{X}_n)^2 / \sigma^2 \sim \chi^2_{n-1}$$

Then it can be shown that Z and U are independent.

So
$$\frac{Z}{\sqrt{(U/(n-1))}} = \frac{\sqrt{n}\,(\bar{X}_n - \mu)}{\sqrt{\{\Sigma(X_i - \bar{X}_n)^2/(n-1)\}}} = \frac{\sqrt{n}\,(\bar{X}_n - \mu)}{s} \tag{6.65}$$

has Student's t-distribution with $(n - 1)$ degrees of freedom.

6.12 *F*-Distribution

This distribution is also of practical interest with problems of inferences about variances and in analysis of variance (ANOVA) to be discussed in subsequent chapters. Hence this distribution is also defined and its properties are discussed in this section.

Definition: If X is a random variable having density given by

$$f(x) = \frac{\Gamma[(m+n)/2]}{\Gamma(m/2)\,\Gamma(n/2)} \left(\frac{m}{n}\right)^{m/2} \frac{x^{(m-2)/2}}{[1+(m/n)x]^{(m+n)/2}}, \quad x > 0 \tag{6.66}$$

then X is defined to have F-distribution with m and n degrees of freedom. The order in which the degrees of freedom are given is of importance and can not be interchanged since the F-distribution changes if the order of degrees of freedom is changed.

F-distribution can also be seen to be distribution of ratio of two Chi-square variables as given in the following result.

Result: Let U be a Chi-square random variable with m degrees of freedom, V be a Chi-square random variable with n degrees of freedom and U and V are independent. Then the random variable X given by

$$X = \frac{U/m}{V/n} \text{ has } F\text{-distribution with } m \text{ and } n \text{ degrees of freedom} \tag{6.67}$$

with pdf in (6.66).

In symbols one can write $X \sim F_{m,n}$. Figure 6.23 gives the graph of pdf of $F_{m,n}$ for various values of m and n.

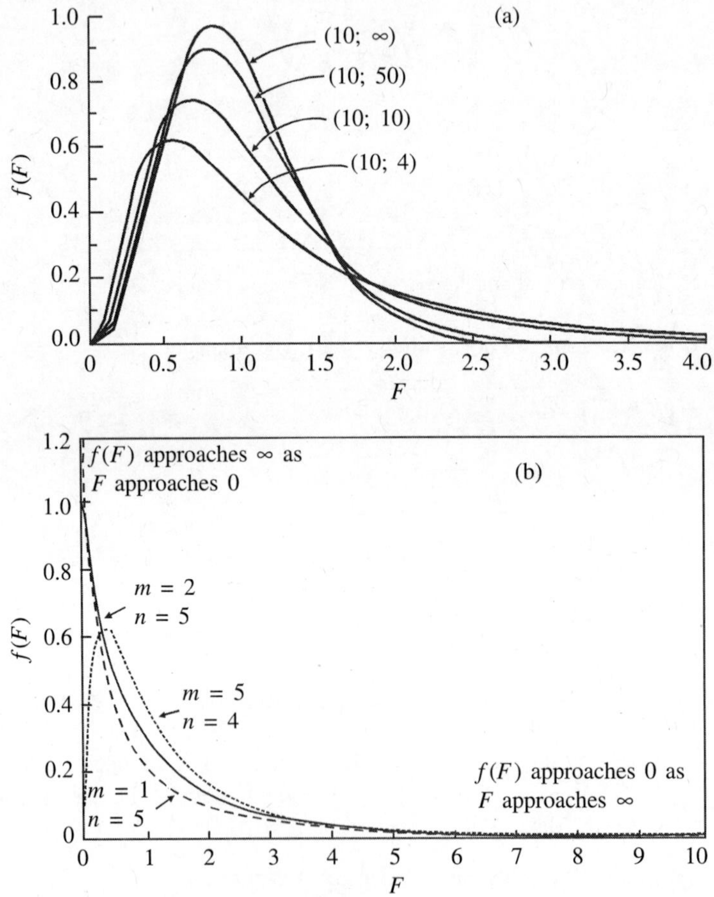

Fig. 6.23. *F*-distribution for various degrees of freedom m and n. (Source: (a) Croxton et al. (1975), p. 603 and (b) Daniel (1987), p. 163.)

6.12.1 Properties of *F*-distribution

If $X \sim F_{m,n}$ then

1. Range of *F*-distribution is $(0, \infty)$.

2. Shape of the pdf is reverse *J* if $m \leq 2$ and bell shaped for $m > 2$. Hence *F*-distribution is positively skewed.

3. $$E(X) = \frac{n}{n-2} \quad \text{for} \quad n > 2 \tag{6.68}$$

and $$V(X) = \frac{2n^2(m+n-2)}{m(n-2)^2(n-4)} \quad \text{for} \quad n > 4 \tag{6.69}$$

4. *Reciprocal Property*: If *X* has *F*-distribution with *m* and *n* degrees of freedom, $Y = 1/X$ has *F*-distribution with *n* and *m* degrees of freedom. This property allows one to tabulate only the upper tail values of $F_{m,\,n}$. The lower tail values can be easily computed using this property.
5. Because of the reciprocal property of *F*, $100 \times p^{th}$ percentile of *X* following $F_{m,\,n}$ distribution is reciprocal of $100(1-p)^{th}$ percentile of $Y(= 1/X)$ having $F_{n,\,m}$ distribution. In symbols

$$F_{m,n,p} = \frac{1}{F_{n,m,(1-p)}} \tag{6.70}$$

where $F_{m,\,n,\,p}$ is $100\, p^{th}$ percentile of *F*-distribution with *m* and *n* degree of freedom (see Fig. 6.24).

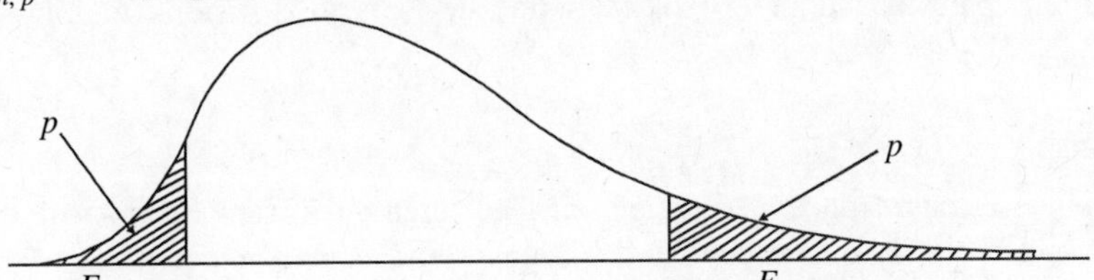

Fig. 6.24. $100\, p^{th}$ and $100\,(1-p)^{th}$ percentile of $F_{m,\,n}$ distribution (source: Hoel (1971), p. 272).

6. The square of a random variable *t*-following Student's *t*-distribution with *n* degrees of freedom has *F*-distribution with 1 and *n* degrees of freedom. In symbols

$$\text{If } t \sim t_n \text{ then } t^2 \sim F_{1,\,n} \tag{6.71}$$

7. If $X_1, X_2, ..., X_{m+1}$ is a random sample from $N(\mu_x, \sigma^2)$ and $Y_1, Y_2, ..., Y_{n+1}$ is a random sample from $N(\mu_y, \sigma^2)$ and if the two samples are independent, then

$$U = \frac{1}{\sigma^2}\sum_{i=1}^{m+1}\left(X_i - \bar{X}\right)^2 \sim \chi_m^2 \tag{6.72}$$

$$V = \frac{1}{\sigma^2}\sum_{j=1}^{n+1}\left(Y_i - \bar{Y}\right)^2 \sim \chi_n^2 \tag{6.73}$$

then *U* and *V* are independent. So

$$F = \frac{U/m}{V/n} = \frac{\sum_{i=1}^{m+1}\left(X_i - \bar{X}\right)^2/m}{\sum_{j=1}^{n+1}\left(Y_j - \bar{Y}\right)^2/n} = \frac{S_x^2}{S_y^2} \tag{6.74}$$

has *F*-distribution with *m* and *n* degrees of freedom.

7. Biostatistical Inference

Last three chapters of this book emphasize definition and real life application of probability theory and random variables. To answer judgmental questions and to take some decisions the distribution of associated random variables are needed to compute the probabilities of relevant values of the random variables.

One important part of the theory of statistics deals with decisions making in presence of uncertainty.

7.1 Inference

Progress in science is based on experimentation. The research worker in most of the physical and social sciences performs an experiment and obtains some data. On the basis of the data certain conclusions are drawn. The conclusions usually go beyond the materials and operations of the particular experiment. In other words, the scientists wish to generalize from a particular experiment to the class of all similar experiments. This sort of extension from a particular to the general may be called *inductive inference.*

An interested reader may supplement his knowledge in these techniques by consulting the books on mathematical statistics or statistical inference, such as Hoel (1971), Meyer (1975), Mood, Graybill and Boes (1974), Rohatgi (1994). The books by Milton and Tsokos (1983) and Steel and Torrie (1981) give several examples of real world problems related to biological experiments.

In such inferences always an element of uncertainty is present. One can not make absolutely certain generalizations. However, if the experiment is performed according to certain principles, the degree of uncertainty can be measured. Statistical theory provides technique for making inductive inferences and for measuring the degree of uncertainty in terms of probability. This chapter deals with some concepts on which these statistical techniques are based.

Before giving the concepts leading to inductive inference a brief idea of another type of inference viz. *deductive inference* which is conclusive, is given below.

Consider the statements

(i) One of the interior angle of a right angle triangle is of 90°.
(ii) Triangle A is a right angle triangle.

If these two statements are correct then

(iii) One of the angles of triangle A is 90°.

This type of inference is called *deductive inference.*
While deductive inference is extremely important, much of the new knowledge in the real world comes from the inductive inference.

7.1.1 Examples of use of Inductive Inference

In this section some problems are considered where inductive inference should be made.

Example 7.1: A bag is full of about 10 million of flower seeds. It is known that seeds produce either white or red flowers only. It is desired to know the proportion of seeds which will give white flowers.
Example 7.2: It is desired to know the average weight of Post Graduate (PG) students of Banaras Hindu University (BHU) in 2007-08.
Example 7.3: It is desired to know if the new drug for treatment of tumour is more effective than existing drug, by administering the two drugs to two groups of tumour induced mice of same age.
Example 7.4: In a country the average number of children born to married women in age group 18 to 24 years was 0.8 in 1980. A demographer feels that in 2000, this figure has reduced. Can the country records of 2000 support or refute this contention?

In all these examples some conclusion is to be drawn for a general group of elements.

The answer to question in Examples 7.1 and 7.2 can be given in two ways. One method is to observe each of the elements of these examples and determine the proportion of the seeds giving white flowers and by measuring weight of the each PG students enrolled. If all the seeds are sown to see the flower colour, there will be left nothing for sale and will involve lot of money and land. Locating each PG student and taking his weight is also very cumbersome task. This method is therefore, infeasible for both the problems of Examples 7.1 and 7.2.
Second method to answer these problems is to select only a few elements from all the elements of interest and observe only the selected ones. Thus only a few seeds (say 30) may be sown and flower colour may be noted to find the proportion of the seeds with white flower. Similarly say 50 PG students may be contacted to measure their weights and calculate the average weight of the 50 PG students.

Both of these methods require properly planned and statistically sound techniques. The conclusion drawn by any one of the two methods are *inductive inference.*

7.2 General Concepts

Before going into statistical techniques of inductive inference some concepts are defined.

Population

Totality or an aggregate of all the objects under study and about which some information is required for a statistical problem is called a *population.* Thus

(a) All the seeds (about 10 million) in the bag,
(b) All the PG students of BHU enrolled in 2007-08,
(c) Mice with induced tumor of a given age, and
(d) All the married women in age group 18-24 years in the country,

may be called a population for respective studies.
Objects in a population can be living, such as people, animals, bacteria, plants etc or non living such as records, seeds, light bulbs of a given company, tables supplied by a shop etc.

Sample

A subset (or part) of all the elements under study is called a *sample*.
50 PG students of BHU in 2007-08 form a sample for Example 7.2.

Complete Enumeration or Census

The first method of collecting desired information on each and every element in the population is called *census* or *complete enumeration*.

Sample Survey

The second method of collecting desired information on only sampled elements is called *sample survey*.

Methods of Sampling

Methods of selecting a sample from a population are dealt in *theory of sampling* and form a more than one semester course for graduates in statistics.

A proper choice of sampling procedure depends on the nature and some known characteristics of the population under study. After a choice has been made the prescribed sampling procedure must be followed correctly.

Random Sampling

The simplest *and quite useful* method of sampling to enable measurement of uncertainty is *simple random sampling*. The sample thus obtained is called a *simple random sample* or a *random sample*. A random sample from any finite population and that from a population which has a density are defined below.

Random Sample from a Finite Population

Let there be a population of finite number (say N) of objects and out of these n objects are to be selected. For this, each object in the finite population is labeled numerically so that each object has a unique identity number.

This step is called *preparing a sampling frame*. Without loss of generality a population of N objects may be represented as 1, 2, ..., N.

Definition: A sample of size n is said to be a *simple random sample* if it satisfies the following conditions

1. All the units in the population have an equal chance of being selected in the sample in a particular draw.
2. All possible samples have an equal chance of being selected.

A *simple random sample* of n objects may be drawn from N objects *with replacement* (called *SRSWR*) or *without replacement* (called *SRSWOR*). To illustrate the method of selection, consider the following random variable:

Let X_i = population value of the random variable X (characteristic under study) when the ith unit is selected in the sample, i = 1, 2, ..., n (7.1)

Probability distribution of the random variables $X_1, X_2, ..., X_n$ depends on the method of selection of sample.

If sampling is SRSWR, at each of the n draws of an object from a population of size N, each of these N objects are given an equal chance of being selected. An *object* may *appear more than once* in the sample. In this case, the random variables X_1, X_2, ..., X_n are independently and identically distributed as

$$P[X_i = j] = 1/N, \quad i = 1, 2, \ldots\ldots, n$$
$$j = 1, 2, \ldots\ldots, N \tag{7.2}$$

In case of simple random sampling without replacement i.e. for SRSWOR, the random variables X_1, X_2, ..., X_n are not independent, because the number of objects from which to select one in sample will be one less at each subsequent draw. Thus the joint probability distribution of the sample is given by

$$P\left[X_1 = j_1, X_2 = j_2, \ldots\ldots, X_n = j_n\right] = \frac{1}{N(N-1)\ldots\ldots(N-n+1)} \tag{7.3}$$

where j_1, j_2, ..., j_n are any n values from the objects 1, 2, ..., N.

It can be shown that the marginal densities of X_1, X_2, ..., X_n in SRSWOR, irrespective of the values taken by other $(n - 1)$ X_i's are same as that for SRSWR.

The random variable X_i is a representation for the numerical value that the ith sampled object or element will assume (i = 1, 2, ..., n).

After the sample is selected the actual values of X_1, X_2, ..., X_n are known and as usual, these observed values are denoted by x_1, x_2, ..., x_n. Some times these observed values are called a random sample, if these are the values of the random sample X_1, X_2, ..., X_n.

Random Sample from Population with Density

Most often the scientific study is related to some numeric characteristic of a population elements, i.e. about a random variable associated to it, say X, which has a pdf.

Definition: Let X have a pdf $f(x, \theta)$ and let X_1, X_2, ..., X_n be observations on a sample of n elements from this population. Then X_1, X_2, ..., X_n are also random variables and have a joint pdf.

If the sample is such that the joint density of X_1, X_2, ..., X_n denoted by $L(x_1, x_2, \ldots\ldots, x_n; \theta)$ is given by

$$L(x_1, x_2, \ldots\ldots, x_n; \theta) = f(x_1, \theta)\, f(x_2, \theta), \ldots\ldots, f(x_n, \theta)$$
$$= \prod_{i=1}^{n} f(x_i, \theta) \tag{7.4}$$

where $f(x_i, \theta)$ is the common density of each X_i, i = 1, 2, ..., n,
then X_1, X_2, ..., X_n is said to be a *random sample* of size n on X.

It may be noted that for a sample to be a random sample, the observations X_1, X_2, ..., X_n must be independent and they must have the same density as X.

In other words we can say that if X_1, X_2, ..., X_n is a random sample, X_i's (i = 1, 2, ..., n) are independently and identically distributed (iid) with pdf $f(x_i, \theta)$.

Method of Selecting a Random Sample from a Finite Population

Suppose a finite population consists of N elements and a random sample of size $n(n < N)$ is to be selected. In order that the selected sample satisfies the conditions (1) and (2) given in the definition, the selection of elements is done by using random number tables (see Appendix A.2).

The *steps* for selection are as follows:

1. Assign a numeric identity number to each unit (or element) in the population i.e. 1, 2, ..., N.
2. Open a page of random number table randomly and select any row and any column in it.
3. If the number N is of K digits, start reading K consecutive digits in the random number table row-wise or column-wise. Those K digits form a random number R.
4. If $R \leq N$, element with identity number R is selected in the sample.
   If $R > N$, find remainder R_m of R/N.
   If $R_m > 0$, element with identity number R_m is selected.
   If $R_m = 0$, choose element with identity N.
5. Read the next K digits row-wise or column-wise (opt the same pattern till the selection of n elements in sample is completed). Repeat steps 4 and 5 till n units are selected in the sample.
6. If sampling is without replacement, a repeated random number R is not selected or considered again but neglected. If sampling is SRSWR, R can be retained so that same element of the population may be repeated in the sample.

Parameter

The constants appearing in the pdf or pmf of a random variable are called *parameters.*

Apart from these constants, appearing in the distribution of random variables, any characteristic of an existing (so finite) population is also called a parameter.

For example, the mean weight, standard deviation of weights, range of heights of PG students of BHU in 2007-08 are parameters. These are obtained by calculations on each of the N (= 8000 for BHU PG students of 2007-08) units of a finite population.

Parameters are usually denoted by θ, are generally unknown and the problem of statistical inference is to infer about unknown parameters. Knowledge of the value(s) of the parameter(s) of pdf or pmf specifies the distribution completely and all probability statements can be made about the associated random variable.

Statistic

A function of observed values in a sample, which does not involve any unknown parameters, is called a *statistic.*

Because it is a function of sample observations, the value of a statistic is known if the sample values are available.

Let $X_1, X_2, ..., X_n$ be a random sample from population, then

$$T(X_1, X_2, ..., X_n) \text{ is a statistic and its value is given by } t = T(x_1, x_2, ..., x_n) \tag{7.5}$$

Here t depends only on the sample values $x_1, x_2, ..., x_n$.

Sampling Distribution

A statistic being a function of sample observations, is likely to have different values, when computed from different samples.

If a large number of samples are drawn from a finite population, and values of statistic(s) are computed from each sample, one could prepare a frequency distribution of the values of statistic(s). This distribution is called *sampling distribution* of the statistic because the variation in values of a statistic arises due to sampling.

If sample is from a population with pdf, theoretically the distribution of T, which is a function of $X_1, X_2, ..., X_n$ can be determined from the joint pdf (or joint pmf) of the random sample. This distribution will be the sampling distribution of T.

The statistical technique of drawing inductive inference uses the sampling distribution of some suitable statistic. Measurement of uncertainty in the inference is also possible from this sampling distribution.

Standard Error

The standard deviation of the statistic obtained from the sampling distribution of a statistic is called the *standard error* of the *statistic*.

Thus in symbols, the standard error (SE) of the statistic T is given by

$$\text{SE}(T) = \sqrt{V(T)} \tag{7.6}$$

After having defined some of the concepts needed for inductive inference, it may be mentioned again that inductive inference or statistical interference is a type of decision making based on probability. It can be seen that a large part of the statistical inference falls in two categories. Either they involve *estimation of parameters* or other characteristics of the probability density function of an associated random variable (selected as a model) or they involve *testing of some* (one or more than one) *hypotheses* about the model (the pdf).

This chapter is giving a brief description of these two types of statistical inference. The next section deals with estimation and the subsequent one the testing of hypothesis.

7.3 Estimation

The **problem of estimation** can be briefly defined as follows: suppose a characteristic of elements in a population can be represented by a random variable X with pdf $f(x, \theta)$. The form of the pdf i.e. the function $f(.,.)$ is known but the value of the parameter θ is not known. If θ were also known the pdf $f(x, \theta)$ would be completely known and there would be no need for drawing inference about X. Further assume that it is possible to obtain the random sample $X_1, X_2, ..., X_n$ on X.

On the basis of the observed values $x_1, x_2, ..., x_n$ of the random sample it is desired to estimate the value of the unknown parameter θ or value of some function, say $\tau(\theta)$ of unknown θ.

7.3.1 Point and Interval Estimation

Two types of estimation procedures are in common use. One is called *point estimation* and the other *interval estimation*.

First procedure which gives a *point estimate*, consists in using a statistic, say $T(x_1, x_2, ..., x_n)$ to estimate the unknown $\tau(\theta)$. The value of the statistic $T(x_1, x_2, ..., x_n)$ is a point estimate obtained from a given sample, whereas the formula of this statistic $T(X_1, X_2, ..., X_n)$ which is a known function of the observations is called the *point estimator*.

The second procedure consists in defining two statistics $l_1(X_1, X_2, ..., X_n)$ and $l_2(X_1, X_2, ..., X_n)$ such that $l_1(X_1, X_2, ..., X_n) < l_2(X_1, X_2, ..., X_n)$ and $(l_1(.), l_2(.))$ is a random interval for which the probability that the unknown $\tau(\theta)$ is included in this interval can be calculated.

The specific interval obtained for a particular sample observations is called an *interval estimate* or *confidence interval* for $\tau(\theta)$ and the formula or the known function $(l_1(.), l_2(.))$ is called an *interval estimator* for $\tau(\theta)$. The probability with which the said random interval is expected to include the unknown $\tau(\theta)$ is the confidence an experimenter can have in the calculated interval, that it includes the unknown parameter in it, hence the name **confidence interval.**

The estimators are the statistics which are the random variables and the estimates are the values from sample observations.

The answers to the questions in Example 7.1 and 7.2 are problems of estimation of unknown proportion (say p) of seeds with white flowers and of unknown average weight (say μ) of PG students of BHU, respectively.

Solution to Example 7.1: Suppose a random sample of n seeds is selected and sown in a field. The number of plants which gave white flowers (= r) out of total plants (n) are counted. Then ***r/n* is a possible estimator** of p. In symbols

$$\hat{p} = \frac{r}{n} \tag{7.7}$$

and if in the experiment $r = 45$ and $n = 225$, the **point estimate** of

$$p = \frac{45}{225} = \frac{1}{5} = .2$$

A possible **interval estimator** of p could be

$$\left(\hat{p} - 1.96\sqrt{\frac{\hat{p}(1-\hat{p})}{n}},\quad \hat{p} + 1.96\sqrt{\frac{\hat{p}(1-\hat{p})}{n}}\right) \tag{7.8}$$

which gives a **95% confidence interval** for p as

$$\left(.2 - 1.96\sqrt{\frac{.2\times.8}{225}},\quad .2 + 1.96\sqrt{\frac{.2\times.8}{225}}\right) \text{ or } (.1477, .2523)$$

Solution to Example 7.2: One possible point estimator for $\tau(\theta) = \mu$ is $(X_1 + X_2 +, \ldots\ldots, + X_n)/n = \bar{X}_n$. In symbols

$$\hat{\mu} = \bar{X}_n \tag{7.9}$$

and $\left(\bar{X}_n - 1.96\frac{S}{\sqrt{n}},\quad \bar{X}_n + 1.96\frac{S}{\sqrt{n}}\right)$ is a possible **interval estimator** for μ, where S is sample standard deviation.

Suppose a random sample of 100 PG students of 2007-08 of BHU was selected and their weights measured. Let the mean of these 100 weights ($\bar{X}_{100}$) be 53 kg. and standard deviation of these 100 weights (S) be 16 kg.

Then the **point estimate** *of* μ is 53 kg and **95% confidence interval** *for* μ is $\left(53 - 1.96\frac{16}{\sqrt{100}},\quad 53 + 1.96\frac{16}{\sqrt{100}}\right)$ or (53 − 3.136, 53 + 3.136) or (49.864, 56.136).

Note: One can never know whether a 95% or even 99% confidence interval contains the unknown parameter p or μ in it or not. It only means that 95% or (99%) of random intervals $(l_1(.), l_2(.))$ trap the true but unknown parameter value in it. As an example see Fig. 7.1.

From the above two examples, it is evident that a statistic is chosen to be used as an estimator of the unknown parameter. The *problem of estimation* has two issues. *First* is to devise *some means* of *obtaining a statistic* to be used as an estimator and *second* to *set and select some criteria* to find the *best estimator.* Some of the commonly used methods to choose a statistic is given in the subsection 7.3.3 and the desirable properties or criteria of an estimator are given in the subsequent subsection 7.3.2.

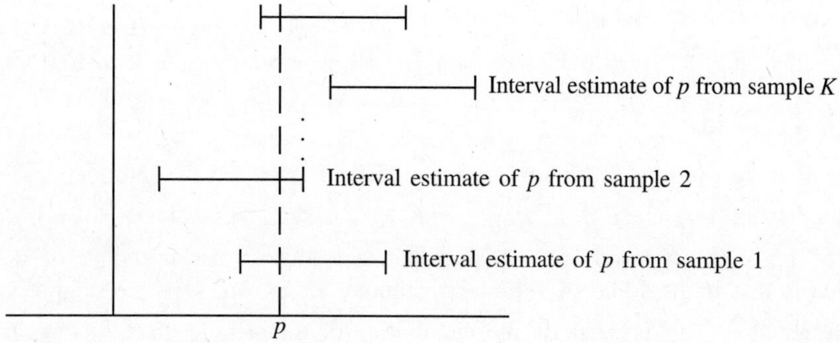

Fig. 7.1. 95% of the intervals calculated from equation (7.8) are expected to contain true value of p (source: Milton and Tsokos (1983), p. 158).

7.3.2 Criteria for a Good Estimator

Previous section considered some methods of obtaining point estimators. It is well known that often there exist more than one possible estimators of a parameter. For example, the mean, median and mode are well known possible estimators of population mean. A question, therefore, arises that which estimator among several should be used to estimate a parameter.

Some criteria or properties, that a good estimator should possess, are defined in this section to decide which one among many is a better estimator. Some of the common estimators will be discussed and the properties possessed by them will be enumerated. In common practice, a good estimator should have the following four properties (i) Unbiasedness (ii) Consistency (iii) Efficiency (iv) Sufficiency.

The definitions of estimator with these properties are given below. Let $\hat{\theta} = T(X_1, X_2, \ldots\ldots, X_n)$ be an estimator of the unknown parameter θ associated with pdf of a random variable X.

(i) Unbiased Estimator

$\hat{\theta}$ is an *unbiased estimator* of θ if

$$E(\hat{\theta}) = \theta \qquad \text{for all } \theta \tag{7.10}$$

(ii) Consistent Estimator

$\hat{\theta}$ is said to be a *consistent estimator* of θ if

$$\lim_{n \to \infty} P\left[\left|\hat{\theta} - \theta\right| > \varepsilon\right] = 0 \text{ for all } \varepsilon > 0 \tag{7.11}$$

An equivalent definition is more convenient and is as follows:

If $$\lim_{n \to \infty} E(\hat{\theta}) = \theta \qquad \text{and} \qquad \lim_{n \to \infty} V(\hat{\theta}) = 0$$

then $\hat{\theta}$ **is** a *consistent estimator* of θ.

(iii) Efficient Estimator

$\hat{\theta}$ is said to be a *more efficient* estimator of θ than any other estimator $\hat{\theta}^*$ if

$$V(\hat{\theta}) < V(\hat{\theta}^*) \tag{7.12}$$

If $V(\hat{\theta}) < V(\hat{\theta}^*)$ for all $\hat{\theta}^*$ then $\hat{\theta}$ is said to be *most efficient* or *best estimator*.
If $V(\hat{\theta})$ is smaller and $\hat{\theta}$ is also unbiased, then $\hat{\theta}$ will be more close to $\boldsymbol{\theta}$ than $\hat{\theta}^*$.

(iv) Sufficient Estimator

An estimator $\hat{\theta}$ is said to be a *sufficient estimator* of θ, if it has all the information about θ, which the whole sample can have about θ. In other words, $\hat{\theta}$ is a *sufficient estimator* of θ if the conditional distribution of $X_1, X_2, ..., X_n$ given $\hat{\theta} = \hat{\theta}_0$ does not depend on θ for any value of $\hat{\theta}$.

The idea is that if the value of sufficient estimator is known, then the sample values can tell nothing more about θ. This is, because the distribution of the sample given $\hat{\theta}$ does not depend on θ, so nothing can be learnt about θ from the distribution which does not depend on θ.

The **sample mean $\bar{X}_n$** is an estimator of mean of a normal population which is **unbiased, consistent, sufficient and efficient** (among median, mode, and first observation X_1).

X_1, the first observation in a random sample from $X \sim N(\mu, \sigma^2)$ is an unbiased estimator of μ, but $\bar{X}_n$ is much efficient than the X_1 since

$$V(\bar{X}_n) = \sigma^2/n < V(X_1) = \sigma^2 \tag{7.13}$$

$\hat{\theta}$ is said to be a *best unbiased estimator* of θ if

$$E(\hat{\theta}) = \theta \text{ and } V(\hat{\theta}) \le V(\hat{\theta}^*) \text{ for all } \hat{\theta}^* \text{ satisfying } E(\hat{\theta}^*) = \theta \tag{7.14}$$

$\hat{\theta}$ is said to be a *best linear unbiased* estimator of θ if

(a) $E(\hat{\theta}) = \theta$

(b) $\hat{\theta} = \sum_{i=1}^{n} a_i x_i$, for constants a_i and

(c) $V(\hat{\theta}) \le V(\hat{\theta}^*)$, where $\hat{\theta}^*$ is any other estimator of θ which satisfies (a) and (b) above.

Some examples of good estimators are given below:

Example 7.5: Suppose that r/n is the proportion of defectives in a sample of size n. Then $r \sim B(n, p)$ and $\hat{p} = \frac{r}{n}$ is an *unbiased estimator* of p.

$V(\hat{p}) = \frac{p(1-p)}{n} \to 0$ as $n \to \infty$. So $\hat{p}$ is also a consistent estimator of p.

Example 7.6: Let X be a random variable with finite mean μ and variance σ^2. Let $\bar{X}_n$ be the mean of the random sample of size n. Then $\bar{X}_n$ is unbiased for μ and is also consistent since $V(\bar{X}_n) = \frac{\sigma^2}{n} \to 0$ as $n \to \infty$.

7.3.3 Methods of Estimation

Some of the commonly used methods of estimation are given in this section.

(a) Method of Moments

Let $f(x, \theta_1, \theta_2, ..., \theta_k)$ be pdf of a random variable X, which has k parameters. Further, let X_1, X_2,

......, X_n be a random sample from this density. μ_r, the rth raw moment (= $E(X^r)$) of X is in general a function of the k parameters. In symbols

$$\mu_r' = h_r(\theta_1, \theta_2, ..., \theta_k) \tag{7.15}$$

One can also find from the sample, the rth raw moment as

$$m_r' = \frac{1}{n}\sum_{i=1}^{n} X_i^r$$

The method of moments consists in forming the k equations

$$m_r' = \mu_r' = h_r[\theta_1, \theta_2, ..., \theta_k] \quad r = 1, 2, ..., k \tag{7.16}$$

and solving these for θ_1, θ_2, ..., θ_k, to get $\hat{\theta}_1, \hat{\theta}_2, ..., \hat{\theta}_k$ as known functions of sample moments. $\hat{\theta}_1, \hat{\theta}_2, ..., \hat{\theta}_k$ are then called *estimators* of θ_1, θ_2, ..., θ_k, respectively, obtained by **method of moments**.

Some examples follow

Example 7.7: Let $X \sim N(\mu, \sigma^2)$ and let X_1, X_2, ..., X_n be a random sample from this distribution Thus, $\mu_r' = m_r' = h_r(\mu, \sigma)$, r = 1, 2 to form the equations (7.16).
It can be seen that

$$\mu_1' = h_1(\mu, \sigma) = \mu \text{ and } \mu_2' = h_2(\mu, \sigma) = \sigma^2 + \mu^2 \tag{7.17}$$

So

$$\mu_1' = \mu = m_1' = \bar{X}_n \text{ and } \mu_2' = \sigma^2 + \mu^2 = \frac{1}{n}\sum_{i=1}^{n} X_i^2 = m_2' \tag{7.18}$$

Hence solving for μ and σ from (7.18), the method of moment estimators are

$$\hat{\mu} = \bar{X}_n \text{ and } \hat{\sigma}^2 = m_2' - \bar{X}_n^2$$

or

$$\hat{\sigma}^2 = \frac{1}{n}\sum_{i=1}^{n} X_i^2 - \bar{X}_n^2 = S^2 \tag{7.19}$$

The sample mean and sample variance are thus moment estimators of μ and σ^2, respectively. (Source: Mood et al. (1974), p. 275.)

Example 7.8: Let X_1, X_2, ..., X_n be a random sample from Poisson distribution with parameter λ. Since it has only one parameter λ, so only one equation to obtain moment estimator of λ will be

$$\mu_1' = m_1' \text{ or } h_1(\lambda) = \lambda = \bar{X}_n$$

Hence sample mean is method of moment estimator of λ. (Source: Mood et al. (1974), p. 275.)

(b) Method of Maximum Likelihood

Let X_1, X_2, ..., X_n be a random sample from X with pdf $f(x, \theta)$. Then the joint density of the random sample denoted by $L(x_1, x_2, ..., x_n, \theta)$, is called the likelihood function. Once a sample has been obtained this may be considered as a function of unknown parameter θ.

$L(x_1, x_2, ..., x_n, \theta)$ gives the *likelihood* that the random variables X_1, X_2, ..., X_n assume a particular value x_1, x_2, ..., x_n. The likelihood is the value of a density function, so for discrete random variables it is a probability if θ is known. Problem of estimation wants to know from which density (i.e. for what value of θ) is the likelihood largest to obtain the observed sample.

In other words it is desired to find the values of θ say $\hat{\theta}$ in the range space of θ, which maximizes the likelihood function. The value $\hat{\theta}$ which maximizes $L(\theta, x_1, x_2, \ldots\ldots, x_n)$ is in general a function of $x_1, x_2, \ldots\ldots, x_n$ say $T(x_1, x_2, \ldots\ldots, x_n)$.

Definition: Maximum likelihood estimator (m.l.e.): Let $L(\theta) = L(\theta, x_1, x_2, \ldots, x_n)$ be the likelihood function of the $X_1, X_2, \ldots\ldots, X_n$. If $\hat{\theta} = T(x_1, x_2, \ldots\ldots, x_n)$ is a function of observations, which maximizes $L(\theta)$, then $\hat{\theta}$ is the *maximum likelihood estimator (m.l.e.) of* θ for the sample $(x_1, x_2, \ldots\ldots, x_n)$. Many likelihood functions satisfy regularity conditions so, m.l.e. can be obtained by solving

$$\frac{d\,L(\theta)}{d\theta} = 0 \tag{7.20}$$

Since $L(\theta)$ and $\log L(\theta)$ have maxima at the same value of θ, so an alternative equation, which is often more convenient, can be used. It is given by

$$\frac{d \log L(\theta)}{d\theta} = 0 \tag{7.21}$$

Since
$$L(\theta) = \prod_{i=1}^{n} f(x_i, \theta)$$

$$\text{Log } L(\theta) = \sum_{i=1}^{n} \log f(x_i, \theta)$$

$$\frac{d \log L(\theta)}{d\theta} = \sum_{i=1}^{n} \frac{d \log f(x_i, \theta)}{d\theta} \tag{7.22}$$

If θ is a vector containing k parameters,

$$L(\underline{\theta}) = L(\theta_1, \theta_2, \ldots\ldots, \theta_k, x_1, x_2, \ldots, x_n)$$

$$= \prod_{i=1}^{n} f(x_i, \theta_1, \theta_2, \ldots\ldots, \theta_k) \tag{7.23}$$

Then the m.l. estimators of the parameters $\theta_1, \theta_2, \ldots\ldots, \theta_k$ are the random variables $\hat{\theta}_1 = T_1(x_1, x_2, \ldots, x_n) \ldots\ldots, \hat{\theta}_k = T_k(x_1, x_2, \ldots\ldots, x_n)$ which maximize $L(\theta_1, \theta_2, \ldots\ldots, \theta_k, x_1, x_2, \ldots\ldots, x_n)$ or $L(\underline{\theta})$. Such a point can also be obtained by solving the following k equations, if certain regularity conditions are satisfied

$$\frac{\partial L(\theta_1, \theta_2, \ldots\ldots, \theta_k)}{\partial \theta_1} = 0$$

$$\frac{\partial L(\theta_1, \theta_2, \ldots\ldots, \theta_k)}{\partial \theta_2} = 0$$

$$\vdots$$

$$\frac{\partial L(\theta_1, \theta_2, \ldots\ldots, \theta_k)}{\partial \theta_k} = 0 \tag{7.24}$$

In place of $L(\underline{\theta})$, $\log L(\underline{\theta})$ may also be used in (7.24).

Example 7.9: Suppose that a random sample of size n is drawn from the Bernoulli distribution

$$f(x, p) = p^x(1 - p)^{1-x} \quad x = 0, 1; \; 0 \le p \le 1$$

Find m.l.e. of p. (Source: Mood et al. (1974), p. 280.)

Solution: Let $X_1, X_2, \ldots\ldots, X_n$ may be a random sample from $f(x, p)$ giving the sequence of observations $x_1, x_2, \ldots\ldots, x_n$. So

$$L(p, \underline{x}) = p^{\sum_{i=1}^{n} x_i} (1-p)^{n - \sum_{i=1}^{n} x_i}$$

$$\text{Log } L(p) = y \log p + (n - y) \log (1 - p)$$

where

$$y = \sum_{i=1}^{n} x_i$$

$$\frac{\partial \log L(p)}{\partial p} = \frac{y}{p} - \frac{(n-y)}{1-p} \tag{7.25}$$

Equating (7.25) to zero and solving for p the m.l.e. of p is obtained as

$$\hat{p} = \frac{y}{n} = \frac{1}{n} \sum_{i=1}^{n} x_i = \bar{x} \tag{7.26}$$

Example 7.10: Suppose $X \sim N(\mu, \sigma^2)$ and it is desired to find maximum likelihood estimator of $\underline{\theta} = \{\mu, \sigma^2\}$. (Source: Mood et al. (1974), p. 281.)

Let $x_1, x_2, \ldots, x_n$ be the observed sample of size n from this distribution. Then

$$L(\underline{\theta}, \underline{x}) = \left(\frac{1}{\sqrt{2\pi}\sigma}\right)^n \exp\left\{-\frac{1}{2\sigma^2} \sum_{i=1}^{n} (x_i - \mu)^2\right\} \tag{7.27}$$

$$\log L(\underline{\theta}, \underline{x}) = \frac{-n}{2} \log(2\pi) - \frac{n}{2} \log \sigma^2 - \frac{1}{2\sigma^2} \sum_{i=1}^{n} (x_i - \mu)^2$$

the m.l.e. is obtained by solving the following

$$\frac{\partial \log L}{\partial \mu} = 0 = +\frac{1}{2\sigma^2} \sum_{i=1}^{n} 2(x_i - \mu)$$

$$\frac{\partial \log L}{\partial \sigma^2} = 0 = -\frac{n}{2\sigma^2} + \frac{1}{2\sigma^4} \sum_{i=1}^{n} (x_i - \mu)^2$$

Solving these two equations, the m.l.e. are

$$\hat{\mu} = \frac{1}{n} \sum_{i=1}^{n} x_i = \bar{x}$$

$$\hat{\sigma}^2 = \frac{1}{n} \sum_{i=1}^{n} (x_i - \bar{x})^2 \tag{7.28}$$

$$= S^2 \text{ (say)}$$

The m.l. estimates of p, μ, σ^2 are $\bar{x}$, $\bar{x}$ and S^2, where $\bar{x}$ and S^2 are sample mean and sample variance.

The m.l. estimators of these parameters will be $\bar{X}_n, \bar{X}_n$ and $\frac{1}{n} \sum_{i=1}^{n} (X_i - \bar{X}_n)^2$, respectively.

Apart from the method of moments and the method of maximum likelihood, there are several other methods also. Some of these are: (i) *Method of least squares* (ii) *the Bayes method* (iii) *the minimum–Chi-square method* and (iv) *the minimum – distance method.*

The method of least squares will be discussed in Chapter 9. For the rest of the methods the interested reader may consult any book on Mathematical Statistics or Statistical Inference e.g. books by Mood et al. (1974 or later edition) or Rohatgi (1994).

7.4 Testing of Hypothesis

This section deals with general principles of second type of inductive or statistical inference, viz. testing of hypothesis. In general **a statistical hypothesis** is *any type of assertion about the pdf of a random variable.*

Thus a statement that the protein content in 10 gm of dry Soyabeans from various plots follows normal distribution is a statistical hypothesis. The number of bacterial count in 1 ml of water sampled from a pond has Poisson distribution with mean 10, is also a statistical hypothesis.

The first of these statistical hypothesis is about the functional form of the pdf. The second of these assumes the form of the distribution and assertion is about only the value of a parameter of the distribution.

As some more examples of the practical situations requiring a test of statistical hypothesis suppose that

(a) an agricultural scientist who developed a hybrid variety of tomato wants to test if the new variety gives higher yield than the existing and commonly used variety and hence he could recommend to the farmers to start using the new hybrid variety.
(b) an entomologist wants to test whether a newly developed germicide is more effective in treating certain infections harming the crop than the one which is commonly used.
(c) a nutritionist wants to know if the new iron tablet suggested by him is effective in increasing haemoglobin level in children suffering from malnutrition.

As an illustration of testing of hypothesis consider the example of hybrid variety of tomato. Suppose the existing variety is known to give an average yield of 10 quintals per hectare. The agricultural scientist may take a decision to recommend the new hybrid variety if he could find from a statistical test that the average yield of the new variety is significantly more than 10 quintals per hectare.

To answer this question or any other similar question one will set up a statistical hypothesis that the average yield of hybrid variety is no more than that of the existing variety.

On the basis of the sample of yields from different plots sown with two varieties the test of the said hypothesis can be conducted either to accept or reject the hypothesis. If the hypothesis is not rejected it implies that the average yields of both the varieties are same. So the new variety can not be recommended.

However, if the hypothesis is rejected, what should be the assertion? This also needs to be formulated before performing the test. This alternative assertion is written as an alternative hypothesis.

The test procedure depends on the formulated alternative hypothesis also. A convenient and simple way of describing tests of hypothesis requires some more terms, notations and concepts related to it. These are defined and described as follows:

(i) A *statistical hypothesis* is denoted by H.
(ii) A statistical hypothesis to be tested is called a *null hypothesis* and is *denoted by* H_0.

(iii) If null hypothesis is rejected the assertion about the parameter(s) which would be made is called *alternative hypothesis* and is *denoted by* H_1.

(iv) If a statistical hypothesis completely specifies the density function of the associated random variable, it is called a *simple hypothesis*. If the density function is not completely specified by the hypothesis, it is called a *composite hypothesis*.
For example, H_0: $\mu = 10$ for a random variable $X \sim N(\mu, 1)$ is a simple hypothesis and H_1: $\mu > 10$ for the same X is a composite hypothesis H_0: $\mu = 10$ for a random variable $Y \sim N(\mu, \sigma^2)$ is a composite hypothesis.

(v) *Test of a statistical hypothesis* is a rule or procedure for deciding whether to reject or accept H_0. Rejection of H_0 implies acceptance of the H_1 and vice versa. It is said that H_0 is tested against an alternative H_1.
In real life problems, one selects a random sample of observations $X_1, X_2, ..., X_n$ and on the basis of the sampled observation values $x_1, x_2, ..., x_n$, a decision is made to reject H_0 or not to reject H_0 (or accept H_0).

(vi) The part of the sample space S which corresponds to rejection of H_0 is called *critical region* and may be denoted by C. The rest of the sample space S-C is called the *acceptance region* and is denoted by A.

If a random sample of size n is drawn to test the hypothesis, the sample space is n-dimensional. For a test of H_0 a statistic $T(x_1, x_2, ..., x_n)$ is computed, (thus reducing the dimension to one) whose distribution is completely known if H_0 is true. Such a statistic is generally used for testing H_0 and is called a *test statistic*.

Some values of this one dimensional space will be the **critical region**. These are the set of unusual or less probable values of T, if H_0 is true.

7.4.1 Two Types of Error

It is seen that test procedure chooses a test statistic and uses some values of the test statistic as critical region. If the value of test statistic computed from the observed random sample falls in the critical region, H_0 is rejected according to the test procedure. Regardless of all the care, there are possibilities of two types of error.

Rejection of H_0 when H_0 is true is an incorrect decision and is called *type I error* and *acceptance of H_0 when H_0 is false* is also an incorrect decision and is called *type II error*.

These two incorrect decisions as well as the two correct decisions are possible and are displayed in the following table.

Table 7.1 Two types of error

Decisions	True situation	
	H_0 True	H_1 True
H_0 rejected	Type I error	Correct decision
H_0 not rejected	Correct decision	Type II error

Type I and Type II errors will depend on the choice of the critical region. So to make its wise choice, there should be some way of measuring the seriousness of the two errors. This is accomplished by using sizes of these two errors, which is given by the probabilities of committing these two errors. Thus

$$\text{Size of type I error} = \alpha = P[\text{Type I error}]$$
$$= P[\text{Reject } H_0/H_0 \text{ is true}] \quad (7.29)$$

and
$$\text{Size of type II error} = \beta = P[\text{Do not reject } H_0/H_1 \text{ is true}] \quad (7.30)$$

Thus α is the probability that sample point falls in critical region when H_0 is true and β is the probability that the sample point falls in non critical region when H_1 is true.

Power of the Test

Probability of rejecting H_0 if H_1 is true (it is a correct decision) is called *power of the test.* It can be seen that

$$\text{Power of the test} = 1 - \beta \quad (7.31)$$

Level of the Significance

Size of the type I error, α is also called *level of significance.*

P-value

The *P*-value for hypothesis test is the probability of obtaining the calculated value of the test – statistic as extreme or more extreme (in appropriate direction depending on H_1) than actually obtained if H_0 is true.

In area of life sciences the term *P*-value is used commonly instead of quoting level of significance.

7.4.2 Procedure of Testing of Hypothesis

After defining concept of the two types of error, principles of determining good tests of hypotheses can be mentioned. Ideally one would like to choose a test of hypothesis which minimizes probabilities of both the errors i.e. both α and β. This is not possible since if one tries to minimize α, β increases and if one tries to minimize β, α increases. So a general principle is to fix value of α, and use a test which has smallest β.

This **principle**, suggested by **Neyman and Pearson**, is to choose a test with size of type II error as small as possible among all the tests with same size of the type I error. In other words, among all tests with same level of significance α, choose one with smallest β or highest power (= $1 - \beta$).

The above principle is used for testing a simple null hypothesis against a simple alternative.

Suppose X is a random variable with pdf $f(x, \theta)$ and it is desired to test H_0: $\theta = \theta_0$ against H_1: $\theta = \theta_1$. Both H_0 and H_1 are simple. For this a random sample $x_1, x_2, ..., x_n$, is selected on X. The test is performed using **simple likelihood ratio** given by

$$\lambda = \frac{L(\theta_0; x_1, x_2, ..., x_n)}{L(\theta_1; x_1, x_2, ..., x_n)} \quad (7.32)$$

The test procedure is to reject H_0 if $\lambda < k$ (7.33)

Another principle may be to minimize sum of the sizes of the two types of error i.e. to choose a test which has minimum ($\alpha + \beta$). This method is not in common use.

Two other tests of simple H_0 against simple H_1 are minimax test and Bayes test. In this book these will not be considered. An interested reader may refer book by Mood and Graybill (1974).

For testing a composite hypothesis it is assumed that a random sample from $f(x, \theta)$ is available. $\theta \varepsilon \underline{\theta}$ and it is desired to test H_0: $\theta \varepsilon \underline{\theta}_0$ against hypothesis H_1: $\theta \varepsilon \underline{\theta}_1$

Where $\underline{\hat{\theta}_0} \subset \underline{\theta}, \underline{\theta_1} \subset \underline{\theta}$ and $\underline{\theta_0} \cap \underline{\theta_1} = \phi$ i.e. $\underline{\theta_0}$ and $\underline{\theta_1}$ are disjoint. For such test of hypothesis **generalized likelihood ratio** is used. It is given by

$$\lambda_n = \lambda(x_1, x_2, \ldots\ldots, x_n) = \frac{L(\underline{\hat{\theta}_0}\, x_1, x_2, \ldots\ldots, x_n)}{L(\underline{\hat{\theta}}, x_1, x_2, \ldots\ldots, x_n)} \tag{7.34}$$

where $\underline{\hat{\theta}_0}$ is m.l.e. of $\underline{\theta}$ when H_0 is true and $\underline{\hat{\theta}}$ is m.l.e. of $\underline{\theta}$.

The **generalized likelihood ratio test procedure** is

$$\text{Reject } H_0 \text{ if } \lambda_n \le \lambda_0 \tag{7.35}$$

where λ_0 is some fixed constant satisfying $0 \le \lambda_0 \le 1$ chosen to fix the size of the test.

Equation (7.34) gives a test statistic as λ_n is independent of any unknown parameter and (7.35) gives the test procedure.

One can note that a useful and intuitive technique for obtaining tests is the following:

Choose some statistic which has different distribution under H_0 and H_1. For example suppose it is desired to test H_0: $\mu = \mu_0$ against H_1: $\mu > \mu_0$, where the random sample $X_1, X_2, \ldots, X_n$ is available on $X \sim N(\mu, 1)$.

The statistic $\bar{X}_n$ is a statistic which has normal distribution with variance $1/n$. If H_0 is true its mean is μ_0 and if H_0 is not true, mean $\bar{X}$ will be larger than μ_0. The statistic $\bar{X}$ has $N(\mu_0, 1/n)$ distribution if H_0 is true and it will have $N(\mu, 1/n)$ distribution (with $\mu > \mu_0$) when H_0 is not true. Hence a reasonable test will be to reject H_0, if $\bar{X}$ is large, i.e. reject H_0 if $\bar{X} \ge k$, where k is determined to fix size of type I error. In symbols, if α is chosen as size of type I error, choose k such that

$$P(\bar{X} > k/H_0 \text{ is true}) = \alpha \tag{7.36}$$

or

$$P\left[\sqrt{n}\,(\bar{X} - \mu_0) > (k - \mu_0)\sqrt{n}\right] = \alpha \tag{7.37}$$

Looking at tables of areas under standard normal curve, numeric values of Z_α can be seen for specified α. Equating

$$Z_\alpha = (k - \mu_0)\sqrt{n} \tag{7.38}$$

k can be determined and test procedure will be to Reject H_0 if $\bar{X} > k$ where k will be determined from (7.38).

These principles are generally used in the problems of tests of significance. In this book theory of testing of various hypotheses will not be discussed any more, but various applications of the theory in practical real life experiments will be considered.

8. Tests of Significance

After having an idea of various concepts of *testing of hypothesis* in the last chapter, the applications of the said procedure are considered in this chapter, all of which may be termed as *test of significance*. This is a type of statistical inference that is closely related to hypothesis testing.

The purpose of test of significance is to aid the clinician, the researcher or the administrator in reaching a decision concerning a population by examining only a sample from the population. As already mentioned the sample must be random to reach a reliable and valid decision.

Researchers are mainly concerned with *research hypotheses*. The research hypothesis is the *conjecture* or *supposition* that motivates research. It may be the result of years of observation by the researcher or his team. For example a physician may have numerous instances in which certain combinations of therapeutic measures were more effective than any one alone. Research problems are taken up from the desire of such health practitioners to determine whether or not their theories or suspicions can be supported by a well planned rigorous scientific investigation.

Research hypotheses lead directly to statistical hypotheses. The null hypothesis (H_0) are often the statistical hypotheses which are set up so that its rejection or not rejection will indicate truth of the researcher's conjecture. The alternative hypothesis (H_1) may be set so that its acceptance (i.e. rejection of null hypothesis) supports the researchers' theory or suspicion.

The statistical hypothesis or the null hypothesis is set in such a way that the distribution of the test statistic becomes known when H_0 is true. Usually null hypothesis is of the form H_0: $\theta = \theta_0$ i.e. it specifies the unknown parameter so that complete distribution of test statistic becomes known and critical region can be chosen in the one dimensional space of a specific size of type I error; α.

In general a test of significance procedure may be formally described as the following steps.

Steps of Test of Significance

1. **About data:** It must be understood, whether the data that form the basis of scientific study and test of significance consists of counts or measurements. This knowledge will determine the particular test to be employed.
2. **Assumptions:** Assumptions about the possible distribution of sampled observations, knowledge of other parameters and independence of samples should be narrated for a proper choice of a testing procedure for the research problem.
3. **Setting of hypotheses:** The null (H_0) and alternative (H_1) hypotheses must be set according to the objective of the research study. The null hypothesis (H_0) is sometimes referred to as a *hypothesis of no difference* and in general is set with a purpose of being rejected. Thus mostly

the complementary conclusion to the desired one is set as null hypothesis and the desired one, sought to be proved by researcher, is set as alternative hypothesis.

If the set H_0 is not rejected it is concluded that the data does not provide sufficient evidence to cause rejection of H_0. One may also conclude that the $\boldsymbol{H_0}$ **may be true**. It is **never** stated **as an absolute truth** i.e. **never that** $\boldsymbol{H_0}$ **is true**. It must be kept in mind that not rejecting H_0 is not a proof of H_0.

4. **Choice of the test statistic:** The nature of the data and the assumptions guide the choice of a proper test statistic. After the choice is made it is computed from the sample observations, so the test statistic is also a random variable and its distribution should be known if H_0 is true. Decision to reject or not to reject H_0 depends on the value of the chosen test statistic.

   A general formula for a test statistic, applicable in many of the tests of significance is given by

$$\text{Test statistic} = \frac{\text{Relevant statistic} - \text{Hypothesized parameter}}{\text{Standard error of the relevant statistic}} \tag{8.1}$$

   For example to test H_0: $\mu = \mu_0$ against H_1: $\mu \neq \mu_0$ the following test statistic may be used

$$Z = \frac{(\bar{X} - \mu_0)}{(\sigma/\sqrt{n})} \tag{8.2}$$

5. **Distribution of the test statistic:** After choosing the test statistic, it is necessary to specify its distribution if H_0 is true and the assumptions are met, so that the critical region is decided.
6. **Test procedure or rule:** The test statistic is one dimensional random variable so takes values on the horizontal axis. Test procedure specifies the two parts of the values of the statistic, one the critical region or rejection region and the other acceptance region.

   Critical region consists of values which are less likely to occur if H_0 is true and acceptance region consists of values which are more likely to occur if H_0 is true. So these can be specified if level of significance α is chosen. The critical region is specified as the sixth step.
7. **Calculation of the test statistic:** Now the test statistic already chosen in step 4 is computed from the observed sample values.
8. **Statistical decision:** Decision to reject the null hypothesis is taken and stated, if the computed value of test statistic lies in the critical region. H_0 is not rejected if the calculated value of the statistic falls in acceptance region.
9. **Conclusion:** If H_0 is rejected the conclusion is given, by clearly indicating that the researcher's, conjecture expressed in H_1 is supported by data or it is not supported. If it is not supported, the conclusion should be **that** $\boldsymbol{H_0}$ **may be true**. It can not be concluded that $\boldsymbol{H_0}$ **is true**.

It may be borne in mind clearly that the outcome of the statistical test is only one piece of evidence to be considered by researcher for taking decision. The statistical decision should not be taken as definitive, but should be considered along with all the other relevant information available to the experimenter.

After these general discussions, specific tests of significance are considered in the following subsections. This chapter deals with comparison of means and variances for one, two, and several populations. These tests assume that distribution of considered random variable is normal or not, sample sizes are large or small.

Books by Steel and Torrie (1981), Daniel (1987), Milton and Tsokos (1983) and Zar (2004) give various examples of biological experiments in which the discussed techniques are applicable.

8.1 One Sample Problems for Testing Mean

In this section testing of mean μ of a random variable X (which is not a count) in a population is considered. X has variance $V(X) = \sigma^2$ and a random sample $X_1, X_2, \ldots\ldots, X_n$ of a size n on X is in hand. Test procedures are considered for the following conditions:

(A) *X* has normal distribution

(a) σ^2 *is known* and *sample size n is large* (i.e. $n > 30$) or *n is small* (i.e. $n \leq 30$)
(b) σ^2 *is not known*

(i) Sample size *n is large* (i.e. $n > 30$)
(ii) Sample size *n is small* (i.e. $n \leq 30$)

(B) *X* has approximately normal distribution or distribution of X is not drastically different from normal

(a) σ^2 is known and n is large.
(b) σ^2 is not known and n is large.

For each of the conditions, which are to be mentioned as assumptions while actually performing the test of significance, the test procedure is given below.

Test Procedures for Various Conditions

A (a) Assumptions $X \sim N(\mu, \sigma^2)$, **σ^2 is known and *n* is large or small.**

Hypothesis H_0: $\mu = \mu_0$ H_1 may be any one of the following three depending on the conjecture of the researcher.
(i) H_1: $\mu \neq \mu_0$ or (ii) H_1: $\mu > \mu_0$ or (iii) H_1: $\mu < \mu_0$.
If H_1 is as in (i) test is called two tail test.
If H_1 is as in (ii) test is called upper tail or right tail test.
If H_1 is as in (iii) test is called lower tail or left tail test.
Test statistic is

$$Z = \frac{\bar{X} - \mu_0}{\sigma/\sqrt{n}} = \frac{\sqrt{n}\left(\bar{X} - \mu_0\right)}{\sigma} \tag{8.3}$$

Distribution of Z: $Z \sim N(0, 1)$ if H_0 is true. In words: the statistic Z follows standard normal distribution, if H_0 is true.
Test procedure or rule or critical region: The level of significance or size of type I error α is to be decided for defining critical or rejection region. Common choice of α is .05. Once α is specified the critical region depends on H_1.

The critical region will be on the two tails of distribution of test statistic, if H_1 is two tailed i.e. (i). It will be upper tail of distribution of Z, for upper tail H_1 i.e. for (ii) and lower tail of the distribution of Z, for lower tail H_1 i.e. for (iii). Figure 8.1 gives the critical region in the said three cases.

In a specific testing problem the appropriate critical region is specified in symbols. For example. Reject H_0, if $|Z| > 1.96$, for a two tail test.
In other words reject H_0 if either $Z < -1.96$ or $Z > 1.96$

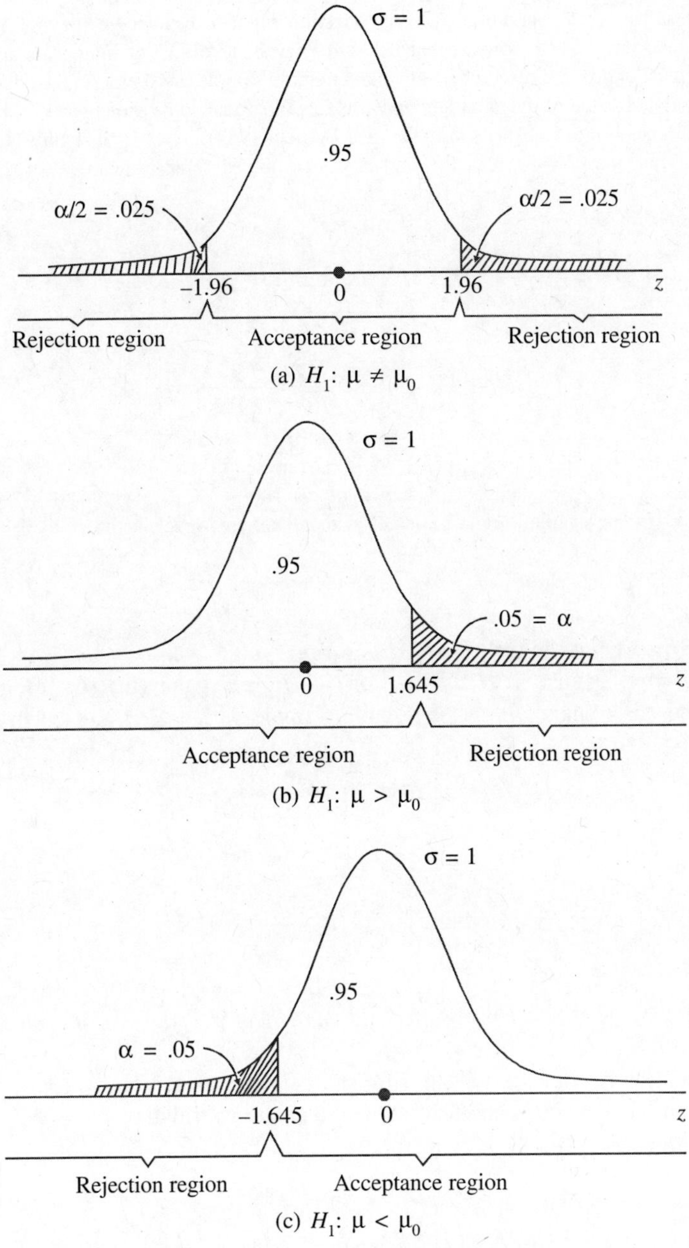

Fig. 8.1. Critical regions of size α = .05 for the three alternatives. (Source: (a) Daniel (1987), p. 196, (b) Daniel, p. 203, (c) Daniel, p. 199)

Computation of test statistic is done from the sampled observations and *Statistical decision* of rejection or acceptance of H_0 is done on the basis of computed value of the test statistic $Z = Z_{cal}$

If Z_{cal} falls in the critical region, H_0 is rejected.

Conclusion is written according to statistical decision reached and the research question to be answered by testing. An example is considered below.

Example 8.1: A researcher is interested to answer the question whether the average enzyme level in a certain human population is different from 25. It is known that the said enzyme level (X) of interest is approximately normally distributed with variance of 45. The researcher selected a random sample of 10 individuals, determined enzyme level on each and the calculated value of the sample mean was 22. What should be the answer to the question of research on the basis of available sample observation (source: Daniel (1987), p. 194, data only)?
Solution: To answer the question the data in hand is the computed sample mean $\bar{X}_{10} = 22$.
Assumption: is that $X \sim N(\mu, 45)$ and $n = 10$ which is small.
Hypothesis to be tested is H_0: $\mu = 25$ against H_1: $\mu \neq 25$

Test statistic to be used is $Z = \dfrac{\bar{X}-\mu_0}{\sigma/\sqrt{n}} \sim N(0, 1)$ if H_0 is true.

$$\textit{Calculated } Z_{\text{cal}} = \frac{22-25}{\sqrt{45/10}} = \frac{-3}{2.1213} = -1.41$$

Critical region is $|Z| > 1.96$ i.e. if $|Z_{\text{cal}}| > 1.96$ H_0 is rejected
Since $|Z_{\text{cal}}| = 1.41$, H_0 is not rejected is the *statistical decision.*
Hence the *conclusion* is: The available data indicates that the enzyme level in the studied population may not be different from 25.

(A) (b) (i)

Assumption: $X \sim N(\mu, \sigma^2)$, **σ^2 is unknown, n is large**
Hypotheses: H_0: $\mu = \mu_0$ against (i) H_1: $\mu \neq \mu_0$ or (ii) H_1: $\mu > \mu_0$ or (iii) H_1: $\mu < \mu_0$
Test statistic: Since σ^2 is unknown and n is large, sample variance S^2 is used in place of σ^2, where

$$S^2 = \frac{1}{n}\sum_{i=1}^{n}(X_i - \bar{X})^2 \tag{8.4}$$

The statistic

$$Z = \frac{\bar{X} - \mu_0}{S/\sqrt{n}} \tag{8.5}$$

is used to test H_0 against H_1.
Distribution of Z is standard normal, if H_0 is true. In symbols $Z \sim N(0, 1)$ if H_0 is true.

The *critical region* or the *test procedure*, the *statistical decision* and *conclusion* will be similar to previous case.

These three and the hypotheses will be similar in all the cases. Hence in all the subsequent cases only the assumptions, null hypothesis, test statistic and the distribution of test statistic, if H_0 is true, will be given.

(A) (b) (ii)

Assumption: $X \sim N(\mu, \sigma^2)$, **σ^2 is unknown, n is small**
Null Hypothesis: H_0: $\mu = \mu_0$ against H_1 as above
Test statistic: If n is small, sample variance S^2 can not be used as an estimate of σ^2. Instead the unbiased estimator (s^2) of σ^2 given by

$$s^2 = \frac{1}{(n-1)}\sum_{i=1}^{n}(X_i - \bar{X})^2 = \frac{nS^2}{(n-1)} \tag{8.6}$$

is used in place of σ^2. The resulting statistic by this change does not have standard normal distribution, but has Student's t-distribution.

The statistic to be used in this case is

$$t = \frac{\bar{X} - \mu_0}{s/\sqrt{n}} = \frac{\sqrt{n}\left(\bar{X} - \mu_0\right)}{s} \tag{8.7}$$

Distribution of the test statistic is Student's t-distribution with $(n - 1)$ degree of freedom (d.f.), if H_0 is true. In symbols

$$t \sim t_{n-1} \text{ if } H_0 \text{ is true} \tag{8.8}$$

Test procedure or rule is based on the Student's t-distribution.

Taking α as the level of significance, the critical regions are given by the following, for the three alternatives.

(a) If H_1: $\mu \neq \mu_0$, reject H_0 if $| t_{cal} | > t_{n-1, \alpha/2}$ (8.9)
In other words reject H_0 if either $t < -t_{n-1, \alpha/2}$ or $t > t_{n-1, \alpha/2}$

(b) If H_1: $\mu > \mu_0$, reject H_0 if $t_{cal} > t_{n-1, \alpha}$ (8.10)

(c) If H_1: $\mu < \mu_0$, reject H_0 if $t_{cal} < -t_{n-1, \alpha}$ (8.11)
$t_{d, \alpha}$ is 100 $(1 - \alpha)$ % point of Student's t-distribution with d d.f.

Figure 8.2 below gives the critical regions for various alternatives.

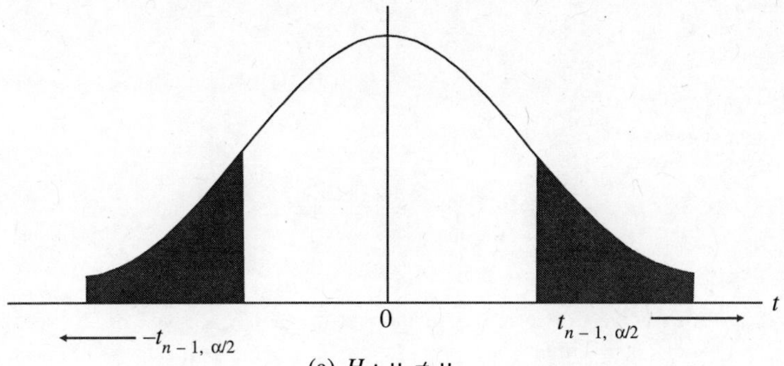

(a) H_1: $\mu \neq \mu_0$

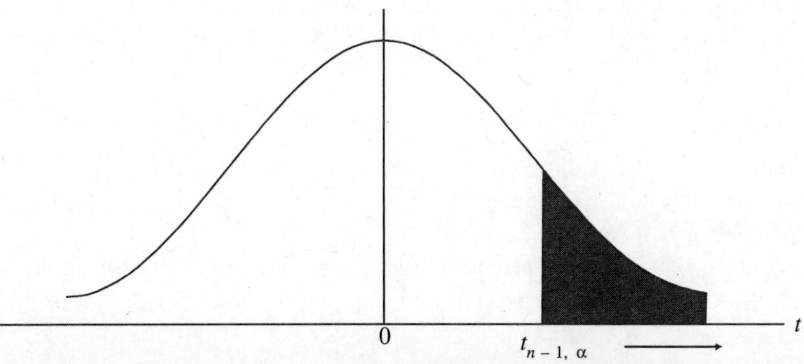

(b) H_1: $\mu > \mu_0$

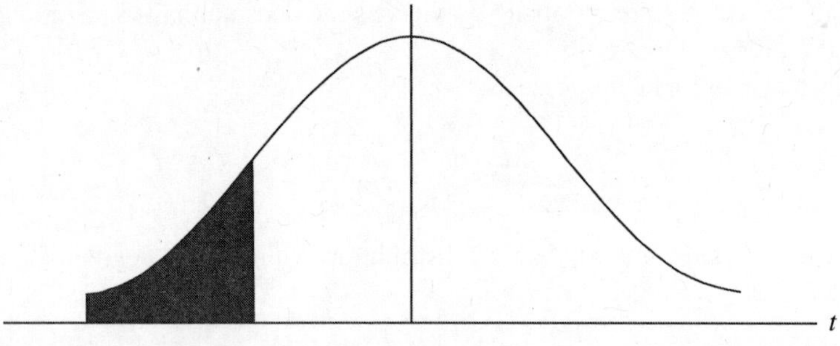

(c) H_1: $\mu < \mu_0$

Fig. 8.2. Critical regions of size α based on Student's *t*-distribution (source: (a), (b) and (c) Milton and Tsokos (1983), p. 191).

Example 8.2: Researchers collected serum amylase values from a random sample of 16 apparently healthy subjects. They want to know if they can conclude that the mean of the said values in the population, from which sample for serum amylase determination, was selected is different from 120. The mean and standard deviation from the observed sample are 96 and 36 units/100 ml (source: Daniel (1987), p. 200 (modified)).

Solution: *Data*: the sample mean $\bar{X}$ = 96 units/100 ml and sample variance s^2 = 36 units/100 ml.

Assumption: The random sample is coming from a normal population $N(\mu, \sigma^2)$ in which σ^2 is unknown and n = 16 is small.

The Hypothesis is H_0: μ = 120 against H_1: $\mu \neq 120$

The *test statistic to test* H_0 is

$$t = \frac{\bar{X} - \mu_0}{s/\sqrt{n}} \sim t_{n-1} \text{ if } H_0 \text{ is true}$$

Computation: The test statistic to be calculated as

$$t = \frac{\bar{X} - \mu_0}{s/\sqrt{n}} = \frac{(96 - 120)\sqrt{16}}{36} = \frac{-8}{3}$$

$$= -2.66$$

Test procedure or rule: Alternative is two tailed so the test procedure is given by

Reject H_0 if $| t_{cal} | > t_{n-1,\, \alpha/2}$ or if either $t_{cal} < -t_{n-1,\, \alpha/2}$ or $t_{cal} > t_{n-1,\, \alpha/2}$, where α is chosen as .05

From the Table A.4 $t_{15,\, .025}$ = 2.131

Since $t_{cal} < -2.131$, it lies in the critical region. So H_0 *is rejected*

Conclusion: The data indicates that mean serum amylase value in the population may be different from 120.

B (a)

Assumptions

X **does not have exact normal distribution** but it is not drastically different from normal. $E(X) = \mu$ and $V(X) = \sigma^2$ **is known and** n **is large**.

Null Hypothesis H_0: $\mu = \mu_0$ against H_1 as above

Test statistic: Since n is large and distribution is not very different from normal, the test statistic used is

$$Z = \frac{\bar{X} - \mu_0}{\sigma/\sqrt{n}} = \frac{\sqrt{n}\,(\bar{X} - \mu)}{\sigma} \tag{8.12}$$

Distribution of the test statistic is standard normal if H_0 is true. In symbols

$$Z \sim N(0, 1) \text{ if } H_0 \text{ is true} \tag{8.13}$$

The test procedure or rule is same as that for cases A (a) and A (b) (i) given in Fig. 8.1. In symbols for $\alpha = .05$

(i) Reject H_0 if $|Z| > 1.96$ for H_1: $\mu \neq \mu_0$ (8.14)
In other words reject H_0 if either $Z < -1.96$ or $Z > 1.96$

(ii) Reject H_0 if $Z > 1.645$ for H_1: $\mu > \mu_0$ (8.15)

(iii) Reject H_0 if $Z < -1.645$ for H_1: $\mu < \mu_0$ (8.16)

B (b)

Assumptions

X does not have exact normal distribution but it is not drastically different from normal. $E(X) = \mu$ and **$V(X) = \sigma^2$ is unknown** and **n is large.**

Null Hypothesis H_0: $\mu = \mu_0$ against H_1 same as above

Test statistic: Since n is large, σ^2 is replaced by sample variance S^2 as distribution of X does not differ drastically from normal. Hence the test statistic is given by

$$Z = \frac{\bar{X} - \mu_0}{S/\sqrt{n}} = \frac{\sqrt{n}\,(\bar{X} - \mu_0)}{S} \tag{8.17}$$

The *distribution of the test statistic* is standard normal if H_0 is true. In symbols

$$Z \sim N(0, 1) \text{ if } H_0 \text{ is true} \tag{8.18}$$

Test procedure or rule is same as that for B (a) i.e. as given in (8.14), (8.15) and (8.16). Some real examples are given which illustrate the above procedures.

Example 8.3: It is desired to know if it could be concluded that mean daily caloric intake in the adult rural population of a developing country is less than 2000. A sample of 500 had a mean of 1985 and a standard deviation of 210. Choose $\alpha = .05$ (source: Daniel (1987), p. 205).

Solution: *Data*: Sample of size 500 gave $\bar{X} = 1985$ and $S = 210$.

Assumption: X, the daily caloric intake has appropriately normal distribution. Variance of X is unknown and sample size $n = 500$ is large. So the problem may be considered of type B (b).

Hypothesis H_0: $\mu = 2000$ against H_1: $\mu < 2000$

Test statistic to be used is

$$Z = \frac{\bar{X} - \mu_0}{S/\sqrt{n}} \sim N(0, 1) \text{ if } H_0 \text{ is true}$$

Test procedure for $\alpha = .05$ is

Reject H_0 if $Z < -1.645$

$$\text{Calculation of } Z = \frac{(1985 - 2000)\sqrt{500}}{210} = -1.597$$

Since calculated Z does not lie in rejection region, H_0 is not rejected. Hence it can not be concluded from the observed sample that average daily caloric intake is less than 2000.

Example 8.4: In a health survey of a certain community, 150 persons were interviewed. One of the information collected was the number of prescriptions each person had filled during the past year. The average number for the sample of 150 people was 5.8 with a standard deviation of 3.1. Does the data indicate that population mean is greater than 5 (source: Daniel (1987), p. 202)?

Solution: *Data*: The number of prescriptions (X) filled for 150 persons in the last year, with mean = 5.8 = $\bar{X}$ and standard deviation S = 3.1

Assumptions: X does not follow normal distribution but is approximately normal, with $E(X) = \mu$ and $V(X) = \sigma^2$ (unknown), n = 150 (large).

Hypothesis: H_0: $\mu = \mu_0 = 5$ against H_1: $\mu > 5$.

Test statistic: Since the sample size is large and σ^2 is unknown the problem is identical to case B (b). So the test statistic to test H_0 is

$$Z = \frac{\bar{X} - \mu_0}{S/\sqrt{n}}$$

Distribution of Z is $N(0, 1)$ if H_0 is true.

Test procedure: Reject H_0 if $Z > 1.645$, choosing $\alpha = .05$

$$\text{Calculated } Z = \frac{(5.8 - 5)\sqrt{150}}{3.1} = 3.2$$

Since calculated Z lies in rejection region, H_0 is rejected.

Hence it is concluded that the data indicates that the mean number of filled up prescriptions in the population is greater than 5.

8.2 Two Sample Problems for Testing Means

This section deals with problem of comparing means of a characteristic in two populations. The null hypothesis is, therefore, set that means do not differ.

From the two populations random samples of sizes n_1 and n_2 are selected and the value of the characteristic X_1 and X_2 are measured in the first and second populations, respectively. $\bar{X}_1$ and $\bar{X}_2$ are the means of random samples from the two populations. Mean and variance of X_1 in first population are μ_1 and σ_1^2 and those in the second population are μ_2 and σ_2^2. Again there will be several cases as given below:

A: X_1 and X_2 have normal distribution and **samples are independently** selected

(a) σ_1^2 ***and*** σ_2^2 ***are known,*** **sample sizes, n_1 and n_2 may be large or small.**
(b) σ_1^2 ***and*** σ_2^2 ***are unknown,*** **n_1 and n_2 are large.**
(c) σ_1^2 and σ_2^2 **are unknown, but** known to be **equal** i.e. $\sigma_1^2 = \sigma_2^2$, **n_1 and n_2 are small.**
(d) $\sigma_1^2 \neq \sigma_2^2$ **are unknown and n_1, n_2 may be small.**

B: X_1 and X_2 have normal distributions. Samples are not independent but paired by an experimental procedure making use of related observations.

Related or paired observations may be obtained in any of the following situations.

1. The same subject may be observed before and after receiving some treatment.
2. Litters of the same sex may be randomly assigned to receive either a treatment or a placebo.
3. Pairs of twins or siblings may be randomly assigned to treatments in such a way that members of a single pair receive different treatments.

4. In comparing two methods of analysis, the material to be analyzed may be equally divided so that one half is analyzed by one method and the second half by the other.
5. Pairs may be formed by matching individuals on some characteristic, which is closely related to the measurement of interest, say, post treatment scores on some test.

In such situations, instead of analyzing observations on each individual, the difference between paired individuals is considered as variable of interest.

C. Distributions of X_1 and X_2 are not exactly normal but it is not drastically different from normal **n_1 and n_2 are large**

(a) **σ_1^2 and σ_2^2 known**
(b) **σ_1^2 and σ_2^2 are unknown.**

Now the items given in the last Section 8.1 will be given for each case stated above.

A (a)

Assumptions: X_1 and X_2 the measured characteristic in the two populations have *normal distribution. Independent random samples* of sizes n_1 and n_2 are available from first and second populations, respectively, with means, μ_1 and μ_2 and **known variances σ_1^2 and σ_2^2, n_1 and n_2 are large or small.**
Hypothesis: H_0: $\mu_1 = \mu_2$ against one of the following
(i) H_1: $\mu_1 \neq \mu_2$ (ii) H_1: $\mu_1 > \mu_2$ (iii) H_1: $\mu_1 < \mu_2$
Test statistic to test H_0 is **same for n_1 and n_2 large or small** given by

$$Z = \frac{\bar{X}_1 - \bar{X}_2}{\sqrt{\left(\sigma_1^2/n_1\right) + \left(\sigma_2^2/n_2\right)}} \tag{8.19}$$

Distribution of the test statistic is standard normal if H_0 is true.
Test procedure or rule is as for case A (a) of one sample problem for the three alternatives.

In symbols, choosing $\alpha = .05$

(i) Reject H_0: $\mu_1 = \mu_2$ against H_1: $\mu_1 \neq \mu_2$ if $|Z| > 1.96$
In other words, reject H_0 if, either $Z < -1.96$ or $Z > 1.96$ (8.20)
(ii) Reject H_0: $\mu_1 = \mu_2$ against H_1: $\mu_1 > \mu_2$ if $Z > 1.645$ (8.21)
(iii) Reject H_0: $\mu_1 = \mu_2$ against H_1: $\mu_1 < \mu_2$ if $Z < -1.645$ (8.22)

Even if n_1 and n_2 are small, the same test statistic and critical regions will be used.

Example 8.5: It is desired to know if the data provides sufficient evidence to indicate a difference in mean serum uric acid levels between normal individuals and individuals with Mongolism. The uric acid level is known to follow normal distribution in both types of individuals with variance 1. The random sample of 12 Mongoloid and 15 normal individuals gave means $\bar{X}_1 = 4.5$ mg/100 ml, $\bar{X}_2 = 3.4$ mg/100 ml (source: Daniel (1987), p. 207).
Solution: *Assumption*: Let X_1 and X_2 be the uric acid levels in the two groups of individuals. Then $X_1 \sim N(\mu_1, 1)$ and $X_2 \sim N(\mu_2, 1)$. Independent samples of sizes 12 and 15 are drawn from Mongoloids and normal individuals, respectively. So the testing problem is of type A (a).
Hypotheses To test H_0: $\mu_1 = \mu_2$ against H_1: $\mu_1 \neq \mu_2$
The test statistic for testing H_0 against H_1 is

$$Z = \frac{\bar{X}_1 - \bar{X}_2}{\sqrt{(\sigma_1^2/n_1) + (\sigma_2^2/n_2)}} \sim N(0, 1) \text{ if } H_0 \text{ is true}$$

The *calculated value* of

$$Z_{cal} = \frac{4.5 - 3.4}{\sqrt{\frac{1}{12} + \frac{1}{15}}} = \frac{1.1}{.39} = 2.82$$

The test procedure is to reject H_0 if $| Z | > 1.96$ or in other words reject H_0 if either $Z < -1.96$ or $Z > 1.96$. The *statistical decision* is therefore to reject H_0 since Z_{cal} belongs to the critical region because $Z_{cal} = 2.82 > 1.96$. The *conclusion*, therefore is that data gives sufficient evidence to indicate that there is difference in average serum uric acid levels in Mongoloids and normal individuals.

A (b)

Assumptions: All the assumptions are same as in A (a) except the following two hold in this case

(i) σ_1^2 and σ_2^2 are not known
(ii) n_1 and n_2 are large

Hypotheses: They are same as in case A (a) of two sample problem just mentioned.
Test statistic: Since n_1 and n_2 are large using property of central limit theorem, $\bar{X}_1$ and $\bar{X}_2$ still have normal distribution, even when the unknown variances are replaced by sample variances to give the test statistic as

$$Z = \frac{\bar{X}_1 - \bar{X}_2}{\sqrt{\frac{S_1^2}{n_1} + \frac{S_2^2}{n_2}}} \qquad (8.23)$$

Distribution of Z is still standard normal if H_0 is true and the test procedure is same as given in last case using critical regions (8.20), (8.21) and (8.22).

Example 8.6: Protoporphyrin levels were measured in two samples of subjects, First was of 50 adult male alcoholics with ring sideroblasts in the bone marrow and second was of 40 apparently non-alcoholic males. The mean and standard deviation of protoporphyrin levels from the two samples were: $\bar{X}_1 = 340$, $\bar{X}_2 = 45$, $S_1 = 250$ and $S_2 = 25$. On the basis of these observations, can one conclude that protoporphyrin level is higher in the alcoholic population than in the non-alcoholic population (source: Daniel (1987), p. 215)?
Solution: As the sample sizes are large and variances are unknown due to central limit theorem the solution will be same whether considered as of case A (b) or of case C.
Assumption: Let X_1 and X_2 be the measured protoporphyrin levels in alcoholic and apparently non-alcoholic adults. Then $E(X_1) = \mu_1$ $V(X_1) = \sigma_1^2$, $E(X_2) = \mu_2$ and $V(X_2) = \sigma_2^2$. X_1 and X_2 have distribution which does not depart much from normal.
Hypothesis H_0: $\mu_1 = \mu_2$ against H_1: $\mu_1 > \mu_2$
Test statistic to be used is

$$Z = \frac{\bar{X}_1 - \bar{X}_2}{\sqrt{\frac{S_1^2}{n_1} + \frac{S_2^2}{n_2}}} \sim N(0, 1) \text{ if } H_0 \text{ is true}$$

$$Z_{cal} = \frac{340 - 250}{\sqrt{\frac{250^2}{50} + \frac{25^2}{40}}} = \frac{90}{35.57} = 2.53$$

Test procedure or rule: At level of significance $\alpha = .05$. Reject H_0 if $Z_{cal} > 1.645$.
Statistical decision is to reject H_0
Hence the conclusion is that average protoporphyrin level in alcoholics is higher than that in non-alcoholic adult.

A (c)

Assumption: All the assumptions are same as in case A (a) except the following

(i) **σ_1^2 and σ_2^2 are unknown but $\sigma_1^2 = \sigma_2^2$**
(ii) n_1 and n_2 are small.

Hypotheses: are same as in last two cases.
Test statistic is given by

$$t = \frac{\bar{X}_1 - \bar{X}_2}{Sp\sqrt{\left(\frac{1}{n_1} + \frac{1}{n_2}\right)}} \tag{8.24}$$

where S_p^2 is variance obtained by pooling the two samples, given by

$$S_p^2 = \frac{\sum_{i=1}^{n_1}\left(X_{1j} - \bar{X}_1\right)^2 + \sum_{j=1}^{n_2}\left(X_{2j} - \bar{X}_2\right)^2}{n_1 + n_2 - 2}$$

$$= \frac{n_1 S_1^2 + n_2 S_2^2}{n_1 + n_2 - 2}$$

S_1^2 and S_2^2 are variances from the first and second samples, respectively.

The distribution of the test statistic if H_0 is true is Student's t, with $(n_1 + n_2 - 2)$ degrees of freedom. In symbol's

$$t \sim t_{n_1+n_2-2}, \text{ if } H_0 \text{ is true} \tag{8.27}$$

The *test procedure or rule* is given by

(i) $|t| > t_{n_1+n_2-2,\alpha/2}$ if H_1: $\mu_1 \neq \mu_2$
In other words reject H_0 if either $t < -t_{n_1+n_2-2,\alpha/2}$ or $t > t_{n_1+n_2-2,\alpha/2}$ (8.28)
(ii) $t > t_{n_1+n_2-2,\alpha}$ if H_1: $\mu_1 > \mu_2$ (8.29)
(iii) $t < -t_{n_1+n_2-2,\alpha}$ if H_1: $\mu_1 < \mu_2$ (8.30)

The rejection region is similar to that given in Fig. 8.2 but degrees of freedom of t is $n_1 + n_2 - 2$.

Example 8.7: A study of angina on rats was conducted on 18 rats with a history of angina. These 18 rats were split randomly into two groups of 9 each. One group was given a placebo and the other an experimental drug FL 113. After controlled exercise on a tread-mill, the recovery time of each rat was determined. The following information is available (source: Milton and Tsokos (1983), p. 249).

	Sample size (n_i)	Mean ($\bar{X}_i$)	Sample variance (s_i^2)
Placebo	9	329 seconds	45 seconds^2
FL 113	9	283 seconds	43 seconds^2

Do the data support the convention that FL 113 will reduce the average recovery time? Recovery time may be assumed to be normal.

Solution: This is a problem of type A (c).

Assumption

Let X_1 and X_2 be the recovery times of rats that are given placebo and FL 113, respectively. Then it is assumed that $X_1 \sim N(\mu_1, \sigma_1^2)$ and $X_2 \sim N(\mu_2, \sigma_2^2)$, $\sigma_1^2 = \sigma_2^2$.

Hypotheses: H_0: $\mu_1 = \mu_2$ against H_1: $\mu_1 > \mu_2$

Level of Significance: $\alpha = .05$ is chosen.

The test-statistic to test H_0 is

$$t = \frac{\bar{X}_1 - \bar{X}_2}{Sp\sqrt{\left(\frac{1}{n_1} + \frac{1}{n_2}\right)}}, \text{ where } S_p^2 = \frac{(n_1 - 1)\, s_1^2 + (n_2 - 1)\, s_2^2}{n_1 + n_2 - 2}$$

The distribution of test statistic is given by

$$t \sim t_{n_1+n_2-2}, \text{ if } H_0 \text{ is true}$$

Choosing level of significance $\alpha = .05$,

The test procedure or rule is Reject H_0 if $t \sim t_{n_1+n_2-2,\,.05}$

Computation

$$S_p^2 = \frac{8(45) + 8(43)}{9 + 9 - 2} = \frac{45 + 43}{2} = 44$$

Hence

$$t_{\text{cal}} = \frac{329 - 283}{\sqrt{44\left(\frac{1}{9} + \frac{1}{9}\right)}} = \frac{46 \times 3}{\sqrt{88}} = 14.71$$

From Table A.4 $t_{9+9-2,\,.05} = t_{16,\,.05} = 1.746$

Statistical decision: Since $t_{\text{cal}} = 14.71 > 1.746$, H_0 is rejected

Hence it can be concluded that the drug FL113 is effective in reducing recovery time.

A (d)

Assumptions: All the assumptions are same as in A (a) except the following

(i) $\sigma_1^2 \neq \sigma_2^2$ **are unknown.**

(ii) n_1, n_2 **may be small.**

This test procedure is also known as **Behrens Fisher Problem.**

Hypotheses: are same as in case A (a) i.e. H_0: $\mu_1 = \mu_2$ against one of the following alternatives.

(i) H_1: $\mu_1 \neq \mu_2$ (ii) H_1: $\mu_1 > \mu_2$ (iii) H_1: $\mu_1 < \mu_2$

Test statistic and the Test procedure in this case is given in section 14.1, eqs. (14.1) to (14.3a).

B. Paired *t*-test

Assumption: Sample can be paired and n pairs result. X_1 and X_2 both follow normal distribution with means μ_1 and μ_2 and variances σ_1^2 and σ_2^2, respectively.

Data: The observations are paired so find the differences between X_1 and X_2 on n pairs to give d_1, d_2, ..., d_n. Find the mean and variances of di's as $\bar{d}$ and s_d^2.

The test statistic to test H_0 is given by

$$t = \frac{\sqrt{n}\,\bar{d}}{s_d} \tag{8.31}$$

where

$$\bar{d} = \frac{1}{n}\sum_{i=1}^{n} d_i$$

and

$$s_d^2 = \frac{1}{n-1}\sum_{i=1}^{n} \left(d_i - \bar{d}\right)^2 \tag{8.32}$$

The distribution of the test statistic is Student's t-distribution with $(n - 1)$ d.f. if H_0 is true. In symbols

$$t \sim t_{n-1} \text{ if } H_0 \text{ is true} \tag{8.33}$$

The test procedure or rule is given by

Reject H_0 if $|t| > t_{n-1,\,\alpha/2}$ if H_1: $\mu_1 \neq \mu_2$.

In other words reject H_0 if either $t < -t_{n-1,\,\alpha/2}$ or $t > t_{n-1,\,\alpha/2}$

Reject H_0 if $t > t_{n-1,\,\alpha}$ if H_1: $\mu_1 > \mu_2$ and

Reject H_0 and if $t < -t_{n-1,\,\alpha}$ if H_1: $\mu_1 < \mu_2$ (8.34)

Example 8.8: To study the effect of physical training on the serum cholesterol level, it was measured on 11 subjects before and after the training program and is given in the following table.

Table 8.1 Serum cholesterol levels before and after training (Source: Milton and Tsokos (1983), p. 264 (data only))

Subjects	Pre-training level (X_1) (mg/dl)	Post-training level (X_2) (mg/dl)
1	182	198
2	232	210
3	191	194
4	200	220
5	148	138
6	249	220
7	276	219
8	213	161
9	241	210
10	480	313
11	262	226

Does the data support the proposition that physical training program effectively reduced the serum cholesterol level? Normality of cholesterol level may be assumed.

Solution: Since the data is on the same subject before and after the training, the paired t-test will be used. Thus the steps for the type of case B is applied.

Assumptions: $X_1 \sim N(\mu_1, \sigma_1^2)$, $X_2 \sim N(\mu_2, \sigma_2^2)$

$n_1 = n_2 = n = 11$

Hypothesis: H_0: $\mu_1 - \mu_2 = 0$ against H_1: $\mu_1 - \mu_2 > 0$

The test statistic to be used is given by

$$t = \frac{\bar{d}}{s_d/\sqrt{n}} \sim t_{n-1} \text{ if } H_0 \text{ is true.}$$

The test procedure or rule or critical region for level of significance $\alpha = .05$ is

Reject H_0 if $t > t_{n-1, .05}$

From the Table A.4 $t_{n-1, .05} = t_{10, .05} = 1.812$

Computation of test statistic is done by finding the differences $d = X_1 - X_2$ and d_i^2 for each subject in the following table.

Subjects (i)	1	2	3	4	5	6	7	8	9	10	11	Total
$d_i = X_{1i} - X_{2i}$	–16	22	–3	–20	10	29	57	52	31	167	36	365
d_i^2	256	484	9	400	100	841	3249	2704	961	27889	1296	38189

So that

$$\bar{d} = \frac{365}{11} = 33.18$$

and

$$s_d^2 = \frac{1}{10}\left[\sum_{i=1}^{11} d_i^2 - 11\,\bar{d}^2\right] = \frac{1}{10}[38189 - 12111.3636]$$

$$= 2607.76$$

$$s_d = \sqrt{2607.76} = 51.07$$

Hence

$$t_{\text{cal}} = \frac{33.18\sqrt{11}}{51.07} = 2.15$$

The *statistical decision* is to reject H_0 since $t_{\text{cal}} = 2.15 > t_{10, .05} = 1.812$.

The conclusion, therefore, is that the observed data indicates that the physical training program is effective in reducing the serum cholesterol level.

C (a)

Assumption: X_1 and X_2 are not exactly normally distributed but do not differ drastically from normal distribution. The sample sizes n_1 and n_2 are large σ_1^2, σ_2^2 are known.

The *test statistic* to be used for testing H_0 is

$$Z = \frac{\bar{X}_1 - \bar{X}_2}{\sqrt{\dfrac{\sigma_1^2}{n_1} + \dfrac{\sigma_2^2}{n_2}}} \tag{8.35}$$

The distribution of the test statistic can be approximated by standard normal due to Central Limit Theorem, since sample sizes are large. Thus in symbols

$$Z \sim N(0, 1) \text{ if } H_0 \text{ is true.} \tag{8.36}$$

The test procedure is as given in case A (a). In symbols

Reject H_0 if $|Z| > 1.96$ if H_1: $\mu_1 \neq \mu_2$,

In other words reject H_0 if either $Z < -1.96$ or $Z > 1.96$,

Reject H_0 if $Z > 1.645$ if H_1: $\mu_1 > \mu_2$ and

Reject H_0 if $Z < -1.645$ if H_1: $\mu_1 < \mu_2$ (8.37)

Example 8.9: A hospital administration wished to know if the population which patronizes the hospital A has a larger mean family income than does the population which patronizes hospital B. The patronizing populations are known to be non normal with σ_1 = Rs. 600 and σ_2 = Rs. 500. Sample of 75 patients admitted to hospital A and 80 patients admitted to hospital B had average weekly family income Rs. 6800 and Rs. 5450.

Do the data indicate that hospital A patients have higher mean family income than those of the hospital B? (Source: Daniel (1987), p. 212.)

Solution: Since sample sizes are large and standard deviations are known, because of Central Limit Theorem the procedure to conduct the test will be same as A (a) or C (a).

Assumption: Let X_1 and X_2 denote the weekly family income of patients of hospital A and B, respectively. Moreover, let $E(X_1) = \mu_1$, $E(X_2) = \mu_2$, $V(X_1) = \sigma_1^2 = 600^2$, $V(X_2) = \sigma_2^2 = 500^2$, $n_1 = 75$, $n_2 = 80$.

Hypothesis H_0: $\mu_1 = \mu_2$ against H_1: $\mu_1 > \mu_2$.

Test statistic for testing H_0 against H_1 is

$$Z = \frac{\bar{X}_1 - \bar{X}_2}{\sqrt{\dfrac{\sigma_1^2}{n_1} + \dfrac{\sigma_2^2}{n_2}}} \sim N(0, 1) \text{ if } H_0 \text{ is true}$$

The test procedure or rule or critical region is (for $\alpha = .05$)

$$Z > 1.645$$

The *computed* value of test statistic is

$$Z_{cal} = \frac{6800 - 5450}{\sqrt{\dfrac{600^2}{75} + \dfrac{500^2}{80}}} = \frac{1350}{100\sqrt{\dfrac{36}{75} + \dfrac{25}{80}}}$$

$$= \frac{1350}{100 \times \sqrt{.7925}} = \frac{1350}{89} = 15.17$$

The *statistical decision* is to reject H_0 since $Z_{cal} = 15.17 > 1.645$.

It may be therefore, *concluded* that mean family income of those patronizing hospital A is more than those patronizing hospital B.

C (b)

Assumptions: σ_1^2 are σ_2^2 **not known** in this case and other assumptions are same as in case in C(a). *The test statistic* to be used replaces σ_1^2 and σ_2^2 by the sample variances S_1^2 and S_2^2 in (8.35). In symbols the test statistic is

$$Z = \frac{\bar{X}_1 - \bar{X}_2}{\sqrt{\dfrac{S_1^2}{n_1} + \dfrac{S_2^2}{n_2}}} \tag{8.38}$$

The distribution of Z is standard normal if H_0 is true as in C(a) and *the test procedure* or rule is also same as in (8.37).

Example 8.10: In dairy lands, Strontium 90 a radioactive element closely related to calcium, can make its way into the milk via the grasses eaten by dairy cows. Then it becomes concentrated in the bones of those who drink this milk. In 1959 a study was conducted to compare mean concentration of Strontium 90 in the bones of the children to those of adults. It was thought that the level in children was higher, due to more intake of milk during their formative years. Is this contention supported by the following data? Approximate normality may be assumed. (Source: Milton and Tsokos (1983), p. 258.)

Children $n_1 = 121$ $\bar{X}_1 = 2.6$ picco curies/gm $S_1 = 1.2$ pc/gm

Adults $n_2 = 61$ $\bar{X}_2 = .4$ picco curies/gm $S_2 = .11$ pc/gm

Solution: Since sample sizes are large and approximate normal distribution of X_1: the Strontium 90 concentration in children and of X_2: the same concentration in adults can be assumed, the solution follows the line of cases A (b) and C (b).

Assumption: Let $E(X_1) = \mu_1$ and $E(X_2) = \mu_2$, $V(X_1) = \sigma_1^2$ $V(X_2) = \sigma_2^2$, σ_1^2 and σ_2^2 unknown and $n_1 = 121$ and $n_2 = 61$, are large.

The hypotheses are H_0: $\mu_1 = \mu_2$ against $H_1 = \mu_1 > \mu_2$

The test statistic to be used is

$$Z = \frac{\bar{X}_1 - \bar{X}_2}{\sqrt{\frac{S_1^2}{n_1} + \frac{S_2^2}{n_2}}} \sim N(0, 1) \text{ if } H_0 \text{ is true}$$

The test procedure or rule or critical region is for $\alpha = .05$

Reject H_0 if $Z > 1.645$

The *computed test statistic* is given by

$$Z_{cal} = \frac{2.6 - .4}{\sqrt{\frac{1.2^2}{121} + \frac{.11^2}{62}}} = \frac{2.2}{.10999} = 20.002$$

The *statistical decision* is to reject H_0, since Z_{cal} (> 1.645) lies in the critical region.

Hence it can be *concluded* that the available data support the contention that the concentration of Sr 90 in bones of children is higher than that in adults.

8.3 One Sample Problems for Testing Variance

The last two sections were concerned with inferences about one or two population means. On some occasions inferences about variances of one population or about variances of two populations are required. In this section, hypothesis testing procedures for a single population variance are described.

Again variance of measured characteristic X (= σ^2) in a population is of interest, and it is desired to conclude if σ^2, has a specified value say σ_0^2. Conclusion is **drawn from a random sample of size *n* on the random variable *X*,** which is **assumed** to have **normal distribution** i.e. $X \sim N(\mu, \sigma^2)$. There are **two cases** of testing problem: (A) when μ is known (B) μ is not known. Both of these cases are considered and all the steps of testing procedure are given below.

(A) *Assumptions:* The random variable of interest X has normal distribution with mean μ and variance σ^2. In symbols: **$X \sim N(\mu, \sigma^2)$, where μ is known.**

The *hypotheses to be tested* H_0: $\sigma^2 = \sigma_0^2$ against any of the three possible alternatives.

(i) H_1: $\sigma^2 \neq \sigma_0^2$ (ii) H_1: $\sigma^2 > \sigma_0^2$ or (iii) H_1: $\sigma^2 < \sigma_0^2$

The test statistic to be used for testing H_0 is

$$\chi^2 = \sum_{i=1}^{n} \frac{(X_i - \mu)^2}{\sigma_0^2} \tag{8.39}$$

The distribution of the test statistic is chi-square distribution with n degree of freedom if H_0 is true. In symbols

$$\chi^2 \sim \chi_n^2 \text{ if } H_0 \text{ is true.}$$

The level of significance is specified as α.

The test procedure or *rule* or *critical region* depends on the alternative against which H_0 is tested. Thus the critical regions are

(i) Reject H_0 if $\chi^2 < \chi^2_{n,\alpha/2}$ or $\chi^2 > \chi^2_{n,1-\alpha/2}$ if H_1: $\sigma^2 \neq \sigma_0^2$ is two sided.

(ii) Reject H_0 if $\chi^2 > \chi^2_{n,1-\alpha}$ if H_1: $\sigma^2 > \sigma_0^2$ is upper tail test.

(iii) Reject H_0 if $\chi^2 < \chi^2_{n,\alpha}$ if H_1: $\sigma^2 < \sigma_0^2$ is lower tail test.

Figure 8.3 below gives the critical regions for the above three alternatives.

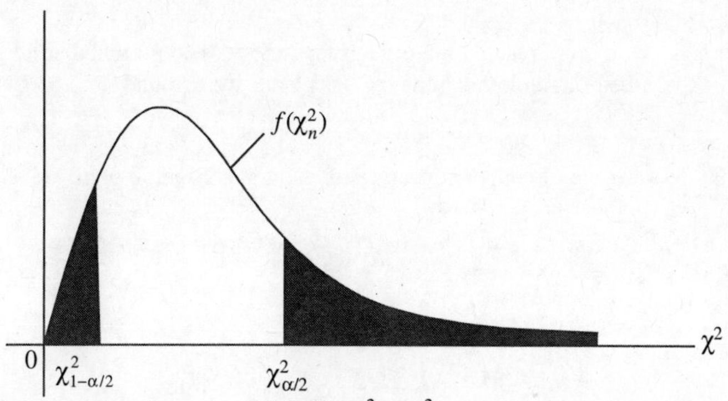

(a) H_1: $\sigma^2 \neq \sigma_0^2$

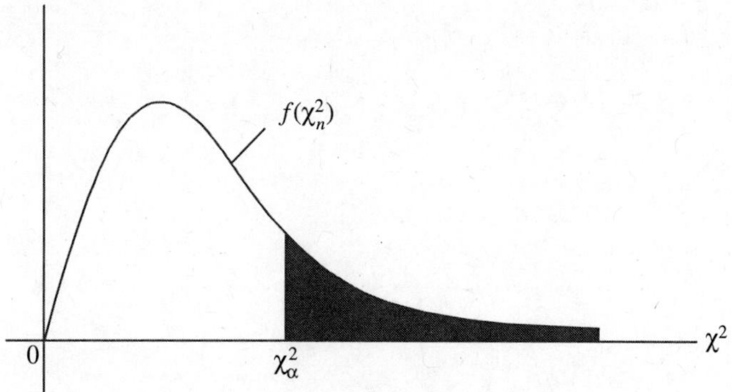

(b) H_1: $\sigma^2 > \sigma_0^2$

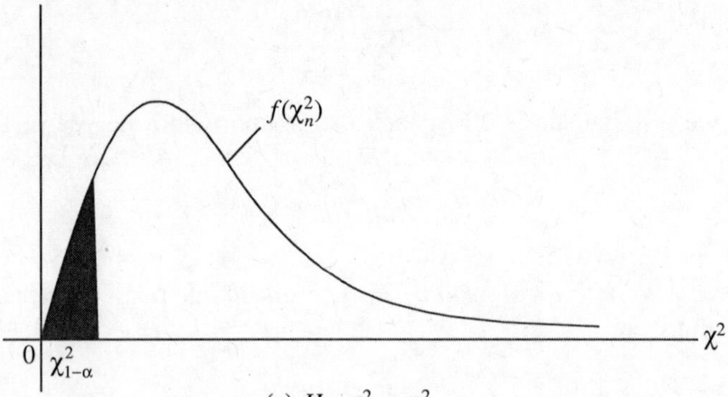

(c) H_1: $\sigma^2 < \sigma_0^2$

Fig. 8.3. Critical regions for test of variance (source of (a), (b), and (c) Milton and Tsokos (1983), p. 204).

The *computation* of test statistic is done using the observed sample values in (8.39).
The *statistical decision and conclusion* are written depending on the calculated value of χ^2 statistic ($= \chi^2_{cal}$).

Example 8.11: Copper concentration in plants found in a river valley *X* had normal distribution with mean 50 ppm, copper concentration in the ashes of random sample of 16 plants were found as:

5 3 34 18 27 8 50 38 43 35 20 70 25 60 19 14

Do the data indicate that the variance of copper concentration is larger than 250 ppm^2? (Source: Milton and Tsokos (1983), p. 202.)

Solution: The problem is of type A in which the *assumptions* are $X \sim N(50, \sigma^2)$ $n = 16$.
The hypotheses are H_0: $\sigma^2 = \sigma_0^2 = 250$ against H_1: $\sigma^2 > 250$
The test statistic to test H_0 is

$$\chi^2 = \sum_{i=1}^{n} \frac{(X_i - \mu)^2}{\sigma_0^2} \sim \chi^2_n \text{ if } H_0 \text{ is true}$$

The level of significance is $\alpha = .05$
The *computation* of the test statistic is as

$$\chi^2 = \sum_{i=1}^{n} \frac{(X_i - \mu)^2}{\sigma_0^2} = \frac{1}{\sigma_0^2}\left(\sum_{i=1}^{n} X_i^2 - 2\mu \sum_{i=1}^{n} X_i + n\mu^2\right)$$

$$= \frac{1}{250}\left[19407 - 2(50)(469) + 16(50^2)\right] = \frac{1}{250}(19407 - 46900 + 40000)$$

$$= \frac{12507}{250} = 50.03$$

The critical region is - Reject H_0 if $\chi^2_{cal} > \chi^2_{16,.95} = 26.3$ (from Table A.5)
The statistical decision is to reject H_0 since $\chi^2_{cal} = 50.03 > 26.3$
The *conclusion*, therefore, is that the data indicates that copper concentration in plants has larger variance than 250 ppm^2.

(B) The *data* is available as for case A, but the value of μ is not known. All other assumptions are same as in A. Thus

Assumptions: **A random sample of size *n*** is available **on *X*,** where $X \sim N(\mu, \sigma^2)$ and μ **is unknown.**

The *hypotheses* are also same $H_0 : \sigma^2 = \sigma_0^2$ against

(a) H_1: $\sigma^2 \neq \sigma_0^2$ (b) H_1: $\sigma^2 > \sigma_0^2$ (c) H_1: $\sigma^2 < \sigma_0^2$

The *test statistic* to be used to test H_0 is

$$\chi^2 = \frac{(n-1)\, s^2}{\sigma_0^2} \tag{8.40}$$

where $$s^2 = \frac{1}{n-1} \sum_{i=1}^{n} (X_i - \bar{X})^2$$

The *distribution of the test statistic* given in the (8.40), if H_0 is true, is Chi-square with $(n - 1)$ degree of freedom. In symbols

$$\chi^2 \sim \chi^2_{n-1} \text{ if } H_0 \text{ is true}$$

The level of significance $\alpha = .05$

The test procedure or rule or critical region is similar to that of case A given in Fig. 8.3 except that the **degrees of freedom** of the Chi-square distribution is $(n - 1)$, in place of *n.* In symbols

(i) Reject H_0 if $\chi^2 < \chi^2_{n-1,\alpha/2}$ or $\chi^2 > \chi^2_{n-1,1-\alpha/2}$ if $H_1: \sigma^2 \neq \sigma_0^2$

(ii) Reject H_0 if $\chi^2 > \chi^2_{n-1,1-\alpha}$ if $H_1: \sigma^2 > \sigma_0^2$

(iii) Reject H_0 if $\chi^2 < \chi^2_{n-1,\alpha}$ if $H_1: \sigma^2 < \sigma_0^2$

The statistical decision and conclusion is written as in the Case A when μ is known.

Example 8.12: A simple random example of 15 nursing students who participated in an experiment took a test to measure manual dexterity. The variance of the sample observations was 1225. Can it be concluded from this data that the population variance is different from 2500 (source: Daniel (1987), p. 228)?

Solution: *Assumptions:* Sample is from a normal population with variance σ^2.

Hypothesis H_0: $\sigma^2 = 2500$ against H_1: $\sigma^2 \neq 2500$

The test statistic to test H_0 is

$$\chi^2 = \frac{(n-1)\, s^2}{\sigma_0^2} \sim \chi^2_{n-1} \text{ if } H_0 \text{ is true}$$

The *level of significance is* $\alpha = .05$

The test procedure or critical region is given by

Reject H_0 if $\chi^2 < \chi^2_{14,.025}$ or if $\chi^2 > \chi^2_{14,.0975}$

The *computed* test statistic is

$$\chi^2_{\text{cal}} = \frac{14 \times 1225}{2500} = 6.86$$

From the Table A.5 $\chi^2_{14,.025} = 5.629$ and $\chi^2_{14,.0975} = 26.118$

Statistical decision is not to reject H_0 since $\chi^2_{cal} > 5.629$ and $\chi^2_{cal} < 26.118$. Thus χ^2_{cal} does not belong to critical region.

The *conclusion* is that the data does not support the contention that the population variance is different from 2500.

8.4 Two Sample Problems for Testing Variances

Comparison of two population means was based on two independent samples from the two normal populations with the two assumptions that (i) σ_1^2 and σ_2^2 are equal and that (ii) σ_1^2 and σ_2^2 are not equal.

It is advisable that before deciding whether to use *t*-test or Behren's Fisher *t*-test for comparing means, the variances of the two populations should be tested for equality.

In this section, therefore, test **of equality of two population variances** is considered. Let a characteristic is measured on two populations. Let X_1 denotes this measure on population 1 and X_2 that on population 2. Moreover let $X_1 \sim N(\mu_1, \sigma_1^2)$ and $X_2 \sim N(\mu_2, \sigma_2^2)$. Samples of sizes n_1 and n_2 are in hand. To test H_0: $\sigma_1^2 = \sigma_2^2$ on the basis of available data the following steps are followed:

(A) *Assumption:* Independent random samples of sizes n_1 and n_2 are available on X_1 and X_2 **where $X_1 \sim N(\mu_1, \sigma_1^2)$ and $X_2 \sim N(\mu_2, \sigma_2^2)$, μ_1 and μ_2 are known.**

Hypothesis to be tested is H_0: $\sigma_1^2 = \sigma_2^2$ against one of the following alternatives

(a) H_1: $\sigma_1^2 \neq \sigma_2^2$ (b) H_1: $\sigma_1^2 > \sigma_2^2$ (c) H_1: $\sigma_1^2 < \sigma_2^2$

The *test statistic* for testing H_0 is

$$F = \frac{\sum_{i=1}^{n_1} (X_{1i} - \mu_1)^2 / n_1}{\sum_{j=1}^{n_2} (X_{2j} - \mu_2)^2 / n_2} \tag{8.41}$$

The *test statistic follows F-distribution* with n_1 and n_2 degrees of freedom, if H_0 is true. In symbols

$$F = \frac{S_1^2}{S_2^2} \sim F_{n1, n2} \text{ if } H_0 \text{ is true} \tag{8.42}$$

The level of significance is to be specified and normally $\alpha = .05$ (8.43)

The test procedure or rule or critical region depends on the alternatives against which H_0 is tested. Thus one of the following rejection regions is used.

(i) Reject H_0 if $F_{cal} < F_{n1,\, n2,\, \alpha/2}$ or $F_{cal} > F_{n1,\, n2,\, 1-\alpha/2}$ if H_1: $\sigma_1^2 \neq \sigma_2^2$

(ii) Reject H_0 if $F_{cal} > F_{n1,\, n2,\, 1-\alpha}$ if H_1: $\sigma_1^2 > \sigma_2^2$

(iii) Reject H_0 if $F_{cal} < F_{n1,\, n2,\, \alpha}$ if H_1: $\sigma_1^2 < \sigma_2^2$ (8.44)

Some times **only upper tail 5% and 1% significant points are available in the *F*-Table.** In such a case **instead of setting H_1: $\sigma_1^2 < \sigma_2^2$ it may be set as $H_1 : \sigma_2^2 > \sigma_1^2$** and the ***F* statistic may also have S_2^2 in the numerator and S_1^2 in the denominator**. Thus for modified H_1, the test statistic should be

$$F_m = \frac{S_2^2}{S_1^2} \sim F_{n2, n1} \text{ if } H_0 \text{ is true} \tag{8.45}$$

and *critical region* becomes

$$\text{Reject } H_0 \text{ if } F_m > F_{n2,\, n1,\, 1-\alpha} \tag{8.46}$$

It is **a general convention** that **H_1 is set with '>' symbol** only and **sample variance from higher variance population is taken as numerator in F-statistic.**

Computation of test statistic may be done using the following

$$S_i^2 = \frac{1}{n_i}\sum_{j=1}^{ni}\left(X_{ij}-\mu_i\right)^2 = \frac{1}{n_i}\left[\sum_{j=1}^{ni} X_{ij}^2 + n_i\mu_i^2 - 2\mu_i\sum_{j=1}^{ni} X_{ij}\right]$$

$$= \frac{1}{n_i}\sum_{j=1}^{ni} X_{ij}^2 + \mu_i^2 - 2\mu_i\,\bar{X}_i \quad (i=1,2) \tag{8.47}$$

Statistical decision and conclusion are written as given in all the tests of significance.

(B) If μ_1 and μ_2 are unknown there are following two differences in the above test procedure.

1. The *test statistic* becomes

$$F = \frac{s_1^2}{s_2^2} \tag{8.48}$$

where,
$$s_i^2 = \frac{1}{n_i - 1}\sum_{j=1}^{ni}\left(X_{ij}-\bar{X}_i\right)^2$$

$$= \frac{1}{(n_i-1)}\left[\sum_{j=1}^{ni} X_{ij}^2 - n_i\,\bar{X}_i^2\right] \quad (i=1,2) \tag{8.49}$$

and

2. The *distribution of the test statistic* is still F-distribution but with degrees of freedom $(n_1 - 1)$, $(n_2 - 1)$. In symbols, now

$$F \sim F_{(n1-1),\,(n2-1)} \text{ if } H_0 \text{ is true}$$

Figure 8.4 gives below the critical regions of the said tests for two alternative hypotheses.

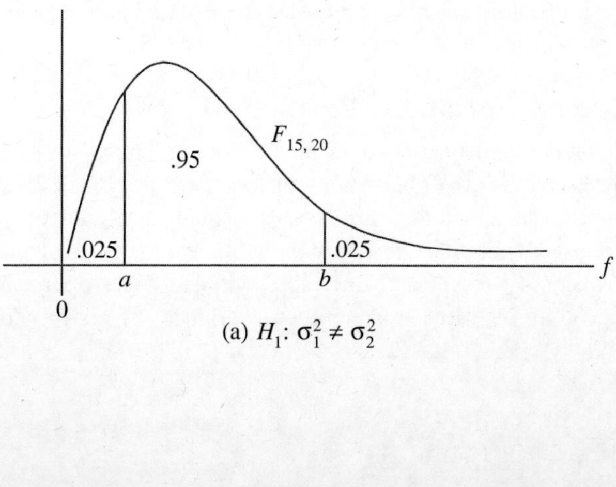

(a) H_1: $\sigma_1^2 \neq \sigma_2^2$

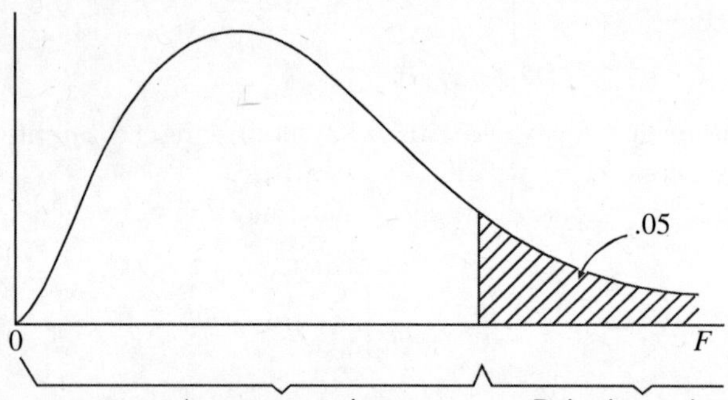

(b) H_1: $\sigma_1^2 > \sigma_2^2$

Fig. 8.4. Critical regions for test of equality of two variances. (Source: (a) Milton and Tsokos (1983), p. 248, (b) Daniel (1987), p. 232.)

Example 8.13: Cardiac index values (lit/minutes/M^2) were analyzed in two groups of patients following prosthetic valve replacement. The sample sizes and variances were as follows: n_1 = 16, s_1^2 = 3.75 n_2 = 10, s_2^2 = 1.8. Do these data provide sufficient evidence to indicate a difference in population variances? Let α = .05, and assume normality of distributions of cardiac index values. (Source: Daniel (1987), p. 232.)
Solution: This problem is of second type and the
Assumptions are: Let X_1 and X_2 denote the cardiac index values in two groups of patients. Then $X_i \sim N(\mu_i, \sigma_i^2)$, i = 1, 2.
Independent random samples of sizes n_1 = 16 and n_2 = 10 are drawn from these two groups.
Hypotheses to be tested are H_0: $\sigma_1^2 = \sigma_2^2$ against H_1: $\sigma_1^2 \neq \sigma_2^2$
The *test statistic* to be used is

$$F = \frac{s_1^2}{s_2^2} \sim F_{n1-1,\,n2-1} \quad \text{if } H_0 \text{ is true}$$

For *the level of significance* α = .05, *the critical region* is
Reject H_0 if $F < F_{n1-1,\,n2-1,\,\alpha/2}$ or $F > F_{n1-1,\,n2-1,\,1-\alpha/2}$
The *calculated* test statistic is

$$F = \frac{3.75}{1.8} = 2.08$$

From the Table A.6 only the upper tail value is seen, since $F > 1$ to give $F_{15,\,9,\,.975}$ = 3.77
Statistical decision is do not reject H_0 since F_{cal} = 2.08 does not lie in the critical region.
The conclusion, therefore, is that the data does not provide sufficient evidence to indicate that there is a difference in variances of cardiac index values in two populations.

8.5 Comparing Several Variances: Bartlett's Test

It has been seen that for comparison of two means of normal populations the test statistic depends on whether the variances, of the two populations differ or do not differ. Generally an assumption is stated before performing a test, that the variances are equal, and then only pooled *t*-test is used.

If it is not known in advance that the two population variances are equal, it is advisable to first check equality of variances by *F*-test and then apply pooled *t*-test or the Behrens's Fisher test for equality of means depending as whether equality of variances is confirmed by test or not.

There are many situations, when more than two populations' means are compared for their equality. Apart from normality of these populations (i.e. of measured characteristic of interest), the commonly applied test, called one-way ANOVA (Analysis of Variance) assumes equality of variances also.

An experimenter, exploring equality of means, should ascertain first equality of variances, unless it is already established from past experience. In this section one such test called **Bartlett's (1937) test of homogeneity of variances** is being discussed, which is based on Neymann Pearson likelihood ratio test (1931).

The data in hand consists of independent random samples of sizes n_i $(i = 1, 2, ..., k)$ from k normal populations on a characteristic X. So X_i is the measured characteristic on i^{th} population such that $X_i \sim N(\mu_i, \sigma_i^2)$. It is desired to test H_0: $\sigma_1^2 = \sigma_2^2 = ...\ \sigma_i^2 = ... = \sigma_k^2$ against an alternative H_1: At least two of σ_i^2 differ $(i = 1, 2, ..., k)$.

From the samples of n_i observations from i^{th} population s_i^2 $(i = 1, 2, ..., k)$ are the estimates of population variances σ_i^2.

For testing homogeneity of variances, steps to be followed are given below.

Assumptions: X_i $(i = 1, 2, ..., k)$ follow normal distribution with means μ_i and variances σ_i^2. In other words, the measured characteristic X on the subjects belonging to k populations, follows normal distributions. If X_i denotes the random variable measuring the characteristic in i^{th} population $(i = 1, 2, ..., k)$ then $X_i \sim N(\mu_i, \sigma_i^2)$.

Data: Independent random samples of n_i observations on X_i are available from i^{th} population. Thus $n_1, n_2, ..., n_k$ observations are available on the k population subjects and estimates of variances are computed as $s_1^2, s_2^2, ..., s_k^2$, s_i^2 being unbiased estimate of σ_i^2 $(i = 1, 2, ..., k)$.

The *hypothesis* to be tested is H_0 against H_1 given above.

The *test statistic* for testing H_0 is

$$\chi^2 = \frac{\left[\left\{\sum_{i=1}^{k} (n_i - 1)\right\} l_n\, s_p^2 - \left\{\sum_{i=1}^{k} (n_i - 1)\, l_n\, s_i^2\right\}\right]}{C.F.} = \frac{B}{C.F.} \qquad (8.50)$$

where $$s_i^2 = \frac{1}{n_i - 1} \sum_{j=1}^{ni} \left(X_{ij} - \bar{X}_i\right)^2 \qquad (8.51a)$$

$$s_p^2 = \frac{\left[\sum_{i=1}^{k} \left\{\sum_{j=1}^{ni} \left(X_{ij} - \bar{X}_i\right)^2\right\}\right]}{\sum_{i=1}^{k} (n_i - 1)} = \text{pooled estimate of common variance} \qquad (8.51b)$$

$$l_n(\cdot) = \log_e(\cdot) \text{ or natural log } (\cdot) \qquad (8.51c)$$

$$C.F. = \text{Correction Factor} = 1 + \frac{1}{3(k-1)}\left[\sum_{i=1}^{k} \frac{1}{(n_i - 1)} - \frac{1}{\sum_{i=1}^{k} (n_i - 1)}\right] \qquad (8.52)$$

If tables of natural log are not in hand, the numerator B of the test statistic in (8.50) can be modified as

$$B = 2.30259\left[\left\{\sum_{i=1}^{k}(n_i - 1)\right\}\log s_p^2 - \left\{\sum_{i=1}^{k}(n_i - 1)\log s_i^2\right\}\right] \tag{8.53}$$

The distribution of the test statistic: The test statistic follows Chi-square distribution with $k - 1$ degree of freedom if H_0 is true. In symbols

$$\chi^2 \sim \chi^2_{k-1} \text{ if } H_0 \text{ is true} \tag{8.54}$$

If *level of significance* is chosen as α, the *test procedure* or *rule* or *critical region* is

$$\text{Reject } H_0 \text{ if } \chi^2 > \chi^2_{k-1,\alpha} \tag{8.55}$$

Computation of test statistic will depend on available data. If the observed sample values are given s_i^2 are computed from (8.51a) and s_p^2 may be obtained using (8.51b).

If however, the sample observations are not given but the values of s_i^2, $i = 1, 2, ..., k$ are given, (8.51a) is skipped. s_p^2 can be calculated by the following expression instead of (8.51b).

$$s_p^2 = \frac{\left[\sum_{i=1}^{k}(n_i - 1)s_i^2\right]}{\sum_{i=1}^{k}(n_i - 1)} \tag{8.56}$$

The *statistical decision and the conclusion* are written according to the computed value of the test statistic.

An example is given below.

Example 8.14: Nineteen pigs were divided into four groups, and each group was given a different feed. Weights of pigs were recorded after a month of feeding as given below. It is desired to test whether the variances of weights of pigs on the four feeds are the same. (Source: Zar (2004), p. 203 (data only).)

Table 8.2 Weights of pigs on four feeds (Source: Zar, p. 203 (data only))

Feed 1	Feed 2	Feed 3	Feed 4
60.8	68.7	102.6	87.9
57.0	67.7	102.1	84.2
65.0	74.0	100.2	83.1
58.6	66.3	96.5	85.7
61.7	69.8		90.3

Solution: The test will be based on the following assumptions.

Assumptions: Let X_1, X_2, X_3, X_4 denotes the weight of pigs on the four feeds respectively. Then $X_i \sim N(\mu_i, \sigma_i^2)$ for $i = 1, 2, 3, 4$. Pigs are randomly allotted and n_1, n_2, n_3, n_4 the number of pigs on the four feeds are 5, 5, 4, 5 respectively.

Hypothesis to be tested is H_0: $\sigma_1^2 = \sigma_2^2 = \sigma_3^2 = \sigma_4^2$ against H_1: at least two variances differ.

The *test statistic* to test H_0 is

$$\chi^2 = \frac{B}{C.F.} \text{ as given in (8.52)}$$

$$\chi^2 \sim \chi^2_{k-1} \text{ if } H_0 \text{ is true}$$

Let the level of significance be α = .05 and so critical region will be:

Reject H_0 if $\chi^2 \sim \chi^2_{k-1,\alpha}$

The computation of s_i^2 will be done using (8.52a) and details are shown in table below.

Table 8.3 Bartlett's test of homogeneity of four variances

Class or group (i)	$n_i - 1$	$\sum_{j=1}^{ni} (X_{ij} - \bar{X}_i)^2$	s_i^2	Log s_i^2	$(n_i - 1) \log s_i^2$	$1/(n_i - 1)$
Feed 1	4	37.57	9.39	.9727	3.8908	0.250
Feed 2	4	34.24	8.56	.9325	3.7300	0.250
Feed 3	3	22.97	7.66	.8842	2.6526	0.333
Feed 4	4	33.55	8.39	.9238	3.6952	0.250
Total	15	128.35			13.9686	1.083

$$s_p^2 = \frac{\left[\sum_{i=1}^{k}\left\{\sum_{j=1}^{ni}(X_{ij} - \bar{X}_i)^2\right\}\right]}{\sum_{i=1}^{k}(n_i - 1)} = \frac{128.35}{15} = 8.56$$

Log s_p^2 = .9325 from (8.53)

$$B = 2.30259 \; [15 \; (.9325) - 13.9686] = .0435$$

From (8.52)

$$C.F. = 1 + \frac{1}{3 \times 3}\left(1.083 - \frac{1}{15}\right) = 1.113$$

Hence

$$\chi^2_{\text{cal}} = \frac{.0435}{1.113} = .0391$$

From Table A.5 $\chi^2_{3,.05} = 7.815$

Hence *Statistical decision* is not to reject H_0 since χ^2_{cal} does not lie in the critical region.

It can, therefore, *be concluded* that the variances of the weights of pigs on four feeds may not be different.

8.6 Comparison of Several Means

Section 8.2 considers testing of difference between two populations means. However, quite often an investigator needs to test equality of means of several (say k) populations. A novice may begin by testing difference between a pair of population from among the k, selecting each pair one by one. The number of possible pairs is $\binom{k}{2}$ which is quite large (= 10) even for k = 5. Moreover, subsequent tests distort α. Hence it is advisable that equality of k-population means be done using a single test. In this section this test is being discussed.

A general problem may be selected as follows: There are k populations, and a response variable X is measured on each individual in the independent random samples of sizes n_1, n_2, ..., n_k. Let X_1, X_2, ..., X_k be the random variables measured on the said k populations, so that X_i, $i = 1, 2, ..., k$ is variable in the ith populations. $X_i \sim N(\mu_i, \sigma_i^2)$. The k populations are studied to determine if these k populations' responses differ as regards one characteristic or one factor or level of a treatment which the k populations, receive. The data collected can be arranged or recorded in the format given in Table 8.4 for testing H_0: $\mu_1 = \mu_2 =$ ......, μ_k against H_1: at least two means differ.

Table 8.4 Data lay out for testing equality of k means

	Groups								
	1	2	3	...	i	...	k		Total
	X_{11}	X_{21}	X_{31}	...	X_{i1}	...	X_{k1}		...
	X_{12}	X_{22}	X_{32}	...	X_{i2}	...	X_{k2}		...
	$\vdots$	$\vdots$	$\vdots$		$\vdots$		$\vdots$		...
	X_{1j}	X_{2j}	X_{3j}	...	X_{ij}	...	X_{kj}		...
	$\vdots$	$\vdots$	$\vdots$		$\vdots$		$\vdots$		$\vdots$
	X_{1n_1}	X_{2n_2}	X_{3n_3}	...	X_{in_i}	...	X_{kn_k}		$\vdots$
Total	$T_{1.}$	$T_{2.}$	$T_{3.}$	...	$T_{i.}$	...	$T_{k.}$		$T_{..}$

Thus X_{ij} denotes the response of jth individual in the ith group. To use the sample observations for comparing k population means, the following calculations are required.

$Ti.$ = Total of responses from sample on i^{th} group or population

$$= \sum_{j=1}^{ni} X_{ij}, \quad i = 1, 2, ..., k \tag{8.57}$$

$$T.. = \text{Grand Total} = \sum_{i=1}^{k} \sum_{j=1}^{ni} X_{ij} = \sum_{i=1}^{k} T_{i.} \tag{8.58}$$

$$\sum_{i=1}^{k} \sum_{j=1}^{ni} X_{ij}^2 = \text{Sum of squares of all the responses} \tag{8.59}$$

$$n = \sum_{i=1}^{k} n_i$$

$$T^2../n = C.F. \text{ (Correction Factor)} \tag{8.60}$$

It may be noted that dot indicates the subscript over which summation has been taken. The following sum of squares (SS) of deviations are also calculated

$$TSS\,(\text{Total sum of squares}) = \sum_{i=1}^{k} \sum_{j=1}^{ni} X_{ij}^2 - C.F. \tag{8.61}$$

$$SSGr\,(\text{Sum of squares due to groups or treatments}) = \sum_{i=j}^{k} T_{i.}^2/n_i - C.F. \tag{8.62}$$

$$SSE \text{ (Residual sum of squares or error sum of squares)} = TSS - SSGr \tag{8.63}$$

Let the level of significance $\alpha = .05$. Mean sum of squares are calculated as

$$MSGr = SSGr/\text{degree of freedom} = SSGr/(k - 1) \tag{8.64}$$

$$MSE = SSE/\text{d.f.} = SSE/(n - k) \tag{8.65}$$

$$F = MSGr/MSE \text{ is the } \textit{statistic to test } H_0 \textit{ against } H_1 \tag{8.66}$$

The result of these computations is shown in the following ANOVA (Analysis of variance) Table:

Table 8.5 ANOVA table for testing k group means

Source of variation	d.f.	Sum of squares (*SS*)	Mean sum of square (*MSS*)	F_{cal}	*F*-Tabulated
Groups	$k - 1$	*SSGr*	$MSGr = SSGr/(k - 1)$	$MSGr/MSE = F$	$F_{k-1, n-k, .05}$
Error	$n - k$	By subtraction $TSS - SSGr = SSE$	$MSE = SSE/(n - k)$		$F_{k-1, n-k, .01}$
Total	$n - 1$	*TSS*			

The *test statistic* $F \sim F_{k-1,\, n-k}$ *if* H_0 *is true* (8.67)

The *critical region* is given by

$$\text{Reject } H_0 \text{ if } F_{cal} > F_{k-1,\, n-k,\, .05} \text{ at 5\% level of significance} \tag{8.68}$$

and

$$\text{Reject } H_0 \text{ if } F_{cal} > F_{k-1,\, n-k,\, .01} \text{ at 1\% level of significance} \tag{8.69}$$

There is a convention to put a single '*' on F_{cal} if (8.68) holds and a double '**' if (8.69) holds. As an illustration of this method an example is considered below.

Example 8.15: In a study of the effect of glucose on insulin release, specimens of pancreatic tissue from experimental animals were randomly assigned to be treated with one of the 5 different stimulants. Later a determination of the amount of insulin released by each tissue was made and is given in table below. Can it be concluded that there is difference among the five treatments with respect to the mean amount of insulin released? (Source: Daniel (1987), p. 276.)

Table 8.6 Amount of insulin released due to 5 stimulants

	Stimulants					Total
	1	2	3	4	5	
	1.53	3.15	3.89	8.18	5.86	
	1.61	3.96	3.68	5.64	5.46	
	3.75	3.59	5.70	7.36	5.69	
	2.89	1.89	5.62	5.33	6.49	
	3.26	1.45	5.79	8.82	7.81	
		1.56	5.33	5.26	9.03	
				7.10	7.49	
					8.98	
Total	13.04	15.60	30.01	47.69	56.81	163.15
n_i	5	6	6	7	8	32

Solution: Let X_1, X_2, X_3, X_4, and X_5 denote the insulin amount released under 1st, 2nd, 3rd, 4th, 5th, treatment and the sizes of the random samples tissues subjected to these 5 treatments or stimulants are $n_1 = 5$, $n_2 = 6$, $n_3 = 6$, $n_4 = 7$, and $n_5 = 8$.
Assumptions: It is assumed that $X_i \sim N(\mu_i, \sigma_i^2)$, $i = 1, 2, ..., 5$ and that $\sigma_1^2 = \sigma_2^2 = \sigma_3^2 = \sigma_4^2 = \sigma_5^2$.
Hypothesis to be tested is H_0: $\mu_1 = \mu_2 = \mu_3 = \mu_4 = \mu_5$ against the alternative H_1: at least two means differ.
Let $\alpha = .05$ be the *level of significance*
The test statistic for testing H_0 against H_1 is

$$F = \frac{MST_r}{MSE} \sim F_{4,27} \text{ if } H_0 \text{ is true}$$

The *critical region* is to reject H_0 if $F_{cal} > F_{4, 27, \alpha}$
The *computation and preparation of ANOVA table*

1. The treatment totals $T_{i.}$ are computed for each of the 5 stimulants and shown in the last row of Table 8.6, above n_i values. Thus

$$T_{1.} = 13.04,\ T_{2.} = 15.60 \ ....T_{5.} = 56.81 \text{ and } T.. = 163.15$$

2. Sum of squares of individual observations $= \sum_{i=1}^{5} \sum_{j=1}^{ni} \left(X_{ij}^2\right) = 994.3529$ and

$$C.F. = \frac{T_{..}^2}{n} = \frac{163.15^2}{32} = 831.81008$$

3. Hence $TSS = 994.3529 - 831.81008$

$$=162.54282$$

$$SST_r = \sum_{i=1}^{k}\left(T_{i.}^2/n_i\right) - C.F$$

$$= \frac{(13.04)^2}{5} + \frac{(15.60)^2}{6} + \frac{(30.01)^2}{7} + \frac{(47.69)^2}{7} + \frac{(56.81)^2}{8} - 831.81008$$

$$= 952.99551 - 831.81008 = 121.18543$$

$$SSE = TSS - SST_r = 162.54282 - 121.18543 = 41.35739$$

$$MST_r = SST_r/(k - 1) = 121.18543/4 = 30.296358$$

$$MSE = SSE/(n - k)$$

$$= 41.35739/27 = 1.53177$$

Table 8.7 ANOVA table for equality of 5 mean insulin release

Sources of variation	d.f.	*SS*	*MSS*	*F*-ratio	Tabulated *F*
Stimulants	4	$SSTr = 121.18543$	$MSTr = 30.296358$	$MSTr/MSE$	$F_{4, 27, .05} = 2.73$
Error	27	$SSE = TSS - SSTr$ $= 41.35739$	$MSE = 1.531776$	$F = 19.78$**	$F_{4, 27, .01} = 4.11$
Total	31	$TSS = 162.54282$			

Since *calculated* value of $F > F_{4,\,27,\,.01} = 4.11$, it is highly significant.
Thus it may be *concluded* that the data indicates that mean amount of insulin released by five stimulants are **highly significantly different**.

Calculation of *P*-values

As already mentioned, researchers in life sciences normally quote the *P*-value after the test of significance. The definition of *P*-value is, therefore refreshed below.

Definition of *P*-value: The *P*-value for hypothesis test is the probability of obtaining the calculated value of the test – statistic as extreme or more extreme (in appropriate direction depending on H_1) than actually obtained if H_0 is true.

Instead of concluding that test statistic is significant or non significant at certain level of significance some researchers prefer to report the exact value of probability of getting the obtained value of the test statistic if H_0 is true. Many researchers quote instead of exact value of this probability (*P*-value) give an upper limit or a lower limit of the *P*-value. Thus they indicate whether calculated value of test statistic is very improbable or rare to obtain if H_0 is true. A high *P*-value indicates possibility of H_0 being true.
Thus there are following two ways of reporting *P*-value of a statistical test.

1. In Example 8.15 the calculated value of *F*-test statistic is 19.78. The *F*-statistic has $F_{4,\,27}$ distribution. One can look the table to find $P[F_{4,\,27,} > 19.78]$. It may not be possible to read it. In such a case one compares with upper tail 5%, 1%, .5% values of $F_{4,\,27}$ distribution and finds that $F_{cal} = 19.78$ is much higher than the said three values 2.73 (5% value), 4.11 (1% value), 4.74 (.5% value). The researcher may, therefore, report that *P*-value is much smaller than .5%. Similar reporting of *P* values may be possible, if *t*-table does not give area for all values of *t*, as for standard normal distribution. Most *t*-tables give some percentage points (99.5%), 99%, 97.5%, 95%, 90%) only, for various degrees of freedom. Hence from actual value of calculated *t* one has to compare various significant values of *t* and accordingly give limit of *P*-value.
2. For normal tests one can find exact *P*-values since one can read the probability for any value of standard normal test statistic.

(a) Consider Example 8.8 which uses *Z*-statistic, following *N*(0, 1) distribution if H_0 is true, to test H_0: $\mu_1 = \mu_2$ against H_1: $\mu_1 > \mu_2$ $Z_{cal} = 15.17$. *P*-value = $P[Z > 15.17] = .00$ so getting $Z = 15.17$ is almost impossible if H_0 is true.
(b) Consider another example in which H_0: $\mu = 25$ against H_1: $\mu \neq 25$. The value of test statistic, $Z_{cal} = -1.41$. Since $Z \sim N(0, 1)$ and test is two sided, in this case *P*-value = $P[Z < -1.41$ or $Z > 1.41] = [.0793 + .0793] = .1586$.

This indicates that it is quite probable to get $Z_{cal} = -1.41$, if H_0 is true.

9. Bivariate and Multivariate Data: Measuring and Testing Relationship

In the last few chapters concern is mainly on univariate random variables. It may be recalled that in Chapter 5, Section 5.3, it was mentioned that in many experimental problems an experimenter may determine values of more than one characteristics on selected subjects or experimental outcomes.

When only two characteristics are observed and measurements (x, y) are taken on these two, one obtains pairs of values of (X, Y) for each individual or observation unit. Such data is called bivariate data. In this manner when two characteristics are measured on each of the n items or units, one gets n pairs of values (X_1, Y_1), (X_2, Y_2), ..., (X_n, Y_n). If more than two say p, $(p > 2)$ characteristics are measured the resulting data is *multivariate data.*

9.1 Simple or Pearson's Product Moment Correlation Coefficient

One of the most common problems with bivariate data is to study quantitatively the influence or effect of one of the variables on the other.

The basic aid for studying the relation between two variables X and Y is to draw a scatter diagram or scatter plot. In a scatter plot each of the n pairs of observations (X_i, Y_i), $i = 1, 2, ..., n$ are plotted on a graph taking X's on X-axis and Y's on the Y-axis. Thus one gets a swarm of points on the graph. As an example see the figure below for the example 9.1 given below.

Example 9.1: A team of professional mental health workers wanted to test usefulness of new instruments for measuring the levels of withdrawn patients response. The team decided to examine the relationship between the scores on new, easy and less time taking test and the existing standardized test which is more expensive and time consuming. Team was interested only in carrying out the analysis for standardized scores between 50 and 100. Scores of 11 patients are given below (source: Daniel (1987), pp. 371-72).

Patient No.	1	2	3	4	5	6	7	8	9	10	11
Scores on new test (X)	50	55	60	65	70	75	80	85	90	95	100
Scores on standardized test (Y)	61	61	59	71	80	76	90	106	98	100	114

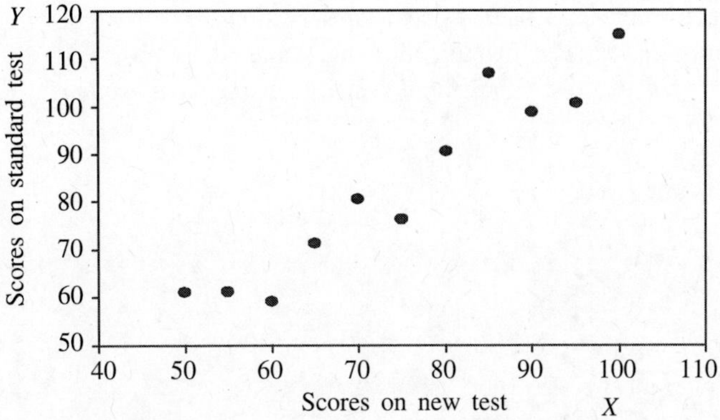

Fig. 9.1. Scatter diagram for data of Example 9.1.

In this figure the pattern of points suggests that with increase in X, Y also increases. Figure 9.2 given below shows another scatter diagram with a different pattern.

Example 9.2: For a study of water loss in *Tribolium confusum*, 9 batches of 25 beetles were weighed, kept at different relative humidities and weighed again after 6 days of starvation. Weight loss in mg (Y) was computed for each batch put on different relative humidity X (source: Sokal and Rohlf (1995), p. 459).

Batches →	1	2	3	4	5	6	7	8	9
Percent relative humidity (X)	0	12	29.5	43	53	62.5	75.5	85	93
Weight loss in mg (Y)	8.98	8.14	6.67	6.08	5.90	5.83	4.68	4.20	3.72

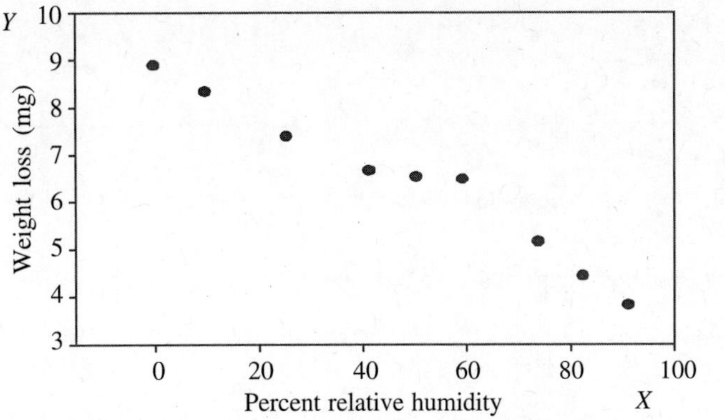

Fig. 9.2. Scatter diagram for data in Example 9.2.

In this figure it is found that for increasing X, Y is decreasing.
Even though the points in the scatter diagram do not lie on a single straight line but it seems the points are around an invisible straight line. There appears to be linear trend in the observations.

The scatter diagram reveals a pattern in the observations and it suggests a possible relationship between the two variables, as seen in the following scatter diagrams. Figure 9.3 shows the feeding rate of the *Didinium* (Y), based on the initial density of the *Paramecium* in the medium (X).

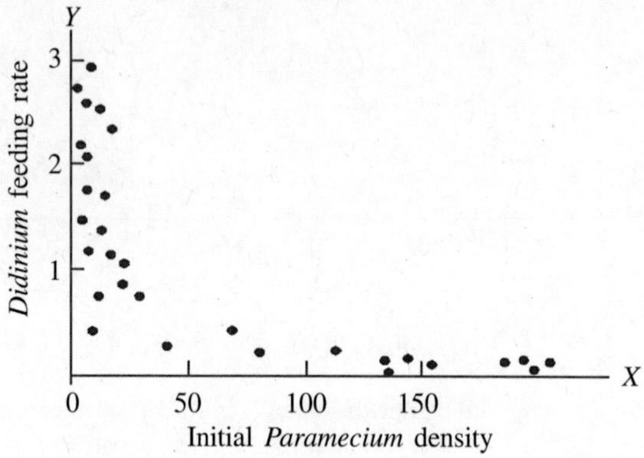

Fig. 9.3. Scatter diagram of initial *Paramecium* density and *Didinium* feeding rate (source: Milton and Tsokos (1983), p. 338).

Some times the scatter diagrams may be of the following types.

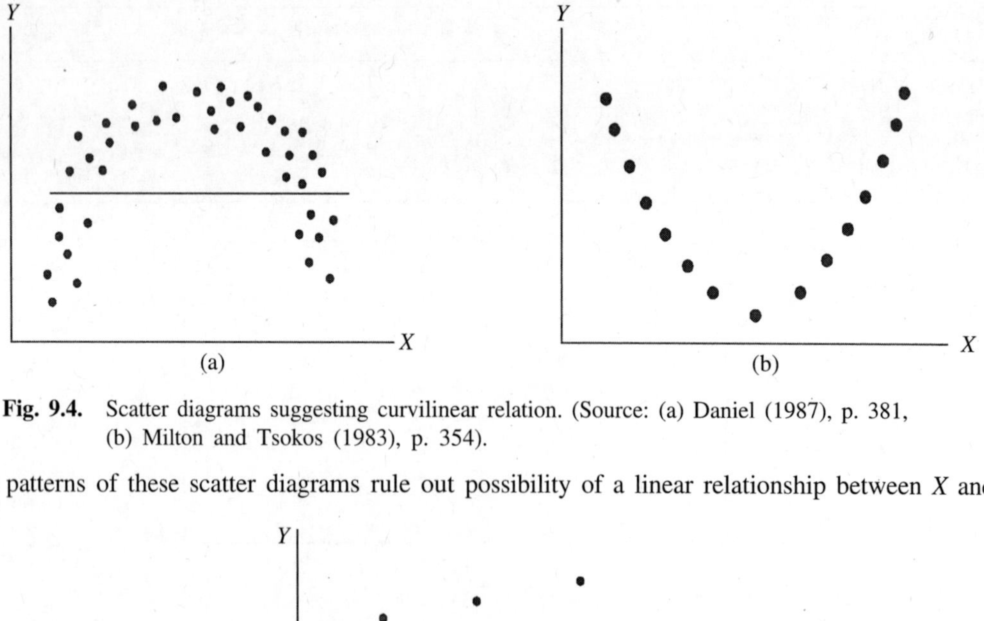

Fig. 9.4. Scatter diagrams suggesting curvilinear relation. (Source: (a) Daniel (1987), p. 381, (b) Milton and Tsokos (1983), p. 354).

The patterns of these scatter diagrams rule out possibility of a linear relationship between X and Y.

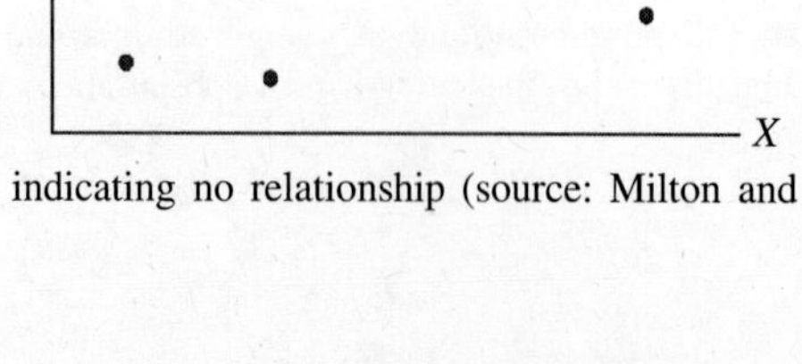

Fig. 9.5. Scatter diagram indicating no relationship (source: Milton and Tsokos (1983), p. 354).

It is seen in Fig. 9.5 that for all the values of *X*, *Y* is taking all its possible values. Similarly for any given values of *Y*, *X* can have values over the whole range of *X*. Thus it is evident that the two variables *X* and *Y* are taking values independently of the values taken by the other variable or they seem unrelated.

Scatter diagram is extremely useful graph for preliminary analysis of bivariate data. It reveals to the eye (i) the presence or absence of any relationship between the two variables (ii) broadly the nature of that relationship whether it is linear or curvilinear.
Three situations may arise after plotting a scatter diagram:

(a) One of the variables say *Y*, depends on the other variable *X* only, and follows a straight line relationship i.e. for a given *X* there is only one value of *Y* and for each unit change in *X*, there is an equal amount of change in *Y*. Then the scatter diagram reduces to a series of points lying on a straight line.
See Fig. 9.6. Real data rarely is of this perfect linear form.
(b) If *Y* is completely independent of *X* and varies due to some other factor, then at each value of *X* all values of *Y* will be possible to be attained. The scatter diagram will have points scattered over the whole of the range of *X* and *Y* as in Fig. 9.5.
(c) In the third situation neither of the above two extreme situations may be present. The values of *Y* do depend on *X* but there is some variability in *Y* (i.e. several values of *Y*) at each value of *X*. See Fig. 9.7.

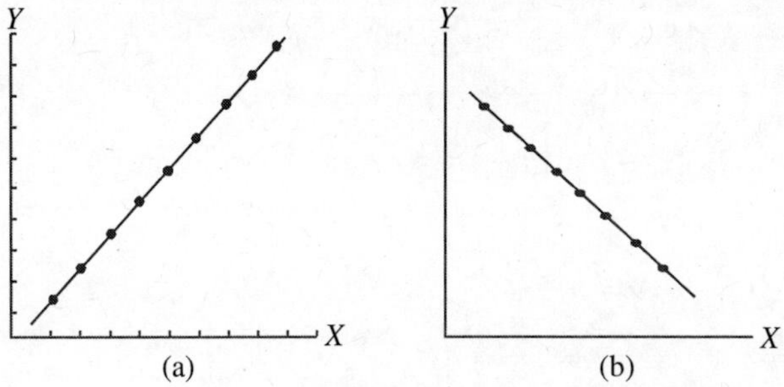

Fig. 9.6. Scatter diagram: All points in a straight line. (Source: (a) Daniel (1987), p. 388, (b) Daniel (1987), p. 403.)

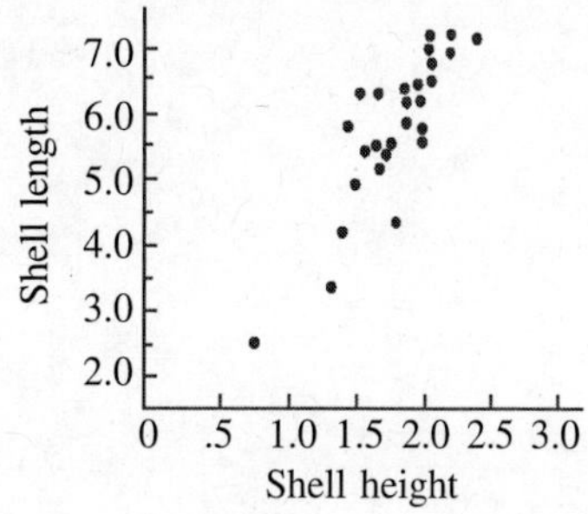

Fig. 9.7. Scatter diagram of shell height and shell length. (Source: Milton and Tsokos (1983), p. 337.)

In this intermediate case, two types of problems are faced (i) to measure the degree of linear relationship between the two variables and (ii) to find the mathematical relationship between the two variables, so that one of them may be predicted from the given values of the other.

Quantitative measurement of the degree of linear relationship between *X* and *Y* is called **problem of correlation** and is discussed in this section.

Finding mathematical relationship between the two variables so that one of them may be predicted from the given values of the other **is called problem of regression**, will be discussed in the next section.

Simple Correlation

As already discussed, through scatter diagram one gets an idea of the relationship between two variables *X* and *Y*. But a quantitative or numerical measure gives a precise, unique value of the degree of the linear relationship between the two variables. **Such a measure,** which is commonly used, is called **simple correlation** or product moment correlation coefficient. As it was suggested by Karl Pearson (1896), it is also called **Pearson's product moment correlation coefficient**. For two variables X and Y it is defined as

$$\rho_{XY} = \frac{Covariance\,(X, Y)}{\sqrt{V(X)\,V(Y)}} \tag{9.1}$$

$$= \frac{E\left[(X-\mu_x)(Y-\mu_y)\right]}{\sigma_x\,\sigma_y} \tag{9.2}$$

This value is the correlation coefficient of the X and Y in the population and the estimate of ρ_{XY} can be obtained as r_{XY} or r from the sampled n pairs of observations, given by

$$r = \frac{\frac{1}{n}\sum_{i=1}^{n}(X_i-\bar{X})(Y_i-\bar{Y})}{\left[\left\{\frac{1}{n}\sum_{i=1}^{n}(X_i-\bar{X})^2\right\}\left\{\frac{1}{n}\sum_{i=1}^{n}(Y_i-\bar{Y})^2\right\}\right]^{1/2}} \tag{9.3}$$

$$= \frac{\sum_{i=1}^{n}(X_i-\bar{X})(Y_i-\bar{Y})}{\left[\sum_{i=1}^{n}(X_i-\bar{X})^2\sum_{i=1}^{n}(Y_i-\bar{Y})^2\right]^{1/2}} = \frac{\sum_{i=1}^{n}(X_i-\bar{X})(Y-\bar{Y})}{n\,S_x\,S_y} \tag{9.4}$$

$$= \frac{\sum_{i=1}^{n}X_iY_i - n\bar{X}\bar{Y}}{\sqrt{\left\{\sum_{i=1}^{n}X_i^2 - n\bar{X}^2\right\}\left\{\sum_{i=1}^{n}Y_i^2 - n\bar{Y}^2\right\}}}$$

$$= \frac{n\sum_{i=1}^{n}X_iY_i - \sum(X_i)\sum(Y_i)}{\sqrt{\left\{n\sum_{i=1}^{n}X_i^2 - \left(\sum_{i=1}^{n}X_i\right)^2\right\}\left\{n\sum_{i=1}^{n}Y_i^2 - \left(\sum_{i=1}^{n}Y_i\right)^2\right\}}} \tag{9.5}$$

Equation (9.4) defines the sample correlation coefficient and (9.5) is a form of the same, which is more convenient for computation. It is evident from (9.5) that the determination of the correlation coefficient from sampled observations requires computation of five sums of (1) X_i (2) Y_i (3) X_i^2 (4) Y_i^2 and (5) X_iY_i.

Note: The definitions of simple correlation coefficient in population and in sample include three measures. Variances and covariances in population and sample are defined here for convenience of the reader as follows:

In population

$$\text{Variance of } X = V(X) = E(X - \mu_x)^2 = \sigma_x^2$$

$$\text{Variance of } Y = V(Y) = E(Y - \mu_y)^2 = \sigma_y^2$$

$$\text{Covariance (X, } Y) = E[(X - \mu_x)(Y - \mu_y)]$$

In sample

$$\text{Variance of } X = V(X) = \left\{\frac{1}{n}\sum_{i=1}^{n}(X_i - \bar{X})^2\right\} = S_x^2$$

$$\text{Variance of } Y = V(Y) = \left\{\frac{1}{n}\sum_{i=1}^{n}(Y_i - \bar{Y})^2\right\} = S_y^2$$

$$\text{Covariance } (X, Y) = \frac{1}{n}\sum_{i=1}^{n}(X_i - \bar{X})(Y_i - \bar{Y}) = S_{xy}$$

Properties of Correlation Coefficient

1. It can be seen that the **correlation coefficient is a pure number** and it **has no unit**. It has unit of neither variable X, nor that of variable Y.
2. The value of r must satisfy the inequality that $-1 \le r \le +1$. Value of $r = \pm 1$ if and only if, the points of the scatter diagram lie on a straight line.
3. If $r = 0$ the scatter diagram will be of the form as shown in Fig. 9.5. If $r = +1$ the scatter diagram will be of the form as given in Fig. 9.8 (a) and for $r = -1$, the scatter diagram will be of the form of Fig. 9.8 (b). For $\rho = \pm 1$ the same forms of the scatter diagrams are obtained.
4. (a) If $r = 0$, X and Y have no linear relationship but they may depend on each other according to some other functional relation (See Fig. 9.4).
   (b) If X and Y are independent, $\rho_{x,y} = 0$ and r is also close to 0.
   (c) If $r = +1$, X and Y have positive relation. Both X and Y increase or decrease together.
   (d) If $r = -1$, X and Y have negative relationship. If X increases Y decreases and vice versa.

Moreover, it may be noted that in correlation studies X and Y both are random variables. An example is given below to explain the computational procedure in detail. It may be seen than X and Y both are random variables. X is not a controlled variable, so with no specified values.

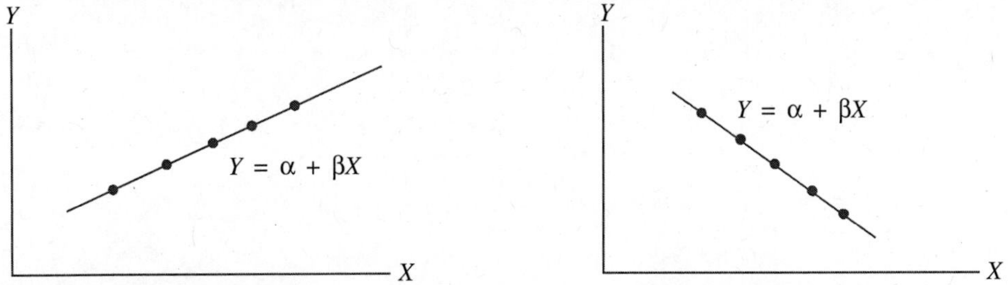

(a) Scatter diagram for $r = +1$ (b) Scatter diagram for $r = -1$

Fig. 9.8. Scatter diagrams for (a) perfect positive (+1) and (b) perfect negative (–1) correction (source of both: Milton and Tsokos, 1983, p. 354).

Example 9.3: Percentage overweight (X) and threshold of nociceptive flexion reflex (Y) is observed on a random sample of 10 individuals. It is desired to estimate ρ_{xy}. The data is given as (Source: Milton and Tsokos (1983), p. 355.)

Individual (i)	1	2	3	4	5	6	7	8	9	10
Percentage overweight (X)	89	90	75	30	51	75	62	45	90	20
Threshold of nociceptive flexion reflex (Y)	2	3	4	4.5	5.5	7	9	13	15	14

Solution: The required sums are obtained from the data and the values are

$$\sum_{i=1}^{10} X_i = 627 \qquad \sum_{i=1}^{10} Y_i = 77 \qquad \sum_{i=1}^{10} X_i^2 = 45,141 \qquad \sum_{i=1}^{10} Y_i^2 = 799.5$$

$$\sum_{i=1}^{10} X_i Y_i = 4461.5$$

Using equation (9.5) $\hat{\rho} = r$ is given by

$$r = \frac{\left\{n \sum_{i=1}^{10} X_i Y_i - \sum_{i=1}^{10} X_i \sum_{i=1}^{10} Y_i\right\}}{\sqrt{10\left\{\sum_{i=1}^{n1} X_i^2 - \left(\sum_{i=1}^{10} X_i\right)^2\right\}\left\{\sum_{i=1}^{10} Y_i^2 - \left(\sum_{i=1}^{10} Y_i\right)^2\right\}}}$$

$$= \frac{10 \times 4461.5 - (627)(77)}{\sqrt{10\{45141 - 627^2\}\{799.5 - 77^2\}}} = -.33$$

The scatter diagram of data is given below in Fig. 9.9.

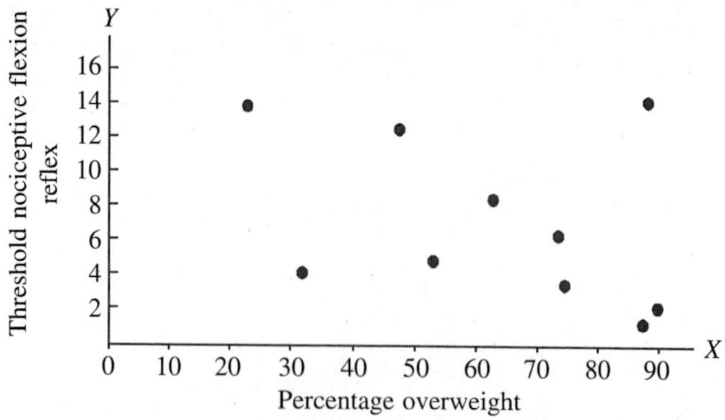

Fig. 9.9. Scatter diagram for data in Example 9.3 (source: Milton and Tsokos (1983), p. 356).

The scatter diagram also shows a linear trend. With increasing X, Y is decreasing. So observed data supports the finding of negative correlation, though not substantially high in magnitude. So to be sure, it will be **advisable to test H_0: $\rho = 0$ against H_1: $\rho < 0$.**

9.2 Simple Linear Regression

It has been already mentioned that if a random sample of n pairs of values of bivariate random variables X and Y is available, the **problem of regression** is **to determine a mathematical relationship between X and Y** in the population of the individuals from which sample was selected. Thus the aim is to determine the equation say

$$Y = f(X) \tag{9.6}$$

where $f(X)$ is some function of X (may be linear or non-linear).

Simple linear regression tries to determine the linear relationship between X and Y in the population i.e.

$$f(X) = A + BX$$

or

$$Y = A + BX \tag{9.7}$$

is chosen as the mathematical relationship.

The linear relation (9.7) becomes known if A and B are known.

Equation (9.7) is called the line of regression of Y on X and the problem of linear regression is to determine the values of A and B from the observed n pairs of sampled values.

The nature of the variable (X, Y) may be of the following types and analysis procedure may vary according to its nature.

(A) X and Y Both Random Variables

Some times values of any of the two variables can not be decided or controlled by the experimenter. Some examples of such situations are:

(a) Height and weight of elephants.
(b) Blood sugar level and blood cholesterol level of University teachers.
(c) Rainfall in June and minimum temperature in January in last ten years.

In all of these examples both the variables are changing randomly and the experimenter can not fix values of any one of the two variables.

(B) X can be Fixed by Experimenter, and only Y is Random

In some situations the experimenter can decide or control values of one of the variable (normally of X) and observes the values taken by the second variable Y, which can not be restricted by the experimenter. Examples of such variables are given below:

(a) Doses of certain insecticides are chosen by agriculture scientist for administration on 100 insects and percentage of deaths are observed.
(b) Children of 5 different ages are selected and their I.Q. determined.
(c) 10 selected amounts of a manure (X) is used on similar plots of particular variety of potato and yield from each plot (Y) is measured.

Whether X can be fixed by experimenter or it can not be fixed but Y is always random. There will always be variations in values of Y even for a given X. **Variation in Y may be due to measurement error**. As an example, though there is a linear formula to determine temperature from Celsius to Fahrenheit scale but through experimental measurements a variation is always found.

Similarly even if humans or animals of a given body weight are selected, height is found to vary. Thus one can conclude that there exists a population of Y for a given X.

Regression analysis, therefore, tries to find the relation between the average in the population of Y's for a given X viz. $E(Y/X)$ and X. One will never find a unique single value of Y for an X in real data. Though the problem of regression has been said to determine the relation $Y = f(X)$, but strictly speaking one determines a relation

$$E\,(Y/X) = f(X) \tag{9.8a}$$

The **problem of linear regression finds a linear relationship between $E(Y/X)$ and X** though the relation is **written as $Y = A + BX$** as in equation (9.7) but strictly speaking it is

$$E\,(Y/X) = A + BX \tag{9.8b}$$

Hence the line of regression predicts expected values of Y for a specified value of X. The values of X should be in a proper range, where the relation (9.8b) holds. Beyond that range of X the data (X, Y) may be far from the regression line (9.8b).

In regression analysis most often one of the variable is such that its values are influenced by the values of the other variable. The variable which is influenced is called *dependent variable* and is usually denoted by Y. It is also called a *regressed or predicted or response variable.*

The other variable which influences the dependent variable Y, is called *independent variable* and is usually denoted by X. X is also called a *regressor or predictor or explanatory variable.*

The term *regression* was first used by Galton (1822-1911) in connection with his research on heredity. It was observed that stature of offsprings reverts back to average stature of population, even though stature of its parents may be tall or short. Galton used the word *regression* for this phenomenon of reverting to population average.

The regression analysis is used to predict the average or expected value of the population of Y for a given value of X. The sampled pairs, as seen in scatter diagram, will not lie exactly on the regression equation but cluster around it. This unknown mathematical relation $Y = f(X)$ is determined from the observed data. The simplest and mathematically easy relation is linear relation. So in this section the method of determining the regression line (9.7) is considered below.

Equation (9.7) or (9.8b) is called the line of regression of Y on X. It specifies that the mean of distribution of Y for a given X varies linearly with X. It is further assumed that the distribution of Y given X is normal and variance of this distribution does not change with value of X. The variance of Y for a given X is denoted by $\sigma^2_{Y,X}$.

Determining the Regression Lines

Regression line $Y = A + BX$ becomes known or can be plotted if values of A and B are known. A and B are population parameters and are usually unknown. These values are determined from the n pairs of observed values (X_i, Y_i), $i = 1, 2, ..., n$.

The method applied for estimating A and B from the observed sample is **method of least square.** This method consists in choosing a straight line among infinite number of possible straight lines such that the sum of squares of deviations of the observed points (X_i, Y_i), from the chosen straight line is minimum.

This procedure of estimating the line of regression (by estimating A and B) is called **fitting straight line** to the data **by least squares.**

One chooses a line

$$Y = a + bX \tag{9.9}$$

such that the sum of squares of vertical distances of all the points (X_i, Y_i) from the line (9.9) is minimum.

The vertical distance of (X_i, Y_i) from (9.9), as also evident from Fig. (9.10), is

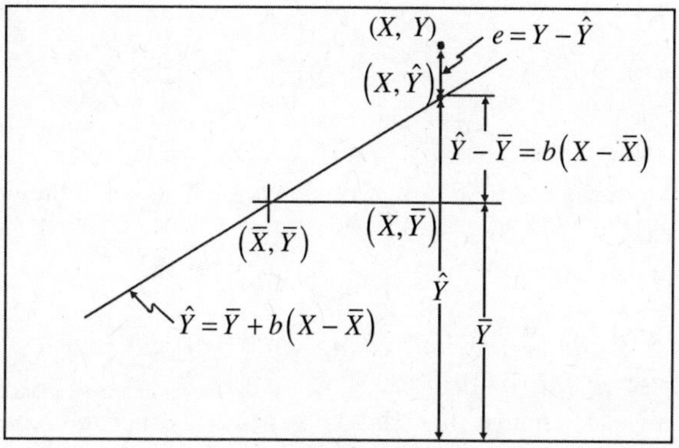

Fig. 9.10. Source of variation in Y in regression of Y on X (source: Steel and Torrie (1981), p. 249).

$$e_i = Y_i - \hat{Y} = Y_i - (a + bX_i) \tag{9.10}$$

The method of least squares chooses a and b such that

$$S^2 = \sum_{i=1}^{n} \left(Y_i - \hat{Y}_i\right)^2 = \sum_{i=1}^{n} \left[Y_i - (a + bX_i)\right]^2 \tag{9.11}$$

is minimum. Using result of calculus, this is achieved by solving the equations

$$\frac{\partial S^2}{\partial a} = 0 \quad \text{and} \quad \frac{\partial S^2}{\partial b} = 0 \tag{9.12}$$

(9.12) gives

$$b = \frac{\sum_{i=1}^{n} (X_i - \bar{X})(Y_i - \bar{Y})}{\sum_{i=1}^{n} (X_i - \bar{X})^2} = \frac{n\sum_{i=1}^{n} X_i Y_i - \sum_{i=1}^{n} X_i \sum_{i=1}^{n} Y_i}{n\sum_{i=1}^{n} \left(X_i^2\right) - \left(\sum_{i=1}^{n} X_i\right)^2} \tag{9.13}$$

$$a = \bar{Y} - b\bar{X} \tag{9.14}$$

a is the intercept of the line of the regression of Y on X, on Y axis i.e. the point where line of regression (9.9) intersects Y-axis.

b is the slope of the line of regression of Y on X with the X-axis.

b is called *the regression coefficient* of Y on X and is some times denoted by $\boldsymbol{b_{yx}}$.

If one tries to minimize the sum of squares of the horizontal distances of points from the straight line in Fig. 9.10, one gets a different straight line, say

$$X = c + dY \tag{9.15}$$

The constants c and d are obtained by exchanging X and Y in (9.13) and (9.14) to get

$$d = \frac{n\sum_{i=1}^{n} X_i Y_i - \sum_{i=1}^{n} X_i \sum_{i=1}^{n} Y_i}{n\sum_{i=1}^{n} Y_i^2 - \left(\sum_{i=1}^{n} Y_i\right)^2} \tag{9.16}$$

and
$$c = \bar{X} - d\bar{Y} \tag{9.17}$$

c is the intercept of the line of the regression of X on Y, on the X-axis i.e. the point where the regression line (9.15) intersects X-axis.
d is the slope of the line of the regression of X on Y, with the Y-axis. d is called *the regression coefficient of X on Y*, and is some times denoted by $\boldsymbol{b_{xy}}$.
$b = b_{yx}$ is the *change in Y variable, per unit change in X variable* and $d = b_{xy}$ is the *change in the X variable per unit change in the Y variable.*

Thus b_{yx} and b_{xy} are rates of change in Y and X, respectively, per unit change in the other variable and have units of Y and X, respectively.

Use of Regression Lines

It is seen that from a bivariate data, one can find out two lines of regression.
The regression line of Y on X of equation (9.7) is obtained by minimizing sum of squares of vertical deviations of data points from regression lines. In other words, the sum of squares of deviations of Y's from $\hat{Y}$'s (the estimated Y's from the regression line) are minimized. This **line of regression of Y on X** is, therefore, **used to estimate or predict the average or expected value of Y for a known value of X.**

The line of the regression of X on Y is obtained by minimizing sum of the squares of deviations of X's from the estimated values of X's from the regression line (9.15). Hence, **the line of regression of X on Y** is **used to predict or estimate the average or expected value of X for a known value of Y.** However, it must be taken care that the line of regression of X on Y should be obtained and used only in appropriate situations. The experimenter should ensure before using this regression line, that X may also logically have dependence on Y. Thus the body weight X and height of animals Y can be logically thought of as having both ways of dependence: body weight may depend on height, whereas height may also depend on body weight. In such situations determining and using both the lines of regression is proper.

Consider data on maximum temperature (X) in summer and death due to heat stroke in a state (Y) during last 10 years. It is very logical to treat heat stroke deaths as dependent variable and maximum summer temperature (X) causing heat stroke deaths as independent variable. So using line of regression of Y on X is proper. But heat stroke deaths, can not affect summer temperature. So it is not proper to determine or use line of regression of X on Y. For such type of data only one line of regression should be determined and used.
Now some examples are considered to obtain the lines of regression.

Example 9.3 (Revisited): The data seems to be of the form that only line of regression of Y on X should be determined. So a and b are determined as

$$b = \frac{n \sum X_i Y_i - \sum_{i=1}^{n} X_i \sum_{i=1}^{n} Y_i}{n \sum X_i^2 - \left(\sum_{i=1}^{n} X_i \right)^2} \quad \text{and } a = \bar{Y} - b\bar{X}$$

From the sums given in Example 9.3

$$b = \frac{10 \times 4461.5 - (627)(77)}{10 \times 4514 - (627)^2} = -.063$$

$$a = 7.7 - (-.063)\ 62.7 = 11.650$$

Hence line of regression is $Y = 11.650 - .063\ X$.
Negative value of b indicates that with increase in value of X, Y will decrease, though the rate of change is quite small.

Example 9.1 (Revisited): Data of this example has two variables X (scores on new test) and Y (scores on standardized test) for which dependency of X on Y is also quite possible as that of Y on X. An experimenter may be interested to predict scores on the other test, if he knows scores on either one of the two tests. Hence, in this example both the lines of regression will be useful.
Hence computation of a, b, c, d will be required and all the following sums will be needed.
One can see from the data of Example 9.1 that

$$\sum_{i=1}^{11} X_i = 825 \quad \sum_{i=1}^{11} Y_i = 916 \quad \sum_{i=1}^{11} X_i^2 = 64625 \quad \sum_{i=1}^{11} Y_i^2 = 80076 \quad \sum_{i=1}^{11} X_i Y_i = 71790$$

From the formula (9.13), (9.14), (9.16) and (9.17), it may be noted that numerator of b and d are same

$$b = \frac{n \sum_{i=1}^{n} X_i Y_i - \sum_{i=1}^{n} X_i \sum_{i=1}^{n} Y_i}{n \sum_{i=1}^{n} X_i^2 - \left(\sum_{i=1}^{n} X_i \right)^2} = \frac{11 \times 71790 - 825 \times 916}{11 \times 64625 - 825^2}$$

$$= \frac{33990}{30250} = 1.1236$$

$$a = \bar{Y} - b\bar{X} = \frac{916 - 1.1236 \times 825}{11} = -.9973$$

$$d = \frac{n \sum_{i=1}^{n} X_i Y_i - \sum_{i=1}^{n} X_i \sum_{i=1}^{n} Y_i}{n \sum_{i=1}^{n} Y_i^2 - \left(\sum_{i=1}^{n} Y_i \right)^2} = \frac{33990}{11 \times 80076 - 916^2}$$

$$= \frac{33990}{41780} = .8135$$

$$c = \bar{X} - d\bar{Y} = \frac{825 - .8135 \times 916}{11} = 7.2576$$

Thus the line of regression of Y on X is

$$Y = -.9973 + 1.1236\ X \text{ and}$$

The line of regression of X on Y is

$$X = 7.2576 + .8135\ Y$$

Plotting Regression Lines

After drawing scatter diagram and determination of regression line(s), these can be plotted on the same by using any one of the following two methods.

Any straight line can be drawn by joining two points which lie on it and extending the line beyond these two points up to the feasible and proper range of X and Y. The two points may be found by

(a) *To draw line of regression of Y on X*, choose two values of X, say x_1 and x_2, x_1 should be near lowest value of observed X and x_2 near highest value of X.

Find
$$y_1 = a + bx_1 \text{ and } y_2 = a + bx_2 \tag{9.18}$$

Plot the two points (x_1, y_1) and (x_2, y_2) on the scatter diagram and join these two points to get the line of regression of Y on X i.e. $Y = a + bX$.

To draw line of regression of X on Y, choose two values of Y say y_1' and y_2'. y_1' should be near the lowest value of observed Y and y_2' should be near the highest observed value of Y.

Now find
$$x_1' = c + d\,y_1' \quad \text{and} \quad x_2' = c + d\,y_2' \tag{9.19}$$

and plot (x_1', y_1') and (x_2', y_2') on the graph and join these two points as mentioned above, to get the line of regression of X on Y.

Either the equations of the lines may be written on the respective lines or it should be indicated which line is regression line of Y on X and which one is line of regression of X on Y.

(b) It can be seen that $(\bar{X}, \bar{Y})$ is a point which **lies on both the regression lines**.

Substituting the value of $a = \bar{Y} - b\bar{X}$ from (9.14) in the equation of line of regression of Y on X given in (9.9), one gets

$$Y = \bar{Y} - b\bar{X} + bX$$

or
$$(Y - \bar{Y}) = b(X - \bar{X}) \tag{9.20}$$

Similarly substituting the value of $c = \bar{X} - d\bar{Y}$ from (9.17) in the equation of the line of regression of X on Y.

$$X = \bar{X} - d\bar{Y} + dY$$

$$(X - \bar{X}) = d(Y - \bar{Y}) \tag{9.21}$$

Now it is clear that $(\bar{X}, \bar{Y})$ is a point which lies on both the lines in (9.20), (9.21) since putting $X = \bar{X}$ and $Y = \bar{Y}$ both the equations, (9.20) and (9.21) are satisfied (both L.H.S. and R.H.S. become zero).

So instead of finding out two points on each of the two regression lines, one can find only one point (x_1, y_1) on line regression of Y on X and one (x_1', y_1'), on the line of regression of X on Y

from equations (9.18), and (9.19). It should be kept in mind that x_1 is not very close to $\overline{X}$ and y_1' is not very close to $\overline{Y}$.

Thus one joins $\left(\overline{X}, \overline{Y}\right)$ and (x_1, y_1) to get line of regression of Y on X and joins $\left(\overline{X}, \overline{Y}\right)$ and (x_1', y_1') to get line of regression of X on Y.

The two lines are extended beyond the said two points on each line, keeping in mind the point mentioned above.

Some More Expressions and Relationships of Regression and Correlation Coefficient

The correlation coefficient and the regression coefficients are defined using the five sums on the data values in the equations (9.5), (9.13) and (9.16). In this part, alternative expressions for these are given and some expressions are given which relate regression coefficients b and d with correlation coefficient r. For doing so let the following sample estimates be refreshed and new one for covariance be introduced.

$$S_x^2 = \frac{1}{n}\sum_{i=1}^{n}\left(X_i - \overline{X}\right)^2 \quad \text{is a sample estimate of } V(X) \tag{9.22}$$

$$S_y^2 = \frac{1}{n}\sum_{i=1}^{n}\left(Y_i - \overline{Y}\right)^2 \quad \text{is a sample estimate of } V(Y) \tag{9.23}$$

$$S_{xy} = \frac{1}{n}\sum_{i=1}^{n}\left(X_i - \overline{X}\right)\left(Y_i - \overline{Y}\right) \quad \text{is a sample estimate of Cov}(X, Y) \tag{9.24}$$

From (9.4)

$$r = \frac{S_{xy}}{S_x S_y} \tag{9.25}$$

From first expression in (9.13) for b, by dividing both numerator and denominator by n^2.

$$b = \frac{S_{xy}}{S_x^2} \tag{9.26}$$

Similarly from (9.16) for d, dividing both numerator and denominator by n^2

$$d = \frac{S_{xy}}{S_y^2} \tag{9.27}$$

From these three equations

$$b \times d = r^2 \tag{9.28}$$

or

$$r = \sqrt{b \times d} \tag{9.29}$$

It is to be noted from (9.25), (9.26) and (9.27) that the signs of r, b and d are same as that of S_{xy}, since the denominators in these three equations are always positive.

If any two of r, b and d are given the third can always be found using the relations (9.28) and (9.29). *Signs of r, b and d will be either all positive or all negative depending on the sign of Covariance* S_{xy}.

Thus as a **continuation of Example 9.1 revisited** one can determine the correlation between the new test scores and standard test scores by the formula.

$$r = \sqrt{b \times d} = \sqrt{1.1236 \times .8135} = .9560$$

Since b and d are positive, sign of r is also positive. Thus r = +.9560, even though mathematically square root of a positive number can have both + and – sign.

9.3 Tests of Correlation

In section 9.1 the definition and computation of simple correlation coefficient has been considered. Once from a sample value of the correlation coefficient r has been obtained, it can not be ensured that the correlation coefficient ρ in the populations is close to r. Moreover, even if r is non zero there may be no correlation between X and Y in population.

Thus after obtaining correlation coefficient from a sample of n pairs (X_i, Y_i), $i = 1, 2, ..., n$, one would like to ensure (a) whether in population $\rho = 0$ or (b) whether in population $\rho = \rho_0$, where ρ_0 is some specified value in the range $-1 \leq \rho \leq 1$.

For these decisions a statistical test will be required.

In this section, therefore, tests of H_0: $\rho = 0$ and of H_0: $\rho = \rho_0$ are considered.

The tests are based on the **assumption** that the bivariate random variable (X, Y) has bivariate normal distribution and the random sample of n pairs is drawn from it.

The distribution of r is approximately normal if population correlation coefficient ρ is zero. When $\rho \neq 0$ the distribution of r is non normal and **a transformation of *r*, called *Z* transformation, is used for testing H_0, given by**

$$Z = \frac{1}{2} \ln \left(\frac{1+r}{1-r} \right) \tag{9.30}$$

Z is approximately normally distributed with

$$E(Z) = \frac{1}{2} \ln \frac{1+\rho}{1-\rho} \text{ and } V(Z) = \frac{1}{n-3} \tag{9.31}$$

(ln = natural log or $\log_e$). In symbols

$$Z \sim N\left(\frac{1}{2} \ln \frac{1+\rho}{1-\rho}, \frac{1}{n-3} \right)$$

A Test of H_0: $\rho = 0$ against H_1: $\rho \neq 0$

Under the assumptions given above it is **simple and correct** to **use *t*-test** statistic given by

$$t = \frac{r\sqrt{n-2}}{1-r} \sim t_{n-2} \quad \text{if } H_0 \text{ is true} \tag{9.32}$$

The **test procedure** for α level of significance

$$\text{Reject } H_0 \text{ if } |t| > t_{n-2,\, \alpha/2} \tag{9.33}$$

where $t_{n-2,\, \alpha/2}$ is upper tail 100 α/2% significance point of t-distribution with n – 2 degrees of freedom.

From the Table A.4 of Student's t-distribution the value is read and calculated value of t of equation (9.32) is compared with it for *statistical decision*.

$$H_0 \text{ is rejected against } H_1 \text{ if either } t_{cal} > t_{n-2,\,\alpha/2} \text{ or } t_{cal} < -t_{n-2,\,\alpha/2} \tag{9.34}$$

Example 9.3 (revisited): As was mentioned above, it is advisable to test H_0: $\rho = 0$ against H_1: $\rho < 0$.
The sample of $n = 10$ pairs of observations gave $r = -.33$.
The test statistic for this test is given by

$$t = \frac{r\sqrt{n-2}}{\sqrt{1-r^2}} \sim t_{n-2} \text{ if } H_0 \text{ is true}$$

The test is lower tail test so for level of significance α, the test procedure is to reject H_0, if $t < -t_{n-2,\,\alpha}$.
If H_0 is rejected it will support significant negative linear relationship.
For $\alpha = .05$ $t_{n-2,\,\alpha} = t_{8,\,.05} = 1.860$, from Table A.4

$$t_{cal} = \frac{-.33\sqrt{8}}{\sqrt{1-.33^2}} = -.9888$$

Since $t_{cal} = -.9888 > -t_{8,\,.05} = -1.860$
H_0 is not rejected
So the test suggests that X and Y may not have any correlation or linear relationship in the population.
$\hat{\rho} = r = -.33$ has already been obtained in Example 9.3

B Test of H_0: $\rho = \rho_0$ against H_1: $\rho \neq \rho_0$

If ρ_0 is not zero one has to look for Z transformation i.e. compute

$$Z = \frac{1}{2}\ln\left(\frac{1+r}{1-r}\right) \text{ and } V(Z) = \frac{1}{n-3}$$

$$\mu_z = \frac{1}{2}\ln\left(\frac{1+\rho_0}{1-\rho_0}\right)$$

Now to test H_0 **use the statistic**

$$Z_t = (Z - \mu_z)\sqrt{n-3} \sim N(0,1) \text{ if } H_0 \text{ is true} \tag{9.35}$$

The test procedure is

$$\text{Reject } H_0 \text{ if } |Z_t| > 1.96 \tag{9.36}$$

In other words, reject H_0 if either $Z_t > 1.96$ or if $Z_t < -1.96$
Table A.19 may be consulted to get Z for a given r.

Example 9.4: A random sample of 28 observations gave correlation coefficient $r = .95$. Can the correlation coefficient in populations be $\rho = .98$? Test the hypothesis at 5% level of significance.
Solution: It is desired to test H_0: $\rho = .98$ against H_1: $\rho \neq .98$
From the Table A.19 one can see that for $r = .95$, $Z = 1.83178$ and for $\rho = .98$ $\mu_z = 2.29756$
Hence the test statistic to be used is

$$Z_t = (Z - \mu_z)\sqrt{25} = 5(1.83178 - 2.29756)$$
$$= 5(-.46578) = -2.32899$$

Since $Z_t < -1.96$, H_0 is rejected

Hence it can be concluded that as indicated by the sampled observations, correlation coefficient in the population may not be .98.

C Hotelling's Test

The above Z transformation suggested by Fisher (1921) is not advisable to use for $n < 25$. For such situations Hotelling (1953) suggested the following transformation. The sample correlation coefficient, r is transformed to

$$Z^* = Z - \frac{3Z + r}{4n} \tag{9.37}$$

$$E\left(Z^*\right) = \xi \text{ where } \xi = \mu_z - \frac{3\mu_z + \rho}{4n} \tag{9.38}$$

and $V\left(Z^*\right) = \frac{1}{n-1}$

Hence *to test* H_0*:* $\rho = \rho_0$ *against* H_1*:* $\rho \neq \rho_0$

Use the statistic

$$Z_H = \left(Z^* - \xi\right)\sqrt{n-1} \sim N(0,1) \text{ if } H_0 \text{ is true}$$

The *test procedure* is same as in the previous case.

$$\text{Reject } H_0 \text{ if } |Z_H| > 1.96$$

In other words, reject H_0 if either $Z_H > 1.96$ or if $Z_H < -1.96$.

Example 9.5: A random sample of 10 observations on (X, Y) gave correlation coefficient $r = .84$. Can it be taken as indicator of population correlation being greater than .80?

Solution: It is desired to test H_0: $\rho = .80$ against H_1: $\rho > .80$

$n = 10$ and (X, Y) is *assumed* to follow bivariate normal distribution.

Since $n < 25$ Hotelling's test should be used.

For $r = .84$, $Z = 1.22117$ and for $\rho = .80$, $\mu_z = 1.09861$

So

$$Z^* = Z - \frac{3Z + r}{4n} = 1.22117 - \frac{3(1.22117) + .84}{4 \times 10}$$

$$= \frac{44.3109}{40} = 1.1078 \text{ and}$$

$$\xi = \mu_z - \frac{3\,\mu_z + \rho}{4n} = 1.09861 - \frac{3(1.09861) + .80}{4 \times 10} = \frac{39.8486}{40} = .9962$$

To test H_0 *the test statistic used* is

$$Z_H = \left(Z^* - \xi\right)\sqrt{n-1} = \sqrt{9}\,(1.1078 - .9962)$$

$$= 3 \times .1116 = 0.3348$$

Choosing $\alpha = 5\%$, the *test procedure* is: Reject H_0 against H_1 if $Z_H > 1.645$

Since calculated Z_H does not lie in the critical region, H_0 is not rejected.

Hence the sample correlation coefficient does not indicate that population correlation coefficient may be larger than .80.

9.4 Tests of Regression Coefficient

Once a regression equation has been obtained, one may need to know whether the regression line adequately describes the relationship between X and Y. The obtained regression line is based on sample observations and one would like to know whether in population the relationship can be adequately described by the estimated regression line.

Thus one is interested to test whether (i) the regression coefficient in population, $\beta = 0$ (ii) regression coefficient in population has some specified non zero value β_0 i.e. $\beta = \beta_0$.

If $\beta = 0$, there is no linear relationship of Y with X and the regression line obtained from the data will have little or no value for prediction.

The form or nature of linear relationship depends on the value of β. Thus if a linear relationship does exist, the nature or form of the relationship may be ascertained by testing if data support that β has a specified value β_0 in the population. Its value indicates if relationship is reverse and its strength. Hence an experimenter may require testing any of the above two hypotheses. Testing of H_0: $\beta = 0$ against H_1: $\beta \neq 0$ is discussed first, and that of H_0: $\beta = \beta_0$ against H_1: $\beta \neq \beta_0$, subsequently.

The tests of regression coefficient are valid if the following assumptions are satisfied by the data.

1. The bivariate random variable (X, Y) follows a bivariate normal distribution so that Y for a given X has normal distribution with mean $E(Y/X)$ and variance $\sigma^2_{Y,X}$.
2. The variance $\sigma^2_{Y,X}$ is same for different values of X.
3. The sampled observations are selected independently i.e. values of (X_i, Y_i) is not affected by value of another observations (X_j, Y_j) $i \neq j$.

A. Test of H_0: $\beta = 0$ against H_1: $\beta \neq 0$ using F-statistic

One method of testing H_0: $\beta = 0$ against H_1: $\beta \neq 0$ depends of F-statistic obtained in ANOVA. The total sum of square is partitioned into two parts- one due to regression and another due to error or residual. It can be seen from the following Fig. 9.10 as well as from the following discussion.

A sample regression line gives estimated or predicted Y for a given X. As evident from Fig. 9.10, deviation of an observed Y from its mean $\bar{Y}$ consists of two parts:

$$Y - \bar{Y} = Y - \hat{Y} + \left(\hat{Y} - \bar{Y}\right) \tag{9.39}$$

or

$$Y - \bar{Y} = \left(\hat{Y} - \bar{Y}\right) + \left(Y - \hat{Y}\right)$$

or

Total deviation = Deviation explained by regression + Unexplained or residual deviation (E) (9.40)

Now it can be seen that minimized sum of square of deviations = Residual sum of square = $SSE = \sum_{i=1}^{n} \left(Y_i - \hat{Y}\right)^2$.

$$= \sum_{i=1}^{n} \left[Y_i - (a + bX_i)\right]^2 = \sum_{i=1}^{n} \left[Y_i - \left(\bar{Y} - b\bar{X} + bX_i\right)\right]^2$$

$$= \sum_{i=1}^{n} \left[\left(Y_i - \bar{Y}\right) - b\left(X_i - \bar{X}\right)\right]^2$$

$$= S_Y^2 - bS_{xy} \tag{9.41}$$

or
$$S_Y^2 = bS_{xy} + SSE$$

or Total sum of square = Variability in Y due to regression on X + Residual or error sum of square

or
$$TSS = SSR + SSE$$

where SSR is sum of square due to regression. Thus one can write these in an ANOVA table and perform a test of H_0: $\beta = 0$ against H_1: $\beta \neq 0$ by the variance ratio statistic

$$F = \frac{SSR/d.f.\ of\ SSR}{SSE/d.f.\ of\ SSE} = \frac{MSR}{MSE} \tag{9.42}$$

where,
$$MSR = SSR/1 \text{ and } MSE = SSE/(n-2) \tag{9.43}$$

If $\beta = 0$, then all the variability in Y will be due to error component and in that case the F-statistic will be close to 1 or insignificant. Otherwise SSR will be large and so F will be large, so that H_0: $\beta = 0$ is rejected.

Table 9.1 ANOVA Table for testing goodness of linear regression

Source of variation	d.f.	Sum of squares: SS	Mean square: MS	F-calculated	F (from Table)
Regression	1	$bS_{xy} = SSR$	$SSR/1 = MSR$	$F_{1,\, n-2} = MSR/MSE$	$F_{1,\, n-2,\, .05}$
Error or residual	$n - 2$	$S_{yy} - bS_{sy} = SSE$	$SSE/(n-2) = MSE$		$F_{1,\, n-2,\, .01}$
Total	$n - 1$	$TSS = S_{yy}$			

Thus rejection of H_0 or significant value of F indicates that data has linear relationship. Hence linear regression in population is supported statistically from the observed sample.
The F-statistic follows $F_{1,\, n-2}$ *distribution,* if H_0 is true. In symbols

$$F \sim F_{1,\, n-2} \text{ if } H_0 \text{ is true} \tag{9.44}$$

The *test procedure* is (for level of significance α)
Reject H_0 against H_1 if $F > F_{1,\, n-2,\, \alpha}$

B. Testing H_0: $\beta = 0$ by t-statistic

Under the assumptions mentioned above, the regression coefficient b is unbiased estimator of population regression coefficient β. The sampling distribution of b is normal with mean and variance given by

$$E(b) = \beta \text{ and } V(b) = \frac{\sigma_{Y/X}^2}{\sum_{i=1}^{n}(X_i - \bar{X})^2}$$

In symbols
$$b \sim N\left(\beta, \frac{\sigma^2_{Y/X}}{\sum_{i=1}^{n}(X_i - \bar{X})^2}\right) \tag{9.45}$$

Thus H_0: $\beta = 0$ against H_1: $\beta \neq 0$ can be tested using (9.45). Normally $\sigma^2_{Y/X}$ is not known. So its estimator $s^2_{y/x}$ from the sample is used, which is given by

$$s^2_{y/x} = \sum_{i=1}^{n}(Y_i - \hat{Y})^2 \Big/ (n-2) = \frac{SSE}{n-2} = \frac{R_0^2}{n-2} \tag{9.46}$$

$$= MSE \text{ from ANOVA Table 9.1}$$

Thus
$$\hat{V}(b) = s_b^2 = \frac{s^2_{y/x}}{\sum_{i=1}^{n}(X_i - \bar{X})^2} \tag{9.47}$$

Thus the *test statistic* to test H_0 is

$$t = \frac{b - o}{s_b} \sim t_{n-2} \text{ if } H_0 \text{ is true} \tag{9.48}$$

So the *test procedure* to test H_0 against H_1 for a choice of significance level α is

Reject H_0 if $|t| > t_{n-2, \alpha/2}$

In other words – Reject H_0 against H_1 if either $t > t_{n-2, \alpha/2}$ or if $t < -t_{n-2, \alpha/2}$ (9.49)

An alternative expression for $s^2_{y/x}$ can also be written as

$$s^2_{y/x} = \left(S_y^2 - bS_{xy}\right) / (n-2) \text{ from (9.41) and (9.46)}$$

$$= \left(S_y^2 - \frac{S_{xy}^2}{S_x^2}\right) \Big/ (n-2) \tag{9.50}$$

C. Testing of H_0: $\beta = \beta_0$ against H_1: $\beta \neq \beta_0$

Using the result of *B* above, to test H_0: $\beta = \beta_0$ against H_1: $\beta \neq \beta_0$ one can easily use the statistic

$$t = \frac{b - \beta_0}{s_b} \sim t_{n-2} \text{ if } H_0 \text{ is true}$$

and *follow the same test procedure* with significance level α and

Reject H_0 if $|t| > t_{n-2, \alpha/2}$

In other words – Reject H_0 against H_1 if either $t > t_{n-2, \alpha/2}$ or if $t < -t_{n-2, \alpha/2}$ (9.51)

As an illustration of the two test procedures, let us consider again an example.

Example 9.6: Height (*Y*) and length (*X*) of 28 shells (each in mm) in sheltered sea shores gave the following results:

$$n = 28,\ \sum_i X_i = 56.6,\ \sum_i Y_i = 151.1,\ \sum_i X_iY_i = 311.96,\ \sum_i X_i^2 = 117.68\ \sum_i Y_i^2 = 832.85$$

It is desired to test if the assumption of linear regression is valid. (Source: Milton and Tsokos (1983), pp. 336, 365.)
Solution: Both the tests i.e. using F-statistic and t-statistic will be used. To test validity of linear regression statistically, a test of H_0: $\beta = 0$ against H_1: $\beta \neq 0$ will be done. For this first find

$$S_x^2 = \frac{\left[n \sum_i X_i^2 - \left(\sum_i X_i\right)^2\right]}{n} = \frac{28(117.68) - 56.6^2}{28} = 3.27$$

$$S_y^2 = \frac{\left[n \sum_i Y_i^2 - \left(\sum_i Y_i\right)^2\right]}{n} = \frac{28(832.85) - 151.1^2}{28} = 17.45$$

$$S_{xy} = \frac{\left[n \sum_i X_iY_i - \sum_i X_i \sum_i Y_i\right]}{n} = \frac{28(311.96) - (56.6)(151.1)}{28} = 6.52$$

Test using *F*-statistic

$$TSS = S_y^2 = 17.45 \quad SSR = bS_{xy} = \frac{S_{xy}^2}{S_x^2} = \frac{6.52^2}{3.27} = 13.00$$

$$SSE = TSS - SSR = 17.45 - 13.00 = 4.45$$

Hence the ANOVA Table is

Table 9.2 ANOVA Table for testing validity of regression between shell height and length

Source of variation	d.f.	*SS*	*MS*	F_{cal}	F_{tab}
Regression	1	13.00	13.00	F = 13.00/.17 = 76.47**	$F_{1,\,26,\,.05} = 4.23$
Residual or error	26	4.45	4.45/26 = .17		$F_{1,\,26,\,.01} = 7.72$
Total	27	17.45			

Since $F_{cal} > F_{1,\,26,\,.01}$, the assumption of linear regression is strongly validated statistically.

Test using *t*-statistic
To test H_0: $\beta = 0$ against H_1: $\beta \neq 0$ use the t-statistic

$$t = \frac{b}{s_b} \sim t_{n-2} \text{ if } H_0 \text{ is true}$$

$$b = \frac{S_{xy}}{S_x^2} = \frac{6.52}{3.27} = 1.99$$

$$s_b^2 = \frac{MSE}{nS_x^2} = \frac{.17}{28 \times 3.27} = .0019$$

$$s_b = .044$$

$$t = \frac{1.99}{.044} = 45.23$$

From Table A.4, $t_{26,.005} = 2.779$

Since calculated t is much larger than 2.779, this test also strongly supports validity of linear regression.

It is quite evident that if required, using t-statistic one can easily test H_0: $\beta = \beta_0$ against H_1: $\beta \neq \beta_0$ (two tailed) or lower tailed or upper tailed test just by replacing the numerator in the t-statistic of equation (9.46) by b-β_0 and use rejection region which is two tailed or lower tailed or upper tailed.

In the next section testing equality or homogeneity of two or more correlation coefficients and that of two or more regression coefficients is considered.

9.5 Testing Homogeneity of Correlation and Regression Coefficients

Some times two or more independent random samples of a bivariate variable are available. It is often a natural question that arises, whether the correlation coefficients in the said populations are equal. If statistically test does not confirm differences between group correlation coefficients, the common correlation coefficient can be determined after pooling observations in the two or more samples.

One can also do similar testing for equality of various groups' regression coefficients, specifically whether all the groups' regression lines have the same slope. If test of equality is not rejected, the two or more groups may be pooled together and the common regression line may be obtained. With this idea, first tests of equality or homogeneity of two or more correlations are considered, and subsequently equality or homogeneity of regression coefficients are considered.

A: Testing Equality of two Correlation Coefficients

Suppose that two independent and sufficiently large random samples of sizes n_1 and n_2 on (X, Y) are available. Let r_1 and r_2 be the simple correlation coefficients computed from the two samples of sizes n_1 and n_2 and let ρ_1 and ρ_2 be correlation coefficients in the two populations (or groups) from which samples are drawn. Then it is desired to test H_0: $\rho_1 = \rho_2$ against H_1: $\rho_1 \neq \rho_2$.

The Z transformations of r_1 and r_2 are done to get Z_1' and Z_2'. In such a case

$$Z_i' \sim N\left(\mu_{zi}, \frac{1}{n_i - 3}\right) \tag{9.51}$$

where Z_i' and μ_{zi} are given in (9.30) and (9.31) as

$$Z_i' = \frac{1}{2}\ln\left(\frac{1+r_i}{1-r_i}\right) \text{ and } \mu_{zi} = \frac{1}{2}\ln\left(\frac{1+\rho_i}{1-\rho_i}\right); \quad i = 1, 2 \tag{9.52}$$

Hence to test H_0 against H_1, large sample normal test is used. Thus the test statistic is

$$Z = \frac{(Z_1' - Z_2')}{\sqrt{\frac{1}{n_1 - 3} + \frac{1}{n_2 - 3}}} \sim N(0, 1) \text{ if } H_0 \text{ is true}$$

The test procedure for $\alpha = .05$, is Reject H_0 if $|Z| > 1.96$
The following example illustrates the testing method.

Example 9.7: Random samples of sizes 22 and 21 were collected on strains 109 and 130 of rubber wood. Correlation coefficient of rubber content of branch and diameter of wood in guayule are obtained as 0.310 and 0.542. Test if the correlation coefficient in strains 109 and 130 do not differ. (Source: Steel and Torrie (1981), p. 280 (data only).)
Solution: Let ρ_1 and ρ_2 be the correlation coefficients in strains 109 and 130 between the rubber content of branch and diameter of wood in guayule. r_1 and r_2 are sample correlation coefficients in the two stations. The computational part is shown in the following table.

Table 9.3 Test of H_0: $\rho_1 = \rho_2$

Strain of Guayule	Sample size	Y	Z'	$1/(n_i - 3)$
109	22	.310	.32055	.0526
130	21	.542	.60698	.0556

The test statistic for testing H_0 is:

$$Z = \frac{(Z_1' - Z_2')}{\sqrt{\frac{1}{n_1 - 3} + \frac{1}{n_2 - 3}}} = \frac{.32055 - .60698}{\sqrt{.0526 + .0556}}$$

$$= \frac{-.28643}{\sqrt{.1082}} = \frac{-.28643}{.329} = -0.87$$

Choosing $\alpha = .05$, the test procedure is to reject H_0 if $|Z| > 1.96$. In other words

Reject H_0 if either $Z > 1.96$ or if $Z < -1.96$

In this case $|Z| = 0.87 < 1.96$. So Z_{cal} is not in critical region. Hence, H_0 is not rejected at 5% level. Therefore, the population correlation coefficients of the strains 109 and 130 may be considered to be equal.

B: Testing Equality of Several (*k*) Correlation Coefficients

Suppose there are k groups in population on which bivariate data has been observed giving sample correlation coefficients $r_1, r_2, ..., r_k$. Let ρ_i, $i = 1, 2, ..., k$ be the population correlation coefficients. It is desired to test

H_0: $\rho_1 = \rho_2 = ...\ \rho_k$ against H_1: at least two ρ's differ.
For testing H_0 *the test statistic* is

$$\chi^2 = \sum_{i=1}^{k} (n_i - 3)(Z_i' - \bar{Z}_\omega)^2 \tag{9.53}$$

where $\bar{Z}_\omega$ is weighted mean of Z_i' given by

$$\bar{Z}_\omega = \frac{\sum_{i=1}^{k} (n_i - 3) Z'_i}{\sum_{i=1}^{k} (n_i - 3)} \tag{9.54}$$

If H_0 is true, *the test statistic has χ^2-distribution with k – 1 degree of freedom.* In Symbols

$$\chi^2 \sim \chi^2_{k-1} \text{ if } H_0 \text{ is true} \tag{9.55}$$

Test procedure is:

$$\text{Reject } H_0 \text{ if } \chi^2 > \chi^2_{k-1,\,\alpha} \tag{9.56}$$

If hypothesis of equality or homogeneity of correlation coefficients is not rejected, the common correlation coefficient in the combined groups may be obtained.

This is done by reversing the *Z*-transformation of $\bar{Z}_\omega$ to get r_{comb} from the pooled observations in all the *k* groups. As an illustration of the method consider the following Example 9.8.

Example 9.8: Random samples of 50 plants of guayule gave the correlation coefficient between percent resin content and percent rubber content in the three strains 405, 407 and 416 as shown in the third column of the table below.

$$\bar{Z}_\omega = \frac{\sum_{i=1}^{k} (n_i - 3) Z'_i}{\sum_{i=1}^{k} (n_i - 3)} = 0.4703$$

Strain	n_i	r_i	Z'_i	$Z'_i - \bar{Z}_\omega$	$(n_i - 3)(Z'_i - \bar{Z}_\omega)^2$
405	50	.362	0.3791	– 0.0912	0.391
407	50	.419	0.4455	– 0.0248	0.029
416	50	.527	0.5860	+ 0.1157	0.629
				Total	1.049

(Source: Steel and Torrie (1981), p. 281)

The test statistic for testing H_0 is

$$\chi^2 = \sum_{i=1}^{k} (n_i - 3)(Z_i - \bar{Z}_\omega)^2 = 1.049$$

For $\alpha = .05, \chi^2_{2,\,.05} = 5.99$ from Table A.5.

Since $\chi^2_{cal} = 1.049 < \chi^2_{2,\,.05} = 5.99$, H_0 can not be rejected

Hence the correlation in the three strains are homogeneous and the common correlation coefficient is obtained by inverse transformation of $\bar{Z}_\omega$ to get common *r* from pooled sample. The inverse transformation (i.e. to read *r* for $Z' = .4703$) gives $r = .4396$.

C: Test of Equality of Two Regression Coefficients

In this section test of equality of two regression coefficients (or slopes of lines of regression) from two independent samples of sizes n_1 and n_2 are considered. In symbols it is desired to test

$$H_0: \beta_1 = \beta_2 \text{ against } H_1: \beta_1 \neq \beta_2 \tag{9.57}$$

when, there are available independent samples from two groups of bivariate data of sizes n_1 and n_2. Thus a prefix 1 is used for 1st sample observations and 2 for 2nd group of observations.

Analogous to Student's t-statistic for equality of two means the test statistic for testing H_0 against H_1 in (9.57) is given by

$$t = \frac{b_1 - b_2}{s_p \sqrt{\frac{1}{S_{x1}^2} + \frac{1}{S_{x2}^2}}} \tag{9.58}$$

where b_1 and b_2 are slopes (or regression coefficients) of the regression lines obtained from n_1 group 1 observations and n_2 group 2 observations.

$$S_{x,j}^2 = \sum_{i=1}^{nj} \left(X_{ij} - \bar{X}_j\right)^2 \qquad j = 1, 2 \tag{9.59}$$

s_p^2 is pooled estimator of common $\sigma_{Y/X}^2$, given by

$$s_p^2 = \frac{1}{(n_1 - 2) + (n_2 - 2)} \left[(\text{Residual } SS)_1 + (\text{Residual } SS)_2\right] \tag{9.60}$$

$(\text{Residual } SS)_j = (SSE)_j \; j = 1, 2$

= Error or residual sum of square of ANOVA Table from j^{th} sample

$$= \left(S_y^2\right)_j - \left(S_{xy}^2\right)_j / \left(S_x^2\right)_j \tag{9.61}$$

This **test also assumes** that $\sigma_{Y/X}^2$ **is same in both the groups.**
If H_0 is true, *the test statistic has Student's t-distribution* with $v = n_1 + n_2 - 4$ d.f. In symbols

$$t \sim t_v \text{ if } H_0 \text{ is true} \tag{9.62}$$

So

$$H_0 \text{ is rejected if } |t| > t_{v,\,\alpha/2} \tag{9.63}$$

In other words: Reject H_0 if either $t > t_{v,\,\alpha/2}$ or if $t < -t_{v,\,\alpha/2}$
If $H_0 : \beta_1 = \beta_2$ *is not rejected*

(a) *The common slope is computed by combining the two samples.*
(b) A test of $H_0 : a_1 = a_2$, equality of intercepts should be performed. *If this test of equality of two intercepts is not rejected, common regression line by combining both the samples is obtained.*

The test of equality of two intercepts is done as follows:
The hypotheses are

$$H_0\text{: } \alpha_1 = \alpha_2 \text{ against } H_1\text{: } \alpha_1 \neq \alpha_2 \tag{9.64}$$

a_1 and a_2 are the intercepts of regression lines from two samples which are statistically proved to have common slope i.e. are parallel.

If hypothesis of equality of intercepts is not rejected, the two lines coincide so a common regression line should be obtained. For testing H_0: $\alpha_1 = \alpha_2$ some more computations are required as given below.

$$\text{Common sum of squares of } X = (S_x^2)_c = (S_x^2)_1 + (S_x^2)_2 \tag{9.65}$$

$$\text{Common sum of cross product } XY = (S_{xy})_c = (S_{xy})_1 + (S_{xy})_2 \quad (9.66)$$

$$\text{Common sum of squares of } Y = (S_y^2)_c = (S_y^2)_1 + (S_y^2)_2 \quad (9.67)$$

$$\text{Common residual sum of square} = (SSE)_c = (S_y^2)_c - \frac{(S_{xy})_c^2}{(S_x^2)_c} \quad (9.68)$$

$$\text{d.f. of common residual } SS = (df)_c = n_1 + n_2 - 3 \quad (9.69)$$

$$\text{So common residual mean square } (MSR)_c = \frac{(SSE)_c}{(df)_c} \quad (9.70)$$

$$\text{Estimate of common } \sigma_{Y/X}^2 \text{ or pooled estimate of common } \sigma_{Y/X}^2 = (s_p^2)_c = (MSR)_c \quad (9.71)$$

$$\text{The combined slope } b_c = (S_{xy})_c/(S_x^2)_c \quad (9.72)$$

Using these, *the test statistic for testing* H_0: $\alpha_1 = \alpha_2$ is

$$t = \frac{(\bar{Y}_1 - \bar{Y}_2) - b_c(\bar{X}_1 - \bar{X}_2)}{\sqrt{(s_p^2)_c\left[\frac{1}{n_1} + \frac{1}{n_2} + \frac{(\bar{X}_1 - \bar{X}_2)^2}{(S_x^2)_c}\right]}} \quad (9.73)$$

If $\quad H_0$: $\alpha_1 = \alpha_2$ is true, $t \sim t_{n1 + n2 - 3}$ (9.74)

So $\quad H_0$ is rejected if $|t| > t_{n1 + n2 - 3,\ \alpha/2}$ (9.75)

Example 9.9: Random samples of 13 and 15 men over 40 years, with two different occupations, were selected to consider regression of systolic blood pressure (Y) on age (X). Regression line of Y on X was obtained for men in two occupations. The computations gave the following results for the two samples:

Sample 1	Sample 2
$n = 13$	$n = 15$
$\bar{X} = 54.65$ yr	$\bar{X} = 56.93$ yr
$\bar{Y} = 170.23$ mm Hg	$\bar{Y} = 162.93$ mm Hg
$n\,S_x^2 = 1012.1923$	$n\,S_x^2 = 1659.4333$
$n\,S_y^2 = 2618.3077$	$n\,S_y^2 = 3848.9333$
$n\,S_{xy} = 1585.3385$	$n\,S_{xy} = 2475.4333$
$b = 1.57$ mm Hg/yr	$b = 1.49$ mm Hg/yr
$a = 84.6$ mm Hg	$a = 78.0$ mm Hg
$SSE = 136.322$	$SSE = 156.244$
Residual d.f. = 11	Residual d.f. = 13

It is desired to test H_0: $\beta_1 = \beta_2$ against H_1: $\beta_1 \neq \beta_2$ and then H_0: $\alpha_1 = \alpha_2$ against H_1: $\alpha_1 \neq \alpha_2$. (Source: Zar (2004), p. 365.)

Solution: (I) For testing of H_0: $\beta_1 = \beta_2$ against H_1: $\beta_1 \neq \beta_2$ test statistic to be used is

$$t = \frac{b_1 - b_2}{s_p\sqrt{\frac{1}{S_{x1}^2} + \frac{1}{S_{x2}^2}}} \sim t_{11+13} \text{ if } H_0 \text{ is true}$$

where
$$s_p^2 = \frac{(SSE)_1 + (SSE)_2}{n_1 - 2 + n_2 - 2} = \frac{136.2230 + 156.2449}{11 + 13}$$
$$= 12.1862$$
$$s_p\sqrt{\frac{1}{S_{x1}^2} + \frac{1}{S_{x2}^2}} = .1392$$
$$t = \frac{1.57 - 1.49}{.1392} = .575$$

From Table A.4, $t_{24,\ .025} = 2.064$

Since $t < 2.064$ H_0 is not rejected

Hence now to test H_0: $\alpha_1 = \alpha_2$ against H_1: $\alpha_1 \neq \alpha_2$, first the combined slopes b_c and combine sums will be needed.

$$\left(S_x^2\right)_c = \left(S_x^2\right)_1 + \left(S_x^2\right)_2 = 1012.1923 + 1659.4333 = 2671.6256$$
$$\left(S_y^2\right)_c = \left(S_y^2\right)_1 + \left(S_y^2\right)_2 = 2618.3077 + 3484.933 = 6467.2410$$
$$\left(S_{xy}\right)_c = \left(S_{xy}\right)_1 + \left(S_{xy}\right)_2 = 1585.3385 + 2475.4333 = 4060.7718$$
$$b_c = \frac{4060.7718}{2671.6256} = 1.520 \text{ mm Hg/yr}$$
$$(SSE)_c = \left(S_y^2\right)_c - \frac{\left(S_{xy}\right)_c^2}{\left(S_x^2\right)_c} = 6467.2410 - \frac{(4060.7718)^2}{2671.6256} = 295.0185$$
$$(df)_c = 13 + 15 - 3 = 25$$
$$\left(s_p^2\right)_c = \frac{295.0185}{25} = 11.8007$$

To test H_0: $\alpha_1 = \alpha_2$ against H_1: $\alpha_1 \neq \alpha_2$, the test statistic is

$$t = \frac{\left(\bar{Y}_1 - \bar{Y}_2\right) - b_c\left(\bar{X}_1 - \bar{X}_2\right)}{\sqrt{\left(s_p^2\right)_c\left[\frac{1}{n_1} + \frac{1}{n_2} + \frac{\left(\bar{X}_1 - \bar{X}_2\right)^2}{\left(S_x^2\right)_c}\right]}} \sim t_{25} \text{ if } H_0 \text{ is true}$$
$$= \frac{(170.23 - 162.93) - 1.520\,(54.65 - 56.93)}{\sqrt{11.8007\left[\frac{1}{13} + \frac{1}{15} + \frac{(54.65 - 56.93)^2}{2671.6256}\right]}} = \frac{10.77}{1.3105} = 8.218$$

From the Table A.4 $t_{25,\ .025} = 2.060$

Since $t_{cal} = 8.218 > 2.060 = t_{25,\ .025}$, null hypothesis of equality of intercepts is rejected. Hence two occupation groups of men do not have common regression line even though the lines of regression of the two groups have common slopes (= b_c). Hence the regression lines of the two groups are

$$Y = 84.6 + 1.52X$$

and
$$Y = 78.0 + 1.52X$$

for the 1st and 2nd groups, respectively.

D: Testing Homogeneity of More than Two Regression Coefficients

Suppose that independent random samples are available from k groups (or populations) on bivariate random variable (X, Y). k regression lines of Y on X have been obtained from the k samples, size of ith sample being n_i, $i = 1, 2, ..., k$.

It is desired to test that the slopes of k regression lines in the k populations $\beta_1, \beta_2, ..., \beta_k$ are equal. In symbols it is desired **to test H_0: $\beta_1 = \beta_2 = ... = \beta_k$ against the alternative H_1: All the β_i's are not equal**. This *test also assumes* that $\sigma^2_{Y/X}$ are same in the k groups i.e.

$$\left(\sigma^2_{Y/X}\right)_1 = \left(\sigma^2_{Y/X}\right)_2 = ... = \left(\sigma^2_{Y/X}\right)_k \tag{9.76}$$

Various sums for obtaining the k regression lines viz. $(S^2_x)_i$, $(S^2_y)_i$, $(S^2_{xy})_i$, $i = 1, 2, ..., k$ are required for comparing the slopes of k regression lines.

The residual or error sum of squares $(SSE)_i$ and corresponding d.f. are also used. Table 9.4 below gives these values and further computations **for obtaining pooled residual or error sum of squares** and S^2_p its **pooled degrees of freedom dfp.**

Combined or summed sums for **each of the k-samples** are **found** to obtain the 'common' residual or error sum of square $(SSE)_c$.

To *test* H_0: $\beta_1 = \beta_2 = ... = \beta_k$ *against* H_1: *All the* β_i'*s are not equal the statistic to be used is*

$$F = \frac{\left[(SSE)_c - (SSE)_p\right] / (k-1)}{(SSE)_p / (n-2k)} \sim F_{k-1, n-2k}, \text{ if } H_0 \text{ is true} \tag{9.77}$$

The **null hypothesis** of equality of slope **is rejected** at α level

$$\text{if} \quad F > F_{k-1, n-2k, \alpha} \tag{9.78}$$

If this hypothesis of homogeneity of slopes is accepted common slope b_c is obtained Using the calculations for common regression as

$$b_c = \frac{\sum_{i=1}^{k} (S_{xy})_i}{\sum_{i=1}^{k} (S^2_x)_i} \tag{9.79}$$

After obtaining common slope b_c, one should also test equality or homogeneity of k intercepts. So that if equality of intercepts is not rejected, a common regression line may be obtained after combining all the k-samples.

Test of H_0: $\alpha_1 = \alpha_2 ...... = \alpha_k$ against H_1: all the intercepts are not equal is performed now.

For testing the null hypothesis of equality of intercepts all the $n = \sum_{i=1}^{k} n_i$ observations are used to compute $(S^2_x)_t$, $(S_{xy})_t$ and $(S^2_y)_t$.

Table 9.4 Calculations for homogeneity of slopes and intercepts of k simple linear regressions (source: Zar (2004), p. 370 (modified))

For	S_x^2	S_{xy}^2	S_y^2	Residual or error sum of square SSE	d.f. of SSE
Sample 1	$(S_x^2)_1$	$(S_{xy})_1$	$(S_y^2)_1$	$(SSE)_1 = (S_y^2)_1 - \frac{(S_{xy})_1^2}{(S_x^2)_1}$	$n_1 - 2$
Sample 2	$(S_x^2)_2$	$(S_{xy})_2$	$(S_y^2)_2$	$(SSE)_2 = (S_y^2)_2 - \frac{(S_{xy})_2^2}{(S_x^2)_2}$	$n_2 - 2$
⋮					⋮
Sample k	$(S_x^2)_k$	$(S_{xy})_k$	$(S_y^2)_k$	$(SSE)_k = (S_y^2)_k - \frac{(S_{xy})_k^2}{(S_x^2)_k}$	$n_k - 2$
Pooled regression				$s_p^2 = (SSE)_p = \sum_{i=1}^{k} (SSE)_i$	$\sum_{i=1}^{k} n_i - 2k = n - 2k$
Common regression				$(S_x^2)_c = \sum_{i=1}^{k} (S_x^2)_i$ $(S_{xy})_c = \sum_{i=1}^{k} (S_{xy})_i$	$(S_y^2)_c = \sum_{i=1}^{k} (S_y^2)_i$ $(SSE)_c = *$ $df_c = \sum_{i=1}^{k} n_i - k - 1$
Total regression	$(S_x^2)_t$	$(S_{xy})_t$	$(S_y^2)_t$	$(SSE)_t = (S_y^2)_t - \frac{(S_{xy})_t^2}{(S_x^2)_t}$	$df_t = \sum_{i=1}^{k} n_i - 2 = n - 2$

$$* = (S_y^2)_c - \frac{(S_{xy})_c^2}{(S_x^2)_c}$$

Thus

$$(S_x^2)_t = \sum_{j=1}^{n} (X_j - \bar{X})^2 \quad \text{where} \quad \bar{X} = \frac{1}{n} \sum_{j=1}^{n} X_j \tag{9.80}$$

In a similar manner all the three sums for 'total' regression in Table 9.4 will be obtained and residual sum of square $(SSE)_t$ will be computed as shown in Table 9.4. Testing of homogeneity of intercepts is done by using ***F*-statistic** given by

$$F = \frac{\left[(SSE)_t - (SSE)_c\right]/(k-1)}{(SSE)_c/(n-k-1)} \sim F_{k-1,\, n-k-1} \text{ if } H_0 \text{ is true} \tag{9.81}$$

If the null hypotheses of equality of regression coefficients or of intercepts are rejected one can **apply multiple comparisons** to find the slopes which differ from which other slopes, and similarly the intercepts which differ from which other intercepts. Multiple comparison will be dealt in Chapter 14. The test procedure for testing of equality of slopes and intercepts is illustrated in the Example 9.10.

Example 9.10: Three independent random samples of 24, 29 and 30 observations from three populations of bivariate random variable (X, Y) were selected. Regression analyses were performed on the three samples and the computed values for testing homogeneity of regression coefficients and intercepts are given in the following Table. The required computations for homogeneity tests are also included in the said Table 9.5.

Table 9.5 Testing homogeneity of regression coefficients and intercepts (source: Zar (2004), p. 371 (modified))

Sample no.	Sample size (n)	S_x^2	S_{xy}^2	S_y^2	b	Residual or error SS	$d.f.$
1	24	**430.14**	**648.97**	**1085.34**	1.51	86.21	22
2	29	**448.65**	**694.36**	**1184.12**	1.55	109.48	27
3	30	**502.31**	**714.33**	**1186.52**	1.42	170.68	28
Pooled regression	$n = 83$					$s_p^2 = (SSE)_p = 366.37$	77
Common regression	$n - 2k = 77$	$(S_x^2)_c = 1381.10$	$(S_{xy})_c = 2057.66$	$(S_y^2)_c = 3435.98$	$b_c = 1.49$	$(SSE)_c = 370.33$	79
Total regression	$n = 83$	**2144.06**	**3169.78**	**5193.48**		$(SSE)_t = 427.10$	81

Note: Bold values are calculated from raw data. All other normal values are computed from bold values.

Solution: First a test of H_0: $\beta_1 = \beta_2 = \beta_3$ will be performed against H_1: all the three β_i'*s* are not equal.

If homogeneity of β_i'*s* is not rejected, a test of homogeneity of intercepts will be performed by testing H_0: $\alpha_1 = \alpha_2 = \alpha_3$ against H_1: all the intercepts are not equal.

*From the data in **Table 9.5***

$$(SSE)_p = \sum_{j=1}^{3} (SSE)_j$$

The $(S_x^2)_c, (S_{xy})_c, (S_y^2)_c$ are totals of three values in corresponding columns

$$(SSE)_c = (S_y^2)_c - \frac{(S_{xy})_c^2}{(S_x^2)_c} = 3435.98 - \frac{(2057)^2}{1381.10} = 370.33$$

$$b_c = \frac{(S_{xy})_c^2}{(S_x^2)_c} = \frac{2057}{1381.10} = 1.49$$

For testing H_0: $\beta_1 = \beta_2 = \beta_3$ against H_1: all the three β_i'*s* are not equal, use the F-statistic given in (9.77). The calculated value of this

$$F = \frac{(370.33 - 366.37)/2}{366.37/77} = 0.42$$

From the Table A.6 $F_{2,\,77,\,.05} \simeq 3.13$, so H_0 is not rejected.

Hence a test of H_0: $\alpha_1 = \alpha_2 = \alpha_3$ against H_1: all the intercepts are not equal is performed by using F-statistic given in (9.81).

$$F = \frac{(427.10 - 370.33)/2}{370.33/79} = 6.06$$

$$\text{since } (SSE)_t = (S_y^2)_t - \frac{(S_{xy})_t^2}{(S_x^2)_t} = 427.10$$

From the F Table A.6 $F_{2,\,79,\,.05} \simeq 3.13$.
Since $F_{cal} > F_{2,\,79,\,.05}$, H_0: the equality of intercepts is rejected.
Hence it is concluded that the three groups have parallel regression lines with common slope 1.49. The intercepts of the three lines are not same.
By multiple comparison technique given in Chapter 14, it will be possible to detect whether all the three intercepts differ or any two are equal.

9.6 Intraclass and Spearman's Rank Correlation

A Intraclass Correlation Coefficient

Pearson's Product moment correlation coefficient has been considered between two definitely defined measured variables designated as X and Y such as age and yield of milk in cows, stature of father and sons, weight and systolic blood pressure, length of shell and its height etc.

But there occur situations mainly in biological studies, when it is not possible to say that which of the pair of observations should be designated as X and the other as Y. For example

1. One wishes to determine relationship between weights of identical twins.
2. It is desired to examine relationship between heights of two brothers.

In both of these there are two variates, but which one the first and which one the second may be resolved by setting some rule. Such as the elder brother first or the taller as first. But by doing so, one would get relation between height of elder and younger brother or the height of taller and shorter brother. It will not give relation between heights of brothers in general.

The relationship of brotherhood or twin ship is in fact symmetrical. There is no relevant character to enable singling out one of the two as first. If A is brother of B, B is brother of A. If A is twin of B, B is twin of A.

Accordingly, each of the subject of a pair may be taken in two ways, (1) A as first and B as second (2) B as first and A as second. Height of brothers or weight of twins will be the first or second value for finding the correlation.

Instead of considering two brothers in a family, one may consider k brothers. For computing correlation coefficient all possible pairs of a family will be $2x\,({}^kC_2) = k(k-1)$ for n families there will be $nk(k-1)$ pairs. Hence these many entries will be required in the table for computation of correlation.

Such a table of paired values is called *intraclass correlation table* and the correlation between two variates of the table is called **intraclass correlation.**

The intraclass correlation table has certain peculiarities. It is not of such a general type as considered for simple correlation.

I. Let the variate values in the ***n*** **families of** ***k*** **members** be as given below.

In the intraclass correlation table x_{11} will be associated as X-variate with each of the $k - 1$ values of family 1, viz. $x_{12}, x_{13}, ..., x_{1k}$. Hence x_{11} appears $k - 1$ times. Similarly, every other value also appears $k - 1$ times.

Hence the sum of X-values in intraclass table = $(k-1)\ \Sigma(x)$ where summation extends on all the nk variate values of Table 9.6.

Table 9.6 Variate values in n families of k members

Family	Variate values
1	$x_{11}, x_{12}, \ldots\ldots, x_{1k}$
2	$x_{21}, x_{22}, \ldots\ldots, x_{2k}$
:	
:	
n	$x_{n1}, x_{n2}, \ldots\ldots, x_{nk}$

Since there are $nk(k-1)$ values in intraclass correlation table, so

$$\bar{X} = \frac{1}{nk(k-1)}(k-1)\sum x = \frac{\sum x}{nk} \tag{9.82}$$

Similarly

$$\bar{Y} = \frac{1}{nk}\sum x \tag{9.83}$$

Thus means of the X and Y variates of intraclass correlation are same.

The variance of $X = \sigma_x^2 = \frac{1}{nk(k-1)}$ sum of $(x-\bar{X})^2$ and since each $(x-\bar{X})$ occurs $(k-1)$ times

$$\sigma_x^2 = \frac{\sum (x-\bar{X})^2}{nk} \tag{9.84}$$

where summation extends over all the values of x.

Similarly

$$\sigma_y^2 = \frac{1}{nk}\sum (x-\bar{Y})^2 = \frac{1}{nk}\sum (x-\bar{X})^2 = \sigma_x^2 \tag{9.85}$$

Hence $\sigma_x = \sigma_y = \sigma$

The intraclass correlation coefficient r_I is given by

$$\sigma^2 r_I = \frac{1}{nk(k-1)}\sum_{j,m}(x_{ij}-\bar{X})(x_{im}-\bar{X}) \tag{9.86}$$

where $\sum\limits_{j,m}$ extends over all possible pairs.

Let

$$\bar{X}_1 = \frac{1}{k}(x_{11} + x_{12} + \ldots\ldots + x_{1k}) \tag{9.87}$$

$$= \text{Mean of first family}$$

Further let

$$\sigma_m^2 = \frac{1}{n}\sum_{i=1}^{n}\left(\bar{X}_i - \bar{X}\right)^2 \tag{9.88}$$

= variance of the means of families about overall mean $\bar{X}$.

Then after simple algebra (9.86) can be written as

$$nk(k-1)\sigma^2 r_I = k^2 n\sigma_m^2 - \sigma^2 nk$$

or

$$\sigma^2\left[1+(k-1)r_I\right] = k\sigma_m^2 \tag{9.89}$$

or

$$r_I = \frac{k\sigma_m^2 - \sigma^2}{(k-1)\sigma^2} \tag{9.90}$$

From (9.89) since σ_m^2 is non negative

$$1 + (k-1)\, r_I \geq 0$$

or

$$r_I > -\frac{1}{(k-1)} \tag{9.91}$$

Thus though **intraclass correlation can vary between [–1, + 1] it can not be less than** $-\frac{1}{(k-1)}$**.**

Example 9.11: In five families of 3 brothers, their heights are as given below (source: Yule and Kendall (1958), p. 276):

Family	Heights		
1	5′ 9″	5′ 10″	5′ 11″
2	5′ 10″	5′ 11″	6′ 0″
3	5′ 11″	6′ 0″	6′ 1″
4	6′ 0″	6′ 1″	6′ 2″
5	6′ 1″	6′ 2″	6′ 3″

Find intraclass correlation coefficient.

Solution: Mean of all the heights $\bar{X}$ = 6′, n = 5, k = 3

$$\sigma^2 = \frac{1}{5\times 3}\left[\textit{sum of } \left(x-\bar{X}\right)^2 \textit{ over all the values}\right]$$

$$= \frac{1}{15}\left[9+4+1+4+1+0+1+0+1+0+1+4+1+4+9\right]$$

$$= \frac{1}{15}\left[40\right] = \frac{8}{3}$$

$$\sigma_m^2 = \frac{1}{5}\sum_{i=1}^{5}\left(\bar{X}_i - \bar{X}\right)^2 = \frac{1}{5}\left[4+1+0+1+4\right] = 2$$

So

$$r_I = \frac{k\sigma_m^2 - \sigma^2}{2\sigma^2} = \frac{3\times 2 - 8/3}{2\times 8/3} = \frac{5}{8}$$

$$= .625$$

II. The above case of equal number of k members per family is not always prevalent. The ***n* families may have unequal number of members**. In this subsection this case is considered. Let k_i be the number of members in the ith family (i = 1, 2, ..., n). The ith family will contribute $(k_i(k_i - 1)$ pairs of values in the intraclass correlation table.

$$\text{Hence the total number of pairs} = \sum_{i=1}^{n} k_i \left(k_i - 1\right) = N \text{ (say)} \tag{9.92}$$

The intra class correlation coefficient in this case is given by

$$N\sigma^2 r_I = \sum_{i=1}^{n} k_i^2 \left(\bar{X}_i - \bar{X}\right)^2 - \sum_{j=1}^{n} \sum{}' \left(x_{ij} - \bar{X}\right)^2 \tag{9.93}$$

where
$$\sigma^2 = \frac{1}{N} \sum{}'' \left\{ \left(k_i - 1\right) \sum_{i=1}^{n} \left(x_{ij} - \bar{X}\right)^2 \right\}$$

Σ' extends over all the members of all i^{th} family and Σ'' extends over all the possible pairs.

B Spearman's Rank Correlation Coefficient

In calculating product moment correlation coefficient, it is necessary that the two variates should be definitely measured. However, in practice there are situations when the qualities can not be measured as numbers of units on an objective scale.

For example, relation between mathematical and numerical ability in a class of students, brightness of different species, colour pattern in parrot and its agility may be required to know.

'Ability', of any kind, is a variate in the sense that it varies from one individual to another. It can be a quantitative variate if some unique way of measuring it is devised. A very common mode of measurement of ability is to allot marks on the abilities of each individual. Thus marks on mathematics and music, of each student may be considered as measures of mathematical and musical ability.

Such method is however, open to many objections, one of which is that if examined by different examiners, the same person may get different marks. A correlation between marks will be therefore, likely to depend to some extent on the examiner.

A method to overcome this difficulty to some extent can be adopted if the students are arranged in **order** of their ability instead of assessing it numerically and then allot to each student a number which indicates his position in the group, according to his ability. One with highest ability given the rank 1 and next the rank 2, and so on. The students are then said to be **ranked**.

A procedure of this kind can be applied to measurable data. Such as, a group of men can be easily ranked according to their heights without actually measuring them.

Charles Spearman (1904) gave first the procedure of rank correlation. If there are a set of individuals ranked according to two different qualities, then *a measure of the degree of relation between the two qualities is given by the rank correlation coefficient* r_s.

The formula for the product moment correlation coefficient may be applied on the ranks X and Y on two qualities to obtain r_s.

If $X_1, X_2, \ldots\ldots, X_n$ and $Y_1, Y_2, \ldots\ldots, Y_n$ are the ranks of n individuals according to quality A and B respectively, the X's and Y's are merely some permutations of first n natural numbers 1, 2, ..., n.

A simpler method exists to find r_s than the procedure of product moment correlation. In this simpler method one has to find the differences $d_k = X_k - Y_k$ in the ranks on two qualities of kth individual (k = 1, 2, ..., n).

Because of the properties of sums of natural numbers the computation of r_s can be done by the formula

$$r_s = \frac{1 - 6 \sum_{k=1}^{n} d_k^2}{n^3 - n} \tag{9.94}$$

The range of r_s is also from –1 to +1.

Example 9.12: The rankings of ten students in mathematics and music are as follows:

Students	1	2	3	4	5	6	7	8	9	10
Mathematics	1	2	3	4	5	6	7	8	9	10
Music	6	5	1	4	2	7	8	10	3	9

What is the rank correlation between students' mathematical and musical abilities? (Source: Yule and Kendall (1958), p. 261.)

Solution: The difference in ranks are

k:	1	2	3	4	5	6	7	8	9	10
d_k:	–5	–3	2	0	3	–1	–1	–2	6	1
d_k^2:	25	9	4	0	9	1	1	4	36	1

$$\sum_{k=1}^{10} d_k^2 = 90$$

Hence $$r_s = 1 - \frac{6 \times 90}{1000 - 10} = 1 - \frac{6 \times 90}{990} = \frac{5}{11} = 0.45$$

Tied Ranks

Some times, it is not possible to distinguish between a number of individuals with almost same quality. In such a case it is customary to average the ranks of these individuals and assign to each this average rank, even if this is fractional.

As an example suppose while ranking 20 individuals one of 20 can be judged best and is assigned rank 1.

There are two individuals which appear next best and it is not possible to say which should be 2nd and which the 3rd. They are, therefore, 'tied' and each is given the rank (2+3)/2 = 2.5.

The next member is ranked as 4. Suppose, there is a tie in the next three members of the group. Each of these three then should be allotted the rank (5 + 6 + 7)/3 = 6. The next member will get the rank 8, and so on.

In such a situation the formula for Spearman's rank correlation should be modified. One could calculate the r_s using the formula for product moment correlation coefficient. But the changes due to tied ranks can be seen (by some arithmetic) to reduce in to the following, to give r_s corrected for ties as

$$(r_s)_c = \frac{1/6\left(n^3 - n\right) - \left(T_x + T_y\right) - \sum_{k=1}^{n} d_k^2}{\left[\left(n^3 - n\right)/6 - 2T_x\right]^{1/2}\left[\left(n^3 - n\right)/6 - 2T_y\right]^{1/2}} \tag{9.95}$$

Here
$$T_x = \sum_j \left(t_j^3 - t_j\right)/12 \tag{9.96}$$

$$T_y = \sum_l \left(t_l^3 - t_l\right)/12 \tag{9.97}$$

t_j are the number of tied values in a group of X and t_1 are the number of tied values in a group of Y. Let the method be illustrated by the following example.

Example 9.13: Two office in charges rank ten subordinates according to their suitability for promotion as follows:

Subordinate	A	B	C	D	E	F	G	H	I	J
In charge 1	1.5	1.5	3	4	6	6	6	8	9.5	9.5
In charge 2	1	2	4	4	4	6	7	8	9	10

Find the Spearman's rank correlation coefficient. (Source: Yule and Kendall (1958), p. 267 (modified)).

Solution: In the ranking by first in charge there are 3 sets of ties. In this 1st is of two ties, 2nd of 3 ties and the third is of 2 ties. i.e. $t_1 = 2$, $t_2 = 3$ and $t_3 = 2$.

Hence
$$T_x = \frac{1}{12}\sum_{j=1}^{3}\left(t_j^3 - t_j\right) = \frac{1}{12}\left[\left(2^3 - 2\right) + \left(3^3 - 3\right) + \left(2^3 - 2\right)\right]$$

$$= \frac{1}{12}\left[6 + 24 + 6\right] = 3$$

In the ranking by 2nd in charge, there is one set of tie of 3 ties.

So
$$T_y = \frac{1}{12}\left(3^3 - 3\right) = 2$$

d_k:	½	–½	–1	0	2	0	–1	0	½	–½
d_k^2:	¼	¼	1	0	4	0	1	0	¼	¼

$$\sum_{k=1}^{0} d_k^2 = 7$$

$$(r_s)_c = \frac{\left(10^3 - 10\right)/6 - (3 + 2) - 7}{\left[\left(10^3 - 10\right)/6 - 2.3\right]^{1/2}\left[\left(10^3 - 10\right)/6 - 2.2\right]^{1/2}}$$

$$= \frac{165 - 12}{\sqrt{(165 - 6)(165 - 4)}} = \frac{153}{\sqrt{159 \times 161}} = .956$$

The rank correlation indicates high degree of agreement between the opinions of the two in charges.

A remark: Various ideas, concepts and examples discussed in this chapter may also be found in books by Yule and Kendall (1958), Milton and Tsokos (1983), Steel and Torrie (1981) and Daniel (1987) etc.

10. Analysis of Categorical Data: Independence and Association

This chapter is concerned with classification of sampled objects or individuals according to the quality they possess. This results in categorical data.

If one observes only presence or absence of certain quality, it results in data in two categories, which are the counts of individuals with and without the presence of an attribute or quality.

At some other time, classification may be in several categories of a single quality. For example, a group of school children may be classified according to their eye colour being: (i) black (ii) brown (iii) blue. Thus data will be count of school children in 3 categories.

A population may be classified according to blood groups AB, A, B, O. So data will result in count of individuals in four categories or cells or classes.

In general there may be count data of one qualitative variable in n cells or categories. Quite often individuals may be classified according to several qualities. For example, a person may be classified as smoker or non-smoker, having or not having coronary disease. Thus in this case there are 2×2 categories.

At other times individuals may be classified in r categories according to a quality. Subsequently individuals of each of the r groups may be classified into c cells or categories according to a second quality. This will result in an $r \times c$ table, also called $r \times c$ contingency table.

The next section deals with inferential problems when data is classified according to one quality giving 2 categories.

The theoretical concepts and numerical examples may also be found in books by Steel and Torrie (1981), Milton and Tsokos (1983) and Zar (2004).

10.1 Two Categories: Estimation and Tests of Proportions

The general situations is that there is a population of interest, on which a trait is being studied and each member is allotted in one of two classes or cells, either having trait or not having it.

Inferences are to be made on a parameter p, the proportion of population with trait.

A. Estimation of p

In this subsection estimation of the proportion p is considered when there are two categories in population and p is the proportion of individuals possessing a certain quality. For estimating p one will draw a random sample of size n from the population. The number of individuals in the sample of size n will be counted (say r) who possess the quality of interest.

Then an estimate of p is $\hat{p}$, which is the observed proportion of individuals in the sample who possess the quality.

Thus
$$\hat{p} = r/n \tag{10.1}$$

$\hat{p}$ is an unbiased estimator of p.

r follows a binomial distribution with parameter p. In symbols $r \sim B(n, p)$.

$$E(\hat{p}) = E(r/n) = p \tag{10.2}$$

$$V(\hat{p}) = p(1-p)/n \tag{10.3}$$

Since p may not be known one can find the estimate of $V(\hat{p})$ by replacing p by $\hat{p}$. Thus

$$\hat{V}(\hat{p}) = \hat{p}(1-\hat{p})/n \tag{10.4}$$

Hence a *point estimator* of p is given by

$$\hat{p} = r/n$$

and estimate of

$$SE(\hat{p}) = \sqrt{\hat{p}(1-\hat{p})/n} \tag{10.5}$$

As binomial distribution tends to normal distribution for large n, a 95% confidence interval for p may be given by

$$\hat{p} \pm Z_{.025}\sqrt{\hat{p}(1-\hat{p})/n} \quad \text{or} \quad \hat{p} \pm 1.96\sqrt{\hat{p}(1-\hat{p})/n} \tag{10.6}$$

If sample size is reasonable this confidence interval is sufficiently good.

B. Tests of Proportion

While performing experiment of the type mentioned in *A* one may be interested in testing whether the population parameter is having some specific value. In other words one is interested in testing H_0: $p = p_0$ on the basis of a random sample of size n in which r possess the trait.

(a) Normal Test

A test of H_0 is based on normal distribution of the test statistic. p is estimated by $\hat{p} = r/n$ and test statistic to test H_0 is

$$Z = \frac{\hat{p} - p_0}{\sqrt{p_0(1-p_0)/n}} \tag{10.7}$$

If n is sufficiently large, *distribution of test statistic is normal* if H_0 is true.

In symbols
$$Z \sim N(0, 1) \text{ if } H_0 \text{ is true} \tag{10.8}$$

The test procedure for various alternatives are as follows:

(a) Suppose H_0: $p = p_0$ is tested against H_1: $p \neq p_0$ at $\alpha = .05$. Reject H_0 if $|Z| > 1.96$. In other words

$$\text{Reject } H_0 \text{ if either } Z > 1.96 \text{ or } Z < -1.96 \tag{10.9}$$

(b) If H_0: $p = p_0$ is tested against H_1: $p > p_0$ at $\alpha = .05$, then

$$\text{Reject } H_0 \text{ if } Z > 1.645 \tag{10.10}$$

(c) If H_0: $p = p_0$ is tested against H_1: $p < p_0$ at $\alpha = .05$, then

$$\text{Reject } H_0 \text{ if } Z < -1.645 \tag{10.11}$$

(b) Chi-square Test

For testing H_0: $p = p_0$ against H_1: $p \neq p_0$ one can also apply Chi-square test. The general form of Chi-square test statistic is

$$\chi^2 = \sum_i \frac{(O_i - E_i)^2}{E_i} \tag{10.12}$$

where O_i = Observed frequency in the ith class, and
E_i = Expected frequency in the ith class if H_0 is true

The number of classes may be two, as in this problem or it may be k, in general.

If there are k classes, n members of sampled population are allotted one of these k classes. Let the allocation results in $n_1, n_2, \ldots n_i, \ldots n_k$ members in 1st, 2nd, …i^{th}, …k^{th} class, respectively. Then

$$n = \sum_{i=1}^{k} n_i \quad \text{and} \quad O_i = n_i \tag{10.13}$$

If p_i = P[a member belongs to i^{th} class] $i = 1, 2, \ldots k$ then hypothesis to be tested is H_0: $p_i = p_{io}$ (i = 1, 2, …k) and

$$E_i = n\, p_i,\ i = 1, 2, \ldots k \tag{10.15}$$

Since each of the observations belong to only one of the k classes

$$\sum_{i=1}^{k} p_i = 1 \quad \text{and} \quad \sum_i^{k} O_i = \sum_i^{k} E_i = n \tag{10.16}$$

Since $(O_i - E_i)$ ($i = 1, 2, \ldots k$) in (10.12) satisfy one relation (10.16) the degree of freedom of χ^2-test statistic is reduced by 1. Thus in general for k classes, the χ^2-test statistic follows Chi-square distribution with $k - 1$ degree of freedom. In symbols

$$\chi^2 \sim \chi^2_{k-1} \text{ if } H_0 \text{ is true} \tag{10.17}$$

If $k = 2$

$$O_1 = r,\ O_2 = (n - r),\ E_1 = np \text{ and } E_2 = n(1 - p) = n - np$$

And the *test statistic has Chi-square distribution with* 1 *degree of freedom.*

The *test procedure* to test H_0: $p = p_0$ against H_1: $p \neq p_0$ at level α is

$$\text{Reject } H_0 \text{ if } \chi^2 > \chi^2_{1,\alpha} \tag{10.18}$$

where $\chi^2_{1,\alpha}$ is upper tail α% significant point of χ^2_1 distribution.

$$H_0: p = p_0 \text{ is equivalent to } H_0: p_1 = p_0, p_2 = 1 - p_0$$

Subsequently it is equivalent to H_0: $\dfrac{np_1}{np_2} = \dfrac{p_0}{1 - p_0} = \dfrac{r_1}{r_2}$ or

$$H_0: E_1: E_2 = r_1: r_2 \tag{10.19}$$

In words, the frequency ratio in two classes are as $r_1: r_2.$

It can be shown by simple algebra, that the test statistic in (10.12) can be written as

$$\chi^2 = \frac{(r_2 n_1 - r_1 n_2)^2}{r_1 r_2 (n_1 + n_2)} = \frac{(r_2 n_1 - r_1 n_2)^2}{r_1 r_2 n} \tag{10.20}$$

for testing H_0: $p = p_0$ in the form (10.19). As the test statistic given in (10.12), this test statistic in (10.20) also has χ_1^2 distribution if H_0 is true. So test procedure is same as (10.18).
These procedures can be illustrated by an example given below.

Example 10.1: Consider a certain F_1 generation of *Drosophila melanogaster* with 35 males and 46 females. It is required to test the hypothesis that the sex ratio is 1 : 1 (source: Steel and Torrie (1981), p. 482).
Solution: In this problem there are only two categories male and female. So $k = 2$.

$$n_1 = 35 \text{ and } n_2 = 46 \quad n = n_1 + n_2 = 81$$

Let p be the proportion of males in F_1 generation of *Drosophila.*

The hypothesis to be tested may be written as

(a) H_0: $p = .5$ against H_1: $p \neq .5$ or
(b) H_0: frequency of females: frequency of males is 1 : 1 against H_1: sex ratio is not 1 : 1

(A) Normal test can be applied for (a) by using

$$Z = \frac{\hat{p} - .5}{\sqrt{(.5 \times .5)/81}} \sim N(0,1) \text{ if } H_0 \text{ is true}$$

Test procedure for $\alpha = .05$ is to reject H_0 if $|Z| > 1.96$

Now $$Z = \frac{35/81 - .5}{.5/9} = -1.22$$

Since calculated Z does not lie in rejection region, H_0 is not rejected.
Hence it may be *concluded* that males and females are in equal proportion in F_1 generation of *Drosophila.*
(B) For applying χ^2 test of H_0: Sex ratio is 1:1 against H_1: sex ratio is not 1:1, the test statistic to be used from (10.20) is

$$\chi^2 = \frac{(r_2 n_1 - r_1 n_2)^2}{r_1 r_2 n} = \frac{(35-46)^2}{1 \times 1 \times 81} = \frac{121}{81} = 1.49$$

From the Table A.5 $\chi^2_{1,\,.05} = 3.84$
Since calculated $\chi^2 < 3.84$, H_0 is not rejected.

Yates Correction for Continuity

The test statistic for testing specified probabilities p_{io} in i^{th} class ($i = 1, 2, \ldots k$) given in equation (10.12), is obtained from categorical frequencies which take only discrete values. Chi-square distribution being continuous, the probability of obtaining a calculated χ^2-value from theoretical χ^2-distribution can be determined only approximately.

However, the approximation is very good one, except when the degree of freedom = 1. Thus in such a case when the χ^2 test statistic has degree of freedom = 1, it is usually recommended that *Yates correction for continuity* (Yates 1934) be applied and χ_c^2, the test statistic corrected for continuity be used for testing. It is given by

$$\chi_c^2 = \sum_{i=1}^{2} \frac{(|O_i - E_i| - .5)^2}{E_i} \tag{10.21}$$

If H_0 involves 1:1 ratio for $k = 2$, it is evident from (10.20) that

$$\chi^2 = \frac{(O_1 - O_2)^2}{n} \tag{10.22}$$

and

$$\chi_c^2 = \frac{(|O_1 - O_2| - 1)^2}{n} \tag{10.23}$$

may be used in place of (10.21) (cf. Zar (1999). It may be noted that (10.22) and (10.23) do not require calculation of E_1 and E_2 and thus rounding errors are avoided in the last two equations.

Some sources suggest that this correction should be applied when degree of freedom = 1 and it should not be applied when degree of freedom > 1. A group suggests that Yates correction should be used when the expected frequency is less than 10. For large sample size it has little effect. It may, therefore, be used when d.f. = 1 regardless of sample size n.

10.2 Testing Independence and Homogeneity in 2 × 2 and $r \times c$ Contingency Table

In this section again a sample of nominal or categorical data is considered. But in this section the qualitative variable of classification is not one as in previous section, but there are *two qualitative or nominal variables*. When each of the two qualitative or nominal variable has two levels or categories of classification, there are in all 2 × 2 = 4 classes. Each observation out of total n observations, belongs to exactly one of the four mutually exclusive classes.

Such type of situation results in a 2 × 2 contingency table which is discussed first in this section.

If instead of two categories of each of the two nominal variables under study, suppose one nominal variable has r groups. Each of the total n observations is assigned to exactly one of r groups. Subsequently observations within each of the r groups are further classified according to c levels of the second variable. In this way the total number of n observations are classified in a table with $r \times c$ cells. Such a table is called $r \times c$ contingency table. This will be discussed after 2 × 2 contingency table in this section.

2 × 2 Contingency Table

Consider an example when a group of 100 class IX students of a school are contacted and the two nominal variables of interest are their sex and whether the courses opted by them are mathematics or non-mathematics.

Thus the students are classified according to *two categories* or *two levels* of the *two nominal variables* resulting in 2 × 2 = 4 classes or categories. For example suppose a quick record of sex and preference for course on a list of 100 students is prepared. It can be arranged in a table given below, which may be called a 2 × 2 contingency table.

Table 10.1 2 × 2 contingency table

Opted course	Male	Female	Total
Mathematics	20	10	30
Non-mathematics	40	30	70
Total	60	40	100

Thus each student falls in *exactly one* of the four *mutually exclusive cells or categories.* Some other examples of observations giving 2 × 2 contingency tables may be

1. In a group of volunteers for a study on new hepatitis vaccine, some were vaccinated and others not vaccinated. After some time the number who caught the hepatitis were counted. It was noted how many caught among the vaccinated and among the unvaccinated volunteers.
   Again each volunteer was either vaccinated or was not vaccinated. Each either caught the hepatitis or did not catch the disease. Thus each one belongs to exactly one of the four classes.
2. A study to consider association between *pepsinogen I* level in blood and presence of peptic ulcer was conducted. Some patients who had ulcer and some who did not have ulcer were examined for *pepsinogen I* level in blood and were classified as level being high or low. Thus giving a 2 × 2 contingency table. Ulcer present and not present with either high or low level of *pepsinogen I,* are the four classes.
   A patient belongs to exactly one of these four mutually exclusive classes.

All these examples give 2 × 2 contingency tables each. To describe the procedure of analysis of these contingency tables some notations are defined below.

Notations: Let n be the total number of observations which are classified in a 2 × 2 contingency table.

Let n_{ij} be the number of observations which belong to i^{th} row and j^{th} column of 2 × 2 contingency table. Then $i = 1, 2$ and $j = 1, 2$.

Further let $n_{i\cdot} = \sum_j n_{ij}$, $n_{\cdot j} = \sum_i n_{ij}$ and $n_{\cdot\cdot} = \sum_i \sum_j n_{ij} = n$

Thus the 2 × 2 contingency Table 10.1 may be written as:

Table 10.2 Notation fro 2 × 2 contingency table

Opted course i	Sex j		Total
	Male	Female	
Mathematics	$20 = n_{11}$	$10 = n_{12}$	$30 = n_{1\cdot}$
Non-mathematics	$40 = n_{21}$	$30 = n_{22}$	$70 = n_{2\cdot}$
Total	$60 = n_{\cdot 1}$	$40 = n_{\cdot 2}$	$100 = n_{\cdot\cdot} = n$

It may be noted that $n_{1.}$ and $n_{2.}$ are the 1st and 2nd row totals and $n_{.1}$ and $n_{.2}$ are the 1st and 2nd column totals. Table of the form 10.2 for any two nominal variables is called a 2 × 2 contingency table.

In this set up there may be two types of hypotheses that may be required to be tested to conclude

(a) Whether the two nominal variables are associated.
(b) Whether the two categories of nominal variable 1 are homogeneous as regards two categories of nominal variable 2.

For the data in Table 10.1, one may be interested to conclude

(a) The proportion of two opted course in school children is not associated with sex.
(b) The proportion of opted courses are same in the female and male school children.

The actual computational procedures, though are very similar, the null hypothesis have different set up. Both the test statistics follow chi-square distribution and the general formula is same as given in (10.2).

The test for (a) is called χ^2-test of independence and that for (b) is called χ^2-test of homogeneity. These two tests may also be required for an $r \times c$ contingency table.

First the-test of independence is considered and next that for homogeneity.

(A) χ^2-test of Independence (for 2 × 2 Contingency Table)

Suppose *the data in hand* on two nominal variables A and B are as shown in Table 10.2.

The hypothesis to be tested can be written as H_0: A and B are independent.

To express this mathematically, let p_{ij} be the probability that an observation belongs to i^{th} row and j^{th} column of Table 10.2, $i = 1, 2$ and $j = 1, 2$.

In other words

$$p_{ij} = P[\text{an observation belongs to } (i, j)^{th} \text{ cell}]\ i = 1, 2, j = 1, 2 \tag{10.24}$$

Further let $$p_{i.} = \sum_j p_{i.j} = P[\text{an observation belongs to } i^{th} \text{ row}]$$

and $$p_{.j} = \sum_i p_{i.j} = P[\text{an observation belongs to } j^{th} \text{ column}] \tag{10.25}$$

$$\sum_i p_{i.} = \sum_j p_{.j} = \sum_i \sum_j p_{ij} = 1 \tag{10.26}$$

Hence the hypothesis of independence of A and B can be written as

$$H_0: p_{ij} = p_{i.} \times p_{.j} \text{ for } i = 1, 2 \text{ and } j = 1, 2$$

Under the null hypothesis $(i, j)^{th}$ cell probability is equal to the product of the probabilities that an observation belongs to i^{th} row and that it belongs to j^{th} column.

The *test statistic* to test H_0 is given by equation (10.12) as

$$\chi^2 = \sum_{i=1}^{2} \sum_{i=1}^{2} \frac{(O_{ij} - E_{ij})^2}{E_{ij}} \tag{10.27}$$

where the expected frequency of $(i, j)^{th}$ cell E_{ij} under H_0 is given by

$$E_{ij} = np_{ij} = n\ p_{i.}\ p_{.j} \tag{10.28}$$

Since actual probabilities p_{ij}, $p_{i.}$ and $p_{.j}$ are not given, their estimates from the data are used. It is logical to use the estimates

$$\hat{p}_{i.} = \frac{n_{i.}}{n_{..}} \text{ and } \hat{p}_{.j} = \frac{n_{.j}}{n_{..}} \tag{10.29}$$

Hence E_{ij} is determined as

$$\hat{E}_{ij} = n\,\hat{p}_{i.}\ \hat{p}_{.j} = n\frac{n_{i.}}{n} \times \frac{n_{.j}}{n}$$

$$= \frac{n_{i\cdot}\, n_{\cdot j}}{n} \tag{10.30}$$

$$= \frac{i^{th} \text{ row total} \times j^{th} \text{ column total}}{\text{Grand total (or total sample size)}} \tag{10.31}$$

Using these the test statistic can be computed. However, it may be noted that all the four E_{ij} for 2 × 2 contingency table need not be computed. It is required to compute only $\hat{E}_{11}$. The other three can be obtained by subtraction since the marginal totals of $\hat{E}_{ij}$ must be maintained. Thus

$$\hat{E}_{12} = n_{1\cdot} - \hat{E}_{11},\ \hat{E}_{21} = n_{\cdot 1} - \hat{E}_{11} \text{ and } \hat{E}_{22} = n_{2\cdot} - \hat{E}_{21} \tag{10.32}$$

The *test statistic follows Chi-square distribution*, if *null hypothesis is true*. The degree of freedom = (2 – 1) (2 – 1) = 1.

In symbols, the test statistic

$$\chi^2 \sim \chi^2_1 \text{ if } H_0 \text{ is true} \tag{10.33}$$

If the *level of significance is α*, the *test procedure* is

$$\text{Reject } H_0 \text{ if } \chi^2 > \chi^2_{1,\,\alpha} \tag{10.34}$$

Large value of χ^2 indicates that expected frequencies do not tally with observed frequencies, if the assumption of independence of two variables is used to obtain expected frequencies.

Now an example of testing independence in a 2 × 2 contingency table is considered by using the data of Table 10.1.

Example 10.2: Considering data of Table 10.1, test the hypothesis that sex of school children is not associated to choice of course.

Solution: The hypothesis to be tested is

H_0: $p_{ij} = p_{i\cdot}\, p_{\cdot j}$ $i = 1, 2, j = 1, 2$ against
H_1: at least one of these relations do not hold

The test statistic requires calculation of $\hat{E}_{11}$, which is

$$\hat{E}_{11} = \frac{n_{1\cdot} \times n_{\cdot 1}}{n} = \frac{30 \times 60}{100} = 18$$

Hence $\hat{E}_{12} = 30 - 18 = 12$

$\hat{E}_{21} = 60 - 18 = 42$ and $E_{22} = 70 - 42 = 28$

The expected frequencies may be included in brackets for ease of computation as shown below

	Sex		
Opted course	Male	Female	Total
Mathematics	20 (18)	10 (12)	30
Non-mathematics	40 (42)	30 (28)	70
Total	60	40	100

Hence

$$\chi^2 = \frac{(20-18)^2}{18} + \frac{(10-12)^2}{12} + \frac{(40-42)^2}{42} + \frac{(30-28)^2}{28}$$

$$= \frac{4}{18} + \frac{4}{12} + \frac{4}{42} + \frac{4}{28} = 4\left(\frac{1}{18} + \frac{1}{12} + \frac{1}{42} + \frac{1}{28}\right)$$

$$= .7936$$

If $\alpha = .05 \quad \chi^2_{1,\ .05} = 3.84$

Since $\chi^2_{cal} < \chi^2_{1,\ .05}$ H_0 is not rejected.

Hence it is concluded that the available data supports the hypothesis of independence of sex and opted course of school children.

Note: Since the degree of freedom of the test statistic is 1, as it was remarked, Yates Correction should be used. However, it is obvious that the corrected χ^2_c will be further decreased from the calculated χ^2. So H_0, even then will not be rejected. Hence the conclusion will not change. So it is not done in this example.

Yates Correction

Since the degree of freedom for a 2 × 2 contingency table is 1, Yates correction should be used by computing

$$\chi^2_c = \sum_{i=1}^{i} \frac{\left[|O_{ij} - E_{ij}| - .5\right]^2}{E_{ij}} \tag{10.35}$$

In a 2 × 2 table there is a computational short cut to evaluate the test statistic. It can be shown algebraically that

$$\chi^2 = \sum_{i=1}^{2}\sum_{j=1}^{2} \frac{\left(O_{ij} - E_{ij}\right)^2}{\hat{E}_{ij}} = \sum_{i}\sum_{j} \frac{\left(n_{ij} - \hat{E}_{ij}\right)^2}{\hat{E}_{ij}}$$

$$= \frac{n\left(n_{11}\,n_{22} - n_{12}\,n_{21}\right)^2}{n_{1\cdot}\,n_{2\cdot}\,n_{\cdot 1}\,n_{\cdot 2}} \tag{10.36}$$

This formula avoids computation of any expected frequency and resulting rounding errors, due to $\hat{E}_{11}$ being possibly often fractional.

A conventional alternative to the formula (10.36), which is easy to remember is given by

$$\chi^2 = \frac{n(ad - bc)^2}{(a+c)\,(b+d)\,(a+b)\,(c+d)} \tag{10.37}$$

when the frequencies in the 2 × 2 table are as

	A		
B	1	2	Total
1	a	b	a + b
2	c	d	c + d
Total	a + c	b + d	a + b + c + d = n

Applying this on the Example 10.2

$$\chi^2 = \frac{(20\times30-10\times40)^2\times100}{60\times40\times30\times70} = \frac{200^2\times100}{60\times40\times30\times70}$$

$$= .7937$$

Which is same as the statistic calculated earlier.

The statistic after applying Yates *correction* is also simpler and is given by

$$\chi_c^2 = \frac{n\left(|\,ad-bc\,|-n/2\right)^2}{(a+c)\,(b+d)\,(a+b)\,(c+d)} \tag{10.38}$$

(B) χ^2-test of Independence (for $r \times c$ Contingency Table)

As already mentioned if the categories or classes or groups with respect to first nominal variable are r and those of second nominal variable are c, classification of n objects in $r \times c$ mutually exclusive cells gives $r \times c$ contingency table.

As an example of $r \times c$ contingency table consider the following Example 10.3.

Example 10.3: For convincing the public to use safety equipment in automobile, a random sample of 1000 accidents is chosen from records. Each accident is classified according to type of safety equipment used and the severity of injury received. The data results in the following table (source: Milton and Tsokos (1983), p. 409).

Type of safety equipment

Extent of injury	Seat belt only	Seat belt and harness	None	Total
None	75	60	65	200
Minor	160	115	175	450
Major	100	65	135	300
Death	15	10	25	50
Total	350	250	400	1000

In this table, the extent of injury has four groups or levels, whereas the second variable the type of safety equipment has three groups, giving a 4 × 3 contingency table.

The public does not give importance to safety equipments as they feel, extent of injury is not associated to use of safety equipment. Thus the experimenter would like to establish that there is an association between the variables. This can be proved if the null hypothesis of independence is rejected by the χ^2 test of independence.

Notation and χ^2-test Procedure for Independence

The notation used for this type of $r \times c$ table are given below to describe the computational procedure It may be noted that

n_{ij}: number of observations in $(i, j)^{th}$ cell
$n_{i\bullet}$: marginal total of ith row: $i = 1, 2, \ldots r$
$n_{\bullet j}$: marginal total of jth column: $j = 1, 2, \ldots c$

$r \times c$ Contingency table

	$j \rightarrow$		Variable A					
Variable B	1	2	3	......	j	....	c	Total
$i \downarrow$								
1	n_{11}	n_{12}	n_{13}	....	n_{1j}	....	n_{1c}	$n_{1\bullet}$
2	n_{21}	n_{22}	n_{23}	....	n_{2j}	....	n_{2c}	$n_{2\bullet}$
3	n_{31}	n_{32}	n_{33}	....	n_{3j}	....	n_{3c}	$n_{3\bullet}$
:	:	:	:	:	:	:	:	:
i	n_{i1}	n_{i2}	n_{i3}	....	n_{ij}	....	n_{ic}	$n_{i\bullet}$
:	:	:	:	:	:	:	:	:
r	n_{r1}	n_{r2}	n_{r3}	....	n_{rj}	....	n_{rc}	$n_{r\bullet}$
Total	$n_{\bullet 1}$	$n_{\bullet 2}$	$n_{\bullet 3}$	....	$n_{\bullet j}$	....	$n_{\bullet c}$	$n_{\bullet\bullet} = n$

Let p_{ij} = P[an observation belongs to $(i, j)^{\text{th}}$ cell]

$p_{i\bullet} = \sum_{j=1}^{c} p_{ij}$ = P[an observation belongs to i^{th} row]

$p_{\bullet j} = \sum_{i=1}^{r} p_{ij}$ = P[an observation belongs to j^{th} column]

The *hypothesis of independence* is again set up as H_0: A and B are independent or

$$H_0\text{: } p_{ij} = p_{i\bullet} \times p_{\bullet j};\ i = 1, 2, \ldots r \quad j = 1, 2, \ldots c \tag{10.39}$$

The *test statistic for testing H_0 against H_1*: The above probability relation does not hold for at least one cell, is similar as that in equation (10.27).

With minor modification it is given by

$$\chi^2 = \sum_{i=1}^{r}\sum_{j=1}^{c} \frac{(O_{ij} - E_{ij})^2}{E_{ij}} \tag{10.40}$$

where, as before $E_{ij} = n\ p_{ij} = n\ p_{i\bullet} \times p_{\bullet j}$ under H_0.

As p_{ij}, $p_{i\bullet}$ and $p_{\bullet j}$ are not known, their estimates from the data are used to find E_{ij}.

Thus $$\hat{p}_{i\bullet} = \frac{n_{i\bullet}}{n_{\bullet\bullet}} \quad \text{and} \quad \hat{p}_{\bullet j} = \frac{n_{\bullet j}}{n} \quad \text{for} \quad i = 1, 2, \ldots r \quad \text{and} \quad j = 1, 2, \ldots c \tag{10.41}$$

So $$\hat{E}_{ij} = \frac{n_{i\bullet}\, n_{\bullet j}}{n}$$

$$= \frac{(i^{th} \text{ row sum}) \times (j^{th} \text{ column sum})}{\text{sample size}} \tag{10.42}$$

The *distribution of the test statistic if H_0 is true* is Chi-square with d.f. = $(r - 1)(c - 1)$. In symbols

$$\chi^2 \sim \chi^2_{(r-1)(c-1)} \text{ if } H_0 \text{ is true} \tag{10.43}$$

If level of significance is chosen as α, the test procedure is

Reject H_0 if $$\chi^2 > \chi^2_{(r-1)(c-1),\alpha} \tag{10.44}$$

Conclusion depends on whether H_0 is rejected or not rejected. If H_0 is rejected variables A and B are associated.

This test procedure for $r \times c$ contingency table is illustrated in Example 10.4 by considering the 4×3 table given in Example 10.3.

Example 10.4: Considering data given in Example 10.3, test whether extent of injury is associated with type of safety equipment used while riding the automobile.

Solution: In the following table the entries in brackets ( ) contain expected frequencies in each cell.

Table 10.3 Test of independence in contingency table

Extent of injury	Type of safety equipments			Total
	Seat belt only	Seat belt and harness	None	
None	75 (70) [0.3571]	60 (50) [2.000]	65 (80) [2.8125]	200
Minor	160 (157.5) [0.0397]	115 (112.5) [0.0556]	175 (180) [0.1389]	450
Major	100 (105) [0.2381]	65 (75) [1.3333]	135 (120) [1.8750]	300
Death	15 (17.5) [0.3571]	10 (12.5) [0.5000]	25 (20) [1.2500]	50
Total	350	250	400	1000

Since the column totals and row totals of expected frequencies should match the observed totals in rows and columns, all the 4×3 cells' expected frequencies need not be calculated. The last row and last column entries of expected frequencies are obtained by subtraction. Only $3 \times 2 = 6$ expected frequencies of first three rows and first two columns are obtained using the formula in equation (10.42).

Thus
$$\hat{E}_{11} = \frac{200 \times 350}{1000} = 70 \quad \hat{E}_{12} = \frac{200 \times 250}{1000} = 50$$

and
$$\hat{E}_{13} = 200 - 70 - 50 = 80$$

Similarly all the values of expected frequencies are calculated and have been included in a bracket ( ) in each cell.

Now the terms $(O_{ij} - \hat{E}_{ij})^2 / \hat{E}_{ij}$ are calculated for each cell and are put in another bracket [ ]. All the values in [ ] are added to obtain the test statistic

$$\chi^2 = \sum_{i=1}^{4} \sum_{j=1}^{3} \frac{(O_{ij} - \hat{E}_{ij})^2}{\hat{E}_{ij}} = 10.9573$$

If H_0 is true, *the* χ^2*-test statistic has* χ^2*-distribution with* $(4 - 1) \times (3 - 1) = 3 \times 2 = 6$ *d.f.* So from the Table A.5, 5% (= α) upper tail value is read as $\chi^2_{6,0.05} = 12.6$.

Since calculated value of $\chi^2 = 10.9573 < 12.6$, H_0 is not rejected.

Hence it may be concluded from the observed data that extent of injury is not associated to safety equipment used by automobile riders.

(C) Testing Homogeneity

Sometimes a number of sets of data are obtained on the same variable. Let the number of data sets be r and let there be c levels of the variable under study. It is generally pre-decided that a definite or fixed number of observations will be taken in each of the r data sets.

Let n_1, n_2, ...n_r be the number of observations (or sample sizes) in the 1st, 2nd, ...r^{th} data set. Each of these data sets is divided into c categories according to the variable. Let n_{ij} be the number of observations of ith data set in j^{th} category or class; (i = 1, 2, ...r and j = 1, 2, ...c). The total number of observations in the various categories of variable are not fixed but are *random*.

These can be arranged in the following table

Table 10.4 Frequency table of r data sets (or r samples)

	Variable categories →							
Data sets	1	2	3	...	j	...	c	Total (Fixed)
1	n_{11}	n_{12}	n_{13}	...	n_{1j}	...	n_{1c}	$n_{1\cdot} = n_1$
2	n_{21}	n_{22}	n_{23}	...	n_{2j}	...	n_{2c}	$n_{2\cdot} = n_2$
3	n_{31}	n_{32}	n_{33}	...	n_{3j}	...	n_{3c}	$n_{3\cdot} = n_3$
⋮	⋮	⋮	⋮	...	⋮	...	⋮	⋮
i	n_{i1}	n_{i2}	n_{i3}	...	n_{ij}	...	n_{ic}	$n_{i\cdot} = n_i$
⋮	⋮	⋮	⋮	...	⋮	...	⋮	⋮
r	n_{r1}	n_{r2}	n_{r3}	...	n_{rj}	...	n_{rc}	$n_{r\cdot} = n_r$
Total	$n_{\cdot 1}$	$n_{\cdot 2}$	$n_{\cdot 3}$	...	$n_{\cdot j}$	...	$n_{\cdot c}$	

Let p_{ij}(i = 1, 2, ...r; j = 1, 2, ..., c) denote the proportion of observations in i^{th} data set which belong to j^{th} category of study variable. Since an observation of ith data set (i = 1, 2, ...r) belongs to exactly one of c mutually exclusive categories of study variable

$$\sum_{j=1}^{c} p_{ij} = 1 \text{ for each } i \tag{10.45}$$

The hypothesis of homogeneity of r data sets implies that in each of the c categories of study variable, none of the r data sets are more prevalent than the other (r – 1) data sets.

In symbols, the *null hypothesis* will be

$$H_0: p_{1j} = p_{2j} = p_{3j} = \ldots = p_{ij} = \ldots = p_{rj} \text{ for } j = 1, 2, \ldots, c \tag{10.46}$$

against the alternative.

H_1: The above equality does not hold for at least one i and j

$$i = 1, 2, \ldots, r;\ j = 1, 2, \ldots, c \tag{10.47}$$

This test is called a *test of homogeneity.*

If the null hypothesis is true, all the data sets are homogeneous, they can be pooled and the estimate of probability that an observation belongs to j^{th} category

$$= \frac{\text{Total observations in category } j}{\text{Total number of observations}} = \frac{n_{\cdot j}}{n} \tag{10.48}$$

where, $n = n_1 + n_2 + \ldots + n_r$; j = 1, 2, ..., c.

Hence the estimated expected frequency of ith data set (i = 1, 2, ..., r) in category j

$$\hat{E}_{ij} = \frac{n_i\, n_{\cdot j}}{n}$$

$$= \frac{(i^{th} \text{ row total}) \, (j^{th} \text{ column total})}{\text{Total number of observations}}; \; j = 1, 2, \ldots, c \tag{10.49}$$

Thus it is seen that the expected frequencies can be computed as for testing independence.

The test statistic for testing H_0 against H_1 is also same as that for test of independence and is given by

$$\chi^2 = \sum_{i=1}^{r} \sum_{j=1}^{c} \frac{\left(O_{ij} - \hat{E}_{ij}\right)^2}{\hat{E}_{ij}} \tag{10.50}$$

The *distribution of the test statistic* if H_0 is true is χ^2 with $(r - 1)(c - 1)$ d.f.

In symbols

$$\chi^2 \sim \chi^2_{(r-1)(c-1)} \text{ if } H_0 \text{ is true} \tag{10.51}$$

The *test procedure with level of significance* α is also same as before, to

$$\text{Reject } H_0 \text{ if } \chi^2 > \chi^2_{(r-1)(c-1),\alpha} \tag{10.52}$$

It may be noted that the null hypothesis is different for testing homogeneity from that for testing independence.

If the hypothesis of homogeneity is not rejected one may decide to combine all the data sets for estimation or other analysis.

As an illustration of the above test procedure consider the following Example 10.5.

Example 10.5: A study was done on the extent of drug usage among Science and Engineering College Students, who were professed users of drug. A random sample of 150 B.Sc./B.Tech. 1st year, 135 of 2nd year and 125 3rd year and 100 M.Sc./M.Tech. Previous students were selected to fill up an anonymous questionnaire indicating the extent of his or her drug use as experimental or casual or moderate to heavy. The result of the study is shown in Table below.

Table 10.5 Extent of drug usage among science and engineering college students (source: Daniel (1987), p. 536)

Class	Extent of drug usage			Total
	Experimental	Casual	Moderate to heavy	
B.Sc./B.Tech. 1st year	57 (63.24)	50 (51.47)	43 (35.29)	150
B.Sc./B.Tech. 2nd year	57 (56.91)	58 (46.35)	20 (31.77)	135
B.Sc./B.Tech. 3rd year	56 (52.70)	45 (42.89)	24 (29.41)	125
M.Sc./M.Tech. (Previous)	45 (42.15)	22 (34.32)	33 (23.53)	100
Total	215	175	120	510

Are these data indicating that the four populations of students are homogeneous with respect to extent of drug use?

Solution: Random samples of sizes $n_1 = 150$, $n_2 = 135$, $n_3 = 125$ and $n_4 = 100$ from four populations are in hand. With respect to 2nd variable, the extent of drug usage these are classified in three categories. Thus there is 4×3 frequency table.

Let $p_{ij} = P$[an observation of ith population belongs to j^{th} category]

$$i = 1, 2, 3, 4; \; j = 1, 2, 3$$

The *hypothesis to be tested* is

$$H_0: p_{1j} = p_{2j} = p_{3j} = p_{4j} \text{ for } j = 1, 2, 3 \text{ against}$$

$$H_1: \text{all the above relations do not hold}$$

The *test statistic for testing* H_0 requires computation of $\hat{E}_{ij} = \frac{n_{i\cdot}\, n_{\cdot j}}{n}$

Thus
$$\hat{E}_{11} = \frac{215}{510} \times 150 = 63.24, \quad \hat{E}_{21} = \frac{215}{510} \times 135 = 56.91$$

$$\ldots \quad \hat{E}_{44} = \frac{120}{510} \times 100 = 23.53$$

The estimated expected frequencies are written in ( ) parentheses of Table 10.5.

Now the χ^2-test statistic can be calculated using equation (10.50) and the value obtained is

$$\chi^2 = 19.4$$

The χ^2-*test statistic has* χ^2-*distribution with* $(4 - 1) \times (3 - 1) = 6$ *d.f.* if H_0 is true.
Thus for testing H_0 at 1% level read $\chi^2_{6,0.01} = 16.812$ from Table A.5.
Since calculated $\chi^2 = 19.4 > 16.812$, H_0 is rejected.
So it is *concluded* that the four populations of students are not homogeneous as regards extent of drug usage.
Note: This test can also be applied if two samples from two populations are available in two categories. The hypothesis of homogeneity in this case is equivalent to H_0: $p_1 = p_2$, where p_1 and p_2 are proportions of observations in one of the two categories of variable in the two populations.

Thus χ^2-test of homogeneity may be treated as an alternative to normal test of equality of two proportions already considered in Section 10.1.

The d.f. for χ^2-test statistic is 1 in this case. So Yates correction should be applied for testing homogeneity in 2×2 table.

Conditions of Validity of χ^2-test

It has been mentioned that the χ^2-test statistic for testing independence or homogeneity are based on discrete data. The limiting distribution of the test statistic is close to the theoretical Chi-square distribution which is continuous. Unless the approximation is quite close to theoretical and continuous distribution the conclusions drawn from the test will not be valid. It has been found that the approximation is quite satisfactory if the expected frequencies E_{ij}'s are not very small.

For many years it was suggested that all the expected frequencies should be at least 5. If the total sample size n is large, the expected frequencies will be ≥ 5. Cochran (1952, 1954) conjectured that even small expected frequency can be tolerated. This was not put in practice for quite some time till Roscoe and Byars (1971) concluded on the basis of extensive empirical analysis, that it appears that *expected frequency as small as 1.0* with $\alpha = 0.05$ or *as small as 2.0* for $\alpha = 0.01$ are acceptable. If $k \geq 3$, $n \geq 10$ *and* $n^2/k \geq 10$, Kochler and Larntz (1981) suggest to use χ^2-tests, if Roscoe and Byars (1971) conditions are also fulfilled. For $k = 2$, however, the requirement of $E_{ij} \geq 5$ should still be followed.

Hence, it is therefore, advisable that n should be taken sufficiently large so that most of expected frequencies are at least ≥ 2.

11. Electronic Data Handling

An *electronic device* that accepts data, stores it, retrieves it when required, processes the data as per instructions to convert it into useful information and communicates it as an output, is called a *computer*. The machine came into being in early 20th century due to the man's search for a fast and accurate calculating device. Very accelerated growth of computer industry has now made it possible that almost every family and almost every educated person owns or has access to a personal computer.

It is now essential in all the disciplines to use computer for proper analysis of relevant data for quick results. Thus knowledge can be deeply explored in no time and results applied for advancement in academic, technical, scientific, agricultural and medical lines. Statistical analysis also requires computers for correct and quick results. The knowledge to use and interact with the machine is essential so that the capabilities of the machine are utilized following correct method purposefully.

In this chapter, some elementary ideas about the structure and features of computer are discussed as data is now electronically handled by computers. How to establish a man machine interaction is described in Section 11.2. Section 11.3 gives some basics of the operating system, describes use of two application softwares and the most useful data transfer platform of computer network.

11.1 Introduction to Computers

As already described, a *computer* is an electronic device which can process information i.e. can store, manipulate and retrieve information as and when required.

The computers may be classified as digital and analogue computer. The modern computers are mostly digital. So the word computer generally accepted is being a *digital computer*, which operates essentially by *counting*. The *analogue computer operates by measuring* rather by counting. This machine has only limited memory and is restricted in the type of calculations it can perform. Thus it can be used for certain specialized engineering and scientific applications.

Components of a Computer System

Basically computer receives information (data and/or instructions) as *input,* it processes this inside another part of the machine called *Central Processing Unit (CPU)* and then presents or gives out the results in a useful form as *output.*

Figure 11.1 gives the basic structure of the computers.

It is evident that a computer is made up of mainly three components

(a) Input Device
(b) Central Processing Unit
(c) Output Device

(a) *Input device* is used for transferring data and/ or instructions to the computer system. Commonly used input devices are keyboard, mouse and magnetic disks (floppy diskettes or compact disks (CD's)).
(b) *Central processing unit*, the CPU as already mentioned is the brain of the computer. This unit has basically three parts.

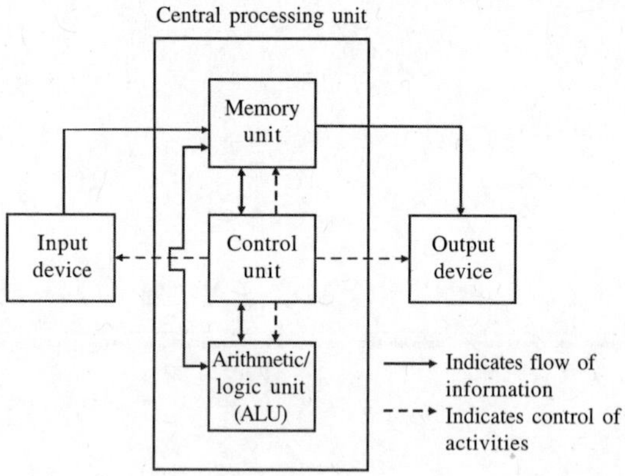

Fig. 11.1. Basic components of the computer (source: Hunt and Shelly (1982), p. 13).

(i) The memory unit
(ii) The control unit
(iii) The Arithmetic/Logical Unit (ALU)

(i) *The memory unit* is used to store the data and instructions inside the computer before the actual processing takes place. Similarly, the results produced by computer are also stored before it is passed out to output device.
(ii) *The control unit* directs and coordinates all the activities in the entire computer system. Instruction for receiving and transferring information from one part of CPU to the other and for communicating the results outside, are all coordinated by this unit. The program instructions are retrieved in proper sequence from the memory and interpreted. Control unit informs the ALU the specific operations to be performed and directs the transfer of any data items to ALU, which is needed for the operation.
(iii) *The ALU* executes the calculations and the comparisons. The arithmetic unit of ALU does all the manipulation of numbers by arithmetical operations of addition (+), subtraction (–), multiplication (×) and division (÷). Comparison and logical operations are performed by the *logical unit* when such instructions are given using logical operators: equal to (=), not equal to (≠), less than (<) and greater than (>). The result of comparison is either 'true' or 'false' and is used to control flow of the program as to which instructions to perform next. Results of comparison are also stored in memory and are used by control unit also, for directing the next steps.

(c) *The output device* communicates the result of processing in computer to the user. The computer monitor with cathode ray tube (CRT) or TFT, printer, magnetic disk (floppy, CD) and graph plotter etc are the out put devices.

The Development of Computers

Evolution of computers took place due to keenness in search of fast and accurate calculating device. The first digital computing instrument named 'abacus' was invented in Asia Minor about 3500 BC. John Napier (1550-1617) during early seventeenth century designed a rod (called Napier's bone) based multiplication calculator. In 1642 Blaize Pascal (1623-1662) invented a device, he called it numerical wheel calculator, which has some automation. It is in the form of a rectangular brass box also called a *Pascaline*. GW Von Leibnitz (1646-1716) extended Pascaline from only addition and subtraction to multiplication and division. In 1820, Charles T. de Colmer (1785-1870) invented a mechanical calculator, called the 'arithometor'. It could perform all the four arithmetic operations.

Charles Babbage (1792-1871) considered to be father of digital computers designed his 'Differential Engine' in 1822 and worked on his 'Analytical Engine'. The former facilitated calculations of the tables while the latter was a general purpose one. Through the 'Analytical Engine', it was recognized that computers could be 'programmed' to perform long sequences of arithmetic and decision operations and printing of results without human intervention.

Babbage proposed to feed instructions and data to the computer on punched cards. For this purpose an electric machine was developed by Herman Hollerith in 1890 for use in US census. In 1896 he established his own company called Tabulating Machine Company (TMC) now known as International Business Machines (IBM).

In 1939-44 Howard H. Aiken developed Mark I, also called *Automatic Sequence Controlled Calculator* which incorporated principles of 'Analytical Engine'. It used the punched paper tape for feeding the program. *Mark I* had speed of about 200 operations per minute. However, it was complex in design and huge in size. Physically it was 51 feet in length and about 8 feet in height, contained $3/4^{th}$ of a million parts and more than 500 miles of wire. It was an electromechanical machine.

In 1946 the first all electronic computer *ENIAC (Electronic Numerical Integrator and Calculator)* was constructed at the Moore School of Engineering of University of Pennsylvania under the leadership of Eckert and Mauchly.It used high speed vacuum tube switching devices. In a single hour ENIAC could accomplish calculations which MARK I would have taken more than one week.

A major break through in the logical design of computers occurred when concepts of *stored program* was proposed by Von Neumann in 1946. This principle was used by Maurice Wilkes of Cambridge University, UK to design the computer called *EDSAC (Electronic Delay Storage Automatic Calculator)* and completed it in 1949.

Commercial production of such electronic computer began in early 1950's. One such early computer of this type was *UNIVAC I* built by Univac Division of Remington Rand in 1951. *EDVAC (Electronic Discrete Variable Automatic Computer)* was also launched during the same period in USA.

The development in computer technology continued and there is no clear cut pattern to demarcate the 'Generations' of computers.

Generations of Computers

The period described above may be called the *first generation* of computers. Table 11.1 gives a summary of the subsequent periods of generations, the basic technology used, some computers of those periods and their distinguishably improved characteristics and applications.

The custom of referring to the computer era in terms of 'generations' came into wide use after 1964. Originally this term was used to distinguish hardware technologies. Since 1968, it has been extended to include in it software also. Since both together make the entire system.

At this point it is worth explaining the terms 'hardware' and 'software'.

The machinery units which comprise all the components of the computer system is called *hardware*. The computer 'hardware' cannot function without a set of instructions, called *software*, that directs the computer to perform a particular task in a particular order. Hardware and Software together form a complete computer system.

Table 11.1 summarizes details of *computer generations*.

Table 11.1 Summary of computer generations

Generation	Period	Electronic components used	Memory	Speed and MTBF*	Software	Brand names	Advantage
First	1946-55	Vacuum tubes	1 Kilobyte (KB) (1024 B)	30 Instructions per second (30 IPS) 1/2 to 1 hour	Machine and assembly languages	ENIAC, EDSAC, UNIVAC, EDVAC	Automatic processing
Second	1956-65	Transistor	128 KB	300 IPS about 10 hours	High level languages, FORTRAN, COBOL, ALGOL, Basic O/S	WHIRLWIND I, IBM 605, IBM 7000, CDC1604, 3600	Smaller in size, fast processing
Third	1966-75	IC chip with SSI and MSI circuit (small scale and medium scale integrated)	> 1 MB (1×10^6 Bytes)	1 K IPS (1×10^3 IPS) about 100 hours	FORTRAN IV, COBOL 68, PL/1, time shared O/S (operating system)	IBM 360, ICL1900, PDP-8, IBM 370/168 CDC 6400, 6600, 7600	More smaller low maintenance cost, more fast, more compact
Fourth	1976-85	Microprocessor with LSI and VLSI Technology (large scale and very large scale integrated)	> 16 MB	> 8 MIPS (8×10^6 IPS) about 1000 hours	FORTRAN 77, Pascal, ADA, COBOL-74 Linking virtual storage text editors	Various company's PC-XT, PC-AT, Main Frame, MINI CYBER 170, HP, HCL	Smaller in size, faster cheapest, minimum maintenance
Fifth	1986-present	Microprocessor with ULSI (Ultra large scale integrated) wireless enabled multimedia over 100 million transistors per chip	1 GB main	> 20 logical instructions per second (LIPS) about 1,20,000 hours	FORTRAN 90/95, C, C++, JAVA, SQL, HTML, Perl, Pascal, Visual Basic etc.	PC-Pentium I, II, III, IV, V, Hp, DELL, WIPRO, COMPAQ etc Supercomputer STAR	Very small in size, very fast, cheap, minimum maintenance

*MTBF is mean time between failures of the processor

Development of Computer Processors

About a decade ago *PENTIUM* processors came on PC's developed by Intel. They launched up to PENTIUM IV dual core last year. By the end of this year quad – and mobile processors are coming up, which will be further greatly reduced in size, will consume significantly less power, will be much speeded up (> 3GHz) in division, video, photo imaging, and for high performance software, will have 50% more Cache. AMD is also bringing out similar processors in competition.

Classification of Computers

Computers are classified as micro, mini, super-mini, midi and maxi based on the CPU speed, word length, main memory size, type and variety of peripherals which can be attached, the software supporting the system and the cost.

The cost of microcomputers or PC's range from Rs. 10 to 30 thousand whereas maxi or super computers cost about several crores of rupees. The Table 11.2 gives some important characteristics of the said five classes of computers.

Table 11.2 Computer classes and their characteristics

Classes	CPU speed	World length	Main memory size	Types of peripherals	Typical machines
1. Micro-PC	500 K IPS	8-16 bits	256 KB-1 MB	Floppy disk, CD, pen, DVD, printer, scanner, FAX, vdeo, hard disks	IBM PC compatibles (India), Apple USA, IBM (USA)
2. Mini	1 MIPS	16 bits	256 KB-4 MB	Hard disks, line printers, magnetic tapes, 4-8 user terminals	HCL, HORIZON, SN23 (India) PDA 11/70 (USA)
3. Super mini	2 MIPS	32 bits	1-8 MB	Exchangeable hard disks, hard disks, line printer, magnetic tapes, 16-32 user terminals	Nordsk (Norway) ND 570 (India) VAX 8400(USA) ICL 2904 (UK)
4. Midi	1-10 MIPS	32-64 bits	2-64 MB	Fast line printers (chain printers) 1000 MB disks high speed tapes. 32-128 user terminals mini computer as a front end processor	VAX 8800, CYBER 180, UNIVAC 1100/ 60, B-7800 (all USA), ACOS 1000, NEC (Japan)
5. Maxi or super mini	100 MIPS	64-96 bits	16-256 MB	Laser printers, very high speed disks, 1000 MB, high speed tapes, midi computer as front end processor	CRAY-2, ETA 10 (USA) NEC super, SX_2 (Japan)

In this section the components, and classification of computers, its brief history and generation have been discussed briefly.

Now in the next section discussion on ways of man machine communication will be included.

Development of Computer Architecture

The computer architecture has also changed along with their generations to cope up with increasing requirements of computer functions. The internal design has been developed so that functioning from *SISD* – single instruction implemented on single data, to *SIMD* – single instruction implemented on multiple data, to *MISD* – multiple instructions implemented on single data (not constructed for machines) to *MIMD* – multiple instructions implemented on multiple data have become possible.

For further reading on these topics one may refer the books by Rajaraman (1995, 2004), Hunt and Shelly (1982), Randell (1973) Rashtriya Computer Saksharta Mission (RCSM)(2001), Bell and Newell (1980), Tannenbaum (1984) and Vigyan Prasar (1999). The books by Hunt and Shelly (1982), Rajaraman (1995, 2004), RCSM (2001) and website cplus.about.com may be referred for the concepts discussed in the next section.

11.2 Man-machine Communication: Binary Code and High Level Languages

Computer is an electronic machine and performs the desired operations after receiving instructions in logical order. A human being can express his requirements in the language spoken by him. He can be obeyed by the person who knows the language in which instructions are given.

In the same way the computer can execute the instructions if it understands these. Being an electronic machine it senses only electromagnetic signals. These signals have only *two states – presence* or *absence*. A switch is *on or off*. A transistor is *conducting or non conducting*. A magnetic surface is *magnetized* or *non-magnetized*. Magnetic poles may be *aligned left to right* ($S \rightarrow N$) *or right to left* ($N \leftarrow S$). An electric pulse or voltage is *present or absent*.

Thus all the information which is needed to be given to computer is transformed or coded as strings of two symbols. The two digits 0 and 1 are conveniently used to express the two possible states. These are known as *bits* derived from *binary digits*.

All the familiar symbols or numbers or words which are used for giving an information to computer are transformed into *combination of bits*, which must be of a unique pattern for each symbol. A set of such combinations should include all the 26 English alphabets, the ten numeric digits 0 to 9, and about 32 special characters – such as punctuation marks, arithmetical symbols and parentheses. It can be seen that there are only $4 = (2 \times 2)$ unique combinations of two bits, namely

$$00 \quad 01 \quad 10 \quad 11 \tag{11.1}$$

If strings of 3 bits are used for a combination, there are following $8 = (2 \times 2 \times 2)$ unique combinations

$$000 \quad 001 \quad 010 \quad 100 \quad 011 \quad 101 \quad 110 \quad 111 \tag{11.2}$$

5 bits will give $2^5 = 32$ combinations, 6 bits will give $2^6 = 64$ combinations, 7 bits will give $2^7 = 128$ characters.

If lower case and upper case alphabets are to be coded in all $26 + 26 + 10 + 32 = 94$ combinations will be required to represent all the alphanumeric characters.

Computer Characters

A string of bits used to represent a character is known as *byte*.

The coding of characters has been standardized to facilitate exchange of recorded information between computers. The most popular standard is ASCII (american standard code for information interchange) code, which uses 7 bits to code each character. It is used in non-IBM computers. Another coding system, known as EBCDIC (extended binary coded decimal interchange code), is used in IBM computers. This code uses 8 bits per character so is capable of representing 256 characters.

It is often required that information in ASCII be used on a computer using EBCDIC and vice versa. A special ready made electronic circuit is used for carrying out this transformation and another circuit for transformation from, EBCDIC to ASCII code.

Though a byte for ASCII could be of 7 bits and that of EBCDIC is of 8 bits, but a *byte is commonly a string of 8 bits.* It can be noted that

1 Byte = 8 bits or one character
1 Kilo Byte (KB) = 1024 Bytes
1 Mega Byte (MB) = 1024 Kilobytes
1 Giga Byte (GB) = 1024 Mega Bytes
1 Tera Byte = 1024 Giga Bytes

Computer 'Words'

Human beings express information in *words*. So the computers also need *words* for exchange of information. A computer *word* is a group of bits, the length of which may vary for different computers. It is arranged differently for alpha numeric information and for numeric data. If a machine has 32 bits words, 8 bit bytes, then a *word* contains 4 characters each of 8 bytes, whereas numeric data uses the entire word to represent a number.

The Computer Languages

The language used by user to pass information to computer is named as *computer language*. Every common language has its own vocabulary or set of words, its grammar and punctuations. So is the case with computer languages. These are with smaller vocabulary and are simpler than natural human languages. However these have to be used with greater precision. The 'grammar' and 'punctuations' are to be strictly followed. A computer can not execute a wrongly written instruction, as it is not able to correct and deduce meaning of an incorrect instruction.

The *computer languages* or *programming languages* are basically of two types

1. Low level languages (a) Machine language and (b) Assembly language
2. High level languages

1.(a) Machine Language

Computer can understand only machine language, which is of binary nature. Structure of machine language is such that a machine instruction has a two part format, as given below in Fig. 11.2.

Operation (Code)	Address (Location)

Fig. 11.2. Form of machine instruction (source: Hunt and Shelly (1982), p. 61).

Both parts are represented internally in the machine's store by a string of binary digits. The set of instruction codes created in conjunction with the computer designer is called *machine language.*It is determined by the actual design of ALU, the control unit, word length and memory size. It is evident that writing program in this language is very difficult.

1.(b) Assembly Language

It is difficult to remember the binary machine codes. Even with a code book or glossary, mistakes can be made while encoding. To overcome these difficulties, the idea of *mnemonics* (or memory aids) or *abbreviated words* was introduced. These are easily identifiable by human brains. Some examples are MULT or MLT, SUB etc. The *mnemonics* are translated to binary patterns in computer, before it can understand the intended instruction and execute it. This act of translation is done by a pre stored program called *assembler*. The program written by user with *mnemonics* is called assembly codes or in assembly language. Since it is also oriented towards basic design of computers, it is also a low level language. Program written for one computer can not be used on another.

2. High Level Languages

The commercial viability and wider use of computer led to the necessity of developing 'High Level' languages by mid 1950's. These are more problem oriented than machine based. These languages use grammatical and mathematical notations similar to every day languages. These are much easier to learn and take much less time to write a program.

The instructions can be given using some commonly used English words and mathematical notations, but with precise rules of syntax and semantics. These languages can be used on almost every computer. However, the program must be translated to machine code for the computer.

Translators are of *two types: interpreter* and *compiler*. An *interpreter* translates a program – one statement at a time and immediately executes it. On the other hand, the *compiler* translates the entire program into lower level language and then executes it.

Types of High Level Languages

There are many high level languages but they can be broadly grouped into four applications areas: (i) scientific (ii) business (iii) special purpose and (iv) interactive.

The choice of a high level programming language depends on the area of applications for which program is to be written and the translator available in computer system.

The most commonly used high level language are FORTRAN, COBOL, BASIC, PASCAL, ALGOL, SNOBOL, LISP, ADA, C. A brief description is given below.

FORTRAN stands for ***For****mula* ***Tran****slation*. It is the oldest high level language developed in 1956-57 by John Backus and his team in IBM. This language was designed to solve the scientific and engineering problems. With evolution of computers and with increasing maturity of users, FORTRAN has also evolved to FORTRAN II and FORTRAN IV. This version was standardized by ANSI (The American National Standard Institute) to FORTRAN 77. The current versions of this language are FORTRAN 90 and FORTRAN 99.

COBOL stands for (***Co****mmon* ***B****usiness* ***O****riented* ***L****anguage*). This grew out of the desire of the data processing professionals for a high level machine independent data processing language, usable with available compiler. Work on the language began in 1959 with strong support from US defence. First report using COBOL was out in 1960. This language was also standardized by ANSI in 1968- and 1974. Currently COBOL is most widely used programming language for business data processing.

BASIC stands for ***B****eginners*' ***A****ll* purpose ***S****ymbolic* ***I****nstruction* ***C****ode*. It was developed by JH Kemeny and TE Kurtz in USA in 1963. This language is designed for a beginner and can be learnt quickly. Most of the computers have the provision of running BASIC program on them and some calculators also.

PASCAL is a programming language named after Blaise Pascal, designed by Niklaus Wirth in 1971, with the primary aim to allow beginners to learn good problem solving and programming method. PASCAL has been standardized by International Standard Organization (ISO) and ANSI. It has facilities to manipulate not only numbers but also vectors, matrices, strings of characters, sets, records, files and lists.

ALGOL stands for ***Algo****rithmic* ***L****anguage*, developed by an international committee in 1960's for scientific computation. It also has many versions and a recent one is ALGOL 68.

SNOBOL stands for ***S****tri****n****g* ***O****riented* ***S****ym****bo****lic* ***L****anguage*. It was designed by a group lead by Griswold in mid 1960s. It has facilities to manipulate strings of characters, such as splitting a string, combining strings, matching strings etc.

LISP stands for ***Lis****t* ***P****rocessing*. It was designed by Mc Carthy in early 1960's. It is used in applications such as game playing, theorem proving and similar non-numeric problem solving requiring logical operations. It is widely used in areas of pattern recognition and artificial intelligence.

ADA has been named in honour of *Lady Ada Augustha Lovelace* who is reputed to have written the first computer program for the mechanical computer Analytic Engine developed by Charles Babbage in 1842 in UK. The language was developed by a group of Honeywell Computer Company team under the leadership of Ichbiah for application in military operations.

C is a general purpose, language developed at Bell Laboratories in early 1970's. It is highly portable and is easy for programming. It can be used with DOS and Unix Operating Systems.

Object Oriented Programming Languages and Systems (OOPS)

Instead of writing program in high level languages, the use of *OOPS* is now very common, since every. thing in the world is an object. The type of object may vary in OOPS but one gets the power to create one's own objects as and when required. OOPS is a programming methodology where each entity is an object. Entities of related data together with routines associated with it are treated as one object of the program.

C^{++}, is currently most favored object oriented programming language. Some more are JAVA, PERL and DOT.net etc. Brief descriptions of some of these are also given.

C^{++} is a general purpose programming language invented in the early 1980s by Bjarne Stroutstrup at Bell Labs. It is similar to C invented in the early 1970s by Dennis Ritchie, but is a safer language than C and includes modern *object oriented programming*.

The purpose of C^{++} is to precisely define a series of operations that a computer can perform to accomplish a task. Any thing that a computer can physically do can be programmed in C^{++}, these may be putting data in a database, or pulling it out, displaying high speed graphics in a game or video, controlling electronic attached to the PC or even playing music and/or sound effects, one can even write software to generate music or composing it.

Apart from these there are some tasks which can be done on C^{++} but not very easily, such as designing GUI screens for applications. There are other languages such as Visual Basic on which this can be easily done.

For using C^{++} one needs its compiler. There are many commercial ones and some can be downloaded free of cost.

Java is a programming language created by James Gosling in June 1991 and originally developed by Sun Microsystems and released in 1995 as a core component of Sun Microsystems' Java platform. The language derives much of its syntax from C and C^{++} but has a simpler object model and fewer low-level facilities. Java applications are typically compiled to byte code that can run on any Java virtual machine (JVM) regardless of computer architecture.

The original and reference implementation Java compilers, virtual machines, and class libraries were developed by Sun from 1995. As of May 2007, in compliance with the specifications of the Java Community Process, Sun made available most of their Java technologies as free software under the GNU General Public License. Others have also developed alternative implementations of these Sun technologies.

The first public implementation was Java 1.0 in 1995. It promised "Write Once, Run Anywhere" (WORA). *Java 2*, and new versions had multiple configurations built for different types of platforms.

J2EE was for enterprise applications *J2ME* was for mobile applications. *J2SE* was the Standard Edition. In 2006, for marketing purposes, new *J2* versions were renamed *Java EE*, *Java ME* and *Java SE*, respectively.

Perl is a dynamic programming language created by Larry Wall working at Unisys, and first released in 1987. The name is "PERL" (for ***P**ractical **E**xtraction and **R**eport **L**anguage*). Perl borrows features from a variety of other languages including C, shell scripting (sh), AWK, sed and Lisp. Perl was widely adopted because it provides powerful text processing facilities without arbitrary data length limits, as were present in many Unix tools at the time. Now used for a wide range of tasks including system administration, web development, network programming, GUI development, and more.

Programming Perl in 1991 was also known as the "Camel Book". Perl 4 went through a series of maintenance releases, culminating in Perl 4.036 in 1993.

Perl 5 was released on October 17, 1994. It was a nearly complete rewrite of the interpreter, and added many new features.

As of 2007, Perl 5 is still being actively maintained. Important features and some essential new language constructs have been added.On December 18, 2007, Perl 5.10.0 was released. It includes notable new features, such as switch statement which brings it closer to Perl 6.

11.3 Working on DOS, Windows, MS Office and Computer Networks

The previous section gave an idea, how a user can communicate with computer through high level languages. A computer user can concentrate on designing an appropriate algorithm for solving a problem. User need not be concerned about the computer's internal structure. An *operating system* is the one that drives the computer machinery. When a computer is tuned on, the BIOS (Basic Input Output System) program comes into play. BIOS resides in the heart of every computer. The BIOS is a series of programs that are needed for all basic low level operations and control of hardware. Process of loading the operating system in memory is called *Booting*.

When a program written in high level language is executed, the following steps are followed.

1. The compiler to translate the program is loaded in the memory.
2. The source program is read and loaded in memory.
3. The source program is compiled into the object code. If there are any syntax errors, appropriate message is printed for user.
4. During compilation, machine language routines for reading and writing required by the program are brought in from a library of routines stored in peripheral store and linked to the object program.
5. The compiler may be taken out of memory to make space for data. The object program is now executed. If any errors are detected during execution, appropriate message is given for user to aid him correct the program.

An operating system manages all these steps. The user thus does not interact with computer hardware only. He interacts with machine + compilers + an operating system. There are various types of operating systems, of which the following are more commonly used:

(1) MS DOS (Microsoft Disk Operating System)
(2) WINDOWS
(3) UNIX
(4) LINUX

In this book some idea of working with MS DOS and how to use WINDOWS, which is currently in almost universal use, is briefly explained. For detailed and deeper knowledge some computer books or the manuals provided by the computer vendor may be consulted (Stevens 1989; Duncan 1988 a, b; Cooper 2001; Schwartz et al. 2001; Dyson and Coleman 2000; MCSE 2007).

DOS and Working with DOS

DOS stands for *disk operating system*. It can take user's English type commands and get it converted into machine language for processing and subsequently returns the processed results to the user in English through an output device. DOS is a link between user and computer. The user must know the command and the function which is performed by the computer while executing the *command*. After switching 'on' the computer, a small bar comes on the screen and blinks. The *small bar* is called a *cursor*. Some times 'A:\>' appears on the left most side of the screen if the DOS for computer is present in floppy disk. It may show 'C:\>' if the DOS is loaded in C: drive.

On computer one works on files. A file contains information on a particular topic. It must be referred while working on computer by a name. So computer may be required to read program from one file say A, read data from another file, say B and after processing the steps desired by user through the program, passes on the result to the third file say C. Thus each file must have a name.

A *file name* has two parts: (i) The primary name and (ii) the secondary name or extension. A dot '.' separates these two parts.

Primary name is of 1 to 8 characters and extension is of 3 or less characters.

Example: BIODATA, STUDENT.DAT, CLUST.DAT

The user has to give commands to computer in a proper order and in correct form. The DOS commands are of two types:

(a) The *internal commands* and (b) *external commands*.

The internal commands are automatically loaded in computer memory and can be executed immediately whereas, the external commands reside separately on hard disk or floppy diskette. They are loaded into memory when called by their specific command names.

Example of internal commands are: DIR, MD, RD, CD, DEL, COPY, TYPE, REN, CLS, etc. Each command has a specific syntax and leads to a specific computer operation.

Syntax of DEL is → DEL file name

Thus DEL RAMA.DAT deletes from computer the file named RAMA.DAT

Syntax of COPY is → COPY Source file name Target file name

COPY MARKS.DAT MARKS9.DAT

Copies the content of file named MARKS.DAT into a file named MARKS9.DAT

A user may consult some book on DOS or manuals which are provided with computers or the manual provided on the DOS disk. Following are some useful books – Stevens 1989; Cooper 2001; Duncan 1988, etc.

Some external commands are: DISKCOPY, CHKDSK, UNDELETE, XCOPY, DOSKEY, FORMAT, UNFORMAT, EDIT, PRINT, SORT, FIND etc.

These commands are used for external jobs to be done. The syntax should be strictly followed.

Syntax of a few are given below and the comprehensive list may be looked into the books and manuals mentioned above.

DISKCOPY A: B: and press < *Enter* > will copy entire content of A: drive into drive B: or DISKCOPY A: B: ↵ (where ↵ indicates press < Enter >).

If DOSKEY utility is loaded in the DOS commands directory, the DOS remembers the commands given by user.
EDIT command is used to create a file
For this the steps are as follows
Step 1 Type Edit ↵
A screen opens up
Step 2 Press ESC and press ALT and F. Now press N.
Step 3 Now type whatever information is required to go in the file.
Step 4 Press ALT F and S keys to save the created file.
Step 5 Computer wants the file name to be entered at a specific bar. So the typed material is saved in the given file name in that bar.
Step 6 To print the created file press ALT F and P.
Step 7 To come out of the created file press ALT F and X.
PRINT command is used to print files in background. The syntax is
PRINT file name ↵ prints the specified file.
Another way to print a file is using TYPE command. The syntax is TYPE file name > prn ↵
Details of other external commands can be seen in the books mentioned earlier.

WINDOWS and Working with Windows

A new family of operating system, after the use of MS DOS was already in practice, was announced by Microsoft in 1983. They named it *Windows.* It was proposed to include all the important features of DOS but with a graphical interface. Apart from various features of DOS many more new features have also been introduced in it.

As the name suggests Windows is an environment in a rectangular structure. A user can open several windows at a time, if he has to handle several files and run several programs.

Windows uses small graphical *icons*–which are pictures to represent documents, application software packages, folders, devices, web pages, and also other computers. Each icon has a text label to aid in the identification of respective functions.

Objects can be selected in many ways to use them for various purposes. There are some short *cut icons* on the desktop which can be selected by a double click of the mouse. Various *menu icons* on the opened Windows can be selected by placing the mouse pointer on it and using a left click. The file or a folder can also be selected either by a double click or by putting mouse pointer on the folder or file name in the list shown on the screen.

With each *menu icons* in a Window, there are listed a set of functions contained on another rectangle which are selected by moving the mouse pointer at it.

With Windows one can easily do the following:

1. Work with files
2. Work on the web
3. Search a file or folder in the computer.
4. Run the computational work program file or other application program.

Windows allows various peripheral devices to be attached and used with computer such as printer, scanner, copier, FAX machine, multimedia hardwares, internet connection etc.

The first of Windows-Windows 1.0 was launched in 1985. Windows family of operating system grew with more and more improved features. Working with larger memories and faster speed, capability for audio visual facilities, web facilities and remote connectivity can now be managed by this family of operating system.

The family *evolved* through:

Windows 2.0 released in 1987
Windows 3.0 released in 1990
Windows 3.1 released in 1992
Windows 3.11 released in 1992
Windows 95 released in 1995
Windows 98 released in 1998
Windows 2000 in 2000 with variants as Windows 2000 Server, Windows 2000 Advanced Server, Windows 2000 Data Center Server and Windows 2000 Professional
Windows XP, Windows XP with SP2, Windows 2008 and Windows Vista the most current one.

A computer system with current WINDOW XP operating system and various other basic application softwares require 64 MB to 4 GB memory, VGA (Video Graphics Array) Monitor, keyboard and mouse, 52 X CD ROM drive, network adapter cards and 80 GB hard disk.

Most of the computer systems with PENTIUM processor have much more speed and memory configurations now.

Most of the desktops and laptops are provided with preloaded WINDOWS 2008 professional or basic with SP2. The latest ones have enormously increased speed and capacity, such as 2.5 Ghz Processor speed., 4Gb ddr3 RAM, 500 Gb hdd sata.

The desktop contains on the monitor screen various icons through which various applications softwares, files, Web etc. can be accessed.

On the left down corner is the *start* menu present. Clicking it a list of options pop up. The list may differ on various versions of Windows. On Windows 95 the items are 1. Shut down 2. Run 3. Help 4. Settings 5. Documents 6. Program

On Window XP SP2 and above, the items are: 1. Shut down 2. Help and Support 3. Default Programs 4. Devices and Printers 5. Control Panel 6. Computer 7. Games 8. Music 9. Pictures 10. Documents 11. Comp.

As explained, selection of any of these can be done by putting mouse–pointer on it, and on clicking the group of items associated with start menu pop up, listed in rectangles.

One can open a file using Windows Explorer, which is accessible by clicking Program menu. Using Windows Explorer one can perform file management quite easily and quickly. For moving or copying a file drag and drop technique by a mouse is even more fast.

The specific methods are all menu driven so most of the users' requirements can be fulfilled by a little practice and trying various options on computer or take help from the help menu or button with each option key.

Moreover, may be, books and manuals may also be needed some times. Apart from the books, sometimes manuals may be consulted or basic books explaining the steps to accomplish various functions may be consulted. Some such references are Microsoft Windows 1995; Rashtriya Computer Sakshartha Mission 2001; MCSE 2007; Dyson and Coleman 2000; Schwartz et al. 2001 etc.

MS Office

Microsoft developed softwares not only for operating system but also application softwares. Some of the application softwares are of common applications, whereas, some are special purpose.

MS Office is an application software which accomplishes almost all generally required day to day office work of personal computer (PC) users. As Windows, it also evolved from its first versions and the MS Office 2003 is in most of the desk tops though MS Office 2007 and 2009 have also come. MS Office has six components with some special functions. These components are: (i) *Word* (ii) *Excel* (iii) *Power Point* (iv) *Access* (v) *Outlook Express* (vi) *Internet Explorer*.

MS Office also has GUI so is quite easy to use after a few hours practice. Using it, saves a lot of working hours in an office or a research lab.

A short description of each component is given below so that an appropriate part of MS Office is tapped by a needy user.

Word is full featured word processing software which can be used to create any kind of letters, memos, reports, newsletters, manuals, web-documents etc in the form of a file with extension 'doc'. The major work area of MS-Word is documentation.

Excel is a spread sheet software to organize, analyse and present the data in graphical way as well as interfacing of a data and graph with other softwares of MS family. It has statistical analysis and data transformation facilities also.

Power Point is a full fledged software for creating presentation slides, (for overhead or LCD projectors) enriched with video, animations, text and graphics and linking facilities with other MS-family.

Access is a database application that enables to store various types of data in a structured database. Access is designed in such a way that existing data of *dbase III*, *Fox Pro* and *Oracle* can be used by it. It can also be used as linker for generating reports.

Outlook Express is a messaging and collaboration client that help to organize and share information on the desktop and communicate with others. Its main goal is to receive and send e-mail and contacting news groups. Other Web based options are also available.

Internet Explorer enables to send and receive e-mail for net users, to browse the internet sites, downloading and uploading. Experts may manage sites, prepare html versions, and java scripting of documents. Also useful for web masters.

This is just an overview of a multifarious application software. It can be learnt also by consulting some guide books or reference books e.g. Dienes 1996; RCSM 2002; Busby and Stultz 2000; Podlin 1999; O' Leary 1999; Bangia 2000; Schwartz 2000; Kettell et al. 2008

As mentioned in Internet Explorer and Outlook Express one of the advancement in computer technology is in Information Technology (IT). It has taken the shape of a separate knowledge area with vast opening for jobs. The base of this technology is computer network, discussed briefly below.

Computer Network

In earlier times the computer was used as 'stand-alone' system fulfilling the requirements of the owner organization only. With increase in the number of computer systems in use, it was realized that it would be advantageous in some situations to use computers from remote points, instead of having several interactive terminals with a computer with time sharing facilities.

The users could be provided the facility of having access to a midi or maxi computer system from their places of work or home without having to go to the computer centre. Such access is being provided by connecting the user's PC's or mini computers by communication lines with the remote powerful computer. Much of the work thus can be done by users at their end. For some facilities which are available on the remote powerful computer, the user may enter the data file and the required program to be executed via communication line and get back the results of processing on his terminal via the same communication line.

If there are a number of computers close together (within 10 km radius), it may be required to be connected, for internal communication. It may be for transferring load of one machine to another, in case of a break down or for exchanging messages between departments. Such a connection of computers is called *Local Area Network* (LAN).

Inter connection of computers may be required between computers at widely dispersed areas, say at different cities of a country or even in different countries. Such a connection of a number of ordinary, mini, midi, and maxi – computers is also known as *computer network.* Network between computers of wider areas beyond the limit of LAN, is called *Wide Area Network* (WAN). The networks accessible from anywhere in the world are called *World Wide Web* or only *Web.*

The communication lines used in computer network results in fast or slow speed of communication. Density of message transfer also affects speed. Communication media for computer network may be the telephone landlines and mobiles (as for BSNL and Reliance network etc.), underground co-axial cables, optic fiber lines and satellite communications.

Some of the *uses of computer networks* are:

1. Users of the network may be allowed to access special library utilities, databases and other special facilities available at any of the computer in the network.
2. Another important application is information retrieval. Some information centres store large amount of data on different subject areas or field observations or experimental findings, technical reports, journals papers, books etc. in an organized fashion and utilizing storage area network (SAN).
3. Computer network of an organization may be used to store files, share program, decentralize the specialized functions and to share the use of expensive and sophisticated peripherals.
4. Railways and air lines networks are useful for the accessing the updated information about available travel facilities, reservation status of passengers etc.
5. Online examinations for admission in various courses and for entering in specialized jobs are also now possible which saves lot of money and time.
6. Person to person or person to organization communication is also possible through computer in the form of 'electronic mail'(e-mail).
7. e-commerce, e-banking, e-learning, web seminar or webinar and video-conferencing etc. are some often used global aids in business, health and academics.

Information transfer may be done by three methods – voice, data, and video. Efficient and cost effective communication links may wish to use even two or all three of the said methods. Requirement of data transfer rates are expanding very rapidly and speedy transfer requires electronic communications through computer network.

Internet

The *Internet* is the vast, worldwide system consisting of people on computers information, computers and communications systems. Internet is so large and complex as well as difficult to comprehend by all. It is an ocean of innovations, information, ideas, and thoughts. Its utilities and applications are infinite. The important internet usage includes e-mail system, the web, the web search engines, the ftp (file transfer protocol), the uploading and downloading of files, telnet (Remote Connections), talk facilities or the Internet Relay Chat (IRC).

The *web* is multilinked information system in the Internet. It is a large system of servers, offers all kinds of information to any one who can access the internet. For using information on the web, one can reach the sources of information through a client program called *browser.* Web is very popular because (i) it is easy to use and (ii) it is easy to create one's own web which can be shared by all the people on net.

On the web information is stored on pages. *Each page* has information as well as links to other pages. Links can be used to jump from one to the other page. The pages may be on any computer on the net. The browser can detect where to find the next page, it contacts the concerned web server, requests for the new page and displayed on the user's screen.

Web search engines (viz. google, medline, pubmed etc.) are used to process searching of information on the web. Nobody yet knows how many web servers are there in the world. Result of search points to *what ever item* fulfills the user's criteria. *ftp* service allows one to copy files from one computer to another computer.

Archive is a tool for ftp activities. There are thousands of anonymous ftp servers in the world and there are a number of archive servers in net, each of which contains a database of almost all the files that are publicly available via anonymous ftp and the addresses of the sites where the requested file can be found. Now ftp can be easily used to download the file.

Telnet (Remote Login) connects to and uses a remote host with some computer system. One needs a users name and password, as for accessing any e-mail account.

Talk facility is private conversation in the net. Communication with other people on the net in real time can be done either by *typing* message *back and forth* or by *voice transmission. IRC* is the public net wide conversation. Within IRC, there are many conversations going on at any time. Many of which may be organized around a particular topic or idea. Each IRC conversation is carried out on a channel. To join an existing conversation, one has to access IRC client and request for a specific channel.

To have an *internet* the following are required

(a) A computer.
(b) Client programs to run on the computer.
(c) A way to connect computer to the net.

Internet can be accessed via a LAN, Telephone and Modem System or ISDN or Data Card through USB port. The *ISP's provide* the following:

(a) Internet addressing
(b) Using the Browser to read article, news, e-books, interact on facebook or blog
(c) Using the Browser to send mail messages.

With passage of time the Internet has become a *magic box*. For more information, an interested reader may consult book by Taunenbaum (1993) and Jain & Aggarwal (2000).

Part II
Advanced

12. Types and Architecture of Studies

Men are continuously observing their surroundings, events and phenomenon intelligently and assimilate it as knowledge. In an attempt to use the knowledge gainfully for improving the life, the assimilated knowledge has to be shared with others with proper justification for any specific conclusion. The observations or the investigations are to be properly conducted for deriving valid conclusions. Collection of observations with a specific purpose or objective may be called an experiment which is the basis of development of scientific knowledge.

This chapter deals with various types of experimental studies. General idea of experiments and planning of an experiment is given in the next Section 12.1. Some basic concepts of design of experiments are discussed in the Section 12.2. and some specific types of experimental and observational studies are described in the last section along with their design issues.

12.1 Planning of Experiments in Lab and in Fields

It is evident that new knowledge is most often obtained by careful collection, analysis and interpretation of data. Considerable thought, therefore, must be given to plan collection of pertinent information so that the desired objective of the study is achieved with minimum time and cost. In this chapter a planned inquiry to obtain new facts or to confirm or deny previously established results will be called an **experiment**. Such inquiry will aid an administrator to take a decision to recommend some variety of crop, some modified procedure in industry, a new pesticide, a new drug for a disease or a new therapy.

Every experiment is set up to provide answers to one or more questions in mind. Accordingly an experiment is conducted to collect observations on experimental units (cf. Cox (1958)). An experiment may be concerned with

(a) Comparing different varieties, procedure, or medical treatments.
(b) Determination of some unknown parameter of the variable under study e.g. average yield of a given variety of the crop, average cost of a given industrial procedure, average time of recovery from a disease etc.

If the purpose of experiment is to recommend a variety or treatment or procedure, it will be easier to conduct an experiment to compare the varieties instead of performing several experiments to find out the average yields of various varieties.

Investigators decide what treatment comparisons are required to provide relevant information. Accordingly an experiment is conducted to measure or to test hypotheses regarding treatment differences under comparable conditions.

Proper experimentation requires proper *formulation* of the *question to be answered* which is relevant to field of inquiry. Experimental procedure is carried out such that it can answer these questions.

From statistical view point, the set of rules is the *experimental procedure* or *experimental design.*

In designing an experiment the *objectives* should be *stated clearly as questions to be answered, hypothesis to be tested and effects to be estimated.*

When the objective of an experiment is comparison of treatments rather than estimation or determination of values of their effects, such experiments are called *comparative.*

The planned investigations, which are not comparative experiments, are concerned with determining the properties of defined sets of things, such as mean diameter of a wire, the number of species of squirrel in a particular area, or the average examination marks of the children in a city who watch television (i) frequently and (ii) infrequently.

The *differences* between the *comparative experiments* and the *planned survey* are

(a) In a comparative experiment, the choice of treatment for each experimental unit is made by experimenter, whereas in the planned survey, the observer has no control at all over what makes one individual fall in one group than the other.
(b) Interesting conclusions can be drawn from planned surveys, even comparisons can be made within similar groups of individuals, e.g. within groups of children of the same age. However, much more cogent conclusions about causal effects can be drawn from comparative experiments than from planned surveys.

Even though all the comparative experiments can not be conducted in a lab but since the experimenter controls assignment of treatments, such experiments will be henceforth referred as experiments in this chapter. This chapter deals with only such type of experiments.

Planning or designing of such experiments is concerned with two issues. The principles of designing apply on:

(a) The choice of (i) treatments to be compared (ii) observations to be made and (iii) the experimental units to be used.
(b) The method of assigning treatments to the experimental units and the decision about how many units should be used.

Part (a) is not discussed in this book as it depends mainly on the specific area of the study i.e. medical, agricultural, industrial etc.

It is appropriate that some terms which have been used, as well as some of those which will be used often in this chapter, are formally defined.

Experiment As already mentioned a well planned inquiry to obtain new facts, to confirm or deny previously established results, is called an *experiment.*

Treatment is the procedure (a manure or variety or medical therapy etc.) whose effect is to be measured and compared with other treatments. Each treatment in the set should be defined carefully.

Experimental unit In this chapter an experimental unit is the unit of material (living or non-living) to which *one application of a treatment* is given.

Some *examples of experimental units* are an animal, a group (pen) of birds, a leaf, a plot of agricultural land, patients etc.

The observations may be taken on a *complete experimental unit* such as an animal on a treatment, or a patient on medical therapy.

Some times *a part from the whole* experimental unit is taken if the experimental unit is quite large, so infeasible to be used as a whole, such as a random sample of 10 (say) leaves from a sprayed plant, or the harvest from 5 square feet of land in the middle of an agricultural plot.

Some times a single unit is very small and inadequate for observational purpose, so a specific number of these (say seedlings) may be used as observation unit.

Experimental Error

Even if the same treatment is given to a group of similar experimental units (of same age, same species) the measurements on them are almost always found to have variation. A part of it may be due to some systematic difference and can be explained. Another part is unexplained and random, called *experimental error.* It does not mean that there is some mistake in the measurements but is due to inherent variability in the experimental units.

Precision or sensitivity of an estimator of mean of n independent observations is given by n/σ^2, where σ^2 is the variance of error due to inherent variation. Hence if σ^2 is large, precision will be small and increase in n will increase the precision or sensitivity.

Precision will be high if standard error (SE) of mean is small. For a large SE (mean), precision will be small.

Requirements of a good experiment are considered now, before some basic principles for part (b) above. Once the treatments, the experimental units and the nature of the observations have been decided, the main requirements are that

(i) *Experimental units* receiving different treatments *should differ* in *no systematic way* from one another. An example below clarifies the idea.

Example 12.1: Let there be an industrial experiment to compare two slightly different processes, A and B. The machinery used are same, but process A is always used in the morning and processes B always in the evening shift. No matter how many lots are processed and compared, it is not possible to separate the differences in the two processes, from the systematic change in the performance due to different operators in the two shifts. Differences in performances of the two processes may also be due to differences in temperature and humidity conditions around the processing machines (source: Goon et al. (2002), p. 65 modified).

(ii) Random errors of estimation should be suitably small with as few experimental units as possible. Absence of systematic error is achieved by randomizations (to be explained shortly). In such a case, the estimate of treatment difference will differ from its true differences only due to random errors.

If σ^2 is the inherent variability of the experimental units, it is already known that precision or sensitivity can be increased for a given σ^2 by increasing n. Large n will involve more time and money to conduct the experiment. Hence a balance has to be maintained while deciding value of n so as to conduct the experiment efficiently with minimum cost.

Another method of imposing precision is to use improved design. It is this approach, which will be discussed (the general principles) in the Section 12.2 and specific points will be given in the last section.

(iii) Range of validity of the conclusions should be quite wide. The conclusions derived from an experiment are obviously related to the particular set of experimental units under the conditions of the experiment. If the conclusions are to be applied to new conditions or units, care should be taken, since additional uncertainty will creep in over and above the standard error of the conducted experiment.

This situation could be avoided by choosing the experimental units from a well defined population of units by a proper sampling procedure.

There should also be some understanding of the reasons of treatment differences. Care should also be taken in designing the experiment to vary the conditions, if possible without inflating the error.

In pure scientific work, best thing is to try to gain thorough insight into some special situations instead of obtaining wide range of conclusions from one experiment.

(iv) The experiment should be simple in design and analysis. If the experiment is to be handled by relatively unskilled people it may be difficult if experiment is complicated. It may be advisable to have few long processes rather than frequent changes.

In scientific work, particularly in preliminary investigations, it will be useful to be able to identify the promising line of inquiry.

Apart from simplicity of design there should also be simple methods of analysis of the results of the experiments. As far as possible such a design should be chosen that analysis is easily feasible.

(v) A proper statistical analysis of the results should be possible without making artificial assumptions.

The experimental data should be of such a nature that the distributional assumptions required for valid treatment comparisons may not appear incorrect. For example if experimental data is binary, assumption of normal distribution may not be correct and routine analysis may result in invalid conclusions.

It is now quite evident that for a good experiment to be conducted, a proper design must be done, so that proper inference could be drawn to answer the questions set in the objectives of the experiment. For more detailed discussion of these ideas, the interested reader may consult books by Cox (1958) and Steel and Torrie (1981). In the next section, the basic principles of design of experiments are given.

12.2 Design of Experiments (DoE)

Design of experiment has been first considered by Sir R.A. Fisher, a renowned geneticist by the famous tea tasting experiment.

Design of experiment is a structured, organized method that is used to determine differences between different factors affecting a process. Experimental design is a strategy to gather knowledge from the experimental data.

Fisher gave an example of experimental design to test a hypothesis that a certain lady can distinguish by flavour alone, whether in the tea cup, the milk or the tea was put first.

Experimental error provides a basis of the confidence an experimenter can place in the inference drawn about the population on the basis of observations on sampled experimental units from the population. The experimental error must, therefore, be measured and controlled. A good design and its appropriate analysis is necessary to provide valid conclusion.

According to Fisher's diagram given below in Fig. 12.1, there are three basic things essential in a good experimental design. These are (a) Replication (b) Randomization and (c) Local control.

Replication

When a treatment appears more than once in an experiments, it is said to be *replicated*. Replication is essential in an experiment since an estimate of experimental error can be obtained only by replication. Unless repeated observations are taken, on a phenomenon that has inherent variability, estimate of variability or experimental error can not be obtained.

Precision of the experiment is improved by increasing replication i.e. n.

If there is only one replicate of the treatments, it is not possible to say whether the observed treatment difference is due to real treatment difference or it is due to inherent variability of the experiment units.

Randomization

A valid or unbiased estimate of experimental error and of differences among treatment means is possible only through randomization. This idea was also given by Fisher in his diagram. Randomization is achieved by using some chance device such as throwing a dice, taking out a card from well shuffled pack, using random number table etc. Random does not mean haphazard. Great care must be taken so that appropriate random methods are used.

Randomization ensures that equal chance is provided to every treatment for being assigned to any experimental unit. Non-random but a selected fashion of applying treatments to experimental units often results in either underestimation or overestimation of experimental error and also making estimates biased. Also this can result in unequal precision of different treatment comparisons.

Example 12.2: Suppose that effects of two different diets for children are to be compared and there are only two children available for the experiment (source: Goon et al. (2002), p. 65).

Suppose the diets are randomly chosen for one child and given to him for a period of one month and the second diet to the second child for the same period. The effect of diets is measured on the two children after a month.

The allocation of diets is random. But the effect of diets will not be same even if the two diets are equally good. It is because of differences due to weight, age, genetic differences and the family background of the two children may not be same, and there may be differences in improvements in the two children.

The difference in improvements may not be entirely because of difference in diets. In order to ensure validity, it is necessary to have several pairs of two children of similar family background. One of two diets is then randomly given to one of the two in a pair for one month.

Thus there are more than one children in each of the two diets and it will be possible to find the valid estimate of the difference in the two diets after eliminating the inherent variability. Experimental error or inherent variability can be measured only because of replication i.e. each diet is received by more than one child.

Thus randomization with replication is necessary to conduct a valid experiment to compare two diets, as shown on left hand side (L.H.S.) of Fig. 12.1. Separate randomization is

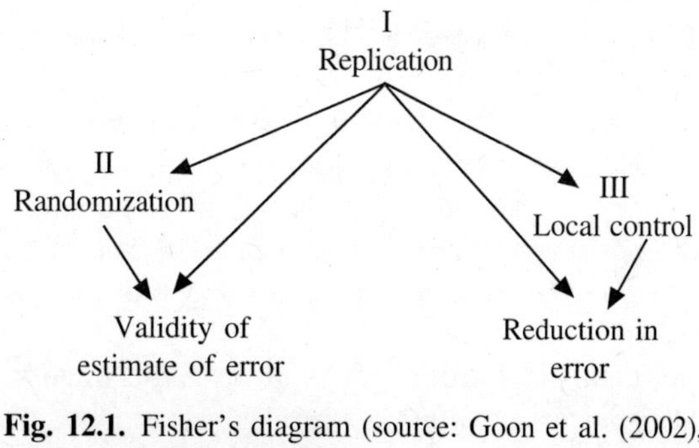

Fig. 12.1. Fisher's diagram (source: Goon et al. (2002), p. 65, modified).

necessary for various replications. The above two principles are essential requirements of a valid experiment, whereas the third, *local control* is a desirable one not indispensable one.

Local Control

The third principle, a desirable one, is called local control. It is also used for further control of error. As shown in the Fig. 12.1, replication is used with local control to reduce error.

This is achieved by restricting the randomization in such a way that a portion of the total variation is subtracted from the error. Thus reducing this variation which is relevant for comparison.

The simplest way to achieve local control is to divide the experimental units in homogeneous groups, or blocks. The variation among the blocks is also subtracted from the error variance thus reducing it.

For a detailed discussions on the principles of design, the reader may go through the book by Fisher (1947).

Another way of controlling error is *use of confounding*. In such designs of experiments the number of treatments is very large and the block size is not equal to the number of treatments but is smaller in size. It is carefully planned in advance according to certain rules, which treatment combination should be confounded and allocation of treatments (less than total combinations) are done accordingly.

This method is adopted since large number of treatments will require large number of experimental units. Most of the times finding large number of homogeneous units is difficult and there are chances of large experimental error.

Number of Replications

It is now obvious that whatever be the source of experimental error, replication of the experiment steadily decreases the error associated with difference of two treatment averages, provided the requirement of the randomization has been incorporated.

From the statistical theory, the rate of reduction of experimental error can be predicted. For this the basic quantity to measure is the *error variance per experimental unit* σ^2.

$$\sigma^2 = E(e^2) \tag{12.1}$$

where e is the error that affects the observation on a single experimental unit. If there are r replications, then

$$V \text{ (difference between two treatment means)} = (2\sigma^2)/r \tag{12.2}$$

$$SE \text{ (difference between two treatment means)} = \sigma\sqrt{2/r} \tag{12.3}$$

The decrease in above SE is proportional to the square root of r.

If $\sigma = 6$, an $SE = 3$ can be detected by 8 replications. If precision or sensitivity is to be increased such that an $SE = 1.5$ can be detected, it will require $r = 2\sigma^2/1.5^2 = 32$ replicates.

However, replication can not be increased beyond certain limit, as it is difficult to find large number of experimental units which are homogeneous. Increased number of replicates increases cost of experiment and observation time. A rule of thumb is to get about 10 degrees of freedom for the experimental error and generally r should not be less than 4.

Importance of Initial Steps of an Experiments

Statisticians are often approached for drawing inferences from the results of the experiment. The inferences which can be drawn depend on the way in which the experiment was conducted. The statisticians, therefore, require to know the detailed description of the experiment and its objectives. It may some times occur to the statistician that no inference can be drawn from the available results or that the possible inferences do not answer the questions the experimenter wanted to answer. It is, therefore, essential that proper care be taken before conducting the experiment, so that an unhappy situation does not arise. A researcher may also consult book by Cochran and Cox (1974) on these ideas.

It is a good practice to make a written draft of the proposed experiment. In general this draft should contain three parts. Namely: (i) a statement of the objectives (ii) a description of the experiment, covering the treatments, to be compared in the experiment, the size of the experiment and experimental material and (iii) an outline of the specific design of the experiment to be used and method of analysis of the results.

Care should be taken to write clearly without any ambiguity all the parts of the draft so that the required time and resources are in hand before starting the experiment. After giving ideas of the experiment and planning of the experiments the next section considers some of the special types of experiments used in medical studies. Any book on clinical trials or epidemiological studies, such as that by Kramer (1988) may be consulted.

12.3 Case-control, Cross Sectional, Longitudinal Studies and Clinical Trials

In this section the studies used for improving the health services are considered. On arrival of a new patient, a medical practitioner utilizes his own experience of treating patients till then, his judgment from the factors known about the patient's conditions and the published results on groups of similar patients. Though the clinician may be aware of the uncertainty about the cure he decides the treatment for the patient he thinks best.

Another approach is epidemiologic. It is group oriented. *Epidemiology* can be defined as the study of disease and other health related phenomenon in groups of persons.

Mutual benefits of the two approaches are now well recognized. Results from a well designed and properly analyzed epidemiological studies are very useful for clinicians in patient care. Evidences from epidemiological studies suggest novel diagnostic and therapeutic options to clinicians.

This section is devoted to some ideas of clinical epidemiologic studies which are concerned with etiology, diagnosis, prognosis, treatment, prevention, analysis of the risks and benefits of various clinical procedures and the evaluation of health care services. The principles and techniques of epidemiology have been extended to study of a variety of diseases, treatments and preventive measures. Development of vaccines, demonstrating association between various risk factors and the resulting diseases, establishing harmful side effects of therapeutic drugs has been possible through epidemiological studies. Many of the controversies and unproven hypotheses in medicine will continue to require epidemiological research. Training and skills in choice of proper experimental design and appropriate statistical analysis of collected data are essential to provide pertinent valid answers. An epidemiological study may be a *descriptive* or *an analytical.*

A descriptive study focuses on describing a *target population* by proper data collection, summarizing it in the form of rates, means, and distributions of one or more attributes in a group of subjects. No association or causes are inferred. *Target population* is specified by its geographic, socio-demographic and clinical attributes. Since the entire target population is rarely available for study, the investigator has to decide a proper study sample.

In *an analytical study*, one or more groups are studied for exploring association between two or more variables, usually a cause- and- effect association. Hence main objective of analytical studies is *causal inference*.

While *planning a descriptive study,* the target population and the variables to be described are unambiguously defined. In *an analytical study* the attributes whose association is being investigated must be clearly specified and the question to be answered from the study must also be properly framed before starting the study.

There are two terms which are commonly used in medical studies viz. *exposure* and *outcome.* A brief description of these two follows.

Exposure is the causal factor or effect that the investigator believes may be responsible for the *outcome* under study. It may be a *natural one*, such as exposure to air, or water pollution to radio-active emissions, to environmental toxicities etc. It may be *intrinsic characteristic* of the subjects, such as sex, race, genotype, family background etc. It may be a *self selected habit* to which the subject has exposed himself, such as drinking, smoking, jogging, morning walk, cycling etc. The fourth type is a treatment prescribed by the subject's clinician or that imposed by investigator himself such as a new drug or a new therapy.

Outcome is the effect that the investigator believes may be caused by exposure. In *observational studies* (also called *surveys*) the investigator is not sure that the groups with different exposures were equally susceptible to the outcome before their exposure. In *analytical studies* or *experiment*, the exposure is assigned by the investigator and the outcome may be death, or some other health or disease state or its prevention or amelioration. Finally it may be a change in quality of life (betterment or worsening).

Through a single research study, multiple exposures and outcomes can also be investigated.

After formulation of a research objective in terms of target population, exposures and outcomes, the *next step* is *planning* the *experiment.* Three things must be considered when plan of an analytical study is decided. These three are (a) Directionality (b) Sample selection and (c) Timing.

Directionality refers to the order in which exposure and outcome are investigated. It may be *forward* i.e. from exposure to outcome; *backward* i.e. from outcome to exposure or *simultaneously* i.e. exposure and outcome being determined at the same point in time, often without knowledge of which actually occurred first.

Sample selection deals with deciding the criteria used to select the subjects for study. It can be based on exposures or on outcomes, or some other criteria.

Timing is concerned with the relation between the study time and exposure and the outcome calendar times i.e. whether it is *historical* in which the exposure and outcome occurred prior to study. It may be *concurrent* i.e. exposure and outcomes both are observed during the study period, of mixed timings. In *mixed timing studies, exposure* may have occurred *in the past* and *outcome* may be *concurrent.*

Types of Analytical Studies

Directionality will be taken as major axis of classification of analytical studies. Directionality is also referred by the terms 'prospective' and 'retrospective' studies. The former has forward directionality and latter, the backward directionality. Analytical studies may be of two types – *observational cohort studies* and *experimental cohort studies.*

Analytical cohort study is in a *forward* direction i.e. from exposure to outcome. These studies may be either *observational* (in which exposure is natural, selected by subject or is prescribed by subjects' clinician) or *experimental* also called *clinical trials* (in these exposure is assigned by investigator). Books by Kramer (1988) and Breslow (2005) may be consulted for these ideas.

12.3.1 Observational Cohort Studies and Longitudinal Studies

Assembling cohort or selecting study subjects can be done by exposure status or by 'other' criteria. Selection of sample by exposure is preferable, especially if exposure is rare, Thus, a sample of sufficient size should be chosen to enable drawing statistically valid conclusions. A discrete number of exposure categories may be formed of the subjects and outcomes may be compared. The non-exposed (or least exposed) group in a cohort study is the *control group* or *control cohort.*

When a new treatment is compared with standard treatment, receiver of the standard treatment are also referred as the *control group.*

Follow up of subjects before, during and after exposure is an important research design feature in cohort studies. It can have marked effect on the results. Time of start of the study is referred to as *zero time.*

Choice of zero time may be the time when exposure begins, or the time when the subjects agree to participate. This is particularly true if the treatments may require a long *waiting period.*

Duration of the follow up should be adequate as it depends on the biological (clinical) relation between the exposure and outcome. The *latent period* (time of development of exposure) of the exposure and *incubation period* of the infection also decide the *duration of the follow up.*

The outcome is usually the occurrence, disappearance, improvement or relief in some health or disease state. In cohort studies often several outcomes are investigated simultaneously. The quantitative expressions of the exposure-outcome association are given below. The analysis and expressions from a cohort study depend on the measurement scale of exposure and outcome.

A. Exposure Categorical and Outcome Continuous

Result of interest in such a study is the mean $\bar{X}$ of the outcome variable in the various exposure groups ($\bar{X}_1$ for exposure group 1, $\bar{X}_2$ for exposure group 2, etc.). Larger the difference in mean outcomes between groups, greater is the exposure-outcome association.

Example 12.3: Consider a hypothetical cohort study of the effect of exposure to asbestos on cardiopulmonary fitness as reflected by the time taken to run a 100 meter race. 100 exposed male asbestos miners and 100 non-exposed healthy male miners of similar age are considered. The exposed and non-exposed miners had the same mean times for running the 100 meter race before the exposure occurred in the miners. Both groups have been observed for 10 years without loss to follow-up. After 10 years, the mean time taken to run the said race by exposed group in 14.4 seconds and 12.2 seconds in the non-exposed (control) group. It is desired to find the degree of association of the outcome with the exposure (source: Kramer (1988), p. 63).

Solution: The first measure of the *degree of exposure outcome association* is

$$A_1 = \bar{X}_2 - \bar{X}_1 = 14.4 - 12.2 = 2.2 \text{ seconds}$$

Another measure is *percent change* from control mean; given by

$$A_2 = \left[\left(\bar{X}_2 - \bar{X}_1\right)/\bar{X}_1\right](100) = \frac{14.4 - 12.2}{12.2} \times 100 = 18\%$$

If there were three exposure groups: non exposed, lightly exposed and highly exposed, the means of the three groups can be obtained as $\bar{X}_1$, $\bar{X}_2$ and $\bar{X}_3$. Now, both $\bar{X}_2$ and $\bar{X}_3$ are compared with $\bar{X}_1$ to obtain both types of measures of association.

B. Exposure Continuous and Outcome Continuous

As an example of such a situation consider Example 12.3 with same outcome of time to run 100 meter race. Suppose the exposure is the number of cigarettes smoked per day.

In this case, the two ways of association measures can be obtained as follows:

1. Categorize the subjects on a continuous scale according to the number of cigarette smoked per day as 0, 1-5, 6-10, ≥ 11. The two association measures can be calculated as in Example 12.3.
2. Another approach of measuring association is to determine linear regression line and simple correlation. Slope of regression line gives the change in outcome per unit change in the exposure.

C. Exposure Continuous and Outcome Categorical

Such a situation may arise, if the number of cigarettes smoked taken as exposure, is treated as continuous, and myocardial infraction (heart-attack) present or absent is dichotomous outcome.

One approach to this type of study will be to compare mean exposures in the two outcome groups. This approach *violates* the principle of *forward directionality* of a cohort study. This approach is an indirect way of comparing outcomes in the differently exposed subjects.

A better approach conforming to the forward dimensionality of a cohort study is to categorize the exposure and then analyze resulting outcomes as given in the following situation, when both exposure and outcomes are categorical.

D. Exposure Categorical and Outcome Categorical

Many cohort studies adopt this format. The simplest of this is in which both exposure and outcome are dichotomous. An example with an appropriate analysis is given in the following Example 12.4. Table 12.1 gives the general 2 × 2 table of results of a cohort study with dichotomous exposure and outcomes where $n = a + c + b + d$.

Table 12.1 Results of cohort study with dichotomous exposure and outcomes

Exposure	Outcome		Total
	O	O^c	
E	a	b	$a + b$
E^c	c	d	$c + d$
Total	$a + c$	$b + d$	n

$$\text{Risk of outcome in exposed } (E) = \frac{a}{a+b} \qquad (12.4)$$

$$\text{Risk of outcome in non-exposed } (E^C) = \frac{c}{c+d} \qquad (12.5)$$

$$\text{Hence relative risk } (RR) = \frac{a/(a+b)}{c/(c+d)} = \frac{a(c+d)}{c(a+b)} \qquad (12.6)$$

$$\text{Attributable risk } (AR) = \frac{a}{a+b} - \frac{c}{c+d} \qquad (12.7)$$

Example 12.4: Cohort study of 200 smokers and 200 non-smokers (controls) for occurrence of myocardial infraction (MI) gave the following results (source: Kramer (1988), p. 65, data only).

Table 12.2 Results of MI on a cohort of smokers and non-smokers

Exposure	MI	No. MI	Total
Smokers	32	168	200
Non-smokers	15	185	200
Total	47	353	400

Determine various risks of MI.

Solution: Risk of MI in smokers $= \dfrac{32}{200} = 16\%$

Risk of MI in non-smokers $= \dfrac{15}{200} = 7.5\%$

Relative risk (RR) $= \dfrac{32/200}{15/200} = 2.13$

Attributable risk (AR) $= \dfrac{32}{200} - \dfrac{15}{200} = \dfrac{17}{200} = 8.5\%$

It may be noted that relative risk (RR) ≥ 0.

If $RR = 1$, it indicates that there is no exposure-outcome association (12.8)

$0 \leq RR < 1 \Rightarrow$ there is negative association

$RR > 1 \Rightarrow$ there is positive association

Use of *RR* and *AR*

The *RR* is the best estimate of strength or magnitude of exposure-outcome association and is useful for causal inference.

The *AR* indicates the frequency with which the outcome can be attributed to exposure in the sample studies and extension to the target population.

AR is more useful for public health purpose.

Etiologic Fraction (*EF*) is another measure for impact of exposure on outcome in target population for which study sample has been taken. It is defined as

$$EF = {(r_T - r_{UE})}\big/{r_T} \tag{12.9}$$

where r_T is risk of outcome in the total target population r_{UE} is risk of outcome among the unexposed.

An algebraically equivalent form is

$$EF = \frac{P_E(RR-1)}{P_E(RR-1)+1}$$

where P_E is prevalence of exposure in target population.

EF can also be interpreted as the proportion of cases of the outcome that would disappear if exposure were eliminated from the target population.

If *RR* can be assumed to remain constant from one population to another, *EF can be used* to compare the proportion of outcome attributable to exposure in setting with different values of prevalence P_E.

Example 12.5: It is understood that maternal smoking doubles the risk of giving birth to an intrauterine growth retarded (IUGR) infant, one-third of IUGR rate can be attributed to maternal smoking in a population in which half of women smoke during pregnancy. (Source: Kramer (1988), p. 67.)

Proof: From the example $RR = 2$, $P_E = 1/2 = 0.5$

So $$EF = \frac{.5(2-1)}{.5(2-1)+1} = \frac{.5}{1.5} = .33 = 1/3$$

If only 1/10th of woman smoke in another population, one eleventh of IUGR rate can be attributed to maternal smoking in this population. This is because for this population:

$$EF_2 = \frac{.1(2-1)}{.1(2-1)+1} = \frac{.1}{1.1} = \frac{1}{11}$$

Hence the above conclusion is correct.

All these *three measures* refer to *fixed cohorts* without losses or addition in it during the follow up period.

In practice, however, cohort does not remain fixed over a passage of time but it changes or that most often cohorts are dynamic.

When there is irregular addition and loss in the cohort, it is advisable to use *person-duration* in the denominator and the resulting rate is called *incidence density (ID).*

Polychotomous ordinal measurement of exposure may also be some times used by investigator. This can give some dose response type effect of exposure on outcome. The risks which are assessed in comparison to the unexposed (or control) group are function of exposures. An example below illustrates the idea.

Example 12.6: Following table gives the result of cohort study of MI and cigarette smoking using three category ordinal scale of exposure (source: Kramer (1988), p. 69).
It is desired to determine the *RR* and *AR*.

Table 12.3 Cohort-study of MI with categorical exposure of smoothing

	Outcome		
Exposure	MI	No. MI	Total
Heavy smokers	20	80	100
Light smoker	12	88	100
Non-smokers	15	185	200
Total	47	353	400

Solution: Risk of MI in non-smokers $= \frac{15}{200} = 7.5\%$

Risk of MI in light smokers $= \frac{12}{100} = 12\%$

$$RR = \frac{12/100}{15/200} = \frac{24}{15} = 1.6$$

$$AR = \frac{12}{100} - \frac{15}{200} = \frac{9}{200} = 4.5\%$$

Risk of MI in heavy smokers $= \frac{20}{100} = 20\%$

$$RR = \frac{20/100}{15/200} = \frac{40}{15} = 2.67$$

$$AR = \frac{20}{100} - \frac{15}{200} = \frac{25}{200} = 12.5\%$$

Thus the *RR* and *AR* are calculated using non-smokers as base, and the change in relative risk with change in exposure is given by 1.6 for light smokers and 2.67 for heavy smokers.

Observational cohort studies are also called *longitudinal studies*. In longitudinal studies subjects are followed over time with continuous or repeated monitoring of risk factors (or exposures) or outcomes or both. These vary enormously in size and complexity. The follow up period may be several decades, such as 1% population of British population that was initially identified in 1971 census by the Office of the Population Censuses and Surveys. Outcomes such as mortality and incidence of cancer have been related to employment status, housing and other variables measured at successive censuses.

Some other longitudinal studies follow up relatively small groups for a few days or weeks, such as firemen actually exposed to noxious fumes may be followed up for a week to identify any effect. A group of unexposed controls are also included for follow up along with the exposed group.

For study of chronic diseases such as cancer, heart diseases and diabetes a large number of people need to be followed up to get sufficient outcomes for statistically meaningful results. For the diseases which have long induction period, the difficulty is further increased.

A *solution* to this is provided by conducting *retrospective studies*. Thus death rates from coronary heart disease could be related to weight at birth and at one year age, provided medical records were available and could be traced out. In long term studies often survival rates are used as outcomes.

12.3.2 Clinical Trials

Already mentioned in the beginning of the last section that a cohort study may be observational or it may use an experimental design. In *experimental cohort studies*, also called *clinical trials*, the exposure is assigned by the investigators. In clinical trials exposure does not occur naturally, neither it is selected by subjects, nor it is prescribed by clinician. This section contains a discussion of clinical trials. In *clinical trials*, the *exposure* is often referred as *treatment*.

In a clinical trial the investigator first defines or identifies the treatment(s) (a drug or some medical device) to be tested. He next decides, with which these are to be compared. This may be one or more existing treatments or a placebo. The existing standard treatment is called a *control treatment*. An inactive control treatment that is indistinguishable from the experimental treatment is called a *placebo*. The investigator must assemble the study sample before the treatments are assigned. The study sample must be such patients who might benefit from the treatments under study.

If the investigator can not obtain enough patients with the specific diseases or condition at his location, investigators at other locations may be contacted. They may obtain the same kind of patients who may be given the same treatment.

For the clinical trial the investigators recruit patients with the predetermined characteristics, administer the treatment(s) and collect data on the patients' health for a pre specified period. These data include some initial conditions of the patients, amount of study drug (dose) given, and his health conditions after a period of time. The investigators send the data collected at their location to the coordinating main investigator. He then analyses the data pooled from various locations using appropriate statistical methods. A clinical trial is always concurrent.

The assignment of treatment is done following principle of randomization to reduce the bias. Such clinical trials are called *randomized clinical trials* (RCT) and are very commonly chosen. Most of the clinical trials compare two medications or devices. However, some trials compare three or four medications, some doses of medication or devices. If the clinical trials are multi-centered the clinical trial design and its objectives are written in a document called a *clinical trial protocol*. This ensures uniformity of procedure and similarity of patients.

The *most common* method of *randomization* makes use of random number tables (already mentioned in Chapter 7) to select the subjects in two groups (for comparing two treatments) or to select in r groups (if r treatments are to be compared). If there are in all N subjects (or patients) N random numbers are selected between 1 to N (as described in Chapter 7, Section 7.2). Order the patients according to their identity number equal to the selected random number. Divide these ordered patients in ($r \geq 2$) parts and assign treatment 1 to 1st part, treatment 2 to 2nd part, …., treatment r to r^{th} part of ordered patients.

The follow-up duration must be sufficiently long to include latent period of the study treatments. The frequency and extent of follow-up visits, examinations, and diagnostic tests should all be documented as standard protocol. It should be adhered to, throughout the clinical trial by all the investigators, in all the centers, if study is multi-centered. Losses in follow up should be minimized to maximize statistical efficiency.

Parallel and Crossover Design

The clinical trials in which one group of subjects receives only one treatment and each of the other groups (one or more) receives (one or more) other treatments, are said to have *parallel design*. Various study groups receive their respective treatments simultaneously in *parallel*.

In a *crossover design* there is no separate comparison group. Each study subject serves as his/her own control since each study subject receives each treatment in series by "crossing over" in sequence from one to the other. Since the same subject receives both the treatments (in a two treatment comparison) (a) a subject is 'counted' twice (b) possibility of variation due to individual subject characteristics is eliminated. Crossover designs are statistically more efficient for treatment comparison.

Proper conduct of crossover clinical trials, requires (i) randomization of treatment sequence (*A*, *B*) versus (*B*, *A*) (ii) time-dependent rather than outcome dependent crossover of treatments and (iii) elimination of (or control for) carry over effects from the first treatment. Analysis of crossover design is like that of matched pairs.

Blinding in Clinical Trials

Apart from controlling inherent variability among subjects by randomization, protection against *information bias* is also necessary. This is achieved by *blinding*. *Subjects blinding* is achieved by providing look-alike, feel-alike, smell-alike, taste-alike placebo, when comparing an experimental treatment with standard treatment or no treatment. Thus the study subjects do not know which treatment is being given to any particular subject (cf. http://en.wikipedia.org)

Double blinding design is used some times by not letting the medico investigator knowing which subject has been given the treatment and which has been given the placebo. If the physician has this knowledge, he is tempted to give extra care to those patients who receive placebo, to compensate its ineffectiveness.

Analysis of Results

Assignment of treatments to groups automatically creates categorical exposure groups. The exposure – outcome association is assessed by comparing the outcomes among various groups. If *outcome is continuous, comparison of means* can be done using appropriate statistical technique.

For *categorical outcomes*, comparison of rates and relative and attributable risks can be obtained.

Efficacy refers to potential effect of treatment under optimal circumstances i.e. whether treatment can have effect on outcome. Efficacy is usually of primary interest to biologists, physiologists and (to some extent) pharmacologists.

An *analysis for efficacy* compares subjects according to treatments actually received rather than the one assigned. It excludes subjects who complied poorly, those who switched over to another treatment and those who withdrew during the trial.

Effectiveness refers to the actual effect of treatment in the "real world" of people who comply poorly, change treatment or become lost to follow up. In other words effectiveness refers to whether treatment has an effect on outcome. Effectiveness is of primary concern of patients and clinicians.

An effectiveness analysis compares all subjects according to their original assigned treatment and hence includes poor compliers, switch-overs and withdrawals.

Ethical issues are of importance in all types of epidemiological studies. The data emerging from observational cohort studies and from clinical trials are often product of a sensitive nature. Though the results are almost always reported in aggregate without disclosing the identities of the subjects, but some subjects are reluctant to be interviewed or examined for purposes other than their own health care. Some subjects object even to their medical records being checked by investigators other than their clinicians. Randomizing and blinding make the aspects of risk compounded and ethical considerations often become more complex. For an RCT to be ethically justiable, none of trial treatments should have proven superior efficacy, when the clinical trial begins.

Moreover the requirements of the randomization and blinding often place the clinicians in the conflicting role of attempting care of patients and of wishing to contribute in advancement of knowledge and through it, of their own career. It is probably rare for a clinician to be nutral concerning treatment of patients. In that case, randomizing his patients can not be allowed by him. It appears difficult to take only 50-50 chance of assigning a new and better drug to half of the patients and for remaining half the existing standard drug.

12.3.3 Case-control Studies

In *cohort studies* (including clinical trials) study subjects are observed in a *forward direction* i.e. from *exposure to outcome.* Association of cause on effect is studied. In *case-control studies* investigator starts with the outcome and looks for prior exposure. The *directionality is backwards*, and inference is about cause from the effect. Case-control studies can therefore, be thought of as the chronological and logical inverse of the cohort-studies. The name "case-control" study has come up from the dichotomous categorization i.e. presence of outcome (case) and absence of outcome (control).

In case-control studies the *subjects are assembled first.* They are *selected* from the target population by either *incident* or *prevalent* outcomes. *Selection from incident outcomes can assemble cases over a period of time. Selection from prevalent outcomes* includes *all subjects with the outcome* at a *given point of time* regardless of when they developed the outcome.

Prevalent cases under represent the cases who died and those who were successfully treated, and those who recovered on their own.

Incident cases include all those newly occurring cases also during the given period of study. These will therefore, *be more representative* of the entire spectrum of disease or any other outcome under study.

Cases should be selected according to predetermined strict objective criteria and should be true representative of cases in the target population. When prevalent cases are used in study, the exposure-outcome association tends to become a measure of association between exposure and non fatal, chronic and incurable cases of the outcome. It is often unsatisfactory in epidemiological studies. Often an association between exposure and entire population spectrum is desired, which can be obtained only by selection from incident cases.

Controls should also be selected according to pre-specified criteria to ensure the absence of outcome ("non-diseased status"). These should be representative of the controls in the target population. It may appear that normal, healthy subjects should be the best control. But it should be selected from the subjects with similar referral, surveillance and other factors which may affect the study sample.

One approach is to select control from the healthy group in the community. Other approach is to select controls from same source (such as a hospital) as that of the cases. If exposure-outcome association measure is similar in both the case-control studies with two selection procedures, the investigator gains more confidence about the result.

The best control group will be the representative sample of subjects free of the outcome, who would have been included as cases if they had developed the outcome.

Exposure should be properly defined taking into account the underlying biological process through which the exposure causes outcome. An exposure of interest should be a *prior one* not the one during the assessment of outcome status. Sex, race, blood groups, inherited deficiency are some of the permanent characteristics that can be presumed to predate the outcome. Contemporary smoking, drinking or dietary habits may not be valid prior habits to cause the study outcome.

The period and duration of exposure should relate to the known and suspected latent period of the exposure. Cancer initiation and promotion are known to require years or decades between exposure to carcinogens and clinical appearance of cancer.

Analysis of Results

Method of analysis will depend on the type of exposure and outcome. If *both are continuous*, linear correlation and linear regression analysis can still be used even though directionality is backwards.

When *outcome is continuous* and *exposure is categorical,* comparison of mean outcomes for various categories is done. Though this analysis seems similar to that in cohort studies, but in case-control studies the groups are not representative of the target population.

Ordinal outcomes are rare in case control studies, but dose response effects can be assessed in this situation also.

If *exposures are continuous* and *outcomes are categorical*, one can have a comparison of mean exposures in various outcome groups. One can not draw an inference about the effect of a given level of exposure, from the mean exposures in cases and controls.

Usual method of analyzing the results of case control study *uses dichotomous outcomes and categorical* (usually dichotomous) *exposures*. Continuous exposures may also be reduced to dichotomous form and the results may be summarized in such cases, as a 2 × 2 contingency table for analyzing the data.

The backward dimensionality imposes a backward method. In general let the contingency table of results of a study be as follows.

Table 12.4 Results of case-control study with dichotomous outcome and exposures

	Case (0)	Controls ($\bar{0}$)	Total
Exposed (*E*)	a	b	$a + b$
Unexposed (*NE*)	c	d	$c + d$
Total	$a + c$	$b + d$	$a + b + c + d = N$

$$\text{Rate of exposure in cases} = \frac{a}{a+c} \quad (12.10)$$

$$\text{Rate of exposure in controls} = \frac{b}{b+d} \quad (12.11)$$

$$\text{Odds of exposure in cases} = \frac{a}{c} \quad (12.12)$$

$$\text{Odds of exposure in controls} = \frac{b}{d}$$

$$\text{Exposure odds ratio} = \text{OR}_e = \frac{a}{c} \bigg/ \frac{b}{d} = \frac{ad}{bc} \quad (12.13)$$

Rates of exposure in cases versus controls are $\frac{a}{a+c}$ and $\frac{b}{b+d}$, respectively, which are comparable as exposure-outcome relationship since two outcome groups: cases and controls are representative of the target population. But these rates do not provide an inference about the risk of exposure in the target population. Use of odds however allows an estimate of relative risk of the exposure in the target population.

Use of Odds

The risks of outcomes in the exposed and the unexposed subjects can provide direct information about the relative risk of exposure in the target population. But these can not be obtained from the data provided by the case-control study. The *odds ratio is a fairly good estimator of the true relative risk of exposure* in the target population, if the outcome is rare.

Sample Odds Ratio and Relative Risk in Target Population

Let the population study was done as case control study and there is no loss to follow up. So every one in the target population can be classified by their exposure status and their outcome status. So at the end of follow up suppose the following 2 × 2 table is obtained.

Table 12.5 Dichotomous exposure and outcome data on a completely follow up population

	Case (0)	Controls ($\bar{0}$)	Total
Exposed (*E*)	*A*	*B*	*A* + *B*
Unexposed (*NE*)	*C*	*D*	*C* + *D*
Total	*A* + *C*	*B* + *D*	*N* = *A* + *B* + *C* + *D*

$$\text{Rate (risk) of outcome among exposed} = \frac{A}{A+B} \tag{12.14}$$

$$\text{Rate (risk) of outcome among un-exposed} = \frac{C}{C+D} \tag{12.15}$$

$$\text{Relative risk } (RR) = \frac{A}{A+B}\Big/\frac{C}{C+D} \tag{12.16}$$

If outcome is rare so that A is much smaller than B and C is much smaller than D, $\frac{A}{A+B} \simeq \frac{A}{B}$ and $\frac{C}{C+D} \simeq \frac{C}{D}$.

$$\text{Therefore, estimated relative risk} = (\widehat{RR}) = \frac{A/B}{C/D} = \frac{AD}{BC} \tag{12.17}$$

Odds of outcomes among exposed = A/B

Odds of outcomes among unexposed = C/D

$$\text{So outcome odds ratio } (OR_O) = \frac{A/B}{C/D} = \frac{AD}{BC} \tag{12.18}$$

Odds of exposure among those with outcome = A/C

Odds of exposure among those without outcome = B/D

$$\text{So exposure odds ratio} = (OR_E) = \frac{A/C}{B/D} = \frac{AD}{BC} \tag{12.19}$$

From equation (12.17), (12.18) and (12.19), it is clear that, $(\widehat{RR})$, OR_E, OR_O *are equivalent.*

If the case-control study is conducted in which sampled cohort is representative of the target population (or when random sampling is adopted) for selection by exposed and unexposed criteria or by case and control criteria, $a/b = A/B$ and $c/d = C/D$. Hence OR_E and OR_O obtained from the Table 12.4 will be $= \frac{ad}{bc}$ and equivalent to $\frac{AD}{BC}$.

An example below gives the complete idea of the algebra above.

Example 12.7: The adjacent Table shows the hypothetical results that would have been obtained, had an entire birth cohort of 3,00,000 from a given community been followed to age 60. It is assumed that there is no loss to follow up due to migration and mortality. Two thirds of this population are tea drinkers (*TD*). Tea drinking is considered as exposure and renal cancer (*RC*) is the considered outcome. It is desired to find the *RR*, *AR*, OR_O, OR_E and $\widehat{RR}$ (source: Kramer (1988), p. 103).

	RC	No RC	Total
Tea drinkers (TD)	400	199600	200000
Non tea drinkers (NTD)	100	99900	100000
Total	500	299500	300000

Solution: Risk of *RC* among *TD* = 400/200000 = .002 (2 per thousand)

Risk of *RC* among *NTD* = 100/100000 = .001 (1 per thousand)

$$\text{Relative risk } (RR) = \frac{400/200000}{100/100000} = 2.00$$

Attributable risk (*AR*) = .002 – .001 = .001 (1 per thousand)

Odds of *RC* among *TD* = 400/199600 = .002004

Odds of *RC* among *NTD* = 100/99900 = .001001

$$\text{Outcome odds Ratio } (OR_O) = \frac{.002004}{.001001} = 2.002$$

Odds of *TD* among subjects with *RC* = 400/100 = 4.00

Odds of *TD* among subjects without *RC* = 199600/99900 = 1.998

$$\text{Exposures odds ratio } (OR_E) = \frac{400/100}{199600/99900} = 2.002$$

Hence $(\widehat{RR}) = OR_O = OR_E = 2.002$.

Remarks

1. A case-*control study is not begun with a representative sample of exposed and non exposed* individuals in target population.
2. The *true rates or risks* which can be calculated are *only the odds of exposure in the cases and controls.*

Interpretation of Odds Ratio

The *OR* is an estimate of *RR*. So *OR* > 1 indicates that exposure is associated with an *increased risk* of developing the outcome. Higher the value of *OR* greater the risk.

Conversely *OR* < 1 indicates that exposure is associated with a *reduced risk* i.e. it *protects* against developing the outcome.

Closer the value of *OR* to zero, greater the protection.

Etiological Fraction (*EF*) from Case Control Study

As in cohort studies, *EF* is given by

$$EF = \frac{P_E(OR-1)}{P_E(OR-1)+1} \tag{12.20}$$

where P_E is prevalence of exposure in the target population and OR is odds ratio in the case-control study.

For the Example 12.7, $P_E = 2/3$. Hence

$$EF = \frac{2/3(2.002-1)}{2/3(2.002-1)+1} = \frac{2/3 \times 1.002}{2/3 \times 1.002 + 1} = \frac{.668}{1.668} = .4004$$

Hence 40% of the renal cancer has developed due to tea drinking.

12.3.4 Cross-sectional Studies

As already indicated *cohort studies* are followed in a *forward direction* i.e. from exposure to outcome. The *case-control* studies are investigated in *backward direction* i.e. from outcome to exposure. Inference in cohort studies are from cause to effect, whereas in case-control studies from effect to cause.

In *cross-sectional studies*, the *exposure and the outcomes*, both are *determined at the same points or cross-section in time*. These studies *are also* called *prevalence studies*. Such studies have many features of case control study but it has a disadvantage that one can not be certain that exposure preceded outcome, so has important implications on causal inference.

Selection of Sample

In cross-sectional studies samples may be selected by exposure or outcome or by some other criteria. Random sampling or some other form of sampling may be used to select a representative sample from the target population.

Cross-sectional studies may profit if sample selection is based on the rarity of exposure or outcome. If exposure is rare statistical efficiency is increased if sample selection is by exposure, and by outcome if outcome is rare.

Outcome

Cross-sectional studies usually consider prevalent outcomes. Fatal cases, dropouts, migrants, or cured cases are not counted. Hence cross-sectional studies are best suited for chronic non fatal ailments. Incident cases could be investigated for simultaneous exposure, even then the design will be said to be cross-sectional.

Exposure is measured at the same point as outcome. It may be noted that same point is an approximation.

If exposure and outcome have been present for some time prior to study, the investigator may find it difficult to ascertain if exposure occurred before the outcome.

If exposures are of permanent attribute like race, sex, genotype, blood type, it can be easily assumed to precede the outcome. For such types of exposures cross-sectional and case- control studies are equivalent.

This is also true if the exposure determined at a particular time might be prevailing in the past also covering the latent period of the outcome, the results of the cross-sectional and case control studies should be similar. The cross-sectional studies, therefore, best suit the outcomes with short latent periods.

Analysis of cross-sectional study data depends on the sample selection method and the types of exposure and outcome variables.

(a) When selection is by exposure, analysis is similar as for cohort studies.
For *continuous-outcomes,* mean outcomes are compared for groups of exposure status.
For *categorical outcomes*, the outcome rates are compared in various exposure groups.
If *incident outcomes are dichotomous*, relative and attributable risks can be calculated.
If the *outcomes* are *prevalent* instead of incident, the relative risk derived from a cross sectional study is often referred as *prevalence rate ratio.*

(b) When sample selection is by outcomes, analysis is similar to that used for case-control study. Usually *outcomes are dichotomized* as case versus control and odds ratios *(OR)* can be calculated. Though it is well known that dichotomous classification of outcomes does not make the cross sectional study a true case-control study, since the exposure is simultaneous in cross-sectional studies.

(c) If the exposure variables are of permanent nature or when latent period of outcome is small, the cross sectional and case control studies are equivalent, therefore, similar analysis can be used.

(d) If the outcomes are rare, odds ratio will be reasonable.

(e) If sample selection is by random procedure, any one the above analyses may be used.

In general, the cohort approach is preferable, since it usually permits a direct comparison of means or rates in various exposure groups. True relative risks can be calculated when the outcome is dichotomous.

"Pseudo-cohort" Cross-sectional Studies

Instead of following one group over time in a cohort study, if different age groups are studied cross-sectionally, this study can look like a cohort study. Data are thus obtained as function of age, but are based on *different subjects* rather than same subjects over a period of time. Such studies (cross-sectional) are called *pseudo-cohort studies* because the data look longitudinal even though they are cross-sectional. An example of this design is insurance company life tables.

12.3.5 Advantages and Disadvantages of Various Studies

After a brief description of design and analysis of observational cohort studies (or longitudinal studies), case-control and cross-sectional studies, this section gives an idea of advantages and disadvantages of these three types of epidemiologic studies in the form of tables (cf. Kramer 1988).

Table 12.6 Advantages and disadvantages of cohort studies

Advantages	Disadvantages
1. Can provide more accurate picture of baseline state.	1. Poorly suited to the study of rare outcomes.
2. Sample selection by exposure essentially for rare outcomes.	2. For long latent period, prolonged follow-up is required.
3. Can minimize losses to follow-up and assess their effect.	3. Careful follow-up, labor intensive and costly.
4. If concurrent, it can protect against contamination and detection bias.	

After listing some advantages and disadvantages of *observational cohort* studies, the following table gives the advantages and disadvantages of *experimental cohort studies or clinical trials.*

Table 12.7 Advantages and disadvantages of clinical trials

Advantages	Disadvantages
1. Reduced potential for confounding bias. 2. Reduced potential for sample distention bias. 3. Conducive to blinding of subjects, observes and care-givers.	1. May be impracticable e.g. to assign any genetic traits, to attempt to direct the mothers to breast feed their new born. 2. May be unethical to provide toxins or carcinogens to subjects. 3. Often expensive since in observational study treatment is not borne by investigator. 4. May reduce generalizability because of possible treatment setting.

A clinical trial may be of two types: (a) Case-control study or (b) Cross-sectional study. Advantages or disadvantages of these as compared to cohort study are now described briefly in the following table.

Table 12.8 Advantages and disadvantages of case-control and cross-sectional studies (compared to observational cohort study)

Case-control studies	Cross-sectional studies
Advantages 1. For rare outcomes these are *more efficient*. 2. The answers of delayed outcomes can also be obtained as soon as exposure histories are obtained. Necessary waiting occurs when incident cases are being accumulated, hence *quicker*. 3. Case-control studies are usually *less costly*, because large number of subjects are not followed up. Disadvantages 1. It has higher potential for sample distortion. 2. Exposure ascertainment more prone to error and bias. 3. Without strict inclusion and exclusion criteria, standardized memory 'probes', interviewer blinding, and unbiased diagnostics, the studies are not 'gold standard'.	Advantages 1. Useful as descriptive epidemiologic studies (a) Disease or other clinical phenomenon can be classified by personal and geographical characteristics and by time. (b) The clinical spectrum (symptoms, signs, laboratory test results and pathologic findings) of a disease can be described well. (c) Prevalence of a variety of diseases and conditions of great importance to a public health can also be considered. 2. Rapid, less expensive analytical studies are possible which can provide early 'clues'. 3. Less prone (than case-control studies) to exposure recall error and bias in situations involving permanent exposure characteristics. Disadvantages 1. Frequent inability to distinguish whether exposure preceded outcome or vice versa, is major limitation unless exposure variable is a permanent attribute. 2. Higher potential for sample distortion bias.

13. Data Collection: Census and Sampling

Purpose of data collection is to obtain information about populations. As already defined earlier, population is a group of units under study. The population may consist of all the fields under a specified crop, persons either of whole country or a particular sector thereof. It could be mammal or animal population in a country or state. One may also be interested in studying bird or insect or pest or fish populations.

There are two basic approaches of data collection. One is called census, which means not only a complete enumeration of the study elements, but also collection of data on each unit of population. The second approach, a more general one, is based on the use of sampling methods and consists of collecting data from only selected units of the population.

13.1 Census of Human Population and Animal Population

Census of human population often called *census*, are now under taken in almost all the countries, generally at ten year intervals. Originally census meant an enumeration at a specified time of all the individuals inhabiting a given geographical area (normally a country). In modern times apart from a count of country men, information about their demographic, social, educational, economical, occupational and ethnic characteristics are also collected. Some censuses publish reports containing some vital rates at national and state level.

Every country has its own planning strategies in various sectors. Accordingly requirement of information of special type, some new requirements or modified form of earlier gathered information may come up. Hence the census procedure has to begin its work much before the reference date of census to give final shape of the coming census activities. It has to work in post census period to compile and process the data for final presentation as census reports. In this period a checking of the collected data is also undertaken by reconfirming a small random sample of data by a second visit.

Census of Human Population in India

Census of India is the largest source of statistics on various characteristics of the people of India. Data has been useful to scholars and researchers, demographers, economists, industrialists, anthropologists, sociologists and statisticians to name a few.

A Brief History

First census in India was held in 1872, which was not synchronized in different parts. From 1881 onwards it is held at every 10 year interval, the last one in 2001. The Indian census procedure is quite reliable and time tested. Due to advent of computer, the data processing has now become quite fast and efficient.

However, till 1951 census, the Census Organization was set up on an ad-hoc basis, though the population census in India is a major administrative function. Census Act was enacted after freedom in 1948 to provide the scheme of conducting population census along with the duties and responsibilities of census officers. In 1949 Government of India established an organization in the Ministry of Home Affairs under the Registrar General and Ex-officio Census Commissioner, India, to initiate steps for developing systematic collection of statistics on the size of population, its growth and its structure. Responsibility of registration of births and deaths was also entrusted to this office. In 2003 it has been assigned the work on a pilot project on issuing Multipurpose National Identity Card (MNIC).

Planning for the Indian Census

India being a very populated country planning of its census starts about 4-5 years of its scheduled date. It has been the practice to have extensive consultations with data users on the data items to be collected and the form of the tables desired to be presented in report.

Since census is an operation which tries to collect all the information at a time, demands from various data users for additional items is quite high. However, the census schedule (form for data collection) can not be too long, so that quality of data can be maintained.

Steps of Census Operations

As a *first step*, the data users' conference is organized about 5 years before census date, to take stock of data items to be collected.

In the *second step*, the Directors of Census Operations and senior officers of State and Union Territory (UT), Directors of Census Operations (i) finalize the strategy to be adopted for census (ii) share their views and experiences gathered about the pre-test of the House list and Household schedule from the House-listing Operation about a year before census date.

At the *third step*, a Technical Advisory Committee consisting of the Registrar General and Census Commissioners, eminent demographers, statisticians and social scientists, discuss the concepts and the revised draft questionnaire to finalize it.

The *fourth step* consists in obtaining approval of Home Minister on the House list and Household schedules.

The *fifth step* is to give adequate publicity among citizens and to request them to cooperate in Census data collection.

The *sixth step* is to start population enumeration, to be completed within a fixed duration of 2-3 weeks. Various government workers and temporarily appointed trained personnel visit house to house to fill up the pertinent information on the census schedules by personal interview of head of household or any other available responsible family member.

The *seventh step* is to compile, enter and analyze the collected data and to conduct post enumeration sample check, to correct the estimates before publication.

Despite all care, there are some unavoidable errors which are found to creep in the census data, such as under enumeration of children and old people, age bias, faulty reporting of income etc.

Proper training of census enumerators and supervision of work done are the ways of controlling errors. Some corrections of the estimates may be done by sample check up by proficient personnel during post census period before publication of report.

Next the special techniques applied for census of animal population are discussed.

Census of Animal Population

As for human population, by convention census of animal population refers to a count of all the animals belonging to the group of interest for a study, in a defined geographic area – a country or a state or a continent (cf. Southwood and Henderson (2000).

For a census of animal population, the study area can be divided into quadrats which may be of different shape and size. It may be a rectangular 10 km^2 area of a forest or 1 m^2 patch of grassland. Then a census of interested animal or insect (on grass patch) can be done on a randomly selected set. Standard statistical methods are used to estimate average of animal or insect population and its variance for the entire area.

Information about animal population is sought for variety of purposes. The objective of the study forms the basis for choice of proper method to be followed. The studies on animal population may be, broadly speaking, classified as *extensive* and *intensive* studies.

Extensive studies are carried over a large area and are often used to provide information on distribution and abundance for conservation and management programs. These may be monitoring of dairy cattle, food animals or birds, national birds, large animals like lion, tiger, elephant etc., of pests, parasites, venomous snakes etc., of aquatic reptiles and fishes etc.

In extensive surveys, normally the total area is divided into convenient subparts often in homogeneous within groups subparts and only a few are sampled for collecting relevant information. A sampled area is often visited once or at most a few times per study period.

Timing of sampling is fixed in relation to life cycle of the animal. For many species only limited stages in the life cycle can be sampled.

Intensive studies involve repeated observation of population of an animal. Usually size of population at various developmental stages are collected so that a life table may be prepared and budget for preserving or monitoring it may be allotted. Factors affecting the population size may be explored to control them beneficially.

If the method of analysis is decided in advance, proper data required for analysis may be collected. In animal population conservation work, the make up of the animal community and changes in its diversity associated with human activities are presently the most frequently desired measures. Special methods are needed for it. Even then difficulties arise because of the impossibility of recording the abundance of all the species within a habitat with equal efficiency.

For animal population the following types of estimates are generally obtained.

1. Absolute and Related Estimates

Only for large animals, that can be easily observed and have small population such as rhino, elephants, whales and some birds, it may be possible to express the population size for a country or state.

However, for most animals their numbers are expressed as a density per unit area or volume or density per unit of habitat (per leaf, per plant, per host). When the density is per habitat, it is necessary that the number of habitat units or its area should also be assessed in the geographic area about which the density will refer.

2. Relative Estimates

Some times the number caught can not be expressed as a density or intensity per area or habitat unit. These estimates can be compared in space and/or time. Such estimates give idea of species distributions, changes in species richness, environmental effect, pattern of animal activity and constitution of a polymorphic population.

For absolute estimates, methods of sampling are not always 100% efficient, whereas the relative estimates can also some time be corrected to give better estimates of density. Thus there is no hard and fast line between methods for absolute and relative estimates.

3. Population Indices

To obtain these the animals themselves are not counted, but their products (e.g. webs, exuviae, tubes, nests, frass etc) or effects (especially plant damage) are recorded.

Population indices and relative estimates of population can sometimes be related to absolute population, (if it is counted at the same time) by regression analysis. If sufficient data is available subsequent estimates of absolute population may be quite good. Such approach is used commonly in fisheries.

Some Methods of Animal Population Census

If *it is possible to count all the animals*, n, within a known area a, then this is *called a census* and the estimate of density D, is given by

$$\hat{D} = \frac{n}{a} \tag{13.1}$$

Counting often requires the observer to move over the census area and use strip transects (long thin quadrats) of length L and width 2ω. Strips can be surveyed by foot, car, boat, aircraft, remote-operated vehicle fitted with camera or submarine.

The counts obtained suffer from errors of bias and tendency of undercounting. It was suggested that photographs or video films be taken to reduce counting errors. This process also suffers from error since all the birds or animals may not be visible for photograph.

Choice of a particular season or time of a day when animals under study concentrate may be made the census time.

Technological detection signals like infrared image, electric signal etc. may be used for some animals.

Some other methods are listed below.

1. Point and Line – Survey Methods

Two such methods are considered here

(a) Indices of abundance using transects are calculated in conservation studies. In these, it is desired to detect any change in abundance rather than absolute count and with minimum cost. The method was suggested by Thomas (1983) based on his studies on a number of species of British butterflies and is applicable on other insects too. His method is

(i) Define the flight area A (or extent of colony) of butterfly by preliminary inspection.
(ii) Mark out a zigzag trail of length L meters, which should either exceed 1000 meters in length or should be long enough to record 40 butterflies.
(iii) Walk the trail and count butterflies say N, seen within an imaginary box 0 to 5 meter ahead, with a fixed width normally of 4 to 6 meters.
(iv) Calculate the index as

$$p = \frac{NA}{L} \tag{13.2}$$

(b) Methods based on Flushing

The number distributed from a known width of habitat are estimated. Under certain circumstances, e.g. in sea bird colonies, all of these may be flushed and an accurate count may be taken. Usually only a part of habitat is disturbed. If this part is a constant proportion of habitats in various localities, the counts give an index of absolute population. It can be changed into absolute population density by a minor mathematical adjustment of proportion and total area.

For birds Hayne (1949) suggested the observer to move along a transect through the habitat and to note the radial distance r_i at which i^{th} bird is flushed. The density is then estimated as

$$\hat{D}_H = \frac{n}{2L}\left(\frac{1}{n}\sum \frac{1}{r_i}\right) \tag{13.3}$$

where n = total number of birds counted
r_i = sight distance of ith bird
L = transect length

(c) The Fourier – Series Estimator

For these methods the following assumptions are made

(i) Objects on the line are always detected.
(ii) The observer does not influence the recorded positions.
(iii) Distances or angles are measured accurately.
(iv) The objects are correctly identified.

The basic procedure is to move along a straight line transect of length L, randomly placed in the field. Record the sighting distance r, (the distance between the observer and the object) and the sight angle θ (the angle of the object from the transect line). If n objects are counted in a strip of length L and width 2ω then the estimated density is

$$\hat{D} = \frac{n}{2\omega L} \tag{13.4}$$

(d) Point Transects

Instead of moving on a line transect, the observer may move to a number of fixed points and record the distance r, of the observed animals. This method is used for bird surveys, where patchy habitats of birds exist. Line transects are inappropriate in such a case, but point transects are often easier to use since observer may find out distance of animal using some locations around the point at known distances from the point transects. Population density is given by

$$\hat{D} = \frac{nb_{(0)}}{2\pi k} \tag{13.5}$$

where n = number of observed animals, k = the number of point transects chosen and $b_{(0)}$ is slope of pdf of detection distances of birds at 0 distance.

Various methods are used to determine $b_{(0)}$. MLE is one such method giving

$$\hat{D} = \frac{n^2}{2\pi k \sum_i r_i^2} \tag{13.6}$$

2. Spatial Distribution and Plotless Density Estimators

These methods are based on the consideration that the animals of a species are randomly distributed. The population density in such a condition can be estimated by *closest individual or distance* method or *nearest neighbor method.*

(a) Closest Individual or Distance Method

A point is selected at random and then, searching around it in concentric rings, the closest animal under study is found and its distance X, from the observation point is measured.

Engeman et al. (1994) call it *ordered distance estimator.* The density estimator is defined as

$$\hat{D} = \frac{sr - 1}{\pi \sum_{i=1}^{s} X_i} \tag{13.7}$$

where r = rank of the distances of the observed animal from the observation point (r = 1 for closest animal), X_i = the distance of closest animal from the ith observation point, i = 1, 2, ... s; and s = the number of random sampling points.

(b) Nearest Neighbor Method

In this method, an animal is selected at random and searching is continued to find another nearest neighbor. The distance between these two, r is measured. All the animals are numbered for identification. Using random number table one of these is selected and is called first animal. Selection is repeated k (an integer) times and distance r_i of the nearest neighbor from i^{th} (i = 1, 2, … k) of these k first animal is determined. Let

$$\bar{r} = \frac{1}{k}\sum_{i=1}^{k} r_i,$$ then the population density is estimated as

$$\hat{D} = \frac{1}{4\bar{r}} \tag{13.8}$$

Census and Sampling

As already mentioned, data collection has two approaches. The first is complete enumeration or census. The second approach is based on the sampling methods. It consists of collecting data on survey items from a few selected units from the population of interest.

Sampling is a scientific and objective procedure of selecting units from the population such that the sample is representative of the whole population of which it is a part. It also provides methods for the estimation of results that would be obtained if data were collected by complete enumeration or census.

The use of sampling in drawing some conclusions about population is possibly as old as civilization. It is used unknowingly in actions and attitudes in day to day life. In scientific study or research also use of sampling is equally prevalent.

There are situations when sampling is the only possible way of gaining information. For example if population is large i.e. number of elements of study interest is a big lot, or experiment to gain information are destructive (such as exploration of blood profile, determination of average life of a branded light bulb etc.), sampling is the only recourse.

Once such a decision is taken next problem is to choose the type of sampling method and to decide the size of sample and other details of data collection.

Normally the first two choices are made so that a balance is achieved between the efficiency of the estimators (which increases with the sample size) and cost involved in a sample survey.

However, in many cases sample survey is conducted in preference to census because of some advantages of sample study over the census. Some of the advantages of sample survey are the following:

(i) There is a *reduction of cost* in terms of money or man hours. Though cost per person in sample survey may be higher for acquiring quality data, but the total survey cost is much smaller than that for complete enumeration.
(ii) Generally in sample survey there is greater scope than in a census. Some enquiries may require highly trained personnel or specialized equipment for data collection. Thus information on large number of questions to be answered may be collected.
(iii) Better quality of data can be obtained than from a census. Very low level workers are generally employed in census, whereas a few highly trained personnel may be employed for a quality data through sample survey.
(iv) A census may be affected by non sampling errors which can not be measured, but sampling errors in a sample survey can be determined if proper survey design is used.
(v) Census can not be used for destructive enumeration. Whereas estimates for the whole population can be obtained by sampling only a few from the population.

Because of various advantages, the sample surveys have been used for seeking information of obvious importance such as on agricultural production and land use, unemployment, size of labour force, industrial production, wholesale and retail prices, health status of people, family income and expenditure, housing and social problems, cost of house building, ages of primary school children, effects of TV on school children, employment of scientists and engineers by industry, years of education, level of education and drop outs, migration, service in armed forces, percentage of tobacco consumers, number of flies in a town, number of people who can fold their tongues, inheritance of genetic debility etc.

Sampling theory is mainly concerned with making sampling methods more efficient. For this it seeks a proper definition and understanding of the nature of the population. It develops methods of sample selection and of estimation. Determination of sample size, the data to be collected and the methods of measurement are also some of the points dealt in sampling theory.

One major difference between sampling survey theory and the classical theory of sampling is that population in survey contains a *finite* number of units. The methods to derive the results are different and they are slightly more complicated when sampling is from finite population. When the sample size is small compared to the size of the population, results for infinite population are valid.

Methods of sampling may be classified as *subjective* and *objective*. *Subjective sampling* depends on the personal judgment or discretion of the sampler. *Objective sampling* adopts the method which is fixed by a rule of sampling theory or is independent of the personal judgment of sampler.

Objective sampling may be *non-probabilistic, probabilistic* or *mixed.*

Non-probabilistic objective sampling follows a rule but there is not any probability attached to mode of selection.

If there is definite preassigned probability of an individuals being selected in sample, the method of sampling is called *probabilistic* and also called *random sampling*. If in particular, the chances of being selected are equal for each individual in the population, it is called *simple random sampling*, already discussed in Section 7.2 of Chapter 7. There are many other methods of sampling, devised for different purposes for different types of populations. In this chapter these are briefly discussed after the next section. Next section contains a brief description of estimator of mean from a simple random sample (SRS) with and without replacement.

13.2 Random Sampling from Theoretical Distribution and from Finite Population

In various scientific studies of real life phenomenon or systems, it is often assumed that the associated random variable has some theoretical distribution. A random sample from this distribution is required to have an insight into the phenomenon or the working of the system. For an empirical study or simulation study, large number of random samples may be needed. This can be very easily done, with the aid of computer, provided the logical steps for generating or obtaining a random sample from the distribution of the associated random variable is known. It may be noted that the definition of the random sample has been given in Chapter 7.

The method of selecting a random sample of size n with or without replacement from a finite population of size N has been already explained in section 7.2 of Chapter 7, hence method of selecting from theoretical distribution is given below.

13.2.1 Selection of Random Sample from a Theoretical Distribution

In this section a basic principle for generating a random sample from a continuous random variable (r.v.) X with pdf $f(x, \theta)$ and from a discrete r.v. X with cdf $F(x)$ is given first. The method is applied to generate random sample from binomial, Poisson, Exponential and Normal random variables.

A Result A commonly used method for selecting a random sample $X_1, X_2, \ldots\ldots X_n$ on a r.v. X with pdf is based on the following *result*:

Let $Y = F(X)$, where $F(x)$ is cumulative density function (cdf) or distribution function of X at x.

Then it is obvious that $Y = F(X)$ is also a random variable, since Y is a function of r.v. X. It can be shown that $Y = F(X)$ has also a uniform distribution on the range $0 \le y \le 1$. In symbols

$$F(X) = Y \sim U(0, 1) \tag{13.9}$$

Hence $F(y) = y$. Therefore, a uniform random variable u generated from $U(0, 1)$ is equated to $y = F(x)$, i.e.

$$F(x) = u \tag{13.10}$$

Solving equation (13.10) for x, one gets a random observation from X as

$$x = F^{-1}(u) \tag{13.11}$$

Thus equating cdf of X with generated observation from $U(0, 1)$ and *solving (13.10) as* $x = F^{-1}(u)$, one gets an observation from *a continuous r.v. X*.

Suppose X *is a discrete* r.v. taking values $x_0, x_1, x_2, \ldots, x_j, \ldots$ *with* pmf $p(x)$ and *cdf* $F(x)$. If a random sample $X_1, X_2, \ldots\ldots, X_n$ is to be selected from X, the following steps are followed. (cf. Karian and Dudewicz (1991)).

Step 1: Generate u_i from $U(0, 1)$
Step 2: If $0 < u_i \le F(x_0)$ choose $X_i = x_0$ and
If $F(x_{j-1}) < u_i \le F(x_j)$; $j = 1, 2$ ......

$$\text{choose } X_i = x_j \tag{13.12}$$

Step 3: Repeat steps 1 and 2 for $i = 1, 2, \ldots\ldots, n$ to get the required sample.

Example 13.1: *Generating a random sample* $X_1, X_2, \ldots, X_n$ from a r.v. $X \sim E(\theta)$ i.e. from *an exponential r.v.* with parameter θ.

Solution: If

$$X \sim E(\theta),\ f(x, \theta) = \frac{1}{\theta} e^{-x/\theta} \quad x > 0,\ \theta > 0 \tag{13.13}$$

and

$$F(x) = 1 - e^{-x/\theta} \tag{13.14}$$

So *the steps* to generate random sample of size n from $E(\theta)$ are

(i) Generate u_i from $U(0, 1)$
(ii) $F(x_i) = u_i \Rightarrow 1 - e^{-x_i/\theta} = u_i$
which on solving gives $x_i = -\theta \log(1 - u_i) = -\theta\, l_n(1 - u_i)$ (13.15)
(iii) Repeat steps (i) and (ii) n times to get $x_1, x_2, \ldots\ldots, x_n$ as the selected random sample of observations from an exponential r.v. X.

Example 13.2: Generating a random sample from $X \sim B(n, p)$.
Solution A: If $X \sim B(n, p)$, the pmf of X is

$$p(x) = {}^nc_x\, p^x (1 - p)^{n - x};\ x = 0, 1, 2, \ldots\ldots, n;\ 0 < p < 1 \tag{13.16}$$

$$p(0) = q^n \text{ and } p(x+1) = \frac{n-x}{x+1}\frac{p}{q}\, p(x),\quad x = 0, 1, 2, \ldots\ldots, (n-1) \tag{13.17}$$

Find

$$F(x) = \sum_{x_j \le x} p(x_j) \tag{13.18}$$

Table 13.1 $p(x)$ and $F(x)$ of binomial random variable $B(n, p)$

x	$p(x)$	$F(x)$
0	$p_{(0)}$	$F_{(0)} = p_0$
1	$p_{(1)}$	$F_{(1)} = p_0 + p_1$
2	$p_{(2)}$	$F_{(2)} = p_0 + p_1 + p_2$
⋮	⋮	⋮
n	$p_{(n)}$	$F_{(n)} = p_0 + p_1 + p_2 + \ldots + p_{(n)} = 1$

Using (13.17) and (13.18) one can prepare a table of x and $F(x)$, as given below.

(i) Now generate u_i from $U(0,1)$ or use random number Table A.18
(ii) If $F(x_{j-1}) < u_i \le F(x_j)$, choose $X_i = x_j$
(iii) Repeat steps (i) and (ii) n times to get observed random sample of size n.

B. Suppose a random sample of size 6 is to be selected from $B(5, 1/3)$. Using equation (13.17) and (13.18) the table of $F(x)$ is prepared as follows, taking $n = 5$ and $p = 1/3$.

Binomial pmf and cdf for B(5, 1/3)

x	$p(x)$	$F(x)$		u_j	Sample x
0	.00001	$F_{(0)} = p_0$	= .00001	u_1 = .75867	5
1	.00045	$F_{(1)} = p_0 + p(1) = F_{(0)} + p_0$	= .00046	u_2 = .85783	5
2	.00810	$F_{(2)} = F_{(1)} + p_{(2)}$	= .00856	u_3 = .41290	5
3	.07290	$F_{(3)} = F_{(2)} + p_{(3)}$	= .08146	u_4 = .05870	3
4	.32805	$F_{(4)} = F_{(3)} + p_{(4)}$	= .40951	u_5 = .82444	5
5	.59049	$F_{(5)} = F_{(4)} + p_{(5)}$	= 1.0000	u_6 = .20247	4

Now using random number table given in Table A.18, suppose 6 random numbers (of 5 digits) are selected starting with 3rd column 9th row and putting a decimal point before the listed number in the 3rd column starting from 9th row. The selected u_i from this table (Table A.18) are shown in the above table. Looking at the values of u_j and $F(x)$ it is found that

$F_{(4)} < u_1 \le F_{(5)}$ so $X_1 = 5$
For u_2, $F_{(4)} < u_2 \le F_{(5)}$ so $X_2 = 5$
For u_3, $F_{(4)} < u_3 \le F_{(5)}$ so $X_3 = 5$
For u_4, $F_{(2)} < u_4 \le F_{(3)}$ so $X_4 = 3$
For u_4, $F_{(4)} < u_5 \le F_{(5)}$ so $X_5 = 5$
For u_6, $F_{(3)} < u_6 \le F_{(4)}$ so $X_6 = 4$

Hence the *random sample* of *size 6* from Binomial distribution with n = 5, and p = 1/3 is 5, 5, 5, 3, 5, 4.

Example 13.3: Generating a random sample of size n from a Poisson random variable X with parameter $\lambda = \nu t$.
Solution: The general method for a discrete distribution is applied first. If $X \sim P(\lambda)$ then the pmf and cdf of X are given by

$$p(x) = \frac{\lambda^x e^{-\lambda}}{x!} \quad x = 0, 1, 2, ... \tag{13.19}$$

$$F(x) = \sum_{x_j \le x} p(x_j) \tag{13.20}$$

Successive terms of $p(x)$ can be obtained from the recurrence formula

$$p(x+1) = \frac{\lambda}{x+1} p(x) \qquad \text{for } x = 0, 1, 2, ...$$

and $\qquad p(0) = e^{-\lambda}$ from (13.19)

Hence the table of $p(x)$ and $F(x)$ is formed as given below:

Steps to generate a random sample of size n from X are:

Step 1: Generate u_j from $U(0, 1)$ or use random number table (Table A.18)
Step 2: If $0 < u_j \le F_{(0)}$ choose $X_j = 0$ and
if $F_{(i-1)} < u_j \le F_{(i)}\, j = 1, 2, ..., n$
Choose $X_j = i$
Step 3: Repeat steps 1 and 2, n times to get a random sample of size n.

Table 13.2 $p(x)$ and $F(x)$ for Poisson random variable $P(\lambda)$

x	$P(x)$	$F(x)$
0	$p(0) = e^{-\lambda}$	$F(0) = p(0)$
1	$p(1) = \frac{\lambda}{2}\, p(0)$	$F(1) = F_{(0)} + p_{(1)}$
2	$p(2) = \frac{\lambda}{3}\, p(1)$	$F(2) = F_{(1)} + p_{(2)}$
3	$p(3) = \frac{\lambda}{4}\, p(2)$	$F(3) = F_{(2)} + p_{(3)}$
4	$p(4) = \frac{\lambda}{5}\, p(3)$	$F(4) = F_{(3)} + p_{(4)}$
⋮	⋮	⋮

Now this procedure is applied to generate a random sample of size 7 from $X \sim P(2)$.

The table below gives the values of $F(x)$ of Poisson distribution with parameter $\lambda = 2$. To find $e^{-\lambda}$ consult Table A.20 to get $p(0) = e^{-2} = .135$. The other values of $p(x)$ may be obtained from the recurrence relation (13.19) and cumulated as shown in Table 13.2 to get $F(x)$. An alternative way is to read Table A.2 for the row containing $\lambda = 2$ to get the $F(x)$ for $x = 0, 1, 2, ..., 8$

x and $F(x)$ for Poisson random variable $X \sim P(2)$

x	$F(x)$	u_j	Sampled x
0	.135	0.48545	2
1	.406	0.75122	3
2	.677	0.92904	4
3	.857	0.69902	3
4	.947	0.94972	5
5	.983	0.19152	1
6	.995	0.36024	1
7	.999		
8	1.000		

Now 7 uniform r.v. are selected using Table A.18. In this example 5 random digits are read from column 7 starting from row 11. The numbers are entered in above table and a decimal point attached to these to make them $U(0, 1)$. Comparing values of u_j with $F(x)$ the values of X in the random sample are 2, 3, 4, 3, 5, 1, 1 as also shown in the above table.

Example 13.3: Method II of Generating random sample of size n from $X \sim P(\lambda)$.

Solution: An alternative and easier method of generating a random sample from Poisson r.v. with parameter λ is based on the relation between Poisson and Exponential r.v.'s (cf. Law and Kelton (1991)). The procedure has the following steps:

Step 1: Find $e^{-\lambda} = p(0)$ from Table A.2 or from Table A.20.

Step 2: Generate $U(0, 1)$ or choose U_i's from random number Table A.18 and find the product $\prod_{i=1}^{j} u_i$.

Step 3: If $\prod_{i=1}^{j} u_i > e^{-\lambda}$ and $\prod_{i=1}^{j+1} U_j < e^{-\lambda}$, select $x_i = j$.

Step 4: Repeat steps 2 to 4, n times to get random sample of size n from $X \sim P(\lambda)$.

Now suppose it is *desired* to select *one random observation* from P (4.2).

First find $e^{-4.2} = .01499$

Now u_i's are selected to find $\prod_{i=1}^{4} u_i = .01807 > e^{-4.2}$ and $\prod_{i=1}^{5} u_i = .01109 < e^{-4.2}$

So the *randomly selected* observation $X_1 = 4$.

Example 13.4: Generating random sample of size n from $N(\mu, \sigma^2)$.
There are several methods for generating random sample from $X \sim N(\mu, \sigma^2)$. Now one can recall that if $X \sim N(\mu, \sigma^2)$ then $Z = (X - \mu)/\sigma \sim N(0, 1)$, i.e. Z has standard normal distribution.

Two methods for generating random sample from standard normal distribution are given below. The first method uses property based on central limit theorem.

Method 1: The following steps are followed to use this method.

Step 1: Generate k $U(0, 1)$ random variables $u_1, u_2, \ldots, u_k$.

Step 2: Find $T = \sum_{j=1}^{k} u_j$.

Step 3: Calculate $Z = (T - k/2)/\sqrt{k/12}$.

Step 4: Repeat steps 1 to 3, n times to get n random observations from $N(0, 1)$

(a) Choose $k = 12$ to get $Z_i = (T - 6)$
(b) If random sample from $N(\mu, \sigma^2)$ is required add one step 3A to calculate $X_i = \sigma Z_i + \mu$ and the random sample of size n will be obtained from normal distribution with mean μ and variance σ^2.
(c) If $n = 48$ is chosen, $Z_i = (T - 24)/2$ are better standard normal variates
Again the transformation $X_i = \sigma Z_i + \mu$ gives random sample from $N(\mu, \sigma^2)$.

Method 2: The following steps give random sample from $N(0, 1)$ (cf. Law and Kelton (1991))

Step 1: Generate U_1 and U_2 as independent $U(0, 1)$.

Step 2: Let $V_i = 2U_i - 1$ for $i = 1, 2$ and let $W = V_1^2 + V_2^2$.

Step 3: If $W > 1$ go to step 1

Otherwise, let $Y = \sqrt{(-2 \ln W)/W}$, $X_1 = V_1 Y$ and $X_2 = V_2 Y$ then X_1, X_2 are iid $N(0, 1)$.

Step 4: Use X_1's for odd calls i.e. for odd sample observations and X_2's for even sample observations.

13.2.2 Random Sampling from a Finite Population

In the last section, methods of generating random sample from some theoretical distributions have been considered. Definition of random sample from a theoretical distribution and finite population has already been given in Chapter 7 and earlier. Methods of generating simple random sample of size n from a finite population of size N have been explained, when random sample is taken with replacement (i.e. SRSWR) or when it is taken without replacement (SRSWOR); in Section 7.2. A more interested reader may consult books on sampling theory such as by Sukhatme and Sukhatme (1976), Cochran(1974), Goon et.al(2002)).

As a ready reference, a definition of random sample from a finite population is given below.

Definition

$X_1, X_2, \ldots, X_n$ will be a random sample from a finite population of size N, if the following conditions are satisfied:

1. Every unit in the population has an equal chance of being selected in the sample.
2. The probability of a specified unit being selected at say, k^{th} draw is same as the "probability of its being selected at the first draw $\left(=\frac{1}{N}\right)$.
3. Every sample of size n has an equal chance of being selected $\{= 1/({}^{N}C_n)\}$ for SRSWOR and $1/(N^2)$ for SRSWR.

The simplest and most commonly used type of probability sampling is simple random sampling. It can be done with replacement (SRSWR) or without replacement (SRSWOR).

It will be assumed, unless otherwise mentioned, that sampling is without replacement. Generally an estimate of mean or total of a characteristic in the population (with unique value for a unit) is desired. In this section some notations are given, which may be used not only in this section but also in other sections dealing with other sampling methods also. The expressions for best estimators of population mean and the variance of the estimators will be given (without proof).

The advantages and uses of various sampling methods will help choosing an appropriate procedure and to determine the estimates and their standard errors.

Notations

N : the number of units in the population or population size

Y : the characteristic under study

Y_i : value of the characteristic for i^{th} unit in the population

$$\bar{Y}_N = \frac{1}{N}\sum_{i=1}^{N} Y_i, \text{ the mean value of the characteristic per unit of the population} \tag{13.20}$$

$$S^2 = \frac{1}{N-1}\sum_{i=1}^{N}\left(Y_i - \bar{Y}_N\right)^2 = \frac{1}{N-1}\left[\sum_{i=1}^{N} Y_i^2 - N\bar{Y}_N^2\right] \tag{13.21}$$

the mean square in the population.

$$V(Y) = \frac{1}{N}\sum_{i=1}^{N}\left(Y_i - \bar{Y}_N\right)^2 = \frac{N-1}{N}S^2 \tag{13.22}$$

the variance of a single observation in the population.

n = the number of units in the sample or the sample size.

$$\bar{Y}_n = \frac{1}{n}\sum_{i=1}^{n} Y_i, \text{ the sample mean} \tag{13.23}$$

and $$s^2 = \frac{1}{(n-1)}\sum_{i=1}^{n}\left(Y_i - \bar{Y}_n\right)^2 = \frac{1}{(n-1)}\left(\sum_{i=1}^{n} Y_i^2 - n\bar{Y}_n^2\right), \text{ the sample mean square} \tag{13.24}$$

Y_i in (13.23) and (13.24) refers to i^{th} sample value (i = 1, 2, ..., n) and is any one of the N values $Y_1, Y_2, \ldots, Y_N$ in the population.

Best Linear Unbiased Estimator of $\bar{Y}_N$

Best linear unbiased estimator (BLUE) of population mean $\bar{Y}_N$ is given by $\bar{Y}_n$. In symbols

$$\hat{\bar{Y}}_N = \bar{Y}_n \tag{13.25}$$

and $$V\left(\bar{Y}_n\right) = \left(\frac{1}{n} - \frac{1}{N}\right)S^2 = \frac{N-n}{N}\frac{S^2}{n} \tag{13.26}$$

It can be seen that

$$E\left(\bar{Y}_n\right) = \bar{Y}_N \text{ so } \bar{Y}_n \text{ is unbiased.} \tag{13.27}$$

and $V\left(\bar{Y}_n\right)$ is smallest among variances of other unbiased linear estimators of $\bar{Y}_N$. So $\bar{Y}_n$ *is BLUE of* $\bar{Y}_N$.

The factor $\frac{N-n}{N}$ is a correction for the finite size of the population and is called *finite population correction* (fpc).

When n is small compared to N, the fpc will be close to one and $V\left(\bar{Y}_n\right)$, the sampling variance of $\bar{Y}_n$ is same as that of sample mean from an infinite population.

The population mean square S^2 is usually unknown. So an estimate of $V\left(\bar{Y}_n\right)$ is given by

$$\hat{V}\left(\bar{Y}_n\right) = \frac{N-n}{N}\frac{s^2}{n} \tag{13.28}$$

and the estimate of SE $\left(\bar{Y}_n\right)$ will be

$$\hat{\text{SE}}\left(\bar{Y}_n\right) = \sqrt{\frac{N-n}{N}}\frac{s}{\sqrt{n}} \tag{13.29}$$

Confidence Interval for Population Mean

If the distribution of Y in population is not very different from normality and sample size n is not very small, $\bar{Y}_n$ is approximately normally distributed with mean $\bar{Y}_N$ and variance $V\left(\bar{Y}_n\right)$ of equation (13.26). A 95% confidence interval (CI) for $\bar{Y}_N$ will be

$$\bar{Y}_n \pm 1.96\sqrt{\frac{N-n}{Nn}}\,S \tag{13.30}$$

If S^2 is unknown, it is replaced by its estimator s^2 and if the conditions for using (13.20) are satisfied, the approximate 95% CI is given by

$$\bar{Y}_n \pm t_{n-1,.025} \sqrt{\frac{N-n}{Nn}}\, s \tag{13.31}$$

Determination of Sample Size for Specified Precision

A statistician has to answer, in planning a sample survey, the size of sample - n to be chosen to achieve a specified *precision* for estimation of mean.

The precision is usually specified in terms of the *permissible margin of error* in the estimate and the coefficient of confidence (= 1 – α) with which one wants to make sure that the estimate is within the permissible margin of error.

Let the permissible error in the estimate of $\bar{Y}_N$ be $\varepsilon \bar{Y}_N$ (say) and desired confidence coefficient is (1 – α). Then one desires to know n, such that

$$P\left[\left|\bar{Y}_n - \bar{Y}_N\right| \geq \varepsilon \bar{Y}_N\right] = \alpha \tag{13.32}$$

Using CI in (13.31) one can see that

$$n = \frac{\dfrac{t^2_{(\infty,\alpha/2)}}{\varepsilon^2}\dfrac{S^2}{\bar{Y}_n^2}}{1 + \dfrac{1}{N}\dfrac{t^2_{(\infty,\alpha/2)}}{\varepsilon^2}\dfrac{S^2}{\bar{Y}_n^2}} \tag{13.33}$$

If S^2 is not known, n could be evaluated more precisely from the following formula

$$n = \frac{\dfrac{t^2_{(n-1,\alpha/2)}}{\varepsilon^2}\dfrac{s^2}{\bar{Y}_n^2}}{1 + \dfrac{1}{N}\dfrac{t^2_{(N-1,\alpha/2)}}{\varepsilon^2}\dfrac{s^2}{\bar{Y}_n^2}} \tag{13.34}$$

In this $t_{(n-1,\ \alpha/2)}$ is not known since n is not known.

So if $t_{(n-1,\ \alpha/2)}$ is replaced by $t_{(\infty,\ \alpha/2)}$ in (13.34), the resulting n is a little underestimated. Since there is likelihood of underestimation of n, the actual n is increased from the calculated n.

Estimation of Population Total

Let $T = N\bar{Y}_N$, the total of Y in the population and it is desired to estimate T. The unbiased estimator is given by

$$\hat{T} = N\bar{Y}_n \tag{13.35}$$

and

$$V(\hat{T}) = N^2 V(\bar{Y}_n)$$

$$= N^2 (N-1)\, S^2/n \text{ for SRSWR}$$

$$= N^2 S^2 \left(\frac{1}{n} - \frac{1}{N} \right) \text{ for SRSWOR} \tag{13.36}$$

Example 13.5: The standard deviation of marks obtained in Zoology by 121 boys is found to be 12.5. Find $\mathrm{SE}\left(\bar{Y}_{10}\right)$ i.e. standard error of estimator of population mean from a random sample of size 10 (a) taken with replacements and (b) taken without replacements.

Solution: Estimator of population mean $\bar{Y}_N$ is sample mean $\bar{Y}_n$.

$$\mathrm{SE}\left(\bar{Y}_n\right) = \sigma/\sqrt{n} = \sqrt{\frac{(N-1)}{n}}\, S \text{ for SRSWR} = \frac{12.5}{\sqrt{10}} = 3.95$$

$$= \sqrt{\frac{N-n}{Nn}}\, S \text{ for SRSWOR}$$

$$= \sqrt{\frac{N-n}{Nn}} \sqrt{\frac{N\sigma^2}{N-1}} = \sqrt{\frac{N-n}{N-1}} \frac{\sigma}{\sqrt{n}} = \sqrt{\left(\frac{121-10}{120}\right)} \frac{12.5}{\sqrt{10}} = 3.80$$

Estimation of Population Proportion

Let p be the proportion of numbers of the population possessing some characteristic. Then an estimator of p is given by

$$\hat{p} = \frac{f}{n} \tag{13.37}$$

where f is the number of persons in the sample of size n who possess the said characteristic

$$V(\hat{p}) = \frac{p(1-p)}{n} \text{ for SRSWR and}$$

$$= \left(\frac{N-n}{N-1}\right) \frac{p(1-p)}{n} \text{ for SRSWOR} \tag{13.38}$$

$$\hat{V}(\hat{p}) = \left(\frac{N-n}{N}\right) \frac{\hat{p}(1-\hat{p})}{n-1} \tag{13.39}$$

Confidence Interval for *p*

If p is neither too small nor too large, $\hat{p}$ will be normally distributed. For small n, the 100(1 – α) percent confidence interval for p is given by

$$\hat{p} \pm t_{\infty, \alpha/2} \sqrt{\left(\frac{N-n}{N}\right) \frac{\hat{p}(1-\hat{p})}{(n-1)}} \tag{13.40}$$

Size of Sample for Specified Precision

Let permissible error in the estimate of p is $\varepsilon\, p$ and the degree of assurance is (1 – α) 100%, then the required sample size is given by

$$n = \frac{\left[\left(t^2_{\infty,\alpha/2}\right)q\right] \Big/ \varepsilon^2 p}{1 + \frac{1}{N}\left\{\frac{\left(t^2_{\infty,\alpha/2}\right)q}{\varepsilon^2 p} - 1\right\}} \tag{13.41}$$

If N is large and ε is not too small, n can be simplified as

$$n = \frac{\left(t^2_{\infty,\alpha/2}\right)q}{\varepsilon^2 p} \tag{13.42}$$

where $q = (1 - p)$.

Example 13.6: Material for the construction of 5000 wells was issued during 1944 in a certain district as part of the "Grow More Food Campaign". List of cultivators to whom it was issued, together with the proposed location of the well, is available. A large part of the material was reported to have been misused by diverting it to other purposes. It is proposed to assess the extent of misuse by a sample spot check. It is proposed to estimate the proportion p of wells actually constructed and used for irrigation. A sample is to be selected by SRSWOR method from the total population of wells for which material was issued. The permissible margin of error in the estimated value is 10% and the degree of assurance desired is 95%. Determine the size of sample for values of p, 0.5 to 0.9 (source: Sukhatme and Sukhatme (1976), p. 19).

Solution: Given that $N = 5000$ $\varepsilon = .1$, $t_{\infty,\ .025} = 1.96$.

As an illustration for $p = .5$ all these values are substituted on r.h.s. of (13.41) to get

$$n = \frac{1.96^2 \times .5/(.01 \times .5)}{1 + \frac{1}{N}\left\{\frac{1.96^2 \times .5}{(.01 \times .5)} - 1\right\}} = 357$$

In a similar way for various values of $p(= .5$ to $.9)$ the values of n are obtained as shown below

p	.5	.6	.7	.8	.9
n	357	244	159	94	42

Since the worst critics do not blame of misuse more than 50% of the issued material, a *sample of 357* out of 5000 proposed wells would be *adequate* for the check.

13.3 Stratified Random Sampling

It is evident from the expressions of variance of estimator of population mean given in equation (13.26), that the precision of $\overline{Y}_n$ depends on sample size n and also on S^2, the measure of variability in the population. For a chosen sample size n, *the precision of* $\overline{Y}_n$ *can be increased by adopting a sampling procedure which is effective for reducing the heterogeneity. The method of stratified random sampling* is one such procedure.

The Procedure of Sampling

The procedure of stratified sampling consists of the following steps:

1. First divide the population of N units in to k classes called *strata* such that N_i units belong to i^{th} strata (i = 1, 2, ..., k) and $\sum_{i=1}^{k} N_i = N$.

   The administrative areas or geographical proximity may be used to *form strata* if sampling is of plots of land or of villages. The plots may also be stratified according to irrigated or non-irrigated. Villages may be stratified according to its population size or area. If households of an urban area are to be sampled, they may be stratified according to the types of residential building as to pucca two or more storied, pucca single storey big, pucca medium and kachcha. With human populations, stratification may be according to people living in institutions (e.g. hostels, hotels, hospitals, prison, religious *Math*, etc.) and those living in houses, since a different approach is needed for sampling from said two groups.

   If business firms are to be studied one may put large firms (criteria may be annual profit, number of workers, etc.) in one strata, medium in the second strata and the small firms in the third strata. The main consideration in forming the strata is that the heterogeneous population should be divided in subpopulations called strata such that there is homogeneity within strata. If each stratum is homogeneous as regards the characteristic under study, the estimator of population mean will be more precise.

2. At the second step it is decided that how many sampling units are to be selected using SRSWOR from various strata. If n_i is the sample size from i^{th} strata, $\sum_{i=1}^{k} n_i = n$ and there are three commonly used methods for allocation of sample in strata:

(a) *Proportional allocation:* In this method the sample size from a strata is proportional to the size of the strata.

In symbols $n_i \propto N_i$ or $n_i = \frac{n}{N} N_i$ (13.42)

(b) *Optimal allocation:* In this method the n_i's are chosen so that the estimator of population mean has a desired degree of precision with minimum possible cost. Another choice may be to achieve maximum precision for a given cost.

If the cost of sampling a unit from i^{th} strata is C_i and these do not differ much so that C_i may be taken as equal, i.e. $C_i = C$, then

$$n_i = n \frac{p_i S_i}{\sum_{i=1}^{k} p_i S_i} \tag{13.43}$$

where $p_i = \frac{N_i}{N}$ and S_i^2 is the mean square in i^{th} strata similar as defined in equation (13.21) for the population of size N and is defined in equation (13.50) for i^{th} stratum. This allocation is also known as Neyman allocation.

(c) *Equal allocation:* In this method equal number of units are selected from each strata. So $n_i = \frac{n}{k}$ for i = 1, 2, …, k.

3. The third step consists in selecting the decided number of n_i units out of known number N_i of units in the ith strata (i = 1, 2, ..., k). Using SRSWOR technique, n_i units are selected from N_i units of ith strata (i = 1, 2, ..., k). This is done for each of the k strata using the procedure described in Section 7.2. $\sum_{i=1}^{k} n_i = n$; also decided before the study starts.

Estimator of Population Mean and its Variance

Let N_i denote the size of ith stratum (i = 1, 2, ... k) or the number of units in it. n_i is the sample size from i^{th} stratum

$$\text{then } \sum_{i=1}^{k} n_i = n, \text{ the total sample size.} \tag{13.44}$$

Further, let Y_{ij} be the jth unit of ith stratum

$$j = 1, 2, ..., N_i \text{ and } i = 1, 2, ..., k.$$

$$\text{and } p_i = \frac{N_i}{N}, \text{ then } \bar{Y}_{N_i} = \frac{1}{N_i}\sum_{j=1}^{N_i} Y_{ij}, \text{ the mean of i}^{\text{th}} \text{ stratum} \tag{13.45}$$

Then the population mean $\bar{Y}_N$ can be written as

$$\bar{Y}_N = \frac{1}{N}\sum_{i=1}^{k} N_i \bar{Y}_{N_i} = \sum_{i=1}^{k} p_i \bar{Y}_{N_i} \tag{13.46}$$

Since the sample of size n_i is drawn from i^{th} stratum by SRSWOR,

$$\bar{Y}_{n_i} = \frac{1}{n_i}\sum_{j=1}^{n_i} Y_{ij} \tag{13.47}$$

the mean of sample from i^{th} stratum, is an unbiased estimator of $\bar{Y}_{N_i}$. Thus, it is natural to define an estimator of $\bar{Y}_N$ as

$$\hat{\bar{Y}}_N = \bar{Y}_{st} = \sum_{i=1}^{k} p_i \bar{Y}_{n_i} \tag{13.48}$$

$\bar{Y}_{st}$ is a weighted mean of the sample means of the strata. The weights are proportional to the strata size.

$\bar{Y}_{st}$ is an unbiased estimator of $\bar{Y}_N$ and its variance is given by:

$$V(\bar{Y}_{st}) = \sum_{i=1}^{k}\left(\frac{1}{n_i} - \frac{1}{N_i}\right) p_i^2 S_i^2 \tag{13.49}$$

where S_i^2 is the mean square of ith subpopulation or i^{th} stratum population, defined as

$$S_i^2 = \frac{1}{N_i - 1}\sum_{j=1}^{N_i}\left(Y_{ij} - \bar{Y}_{N_i}\right)^2 \tag{13.50}$$

The standard error of $\bar{Y}_{st}$ is given by

$$\text{SE}\left(\bar{Y}_{st}\right) = \sqrt{V\left(\bar{Y}_{st}\right)} \text{ and} \tag{13.51}$$

the estimator of

$$\text{SE}\left(\bar{Y}_{st}\right) = \sqrt{\sum_{i=1}^{k}\left(\frac{1}{n_i} - \frac{1}{N_i}\right) p_i^2 s_i^2} \tag{13.52}$$

where $s_i^2 = \frac{1}{n_i - 1}\sum_{j=1}^{n_i}\left(Y_{ij} - \bar{Y}_{n_i}\right)^2$, the mean square of ith strata sample (13.53)

It is now evident that the $V\left(\bar{Y}_{st}\right)$ will be smaller if S_i^2 is small i.e. the units in a strata are homogeneous.

Stratified Random Sampling and Simple Random Sampling

A brief comparison of the two sampling procedures is as follows:

1. Stratified random sampling enables to provide the desired representation of various strata in the sample. In simple random sampling or unrestricted random sampling, adequate representation from all the strata can not always be ensured. The sample selected by SRSWOR may be so distributed among strata, that some may be over represented while some may not be represented at all.
2. More the strata differ in their population means, larger the gain in precision of stratified sampling with proportional allocation over the simple random sampling.
3. More the strata differ in strata mean square as well as in strata means, larger the gain in precision of stratified sampling with optimum allocation or Neyman allocation over the simple random sampling.

13.4 Cluster Sampling and Area Sampling

Simple random sampling and stratified random sampling presuppose that population consists of finite number of distinct, identifiable units called *sampling units*. The smallest units in which a population can be divided are called *elements* of the population and *groups of elements are* called *clusters*.

For some populations the list of elements may not be available so sampling of elements is not feasible. A list of city sectors, a list of villages, a list of houses, however, may be available. Such group of elements or clusters, therefore, may be considered as sampling unit. This procedure of sampling is called *cluster sampling*.

This procedure is valid if every unit or element of the population belongs to one and only one cluster. Under this condition the total number of clusters will cover all the units of the population without any duplication or omission.

Area Sampling is a sampling procedure in which maps, rather than lists or registers serve as the sampling frame. Total area under investigation is divided into smaller sub-areas. A random sample of these sub-areas are selected. Each of the selected sub-areas is then fully inspected and enumerated.

The cluster or area sampling is *widely used* because of (a) non-availability of list of elements and (b) due to saving of time and money. A SRSWOR of elements may lead to scatter of sampled elements over the entire area, whereas sampling of clusters will leave the unselected clusters (areas) from being visited by investigating staff.

For a good representation of the elements in population, the size of the cluster should be small. Smaller the cluster or area it is expected that more accurate will be the estimate.

Estimator of Population Mean and its Variance

The case of equal clusters is considered and let the population be composed of N clusters of M elements each. A random sample of n clusters is chosen from N clusters by SRSWOR.

Let Y_{ij} be the value of characteristic under study of jth element in i^{th} cluster $j = 1, 2, ..., M$ and $i = 1, 2, ..., N$.

$$\bar{Y}_{i\cdot} = \frac{1}{M}\sum_{j=1}^{M} Y_{ij}, \text{ the mean of the } i^{th} \text{ cluster} \tag{13.54}$$

$$\bar{\bar{Y}}_{N\cdot} = \frac{1}{N}\sum_{i=1}^{N} \bar{Y}_{i\cdot}, \text{ the mean of cluster means in population} \tag{13.55}$$

$$\bar{Y}_{\cdot\cdot} = \frac{1}{NM}\sum_{i=1}^{N}\sum_{j=1}^{M} Y_{ij}, \text{ the mean of characteristic in population} \tag{13.56}$$

It is evident that $\bar{\bar{Y}}_{N.} = \bar{Y}_{..}$

Further, let

$$S_i^2 = \frac{1}{M-1}\sum_{j=1}^{M}\left(Y_{ij} - \bar{Y}_{i\cdot}\right)^2, \text{ the mean square of elements of } i^{th} \text{ cluster, } i = 1, 2, \ldots, N \tag{13.57}$$

$$S_b^2 = \frac{1}{N-1}\sum_{i=1}^{N}\left(\bar{Y}_{i\cdot} - \bar{\bar{Y}}_{N\cdot}\right)^2, \text{ the mean square between cluster means in the population} \tag{13.58}$$

The *estimator of* $\bar{\bar{Y}}_{N\cdot}\,(=\bar{Y}_{\cdot\cdot})$ is $\bar{\bar{Y}}_{n\cdot}$ given by

$$\bar{\bar{Y}}_{n\cdot} = \frac{1}{n}\sum_{i=1}^{n} \bar{Y}_{i\cdot}, \text{ mean of sampled } n \text{ clusters} \tag{13.59}$$

$\bar{\bar{Y}}_{n\cdot}$ is unbiased estimator of $\bar{\bar{Y}}_{N\cdot}$ and

$$V\left(\bar{\bar{Y}}_{n\cdot}\right) = \frac{N-n}{N_n} S_b^2 \tag{13.60}$$

13.5 Systematic Sampling

In simple random sampling and cluster sampling, all the sampling units (elements or cluster of elements) are selected with the help of random numbers. *Systematic sampling* is a procedure in which only the first unit of the sample is selected with the help of random numbers, the *rest are selected* according to a *predetermined system.*

Commonly followed pattern in systematic sample is very simple and involves a regular pre-specified spacing of the elements or units of population.

Method of Systematic Sampling

Suppose a population consists of N units numbered from 1 to N and $N = k\,n$ i.e. N is a product of sample size n and an integer k. Now a random number $i \le k$ is selected. The systematic sampling method consists of choosing every k^{th} unit of the population starting from i^{th} unit. Thus the sample will consist of n units with identification numbers $i, i + k, i + 2k, ..., i + (n - 1)\,k$. The sample consisting of these units is called a *systematic sample*.

Some examples of systematic sampling are: (a) every k^{th} strip of a forest (b) every plot in an administrative block, k km. apart (c) every k^{th} village in a list of villages of a district (d) every k^{th} household in a city (e) every k^{th} hour production in a machine (f) number of fishing boats coming to sea coast every k^{th} time interval of 20 minutes. If the ordering of units in the population are not related to the characteristic under study, the systematic sample is equivalent to an SRSWOR.

This method is quite *extensively used* due to its low cost and simplicity of selection procedure. Because of the regular spacing of units to be selected in systematic sample, it is free from the error of missing some part of the population in sample.

Resemblance with Stratified and Cluster Sampling

Since one unit is selected from each of the n strata of k consecutive sampling units, it has some resemblance with stratified sampling. However, in systematic sampling a definite unit from each of the n strata will be selected, whereas in stratified sampling the units from each strata are randomly selected. Moreover, random selection from a stratum is independent of the selection from other strata.

The systematic sampling procedure is more similar to cluster sampling than to stratified sampling. It may be evident from the table below that a systematic sample is equivalent to a sample of one cluster out of k clusters of n units each.

Table 13.3 Composition of population elements equivalent to k clusters and n strata

Stratum number	Cluster numbers					
	1	2	......	i	......	k
1	1	2	......	i	......	$k = k$
2	$1 + k$	$2 + k$	......	$i + k$	......	$k + k = 2k$
3	$1 + 2k$	$2 + 2k$	......	$i + 2k$	......	$k + 2k = 3k$
.	.	.		.		.
.	.	.		.		.
.	.	.		.		.
j	$1 + (j - 1)k$	$2 + (j - 1)k$	......	$i + (j - 1)k$	......	$k + (j - 1)k = jk$
.	.	.		.		.
.	.	.		.		.
.	.	.		.		.
n	$1 + (n - 1)k$	$2 + (n - 1)k$	......	$i + (n - 1)k$	......	$k + (n - 1)k = nk$
Mean	$\bar{Y}_{1\cdot}$	$\bar{Y}_{2\cdot}$	......	$\bar{Y}_{i\cdot}$	......	$\bar{Y}_{k\cdot}$

Estimator of Population Mean and its Variance

Since a systematic sample consists of elements with serial numbers in i^{th} column, if random number i is the random number selected between 1 to k, the mean of the systematic sample is given by

$$\bar{Y}_{sy} = \frac{1}{n}\sum_{j=1}^{n} Y_{i+(j-1)k}$$

and
$$V\left(\bar{Y}_{sy}\right) = \frac{1}{k}\sum_{i=1}^{k}\left(\bar{Y}_{i\cdot} - \bar{Y}_N\right)^2 = \frac{k-1}{k} S_i^2 \tag{13.61}$$

where $\bar{Y}_N$ is population mean and S_c^2 is the mean square of column means given by

$$S_c^2 = \frac{1}{k-1}\sum_{i=1}^{k}\left(\bar{Y}_{i\cdot} - \bar{Y}_{\cdot\cdot}\right)^2 \tag{13.62}$$

Systematic Sampling when N is not a Multiple of n

The sampling method and estimator described above assume that the population size $N = nk$, i.e. N is a multiple of n, the sample size. Situations may arise when the size of population may not satisfy this condition:

Sample size may vary from one choice of i to the other choice, from the same population. In some other situations the population size N may be a prime number say, $N = 23$. Suppose it is divided in 5 groups then there will be 3 systematic samples for which $n = 5$ can be obtained but there will be 2 for which $n = 4$ can be possible.

This fact introduces some disturbance in the simple expressions for estimator and its variance. If $n > 50$, this disturbance is very small and can be ignored.

Interested readers may consult discussion given by Sukhatme and Sukhatme (1976).

Sampling Method

In case, $N \neq nk$, find the integer k nearest to N/n. Now select a random number between 1 to N. Include in the systematic sample the unit with this serial number and every k^{th} unit thereafter. The population units are arranged in a cycle of 1 to N before sampling and sampling is continued till the desired sample size is reached. This type of sample is called *circular systematic sample*.

Example 13.7: Consider a population of eight households a, b, c, d, e, f, g and h. Suppose a sample of size 3 is needed. Enumerate all possible systematic samples (source: Goon et al. (2002), p. 211).

Solution: In this case $N = 8$ and $n = 3$. $N/n = 2\frac{2}{3}$ is not an integer. So the method of *circular systematic sampling* is to be used. So nearest integer to N/n is taken as $k = 3$.

The households in a circular arrangement will be as shown in the figure below:

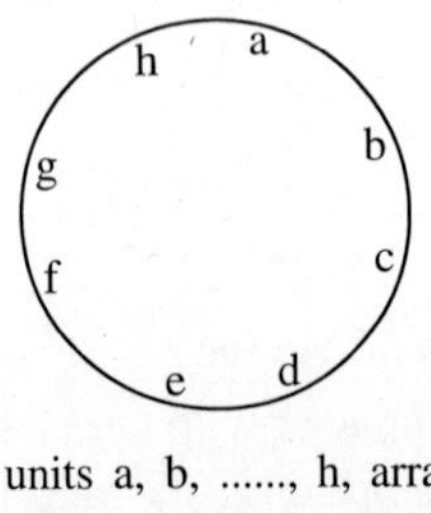

Fig. 13.1. The sampling units a, b, ..., h, arranged in a cyclic fashion.

Now a random number between 1 to 8 (= N) is selected. Suppose selected number is 4. So the 4th unit starting from 1st unit a is selected as the first unit in sample. It is d. The second unit will be 3rd from d, i.e. g. Third unit from g is b which will be the 3rd unit of the sample. Hence *the selected systematic sample of size* $\boldsymbol{n}$ = 3 *will be* d g b.

13.6. Two-stage and Multistage Sampling

There are situations when each sampling unit in the population can be divided into a number of smaller units or elements. For example a set of N clusters exist, in a population of size NM and a sample of n clusters is selected. Normal procedure of cluster sampling selects all the M units of the selected clusters. If the units within a cluster give similar results, it is uneconomical to measure all the elements of the selected clusters. It is advisable in such a situation to select and measure a sample of elements from the selected units. This *technique* of measuring only the chosen pre-specified number of elements from each selected cluster is called *sub-sampling*.

It is also known as *two-stage-sampling* since the sample is *selected in two-steps*. The units (clusters) selected at first step are called the first stage (or primary) sampling units (fsu or psu). The elements or group of elements which form the *units of sampling at the second stage* are called *sub-units or second-stage units* (ssu).

Examples:

1. Cities of a state may be psu and wards in the cities may be ssu.
2. Districts of a country may be fsu and villages may be the ssu.
3. Villages of a district may be fsu and the households in the villages may be ssu.

Multistage Sampling

The above procedure of two-stage sampling may be generalized by adding further stages to have three-stage or four-stage sampling. The method with more than two stages is called *multi-stage sampling*. For *example* to get an estimate of average sugarcane yield per acre of land in UP, the districts may be taken as fsu; the villages in districts as the ssu and the plots of villages as the third or last stage units. The sugarcane crop estimate per acre of the plots will be the collected sample observations.

Multi-stage sampling provides flexibility in sampling procedure. It enables existing divisions and subdivisions to be used as sampling units.

The second-stage frame is required only for the selected fsu. Similarly for the third and subsequent stages frames are needed only for the selected subunits.

This procedure leads to enormous saving in operational cost, for a large area survey, especially if it includes under-developed pockets. However, multi-stage sampling is less efficient than some suitable single-stage sampling method if it is practicable.

Estimation of Population Mean and Variance of Estimator

The method is *given only for two-stage sampling with equal fsu.*

Let there be N fsu's numbered 1, 2, ..., N; each having M ssu's. From the N fsu let n be selected by SRSWOR and from each of the n fsu's consisting of M ssu's let m be selected again by SRSWOR.

Let Y_{ij} be the value of characteristic on j^{th} ssu in i^{th} fsu (i = 1, 2, ... N and j = 1, 2, ..., M).

$$\text{Then } \bar{Y} = \frac{1}{NM}\sum_{i=1}^{N}\sum_{j=1}^{M} y_{ij}$$

Denote by $\bar{Y}_i$, i = 1, 2, ..., n; the mean of ith fsu $= \dfrac{1}{m}\sum_{j=1}^{m} Y_{ij}$ (13.63)

The estimate of population mean $\bar{Y}$ is given by

$$\hat{\bar{Y}} = \bar{Y}_{nm} = \frac{1}{nm}\sum_{i=1}^{n}\sum_{j=1}^{m} Y_{ij} = \frac{1}{n}\sum_{i=1}^{n} \bar{Y}_i \tag{13.64}$$

$\bar{Y}_{nm}$ is the mean per ssu in the sample is an unbiased estimate of $\bar{Y}$.

$$V(\bar{Y}_{nm}) = \left(\frac{1}{n} - \frac{1}{N}\right)S_b^2 + \frac{1}{n}\left(\frac{1}{m} - \frac{1}{M}\right)\bar{S}_w^2 \tag{13.65}$$

where, $$S_b^2 = \frac{1}{N-1}\sum_{i=1}^{N} (\bar{Y}_i - \bar{Y})^2 \text{ and} \tag{13.66}$$

$$S_w^2 = \frac{1}{N}\sum_{i=1}^{N} S_i^2 = \frac{1}{N}\sum_{i=1}^{N} \frac{1}{M-1}\sum_{j=1}^{M} \left(Y_{ij} - \bar{Y}_i\right)^2 \tag{13.67}$$

Some Remarks

(a) If $m = M$, the two-stage sampling is identical to cluster sampling.
(b) If $n = N$, the two-stage sampling is equivalent to stratified sampling.

Similarity with Stratified Sampling and Cluster Sampling

When $n = N$ i.e. every fsu is sampled then the procedure is equivalent to stratified sampling with fsu's as strata and from each strata m units being sampled.

If $m = M$, the selected fsu's are completely enumerated. The procedure is then equivalent to a cluster sampling.

Estimation of the Variance of Estimator of Population Mean

$\bar{Y}_{nm}$ is the estimator of population mean. To estimate the variance of $\bar{Y}_{nm}$, given in equation (13.65) define the following

$$s_b^2 = \frac{1}{n-1}\sum_{i=1}^{n} (\bar{Y}_i - \bar{Y}_{nm})^2 \tag{13.68}$$

$$s_i^2 = \frac{1}{m-1}\sum_{j=1}^{m} \left(Y_{ij} - \bar{Y}_i\right)^2 \text{ and} \tag{13.69}$$

$$s_w^2 = \frac{1}{n}\sum_{i=1}^{n} s_i^2 \tag{13.70}$$

Then the estimate of $V\left(\bar{Y}_{nm}\right)$ is given by

$$\text{Est. of } V\left(\bar{Y}_{nm}\right) = \left(\frac{1}{n} - \frac{1}{N}\right)s_b^2 + \frac{1}{N}\left(\frac{1}{m} - \frac{1}{M}\right)s_w^{-2} \tag{13.71}$$

Three stage sampling materialises if the process of sub-sampling is again conducted by selecting a random sample from the elements of second stage units. A complete enumeration of selected second stage units is not done. The details of the procedure, the estimator of population mean, its variance and estimator of this variance may be consulted from the books of Sukhatme and Sukhatme (1976) and Cochran (1974).

13.7 Purposive or Judgement Sampling

In *purposive sampling* population units are selected in the sample with a definite purpose in mind. So the units which are of one or more specific predefined groups have some possible property of study interest. This method of sampling is more likely to give a typical or representative sample of study interest.

For example an experimenter engaged in market research for a new household product might be looking for educated females between 30-40 years of age. The experimenter may approach such selected women and ask if they are willing to participate. Thus the targeted group can be reached quickly but the selection procedure is influenced by the personal judgement of those responsible for selection.

The procedure of purposive sampling is not objective; neither is it based on the theory of probability. So the method, in most cases may involve some bias of unknown magnitude. Moreover, this method can not provide an estimate of the standard error involved.

The purposive sampling method though may tell more about the mean of the population, but is likely to give a wrong idea of the variability in population. It is because the experimenter has deliberately chosen values near the mean. It is only in a highly restricted class of situations, which rarely materialize in practice, the purposive sampling may give an efficient estimator.

13.8 Snowball Sampling

Snowball sampling is a technique for developing a study sample where a few existing study subjects recruit future subjects from among their acquaintances. *The sample thus appears to grow like a rolling snowball.* As the sample builds up, enough data accumulates for use to draw conclusions for research purpose. The method consists in locating information-rich key informants. These are subsequently asked whether they know of anybody with the characteristic(s) under study.

As an example if a sample of vegetarians or cycle-riders or people with a particular disability or people who support a particular political party, etc., the few in initial contact of experimenter may provide knowledge of other units with the same characteristic(s) or through a support group consisting of other units.

Snowball sampling is not a stand-alone tool, but is a way of selecting participants and then using other tools such as personal visit or mailed enquiry. They may be requested to provide whereabouts of other units and help locating them and to motivate them to give necessary information.

Uses

1. It identifies the people with particular knowledge, skill or characteristics, which may be required to form a committee and/or consultative process.

2. Snowball sampling uses the recommendations of the informants to find people with the specific range of skill, that has been determined as being useful.
3. Snowball sampling makes use of community knowledge about those who have skills in particular areas.
4. This method is used in hidden populations which are difficult to access e.g. drug users or commercial prostitutes or homeless people.
5. Respondent-driven sampling enables to estimate about social network connecting hidden population.

Drawbacks of Snowball Sampling

1. Sample elements are not selected from a sampling frame, so the sample is subject to numerous biases. Thus people who have many friends are more likely to be included in the sample.
2. A variation of snowball sampling called respondent-driven sampling allows researchers to make asymptotically unbiased estimates only under certain conditions.
3. Since it is based on referrals from initial subjects, it is quite unlikely that the sample will be a good section of population.

13.9 Probability Sampling

As is evident in the last two sections, that the non-probability sampling procedures do not always provide a representative sample of the population, and also do not provide reliable estimates of the standard error.

With probabilistic sampling, the odds or the probability that the population is represented in sample is known. Thus one is able to estimate the standard error of the statistic and the confidence interval of the population mean.

It is for this reason that probability sampling is preferred over non-probability sampling as the former is accurate and rigorous.

The simple probability sampling methods are given in the first six sections. The basis of these is SRS discussed in Section 13.1. In this method the selection probabilities are equal for the population elements or the groups of elements (i.e. clusters).

However, if the units vary considerably in size (as are the household sizes, area and population of a city or village etc.), the SRS does not take into account the importance of larger units. There are methods that make use of this ancillary information for better estimates of study characteristic.

Thus it is considered that a more representative sample is obtained if the sample is selected with varying selection probability. One such method is to select the sample units with probability proportional to size of the units.

There are various methods of estimation and various procedures of sampling with varying probability of selection of a unit.

The interested reader may consult Sukhatme and Sukhatme (1976), Cochran (1974), Goon et al. (2002), Murthy (1967), Raj (1968) and Pattan (1990) for all the detailed discussions.

14. Analysis of Data: With Violated Assumptions and from Complex Designs

The general concepts of biostatical inference given in Chapters 7, 8, 9 and 10, deal with analysis of data, methods of testing of hypotheses and estimating unknown parameter(s). Specific methods are applicable depending on the nature of the data i.e. the distribution of the variable on which observations are available. It is, therefore, necessary while choosing a method of estimation or testing of hypothesis, that the experimenter should make sure that the available data satisfies all the conditions under which the chosen method is applicable.

Sometimes one may find that all the conditions desired for a method of testing or estimation are not satisfied. Many statistical procedures require normal distribution of the sampled observations. Equality of variances is another assumption apart from normality which should be satisfied while comparing two or more means from small samples.

It is, therefore, advisable that a test of equality of two variances should be conducted before applying the Student's *t*-test for testing equality of two means. If the test rejects equality of two variances, the Student's *t*-test of equality of two means (given in Section 8.2 equation 8.24) can not be used.

As the title of the chapter indicates, a procedure which can be used for testing equality of two means when the assumption of equal variances is violated is described below.

Comparison of several means has been dealt in Section 8.6 for the simplest design of experiment (name not given there). There are other designs also which are used to conduct experiments for testing equality of means. These are more complex but essential for answering a question under consideration. Some such designs, the conditions when they should be used and the methods of analysis of such designs is discussed in subsequent sections.

14.1 Comparison of Two Means when Variances are Unequal

This problem was first considered by Behrens (1929) and Fisher (1934). A test of hypothesis H_0: $\mu_1 = \mu_2$ against H_1: $\mu_1 (> = <) \mu_2$ when it is known that the two random samples are drawn from two normal populations with unequal variances, is therefore, also called *Behrens-Fisher Problem*. A biological researcher often faces this problem since the two populations from which samples have been obtained may not have normal distribution with equal variances. Numerous studies e.g. Boneau (1960), Box (1953), Cochran (1947), Posten et al. (1982); Srivastava (1958); have shown that if sample sizes are equal or nearly equal and when hypothesis to be tested is a two tail test, *t*-test may still be used because of its robustness.

Some more notable studies on this problem are by Dixon and Massey (1969), Cochran and Cox (1957), Gill (1971) and Satterthwaite (1946).

It is also suggested that if the ratio of two variances is less than or equal to 1.4 (the larger variance in numerator), Student's *t*-test may be applied.

However, for small samples from skewed populations one tail test is less powerful. Actual type I error is also not equal to α, if variances are not equal. The said Behrens-Fisher problem has been solved by various authors after Behrens.
Some of these are described in this section.

I One of the easiest and reliable method, suggested by **Smith (1936) and Satterthwaite (1946)** is given here. The assumptions, the hypothesis and the testing procedures are formally given below.

Assumptions X_1 and X_2, the measured characteristics in the two populations have normal distributions with means μ_1 and μ_2 and variances σ_1^2 and σ_2^2, respectively. From these two populations samples of sizes n_1 and n_2 are selected. $\bar{X}_1$ and $\bar{X}_2$ are the sample means. $\sigma_1^2 \neq \sigma_2^2$ are unknown and n_1 and n_2 may be small.
Hypothesis H_0: $\mu_1 = \mu_2$ against one of the following

(i) H_1: $\mu_1 \neq \mu_2$
(ii) H_1: $\mu_1 > \mu_2$
(iii) H_1: $\mu_1 < \mu_2$

Test statistic in this case is given by

$$t_B = \frac{\bar{X}_1 - \bar{X}_2}{\sqrt{\left(\frac{s_1^2}{n_1} + \frac{s_2^2}{n_2}\right)}} \tag{14.1}$$

where, s_1^2 and s_2^2 are unbiased estimators of σ_1^2 and σ_2^2 with divisors $n_1 - 1$ and $n_2 - 2$, respectively.

The distribution of the test statistic, if H_0 is true, is Student's t with ν degree of freedom. In symbols

$$t_B \sim t_\nu \text{ if } H_0 \text{ is true} \tag{14.2}$$

where

$$\nu = \frac{\left(\frac{s_1^2}{n_1} + \frac{s_2^2}{n_2}\right)^2}{\left[\frac{\left(\frac{s_1^2}{n_1}\right)^2}{(n_1 - 1)} + \frac{\left(\frac{s_2^2}{n_2}\right)^2}{(n_2 - 1)}\right]} \tag{14.3}$$

ν is normally fractional, if computed from (14.3).
The test procedure or rule is given by the following.

(i) Reject H_0 if $|t_B| > t_{\nu,\ \alpha/2}$ for H_1: $\mu_1 \neq \mu_2$. In other words
Reject H_0 if either $t_B > t_{\nu,\ \alpha/2}$ or $t_B < -t_{\nu,\ \alpha/2}$

(ii) Reject H_0 if $t_B > t_{\nu, \alpha}$ for H_1: $\mu_1 > \mu_2$ and
(iii) Reject H_0 if $t_B < -t_{\nu, \alpha}$ for H_1: $\mu_1 < \mu_2$ (14.3a)

To determine significance points of this Student's *t*-distribution with fractional degrees of freedom the nearest smallest integer may be used as degrees of freedom.

Example 14.1: A study of energy requirements for growth and maintenance of nestling house martins was conducted. The following summary statistics were obtained on the normal variable X, the number of kilocalories per gram per hour required per bird (source: Milton and Tsokos (1983), p. 257):

Incubating adults	Pre-breeding adults
$n_1 = 57$	$n_2 = 12$
$\bar{X}_1 = .0167$ k cal/(gm) (hr)	$\bar{X}_2 = .0144$ kcal/(gm) (hr)
$s_1 = .0042$	$s_2 = .0024$

On the basis of the available data can it be concluded that mean energy requirement for incubating bird is higher than that for pre-breeding bird? From the past experience it is known that the variances of energy requirements for two groups of birds are different.

Solution: *Assumptions:* Let $\bar{X}_1$ and $\bar{X}_2$ be the energy requirements of the two groups of birds. Then $X_1 \sim N(\mu_1, \sigma_1^2)$ and $X_2 \sim N(\mu_2, \sigma_2^2)$, where $\sigma_1^2 \neq \sigma_2^2$. So the problem is of type A(d) of Section 8.2 of Chapter 8 known as Behrens-Fisher Problem. The test procedure for this, given above, is applied below.

Hypothesis: H_0: $\mu_1 = \mu_2$ against H_1: $\mu_1 > \mu_2$

The test statistic to be used is

$$t_B = \frac{\bar{X}_1 - \bar{X}_2}{\sqrt{\left(\frac{s_1^2}{n_1} + \frac{s_2^2}{n_2}\right)}} \sim t_\nu \text{ if } H_0 \text{ is true}$$

$$\nu = \frac{\left(\frac{s_1^2}{n_1} + \frac{s_2^2}{n_2}\right)^2}{\left[\frac{\left(\frac{s_1^2}{n_1}\right)^2}{(n_1 - 1)} + \frac{\left(\frac{s_2^2}{n_2}\right)^2}{(n_2 - 1)}\right]}$$

The test procedure or critical region for level of significance $\alpha = .05$ is

$$t_B > t_{\nu, .05}$$

If ν is fractional, nearest lower integer is taken as degree of freedom

Computation of statistic

$$t_{cal} = \frac{.0167 - .0144}{\sqrt{\frac{(.0042)^2}{57} + \frac{(.0024)^2}{12}}} = \frac{.0023}{.000889} = 2.59$$

$$\nu = \left[\frac{.0042^2}{57} + \frac{.0024^2}{12}\right]^2 \Bigg/ \left\{\frac{1}{56}\left(\frac{.0042^2}{57}\right)^2 + \frac{1}{11}\left(\frac{.0024^2}{12}\right)^2\right\}$$

$$= \frac{62.315\times10^{-14}}{.1710\times10^{-14}+2.0945\times10^{-14}} = 27.5$$

So ν is chosen as 27 to see the significant value from Table A.4 as $t_{27,\ .05} = 1.703$

Statistical decision is to reject H_0 since $t_{cal} = 2.59 > 1.703$.

Conclusion at 5% level of significance, therefore, is that according to data, energy requirement for incubating bird is higher than that for pre-breeding bird.

This test is also called "Welch's approximate *t-test*" and is also referred as *Smith, Welch, Satterthwaite t-test.*

II Welch (1938) suggested that replacing $n_1 - 1$ and $n_2 - 1$ by $n_1 - 3$ and $n_2 - 3$, respectively, in the expression of degree of freedom (d.f.) in equation (14.3) gives an improved test.

III Welch and Dixon and Massey (1969) have shown that it would be more appropriate to use the following formula for d.f. of Welch's approximate *t*-test.

$$\nu = \frac{\left(\dfrac{s_1^2}{n_1}+\dfrac{s_2^2}{n_2}\right)^2}{\left\{\dfrac{s_1^4}{n_1^2(n_1+1)}+\dfrac{s_2^4}{n_2^2(n_2+1)}\right\}} - 2 \qquad (14.4)$$

IV Neyman's Approach Without loss of generality let $n_1 \le n_2$. Neyman (1935) suggested to pair the observations X_{1j} (j = 1, 2, ..., n_1) of the first sample with n_1, of the n_2 observations in the second sample, selected randomly. Thus let X_{1j} is paired with X'_{2j}, where X'_{2j} is one of the values X_{21}, X_{22}......X_{2n_2} selected by SRSWOR.

$$\text{Define } d_j = (X_{1j} - X'_{2j}),\ j = 1, 2, \ldots\ldots, n_1 \qquad (14.5)$$

Then use the paired *t*-test of H_0: $\mu_1 = \mu_2$ using

$$t = \frac{\sqrt{n}\,\bar{d}}{s_d}.\ \text{Where } s_d^2 = \frac{1}{(n_1-1)}\sum_j^{n_1}(d_i-\bar{d})^2 \text{ and } \bar{d} = \frac{1}{n_1}\sum_{j=1}^{n1} d_j \qquad (14.6)$$

t has Student's *t*-distribution with $n_1 - 1$ d.f. if H_0 is true.

V Scheffe's Approach Scheffe's (1970) approach is a variant of Neyman's approach. He also suggested pairing of the n_1 observations of the first sample with n_1 observations of the second sample. But his procedure is based on the d_j's defined as

$$d_j = X_{1j} - \sqrt{\frac{n_1}{n_2}}X'_{2j} + \frac{1}{\sqrt{n_1 n_2}}\sum_{j=1}^{n1}X'_{2j} - \frac{1}{n_2}\sum_{j=1}^{n2}X_{2j},\quad j = 1, 2, \ldots\ldots, n_1 \qquad (14.7)$$

It can be seen that d_j's are normally distributed, with

$$E(d_j) = \mu_1 - \mu_2 \text{ for j = 1, 2, ..., } n_1 \qquad (14.8)$$

$$V(d_j) = \sigma_1^2 + (n+1)\sigma_2^2/n_2 \text{ for } j = 1, 2, \ldots\ldots, n_1 \qquad (14.9)$$

and $$\text{Cov}(d_j, d_k) = 0 \text{ for all } j \text{ and } k\ (j \ne k) = 1, 2, \ldots\ldots, n_1 \qquad (14.10)$$

Again to test H_0: $\mu_1 = \mu_2$, the test statistic used is

$$t = \frac{\sqrt{n_1}\,\bar{d}}{s_d} \tag{14.11}$$

where $\bar{d}$ and s_d are obtained using the same formula as given in (14.6) but d_j is from (14.7)

Again this t has Student's t-distribution with $n_1 - 1$ d.f. if H_0 is true (14.12)

So the critical regions are chosen as indicated in Section 8.6 and also given above in (14.3a). Scheffe has shown that among all tests based on Student's t-distribution with n_1-1 d.f. his test is most powerful.

14.1.1 Comparison of Several Means and Completely Randomized Design

Problem of comparing several means considered in Section 8.6 is solved by one way ANOVA. The design of experiment for this problem is called *completely randomized design* (CRD) since n_i observations (or replications) under i^{th} group (or treatment); $i = 1, 2, ..., k$ are selected randomly out of $n\left(=\sum_{i=1}^{n} n_i\right)$. For randomization, random number tables are used. The experimental units are given a serial number 1 to $n\left(=\sum_{i=1}^{n} n_i\right)$ and suppose there are k treatments to be compared. Various treatments are also given numeric values 1, 2, ..., k for their identification.

Following the procedure given in Section 7.2 of Chapter 7, n random numbers are selected and noted down. The n_1 experimental units with serial numbers equal to the first n_1 random numbers selected and noted, are assigned treatment number 1. The next treatment number 2 is assigned to the n_2 experimental units with serial numbers equal to the $(n_1 + 1)^{th}$, $(n_1 + 2)^{th}$, ..., $(n_1 + n_2)^{th}$ random numbers already noted. This process is repeated for $n_3, n_4, ..., n_k$ experimental units also, with serial numbers equal to the noted random numbers in order. As an illustration of layout of experimental units in CRD consider the following example.

Example 14.2: Suppose that three treatments are to be tested using CRD with number of replications 5, 4 and 3. Thus in all 12 experimental units (= 5 + 4 + 3) will be required. These are numbered 1, 2, ..., 12. The 12 random numbers are now selected between 1 to 12 using random number table. Suppose the selected 12 random numbers are

1 8 9 7 5 11 4 10 6 2 3 12

The 5(= n_1) experimental units with serial numbers 1, 8, 9, 7, 5 are assigned the treatment 1.
The 4(= n_2) experimental units with serial numbers equal to next 4 selected random numbers, viz. 11, 4, 10, 6 are assigned treatment 2 and the experimental units with serial numbers 2, 3, and 12 are assigned the treatment 3. This type of layout of the design of experiment is called the CRD.

The analysis of results of such design is given in Section 8.6 of Chapter 8. The analysis is based on the mathematical model of the observations. The basic model of any observation and of observations from experiments with some specific designs are given below. These are useful to determine the technique of analysis of resulting observed data. The models allow to explain the observable facts and also to predict possibly unobserved events.

Linear Additive Model

A common model for describing the make up of an observation assumes that it consists of two parts, a true part or mean and an error. In symbols the simplest linear additive model is given by

$$Y_i = \mu + e_i \tag{14.13}$$

where Y_i, (i = 1, 2, ..., n) is i^{th} observation, μ is the true value or mean and e_i is the error, and both are unknown.

The mean or true value μ may be a single parameter or a linear function of several parameters. This model is fundamental to *analysis of variance.*

Model for CRD or one way ANOVA

The following model is used for ANOVA of CRD observations:

Observation from j^{th} replicate under i^{th} treatment = General effect + Effect due to i^{th} treatment + Random error component

In symbols

$$y_{ij} = \mu + \tau_i + e_{ij} \qquad i = 1, 2, ..., k, \quad j = 1, 2, ..., n_i \tag{14.14}$$

μ and τ_i are a set of constants with $\sum_{i=1}^{k} n_i\tau_i = 0$ and e_{ij}'s are random error components, independently normally distributed with mean zero and variance σ^2, i.e. $e_{ij} \sim N(0, \sigma^2)$.

If $n_i = r$ i.e. there are equal number of replicates under various treatments, τ_i satisfy the relation $\sum_i \tau_i = 0$.

In either case the hypothesis to be tested is H_0: $\tau_1 = \tau_2 = ... = \tau_k = 0$ against H_1: all $\tau_i's$ are not equal to zero.

Fixed Effect Model or Model I

If all the effects $\tau_i's$ are fixed but unknown parameters, they constitute a finite population. The model then is called *fixed effect model.* $\tau_i's$ and σ^2 are the parameters of interest in such a model.

Random Effect Model or Model II

A model in which the $\tau's$ are random variables, is called a *random effect model.* The $\tau's$ under consideration in the experiment are a random sample from a population of $\tau's$ for which mean is zero and variance σ_t^2. σ_t^2 is also a parameter of interest other than σ^2.

This model is also called a *variance-components model.* In a fixed effect model, a repetition of the experiment would consider the same set of $\tau's$ in all the repetitions.

For the random effect model, a repetition of the experiment would consider a new set of $\tau's$ from the same population of $\tau's$. After sufficient repetitions, the whole population of $\tau's$ will appear in some experiments. In this set up the investigator is interested in σ_t^2 and not in a specific set of $\tau_i's$.

For a fixed effect model, inference is drawn about the particular treatments, whereas for random effect model inference is drawn about the population of treatments.

An *agricultural experiment* to test effects of a set of *fertilizers* has a *fixed effects model,* while an *experiment* that examines the *genetic variations among a large population of dairy cows on milk* production has a *random effect model.*

If the model for CRD is written as

$$Y_{ij} = \mu_i + e_{ij} \tag{14.15}$$

where $\mu_i = \mu + \tau_i$, μ_i can be estimated but not μ and τ_i, as different constraints on τ_i's will lead to different estimates.

Advantages of CRD

CRD is the simplest design and has the following advantages:

1. Design is flexible in the sense that any number of treatments and replicates can be used, subject to availability of experimental units.
2. The number of replicates can vary from one treatment to the other.
3. Statistical analysis is simple for unequal replicates and also when data on some experimental units or on entire treatment are missing.
4. The error degree of freedom is maximum so CRD improves precision of experiment.

Disadvantages of CRD

1. The main disadvantage is that this design does not utilize the principle of local control. Since randomization is unrestricted, the experimental error includes the entire variation among the experimental units except the part due to treatments.
2. If the experimental units are not homogeneous CRD should not be used.

14.2 Randomized Block Design

If the experimental units are not homogeneous, the variation among the units vitiates the test of significance of the treatment effects. The simplest design which takes care of the variability among the experimental units is the *randomized block design* (RBD) also called *randomized complete block design.* This design uses all the three principles of good design suggested by Fisher.

There are situations when it is known or can be anticipated that certain experimental units will behave differently even if given the same treatment. For example in agricultural experiments one may know that there is a fertility gradient. Likewise heavier animals in the same age group may show different rate of weight gain than lighter animals. In such situations, it is often possible to group the units in such a way that units within a group are more or less homogeneous. Thus the variation between groups of units or blocks can be measured and can be removed from the experimental error.

Thus RBD is used when the experimental units can be grouped meaningfully. The number of units in a group called *block*, is equal to the number of treatments to be tested. The units in a group or block are as uniform as possible, so that the observed differences in a block will be largely due to treatments. Variability among units of different blocks will be larger than that among units within a block.

Experimental Layout of RBD

Suppose effects of s treatments are to be compared. Each treatment is being replicated say r times. Then in all $n = rs$ experimental units will be needed. Lay out of RBD consists of two steps.

First step consists in dividing the $n(= rs)$ experimental units into r groups with almost s homogeneous units in each group (or block). The number of blocks r is equal to the number of replications of each treatment.

Second step is to assign the s treatments: $t_1, t_2, ..., t_s$, at random to the s units of each block. Random selection of units for assignment of treatment is done for each block afresh. Thus r sets of numbers 1, 2, ..., s are selected randomly and the treatments $t_1, t_2, ..., t_s$ are applied to the units with identification numbers same as the selected random numbers, in each of the r sets.

As an illustration of this procedure consider the following example.

Example 14.3: Obtain a layout of an RBD with 5 treatments each replicated 3 times.
Solution: In this RBD $s = 5$ and $r = 3$. So $n = 15$ (= 3 × 5) units are required. These are first grouped into 3 blocks of 5 homogeneous units. The units in a block are assigned a serial number from 1, 2, ..., 5. Now as in CRD a random permutation of numbers 1, 2, 3, 4, 5 is obtained as explained in Section 7.2. Suppose the first set of random permutation is 4, 2, 1, 5, 3 for block 1. Then treatment t_1 is applied to unit 4, t_2 to unit 2, t_3 to unit 1, t_4 to unit 5 and t_5 to unit 3 of block 1. Another random permutation of numbers 1 to 5 is obtained for block 2 and a third random permutation for block 3. The treatments $t_1, t_2, ..., t_5$ are assigned to the units of the blocks 2 and 3, as for block 1.

Model of observations from RBD is given by

Observation for the ith treatment in the jth block = General effect + j^{th} block effect + i^{th} treatment effect + Random error part (14.16)

In symbols

$$Y_{ij} = \mu + \beta_j + \tau_i + e_{ij} \tag{14.17}$$

$$i = 1, 2, ..., s; j = 1, 2, ..., r$$

μ, β_j and τ_i are constants such that $\sum_j \beta_j = 0 = \sum_i \tau_i$ and e_{ij} are each independently normally distributed with mean zero and variance σ_e^2.

It is desired to test H_0: $\tau_1 = \tau_2 = ... = \tau_s = 0$ against the alternative H_1: all the τ_s' are not equal to zero.

Analysis

The following notations are used to find out various sums of squares. Y_{ij} is observation from j^{th} block on i^{th} treatment. Then

$$Y_{i\bullet} = \sum_{j=1}^{r} Y_{ij} = i^{th} \text{ treatment total}$$

$$= \text{total of observations which received } i^{th} \text{ treatment in all the } r \text{ blocks} \tag{14.18}$$

$$Y_{\bullet j} = \sum_{i=1}^{s} Y_{ij} = j^{th} \text{ block total}$$

$$= \text{total of observations in } j^{th} \text{ block which received all the } s \text{ treatments} \tag{14.19}$$

$$Y_{\bullet\bullet} = \sum_{i=1}^{s}\sum_{j=1}^{r} Y_{ij} = \sum_i Y_{i\bullet} = \sum_j Y_{\bullet j} \tag{14.20}$$

and $$\text{CF} = \frac{Y_{\bullet\bullet}^2}{n} \text{ is the correction factor} \tag{14.21}$$

Now various sums of squares are obtained as

$$\text{Total sum of square (TSS)} = \sum_{i=1}^{s}\sum_{j=1}^{r} Y_{ij}^2 - \text{CF} \tag{14.22}$$

$$\text{Sum of square due to blocks (SSB)} = \frac{\left(\sum_{j=1}^{r} Y_{\bullet j}^2\right)}{s} - \text{CF} \tag{14.23}$$

$$\text{Sum of square due to treatments (SST)} = \frac{1}{r}\sum_{i=1}^{s} Y_{i\bullet}^2 - \text{CF} \tag{14.24}$$

$$\text{Sum of square due to error (SSE)} = \sum_i\sum_j \left(Y_{ij} - Y_{i\bullet} - Y_{\bullet j} + Y_{\bullet\bullet}\right)^2$$

$$= \text{TSS} - \text{SSB} - \text{SST} \tag{14.25}$$

It can be seen from (14.25) that

$$\text{TSS} = \text{SSB} + \text{SST} + \text{SSE} \tag{14.26}$$

A similar partitioning of the d.f. is also possible as

$$rs - 1 = (r - 1) + (s - 1) + (r - 1)(s - 1)$$

$$\text{d.f. of TSS} = \text{d.f. of SSB} + \text{d.f. of SST} + \text{d.f. of SSE} \tag{14.27}$$

It can be seen that SSE can be obtained by subtraction using (14.25) and d.f. of SSE can also be obtained by subtraction from (14.27).

The mean squares (MS) are obtained from corresponding SS by dividing it by its d.f. Thus

$$\text{MSB} = \frac{\text{SSB}}{(r-1)}; \text{MST} = \frac{\text{SST}}{(s-1)} \text{ and MSE} = \frac{\text{SSE}}{(r-1)(s-1)} \tag{14.28}$$

The analysis can be summarized in a table given below. It is called an analysis of variance (ANOVA) table.

Table 14.1 ANOVA table for RBD

Source of variation	d.f.	SS	MS	F	Tabulated F 5%	Tabulated F 1%
Blocks	$r - 1$	SSB	MSB			
Treatments	$s - 1$	SST	MST	MST/MSE = F		
Error	$(r - 1)(s - 1)$ (by subtraction)	SSE (by subtraction)	MSE			
Total	$rs - 1$	TSS				

F = MST/MSE follows F-distribution with $(s - 1)$ and $(r - 1)(s - 1)$ d.f. if H_0 is true
In symbols $F \sim F_{(s-1),\ (s-1)\ (r-1)}$, if H_0 is true.

So in the last column upper tail 5% and 1% points of $F_{(s-1),\ (r-1)\ (s-1)}$ are written from the F-table and may be denoted by $F_{(s-1),\ (r-1)\ (s-1),\ .05}$ and $F_{(s-1),\ (r-1)\ (s-1),\ .01}$.

If calculated $F > F_{(s-1),\ (r-1)\ (s-1),\ .01}$ the treatments differ highly significantly and it may be indicated by putting ** over F.

If $F_{(s-1),\ (r-1)\ (s-1),\ .05} < F < F_{(s-1),\ (r-1)\ (s-1),\ .01}$, the treatments differ significantly and it may be indicated by putting * over F in the ANOVA table.

An example given below illustrates the method of analyzing results of RBD.

Example 14.4: An experiment is conducted to compare the energy requirements of three physical activities: running, walking and bicycle riding. The variable of interest is Y, the number of kilocalories spent per kilometer travelled. To do so, eight subjects were selected and each was asked to run, walk and bicycle a measured distance and the number of kilocalories spent per kilometer were determined for each. The activities are run in random order with time of recovery between activities. Each individual thus serves as a block. Table 14.2 gives Y, the number of kilocalories spent by the 8 subjects on the 3 activities. Perform the analysis of variance and give the conclusions (source: Milton and Tsokos (1983), p. 309).

Table 14.2 Energy spent by 8 subjects on 3 activities

Subjects	Treatments (activities)				
(blocks)	Running	Walking	Bicycling	Total ($Y_{i\cdot}$)	$Y_{i\cdot}^2$
1	1.4	1.1	0.7	3.2	10.24
2	1.5	1.2	0.8	3.5	12.25
3	1.8	1.3	0.7	3.8	14.44
4	1.7	1.3	0.8	3.8	14.44
5	1.6	1.7	0.1	2.4	5.76
6	1.5	1.2	0.7	3.4	11.56
7	1.7	1.1	0.4	3.2	10.24
8	2.0	1.3	0.6	3.9	15.21
Total ($Y_{\cdot j}$)	13.2	9.2	4.8	27.2 (= $Y_{\cdot\cdot}$)	94.14
$Y_{\cdot j}^2$	174.24	84.64	23.04	281.92	

Solution: For this design (RBD) the number of replicates or blocks = 8 = r and the number of treatments (activities) = 3 = s

$$n = r\ s = 24$$

The block totals $Y_{i\cdot}$ are obtained and shown as row totals. The treatment totals $Y_{\cdot j}$ are obtained by adding the three column values and shown in the last row

$$\sum_{i=1}^{8}\sum_{j=1}^{3} Y_{ij}^2 = 36.18 \quad \sum_{i=1}^{8} Y_{i\cdot}^2 = 94.14 \quad \sum_{j=1}^{3} Y_{\cdot j}^2 = 281.92$$

$$\text{CF} = Y_{\cdot\cdot}^2/n = \frac{27.2^2}{24} = 30.83$$

$$\text{TSS} = \sum_{i=1}^{8}\sum_{j=1}^{3} Y_{ij}^2 - \text{CF} = 36.18 - 30.83 = 5.35$$

$$\text{SSB} = \frac{1}{s}\sum_{i=1}^{r} Y_{i\bullet}^2 - \text{CF}$$

$$= \frac{94.14}{3} - 30.83 = 31.38 - 30.83 = .55$$

$$\text{SST} = \frac{1}{r}\sum_{j=1}^{s} Y_{\bullet j}^2 - \text{CF}$$

$$= \frac{281.92}{8} - 30.83 = 35.24 - 30.83 = 4.41$$

$$\text{SSE} = \text{TSS} - \text{SSB} - \text{SST}$$

$$= 5.35 - .55 - 4.41 = 0.39$$

It is desired to test the hypothesis H_0: $\tau_1 = \tau_2 = \tau_3 = 0$ against H_1: all the τ_s' are not equal to zero.

One may also be interested in testing whether individuals (blocks) differ in average energy spent on the three activities. Then one may also test H_0' : $\beta_1 = \beta_2 = \beta_3 = \beta_4 = \beta_5 = \beta_6 = \beta_7 = \beta_8 = 0$ against H_1': all the $\beta's$ are not equal to zero.

This can be tested in the following ANOVA table.

Table 14.3 ANOVA table for Example 14.4 on RBD

Source of variation	d.f.	SS	MS	*F*	Tabulated *F*
Blocks	7	0.55	.55/7 = 0.079	.079/.028 = 2.82*	$F_{7,14,.05} = 2.76$ $F_{7,14,.01} = 4.28$
Treatments	2	4.41	4.41/2 = 2.205	2.205/.028 = 78.75**	$F_{2,14,.05} = 3.74$ $F_{2,14,.01} = 6.51$
Error	14	0.39	.39/14 = 0.028		
Total	24-1	5.35			

Since calculated *F* for blocks is greater than $F_{7,14,.05}$ but less than $F_{7,14,.01}$ one * is put over it, indicating that the individuals differ significantly in the average energy spent on the three activities. So *P*-value for differences in individuals is between .01 to .05.

The calculated *F* for treatment is greater than $F_{2,14,.01,}$ so ** is put over it indicating that the 3 activities differ highly significantly in their average energy requirements. *P*-value is less than .01.

Test of Interaction between Blocks and Treatments

The ANOVA above presumes that there is no interaction between blocks and treatments. If there appears to be some differences in the effect of treatments across blocks or vice versa i.e. that there is an interaction between treatments and blocks, it needs to be tested differently. It requires more replications in each of the *rs* cells. With a single observation per cell, one can not find the sum of square due to interaction or between cell sum of squares. If there are, say, *m* observations in each of the *rs* cells and it can not be presumed that interaction is absent then the model will be

$$Y_{ijk} = \mu + \tau_i + \beta_j + (\tau\beta)_{ij} + e_{ijk} \tag{14.29}$$

where $i = 1, 2, ..., r$; $j = 1, 2, ..., s$ and $k = 1, 2, ..., m$; $(\tau\beta)_{ij}$ is interaction effect of ith treatment and j^{th} block effect on the observations. Moreover, $\sum_i (\tau\beta)_{ij} = 0$ for all *j* and $\sum_j (\tau\beta)_{ij} = 0$ for all *i*

Now in all $rsm = n$ experimental units will be required for the experiment. With this experiment it is possible to test the following three hypotheses.

$$H_{01} : \tau_1 = \tau_2 = ... = \tau_s = 0 \text{ against } H_{11} : \text{all the } \tau\text{'s are not equal to zero} \quad (14.30)$$

$$H_{02} : \beta_1 = \beta_2 = ... = \beta_r = 0 \text{ against } H_{12} : \text{all the } \beta\text{'s are not equal to zero} \quad (14.31)$$

and

$$H_{03} : (\tau\beta)_{11} = (\tau\beta)_{12} = ... = (\tau\beta)_{rs} = 0 \text{ against}$$

$$H_{13} : \text{all the } (\tau\beta)\text{'s are not equal to zero} \quad (14.32)$$

Again the sums of squares of deviations will be needed to perform these tests and the following notations will be used to give the expressions for various SS.

$$Y_{ij\bullet} = \sum_{k=1}^{m} Y_{ijk} = i, j^{\text{th}} \text{ cell total, there will be } rs \text{ totals in all} \quad (14.33)$$

$$Y_{i\bullet\bullet} = \sum_{j=1}^{r}\sum_{k=1}^{m} Y_{ijk} = i^{\text{th}} \text{ treatment total, } s \text{ totals in all}$$

$$= \sum_{j} Y_{ij\bullet} \quad (14.34)$$

$$Y_{\bullet j\bullet} = \sum_{i=1}^{s}\sum_{k=1}^{m} Y_{ijk} = \sum_{i=1}^{s} Y_{ij\bullet} = j^{\text{th}} \text{ block total, } r \text{ totals in all} \quad (14.35)$$

$$Y_{\bullet\bullet\bullet} = \sum_{i=1}^{s}\sum_{j=1}^{r}\sum_{k=1}^{m} Y_{ijk} = Grand\ total = \sum_{i=1}^{s} Y_{i\bullet\bullet} = \sum_{j=1}^{r} Y_{\bullet j\bullet} \quad (14.36)$$

$$\text{CF} = Y_{\bullet\bullet\bullet}^2/rsm = Correction\ factor \quad (14.37)$$

The ANOVA is based on the relation

Total SS (TSS) = Sum of squares due to treatment + Sum of squares due to blocks + Sum of squares due to interaction + Sum of squares due to errors

or

$$\text{TSS} = \text{SST} + \text{SSB} + \text{SSI} + \text{SSE} \quad (14.38)$$

$$\text{SS between cells} = \frac{1}{m}\sum_{i=1}^{s}\sum_{j=1}^{r} Y_{ij\bullet}^2 - \text{CF} \quad (\text{with d.f.} = rs - 1) \quad (14.39)$$

$$\text{SST} = \sum_{i=1}^{s} Y_{i\bullet\bullet}^2/rm - \text{CF} \quad (\text{with d.f.} = s - 1) \quad (14.40)$$

$$\text{SSB} = \sum_{j=1}^{r} Y_{\bullet j\bullet}^2/sm - \text{CF} \quad (\text{with d.f.} = r - 1) \quad (14.41)$$

$$\text{SSI} = \text{SS between cells} - \text{SSB} - \text{SST}$$

$$\text{with d.f.} = (rs - 1) - (s - 1) - (r - 1) = (r - 1)(s - 1) \quad (14.42)$$

$$\text{SSE} = \text{TSS} - \text{SS between cells} = \text{SS within cells}$$

$$\text{with d.f.} = rs\,m - 1 - (rs - 1) = (m - 1)rs \qquad (14.43)$$

The ANOVA table can now be formed to test the said three hypotheses remembering that MSS = SS/d.f.

Table 14.4 ANOVA table for RBD with m observations per cell

Source of variation	d.f.	SS	MSS = SS/df	F (calculated)	Tabulated F upper tail 5% and 1%
Blocks	$r - 1$	SSB	MSB	F_B = MSB/MSE	$F_{(r-1),\, rs(m-1),\, .05}$ $F_{(r-1),\, rs(m-1),\, .01}$
Treatments	$s - 1$	SST	MST	F_T = MST/MSE	$F_{s-1,\, rs(m-1),\, .05}$ $F_{s-1,\, rs(m-1),\, .01}$
Interactions ($\tau \times \beta$)	$(r - 1)(s - 1)$	SSI = SS between cell – SSB – SST (by subtraction)	MSI	F_I = MSI/MSE	$F_{(r-1)(s-1),\, rs(m-1),\, .05}$ $F_{(r-1)(s-1),\, rs(m-1),\, .01}$
Between cells	$rs - 1$	SS between cells			
Error or within cells	$rs(m - 1)$	SSE = TSS – SS between cells (by subtraction)	MSE		
Total	rsm-1	TSS			

If calculated F is greater than the 5% upper tail tabulated value of F, but is the less than the 1% upper tail tabulated value of F, the corresponding differences are significant and it is indicated by putting a * on the relevant F.

If calculated F is greater than the 1% upper tail tabulated value of F, the corresponding differences are highly significant and it is indicated by putting a ** over the relevant F. Example 14.5 given below illustrates the method of analysis of such RBD with $m = s$ observations per cell.

Example 14.5: The data given below are plasma calcium concentration (in mg/100 ml) of birds of both sexes. Half of the birds of each sex are being treated with the hormone and half not treated with the hormone.
Perform ANOVA and test the hypotheses that the average plasma calcium concentration of birds in the sampled population is not affected by (i) Hormonal treatment (ii) Sex of the bird and (iii) Sex and hormone treatment combinations (source: Zar (2004), p. 233).

No hormone treatment		Hormone treatment	
Female	Male	Female	Male
16.5	14.5	39.1	32.0
18.4	11.0	26.2	23.8
12.7	10.8	21.3	28.8
14.0	14.3	35.8	25.0
12.8	10.0	40.2	29.3

Solution: Given data has

s = the number of treatments = 2

r = the number of second factor viz. sex (= blocks) = 2

m = the number of replications in each of rs(= 4) cells = 5

$n = rsm = 2 \times 2 \times 5 = 20$ experimental units in all

It is desired to test the following hypotheses

H_{01}: There is no effect of hormone treatment on mean plasma calcium concentration i.e. $\tau_1 = \tau_2$ against H_{11}: $\tau_1 \neq \tau_2$

H_{02}: Male and female (i.e. sex of) birds have no difference in mean plasma calcium concentration i.e. $\beta_1 = \beta_2$ against H_{12}: $\beta_1 \neq \beta_2$ and

H_{03}: There is no interaction between sex and hormone treatment on the mean plasma calcium concentration of birds against H_{13}: All $(\tau\beta)$'s are not equal to zero.

There are 2 × 2 = 4 treatment and block combinations so in all 4 cells in RBD shown as 4 columns of table in Example 14.5. The four cell totals are

$$Y_{11\bullet} = \sum_{k=1}^{5} Y_{11k} = 74.4 \qquad Y_{12\bullet} = \sum_{k=1}^{5} Y_{12k} = 60.6$$

$$Y_{21\bullet} = \sum_{k=1}^{5} Y_{21k} = 162.6 \qquad Y_{22\bullet} = \sum_{k=1}^{5} Y_{22k} = 138.9$$

The two treatment totals are

$$Y_{1\bullet\bullet} = \sum_{j=1}^{2}\sum_{k=1}^{5} Y_{ijk} = \sum_{j=1}^{2} Y_{1j\bullet} = Y_{11\bullet} + Y_{12\bullet} = 135.0$$

$$Y_{2\bullet\bullet} = Y_{21\bullet} + Y_{22\bullet} = 301.5$$

The two sex (blocks) totals are

$$Y_{\bullet 1\bullet} = \sum_{i=1}^{2}\sum_{k=1}^{5} Y_{i1k} = \sum_{i=1}^{2} Y_{i1\bullet} = Y_{11\bullet} + Y_{21\bullet} = 74.4 + 162.6 = 237.0$$

$$Y_{\bullet 2\bullet} = Y_{12\bullet} + Y_{22\bullet} = 60.6 + 138.9 = 199.5$$

$$Y_{\bullet\bullet\bullet} = Y_{1\bullet\bullet} + Y_{2\bullet\bullet} = 135.0 + 301.5 = 436.5 = Y_{\bullet 1\bullet} + Y_{\bullet 2\bullet} = 237.0 + 199.5$$

$$\text{CF} = Y_{\bullet\bullet\bullet}^2 / rsm = \frac{436.5^2}{20} = 9526.6125$$

$$\sum_i\sum_j\sum_k Y_{ijk}^2 = 11354.31$$

$$\text{TSS} = \sum_i\sum_j\sum_k Y_{ijk}^2 - \text{CF} = 11354.31 - 9526.6125 = 1827.6975$$

d.f. of TSS = $n - 1 = 19$

$$\text{SS between cells} = \frac{1}{5}\left[\sum_i\sum_j Y_{ij\bullet}^2\right] - \text{CF}$$

$$= \frac{[74.4^2 + 162.6^2 + 60.6^2 + 138.9^2]}{5} - 9526.6125$$

$$= 1461.3255$$

d.f. of SS between cells = $rs - 1 = 3$

SSE = TSS − SS between cells = 1827.6973 − 1461.3255 = 366.3720 d.f. of SSE = $rs(m-1) = 2 \times 2\ (4) = 16$

$$SST = \frac{1}{rm}\sum_{i=1}^{s} Y_{i\bullet\bullet}^2 - CF = \sum_{i=1}^{2} \frac{(\text{Treatment total})^2}{\text{Number of values in each total}} - CF$$

$$= \frac{(135.0^2 + 301.5^2)}{2 \times 5} - 9526.6125 = 1386.1125$$

d.f. of SST = $s - 1 = 1$

$$SSB = \frac{1}{sm}\sum_{j=1}^{r} Y_{\bullet j\bullet}^2 - CF = \sum_{j=1}^{2} \frac{(\text{Block or sex total})^2}{\text{Number of values in each total}} - CF$$

$$= \frac{(237.0^2 + 199.5^2)}{2 \times 5} - CF = 70.3125$$

d.f. of SSB = $r - 1 = 1$

SSI = SS between cells – SSB – SST

= 1461.3255 – 70.3125 – 1386.1125 = 4.9005

d.f. of SSI = $(r-1)(s-1) = 1$

ANOVA table for Example 14.5 on RBD 5 observations per cell

Source of variation	d.f.	SS	MSS = SS/df	F(calculated)	Upper tail F value 5%	and 1%
Sex (blocks)	1	70.3125	70.3125	MSB/MSE = 3.07	$F_{1,16,.05}$ = 4.49	$F_{1,16,.01}$ = 8.53
Hormone (treatment)	1	1386.1125	1386.1125	MST/MSE = 60.5**	$F_{1,16,.05}$ = 4.49	$F_{1,16,.01}$ = 8.53
Sex × hormone interaction	1 (by subtract.)	4.9005 (by subtract.)	4.9005	MSI/MSE = 0.214	$F_{1,16,.05}$ = 4.49	$F_{1,16,.01}$ = 8.53
Between cells	3	1461.3225				
Error or within cells	16 (by subtraction)	366.3750	22.8982			
Total	19	1827.6975				

F values in the ANOVA table indicate that H_{02} and H_{03} can not be rejected. P values for both are more than .05. H_{01} is rejected since the F calculated for treatment is highly significant. P-value < .01 because calculated F for treatment is very much larger than 1% upper tail value of $F_{1,16}$.

On the basis of these one can conclude that the hormone treatment has highly significant effect on average plasma calcium concentration in birds. The sex of birds does not have effect on average plasma calcium concentration in birds and sex of birds has no interaction with hormone treatment as regards average plasma calcium concentration in birds.

A Remark

The procedure described above requires equal replications in each of the *rs* cells. In practice there are often the situations when the number of replicates in each cell may not be equal due to missing of data or loss of experimental unit(s) after the experiment, even though initially the experiment was designed with equal replicates in each cell.

Thus one may end up after the experiment, in an RBD with unequal replicates in all the cells. In such a situation the analysis procedure is not as described above, but is a bit more complicated. Analysis of RBD with unequal replicates is not discussed in this book. The interested reader may refer to Zar (2004), Rao (1973), and Yun et al. (2002).

Advantages and Disadvantages of RBD

Advantages

1. It is quite flexible. It is applicable to a moderate number of treatments.
2. If extra replications are necessary for some treatments, these may be applied to more than one unit but to same number of units per block, preferably.
3. Since variability among replicates can be eliminated from experimental error, it is not necessary to use continuous blocks.
4. It also enables to use different techniques to different blocks but it should be same within a block.
5. The analysis is straight forward and remains so, even if data from entire blocks or treatments is missing due to some accident.
6. If data from individual units is missing, one can use Yates's missing plot techniques (see Section 14.8).
7. By grouping the units into blocks, greater precision can be obtained than the CRD.
8. No other design has been used so often as the RBD. If satisfactory results can be obtained using this design, one should not use other complex designs.

Disadvantages

1. If the blocks are not of homogeneous experimental units, a large error term will result.
2. Usually in field experiments, an increase in number of treatments results in large block size. So it is quite likely that blocks may contain heterogeneous units, thus increasing the experimental error.

14.3 Latin Square Design (LS_qD)

The principle of local control to reduce experimental error was used in the RBD by grouping the units in one way, i.e. in blocks containing homogeneous experimental units. This can be carried one step forward by grouping according to two sources of variation among the units. For example in agricultural field experiments, the layout is usually a square, thus allowing to remove from error, the variation along rows as well as along columns, a design called Latin Square Design (LS_qD) may be used. In such experiment fertility variation in two directions at right angles can be controlled simultaneously.

In LS_qD, the *m* treatments are arranged in rows and columns such that each treatment occurs once and only once in each row and in each column. Each row like each column is a complete block. Analysis of data from LS_qD allows to remove variability between rows and columns from SSE. Here rows may be considered as blocks and column are positions in the blocks. If *m* treatments are to be compared, in all $m \times m$ experimental units are needed grouped in *m* rows and *m* columns.

For m = 4 treatments, the arrangement might be

A	D	C	B
B	C	A	D
D	A	B	C
C	B	D	A

Layout

Allotment of treatment (denoted by capital letter) in various cells in the rows and columns following the principle of LS_qD can be done in a large number of ways. The way it is to be done in a layout must be determined randomly.

Latin square with m letters A, B, C..... in the natural order occurring in the first row and first column is called a *standard square* (or *square in canonical form*). Thus the standard square for the 4 treatments A, B, C, D is

A	B	C	D
B	C	D	A
C	D	A	B
D	A	B	C

From a standard $m \times m$ Latin square, one may obtain $m!$ $(m - 1)!$ different LS_qD's. For this, all the orders of the $(m - 1)$ rows, keeping the first row unchanged, are obtained. Then permutations of all the m columns are done. In this way a large number of LS_qD's can be formed.

Randomization in Latin square consists of choosing a square at random from all possible Latin squares. Fisher and Yates (1950) have given detailed instruction and tables of standard LS_qD's given for 4 × 4 to 12 × 12 squares, whereas Saxena (1950) has enumerated all the different squares which would be found in the reference given by him.

Analysis

There are only m^2 observations in the $m \times m$ LS_qD's, but it can be denoted by Y_{ijk}; where i and j are the row and column from which the observation has been taken and k is the identification of the treatment received by this $(i, j)^{th}$ cell unit. Thus i and j = 1, 2, ..., m and k is one of the m letters A, B, C, ... The model for this observation is given by

$$Y_{ijk} = \mu + \alpha_i + \beta_j + \tau_k + e_{ijk}$$

with $$\sum_i \alpha_i = \sum_j \beta_j = \sum_k \tau_k = 0 \text{ and } e_{ijk} \text{ are iid. } N\left(0, \sigma_e^2\right) \tag{14.44}$$

Hypothesis to be tested is H_0: $\tau_1 = \tau_2 = ... = \tau_k = 0$ against
H_1: all the τ's are not equal to zero.

The analysis given below is applicable to the model (14.44) given above. This model assumes that there is no interaction between the rows, columns or treatments. If the analyst is not confident to assume absence of interaction, this analysis is not valid.

Let $$Y_{..} = \sum_{i=1}^{m} \sum_{j=1}^{m} Y_{ijk} = \text{sum of } m^2 \text{ observations} \tag{14.45}$$

$$Y_{i\bullet} = \sum_{j=1}^{m} Y_{ijk} = \text{sum of } m \text{ observations in } i^{\text{th}} \text{ row } i = 1, 2, \ldots\ldots m \tag{14.46}$$

$$Y_{\bullet j} = \sum_{i=1}^{m} Y_{ijk} = \text{sum of } m \text{ observations in } j^{\text{th}} \text{ column } j = 1, 2, \ldots\ldots m \tag{14.47}$$

$$Y_{\bullet\bullet k} = \sum_{ij} Y_{ijk} = \text{sum of } m \text{ cell values from the Latin square where the } k^{\text{th}} \text{ treatment has been applied } k = 1, 2, \ldots\ldots, m \tag{14.48}$$

$$\text{Also find SS of } i^{\text{th}} \text{ row elements} = \sum_{j} Y_{ijk}^2 \tag{A}$$

$$\text{And SS of } j^{\text{th}} \text{ column elements} = \sum_{i} Y_{ijk}^2 \tag{B}$$

Find sum of (A) for all the m rows and sum of (B) for all the m columns. These two sums of m quantities must match and is SS of all the m^2 values in Latin square to be used to find TSS.

$$\text{CF} = Y_{\bullet\bullet}^2 / m^2 \tag{14.49}$$

$$\text{SS due to rows SSR} = \frac{\left(\sum_{i=1}^{m} Y_{i\bullet}^2\right)}{m} - \text{CF} \tag{14.50}$$

$$\text{SS due to column SSC} = \frac{\left(\sum_{j=1}^{m} Y_{\bullet j}^2\right)}{m} - \text{CF} \tag{14.51}$$

$$\text{SS due to treatment SST} = \frac{\left(\sum_{k=1}^{m} Y_{\bullet\bullet k}^2\right)}{m} - \text{CF} \tag{14.52}$$

$$\text{Total SS} = \text{TSS} = \sum_{i=1}^{m} \sum_{j=1}^{m} Y_{ijk}^2 - \text{CF} \tag{14.53}$$

$$\text{SSE} = \text{TSS} - \text{SSR} - \text{SSC} - \text{SST} \tag{14.54}$$

Thus one can form the ANOVA table to test H_0 as given in Table 14.5.

The test procedure is to compare the calculated values F_1, F_2 and F_3, with the 5% and 1% tabulated F values. *Conclusions* are drawn as indicated for RBD.

Table 14.5 ANOVA table for latin square design

Source of variation	d.f.	SS	MSS	F	Tabulated F 5%	1%
Rows	$m-1$	SSR	MSR = SSR/$(m-1)$	F_1 = MSR/MSE	$F_{(m-1),(m-1)(m-2),.05}$	$F_{(m-1),(m-1)(m-2),.01}$
Column	$m-1$	SSC	MSC = SSC/$(m-1)$	F_2 = MSC/MSE	$F_{(m-1),(m-1)(m-2),.05}$	$F_{(m-1),(m-1)(m-2),.01}$
Treatments	$m-1$	SST	MST = SST/$(m-1)$	F_3 = MST/MSE	$F_{(m-1),(m-1)(m-2),.05}$	$F_{(m-1),(m-1)(m-2),.01}$
Error	$(m-1)\times(m-2)$	SSE (by subtraction)	MSE = SSE/$(m-1)(m-2)$			
Total	m^2-1	TSS				

The example given below illustrates the procedure explained above.

Example 14.6: The following table shows the yields in kg per plot (42 meter^2) of wheat in 4 × 4 Latin square. The four varieties of wheat are A: Baladi, B: Mokhtar, C: Giza 139 and D: Thatcher.

Perform the ANOVA and test the hypothesis if there are differences in average yield of wheat of the four varieties (source: Steel and Torrie (1984), p. 223).

Yields (in kg per plot of 42 meter^2) of 4 varieties of wheat in 4 × 4 Latin square

Rows	Columns 1	2	3	4	Total ($Y_{i.}$)	$\sum_j Y_{ijk}^2$
1	C = 10.5	D = 7.7	B = 12.0	A = 13.2	43.4	487.78
2	B = 11.1	A = 12.0	C = 10.3	D = 7.5	40.9	429.55
3	D = 5.8	C = 12.2	A = 11.2	B = 13.7	42.9	495.61
4	A = 11.6	B = 12.3	D = 5.9	C = 13.2	40.0	424.70
$Y_{.j}$	39.0	44.2	39.4	44.6	$Y_{..}$ = 167.2	
$\sum_i Y_{ijk}^2$	401.66	503.42	410.34	522.22		$\sum_i\sum_j Y_{ijk}^2 = 1837.64$

In this case the 4 varieties (= treatments) of wheat are evaluated in 4 × 4 Latin square. So $m = 4$. The hypothesis to be tested is H_0: $\tau_1 = \tau_2 = \tau_3 = \tau_4 = 0$ against H_1: all the four τ's are not equal to zero.

The row totals, column totals, 4 rows and 4 column elements' SS are obtained and shown in the 2 column and 2 row margins.

Now various SS for ANOVA table are obtained as

$$Y_{..} = 167.2 \quad \text{so CF} = Y_{..}^2/m^2 = \frac{167.2^2}{4\times 4} = 1747.24$$

The treatment totals are $Y_{..1}$ = Total for A = 48.0, $Y_{..2}$ = Total for B = 49.1, $Y_{..3}$ = Total for C = 43.2, $Y_{..4}$ = Total for D = 26.9

$$\text{TSS} = \sum_{i=1}^{m}\sum_{j=1}^{m} Y_{ijk}^2 - \text{CF} = 1837.64 - 1747.24 = 90.40$$

with degree of freedom = $4^2 - 1 = 15$

$$SSR = \frac{\left(\sum_{i=1}^{4} Y_{i\cdot}^2\right)}{m} - CF = \frac{\left(43.4^2 + 40.9^2 + 42.9^2 + 40.0^2\right)}{4} - 1747.24 = 1.95$$

with d.f. 4 – 1 = 3

$$SSC = \frac{\left(\sum_{j=1}^{4} Y_{\cdot j}^2\right)}{m} - CF = \frac{\left(39.0^2 + 44.4^2 + 39.4^2 + 44.6^2\right)}{4} - 1747.24 = 6.80$$

with d.f. 4 – 1 = 3

$$SST = \frac{\left(\sum_{k=1}^{4} Y_{\cdot\cdot k}^2\right)}{m} - CF = \frac{\left(48.0^2 + 49.1^2 + 43.2^2 + 26.9^2\right)}{4} - 1747.24 = 78.93$$

with d.f. 4 – 1 = 3

SSE = TSS – (SSR + SSC + SST) = 90.40 – (1.95 + 6.80 + 78.93) = 2.72

with d.f. = (4 – 1) (4 – 2) = 6

Hence ANOVA table for testing H_0 is obtained as

ANOVA table for Example 14.6 on Latin square

Sources of variation	d.f.	SS	MS	F	Tabulated F 5%	Tabulated F 1%
Rows	3	SSR = 1.95	MSR = .65	F_1 = 1.44	$F_{3,6,.05}$ = 4.76,	$F_{3,6,.01}$ = 9.78
Columns	3	SSC = 6.80	MSC = 2.27	F_2 = 5.04*	$F_{3,6,.05}$ = 4.76,	$F_{3,6,.01}$ = 9.78
Treatments	3	SST = 78.93	MST = 26.31	F_3 = 58.47**	$F_{3,6,.05}$ = 4.76,	$F_{3,6,.01}$ = 9.78
Error	6	SSE = 2.72 (by subtraction)	MSE = .45			
Total	15	TSS = 90.40				

From the ANOVA table it is concluded that:

1. Rows do not differ significantly
2. Columns differ significantly .01 < *P*-value < .05
3. The varieties of wheat are very highly significantly different *P*-value < .01

Advantages and Disadvantages of Latin Square Design

1. LS_qD is suitable only in special cases where the land has fertility variation in two directions at 90°. It can also be used in animal experiments where it can be employed to control simultaneously two factors contributing to experimental error. For example litter and body weight differences may be controlled in trials with guinea pigs by assigning them to rows and columns of Latin square.

2. LS_qD requires as many replications as there are treatments. So it is suitable mainly for 4 to 8 treatments.
3. For small number of treatments the number of replications is found to be inadequate, since it results in very few d.f. for error.
4. If there is attack of some disease or pest, so that data on some blocks or cells is missing, the analysis becomes quite complicated.

14.4 Factorial Analysis

The previous sections dealt with testing of a single factor e.g. different varieties or different manures or different diets or different doses of a medical drug; using different designs. A *factor* is some kind of treatment. In a factorial experiment any one factor may supply several treatments. Thus if diet is a factor, several diets will be different levels of this factor. If photoperiod is a factor, the experiment will subject the study animals to different photoperiods. If baking temperature is a factor, baking will be done at different temperatures.

The term *level* refers to the several treatments within any factor. In agricultural experiment different amounts of a fertilizer may be the levels of the factor: fertilizer. The number of levels and factors in a single experiment depends on the practical situations only.

There are situations when two or more factors are needed to be varied simultaneously. The experimenter may be interested in finding the most suitable level of photoperiod and that of temperature. The method of experimenting with one factor first and then on the other i.e. performing two simple experiments with one factor and next with the other will give the optimum levels of the two. These will be correct if the two factors act independently of each other. In many situations e.g. diet and medication, photoperiod and temperature, the two factors do not vary independently. The same amount of change in levels of one factor, may have different effects on the other factors. In such situations technically it can be said that the two factors are not having additive effects, but they interact with each other.

An effective approach, to answer if the two factors actually interact with each other or are independent, is to investigate the effects of two factors together in a single experiment by comparing all possible combinations of the levels of both the factors.

If five varieties of a crop are compared using three different fertilizer combinations, in a single experiment, it is called a 5 × 3 *factorial experiment.*

Thus a *factorial experiment* is one in which the set of treatments consists of all possible combinations of the levels of several factors. When information is wanted on some factor for which recommendations are to be made over a wide range of conditions, factorial experiments are of a greater utility.

If there are n factors each at 2 (or 3) levels, the single experiment to compare all the combinations is called 2^n (or 3^n) factorial experiment. Here n is a positive integer greater than or equal to 2 i.e. $n \geq 2$.

The treatment combinations may be compared in an experiment using CRD or an RBD or an LS_qD. For a 2^2 factorial experiment, the number of treatment combinations will be 4. If RBD of r blocks is used for this, $s = 4$ and analysis will be done as indicated in Section 14.2. If LS_qD is used taking $m = 4$, the analysis will be done as indicated in Section 14.3. But in a factorial experiment one is interested in separate tests for the *main effects* of the factors and their interaction. This is performed by partitioning the SST with 3 d.f. in *three orthogonal* components each carrying a single d.f. Each of these is associated with either one of two main effects or with the interaction.

In this chapter the simplest 2^2 factorial experiment on RBD is considered first, giving the notations, definition and orthogonal partition of the treatment sum of squares. These can be generalized to higher order i.e. for $n > 2$ to find the SS for main effects and interaction. The subsection below describes these. Example of a 2^2 factorial experiment on CRD is also given.

In the subsequent subsection, analysis of a $p \times q$ factorial experiment on RBD is considered.

14.4.1 2^2 Factorial Experiment

Let there be two factors A and B, each at 2 levels. To develop the method of analysis for a 2^n factorial experiment using orthogonal partition of SST, the notations and definitions given by Yates (1933) are described below.

A factor is denoted by capital letter (A and B for 2^2 factorial experiment). The corresponding small letters denote the second level of the factors. Thus *a* and *b* denote the *second levels* of factors A and B, respectively. *First levels* of the factors are signified by absence of the corresponding small letters. With this convention, if there are 2 factors A and B, each at 2 levels, the $2 \times 2 = 4$ treatment combinations will be denoted by

(1): A and B both at the first level
(a): A at the second level and B at the first level
(b): A at the first level at B at the second level
(ab): A and B both at the second levels

Main Effect and Interaction Effect

The symbols [a] and (a) denote the total and mean, respectively, of all the observations receiving treatment combination a. The letters A, B and AB when they refer to numbers, denote the main effects of A and B, and the interaction of A and B.

Effect of changing A from its first level to second level when B is at first level = $(a) - (1)$ (14.53)

Effect of changing A from its first level to second level, when B is at second level = $(ab) - (b)$ (14.54)

These two are called the *simple effects* of the factor A. If the two factors A and B are independent in their effects, the above two simple effects (14.53) and (14.54) are expected to be equal. *The average of these two simple effects is defined as the main effect of A*. Thus the main effect of factor A is

$$= \frac{1}{2}\{(ab) - (b) + (a) - (1)\} = \frac{1}{2}(a-1)(b+1) \tag{14.55}$$

RHS of (14.55) is to be expanded algebraically and enclosed in ( ). Thus the enclosed letter combinations are to be replaced by corresponding treatment means.

If the above two factors A and B are not independent then the above two simple effects of A are not same. *One half of the difference of the first simple effect from the second is taken as a measure of this dependence or interaction.* Thus the *two factor interaction (or the first order interaction)* between A and B is

$$\text{AB} = \frac{1}{2}\{(ab) - (b) - (a) + (1)\} = \frac{1}{2}(a-1)(b-1) \tag{14.56}$$

As for the expression in (14.55) the treatment combinations after algebraic expression are to be replaced by corresponding treatment means.

In a similar manner the two simple effects of the factor B are $(b) - (1)$ and $(ab) - (a)$. So the main effect of B and interaction BA is given by

$$B = \frac{1}{2}\{(ab) - (a) + (b) - (1)\} = \frac{1}{2}(a+1)(b-1) \tag{14.57}$$

and the interaction

$$BA = \frac{1}{2}\{(ab) - (a) - (b) + (1)\} = \frac{1}{2}(a-1)(b-1) \tag{14.58}$$

It can be seen that AB and BA are same in (14.56) and (14.58). Hence interaction of factors does not depend on the order of factors.

Contrasts and Orthogonal Contrasts

From (14.55), (14.56) and (14.57), it can be seen that the main effects and the interaction effects are linear functions of treatment means such that sum of coefficients in the linear functions are all zero. Such *linear function* of the treatment means, where *the sum of coefficients* of means *sum to zero* is called a contrast (or comparison) of the treatment means.

Moreover, it can be seen that the sum of the products of the corresponding coefficients in the contrasts A and AB

$$= \frac{1}{2} \times \frac{1}{2} + \left(-\frac{1}{2}\right)\left(-\frac{1}{2}\right) + \left(\frac{1}{2}\right)\left(-\frac{1}{2}\right) + \left(-\frac{1}{2}\right)\left(\frac{1}{2}\right) = 0 \tag{14.59}$$

Such a pair of contrasts are said to be *orthogonal contrasts.* It can be seen that *B is also orthogonal* to A and (AB). Hence A, B and AB form a set of orthogonal contrasts of treatments.

The following table which gives only the signs and the divisor can be very easily used for defining the main effects and the interaction effects. In this the general mean

$$M = \frac{1}{4}\{(ab) + (a) + (b) + (1)\} \tag{14.60}$$

The following *rules* are followed *to write down the factors*:

Table 14.6 Sign and divisors to write A, B and AB

Effect	Treatment means				Divisor
	(1)	(a)	(b)	(ab)	
M	+	+	+	+	4
A	–	+	–	+	2
B	–	–	+	+	2
AB	+	–	–	+	2

(i) Give + sign to all treatment combinations to find M.

(ii) To each treatment mean give a + sign if the small letter of the corresponding main factor is present.

(iii) The signs of two factor interaction are obtained by multiplying the corresponding signs of main effects (multiplication of two identical sign will be + and two opposite signs will be – in the interaction).

Analysis of Factorial Experiments

Analysis of the factorial experiments will depend, to some extent, on the design which is used for this experiment. This experiment may be performed using CRD or RBD or LS_qD. Most of the times RBD is used for this experiment with r blocks. So each of the 4 treatments is replicated r times.

An experimenter will be interested to test significance of the main effects and the interaction effects of the factors. It is now clear that the factorial effects : main and interaction are orthogonal contrast of the treatment means.

Tests of factorial effects (main or interaction) is done by computing the

$$F\text{-statistic} = \frac{\text{MS due to the factorial effect}}{\text{MSE}} \tag{14.61}$$

where each of the factorial effect has 1 d.f. and MSE is the error MS from the ANOVA table appropriate for the design used.

The factorial effects and their SS are obtained from treatment totals rather than from their means. Thus using Table 14.6 and enclosing treatment symbols in [ ] to denote corresponding treatment total, the factorial effect totals are first obtained from the following.

$$[\text{A}] = [ab] - [b] + [a] - [1] \tag{14.62}$$

$$[\text{B}] = [ab] + [b] - [a] - [1] \tag{14.63}$$

$$[\text{AB}] = [ab] - [b] - [a] + [1] \tag{14.64}$$

Next the sum of squares due to each factorial effect is obtained as follows.
Since each treatment is replicated r times, from (14.55) and (14.62)

Main effect of $\quad \text{A} = [A]/2r \quad$ (14.65)

Similarly main effect of $\quad \text{B} = [B]/2r$ and $\quad$ (14.66)

Interaction $\quad \text{AB} = [AB]/2r \quad$ (14.67)

The $\quad \text{SS due to a factorial effect} = r \times (\text{factorial effect})^2 \quad$ (14.68)

Hence $\quad \text{SS due to main effect of A} = [A]^2/(4r) \quad$ (14.69)

$\text{SS due to main effect of B} = [B]^2/(4r) \quad$ (14.70)

$\text{SS due to interaction AB} = [AB]^2/(4r) \quad$ (14.71)

each with 1 d.f.

Hence the ANOVA Table for a 2^2 factorial experiment in r randomized blocks will be obtained after computing TSS, SSB SSE apart from the above 3 SS.

Now as described in previous Section 14.2 calculate

$$G = \sum_{i=1}^{4}\sum_{j=1}^{r} X_{ij} = \text{Sum of all the } 4r \text{ observations}$$

$$\text{TSS} = \sum_{i=1}^{4}\sum_{j=1}^{r} X_{ij}^2 - G^2/(4r) = \text{Sum of squares of all the } 4r \text{ observations } - G^2/(4r)$$

$$\text{SSB} = \frac{1}{4}\sum_{j=1}^{r} X_{\bullet j}^2 - G^2/(4r)$$

where $X_{.j}$ is j^{th} block total ($j = 1, 2, ..., r$)

$$\text{SSE} = \text{TSS} - \text{SSB} - \text{SS main effect of A} - \text{SS main effect of B} - \text{SS due to interaction AB}$$

The following gives the ANOVA table

Table 14.7 ANOVA table for 2^2 factorial experiment in r RBD

Source of variation	d.f.	SS	MS = SS/d.f.	F	
				Calculated	Tabulated
Blocks	$r - 1$	SSB	SSB/(r – 1) = MSBl	MSBl/MSE	$F_{r-1, 3(r-1), .05}$
Main effect A	1	$[A]^2/(4r)$	MSA	MSA/MSE	$F_{1, 3(r-1), .05}$ $F_{1, 3(r-1), .01}$
Main effect B	1	$[B]^2/(4r)$	MSB	MSB/MSE	$F_{1, 3(r-1), .05}$ $F_{1, 3(r-1), .01}$
Interaction A × B	1	$[AB]^2/(4r)$	MSAB	MSAB/MSE	$F_{1, 3(r-1), .05}$ $F_{1, 3(r-1), .01}$
Error	$3(r - 1)$	SSE (by subtraction)	MSE		
Total	$4r - 1$	TSS			

The hypotheses of no block effect, no main effects and no interaction effect are rejected, if

$$F_{\text{cal}} > F_{\text{tabulated}}$$

The following example illustrates the procedure described above.

Example 14.7: A 2^2 factorial experiment in 6 randomized blocks was conducted in order to obtain an idea of the interaction: spacing (S) × number of seedling per hole (N), along with the effects of different spacings (S) and different numbers of seedlings per hole (N), while adopting Japanese method of cultivation. The levels of the two factors are:

S : 20 cm spacing in between and
: 25 cm spacing in between
N : 3 seedlings per hole
: 4 seedlings per hole

The field plan and yield of dry *Aman* paddy (in kg) for each plot is given below

Block 1				Block 2			
(1) 117	*s* 106	*ns* 109	*n* 114	*ns* 114	(1) 120	*s* 117	*n* 114

Block 3				Block 4			
(1) 111	*n* 117	*s* 114	*ns* 106	*ns* 93	*n* 121	*s* 112	(1) 108

Block 5				Block 6			
ns 75	*s* 97	(1) 73	*n* 38	*n* 58	(1) 81	*ns* 105	*s* 117

Analyze the data to find out if there are any significant treatment effects – main or interaction. (Source: Goon et al. (2002), p. 94.)

Solution: The six block totals are $B_1 = 446$, $B_2 = 465$, $B_3 = 448$, $B_4 = 434$, $B_5 = 283$ and $B_6 = 361$

The treatment totals (from 6 blocks) and factorial effects are computed in the following table.

Treatment combinations	Treatment totals	Factorial effects total [ ]	Main and interaction effects = [ ]/2*r*	SS due to effects [ ]²/4*r*
(1)	[1] = 610	Grand total *G* = 2442		
n	[*n*] = 562	[N] = [*ns*] – [*s*] + [*n*] – [1] = –104	–8.667 = N	450.667
s	[*s*] = 663	[S] = [*ns*] + [*s*] – [*n*] – [1] = 98	8.167 = S	400.167
ns	[*ns*] = 607	[NS] = [*ns*] – [*s*] – [*n*] + [1] = –8	–0.667 = NS	2.667
	Total = *G* = 2442			Total = SST = 853.5

$$\sum_{i=1}^{4}\sum_{j=1}^{6} X_{ij}^2 = \text{Total raw SS} = 259{,}024.00$$

Now perform analysis of RBD

Correction factor (CF) = $G^2/24 = (2442)^2/24 = 248{,}473.5$

So TSS = Total raw SS – CF = 259024.00 – 248,473.5 = 10, 550.5

$$\text{SSB} = \sum_{j=1}^{6} B_j^2/4 - \text{CF}$$

$$= \frac{1}{4}\left[446^2 + 465^2 + 448^2 + 434^2 + 283^2 + 361^2\right] - 248{,}473.5$$

$$= 253652.75 - 248{,}473.5 = 5179.25$$

SSE = TSS – SSB – SS due to N – SS due to S – SS due to NS

= 10550.5 – 5179.25 – 450.667 – 400.167 – 2.667 = 4517.75

As a cross check one can compute SSE using

SSE = TSS – SSB – SST

where $$\text{SST} = \sum_{i=1}^{4} T_i^2/6 - \text{CF}$$

$$= \frac{1}{6}\left[610^2 + 562^2 + 663^2 + 607^2\right] - 248{,}473.5$$

= Total of last column of above table

Now the ANOVA table can be formed as Table 14.8.

Table 14.8 ANOVA table for Example 14.7 of 2^2 factorial experiment

Source of variation	d.f.	SS	MS = SS/d.f.	F = MS../MSE	Tabulated F	
					5%	1%
Blocks	5	5179.25	1035.85		$F_{1,15,.05}$	$F_{1,15,.01}$
N	1	450.667	450.667	1.496	4.54	8.68
S	1	400.167	400.167	1.329	4.54	8.68
NS	1	2.667	2.667	0.009	4.54	8.68
Error	15	4517.75	301.183			
Total	23	10,550.5				

Conclusion

Since all the three calculated values of F are less than 5% tabulated value of $F_{1,15}$, there are no significant main or interaction effects.

The method explained above and the illustration are based on the algebraic expressions for the main and interaction effect totals. Yates has given an algorithmic method of obtaining various effect totals for any 2^n – experiment without use of algebraic expressions for these. Book by Goon et al. (2002) may be consulted by interested reader.

14.4.2 $p \times q$ Factorial Experiment

The previous section considered testing of main and interaction effects of two factors. The two factors had the equal number of levels (n). In this section the situations are considered when the two factors may have unequal number of levels.

Let the two factors under consideration be A and B at p and q levels, respectively. Thus there will be in all $p \times q$ treatment combinations. The $(i, j)^{\text{th}}$ combination will be $a_i b_j (i = 1, 2, \ldots\ldots, p$ and $j = 1, 2, \ldots\ldots, q)$. Let there be n observations which receive $(i, j)^{\text{th}}$ treatment combination. Thus in all there will be npq observations in the whole experiment.

Let Y_{ijk} denote the kth of the n observations which receive the treatment combinations $a_i b_j$. The experiment may be arranged in a completely randomized block design (with n blocks) or in a completely randomized design with n observations in the $(i, j)^{\text{th}}$ cell with treatment combination $a_i b_j$.

Let us consider the $p \times q$ factorial experiment conducted in n randomized blocks. So there will be n replications of each of the pq treatments.

For analysis of such experiment the following calculations will be required for finding SS for effects of block, factor A, factor B, interaction AB and due to error.

$\sum_{k=1}^{n} Y_{ijk} = T_{ij}$, the total of observations receiving $(i, j)^{th}$ treatment combinations $a_i b_j$.

$$i = 1, 2, \ldots\ldots, p \quad j = 1, 2, \ldots\ldots\ldots\ldots\ldots, q$$

$\sum_{j=1}^{q} T_{ij} = A_i$, total of nq observations for i^{th} level of A; $i = 1, 2, \ldots\ldots, p$

$\sum_{i=1}^{p} T_{ij} = B_j$, total of np observations for j^{th} level of B; $j = 1, 2, \ldots\ldots, q$

$\sum_{i=1}^{p} A_i = \sum_{j=1}^{q} B_j = G$, the grand total of all the npq observations

$\sum_i \sum_j \sum_k Y_{ijk}^2$ = raw sum of squares of all the npq observations

$G^2 / (npq)$ = Correction Factor (CF)

Bl_r = Total of pq observations in r^{th} block $r = 1, 2, \ldots\ldots, n$

After these computations the SS for ANOVA table can be obtained as:

$$\text{TSS} = \sum_i \sum_j \sum_k Y_{ijk}^2 - \text{CF} \quad \text{with } (npq - 1) \text{ d.f.} \tag{14.72}$$

$$\text{SSBl} = \frac{1}{pq} \sum_{r=1}^{n} Bl_r^2 - \text{CF} \quad \text{with } (n - 1) \text{ d.f.} \tag{14.73}$$

$$\text{SSA} = \frac{1}{nq} \sum_{i=1}^{p} A_i^2 - \text{CF} \quad \text{with } (p - 1) \text{ d.f.} \tag{14.74}$$

$$\text{SSB} = \frac{1}{np} \sum_{j=1}^{q} B_j^2 - \text{CF} \quad \text{with } (q - 1) \text{ d.f.} \tag{14.75}$$

$$\text{SSAB} = \sum_i \sum_j T_{ij}^2 / n - \text{CF} - \text{SSA} - \text{SSB} \quad \text{with } (p - 1)(q - 1) \text{ d.f.} \tag{14.76}$$

$$\text{SSE} = \text{TSS} - \text{SSBl} - \text{SSA} - \text{SSB} - \text{SSAB} \text{ with } (n - 1)(pq - 1) \text{ d.f.} \tag{14.77}$$

Now the ANOVA table may be prepared for testing effects of factor A, factor B and interaction AB as given below.

If the calculated F value exceeds the upper tail value of F with d.f. shown in Table 14.9, the corresponding main effects or interaction effect are significant.

Table 14.9 ANOVA table for $p \times q$ factorial experiment in n RBD

Source of variation	d.f.	SS	MS = SS/d.f.	F = MS../MSE	5% 1% upper tail values of F with d.f.
Blocks	$n-1$	SSBl	MSBl	MSBl/MSE	$F_{(n-1),(n-1)(pq-1)}$
A	$p-1$	SSA	MSA	MSA/MSE	$F_{(p-1),(n-1)(pq-1)}$
B	$q-1$	SSB	MSB	MSB/MSE	$F_{(q-1),(n-1)(pq-1)}$
AB	$(p-1)(q-1)$	SSAB	MSAB	MSAB/MSE	$F_{(p-1)(q-1),(n-1)(pq-1)}$
Error	$(n-1)(pq-1)$	SSE (by subtraction)	MSE		
Total	npq-1	TSS			

It is desirable that first the significance of interaction effects be tested. If it is found significant, results of the ANOVA Table 14.9 are final. However, if interaction is found insignificant, the SSE and SSAB are added together to get SSE_p. The new pooled mean square error is obtained as

$$MSE_p = (SSE + SSAB)/(\text{d.f. of SSE} + \text{d.f. of SSAB})$$
$$= SSE_p/(\text{d.f. of SSE} + \text{d.f. of SSAB})$$

The new ANOVA table with new MSE_p instead of MSE of Table 14.9 is prepared, which has only SSBl, SSA, SSB and SSE_p. F values are obtained by dividing the MS by MSE_p, which have new d.f. of SSE_p as denominator d.f. In other words, new denominator d.f. in MSE_p is equal to $n(pq-1)-(p-1)-(q-1)$.

The Example 14.8 illustrates this procedure.

Example 14.8: An experiment was laid out in a randomized block design with 6 replicates to test the three varieties of wheat (C518, C591 and C520) with application of four doses of ammonium sulphate (n_0 = 0 lb N per acre, n_1 = 20 lb N per acre, n_2 = 40 lb N per acre and n_3 = 60 lb N per acre). It is desired to compare different varieties and response to application of ammonium sulphate (source: Panse and Sukhatme (1957), p. 167).

There were 3 × 4 = 12 treatment combinations. So each block consisted of 12 plots of 1/50 acre. Yields for each of the 6 × 12 = 72 plots were recorded and are summarized in the given table of treatment totals.

Treatment totals from 6 plots in lbs

Varieties	Doses of N				Total
	n_0	n_1	n_2	n_3	
v_1 = C518	89.6	98.3	97.2	93.6	378.7
v_2 = C591	87.0	96.7	104.9	108.6	397.2
v_3 = C520	89.3	100.0	111.0	108.2	408.5
Total	265.9	295.0	313.1	310.4	1184.4

Block totals are: Bl_1 = 200.2 Bl_2 = 180.4 Bl_3 = 219.6 Bl_4 = 193.0 Bl_5 = 197.0 Bl_6 = 194.2

Solution: Here p = 3, q = 4 and n = 6

T_{ij}'s are the 12 entries in the body of the table above

Varieties totals Vi's; i = 1, 2, 3 are in the last column and

Dose totals Nj's, j = 1, 2, 3, 4 are in the last row of above table

$$\sum_{i=1}^{3} V_i = \sum_{j=1}^{4} N_j = G = 1184.4$$

$$CF = G^2/npq = 1184.4^2/(6 \times 3 \times 4) = 14{,}02{,}803.36/72 = 19{,}810.88$$

Hence

$$\text{TSS} = \sum_i \sum_j \sum_k Y_{ijk}^2 - \text{CF} = 19810.88 - 19483.38 = 327.50$$

$$\text{SSB}_1 = \frac{1}{pq} \sum_{r=1}^{n} Bl_r^2 - \text{CF}$$

$$= \frac{1}{12}\left[200.2^2 + 180.4^2 + 219.6^2 + 193.0^2 + 197.0^2 + 194.2^2\right] - 19{,}483.38$$

$$= 19{,}551.61 - 19{,}483.38 = 68.29$$

$$\text{SSV} = \frac{1}{nq} \sum_{i=1}^{p} A_i^2 - \text{CF} = \frac{1}{24}[378.7^2 + 397.2^2 + 408.5^2] - 19{,}483.38$$

$$= 18.86$$

Here $p = 3$, $q = 4$ and $n = 6$

$$\text{SSN} = \frac{1}{18}\left[265.9^2 + 295.0^2 + 313.1^2 + 310.4^2\right] - 19{,}483.38 = 78.15$$

$$\text{SSVN} = \sum_i \sum_j T_{ij}^2 / n - \text{CF} - \text{SSV} - \text{SSN}$$

$$= \frac{1}{6}\left[89.6^2 + 98.3^2 + 97.2^2 + 93.6^2 + 87.0^2 + \ldots\ldots 108.6^2 + 89.3^2 + \ldots\ldots 108.2^2\right]$$

$$-19{,}483.38 - 18.86 - 78.15 = 120.03 - 18.86 - 78.15 = 23.02$$

$$\text{SSE} = TSS - SSBl - SSV - SSN - SSVN$$

$$= 327.50 - 68.29 - 18.86 - 78.15 - 23.02$$

$$= 139.18$$

Table 14.10 ANOVA table for 3 × 4 factorial experiment of Example 14.8

Source of variation	d.f.	SS	MS = SS/d.f.	F = MS../MSE_p	Upper tail values of F	
					5%	1%
Blocks	5	68.29	13.66	5.40**	2.36	3.33
V	2	18.86	9.43	3.72*	3.4	4.97
N	3	78.15	26.05	10.30**	2.75	4.12
VN	6	23.02	3.83	1.51	2.24	3.11
Error	55	139.18	2.53			
Total	71	327.50				

ANOVA indicates that interaction of varieties and doses of N is not significant. The effect of doses is highly significant. The varieties of wheat also have significant effect. Since interaction of varieties and doses is not significant, the SSVN is added with SSE to obtain pooled SSE_p = 162.20 with 61 d.f.

The revised MSE_p = 162.20/61 = 2.66 and the revised ANOVA table is obtained as follows.

Table 14.11 Revised ANOVA table for Example 14.8

Source of variation	d.f.	SS	MS	F	Upper tail values of F	
					5%	1%
Blocks	5	68.29	13.66	5.13**	2.37	3.34
V	2	18.86	9.43	3.55*	3.15	4.98
N	3	78.15	26.05	9.79**	2.76	4.13
Error (pooled)	61	162.20	2.66			
Total	71	327.50				

However, the results of revised ANOVA about blocks, varieties and doses of N remain as earlier.

14.5 Nested Designs

The designs of experiment considered in the last section contain each level of one factor with all the levels of the second factor.

In some designs, one may have some levels of one factor occurring in combination with levels of one or more factors, while other levels of this factor occur in combination with others (cf. Steel and Torrie (1984), Sokal and Rohlf (1995) and Zar (2003)).

Each major group or class is subdivided into randomly chosen subclass.

For example if an experiment with drosophila is conducted using four mediums: A, B, C, D. Three containers may be filled with each of the four mediums. A sample of drosophila may be reared in each of these 12 jars.

In this type of experiment, mediums are the four groups, the three jars with a medium are the three subgroups. The items in the subgroups are the individual drosophila in the 12 jars.

If only one container with a medium would have been used, it would not be possible to detect if there is an unintentional difference in various parts of the medium or it is due to the morphological differences among the experimental drosophila. With two or more jars it is possible to separate out these two differences.

With the presence of subgroups (3 containers of a medium) it is possible to test if the differences among media is greater than that observed between containers or not.

Such type of design is called a *nested design* since the subgroups are nested within the higher level (groups) of classification. The design is also called *hierarchical design.*

The ANOVA of such design may be called nested ANOVA. The subgroup level of nested ANOVA is always random effect or Model II. The higher level of grouping may be with fixed effect or Model I or with random effect or Model II. If it is Model II, the pure *Model II nested ANOVA is* done. If higher level is Model I or fixed effect, the analysis is *mixed model nested ANOVA.*

Model of Nested Design

The linear model for a *two-level Model II nested design* is given by

$$Y_{ijk} = \mu + A_i + B_{ij} + e_{ijk}; \quad \begin{aligned} i &= 1, 2, \ldots, a \\ j &= 1, 2, \ldots, b \\ k &= 1, 2, \ldots, n \end{aligned} \tag{14.78}$$

where Y_{ijk} is k^{th} observation in the jth subgroup of ith group
μ is the mean of the population.

A_i is the random contribution of the ith group of higher level A
B_{ij} is the random contribution of the jth subgroup (level B) of ith group of A
e_{ijk} is error term of kth item in the jth subgroup of the ith group.
It is *assumed* that A_i, B_{ij} and e_{ijk} are normally distributed with mean 0 and variances σ_A^2, $\sigma_{B\subset A}^2$ and σ^2, respectively. $\sigma_{B\subset A}^2$ is variance of level B within level A.

When the design is *mixed model nested* with highest level as fixed effect, the model is given by

$$Y_{ijk} = \mu + \alpha_i + B_{ij} + e_{ijk} \tag{14.79}$$

$$i = 1, 2, ..., a; j = 1, 2, ..., b, k = 1, 2, ..., n.$$

All the terms in equation (14.79) are same as those in (14.78) except that α_i is a fixed effect of higher level i.

Both the models can be extended to further levels of C, D ... with extra subscripts *l*, *m*, etc.
Analysis of the nested design is given below for *two-level pure Model II.*
The layout of the experiment will be as follows:

A_1	A_2	...	A_i	A_a
B_{11} B_{12}... B_{1j}... B_{1b}	B_{21} B_{22} ... B_{2j} .... B_{2b}	...	B_{ii} B_{i2} ... B_{ij} ... B_{ib}	... B_{a1} B_{a2} ... B_{aj} ... B_{ab}
Y_{111} Y_{121}... Y_{1j1} Y_{1b1}	Y_{211} Y_{221} ... Y_{2j1} ... Y_{2b1}	...	Y_{i11} Y_{i21} ... Y_{ij1} ... Y_{ib1}	... Y_{a11} Y_{a21}... Y_{aj1} ... Y_{ab1}
. . . .	. . . .		. . . .	. . . .
. . . .	. . . .	...	. . . .	. . . .
. . . .	. . . .		. . . .	. . . .
Y_{11n} Y_{12n}... Y_{1jn}... Y_{1bn}	Y_{21n} Y_{22n}... Y_{2jn}... Y_{2bn}	...	Y_{i1n} Y_{i2n}... Y_{ijn}... Y_{ibn}	... Y_{a1n} Y_{a2n}... Y_{ajn}...Y_{abn}

The following quantities are to be computed:

$$ij^{th} \text{ subgroup mean } \quad \bar{Y}_{Bij} = \frac{1}{n}\sum_{k=1}^{n} Y_{ijk} \quad i = 1, 2, ..., a, \; j = 1, 2, ..., b \tag{14.80}$$

$$i^{th} \text{ group mean } \quad \bar{Y}_{A_i} = \frac{1}{nb}\sum_{j=1}^{b}\sum_{k=1}^{n} Y_{ijk} = \frac{1}{b}\sum_{j=1}^{b} \bar{Y}_{Bij} \quad i = 1, 2, ..., a \tag{14.81}$$

$$\text{Grand total } \quad G = \sum_{i=1}^{a}\sum_{j=1}^{b}\sum_{k=1}^{n} Y_{ijk} \tag{14.82}$$

I Grand mean $\bar{\bar{Y}} = \dfrac{G}{abn}$ (14.83)

II SSA = Sum of square among groups = $nb\sum_{i=1}^{a}\left(\bar{Y}_{A_i} - \bar{\bar{Y}}\right)^2$ (14.84)

III SSB = SS among subgroups within groups

$$= n\sum_{i=1}^{a}\sum_{j=1}^{b}\left(\bar{Y}_{B_{ij}} - \bar{Y}_{A_i}\right)^2 \tag{14.85}$$

IV SSE = SS within subgroups or error SS

$$= n\sum_{i=1}^{a}\sum_{j=1}^{b}\sum_{k=1}^{n}\left(Y_{ijk} - \bar{Y}_{B_{ij}}\right)^2 \tag{14.86}$$

Now the ANOVA table for Model II nested design can be formed as

Table 14.12 Hierarchical or nested design ANOVA (model II)

Source of variation	d.f.	SS	MSS = SS/d.f.	F	Expected MSS
Groups	$a - 1$	SSA	MSA = SSA/$(a - 1)$	MSA/MSB = F_A	$\sigma^2 + n\sigma^2_{B\subset A} + nb\sigma^2_A$
Subgroups within groups	$a(b - 1)$	SSB	MSB = SSB/$a(b - 1)$	MSB/MSE = F_B	$\sigma^2 + n\sigma^2_{B\subseteq A}$
Within subgroups or error	$ab(n - 1)$	SSE	MSE = SSE/$ab(n - 1)$		σ^2
Total	$abn - 1$				

Now the calculated F values are used to test the following hypotheses:

1. H_{01}: There is no more variation among groups of A than in the subgroups within groups, against
   H_{11}: Variation among groups of A is more than in the subgroups within groups
2. H_{02}: There are no more variation among subgroups than within subgroups, against
   H_{12}: Variation among subgroups is more than that within subgroups.

$$F_A \sim F_{(a-1),\, a(b-1)} \text{ and } F_B \sim F_{a(b-1),\, ab(n-1)}, \text{ if } H_{01} \text{ and } H_{02} \text{ are true, respectively} \tag{14.87}$$

So reject at 100 α% level of significance

(i) H_{01} if $F_A \geq F_{\alpha,\,(a-1),\, a(b-1)}$ and
(ii) H_{02} if $F_B \geq F_{\alpha,\, a(b-1),\, ab(n-1)}$ (14.88)

A numerical example has been given by Sokal and Rohlf (1995) which studies the phenotypic variation of *Aedes in* at two sampling levels. Rohlf obtained field-caught said mosquito pupae and reared. Of these 12 female pupae were divided in three rearing cages at random, four in each cage.

It was desired to see whether the differences among cages will add variance to the variation among females, which may be due to female mosquitoes' genetic and ecological differences. Length of the left wing was measured twice by a measurer in random order to avoid biased measurements.

This was used as the measurement variable Y.

The cages were groups and females within cages were the subgroups.

The Model II nested ANOVA has been applied and they find that variation among cages (groups) over variation among females (subgroups) is not significant. But variation among females (within cages) is very highly significant.

The interested reader may go through the numerical computations given by them.

14.6 BIBD and PBIBD

When the number of treatments in an experiment increases, the number of experimental units required for a replication increases. If the design is RBD one replication is usually the block. If all the treatments are included in a block the design is *complete block design* since a block is one replication of all the treatments. Large block may result in an increase in experimental error.

Designs are available, where the complete block is subdivided into a number of incomplete blocks, such that each incomplete block contains only a part of all the treatment combinations.

Subdivision of treatments into incomplete blocks is done according to certain rules, so that the experimental error can be estimated among the units within the incomplete blocks. By this subdivision if experimental units within an incomplete block are more homogeneous than the incomplete blocks within a replication, precision is increased.

Confounding

If 2^2 or 2^3 factorial experiment is to be conducted the number of treatments are not large to necessitate incomplete blocks. But for 2^4 factorial (or higher) experiment it is often not advisable to adopt a complete randomized block design. Thus blocks of size less than the number of treatments are taken and a replication needs more than one block. Different groups of treatments are allotted to the blocks. For a 2^4 factorial experiment if block of size 8 is used, 2 blocks form a replication and 8 treatments out of 16 are allotted to one block and the rest 8 to the second block of the replication. It has been seen that an interaction effect or main effect contrasts are divided into two groups, one group (half of total treatment combinations) has + sign and the second group (other half of the total treatment combinations) has – sign.

In such a case, the contrast between the two block totals represents one of the main or interaction effect. Thus this effect, therefore, is mixed up with block effect and can not be separated. In this case this effect is *confounded* with the block effect.

Normally one of the higher order interaction is confounded instead of lower order ones. This device of reducing the block size by making one or more interaction contrasts identical with block contrast is known as *confounding*.

An illustration is given by considering a 2^3 factorial experiment, though normally this experiment can be arranged in complete RBD. Simplicity of the explanation using this experiment is reason for considering 2^3 factorial experiment with 3 factors A, B, C, each at 2 levels.

Highest order interaction is ABC and is given by the contrast:

$$(abc) + (a) + (b) + (c) - (ab) - (ac) - (bc) - (1) \tag{14.89}$$

Suppose that each replicate is divided into 2 blocks of 4 units each, such that one block of a replicate contains (*abc*), (*a*), (*b*) and (*c*) and the other block contains (*ab*), (*ac*), (*bc*) and (1). With 3 replicates before randomization the plan would be as given below.

	Rep I		Rep II		Rep III	
Blocks	1	2	3	4	5	6
	abc	*ab*	*abc*	*ab*	*abc*	*ab*
	a	*ac*	*a*	*ac*	*a*	*ac*
	b	*bc*	*b*	*bc*	*b*	*bc*
	c	(1)	*c*	(1)	*c*	(1)

Fig. 14.1. 2^3 factorial experiment in blocks of 4 units with ABC confounded (source: Cochran and Cox (1957), p. 183).

In all the three replicates block comparisons are the interaction ABC. In such a case ABC is completely *confounded* with blocks. The other six factorial effects A, B, C, AB, AC, BC can all be determined, as all are orthogonal to the block totals. Since two of four treatment combinations of a block have + ve sign and two have – ve sign, block effect has no effect on the said six factorial effects.

Partial Confounding in 2^3 Factorial Experiment

Any of the seven factorial effects may be confounded with blocks as ABC in above design. The rest six effects will then be free from block effects. This fact enables to spread confounding in an experiment among several factorial effects. Figure 14.2 gives a plan in which ABC, AC and BC are confounded in replication I, II and III respectively.

	Rep I		Rep II		Rep III	
Blocks	1	2	3	4	5	6
	abc	*ab*	*abc*	*ab*	*abc*	*ab*
	a	*ac*	*ac*	*bc*	*bc*	*ac*
	b	*bc*	*b*	*a*	*a*	*b*
	c	(1)	(1)	*c*	(1)	*c*
	ABC		AC		BC	

Fig. 14.2. 2^3 factorial experiment in blocks of 4 units with partial confounding of ABC, AC and BC (source: Cochran and Cox (1957), p. 185).

The difference between block totals in replication I represents the ABC interaction. In replication II, the treatment combinations in blocks are different from that of replication I. The difference of block totals in replication II represents the AC interaction. The difference in block totals in replication III represents the BC interaction.

The effects A, B, C and AB are entirely free from block effects. Though ABC is confounded in replicate I, it can be estimated from other two replicates since in the replicates II and III, ABC effect is orthogonal to blocks.

Similarly AC is confounded with blocks in replicate II and it can be estimated from replicates I and III. BC is confounded in replication III and can be estimated from replicates I and II. In this design ABC, AC and BC are partially confounded.

In general suppose in 2^n factorial experiment if a replicate of size 2^n units is divided into blocks of 2^{n-k} units, 2^k blocks will be required in the replicate. Then one can select any k factorial effects to be confounded, provided none of these are *generalized interaction* of any group of others in the selected k.

A generalized interaction of two chosen factorial effects is obtained by combining all the letters denoting these two effects and canceling all the letters that enter twice. For example generalized interaction of ABCDE and BCDE is $AB^1 C^2 D^3 E^4 B^1 C^2 D^3 E^4 = A$ (after canceling letters with same superscripts. After properly deciding k factorial effects to be confounded a further $(2^k - k - 1)$ effects are also automatically confounded, which are the generalized interactions of the group of selected k effects.

For a higher order factorial experiments the detailed methods of complete and partial confounding may be found in the books by Cochran and Cox (1957) or by Das and Giri (1979).

Incomplete Block Design

If in a block the number of units or plots is less than the number of treatments, the block is said to be incomplete and the design is said to be *incomplete block design*.

Reduction in block size leads to confounding of some factorial effects. The incomplete block designs (IBD) may be *balanced or partially balanced*. These were introduced by Yates (1936) in connection with plant breeding and selection experiments.

Let there be t treatments and k ($< t$) be number of plots or units in a block. Then k is also called the block size of the design.

To ensure equal or nearly equal precision of the comparisons of different treatment pairs, the treatments are so allotted in different blocks that each pair of treatments has the same number of replications and each treatment also has the same number of replications.

14.6.1 BIBD (Balanced Incomplete Block Design)

When the number of replications of all pairs of treatments in an IBD is the same, then the emerging important series of designs are known as balanced incomplete block designs (BIBD). A proper definition may be as follows.

An IBD with t treatments distributed over b blocks, each of size k ($< t$) such that *each treatment occurs in r blocks, no treatment occurs more than once* in a block and *each pair of treatments occur together in λ blocks*, is called a balanced incomplete block design (BIBD). The constants t, b, k, r and λ are called the parameters of the design. Two examples of BIBD are given in the following two figures.

Treatments are denoted by numbers 1, 2, ..., 9.

Blocks	Rep I				Rep II				Rep III				Rep IV		
(1)	1	2	3	(4)	1	4	7	(7)	1	5	9	(10)	1	8	6
(2)	4	5	6	(5)	2	5	8	(8)	7	2	6	(11)	4	2	9
(3)	7	8	9	(6)	3	6	9	(9)	4	8	3	(12)	7	5	3

Fig. 14.3. BIBD with parameters $t = 9$, $b = 12$, $k = 3$, $r = 4$ and $\lambda = 1$ (source: Cochran and Cox (1957), p. 376).

It can be noted that these parameters satisfy the following relations

$$tr = bk = \text{total number of plots in the design} \quad (14.90)$$

$$\lambda(t - 1) = r(k - 1) = \text{total number of replications of all possible treatment pairs in each of which a particular treatment is always present} \quad (14.91)$$

Following is an example of BIBD with 7 treatments and 7 blocks of 3 units each:

Blocks															
(1)	1	2	4	(3)	3	4	6	(5)	1	5	6	(7)	1	3	7
(2)	2	3	5	(4)	4	5	7	(6)	2	6	7				

Fig. 14.4. BIBD for 7 treatments in blocks of 3 units (source: Cochran and Cox (1957), p. 377).

It can be observed that in this BIBD, the parameters of the design are $t = 7$, $b = 7$, $k = 3$, $r = 3$ and $\lambda = 1$ and the two relations (14.90) and (14.91) are satisfied by these parameters.

However, it is worth noting, as proved by Fisher (1940), that even though parameters satisfying relations (14.90) and (14.91) can be determined but, a corresponding design may not be available. He proved that no BIBD with $b < t$ is possible.

For the method of construction and analysis of BIBD the reader is referred to the books by Das and Giri (1979) and Cochran and Cox (1957).

In the next subsection, the partially balanced incomplete block designs are considered.

14.6.2 PBIBD (Partially Balanced Incomplete Block Design)

It has been seen that a balanced design can not be constructed for any number of treatments (t) and any block size (k). The minimum number of replications is fixed by t and k. In most cases the number of replications required by a BIBD is too large.

In order to allow for more freedom for choice of replicates, designs which lack complete symmetry of the BIBD must be used. Yates (1936) suggested one type of designs which he called *lattice designs*, whereas Bose and Nair (1939) evolved another type of IBD's which they called *partially balanced incomplete block design* (PBIBD).

Lattice designs have some restrictions such as number of treatments t should be a perfect square in a *square lattice design*. Likewise if t is very large and is cube of number of units in block (k) i.e. $t = k^3$, the set of designs are called *cubic lattices*. Cubic lattice design is suitable for three factors.

These designs are a type of confounded symmetrical factorial designs, with a drastic reduction in block size and also number of replications. The reductions lead to another group of IBD's.

Statistical analysis of the PBIBD is more complicated than that of BIBD. When the variation among blocks is large, some pairs of treatments are more precisely compared than others, and several different standard errors may have to be computed for tests of significance.

Bose and Nair (1939) tried to develop designs (PBIBD) such that analysis for these should not be unduly difficult. For this the PBIBD should satisfy certain properties.

The simplest PBIBD are those with *two associates*. An *m-associate* PBIBD is defined by Bose and Nair (1939) as follows:

A set of t treatments distributed in b blocks of size k each is said to form a PBIBD with m-associate class if

1. every treatment occurs in r blocks
2. a pair of treatment occurs together in λ_1 or λ_2, or .... or λ_m blocks
3. given a treatment say θ, each of n_i treatments occur with θ in λ_i blocks ($i = 1, 2, \ldots, m$) so that

   $$\sum_{i=1}^{m} n_i = t - 1 \text{ and } \Sigma n_i \lambda_i = r(k - 1)$$

   Relative to θ, the remaining treatments are grouped into m groups such that the ith group contains n_i treatments, each of which occurs with θ in λ_i blocks. These n_i treatments are said to be the ith associates of θ.
4. Given two treatments θ and ϕ, which are i^{th} associates, let the number of treatments common between the jth associates of θ and k^{th} associates of ϕ be denoted by p^i_{jk}.
   Then the PBIBD further requires that p^i_{jk} is constant for various choices of θ and ϕ; as long as they are i^{th} associates.

The following figure gives an example of PBIBD which is a 3 × 3 triple lattice.
In this $t = 9$, $b = 9$, $k = 3$, $\lambda_1 = 1$ and $\lambda_2 = 0$, $r = 3$

Blocks		Rep I				Rep II				Rep III	
(1)	1	2	3	(4)	1	4	7	(7)	1	5	9
(2)	4	5	6	(5)	2	5	8	(8)	7	2	6
(3)	7	8	9	(6)	3	6	9	(9)	4	8	3

Fig. 14.5. PBIBD which is a 3 × 3 triple lattice (source: Cochran and Cox (1957), p. 453).

It can be noted that the following properties are satisfied by this PBIBD.

1. Every block contains 3 units. Every treatment is occurring 3 times and no treatment appears more than once in a block.
2. Consider treatment 1. The other treatments that appear in the same block are treatments 2 and 3 in block (1); 4 and 7 in block (4) and 5 and 9 in block (7).
   There are two other treatments 6 and 8 which do not appear in the same block as treatment 1. Similarly, if one starts with any other treatment, it will be found that six treatments are in the same block with it and two that are not.
   Pairs of treatments that occur together in the same block are called *first associates* and pairs that do not, are called *second associates.* This design is said to have *two associate classes* with $\lambda_1 = 1$ and $\lambda_2 = 0$.
3. Consider a pair of treatments that are first associates say 1 and 2. Table 14.14 below gives the relationships of the other treatments with these two. Treatments 3, 4, 5, 7 and 9 are first associates of 1 and 6 and 8 are second associates. For treatment 2; the treatments 3, 5, 6, 7 and 8 are first associates and 4 and 9 are second associates.

Table 14.14 Relationships of other treatments with treatments 1 and 2 (source: Cochran and Cox (1957), p. 454)

		Relation to 1	
		First associates	Second associates
	First associates	3, 5, 7	6, 8
Relation to 2	Second associates	4, 9	None

The numbers of other treatments which fall in the four cells of above table are

$$\begin{matrix} 3 & 2 \\ 2 & 0 \end{matrix} \qquad (14.92)$$

These four numbers will remain same, no matter which pair of 1st associates are considered. With 3 and 9 also the same set of 4 numbers will be obtained.
One can consider treatments 1 and 6 which are second associates. The table below gives the determination of the above four numbers.

Table 14.15 Relationships of other treatments with 1 and 6 which are second associates

		Relation to 1	
		First associates	Second associates
	First associates	2, 3, 4, 5, 7, 9	None
Relation to 6	Second associates	None	8

The four numbers as (14.92) are

$$\begin{matrix} 6 & 0 \\ 0 & 1 \end{matrix} \qquad (14.93)$$

Again with any other pairs which are second associates, same set of four numbers will be obtained.

It is thus observed for PBIBD of Fig. 14.5 that $p_{11}^1 = 3$, $p_{12}^1 = 2 = p_{21}^1$ and $p_{22}^1 = 0$ for *all the pairs* of treatments *which are first associates*. Similarly $p_{11}^2 = 6$, $p_{12}^2 = 0 = p_{21}^2$ and $p_{22}^2 = 1$ for *all the pairs* of treatments *which are second associates*.

This property makes the statistical analysis simpler. It guaranties that only two standard errors are required for testing difference in treatment pairs. In practice these two standard errors are also often very close so that a single error suffices.

The user of PBIBD should be aware of the difficulties that may arise when 'user made' designs are used without ascertaining relationship between factorial effects and block totals. Although the number of designs is too great, some of the plans are given by Cochran and Cox (1957). For construction of designs with $k = 2$, interested readers may consult Clatworthy (1955).

The method of analysis of PBIBD is described in the books by Cochran and Cox (1957) and Das and Giri (1979).

14.7 Multiple Comparisons

In the last sections ANOVA has been described for various designs and the purpose is to test differences in treatment means. A general test in ANOVA is of H_0: there is no difference in treatment means against H_1: some treatment means differ.

Once ANOVA has been performed the result is either (a) H_0 is not rejected or (b) H_0 is rejected.

If situation (a) is reached, based on the available data, any differences among the treatment means can not be detected. In this case the analysis of the data is complete.

However, if result (b) is obtained from ANOVA, the conclusion is that there are differences among the treatment means. In this case it is natural to continue the investigation to try to explore where the differences lie.

Many times the tests for exploring the differences in treatment means are chosen and designed after the results of the experiment. These tests are suggested by the data. They are not planned before the experiment is conducted. A planned comparison names the treatments in advance whereas an unplanned comparison tests the "treatments suggested by the data".

When so little is known about the nature of treatments that one hesitates to propose specific comparisons, then the techniques suitable for unplanned comparisons or testing of effects suggested by data should be taken up.

In this section, first consider the ANOVA of one-way classification to compare k treatment means for significant differences. Case of equal replications of treatments will be described in detail and some method for unequal replications will also be considered.

There are several statistical methods for detecting the differences among k means once the hypothesis of equality has been rejected. There is no agreement as to the best procedure which may be routinely employed.

Some of the commonly used methods along with their comparative merits and demerits are described in the following subsection.

14.7.1 Equal Number of Replications or Equal Sample Sizes

Let us consider a one-way ANOVA conducted on CRD with n observations on each of the k treatments.

In such a case, if ANOVA rejects equality of k treatment means i.e. $\mu_1 = \mu_2 = \ldots = \mu_k$, a test of equality of two treatments means i.e. H_0: $\mu_i = \mu_j$, where $i \neq j = 1, 2, \ldots, k$; is the next step.

Least Significant Difference

Two sample t-test is one option for the second step.

Thus a two sided test of equality of μ_i and μ_j will declare the difference between μ_i and μ_j as significant at α level if

$$\frac{\left|\overline{y_i} - \overline{y_j}\right|}{\sqrt{\frac{2}{n}\text{MSE}}} \geq t_{k(n-1),\alpha/2} \tag{14.94}$$

or

$$\left|\overline{y_i} - \overline{y_j}\right| \geq t_{k(n-1),\alpha/2}\left(\sqrt{\frac{2}{n}\text{MSE}}\right) \tag{14.95}$$

MSE is obtained from ANOVA table of one-way classification and has k $(n - 1)$ d.f. Hence the upper tail 100 $\alpha/2\%$ critical value of t-distribution (from Table A.4) with k $(n - 1)$ d.f. is to be used, as shown in above two equations. The quantity $t_{k(n-1),\ \alpha/2}\left(\sqrt{\frac{2}{n}\text{MSE}}\right)$ is the smallest difference between any two of the k means which will be significant at 100 $\alpha\%$. Hence this quantity is called the least significant difference or LSD. It is calculated once, after MSE has been obtained from ANOVA. This single calculation is sufficient to detect significance of difference between all the treatment means. As an illustration consider following example (source: Steel and Torrie (1981), p. 139).

Example 14.9: Consider the test of nitrogen (N) content in mg of red clover plants inoculated with five cultures of *Rhizobium trifolii* and a composite of five *Rhizobium meliloti* strains. The experiment used CRD with five replicates under each treatment gave the following results (Source: Steel and Torrie (1981), p. 139):

Treatments	3DOK1	3DOK5	3DOK4	3DOK7	3DOK13	Composite
Mean N content from 5 observations	28.8	24.0	14.6	19.9	19.9	18.7

Here $k = 6$ and $n = 5$. The ANOVA table is

Table 14.16 ANOVA table for 6 Rhizobium cultures

Source of variation	d.f.	SS	MS	F	$F_{5,\,24}$	
					5%	1%
Culture	5	847.05	160.41	14.37**	2.62	3.90
Error	24	282.93	11.79			
Total	29	1129.98				

ANOVA table indicates that the 6 cultures have highly significant differences in resulting mean nitrogen content.

When 6 treatments means are detected significantly different one may be interested to check which pairs differ significantly. There may be multitude of tests listed below which may be done

1. Which 5 (total ${}^6c_5$) means are equal and the sixth differs?
2. Which 4 (total ${}^6c_4$) means are equal and other two differ?
3. Which 3 (total ${}^6c_3$) means are equal and the other three differ?
4. Which 2 (total ${}^6c_2$) means are equal and other four differ?

With these tests there will be associated a type I error i.e. wrongly rejecting equality of

1. All 6 treatments
2. 5 treatments out of 6
3. 4 treatments out of 6
4. 3 treatments out of 6
5. 2 treatments out of 6

These type I errors may be denoted as

6-treatment type I error, *5-treatment type I error,*
4-treatment type I error, *3-treatment type I error, and*
2-treatment type I error.

Pearson and Hartley (1942, 1943) have shown that even in the comparison of largest mean with smallest mean in a set of v-treatment means (for $v > 2$) the type I error is larger than chosen 5% value. It depends on the values of k and n. Hence test using the LSD with $v > 2$ is not advisable.
All possible paired comparisons, i.e. comparisons of all possible pairs of means may be done according to following steps

1. Determine LSD = $t_{k(n-1),\,.025}\sqrt{\dfrac{2\text{MSE}}{n}}$

   i.e. LSD = $t_{k(n-1),\,.025}\sqrt{\dfrac{2(11.79)}{5}} = 2.064\sqrt{\dfrac{2(11.79)}{5}} = 4.5$ mg
2. Rank means from smallest to largest, or the reverse.
   In this example they are in ascending order. Number in ( ) is their rank.

3DOK13	3DOK4	Composite	3DOK7	3DOK5	3DOK1
13.3 (1)	14.6 (2)	18.7 (3)	19.9 (4)	24.0 (5)	28.8 (6)

3. As a space-saving presentation list the differences as in Table 14.17 and put an '*' to indicate it is "significant" and leave unmarked if the difference is "not significant".

Table 14.17 Differences between mean nitrogen content under 6 rhizobium cultures

Ranked means	(6)	(5)	(4)	(3)	(2)
(1)	15.5*	10.7*	6.6*	5.4*	1.3
(2)	14.2*	9.4*	5.3*	4.1	
(3)	10.1*	5.3*	1.2		
(4)	8.9*	4.1			
(5)	4.8*				

Scheffe's Method or S-Method

Let

$$\sum_{i=1}^{k} C_i \mu_i \text{ be a contrast of } \mu_i\text{'s i.e. } \sum_{i=1}^{k} C_i = 0 \tag{14.96}$$

This method allows to test any contrast of μ_i's (i = 1, 2, ..., k) satisfying (14.96). Define $S = kF_{\alpha,k,(n-k)}$, where $F_{\alpha,k,(n-k)}$ is upper 100 α% point of

$$F\text{-distribution with } k \text{ and } (n-k) \text{ d.f.} \tag{14.97}$$

The contrast $\sum_{i=1}^{k} C_i \mu_i$ is estimated by $\sum_{i=1}^{k} C_i \bar{y}_i$ and estimate of its variance is given by

$$\hat{V}\left(\sum_{i=1}^{k} C_i \bar{y}_i\right) = s^2 \sum_{i=1}^{k} C_i^2 \tag{14.98}$$

where s^2 is MSE from the ANOVA table with $(n - k)$ d.f.

According to Scheffe (1961), the contrast $\sum_{i=1}^{k} C_i \mu_i$ is significantly different from 0, if

$$\left|\sum_{i=1}^{k} C_i \bar{Y}_i\right| > S\sqrt{\hat{V}\left(\sum_{i=1}^{k} C_i \mu_i\right)} = Ss\sqrt{\sum_{i=1}^{k} C_i^2} \tag{14.99}$$

Whenever a hypothesis of equality of means is rejected by F-test in ANOVA, S-test may be applied for various contrasts of μ_i's. Scheffe suggests to apply S-method choosing α = .10 instead of .05.

Newman – Keul's Test

This test is also known as only Keul's (1952) test or Student-Newman-Keul's test or SNK test, as each of three contributed to this test. Student (1927) suggested that use of range should be made as a procedure to reject divergent samples in routine analysis. Newman (1939) made use of Students' ideas. The steps of SNK test are as follows:

Step 1: The observed treatment means are ranked in ascending order as

$$\bar{Y}_{(1)}, \bar{Y}_{(2)} \ldots \bar{Y}_{(k)}$$

Step 2: Compare the maximum and minimum means by calculating

$$q = \sqrt{n}\left(\bar{Y}_{(k)} - \bar{Y}_{(1)}\right) / \sqrt{MSE} \tag{14.100}$$

where MSE is obtained from ANOVA table.

If $q > q_{\alpha,\, k(n-1),\, k}$, the maximum difference or range is significant. $q_{\alpha,\, k(n-1),\, k}$ is 100 α% upper tail significant point of studentized range (given in Table A.8) with d.f. = k $(n - 1)$. So calculate

$$W_k = q_{\alpha, k(n-1), k} \sqrt{\frac{\text{MSE}}{n}} \tag{14.101}$$

If $\bar{y}_{(k)} - \bar{y}_{(1)} > W_k$ go to step 3, otherwise stop, since the set of means is found homogeneous.

Step 3: Calculate $W_{k-1} = q_{\alpha,\, k(n-1),\, (k-1)} \sqrt{\frac{\text{MSE}}{n}}$ and compare with it $\bar{Y}_{(k)} - \bar{Y}_{(2)}$ and $\bar{Y}_{(k-1)} - \bar{Y}_{(1)}$ i.e. range of $k - 1$ means

If the two ranges are significant continue testing for range of $(k - 2)$, $(k - 3)$, ..., 3, 2 means. Stop at any stage when difference (or range) is non-significant.

The SNK multiple range test compares the range of means in the subset at 100 α% level.

Example 14.10: Consider again the data of Example 14.9 and let the SNK – test be employed.
Solution: In this problem $k = 6$, $n = 5$ so d.f. of MSE = 24 and MSE = 11.79.

Step 1: The ranked treatment means are

13.3 (1)   14.6 (2)   18.7 (3)   19.9 (4)   24.0 (5)   28.8 (6)

Step 2: $\sqrt{\frac{\text{MSE}}{n}} = \sqrt{\frac{11.79}{5}} = 1.54$

$$W_6 = q_{.05,24,6} \sqrt{\frac{\text{MSE}}{n}} = q_{0.05,24,6}(1.54) = 4.37 \times 1.54 = 6.72$$

$\bar{Y}_{(6)} - \bar{Y}_{(1)} = 15.54^*$ is significant. So testing continues

Step 3: Using Table A.8, find

$$W_5 = q_{0.05,\ 24,5}\ (1.54) = 4.17 \times 1.54 = 6.42$$
$$W_4 = q_{0.05,\ 24,4}\ (1.54) = 3.90 \times 1.54 = 6.01$$
$$W_3 = q_{0.05,\ 24,3}\ (1.54) = 3.53 \times 1.54 = 5.44$$
$$W_2 = q_{0.05,\ 24,2}\ (1.54) = 2.92 \times 1.54 = 4.50 = \text{LSD}$$

(i) Again consider Table 14.17

W_5 is compared with

$\bar{Y}_{(6)} - \bar{Y}_{(2)} = 14.2^*$; $\bar{Y}_{(5)} - \bar{Y}_{(1)} = 10.7^*$ Since they are significant.

(ii) W_4 is compared with

$\bar{Y}_{(6)} - \bar{Y}_{(3)} = 10.1^*$; $\bar{Y}_{(5)} - \bar{Y}_{(2)} = 9.4^*$; $\bar{Y}_{(4)} - \bar{Y}_{(1)} = 6.6^*$ All three are significant.

(iii) So W_3 is compared with

$\bar{Y}_{(6)} - \bar{Y}_{(4)} = 8.9^*$; $\bar{Y}_{(5)} - \bar{Y}_{(3)} = 5.3$; $\bar{Y}_{(4)} - \bar{Y}_{(2)} = 5.3$; $\bar{Y}_{(3)} - \bar{Y}_{(1)} = 5.4$.

Only $\bar{Y}_{(6)}$ and $\bar{Y}_{(4)}$ differ significantly.

(iv) So W_2 is compared for $\bar{Y}_{(6)} - \bar{Y}_{(5)} = 4.8^*$

Hence overall it may be concluded that

(1)	(2)	(3)	(4)	(5)	(6)
13.3	14.6	18.7	19.9	24.0	28.8

The means which have lines below are not significantly different.

Tukey's W-test

Tukey (1951, 1953) proposed a multiple range test (MRT) similar in application to LSD test. He suggested several variations of the test. One of them is "honestly significant difference" (*hsd*) and sometimes, referred as the "wholly significant difference" test or *W*-test. It considers pair-wise comparison of treatment means and uses studentized range.

For k treatments $^kc_2 = k(k-1)/2$ pairs of treatment means can be compared.

For Tukey's test
$$hsd = q_{\alpha,k(n-1),k} \times \sqrt{\frac{\text{MSE}}{n}} = W \tag{14.102}$$

is computed.

If two means or groups of means differ by more than the *hsd*, they are said to differ. The procedure consists in

(i) comparing the largest with smallest mean. Then the largest against next smallest, and so on, until the largest is compared with the second largest.
(ii) then one compares second largest with smallest, second largest with next smallest, and so on.

For $k = 6$, first the means are ranked in ascending order

The sequence of comparisons will be

(6) with (1),	(6) with (2),	(6) with (3),	(6) with (4),	(6) with (5)
(5) with (1),	(5) with (2),	(5) with (3),	(5) with (4)	
(4) with (1),	(4) with (2),	(4) with (3),	(3) with (1),	(3) with (2)

and (2) with (1)

(ii) If no difference is found between two means, then it is concluded that no difference exists between any means enclosed by these two, and such differences are not tested.

For the *Rhizobium* data $k = 6$, $n = 5$, MSE = 11.79. The ranked means are

(1)	(2)	(3)	(4)	(5)	(6)
13.3	14.6	18.7	19.9	24.0	28.8

From Table A.8 $W = q_{0.05,24,6}\sqrt{\frac{11.79}{5}} = 4.37 \times 1.54 = 6.7$

Considering Table 14.17

(6) – (1) > W, (6) – (2) > W, (6) – (3) > W, (6) – (4) > W, but (6) – (5) < W

$\Rightarrow$ (6) and (5) may be considered equal

(5) – (1) > W, (5) – (2) > W but (5) – (3) < W. Hence (5), (4), (3) are homogeneous

(4) – (1) < W so (4), (3), (2), (1) are homogeneous.

This result can be presented by underscoring as follows.

(1) (2) (3) (4) (5) (6)

Duncan's New Multiple – Range Test

Duncan (1955) developed a *new multiple range test* (MRT) after his earlier MRT test given in 1951. The new MRT combines the simplicity of the SNK test and power advantages of MRT. The test is quite popular. The procedure is essentially same as SNK test except that the critical values are used from a different Table A.7.

Various steps of this test are illustrated in the following example.

Example 14.11: Considering again *Rhizobium* data $k = 6$, $n = 5$, MSE = 11.79

Step 1: Choose $\alpha = 0.05$ and find $\sqrt{\frac{\text{MSE}}{n}} = \sqrt{\frac{11.79}{5}} = 1.54$

Step 2: Calculate the *least significant ranges* (according to Duncan)

$$R_p = q_{\alpha, k(n-1), p} \sqrt{\frac{\text{MSE}}{n}}; p = 2, 3, \ldots, k \tag{14.103}$$

using Table A.7, as given below:

p	2	3	4	5	6
$q_{0.05, 24, p}$	2.92	3.07	3.15	3.22	3.28
R_p	4.5	4.7	4.9	5.0	5.1

Looking into Table 14.17, it can be seen that results of this test are same as that from LSD. This will not be so always with all the data sets.

Results can be reproduced using underscores

(1) (2) (3) (4) (5) (6)

Comparison of All the Treatment Means with a Control: Dunnett's Test

Objective of an experiment is sometimes to determine the treatments which are different or better than some standard or control. Dunnett (1985) has given a procedure for such testing. It differs from the multiple comparison approaches of Tukey's and S-N-K tests. The investigator is not interested in all possible comparisons of pairs of treatment means but only in $k - 1$ comparisons with 'control'. Dunnett's procedure uses a single value for testing differences of each treatment means and control. Tables for one sided and two sided critical values of $\alpha = 0.05$ and $\alpha = 0.01$ are available. For obtaining critical value, $k - 1$, α, d.f. of MSE from ANOVA, and whether one sided or two sided test is to be used, should be known. The significant difference given by

$$d = td_{\alpha,\ k(n-1),\ k-1}\sqrt{\frac{2\text{MSE}}{n}} \text{ may be calculated using Table A.9(A) or A.9(B)} \qquad (14.104)$$

where *td* is Dunnett's (1985) statistic for comparison of means.

Example 14.12: Consider again the *Rhizobium* data and suppose comparison is needed to be done with the composite culture, which gave mean 18.7 and was ranked (3).
In this case $k = 6$, $n = 5$, d.f. of MSE = 6 × 4 = 24, MSE = 11.79 and test should be two-tailed
From the Table A.9(B) $td_{.05,\ 24,\ 5} = 2.76$

So $d = 2.76\sqrt{\frac{2(11.79)}{5}} = 5.99$ mg

Looking at Table 14.17 the differences from treatment ranked (3) are compared with calculated $d = 5.99$. Only the treatment 3DOK1 ranked (6) is significantly different from the composite.

In the next subsection the modifications in above procedures for unequal replications of different treatments are considered.

14.7.2 Unequal Number of Replications or Unequal Sample Sizes

If the number of replications of treatments are not equal but i^{th} treatment has n_i replications ($i = 1, 2, \ldots, k$), there will be slight modification in the significant difference to be computed. It is because the $SE\left(\bar{Y}_i - \bar{Y}_j\right)$ will be different from that for equal replications. In this section for one way ANOVA for k treatments, it is assumed that i^{th} treatment is replicated n_i times ($i = 1, 2, \ldots, k$) and $\sum_{i=1}^{k} n_i = N$.

Hence the d.f. of MSE will be $(N - k)$.
The LSD under this condition will be given by

$$\text{LSD} = t_{\alpha/2,(N-k)}\sqrt{\text{MSE}\left(\frac{1}{n_i}+\frac{1}{n_j}\right)} \qquad (14.105)$$

for comparing i^{th} and j^{th} treatments.

SNK Test for Unequal Replications

The value W_k for comparison between i^{th} and j^{th} treatment will be

$$W_k = q_{\alpha,N-k,k}\sqrt{\frac{\text{MSE}}{2}\left(\frac{1}{n_i}+\frac{1}{n_j}\right)} \qquad (14.106)$$

Tukey's Test for Unequal Replications

Kramer (1956) and Tukey (1953) gave the modified *hsd* for comparison of ith and j^{th} treatments. It is given by

$$hsd = q_{\alpha,N-k,k}\sqrt{\frac{\text{MSE}}{2}\left(\frac{1}{n_i}+\frac{1}{n_j}\right)} \qquad (14.107)$$

Duncan's New Multiple Range Test for Unequal Replications

For comparing i^{th} and j^{th} treatments the value of R_p will be modified as

$$R_p = q_{\alpha, N-k, p}\sqrt{\frac{\text{MSE}}{2}\left(\frac{1}{n_i}+\frac{1}{n_j}\right)} \tag{14.108}$$

Dunnett's Test for Unequal Replications

For comparing i^{th} treatment with control which is replicated n_c times calculate

$$d = td_{\alpha, N-k, k-1}\sqrt{\text{MSE}\left(\frac{1}{n_i}+\frac{1}{n_c}\right)} \tag{14.109}$$

14.7.3 Multiple Comparison in Two Factor ANOVA

Suppose a two-factor ANOVA on n RBD reveals a significant difference among levels of fixed effect factor with more than two levels. One is then interested in determining between which levels of this factor the difference(s) occur(s). This may be done using any of the methods described in the subsection 14.7.1.

It is assumed that there are p levels of factor A and q levels of factor B, and there are n units or observations under each treatment combination $a_i\ b_j$ ($i = 1, 2, ..., p$; $j = 1, 2, ..., q$).

There are qn observations in each level of factor A and pn observations in each level of factor B. MSE with d.f. $(n - 1)(pq - 1)$ is used from the ANOVA table in various test procedures.

If there is only one observation in this experiment the MSE with $(p - 1)(q - 1)$ d.f. is used for multiple comparison.

If it is desired to compare a control with each of the other level means, Dunnett's test can be used. The MSE from this ANOVA replaces the MSE in equation (14.104) and $k(n - 1)$ is replaced by the d.f. of MSE of 2-way ANOVA.

If interaction AB is detected significant from ANOVA it is not meaningful to test differences between effects of various levels of either of the two factors.

It may be more appropriate to perform multiple comparison to explore significantly differing cell means. The cell means based on n observations in each cell may be compared using the above mentioned procedures. The MSE from ANOVA table with d.f. $(n - 1)(pq - 1)$ is used in various testing procedures.

Multiple Comparison in RBD

If the s treatments are tested using r blocks, the multiple comparison of the means of levels of fixed effect factor can again be done using any of the above mentioned procedures. Unless it is assumed that there is no interaction effect or it is found that interaction effect is non-significant, block means are not desirable to be compared.

If there are m observations in each of the rs cells, the MSE from ANOVA table is used with d.f. $= rs(m - 1)$.

If there is only one observation per cell i.e. $m = 1$, interaction can not be tested and MSE from ANOVA table of RBD is used with d.f. $= (r - 1)(s - 1)$.

15. Non-Parametric Methods I: One Sample Tests

The problems of *estimation and tests of hypotheses* discussed in this book so far, can not be justified unless there is knowledge of probability distribution of the random variable from which sample has been obtained. The inferential procedures described depend on the *assumptions about form of the distribution* and some times *also about the specific values of some parameters*. These techniques may be called *classical-statistical inference* techniques. As they mostly relate to either estimation or tests of the parameters of probability distribution, they are also called *parametric methods*.

However, in a real-world problem, every thing does not come packaged with labels of population of origin. There are alternative techniques available for such situations, which are discussed in this and subsequent two chapters. These have been classified as non-parametric and distribution-free methods.

The distribution-free methods, whether testing or estimation, are based on functions of the sample observations, whose distributions do not depend on the specific cdf or pdf of the population from which the sample was drawn. Thus the *assumptions regarding the underlying population are not necessary*.

On the other hand, the term *non-parametric test* implies test of a hypothesis which is not a statement about parameter values. The permissible statement then will depend on the accepted definition of the parameter. A statement about form of the distribution or some other characteristics of distribution may be tested by non-parametric method.

Though the *distribution-free methods and non-parametric methods* are not synonymous, the former relates to distribution of the test statistic and the latter to the type of hypothesis to be tested, but both the terms are often used interchangeably. In this chapter both types of procedures are considered under the heading non-parametric methods. The methods described in this are testing procedures when there are very general assumptions about the distribution and the test statistic has a distribution which can be determined without knowledge of specific density in population except mostly that it is continuous.

The non-parametric tests mostly relate to tests of location and that of spread. Such hypotheses are often analogous to parametric hypotheses.

There are some advantages and disadvantages of non-parametric methods as listed below:

Advantages

1. Can be used for testing of hypotheses which are not statements about population parameter values. For example χ^2-tests of independence already discussed in Section 10.2 and tests of goodness of fit to be described in the next section.
2. Can be used when the form of the sampled population is not known.

3. Procedures are computationally easier and can be applied quickly.
4. These methods can be applied when the data in hand consists of ranking or classification. In other words if the data is not measured or quantitative, non-parametric methods are applicable.

Disadvantages

1. For large data some non-parametric tests may be laborious.

In this way one can see that non-parametric methods have many advantages especially if the distribution of the variable in population is not known. In this chapter non-parametric test procedures are considered when only one sample is in hand. In the next two sections non-parametric tests of hypotheses have been discussed which are not statement about any parameter. Of the considered tests, two are test of goodness of fit of data to a specified distribution. In the last two sections distribution free tests are considered about the location parameter.

All these tests are performed using random sample $X_1, X_2, \ldots, X_n$ on a random variable X.

15.1 χ^2-Test of Goodness of Fit

Purpose of tests of goodness of fit is to test the null hypothesis that a given set of random observations is drawn from or 'fits' a specified probability distribution against the alternative that it does not 'fit'. In symbols it is desired to test

$$H_0: F(x) = F_0(x) \text{ against } H_1: F(x) \neq F_0(x) \tag{15.1}$$

where $F(x)$ is cdf of X and $F_0(x)$ is the specified cdf.

χ^2-test is applied when sample size n is large. The following steps are followed to test goodness of fit.

Step 1: Group the sampled data in k classes such that O_i is *observed frequency* of ith class and l_{i-1} and l_i are the class boundaries. The prepared frequency table will be as follows.

Table 15.1 Frequency table of sampled data

Class	Class boundaries	Observed frequencies	p_i	$E_i = np_i$
1	$l_0 - l_1$	O_1	p_1	E_1
2	$l_1 - l_2$	O_2	p_2	E_2
.	.	.	.	.
.	.	.	.	.
.	.	.	.	.
i	$l_{i-1} - l_i$	O_i	p_i	E_i
.	.	.	.	.
.	.	.	.	.
.	.	.	.	.
k	$l_{k-1} - l_k$	O_k	p_k	E_k
	Total	n	1.0	n

Step 2: The test statistic to test H_o is

$$\chi^2 = \sum_{i=1}^{k} \frac{(O_i - E_i)^2}{E_i} \tag{15.2}$$

where $E_i = n\, p_i$: the *expected frequency* of the ith class (i, = 1, 2, ..., k) if the sampled observations follow the specified distribution $F_o(x)$

(a) p_i = P [a sampled observation belongs to i^{th} class]

$$= \int_{l_{i-1}}^{l_i} f_0(x)\,dx = F_0(l_i) - F_0(l_{i-1}), \text{ if } X \text{ is continuous} \quad (15.3)$$

$f_0(x)$ is the pdf of the specified distribution of X if null hypothesis is true or in other words, the distribution which is desired to be seen, if it fits the data and i = 1, 2, ..., k.
The value of p_i can be found either analytically, if $F(x)$ is a closed form expression and parameters are known or from the area tables (as for normal distribution).

(b) If X is discrete $p_i = F_0(l_i) - F_0(l_{i-1})$, i = 1, 2, ..., k. (15.4)
$F(l_i)$ can be either read from the tables or can be calculated using recurrence formula for probabilities (for fitting binomial and Poisson distributions). An example below illustrates the process.

(c) If the parameters of $F_0(x)$ are not given, they are estimated from the observed sample. Thus for testing

(i) **$F_0(x)$ is Normal $N(\mu, \sigma^2)$**: μ is estimated by $\hat{\mu} = \overline{X}_n$, using formula for grouped data

$$\overline{X}_n = \frac{1}{n}\sum_{i=1}^{k} O_i X_i \quad (15.5)$$

where X_i is mid point or classmark of ith class. σ^2 is estimated by $\hat{\sigma}^2 = S^2$, the sample variance given by

$$S^2 = \frac{1}{n}\sum_{i=1}^{k} O_i\left(X_i - \overline{X}_n\right)^2 \quad (15.6)$$

(ii) **$F_0(x)$ is Binomial $B(n, p)$**, the parameter p is estimated from the property that mean $E(X) = np$. So p is estimated by the relation

$$n\hat{p} = \overline{X}_n \text{ as } \hat{p} = \frac{\overline{X}_n}{n} \quad (15.7)$$

$\overline{X}_n$ is calculated from the frequency table, when X is discrete.

(iii) **$F_0(x)$ is Poisson $P(\lambda)$**, the parameter λ is estimated from the property that $E(X) = \lambda$. So

$$\hat{\lambda} = \overline{X}_n \quad (15.8)$$

(iv) **$F_0(x)$ is exponential $E(\theta)$**, the parameter θ is estimated from the property that $E(X) = \theta$. So

$$\hat{\theta} = \overline{X}_n \quad (15.9)$$

In this case $$F_0(x) = 1 - e^{-x/\theta} \quad (15.10)$$

and θ is replaced by $\overline{X}_n$, the sample mean and p_i can be calculated using equation (15.4). Values of p_i and E_i are noted as in Table 15.1.

Step 3: If H_0 is true the test statistic

$$\chi^2 \sim \chi^2_{k-1-r-s} \tag{15.11}$$

where r = number of estimated parameters

s = number of classes pooled so that $E_i \geq 5$ for all i

Only adjacent classes are pooled, and these are either at the lower end or upper end or at both ends, depending on the values of E_i in the end classes.

Since almost all the distributions have low probabilities at the tails, the E_i's for these classes are also some times quite small. Hence pooling of adjacent classes at one or both ends are required.

The χ^2-test statistic given in equation (15.3) thus follows χ^2-distribution with d.f. = $k - 1 - r - s$. Table 15.1 shows the method of computation of χ^2-test statistic.

Step 4: The test procedure consists of

$$\text{Reject } H_0 \text{ if } \chi^2 \geq \chi^2_{k-1-r-s,\alpha} \tag{15.12}$$

So from the Table A.5 of χ^2-distribution 100 α% significant point of χ^2-distribution with $k - 1 - r - s$ d.f. is read and compared with the calculated value of χ^2-test statistic.

Step 5: If calculated value of statistic is less than the 100 α% significant point of $\chi^2_{k-1-r-s}$, H_0 is not rejected and it is concluded that the population from which sample is drawn may have the specified distribution $F_0(x)$.

Example 15.1: A new drug is being tested for possible use in relieving nausea associated with motion sickness. The variable studied is Y, the time in minutes needed to obtain relief. Test Y for normality, based on the following 40 observations (source: Miltons and Tsokos (1983), p. 387).

2.2	1.5	3.4	4.4	3.9
3.4	4.3	3.7	2.6	4.2
2.5	3.1	3.2	1.9	3.7
3.3	3.8	3.9	3.5	3.1
4.8	4.7	3.4	3.2	3.3
3.0	3.1	4.5	3.8	4.1
3.7	3.5	3.3	2.9	3.0
4.1	3.1	3.6	3.2	2.6

Solution: The hypothesis to be tested is H_0: $Y \sim N(\mu, \sigma^2)$ against H_1: $Y \nsim N(\mu, \sigma^2)$. In other words H_0: Y follows normal distribution against H_1: Y does not follow normal distribution.

Sample size n = 40. So χ^2-test is applicable and as described above, the following steps are followed.

Step 1: *Preparation of frequency table.* The method of preparing frequency table given in Chapter 2 is followed and number of classes k = 6 and class width h = .6 is chosen. Since the minimum of observed Y = 1.5 and maximum of observed Y = 4.8, the lowest lower class boundary is chosen as 1.45 so that highest upper class boundary is 1.45 + 6 × .6 = 5.05. The frequency table is as shown in Table 15.2.

Since normal distribution is to be fitted, see (*c*)(*i*) of step 2. The parameters μ and σ^2 are not known, so first these are estimated from the above prepared frequency Table 15.2.

The estimate $$\hat{\mu} = \bar{Y} = \frac{1}{40}\sum_{i=1}^{6} O_i Y_i = 3.385$$

Table 15.2 Frequency table of time to obtain relief

Classes	Class boundaries	O_i	Y_i	z_u	$\Phi(z_u)$	p_i	$E_i = 40 \times p_i$	$O_i - E_i$	$(O_i - E_i)^2$	$(O_i - E_i)^2/E_i$
1	1.45-2.05	2	1.75	−1.79	.0367	.0367	*1.47* |			
2	2.05-2.65	4	2.35	−.99	.1611	.1244	*4.98* |	−.45	.2025	.0314
3	2.65-3.25	10	2.95	−.18	.4286	.2675	10.70	−.70	.49	.0458
4	3.25-3.85	14	3.55	.63	.7324	.3038	12.15	1.85	3.4225	.2817
5	3.85-4.45	7	4.15	1.43	.9222	.1893	7.59	−.59	.3481	.0459
6	4.45-5.05	3	4.75	∞	1.0000	.0778	3.11	−.11	.0121	.0039
	Total	40				1.000	40			.4087

Note 1: z_u = standardized upper boundaries.
Note 2: Italic frequency classes are pooled together.

and
$$\hat{\sigma}^2 = S^2 = \frac{1}{40}\sum_{i=1}^{6} O_i Y_i^2 - \bar{Y}^2 = .5536$$

so
$$\hat{\sigma} = S = \sqrt{.5536} = .744$$

Step 2: The test statistic requires p_i to be obtained next. For the first class the lower boundary is taken as $-\infty$, since range of normal distribution is $(-\infty, \infty)$. For the last class the upper boundary is ∞. Hence

$$p_1 = P[-\infty < Y < 2.05] = P\left[-\infty < \frac{Y-3.385}{.744} < \frac{2.05-3.385}{.744}\right] = P[Z < -1.79]$$

$$= \Phi(-1.79) = .0367$$

$$p_2 = P[2.05 < Y < 2.65] = P\left[\frac{2.05-3.385}{.744} < \frac{Y-3.385}{.744} < \frac{2.65-3.385}{.744}\right]$$

$$= P[-1.79 < Z < -.99] = \Phi(-.99) - \Phi(-1.79) = .1611 - .0367 = .1244$$

Thus the standardized upper boundaries z_u are obtained for each class, which is ∞ for the 6th (or last) class, and are shown in Table 15.2.

$\Phi(z_u)$ are read from the normal area table (from Table A.3) and p_i's are obtained by differencing the two consecutive values and are entered in the Table 15.2. E_i's are obtained by multiplying p_i with 40 and shown in the table. $E_1 = 1.47$. So classes 1 and 2 are pooled to get $E_2^p = 6.45$ and $O_2^p = 6$.

Now two more columns are included in Table 15.2 for $O_i - E_i$, for $(O_i - E_i)^2$ and $(O_i - E_i)^2/E_i$. The test statistic now can be obtained as

$$\chi^2 = \sum_{i=1}^{5} (O_i - E_i)^2 / E_i = .4087$$

Step 3: $\chi^2 \sim \chi^2_{6-1-2-1}$ if H_0 is true.

Hence from Table A.5 for $\alpha = .05$, $\chi^2_{2,.05} = 5.99$
P value for calculated χ^2-statistic is between .75 to .90.

Step 4: H_0 is not rejected, and it can be concluded that the data has come from a normal distribution.

Example 15.2: Binomial Distribution: A study was planned to determine patient acceptance of a new pain reliever. 100 physicians each selected a sample of 25 patients to participate in the study. After trying the new pain reliever for a specified period of time, each patient was asked whether the new pain reliever was better than that used regularly in the past. The results of the study given below indicate the frequency of various number of patients out of 25 preferring new pain reliever (source: Daniel (1987), p. 539).

It is desired to test whether the data confirms the hypothesis that the data are drawn from a binomial distribution.

Table 15.3 Results of study on new pain reliever

i	Number of patients out of 25 preferring new pain reliever x	Number of doctors reporting these O_i	$O_i x_i$	p_i	$E_i = 100\ p_i$
1	0	5	0	.0038	.38
2	1	6	6	.0238	2.38
3	2	8	16	.0714	7.14
.	3	10	30	.1369	13.69
.	4	10	40	.1882	18.82
.	5	15	75	.1976	19.76
i	6	17	102	.1647	16.47
.	7	10	70	.1118	11.18
.	8	10	80	.0629	6.29
.	9	9	81	.0297	2.97
11	10 or more	0	0	.0092	.92
	Total	100	500	1.0000	

Solution: If X is the number of patients preferring new drug, it is desired to test if X follows binomial distribution. Each doctor has conducted test on 25 patients and recorded whether they preferred or not the new drug.
So H_0: $X \sim B(n, p)$ against H_1: $X \nsim B(n, p)$
where $n = 25$ and p is to be estimated from the data itself.

As indicated in Step 2 (c) (ii), estimate of $p = \hat{p} = \dfrac{\bar{X}}{n}$

From Table 15.3, to find $\bar{X}$ the formula used is

$$\bar{X} = \frac{1}{100} \sum_{allx} O_i x_i$$

To calculate it the product $O_i\, x_i$ is listed in 3rd column of Table 15.3.

So
$$\bar{X} = \frac{500}{100} = 5 \text{ and } \hat{p} = \frac{5}{25} = .20$$

Now
$$p_i = P[X = x_i] = \binom{25}{x_i} (.2)^{x_i} (.8)^{25-x_i} \text{ for } x_i = 0, 1, 2, ..., 25 \qquad (15.11)$$

Now instead of evaluating p_i for each x_i = 0, 1, 2, …, 9 from the formula (15.11) the recurrence relation is used.

For the last class with x_i = 10 or more, p_i is obtained by subtraction since $\sum_{x_i=0}^{25} p_i = 1$.

$p_i = P[X = 0] = (.8)^{25} = .0038$ and $p(x+1) = \left(\dfrac{n-x}{x+1}\right)\dfrac{p}{q}\, p(x)$, each subsequent 8 probabilities can be determined as

$$p_2 = P[X=1] = \frac{n-0}{0+1}\frac{.2}{.8}\, p(0) \qquad p_3 = P[X=2] = \frac{n-1}{1+1}\frac{.2}{.8}\, p(1)\cdots\cdots$$

$$p_{10} = P[X=9] = \frac{n-8}{8+1}\frac{.2}{.8}\, p(8) \text{ and } p_{11} = 1 - \sum_{i=1}^{10} p_i$$

The values thus computed are noted in the 4th column.

Values of E_i are obtained by multiplying p_i by total frequency (=100) and shown in the fifth column.

It is seen that expected frequencies in the 1st and 2nd classes and those in 10th and 11th (i.e. in the first two and last two) classes are less than 5. Pooling the first two (= 2.76) and the last two (= 3.89) the pooled expected frequencies become 2.76 and 3.89. As suggested by Cochran (1952, 1954) for unimodal distributions $E_i \geq 1$ can also be used. Hence no further pooling is done.

Hence r: number of estimated parameter = 1 and

s: number classes pooled = 2.

So the test statistic

$$\chi^2 = \sum_{i=1}^{9} (O_i - E_i)^2 / E_i \sim \chi^2_{9-1-2} \text{ if } H_o \text{ is true}$$

Table below shows calculation of χ^2-test statistic.

Classes including pooled ones	O_i	E_i	O_i^2/E_i
1 & 2	11	2.76	43.8406
3	8	7.14	8.9636
4	10	13.69	7.3046
5	10	18.82	5.3135
6	15	19.76	11.3866
7	17	16.47	17.5471
8	10	11.18	8.9445
9	10	6.29	15.8983
10 & 11	9	3.89	20.8226
Total	100	100	140.0214

The χ^2-test statistic given in equation (15.3) can also be written as

$$\chi^2 = \sum_{i=1}^{k} O_i^2 / E_i - n \tag{15.12}$$

where n is total frequency or total sample size (= 100 in this example). To use this formula in this example the 4th column contains values of O_i^2/E_i.

Hence $\chi^2 = 140.0214 - 100 = 40.0214$

From Table A.5 $\chi^2_{6,.05} = 12.5916$ and $\chi^2_{6,.01} = 16.8119$

Since calculated $\chi^2 > \chi^2_{6,\,.01}$, H_0 is rejected. So it is concluded that the data does not come from a binomial distribution.

Example 15.3 Poisson Distribution: A hospital administrator wishes to test the null hypothesis that emergency admissions follow a Poisson distribution with $\lambda = 3$. The number of emergency admissions were observed over a period of 90 days. The data obtained are grouped in a frequency table given below (source: Daniel (1987), p. 541):

Table 15.4 Frequency table of number of emergency admissions

Classes (i)	Number of emergency admissions in a day (x)	Number of days said admissions occurred (O_i)	p_i	E_i	O_i^2/E_i
1	0	5	.0498	4.4820	5.5779
2	1	14	.1494	13.4460	14.5768
3	2	15	.2241	20.690	11.1527
4	3	23	.2241	20.1690	26.2284
5	4	16	.1681	715.1290	16.9211
6	5	9	.1009	9.0810	8.9197
7	6	3	.0505	4.545	1.9802
8	7	3	.0216	1.9440	4.5135
9	8	1	.0081	.7290	
10	9	1 $O_i^p = 2$	.0027	.2430 (E_i^p =1.035)	3.8647
11	10 or more	0	.0007	.0630	
	Total	90	1.0000	90.0000	93.7380

Solution: Number of observations = 90 = n. X: number of emergency admissions in a day. It is desired to test if $X \sim P(3)$ i.e. X follows Poisson distribution with parameter $\lambda = 3$. Hence the null hypothesis is H_0: $X \sim P(3)$ against H_1: $X \nsim P(3)$. The alternative hypothesis is that X does not follow Poisson distribution with $\lambda = 3$.

The procedure to find E_i given in Step 2 (c) (iii) is applied. So that

$$p(x) = \frac{e^{-\lambda}\lambda^x}{x!}, \quad x = 0, 1, 2, \ldots\ldots \infty$$

are found for x = 0, 1, 2, ... 9 and 10 and more, using the recurrence relation

$$p(x+1) = \frac{\lambda}{x+1}\, p(x). \qquad x = 0, 1, 2, 3, \ldots\ldots$$

$p(0)$ is obtained first to start the recurrence relation.

$$p(0) = e^{-\lambda} = e^{-3} = .0498$$

$p(1) = \dfrac{\lambda}{1}\, p(0) = 3 \times .0498 = .1494$. Using the recurrence relation $p(x+1) = \dfrac{\lambda}{x+1}\, p(x)$, $p(2)$, $p(3)$ ..., $p(9)$ are computed and the last probability for $P[X \geq 10]$ is obtained by subtraction i.e.

$$P[P \geq 10] = 1.00 - \sum_{x=0}^{9} P[X = x] = 1.00 - (p(0) + p(1) + \ldots + p(9))$$

$$= .0007$$

$p(x)$'s are entered in 4th column of Table 15.4.

E_i, the expected frequencies in 5th column are obtained from the formula

$$E_i = np(i) = 90p(i) \text{ for } i = 1, 2, ..., 11$$

The test statistic to test H_0: $X \sim P(3)$ against H_1: $X \nsim P(3)$ (in words H_0: X follows Poisson distribution with parameter $\lambda = 3$ against H_1: X does not follow Poisson distribution with parameter $\lambda = 3$) is given by

$$\chi^2 = \sum_{i=1}^{11} \left(O_i^2/E_i\right) - n = \sum_{i=1}^{11} \left(O_i^2/E_i\right) - 90 = 3.738$$

It is found that E_i's in the last three classes are very small (<1).

Pooling the last three classes E_i^p in the 9th pooled class becomes 1.035 which is >1. Hence the test statistic has χ^2_{11-1-2} distribution if H_0 is true. Here λ is given so $r = 0$ and $s = 2$.

So looking at χ^2_8 upper tail significant points $\chi^2_{8,\ .05} = 15.507$ from Table A.5.

Since calculated χ^2 is less that $\chi^2_{8,\ .05}$, H_0 can not be rejected. Hence it can be concluded that the observed data on emergency admissions in hospital follows Poisson distribution with parameter $\lambda = 3$.

In a similar manner χ^2-test of goodness of fit can be used for any distribution to a large sample. Having a large sample is not always possible. Hence in the next section a test is given which can be used for small samples also.

15.2 Kolmogorov-Smirnov Test

As mentioned above let $X_1, X_2, ..., X_n$ be a random sample of size n, which may be small ($n < 30$) from a random variable with pdf $f(x, \theta)$ and cdf $F(x)$. It is desired to test if the random sample is drawn from a specified distribution with cdf $F_0(x)$. In symbols H_0: $F(x) = F_0(x)$ against H_1: $F(x) \neq F_0(x)$.

To test H_0 using Kolmogorov-Smirnov test (also called K-S test), one has to find empirical distribution for various values of x.

Empirical distribution function $F_n(x)$ is defined as:

$$F_n(x) = \frac{\text{Number of observations} \leq x}{\text{Total number of observations}} \tag{15.13}$$

$$= \frac{\text{Number of } X_i\text{'s} \leq x}{n} \quad \text{for all } x$$

The K-S test statistic to test H_0 against H_1 is defined as

$$D_n = \text{Sup } \{|F_n(x) - F_0(x)|\} \tag{15.14}$$

where Sup $\{A\}$ is the smallest value which is greater than or equal to all the members of the set $\{A\}$.

When the sample observations are given, one finds the range of observed values. Values of x are chosen in this range.

Now for the chosen values of x, the $F_n(x)$ is computed by counting the number of values in sample which are smaller than or equal to x.

For ease of computation one can arrange the observations in a table in ascending order. Chosen values of x are noted in the second column of the table. Now it is easy to count the number of observations $\leq x$.

Also find the $F_0(x)$, which is the value of cdf of specified distribution at x. If the parameters of this distribution are not known, they are estimated from the sample and are used to replace the unknown parameters to find $F_0(x)$. The differences of $F_n(x)$ and $F_0(x)$ are found and their signs are neglected.

The largest value of these differences is D_n
Choose level of significance = α
The *test-procedure* is to reject H_0 if

$$D_n \geq D_{n,\,\alpha} \tag{15.15}$$

where $D_{n,\,\alpha}$ is 100 α% significant point of *K-S* statistic given in Table A.10.

Example 15.4: Vertical locations of 15 moths on a 25 meter tree trunk were recorded as heights above the ground, *X*. It is desired to test if the observed heights may be regarded as coming from a distribution of *X* which is uniform between 0 to 25 meters (source: Zar (2004), p. 478).

The heights arranged in ascending order of magnitude are as 1.4, 2.6, 3.3, 4.2, 4.7, 5.6, 5.6, 6.4, 7.7, 9.3, 10.6, 11.5, 12.4, 18.6, 22.3.

Solution: From the observed data, $n = 15$. Let *X*: height of location of moth. It is desired to test H_0: *X* follows uniform distribution between 0 to 25 against H_1: *X* does not follow uniform distribution on the range 0 to 25.

In symbols H_0: $X \sim R(0, 25)$ H_1: $X \nsim R(0, 25)$.

If $X \sim R(0, 25)$ $f_0(x) = \dfrac{1}{25}$, $0 < x < 25$ and $F_0(x) = \dfrac{x}{25}$

Let *x* be chosen 3, 5, 8, 13, 15, 18, 23 and 25. Values of *x*, $F_n(x)$ $F_0(x)$ and $|F_n(x) - F_0(x)|$ are given below to determine the *K-S* test statistic given by

$$D_n = \text{Sup } \{|F_n(x) - F_0(x)|\}$$

It is seen that $D_n = .3467$

From the Table A.10 in Appendix, $D_{n,\,\alpha} = D_{15,\,.05} = .34$

x	$F_n(x)$	$F_0(x)$	$\|F_n(x) - F_0(x)\|$
3	2/15 = 0.1333	3/25 = 0.1200	0.0133
5	5/15 = 0.3333	5/25 = 0.2000	0.1333
8	9/15 = 0.6000	8/25 = 0.3200	0.2800
13	13/15 = 0.8667	13/25 = 0.5200	0.3467
15	13/15 = 0.8667	15/25 = 0.6000	0.2667
18	13/15 = 0.8667	18/25 = 0.7200	0.1467
23	15/15 = 1.00	23/25 = 0.9200	0.0800
25	15/15 = 1.00	25/25 = 1.0000	0.0000

Since $D_n > D_{n,\,\alpha}$, H_0 is rejected.
Hence it is concluded that the moth locations are not uniformly distributed from ground to 25 meter on tree trunk.

(A) Alternative *K-S* Test

Instead of using D_n given in (15.14) it is advised to use $D_n^* = \max\{D_n^+, D_n^-\}$ where

$$D_n^+ = \max_{1 \leq i \leq n} \left\{\frac{i}{n} - F(X_{(i)})\right\}$$

and

$$D_n^- = \max_{1 \leq i \leq n} \left[F(X_{(i)}) - \frac{i-1}{n}\right] \tag{15.16}$$

$$\text{Determine } D_n^* = \max\{D_n^+, D_n^-\}$$

$$H_0 \text{ is rejected if } D_n^* > D_{n,\alpha}^* \tag{15.17}$$

Table A.10 gives also significance points $D^*_{n,\alpha}$ of D^*_n and can be read for various n and α.

Both of these *K-S* tests require tables for significant points of *K-S* statistic. The following alternative does not require large tables.

(B) Second Alternative *K-S* Test

Owen (1962) and Stephens (1974) devised following approximations for the significant points which do not depend on the sample size. So large tables are not necessary to perform the test. 5% and 1% values are given for three situations regarding the form of the distribution and whether the parameters are known or unknown. The *K-S* statistic also differs for these cases as given below.

(I) The hypothesis to be tested is $H_0: X \sim N(\mu, \sigma^2)$ against $H_1: X \nsim N(\mu, \sigma^2)$ when both the parameters are unknown.
The values of μ and σ^2 to determine $F_0(x)$ are replaced by $\bar{X}$ and s^2, the sample mean and unbiased estimator of σ^2.

Thus using these estimates, the estimated $F_0(x) = \hat{F}_0(x)$ are obtained for the chosen values of x from the standard normal area table. Now D_n is computed as in (15.14) and the *modified K-S test statistic* is given by

$$T_{KS1} = \left(\sqrt{n} - 0.01 + \frac{0.85}{\sqrt{n}}\right) D_n \qquad (15.18)$$

and the *test procedure* is

$$\text{Reject } H_0 \text{ if } T_{KS1} > C_\alpha \qquad (15.19)$$

where $$C_{.05} = .895 \text{ and } C_{.01} = 1.035 \qquad (15.20)$$

(II) $H_0: X \sim E(\theta)$ against $H_1: X \nsim E(\theta)$ when θ is unknown.
In this case the $F_0(x)$ is estimated by using $\hat{\theta} = \bar{X}_n$ in place of θ in

$$F_0(x) = 1 - e^{-x/\theta}. \text{ Thus } \hat{F}(x) = 1 - e^{-x/\bar{X}_n} \qquad (15.21)$$

and the values of D_n are obtained as in (15.14) viz.

$$D_n = \text{Sup}\left\{\left| F_n(x) - \hat{F}_0(x) \right|\right\} \qquad (15.22)$$

The *modified K-S test-statistic* now is

$$T_{KS2} = \left(D_n - \frac{0.2}{n}\right)\left(\sqrt{n} + 0.26 + \frac{0.5}{\sqrt{n}}\right) \qquad (15.23)$$

and the *test procedure* is

$$\text{Reject } H_0 \text{ if } T_{KS2} \geq C'_\alpha \qquad (15.24)$$

where $$C'_\alpha = 1.096 \text{ for } \alpha = .05 \text{ and } C'_\alpha = 1.308 \text{ for } \alpha = .01 \qquad (15.25)$$

(III) If all the parameters are known to test $H_0: X \sim F(x)$ against $H_1: X \nsim F(x)$, the value of D_n is obtained as in (15.14). The *modified K-S test statistic* is given by

$$T_{KS3} = \left(\sqrt{n} + 0.12 + \frac{0.11}{\sqrt{n}}\right) D_n \tag{15.26}$$

and the *test procedure is*

$$\text{Reject } H_0 \text{ if } T_{KS3} \geq C''_\alpha \tag{15.27}$$

where $C''_\alpha = 1.358$ for $\alpha = .05$ and $C''_\alpha = 1.628$ for $\alpha = .01$ (15.28)

These values of significant points may again be presented in a table for a ready reference.

Table 15.5 Significant points for modified *K-S* tests (source: Law and Kelton (1991))

	Testing goodness of fitting		$\alpha = .05$	$\alpha = .01$
I	$N(\mu, \sigma^2)$, μ, σ^2 both unknown	C_α	.895	1.035
II	$E(\theta)$, θ unknown	C'_α	1.096	1.308
III	General $F(x)$, when all parameters are known	C''_α	1.358	1.628

15.3 Sign Test

It is a distribution free test. It is desired to test if the sample has come from a specified distribution with any form. It is assumed that the distribution is continuous. For such a distribution median is a natural measure of location and has a desirable property that, whatever be the form of the distribution, the probability that a sample value exceeds population median = ½.

Thus on the basis of an observed sample one may devise a test whether the sample observations are from a population with specified median ξ. For this test suppose a random sample of size n: X_1, X_2, ..., X_n is available and a test of

$$H_0: \xi = \xi_0 \text{ against } H_1: \xi > \xi_0 \tag{15.28}$$

is desired. In other words, it is desired to test if the sampled data has come from a population with some specified median or with a higher median.

If H_0 is true $P[X_i - \xi_0] > 0 = ½$, $i = 1, 2, ..., n$. (15.29)

Let
$$Z_i = \begin{cases} 1 & \text{if } X_i - \xi_0 > 0 \\ 0 & \text{otherwise} \end{cases} \tag{15.30}$$

Since Z_i's are independent with only two values 0, 1 with constant probability $p = ½$

$$U = \sum_{i=1}^{n} Z_i \sim B(n, 1/2) \quad \text{if } H_0 \text{ is true} \tag{15.31}$$

$E(U) = n/2$ if H_0 is true. If H_1 holds $E(U)$ will be larger than $E(U)$ under H_0 (= $n/2$).

Hence right tail of the distribution of U is critical or rejection region.

So for small n, rejection region should be determined from binomial table so that upper tail cumulated probability is just not less than chosen value of α.

However, for $p = ½$, even for small n binomial distribution is close to normal. So for using a normal approximation, a small sample correction factor (= ±1/2) is employed.

If left tail is used for critical region (i.e. H_1: $\xi < \xi_0$) correction factor (*CF*) is + ½ . If right tail is to be used, CF = – ½.

Thus to test H_0 against H_1 given in (15.28), the test statistic used is

$$Z = \frac{(U - 1/2) - n/2}{\sqrt{n/4}} \tag{15.32}$$

If H_0 is true $Z \sim N(0, 1)$, so for $\alpha = .05$, the test procedure is

$$\text{Reject } H_0 \text{ if } Z > 1.645 \tag{15.33}$$

An easy way of counting when $Z_i = 1$, is to record only the signs of the differences $X_i - \xi_0$. Thus it is not necessary to find Z_i's. Thus one will get a sequence of + and – signs. The total number of + signs gives the value of U. The following example explains an application of sign test.

Example 15.5: A study of growth based on data gathered from 1971 to 1974 indicates that the median height of men in the 18 to 24 years old group was 69.7 inches. It is felt that as a result of improvement in health care and nutrition; the median height of this age group has increased (source: Milton and Tsokos (1983), p. 415).

Does the following data corroborate this?

Height of 20 men of 18 to 24 years in 1974

70.2	65.8	78.0	74.4
71.4	67.3	72.6	67.5
69.8	71.4	69.9	70.5
67.2	72.1	73.5	70.9
70.1	73.2	65.2	76.0

Solution: It is desired to test

$$H_0: \xi = 69.7 = \xi_0 \text{ against } H_1: \xi > 69.7$$

on the basis of 20 heights. So $n = 20$. The signs of $X_i - \xi = X_i - 69.7$ are as follows

+	–	+	+
+	–	+	–
+	+	+	+
–	+	+	+
+	+	–	+

So the number of positive signs $U = 15$ and the test statistic for right tail critical region is

$$Z = \frac{(U - 1/2) - 20/2}{\sqrt{20/4}} = \frac{15 - 1/2 - 10}{\sqrt{5}} = \frac{4.5}{\sqrt{5}} = 2.018$$

The test procedure is to reject H_0 at level of significance $\alpha = .05$, if

$$Z > 1.645$$

Since calculated $Z = 2.018 > 1.645$, H_0 is rejected.

Hence the data justifies the contention that height of men of age group 18 to 24 years has increased in 1974 from that in 1971.

15.4 Wilcoxon Signed Rank Test

Another test of location of a population from which a random sample $X_1, X_2, ..., X_n$ is available is considered in this section. The test was developed by F. Wilcoxon in 1945. This test is also used for a continuous random variable X to test the null hypothesis that ξ, the median of X has some specified value M_0. In symbols H_0: $\xi = M_0$.

(a) against H_1: $\xi \neq M_0$ or
(b) against H_1: $\xi > M_0$ or
(c) against H_1: $\xi < M_0$ (15.34)

As the name indicates the test procedure involves ranking and also attaches a sign to the ranks. The steps are as follows.

Step 1: The ranks R_i are given to the n absolute differences $| X_i - M_0 |$, $i = 1, 2, ..., n$.
Step 2: Subsequently signs are also attached to the ranks. A positive sign if the difference is positive i.e. $X_i > M_0$ and a negative sign otherwise.

If there is a tie in the difference, average rank is assigned to all the tied ranks.

Step 3: Now find out sum of the positive and negative ranks, given by

$$W_+ = \sum_{\text{positive ranks}} R_i \quad \text{and} \quad W_- = \sum_{\text{negative ranks}} R_i \tag{15.35}$$

Step 4: Determine $W = \min \{W_+, W_-\}$ (15.36)

The test procedure. If H_0 is true W_+ and W_- should be almost equal. Hence a two sided test is used for the alternative (a). For various values of level of significance α and sample size n, the one sided and two sided critical values are given in Table A.11.

(a) Hence to test H_0: $\xi = M_0$ against H_1: $\xi \neq M_0$

$$\text{Reject } H_0 \text{ if } W \leq W_{\alpha/2,\, n} \tag{15.37}$$

(b) If H_1: $\xi > M_0$, W_+ will be large and W_- will be small. So $W = W_-$ and so to test H_0: $\xi = M_0$ against H_1: $\xi > M_0$

$$\text{Reject } H_0 \text{ if } W \leq W_{\alpha,\, n} \tag{15.38}$$

(c) If H_1: $\xi < M_0$, W_+ will be small and W_- will be large. Hence $W = W_+$ and to test H_0: $\xi = M_0$ against H_1: $\xi < M_0$.

$$\text{Reject } H_0 \text{ if } W \leq W_{\alpha,\, n} \tag{15.39}$$

An example given below illustrates this test procedure.

Example 15.6: Suppose the median survival times of patients with acute mylogenous leukemia after achieving complete remission from conventional treatment is 21 months. The survival times of 10 patients who received the new treatment are as follows (source: Milton and Tsokos (1983), p. 417):

24.1	25.8	20.5	20.9	27.3
21.5	20.1	28.9	19.2	26.3

Do the data indicate that median survival time of patients treated with new drug has increased?

Solution: Let X: survival time of patients treated with new drug, and ξ be the median of X. Given that $n = 10$, it is desired to test H_0: $\xi = 21$ against H_1: $\xi > 21$

The procedure is shown in the following table.

i	1	2	3	4	5	6	7	8	9	10
X_i	24.1	25.8	20.5	20.9	27.3	21.5	20.1	28.9	19.2	26.3
$X_i - 21$	3.1	4.8	–.5	–.1	6.3	.5	–.9	7.9	–1.8	5.3
$\|X_i - 21\|$	3.1	4.8	.5	.1	6.3	.5	.9	7.9	1.8	5.3
R_i	6	7	1.5	1	9	1.5	4	10	5	8
Signed rank	6	7	–1.5	–1	9	1.5	–4	10	–5	8

As is evident, the difference between the observations and the specified median, have been obtained and noted. The absolute values of the differences are noted in the 3rd row of the above table. The ranks R_i of the absolute differences are given in 4th row and the positive sign is given to the values of R_i if $(X_i - 21)$ is +ve (or if $X_i - 21 \geq 0$) and a negative sign otherwise. The signed ranks are shown in the 5th row. Now obtain

$$W_+ = \text{Sum of positive (+ve) ranks} = 42.5 \text{ and}$$
$$W_- = \text{Sum of negative (–ve) ranks} = 11.5$$

Now the Wilcoxon signed rank test statistic to test H_0 against H_1 is

$$W = \min \{W_+, W_-\} = \min \{42.5, 11.5\} = 11.5.$$

The test procedure is to reject H_0 if (for $\alpha = .05$) $W \leq W_{\alpha, n} = W_{.05,10}$
From Table A.11, the 5% critical value of test statistic is $W_{.05,\ 10} = 11$
Since calculated $W = 11.5 > 11 = W_{.05,10}$ H_0 can not be rejected.

So it can not be concluded that the new treatment has increased the survival time of the patients of myelogenous leukemia.

16. Non-Parametric Methods II: Two Sample Tests

Let $f_1(x)$ and $f_2(y)$ be two continuous pdf's and there are available random samples $X_1, X_2, ..., X_{n1}$ and $Y_1, Y_2, \ldots, Y_{n2}$ on X and Y, respectively. On the basis of these samples it is desired to test the hypothesis that the distributions of X and Y are same. In symbols it is desired to test H_0: $f_1(x) = f_2(x)$, for all x. In this chapter several two sample non-parametric tests are discussed.

16.1 Sign Test for Two Samples

It has been already mentioned that median being a very convenient measure of location of a distribution having a convenient property, this test also compares medians of the two random variables X and Y. Let ξ_X and ξ_Y be the medians of X and Y, respectively. If the medians ξ_X and ξ_Y are equal, the differences of paired observations on X and Y have median equal to zero.

Thus this test requires paired observations on the random variables X and Y. Let there be n pairs of sampled observations $(X_1, Y_1), (X_2, Y_2), ..., (X_n, Y_n)$ on the random variable (X, Y). Hence the hypothesis required to be tested may be written as

$$H_0: \xi_{X-Y} = 0 \text{ against (a) } H_1: \xi_{X-Y} < 0 \text{ or (b) } H_1: \xi_{X-Y} > 0 \tag{16.1}$$

where ξ_{X-Y} is median of $X - Y$.

Now this test procedure becomes identical to the procedure given in Section 15.3 of Chapter 15 if Z_i is defined as

$$Z_i = \begin{cases} 1 & \text{if } X_i - Y_i \geq 0 \\ 0 & \text{if } X_i - Y_i < 0 \end{cases} \tag{16.2}$$

When H_0 is true, probability is ½ that an X value exceeds the Y value and also that a Y value exceeds X value. Hence again the sum $U = \sum_{i=1}^{n} Z_i$ has binomial distribution with n independent trials with probability p = ½. In symbols

$$U \sim B(n, p) \text{ if } H_0 \text{ is true} \tag{16.3}$$

The test statistic and critical regions are same as that given in Section 15.3.

(a) If H_1: $\xi_{X-Y} < 0$, the test statistic is

$$Z = \frac{\left(U + \frac{1}{2}\right) - n/2}{\sqrt{n/4}} \sim N(0, 1) \text{ if } H_0 \text{ is true} \tag{16.4}$$

and H_0 is rejected at 5% level if $Z < -1.645$. (16.5)

(b) If H_1: $\xi_{X-Y} > 0$, the test statistic to be used is

$$Z = \frac{\left(U - \frac{1}{2}\right) - n/2}{\sqrt{n/4}} \sim N(0, 1) \text{ if } H_0 \text{ is true} \tag{16.6}$$

and H_0 is rejected at $\alpha = .05$ if $Z > 1.645$. (16.7)

Example 16.1: Twelve pairs of patients of a dental clinic matching in age, sex, intelligence and initial oral hygiene scores were selected. One member of each pair received instructions on oral hygiene matters. Six months later all the 24 subjects were examined by a dentist unaware of which subjects received instructions, assigned scores on oral hygiene to them. Low score indicates better hygienic condition. Data on oral hygiene scores of subjects who received instructions (X) and of subjects who did not receive instructions (Y_i) are as follows (source: Daniel (1987), p. 590):

Pair number (i)	1	2	3	4	5	6	7	8	9	10	11	12
X_i	1.5	2.0	3.5	3.0	3.5	2.5	2.0	1.5	1.5	2.0	3.0	2.0
Y_i	2.0	2.0	4.0	2.5	4.0	3.0	3.5	3.0	2.5	2.5	2.5	2.5

It is desired to test if the imparted instructions have improved oral hygienic conditions.

Solution: There are 12 pairs of observations, so $n = 12$.

Subjects with higher oral hygienic condition will get lower score. Hence $X - Y$ should be mostly negative or that the median of $X - Y$ $\xi_{X-Y} < 0$. Thus it is required to test H_0: $\xi_{X-Y} = 0$ against H_1: $\xi_{X-Y} < 0$.

The signs of the differences $X_i - Y_i$ are therefore, found as

– 0 – + – – – – – – + –

and $U = 2$, the number of positive signs out of 11 signs.

Since H_1 is left tailed the critical region is also left tailed. So the test statistic to test H_0 against H_1 is

$$Z = \frac{\left(U + \frac{1}{2}\right) - n/2}{\sqrt{n/4}} \sim N(0, 1) \quad \text{if } H_0 \text{ is true}$$

and H_0 is rejected if $Z < -1.645$.

Since there is one tie 2nd observation (pair) is discarded and $n = 11$ for calculating Z. Thus

$$Z = \frac{\left(2 + \frac{1}{2}\right) - 11/2}{\sqrt{11/4}} = \frac{2.5 - 5.5}{1.658} = -1.809$$

Since calculated $Z < -1.645$, H_0 is rejected at $\alpha = .05$ significance level.

Hence it can be concluded that the imparted instructions have improved the oral hygiene.

A Remark: It may be noted that if there are ties in any pair, either such pairs may be discarded or half of such ties are given + sign and half are assigned – sign.

16.2 Median Test

If there are two independent samples from two continuous distributions, Mood (1950) suggested this test for the hypothesis that both the distributions have *same median*. The procedure is simple if the alternative hypothesis is that the *medians* of the two distributions *are unequal*. In symbols, the median test is used to test

H_0: Medians of the two distributions are equal against

H_1: Medians of the two distributions are not equal

Let $X_1, X_2, ..., X_{n1}$ and $Y_1, Y_2, ..., Y_{n2}$ be two independent samples from random variables X and Y, respectively, which have medians ξ_X and ξ_Y. In symbols, it is desired to test

$$H_0: \xi_X = \xi_Y \text{ against } H_1: \xi_X \neq \xi_Y \tag{16.8}$$

The steps for performing the median test for testing H_0 against H_1 in (16.8) are as follows:

Step 1: Combine the two samples and arrange the $n_1 + n_2$ observations in ascending order of their magnitude.

Step 2: Give ranks to the ordered observations. Rank 1 to the smallest and $n_1 + n_2$ to the highest. If there are ties, average rank is given to each observation in tie. e.g. let 12, 15, 15, 15, 17, 19 are the ordered 6 observations. Ranks given to these are 1, 3, 3, 3, 5, 6. Thus one can see that the rank given to 15 is $= \frac{2+3+4}{3} = 3$ to each of the three tied values 15.

Step 3: Find median of the combined sample = M (say).

Step 4: For each of the two samples find the number of observations greater than the combined median M.

Thus let $\quad r_1$: number of X's > M

and $\quad r_2$: number of Y's > M (16.9)

Let a be total number of observations < M, then

$$a = \begin{cases} \dfrac{n_1 + n_2 + 1}{2} & \text{if } n_1 + n_2 \text{ is odd} \\ \dfrac{n_1 + n_2}{2} & \text{if } n_1 + n_2 \text{ is even} \end{cases} \tag{16.10}$$

If samples come from distributions with different medians, i.e. ξ_X and ξ_Y are different, large number of first sample observations X_i's will lie on one side of combined median M.

If the two samples come from distributions with same median (say ξ) about 50% observations of each sample will lie on either side of M.

Step 5: If n_1 and $n_2 > 10$, prepare a 2 × 2 contingency table as given below.

Step 6: Apply χ^2-test of independence on the above 2 × 2 contingency Table 16.1.

Table 16.1 2 × 2 contingency table for median test

	Sample 1	Sample 2	Total
Number of observations greater than M	r_1	r_2	$n_1 + n_2 - a$
Number of observations less than M	$n_1 - r_1$	$n_2 - r_2$	a
Total	n_1	n_2	$n_1 + n_2$

As already discussed in Section 10.2 of Chapter 10, the χ^2-test statistic is

$$\chi^2 = \left[\sum_{i=1}^{2}\sum_{j=1}^{2} O_{ij}^2 / E_{ij} - (n_1 + n_2)\right] \sim \chi_1^2, \quad \text{if } H_0 \text{ is true} \tag{16.11}$$

where E_{ij} = expected frequency if independence exists.

$$= (n_{i.} \times n_{.j})/(n_1 + n_2) \text{ for } i, j = 1, 2 \tag{16.12}$$

As evident from (16.11), the χ^2-test statistic has only 1 d.f., only one E_{ij} needs to be computed and the rest three of the four E_{ij}'s can be obtained by subtraction.

So one can obtain only
$$E_{21} = \frac{n_1 \times a}{n_1 + n_2} \tag{16.13}$$

and other three are obtained by subtraction.

Now the test-statistic can be calculated.

Step 7: The test procedure for testing H_0 against H_1 of (16.8) at level of significance 100 α % is:

$$\text{Reject } H_0 \text{ if } \chi^2_{cal} > \chi^2_{1,\alpha} \tag{16.14}$$

Step 8: However, if n_1 and n_2 are small χ^2-test for 2 × 2 contingency table can not be applied. In this case, the following probability is calculated, using hyper geometric distribution. It is the probability of obtaining r_1 and r_2 as given in contingency table

$$P[r_1, r_2] = \binom{n_1}{r_1}\binom{n_2}{r_2} \bigg/ \binom{n_1 + n_2}{a} \tag{16.15}$$

$$= \frac{n_1!\,n_2!\,a!\,(n_1 + n_2 - a)!}{r_1!\,r_2!\,(n_1 - r_1)!\,(n_2 - r_2)!\,(n_1 + n_2)!} \tag{16.16}$$

If the value of $P[r_1, r_2] < .05$, H_0 is rejected. (16.17)

Thus it is evident that even if the two samples are not related (as required in Sign test), but are independent, median test can be used for testing equality of medians of two populations from which the two samples have been drawn.

Example 16.2: Members of a random sample of 12 male students from a rural junior high school and an independent random sample of students from an urban junior high school were given a test to measure their level of mental health. The data are given in the following table (source: Daniel (1987), p. 595).

Mental Health Scores of Junior High School Boys

Rural	1	2	3	4	5	6	7	8	9	10	11	12
Scores (X)	29	50	43	22	42	47	42	32	50	37	34	31
Urban	1	2	3	4	5	6	7	8	9	10	11	12
Scores (Y)	35	26	27	21	27	38	23	25	25	27	45	46
Urban	13	14	15	16								
Scores (Y)	33	26	46	41								

It is desired to test if rural and urban male junior high school students differ in their mental health level.

Solution: From the data $n_1 = 12$ and $n_2 = 16$ and let X be scores on mental health of rural students and Y that of urban students of junior high school, with medians ξ_X and ξ_Y, respectively.

The problem requires a test of H_0: $\xi_X = \xi_Y$ against H_1: $\xi_X \neq \xi_Y$.

Since n_1 and n_2 are > 10, the 2 × 2 contingency table for testing H_0 against H_1 is constructed. Various steps given above are followed as shown below:

The two samples merged together to get the combined sample of size ($n_1 + n_2 = 16 + 12 =$) 28 is ordered according to increasing values. The ordered samples will be as given below in Table 16.2.

Table 16.2 Ordered combined sample of 28 observations

i	Sample 1 X (rural)	Sample 2 Y (urban)	Ranks	*i*	Sample 1 X (rural)	Sample 2 Y (urban)	Ranks
1		21	1 (Y)	15	34		15 (X)
2	22		2 (X)	16		35	16 (Y)
3		23	3 (Y)	17	37		17 (X)
4		25	4.5 (Y)	18		38	18 (Y)
5		25	4.5 (Y)	19		41	19 (Y)
6		26	6.5 (Y)	20	42		20.5 (X)
7		26	6.5 (Y)	21	42		20.5 (X)
8		27	9 (Y)	22	43		22 (X)
9		27	9 (Y)	23		45	23 (Y)
10		27	9 (Y)	24		46	24.5 (Y)
11	29		11 (X)	25		46	24.5 (Y)
12	31		12 (X)	26	47		26 (X)
13	32		13 (X)	27	50		27.5 (X)
14		33	14 (Y)	28	50		27.5 (X)

Since $n_1 + n_2 = 28$ is even number, so the median of combined sample M is given by

$$M = (14^{\text{th}} \text{ value} + 15^{\text{th}} \text{ value})/2 = (33 + 34)/2 = 33.5$$

Now looking at the ordered combined sample observations

Number of X's > M (= 33.5) = 8

Number of Y's > M (= 33.5) = 6

Now the 2 × 2 contingency table can be formed, since $n_1, n_2 > 10$.

Contingency Table for Median Test

	Rural	Urban	Total
Number of scores > M (= 33.5)	8 (6)	6 (8)	14
Number of scores < M (= 33.5)	4 (6)	10 (8)	14
Total	12	16	28

$$E_{21} = \frac{n_1 \times a}{n_1 + n_2} = \frac{12 \times 14}{28} = 6 \text{ so } E_{11} = n_1 - E_{21} = 6$$

E_{12} and E_{22} are also = 8 by subtraction.

Now the χ^2-test statistic given by

$$\chi^2 = \left[\sum_{i=1}^{2}\sum_{j=1}^{2} O_{ij}^2/E_{ij} - (n_1 + n_2)\right] \sim \chi_1^2, \text{ if } H_0 \text{ is true}\cdot$$

$$\chi^2 = 8^2/6 + 6^2/8 + 4^2/6 + 10^2/8 - (12 + 16)$$

$$= 30.3334 - 28 = 2.3334$$

5% upper tail significant point of χ^2-distribution with 1 d.f. = $\chi^2_{1,\ .05}$ = 3.841 from Table A.5.

Since $\chi^2_{\text{cal}} = 2.334 < X^2_{1,\ .05} = 3.841$, $P > .05$. So H_0 *is not rejected.*

Hence it is *concluded* that the two samples might be coming from populations with same median. In other words the mental health of junior high school male students from rural and urban area may have same median scores.

16.3 Wald-Wolfowitz Runs Test

As in the last section let there be two independent random samples $X_1, X_2, ..., X_{n_1}$ and $Y_1, Y_2, ..., Y_{n_2}$ on two continuous random variables X and Y, respectively, with pdf's $f_1(x)$ and $f_2(y)$. It is again required to test that X and Y have the same distribution, i.e.

$$H_0: f_1(x) = f_2(x) \text{ for all } x. \tag{16.18}$$

Wald and Wolfowitz (1940) suggested a test which is based on runs of the X and Y observations in the combined and ordered sample of $n_1 + n_2$ observed values.

Definition of Run

A run is a sequence of identical letters preceded and followed by a different letter or no letter.

In order to perform the Wald-Wolfowitz run test, the two samples are combined and the values are arranged in ascending order of their magnitudes as in the median test. A sequence of letters X and Y is prepared by writing X or Y against the combined sample values, keeping track of which observations are from X sample and which are from Y sample.

Theoretically, there can not be a tie, distributions of X and Y being continuous. But in practice there are found ties within samples and some times also across the sample. Namely a value in X sample may be equal to one or more values of Y sample or even of X sample.

If there are ties within sample, there will be no effect on the number of runs, since the ties will bear the same letter X or Y. However, if there are ties across the sample in one tie also, there will be more than one run since both X and Y will appear against tied values across the sample. There is a different and more complex way to handle such case.

The simple case will be discussed here. Let $n_1 = 4$ and $n_2 = 5$, the ordered combined sample may result in the letter combinations as: $X\ Y\ Y\ X\ X\ Y\ X\ Y\ Y$.

This indicates that in the combined ordered sample the smallest observation is an X, the second smallest is Y ...., and the largest observation is a Y value.

If X and Y populations are homogeneous, it is expected that the combined sample of $n_1 + n_2$ observations will be a random sample of size $n_1 + n_2$ from the common population. The X and Y values will then be well mixed and so a large number of runs will be expected. Thus the total number of runs in the combined ordered sample indicates the degree of mixing of X and Y populations.

Too few runs will come up from the two samples from two distinct populations. For example if X's are stochastically smaller than Y's, all the X's will come before all the Y's in the combined samples. Thus there will be only two runs. The same number of only two runs will occur if Y's are stochastically smaller than X's.

The test based on total number of runs, therefore, can not distinguish between these two situations, i.e. whether $\xi_X < \xi_Y$ or $\xi_X > \xi_Y$. This test is thus appropriate for the alternative which is a general one i.e. is two sided H_1: $\xi_X \neq \xi_Y$. It is evident that small value of total number of runs indicate departure from the null hypothesis. Thus this test is, therefore, used for testing.

$$H_0: \xi_X = \xi_Y \text{ against } H_1: \xi_X \neq \xi_Y \tag{16.19}$$

and *the test procedure* is

(a) Combine the two samples and order the combined sample values and assign the letter X or Y to the ordered combined sample values according to it being from sample 1 or from sample 2.
(b) Count the total number of runs = d.
(c) Reject H_0 if $d < d_\alpha$ for level of significance α, where d_α can be obtained from the Table A.12 given in Appendix.

Test Statistic for Large Samples

If n_1, n_2 > 10, *normal approximation of the distribution of d* may be used to test H_0. It can be seen that for large n_1, n_2 the distribution of d is approximately normal (cf. Mood, Graybill and Boes (1974)). If H_0 is true

$$E(d) = \frac{2n_1 n_2}{n_1 + n_2} + 1 \tag{16.20}$$

and

$$V(d) = \frac{2n_1 n_2 (2n_1 n_2 - n_1 - n_2)}{(n_1 + n_2)^2 (n_1 + n_2 - 1)} \tag{16.21}$$

Hence the statistic to test H_0 against H_1 will be

$$Z = \frac{d - E(d)}{\sqrt{V(d)}} \sim N(0, 1), \quad \text{if } H_0 \text{ is true} \tag{16.22}$$

and for α = .05 *the test procedure* is Reject H_0 if $Z < -1.645$ (16.23)

A Remark

The Wald-Wolfowitz run test is applicable to detect differences in shape as well as differences in location.

Example 16.3: Consider again the data on mental health scores of male students of junior high school from rural area (X) and urban area (Y) given in Example 16.2. It is desired to test that the distributions of X and Y are same using Wald-Wolfowitz test.

Solution: Wald-Wolfowitz run test is applicable to detect the differences in locations (or medians) and also differences in shapes of the two distributions of X and Y. Hence this run test is applied on the same data to test

$$H_0: F_X(x) = F_Y(x) \text{ against } H_1: F_X(x) \neq F_Y(x)$$

The combined sample of rural and urban students are ranked according to their mental health scores in the solution of Example 16.2. The same is used to attach the letter X or Y to these combined and ordered observations. X to first sample (rural) observations and Y to the urban students' scores. The resulting sequence is as given below (see Table 16.2).

Y X Y Y Y Y Y Y Y Y X X X Y X Y X Y Y X X X Y Y Y X X X

Total number of runs = d = 12 in 12 + 16 = 28 observations.
From the Table A.12 for n_1 = 12 and n_2 = 16 the 5% critical value = 9.

Since $d = 12 > d_{.05,\ 12,\ 16} = 9$, H_0 is not rejected.

Since n_1, n_2 both are greater than 10, the normal approximation can also be used. The test statistic requires

$$E(d) = \frac{2n_1n_2}{n_1+n_2} + 1 = \frac{2\times12\times16}{12+16} + 1 = 13.7 + 1 = 14.7$$

and

$$V(d) = \frac{2n_1n_2(2n_1n_2 - n_1 - n_2)}{(n_1+n_2)^2(n_1+n_2-1)} = \frac{2\times12\times16(2\times12\times16-12-16)}{28^2\times(12+16-1)} = 6.456$$

So

$$Z = \frac{12-14.7}{\sqrt{6.456}} = \frac{-2.7}{2.54} = -1.063$$

Since $Z > -1.645$ H_0 is not rejected.

Hence it can be concluded that mental health scores of rural and urban male students of junior high school appear to have common distribution.

16.4 Wilcoxon Signed Rank Test

This test is also applicable if paired sample of observations are available on random variable (X, Y). X and Y have continuous distributions $F_X(x)$ and $F_Y(y)$, respectively, as was available for application of Sign test.

The Sign test has the *advantages* that (i) it is easy to apply (ii) homogeneity of variances of X and Y is not required and (iii) it is not affected by minor recording errors.

Sign test has the following *disadvantages* also: (i) Magnitude of differences $X_i - Y_i$ are not given any importance. Thus a lot of information is lost.

(ii) If $n \leq 6$ test is not good and (iii) Handling of ties is subjective and may reduce sample size by discarding these.

Some modifications of the sign test by considering relative magnitudes of the differences as well as the signs of these differences, have been suggested by Wilcoxon (1945) in his signed rank test.

The differences $(X_i - Y_i)$, $i = 1, 2, ..., n$ are obtained as in sign test.

It is assumed that $X - Y$ has a continuous distribution with median ξ and it is desired to test if the medians of X and Y are same so that median of $X - Y$ should be 0. Hence the hypothesis to be tested is H_0: $\xi = 0$ against (a) H_1: $\xi > 0$ or (b) H_1: $\xi < 0$ or (c) H_1: $\xi \neq 0$. (16.24)

To perform the Wilcoxon signed rank test the steps given below are followed:

Step 1: Find the differences $(X_i - Y_i)$, $i = 1, 2, ..., n$.

Step 2: Rank the differences $(X_i - Y_i)$ without regard to sign. In other words rank the differences according to magnitudes or rank the absolute differences.

Step 3: Assign signs of original differences $(X_i - Y_i)$ to the ranks.

Step 4: Compute the following sums of ranks.

$$W_+ = \text{sum of positive ranks and } W_- = \text{sum of negative ranks} \tag{16.25}$$

It can be seen that (cf. Gibbons (1971), Steel and Torrie (1981))

$$W_+ + W_- = \frac{n(n+1)}{2} \tag{16.26}$$

Large values of W_+ correspond to small values of W_- and vice-versa.

Step 5: *Test procedure*

For a level of significance α the test procedure is as follows

(a) Reject H_0: $\xi = 0$ against H_1: $\xi > 0$, if $W_- \le W_{\alpha,\, n}$ (16.27)
(b) Reject H_0: $\xi = 0$ against H_1: $\xi < 0$, if $W_+ \le W_{\alpha,\, n}$ (16.28)
(c) Reject H_0: $\xi = 0$ against H_1: $\xi \ne 0$,
if either $W_+ \le W_{\alpha/2,\, n}$ or $W_- < W_{\alpha/2,\, n}$ (16.29)

The critical points $W_{\alpha,\, n}$ and $W_{\alpha/2,\, n}$ may be obtained from Table A.13.

Example 16.4: Consider again the Example 16.1, giving data on oral hygiene scores of 12 pairs of patients of dental clinic. 12 of these are on the patients who received instructions on oral hygiene matters (*X*) and other 12 on the patients who did not receive instructions (*Y*). Apply Wilcoxon-signed rank test to test the hypothesis that median of the differences (*X* – *Y*) is zero against the alternative that median of (*X* – *Y*) is negative i.e. instructions lowered the scores indicating that oral hygiene level of those patients who received instructions has increased.
Solution: Let the differences *X* – *Y* in the given 12 pairs be obtained and given below:

Pair no. (*i*)	1	2	3	4	5	6	7	8	9	10	11	12
$X_i - Y_i$	–.5	0	–.5	.5	–.5	–.5	–1.5	–1.5	–1	–.5	.5	–.5
Rank R_i	5.5	1	5.5	5.5	5.5	5.5	11.5	11.5	10	5.5	5.5	5.5
Signed ranks	–5.5	1	–5.5	5.5	–5.5	–5.5	–11.5	–11.5	–10	–5.5	5.5	–5.5

Hence $W_+ = 12$ and $W_- = 66$

The hypothesis to be tested is H_0: $\xi = 0$ against H_1: $\xi < 0$, where ξ = median of $X - Y$

Looking at (16.28) of Step 5, the test procedure is

Reject H_0 against H_1 if $W_+ \le W_{\alpha,\, 12}$

For $\alpha = .05$ $W_{.05,\, 12} = 17$, from Table A.13.

Since $W_+ = 12 < 17 = W_{.05,\, 12}$ H_0 is rejected

Hence it can be concluded that median of $X - Y$ is negative. In other words, the patients who received instructions have higher level of oral hygiene.

Large Sample

If *n* is large tables of significant points are generally not available. Since W_+ is sum of *n* independent random variables, a generalization of central limit theorem enables to use normal approximation of distribution of W_+ with

$$E(W_+) = \frac{n(n+1)}{4} \text{ and } V(W_+) = \frac{n(n+1)(2n+1)}{24} \tag{16.30}$$

Hence

$$Z = \frac{W_+ - \dfrac{n(n+1)}{4}}{\sqrt{\dfrac{(n(n+1)(2n+1)}{24}}} \sim N(0, 1) \text{ if } H_0 \text{ is true}$$

or

$$Z = \frac{4W_+ - n(n+1)}{\sqrt{2n(n+1)(2n+1)/3}} \sim N(0, 1) \text{ if } H_0 \text{ is true} \tag{16.31}$$

For $n \ge 15$ test statistic *Z* can be used and the H_0: $\xi = 0$ is rejected at level of significance $\alpha = .05$

(a) against H_1: $\xi > 0$ if $Z \geq 1.645$
(b) against H_1: $\xi < 0$ if $Z \leq -1.645$ and
(c) against H_1: $\xi \neq 0$ if $|Z| \geq 1.96$

16.5 Wilcoxon-Mann-Whitney *U*-Test

This test is also applicable when the samples from *X* and *Y* are independently drawn, as is the Wald-Wolfowitz run test.

Wilcoxon (1945) developed this test for two independent samples of equal sizes i.e. $n_1 = n_2$. Mann and Whitney (1947) extended this test to case when samples are of unequal sizes i.e. $n_1 \neq n_2$. The ranks of the *Y*'s in relation to those of *X*'s are considered to give an idea of the relationship of the populations of *X* and *Y*. Wald-Wolfowitz test uses the total number of runs, whereas this test uses the ranks of *X*'s and *Y*'s in the combined ordered sample of $n_1 + n_2$ observations.

Mann-Whitney *U*-statistic is the number of times a *Y* precedes an *X* in the combined ordered sample of $n_1 + n_2$ observations.

Consider that two independent samples $X_1, X_2, ..., X_{n1}$ and $Y_1, Y_2, ..., Y_{n_2}$ is available from *X* and *Y* having continuous distributions with medians ξ_X and ξ_Y. To conclude that *X* and *Y* have identical distributions, the equality of medians is set as the null hypothesis. Thus in symbols

$$H_0: \xi_x = \xi_y \tag{16.32}$$

The test procedure is as follows:

Step 1: Combine the two samples on *X* and *Y*, arrange them in ascending order of magnitude.

Step 2: Rank the observations in the combined sample keeping track of which observation is from sample 1 and which from sample 2.

Step 3: Find the following sums of ranks

R_1 = Sum of ranks of observations in Sample 1 (of smaller sample of size n_1).

R_2 = Sum of ranks of observations in Sample 2 (of size n_2).

Step 4: Determine

$$U_1 = n_1 n_2 + ½\, n_1(n_1 + 1) - R_1 \tag{16.33}$$

$$U_2 = n_1 n_2 + ½\, n_2 (n_2 + 1) - R_2 = n_1 n_2 - U_1 \tag{16.34}$$

and

$$U = \min (U_1, U_2) \tag{16.35}$$

H_0 is rejected against

(a) $H_1 : \xi_X \neq \xi_Y$ if $U < U_{\alpha, n_1, n_2}$ (16.36)

(b) $H_1 : \xi_X > \xi_Y$ if $U_1 < U_{\alpha, n_1, n_2}$ (16.37)

(c) $H_1 : \xi_X < \xi_Y$ if $U_2 < U_{\alpha, n_1, n_2}$ (16.38)

The significant points, for a specified level of significance are noted from Table A.14 A and B.

If n_1, n_2 are large i.e. $n_1, n_2 > 20$, normal approximation due to central limit theorem may be used with

$$E(U) = \frac{1}{2} n_1 n_2 \quad \text{and} \quad V(U) = \frac{1}{12} n_1 n_2 \left(n_1 + n_2 + 1\right), \text{ if } H_0 \text{ is true} \tag{16.39}$$

To test H_0 use the statistic

$$Z = \frac{U - E(U)}{\sqrt{V(U)}} \sim N(0, 1) \text{ if } H_0 \text{ is true} \tag{16.40}$$

$$= \frac{U - \frac{n_1 n_2}{2}}{\sqrt{\frac{n_1 n_2 (n_1 + n_2 + 1)}{12}}} = \frac{2U - n_1 n_2}{\sqrt{n_1 n_2 (n_1 + n_2 + 1)/3}} \tag{16.41}$$

Reject H_0 if $| Z | > 1.96$ for $\alpha = .05$ and two sided alternative (16.42)

Example 16.5: Consider again the data of Example 16.2 on the mental health score of male students of junior high school from rural (X) and urban (Y) area having 12 and 16 observations, respectively.

It is again desired to test that the male students of rural and urban junior high schools have identical distributions, using Wilcoxon-Mann-Whitney U-test.

Solution: The given data has $n_1 = 12$ and $n_2 = 16$

The combined ordered 28 observations are shown in Table 16.2.

R_1: Sum of ranks of Sample 1 observations

$= 2 + 11 + 12 + 13 + 15 + 17 + 20.5 + 20.5 + 22 + 26 + 27.5 + 27.5 = 214$

R_2: Sum of ranks of Sample 2 observations

$= 1 + 3 + 4.5 + 4.5 + 6.5 + 6.5 + 9 + 9 + 9 + 14 + 16 + 18 + 19 + 23 + 24.5 + 24.5 = 192$

U_1: $n_1 n_2 + ½ n_1 (n_1 + 1) - R_1 = 12 \times 16 + ½ (12 \times 13) - 214 = 192 + 78 - 214 = 56$

U_2: $n_1 n_2 + ½ n_2 (n_2 + 1) - R_2 = 12 \times 16 + ½ (16 \times 17) - 192 = 192 + 136 - 192 = 136$

The hypothesis to be tested is H_0: $\xi_X = \xi_Y$ against H_1: $\xi_X \neq \xi_Y$

So $U = \min (U_1, U_2) = \min (56, 136) = 56$ is used.

The test procedure for $\alpha = .05$ is

Reject H_0 if $U < U_{.05,\ 12,\ 16}$

From Table A.14B $U_{.05,\ 12,\ 16} = 53$

Since $U = 56 > U_{.05,\ 12,\ 16} = 53$, H_0 is not rejected.

So it is concluded that the rural and urban male students of junior high school may be said to have identical distributions.

16.6 Kolmogorov-Smirnov Two Sample Test

Kolmogorov-Smirnov test has been used in Chapter 15 to test if a random sample of size n might have been drawn from some specified distribution with cdf $F_0(x, \theta)$.

This one-sample test can also be extended to the two sample problem. Thus suppose two independent random samples $X_1, X_2, ..., X_{n1}$ and $Y_1, Y_2, ..., Y_{n_2}$ are available from continuous distributions with cdf's F_X and F_Y.

It is desired to test if the two samples have come from identical distributions. In symbols the null hypothesis is

$$H_0: F_X(x) = F_Y(x) \tag{16.43}$$

As in one sample Kolmogorov-Smirnov test, the two sample test statistic is based on the empirical distributions from the two samples. Since the maximum of the absolute difference is used as test statistic, the Kolmogorov-Smirnov (K-S) two sample test is appropriate for two sided alternative H_1: $F_X(x) \neq F_Y(x)$.

Thus the *K-S* two sample test is applied for testing

$$H_0: F_X(x) = F_Y(x) \text{ against } H_1: F_X(x) \neq F_Y(x) \quad (16.44)$$

The steps for performing the test are:

Step 1: Order the observations in the two samples

$$\text{as } X_{(1)}, X_{(2)}, ..., X_{(n_1)} \text{ and } Y_{(1)}, Y_{(2)}, ..., Y_{(n_2)} \quad (16.45)$$

Step 2: Find minimum and the maximum of all the $n_1 + n_2$ sample values and choose some 7-10 values of x in the range of combined samples including the minimum and maximum values or close to these.

Step 3: Determine the empirical distributions from two samples $F_{n_1}(x)$ and $F_{n_2}(x)$ given by

$$F_{n_1}(x) = [\text{Number of observations in first sample} \leq x]/n_1$$

$$= \frac{[\text{Number of } X_i'\text{s} \leq x]}{n_1} \quad (16.46)$$

$$F_{n_2}(x) = [\text{Number of observations in second sample} \leq x]/n_2$$

$$= \frac{[\text{Number of } Y_j'\text{s} \leq x]}{n_2} \quad (16.47)$$

Step 4: The *K-S* test statistic is computed as

$$D_{n_1,n_2} = \max_x \left| F_{n_1}(x) - F_{n_2}(x) \right| \quad (16.48)$$

Step 5: The test procedure is Reject H_0: $F_x(x) = F_Y(x)$ against H_1: $F_X(x) \neq F_Y(x)$ if

$$D_{n_1,n_2} \geq D_{n_1,n_2,\alpha} \quad (16.49)$$

where $D_{n_1,n_2,\alpha}$ is 100 α% significant point of the Kolmogorov-Smirnov two sample test statistic given in Table A.15.

Table A.15 (A) is given for equal sample sizes $n_1 = n_2$ and Table A.15 (B) is for unequal sample sizes $n_1 \neq n_2$.

Example 16.6: Consider again Example 16.2. Two independent samples of mental health scores of junior high school male students from rural and urban area are ordered in Table 16.2. It is desired to use *K-S* test to detect if the two sets of scores may be considered to have a common distribution.

Solution: Sample sizes $n_1 = 12$ and $n_2 = 16$. The null hypothesis to be tested is

$$H_0: F_X(x) = F_Y(x) \text{ against } H_1: F_X(x) \neq F_Y(x).$$

The ordered samples given in Table 16.2 are reproduced below with chosen values of x's.

Computations of *K-S* Two Sample Test Statistic for Example 16.6

$X_{(i)}$	$Y_{(j)}$	x	$F_{n_1}(x)$	$F_{n_2}(x)$	$\lvert F_{n_1}(x) - F_{n_2}(x) \rvert$
22	21	21	0	1/16	1/16
29	23	25	1/12	4/16	2/12
32	25	30	2/12	9/16	19/48**
34	26	35	4/12	11/16	17/48
37	26	40	6/12	12/16	1/4
42	27	45	9/12	14/16	1/8
42	27	50	12/12	16/16	0
43	27				
47	33				
60	35				
50	38				
	41				
	45				
	46				
	46				

The table shows the computation of the absolute differences $\left|F_{n_1}(x) - F_{n_2}(x)\right|$. The maximum absolute difference i.e. the *K-S* statistic

$$D_{n_1,n_2} = D_{12,16} = 19/48$$

From the Table A.15 (B) $D_{n_1,n_2,.05} = D_{12,16,.05} = 7/16\,(= 21/48)$

Since $D_{n_1,n_2} = D_{12,16} = 19/48 < D_{12,16,.05} = 21/48$, H_0 is not rejected.

Hence, it may be concluded that the mental health scores of rural and urban male junior high school students might have identical distributions.

17. Non-Parametric Methods III: k-Sample Tests

Chapter 16 dealt with non-parametric distribution free methods for comparison of two samples. This chapter considers comparison of k-samples. Such a problem has been already discussed in Section 8.6, when k-samples were available from normal or approximately normal distributions with common variance. Thus equality of means implied equality of distributions.

In this chapter an assumption about the distributions from which the k-independent samples are obtained is simply of continuity of the distributions. Tests discussed here are for the equality of the distributions. Any assumption about the form of the distribution is not made. Thus the problem dealt in this chapter may be stated as follows:

k-independent random samples of sizes n_1, n_2, ..., n_i, ..., n_k are in hand from k continuous populations with cdf's $F_1(x)$, $F_2(x)$, ..., $F_i(x)$, ..., $F_k(x)$, and it is desired to test if the k-samples are from a common probability distribution. In symbols, the null hypothesis is

$$H_0 : F_1(x) = F_2(x) = ... = F_i(x) = ... F_k(x) \text{ for all } x \tag{17.1}$$

against

$$H_1 : \text{All the equalities in } H_0 \text{ do not hold.} \tag{17.2}$$

Also

$$\sum_{i=1}^{k} n_i = N \tag{17.3}$$

Extension of the two tests given in previous chapter and some more tests are also included in this chapter.

17.1 Median Test for k-Samples

If the null hypothesis (17.1) is true the k-samples may be considered to have come from a common-population and thus will have a common measure of location, median. Hence the test procedure consists of the following steps.

Step 1: Combine all the samples and arrange them in ascending order of their magnitudes.

Step 2: Find median of the combined sample M. It will be given by

$$M = \begin{cases} \dfrac{N+1}{2}\text{th observation in combined sample} & \text{if } N \text{ is odd, and} \\ [N/2\text{th value} + (N/2+1)\text{th value}]/2 & \text{if } N \text{ is even} \end{cases} \tag{17.4}$$

Step 3: Prepare a $2 \times k$ table showing O_i, the number of observations in i^{th} sample less than or equal to combined median M.

In the second row of this table, the number of observations greater than M in i^{th} sample ($= n_i - O_i$) are entered. The table obtained will be of the form given below.

Table 17.1 $2 \times k$ table for k-sample median test

Sample No.	1	2	3	....	i	....	k	Total
Number of observations $\le M$	O_1	O_2	O_3	....	O_i	....	O_k	t
Number of observations $> M$	$n_1 - O_1$	$n_2 - O_2$	$n_3 - O_3$	....	$n_i - O_i$	....	$n_k - O_k$	$N - t$
Total	n_1	n_2	n_3	....	n_i	....	n_k	N

Step 4: One can find the probability of obtaining the above number of observations in various samples less than M by the formula

$$P(O_1, O_2, ..., O_k/t) = \frac{\binom{n_1}{O_1}\binom{n_2}{O_2}......\binom{n_k}{O_k}}{\binom{N}{t}} \tag{17.5}$$

H_0 is rejected if calculated probability

$$P(O_1, O_2, ..., O_k/t) < .05 \tag{17.6}$$

Computations are not simple so an *alternative procedure* may be applied (cf. Gibbons (1971)) using χ^2-test in a $2 \times k$ contingency table, even if N is as small as 25 (i.e. $N \ge 25$) and each sample size is as small as 5 (i.e. $n_i \ge 5$ for $i = 1, 2, ..., k$).

The χ^2-test statistic to test H_0 against H_1 given in (17.1) and (17.2) is given by

$$\chi^2 = \sum_{i=1}^{k}\sum_{j=1}^{2}\frac{(f_{ij} - e_{ij})^2}{e_{ij}} = \sum_{i=1}^{k}\sum_{j=1}^{2} f_{ij}^2 \Big/ e_{ij} - N \tag{17.7}$$

where

$$f_{i1} = O_i,\ i = 1, 2, ..., k \text{ and}$$
$$f_{i2} = n_i - O_i,\ i = 1, 2, ..., k \tag{17.8}$$

If H_0 is true the test statistic $\chi^2 \sim \chi^2_{k-1}$.

So H_0 is rejected at 100 α% level of significance if

$$\chi^2 \ge \chi^2_{k-1,\alpha} \tag{17.9}$$

The following example illustrates this test procedure.

Example 17.1: A biologist counted number of specific type of moth captured in traps in five different woodlands. The results are as follows:

Woodland 1	Woodland 2	Woodland 3	Woodland 4	Woodland 5
2	8	34	13	52
15	12	45	5	35
23	60	65	32	60
50	18	90	19	82
40	22	65	25	22
	43	50		13
		49		

Do the data indicate that population structure of the studied moth differ in the 5 woodlands?

Solution: It is desired to test if the distributions of moth populations differ in the 5 woodlands. For this median test for $k = 5$ samples, will be applied to test the null hypothesis

$$H_0: F_1(x) = F_2(x) = F_3(x) = F_4(x) = F_5(x) \text{ against}$$

$$H_1: \text{All the equalities do not hold.}$$

Step 1: Combining the samples of sizes $n_1 = 5$, $n_2 = 6$, $n_3 = 7$, $n_4 = 5$ and $n_5 = 6$, $N = 5 + 6 + 7 + 5 + 6 = 29$.

Step 2: Ranking the observations in ascending order the median of the combined sample is value of $\frac{29+1}{2}^{\text{th}}$ or 15th observation.

The ordered observations are:

1	2	3	4	5	6	7	8	9	10
2	5	8	12	13	13	15	18	19	22
22	23	25	32	34	35	40	43	45	49
50	50	52	60	60	65	65	82	90	

Hence median of combined sample

$$M = 34$$

Now the 2 × 5 table for median test is obtained as:

Computations for median test for Example 17.1

Woodlands →	1	2	3	4	5	Total
Number of observations ≤ 34	3(2.59)	4(3.10)	1(3.62)	5(2.59)	2(3.10)	15
Number of observations > 34	2(2.41)	2(2.90)	6(3.38)	0(2.41)	4(2.90)	14
Total	5	6	7	5	6	29

Note: The expected frequencies are shown in parenthesis

$$e_{1j} = n_j \, {}^{15}\!/_{29}, \; j = 1, 2, 3, 4$$

and e_{2j}'s $j = 1, 2, ..., 5$ are obtained by subtraction.

The test statistic is given by

$$\chi^2 = \sum_{i=1}^{2}\sum_{j=1}^{5} {f_{ij}^2}\Big/{e_{ij}} - N = {3^2}\!/_{2.59} + {4^2}\!/_{3.10} + ... + {4^2}\!/_{2.90} - 29$$

$$= 38.5756 - 29 = 9.5756$$

The test procedure is to reject H_0 at $\alpha = .05$ if

$$\chi^2 \geq \chi^2_{5-1,\,.05}$$

From the Table A.5 $\chi^2_{4,\,.05} = 9.488$ $\chi^2_{4,\,.01} = 11.143$

Since $\chi^2_{cal} = 9.5756 > \chi^2_{4,\,.05} = 9.488$, H_0 is rejected at 5% level of significance but at 1% level H_0 is not rejected. It may be, therefore, concluded that the distributions of moth population differ significantly but not very significantly.

17.2 Kruskal-Wallis *k*-Sample Test

Under the conditions mentioned above, when *k*-independent samples are available and it is desired to test that they have been drawn from identical populations, Kruskal-Wallis test also can be applied to test H_0. Just as the one-way ANOVA tests equality of *k* means, this test also tests equality of medians of the *k*-populations from which the independent samples of sizes $n_1, n_2, ..., n_i, ..., n_k$ are available. Kruskal-Wallis test uses the ranks of the observations in the combined sample of *N* observations, since if H_0 is true all of these *N* observations belong to a common population. A track is kept on the sample from which the observation is being ranked. The ANOVA requires approximate normality (if not fully) and homogeneity of variances. Kruskal-Wallis test can be applied if data are from any continuous distribution in population.

Since there are *N* observations in the combined sample, the *sum of ranks* will be sum of first *N* natural numbers i.e.

$$1 + 2 + ... + N = N(N + 1)/2 \tag{17.10}$$

If the *k*-samples are drawn from identical populations, these ranks will be divided in proportion to the sizes of the *k*-samples. Thus for i^{th} sample of size n_i, the expected sum of ranks will be

$$\frac{n_i}{N}\frac{N(N+1)}{2} = \frac{n_i(N+1)}{2} \tag{17.11}$$

If R_i is the actual sum of ranks of ith sample observations, a reasonable test to detect differences in the *k*-population locations will be

$$S = \sum_{i=1}^{k}\left[R_i - \frac{n_i(N+1)}{2}\right]^2 \tag{17.12}$$

Large values of *S* lead to rejection of null hypothesis H_0 against H_1 given in equations (17.1) and (17.2).

Rijkoort (1952) gave tables of exact probabilities of *S* for $k = 3, 4, 5$ but only for small and equal n_i. For some larger values of equal sample sizes critical values are also given.

However, Kruskal-Wallis (1952) gave an alternative test criteria, which is given by

$$H = \frac{12}{N(N+1)}\sum_{i=1}^{k}\frac{1}{n_i}\left[R_i - \frac{n_i(N+1)}{2}\right]^2$$

$$= \frac{12}{N(N+1)}\sum_{i=1}^{k}\frac{R_i^2}{n_i} - 3(N+1) \tag{17.13}$$

If no n_i is very small, H has approximately a χ^2-distribution with $(k - 1)$ d.f. if H_0 is true (cf. Kruskal (1952)). In symbols

$$H \sim \chi^2_{k-1}, \text{ if } H_0 \text{ is true} \tag{17.14}$$

and the *test procedure* for level of significance α is

$$\text{Reject } H_0 \text{ if } H > \chi^2_{k-1,\,\alpha} \tag{17.15}$$

Exact probabilities for H have been obtained by Kruskal-Wallis (1952) for $k = 3$ and $n_i \le 5$. Daniel (1987) has reproduced the critical values of H for various of α, given by them.

The following example applies Kruskal-Wallis test procedure on the data on population of a specific moth in 5 woodlands of Example 17.1.

Example 17.2: Consider again data of Example 17.1 and apply Kruskal-Wallis test to conclude if the 5 samples may have come from an identical population.

Solution: Again it is desired to test H_0: $F_1(x) = F_2(x) \ldots = F_5(x)$ against H_1: all the equalities do not hold.

The samples are of sizes $n_1 = 5$, $n_2 = 6$, $n_3 = 7$, $n_4 = 5$ and $n_5 = 6$ so $N = 29$.

The combined ordered observations are again reproduced indicating the rank and the sample number in parenthesis.

It may be noted that the average rank is to be assigned in case of ties.

The ordered observations in combined sample

	1	2	3	4	5	6	7	8	9	10
Values	2	5	8	12	13	13	15	18	19	22
Ranks (sample number)	1(1)	2(4)	3(2)	4(2)	5.5(4)	5.5(5)	7(1)	8(2)	9(4)	10.5(2)
Values	22	23	25	32	34	35	40	43	45	49
Ranks (sample number)	10.5(5)	12(1)	13(4)	14(4)	15(3)	16(5)	17(1)	18(2)	19(3)	20(3)
Values	50	50	52	60	60	65	65	82	90	
Ranks (sample number)	21.5(1)	21.5(3)	23(5)	24.5(2)	24.5(5)	26.5(3)	26.5(3)	28(5)	29(3)	

Ranks of observations in various samples

Sample 1	Sample 2	Sample 3	Sample 4	Sample 5
1	3	15	2	5.5
7	4	19	5.5	10.5
12	8	20	9	16
17	10.5	21.5	13	23
21.5	18	26.5	14	24.5
	24.5	26.5		28
		29		
$R_1 = 58.5$	$R_2 = 68$	$R_3 = 157.5$	$R_4 = 43.5$	$R_5 = 107.5$

The Kruskal-Wallis test statistic is given by

$$H = \frac{12}{N(N+1)} \sum_{i=1}^{k} \frac{R_i^2}{n_i} - 3(N+1) = \frac{12}{29(30)} \left[\frac{58.5^2}{5} + \frac{68^2}{6} + \frac{157.5^2}{7} + \frac{43.5^2}{5} + \frac{107.5^2}{6} \right] - 3(30)$$

$$H = \frac{2}{145}[7280.8944] - 90 = 100.4261 - 90$$

$$= 10.4261$$

From Table A.5 of χ^2, 5% upper tail significant point

$$\chi^2_{5-1,\ .05} = 9.49$$

Since $H > 9.49$, H_0 is rejected at 5% level of significance.
Hence the 5 Woodlands' population of moths may not be considered as being identical.

17.3 Friedman's Test for RBD

The previous two tests were applied when there were k-independent samples from each of the k continuous distributions $F_1(x)$, $F_2(x)$, ..., $F_k(x)$. In this and the next subsections the available data can be presented as a two-way layout of r rows and s columns, with only one entry in each of the rs cells. In the previous sections, if the k-independent samples were of the same size say n, then there may be again similar $k \times n$ layout as of rs cells. But in the $k \times n$ data, the variation is only due to different populations. There is no other factor affecting the data. Hence the two tests mentioned in the previous sections may be considered to be applicable when data is in the form of one-way ANOVA.

The data considered in this and subsequent sections is in the form of two-way ANOVA problem. However, when the assumptions of techniques of ANOVA discussed in Chapter 14 do not hold, non-parametric methods only can be used.

In this layout of r rows and s columns, the rows indicate the blocks or subjects or sample numbers and the columns represent different treatments. The observations in various rows or blocks may be considered to be independent but the observations in various columns can not be considered as independent since they receive different treatments.

Friedman (1937, 1940) suggested a test of the hypothesis that the s treatment effects are equal, which does not require the assumptions of ANOVA to be satisfied by the data.

Instead of using the observed value in the ith block ($i = 1, 2, ..., r$) he suggested to use the ranks of the observations in each of the blocks (or rows).

As there are s treatments, so s observations in each of the r blocks (or rows) assigned ranks in each block are 1, 2, ..., s according to the magnitude of the observed values, compared to the other observations in the same block.

Let R_{ij} denote the rank of jth observation in i^{th} block ($i = 1, 2, ..., r$; $j = 1, 2, ..., s$) then R_{ij} is the rank of treatment number j in i^{th} block.

$R_{i1}, R_{i2}, ..., R_{is}$ is a permutation of integers 1, 2, ..., s and $R_{1j}, R_{2j}, ..., R_{rj}$ are the ranks of jth treatment in various blocks.

Let $$R_{\cdot j} = \sum_{i=1}^{r} R_{ij} = \text{sum of ranks of } j^{th} \text{ treatment} \qquad (17.16)$$

If all the s treatment effects are same sum of ranks of jth treatment will be expected to be equal to average of the ranks of s treatments.

$$\text{Sum of ranks in each row} = \frac{s(s+1)}{2} \qquad (17.17)$$

$$\text{Hence sum of ranks in all the rows} = \frac{rs(s+1)}{2} \qquad (17.18)$$

and average of ranks of s treatments from r rows = sum of ranks in all the r rows

$$= \frac{rs(s+1)}{s.2} = \frac{r(s+1)}{2} \tag{17.19}$$

Hence in order to test the hypothesis
H_0: s treatment effects are equal against
H_1: at least one treatment effect differs from others
a statistic S defined as

$$S = \sum_{j=1}^{s} \left[R._j - \frac{r(s+1)}{2} \right]^2 \tag{17.20}$$

may be used.
It has been shown (cf. Gibbons (1971)), that the following statistic F can be used to test H_0 against H_1.

$$F = \frac{12S}{rs(s+1)} = \frac{12\sum_{j=1}^{s} R_{\bullet j}^2}{rs(s+1)} - 3r(s+1) \tag{17.21}$$

If H_0 is true F has a χ^2-distribution with $(s - 1)$ d.f.
In symbols, if H_0 is true

$$F \sim \chi^2_{s-1} \tag{17.22}$$

Large values of F indicate departure from null hypothesis.
Hence the *test procedure* for a specified level of significance α is

$$\text{Reject } H_0 \text{ if } F > \chi^2_{s-1,\, \alpha} \tag{17.23}$$

Use of the Friedman test is illustrated in the following example.

Example 17.3: Effects of aromatic hydrocarbon content, concentration of nickel and vanadium in crude oil on the embryonic development of mallards are studied. Ducks from six clutches (nests) to be used as blocks, are used in the experiment. Four eggs are randomly selected from each clutch and assigned at random one of the following four treatments (source: Milton and Tsokos (1983), p. 439):

1. No crude oil applied (control)
2. 1 μl crude oil
3. 1 μl crude oil with 700 ppm vanadium
4. 1 μl crude oil with 700 ppm nickel

After allowing the eggs to develop for 18 days, the weights in gms of the eggs are determined. The results of the experiment are in the given table.

Weights of eggs, 18 days after the treatment

Blocks (Clutches)	Treatments			
	1	2	3	4
1	16.77 (4)	14.34 (2)	16.08 (3)	14.29 (1)
2	15.61 (4)	11.92 (1)	13.22 (2)	13.95 (3)
3	14.46 (4)	14.45 (3)	11.72 (1)	13.59 (2)
4	13.08 (3)	16.11 (4)	10.18 (1)	12.22 (2)
5	14.47 (4)	12.85 (1)	13.52 (3)	13.22 (2)
6	12.01 (2)	16.13 (4)	12.68 (3)	10.73 (1)

Use the Friedman's procedure to test if the four treatments affect embryonic development differently.

Solution: The experimental data is from 6 blocks and 4 treatments. So $r = 6$ and $s = 4$. To apply Friedman's test the following steps are followed.

The null hypothesis H_0: All the four treatments have equal effects against H_1: At least one treatment differs.

Step 1: Rank the observed weights in each of the six blocks in increasing order of magnitude. The ranks are included in parentheses besides the values of weights.
Step 2: Find the sum of ranks of each of the four treatments to get

$$
\begin{aligned}
R_{\bullet 1} &= \text{Sum of ranks of treatment 1} \\
&= 4 + 4 + 4 + 3 + 4 + 2 = 21 \\
R_{\bullet 2} &= 2 + 1 + 3 + 4 + 1 + 4 = 15 \\
R_{\bullet 3} &= 3 + 2 + 1 + 1 + 3 + 3 = 13 \\
R_{\bullet 4} &= 1 + 3 + 2 + 2 + 2 + 1 = 11
\end{aligned}
$$

Step 3: Compute the statistic F given by

$$
\begin{aligned}
F &= \frac{12\sum_{j=1}^{4} R_{\bullet j}^2}{rs(s+1)} - 3r(s+1) \\
&= \frac{12(21^2 + 15^2 + 13^2 + 11^2)}{6 \times 4(4+1)} - 3 \times 6(4+1) \\
&= \frac{956}{10} - 90 = 5.6
\end{aligned}
$$

Step 4: If H_0 is true $F \sim \chi^2_{s-1}$ and the test procedure is

$$\text{Reject } H_0 \text{ if } F > \chi^2_{s-1,\, \alpha}$$

Choosing $\alpha = .05$ the upper tail 5% critical point of χ^2 distribution with $s - 1 = 3$ d.f. is read from Table A.5. It is seen that

$$\chi^2_{3,\, .05} = 7.81$$

Since $F = 5.6 < \chi^2_{3,\, .05} = 7.81$ H_0 is not rejected.
Thus data does not indicate that there is any difference in the four treatment effects.

A Remark

Friedman's test can also be applied when the data is given in the form of ranks and it may be considered to be non-parametric analogue of the classical ANOVA of RBD.

17.4 Median Test for Two-Way Classification

This test is also applicable to a two-way classified data and it is *desired to test* whether the *treatments effects are same* on the s plots of each block. The *distributions* are *assumed continuous* and identical. Thus null hypothesis is

H_0: The s treatments effects are equal against

H_1: At least one treatment effect differs.

The test procedure is an extension of the two-sample median test. This testing procedure also prepares a $2 \times s$ contingency table showing the number of observations above and below the median. The *medians are obtained separately for each block.* The blocks are not combined to obtain a common median as was done for two-sample median test.
Thus the following steps are followed for this median test for two-way classification:
Step 1: Find out the median of the s observations in a block for each of the r blocks.

Step 2: Put a + sign near the observations if it is above the block median and – sign if it is below median.

Step 3: Prepare a 2 × s contingency table showing the number of + and – signs for the s treatments.

Step 4: Apply χ^2-test as on a contingency table using the test statistic

$$\chi^2 = \sum_{i=1}^{2}\sum_{j=1}^{s}\left[\left(O_{ij} - E_{ij}\right)^2 / E_{ij}\right]$$

$$= \sum_{i=1}^{2}\sum_{j=1}^{s}\left(O_{ij}^2 / E_{ij}\right) - rs \qquad (17.24)$$

The test statistic χ^2 in (17.24) follows χ^2 distribution with $(s - 1)$ d.f. In symbols $\chi^2 \sim \chi^2_{s-1}$ if H_0 is true.

So the test procedure for a specified α is:

$$\text{Reject } H_0 \text{ if } \chi^2 > \chi^2_{s-1,\,\alpha} \qquad (17.25)$$

This procedure is illustrated using data given in Example 17.3.

Example 17.4: Apply median test on the 2-way classified data of Example 17.3.

Solution: For convenience the data is reproduced below:

Step 1: Since there are four treatments, so four observations are in each of the six blocks. The median for each

Weights of eggs, 18 days after the treatment

Block clutches	Treatments 1	2	3	4	Block medians
1	16.77+	14.34–	16.08+	14.29–	15.21
2	15.71+	11.92–	13.22–	13.95+	13.59
3	14.46+	14.45+	11.72–	13.59–	14.02
4	13.08+	16.11+	10.18–	12.22–	12.65
5	14.47+	12.85–	13.52+	13.22–	13.37
6	12.01–	16.13+	12.68+	10.73–	12.35

block (since s = 4) will be

$$= \frac{\text{Value of 2nd observation + Value of 3rd observation}}{2}$$

after ranking the observations of each block. The ranks are already entered in parentheses in the table of Example 17.3.

$$\text{Hence median for 1}^{\text{st}}\text{ block} = M_{e1} = \frac{14.34 + 16.08}{2} = \frac{30.42}{2} = 15.21$$

Similarly all the six medians can be obtained as

$$\text{For 2}^{\text{nd}}\text{ block} = M_{e2} = \frac{13.22 + 13.95}{2} = 13.59$$

$$\text{For 3}^{\text{rd}}\text{ block} = M_{e3} = \frac{13.59 + 14.45}{2} = \frac{28.04}{2} = 14.02$$

$$\text{For } 4^{\text{th}} \text{ block} = M_{e4} = \frac{12.22 + 13.08}{2} = \frac{25.30}{2} = 12.65$$

$$\text{For } 5^{\text{th}} \text{ block} = M_{e5} = \frac{13.22 + 13.52}{2} = \frac{26.74}{2} = 13.37$$

$$\text{For } 6^{\text{th}} \text{ block} = M_{e6} = \frac{12.01 + 12.68}{2} = \frac{24.69}{2} = 12.35$$

Values of medians are entered in the last column of the table above.

Step 2: If the observations of a block are larger than block median + sign is written besides it. – sign is written otherwise.

Step 3: The 2 × 4 contingency table is constructed as

Observations	Treatments				
	1	2	3	4	Total
Above median	5(3)	3(3)	3(3)	1(3)	12
Below median	1(3)	3(3)	3(3)	5(3)	12
Total	6	6	6	6	24

Step 4: For computation of test statistic the expected frequencies E_{ij} are to be computed using the formula

$$E_{ij} = \frac{n_{i.} \times n_{.j}}{n_{..}} = \frac{i^{th} \text{ row total} \times j^{th} \text{ column total}}{\text{grand total}}$$

Only three E_{ij}'s i.e. E_{11}, E_{12}, E_{13} will be calculated using this formula. The rest can be obtained by subtraction.

$$E_{11} = \frac{12 \times 6}{24} = 3,\ E_{12} = \frac{12 \times 6}{24} = 3,\ E_{13} = \frac{12 \times 6}{6} = 3 \text{ Hence}$$

$$E_{14} = 3\ E_{21} = 3\ E_{22} = 3\ E_{23} = 3 \text{ and } E_{24} = 3.$$

The expected frequencies are entered in the parentheses besides the observed cell frequencies.

Step 4: Now the test statistic is obtained as

$$\chi^2 = \sum_{i=1}^{2}\sum_{j=1}^{4}\left(O_{ij} - E_{ij}\right)^2 / E_{ij} = \frac{(5-3)^2}{3} + \frac{(3-3)^2}{3} + \frac{(3-3)^2}{3} + \frac{(1-3)^2}{3} + \frac{(1-3)^2}{3} + \frac{(3-3)^2}{3} + \frac{(3-3)^2}{3} + \frac{(5-3)^2}{3}$$

$$= \frac{1}{3}[4 + 0 + 0 + 4 + 4 + 0 + 0 + 4] = \frac{16}{3} = 5.333$$

The χ^2 statistic has χ^2 distribution with 4 – 1 = 3 d.f.

From Table A.5, 5% upper tail significant point of χ^2_3 distribution is $\chi^2_{3,\ .05} = 7.81$

Since calculated $\chi^2 = 5.333 < \chi^2_{3,\ .05} = 7.81$

H_0 is not rejected. Hence the data does not indicate that treatment effects differ.

17.5 Olmstead-Tukey Corner (or Quadrant Sum) Test of Association

The above two tests deal with two way classified data and may be considered as non-parametric anologues of the ANOVA of RBD.

For quantitative bivariate data on (*X*, *Y*), the simple (or product moment) correlation coefficient measures the degree of linear relationship between *X* and *Y*. If *X* and *Y* are related but relationship is non-linear, for some range of values of X sample correlation r may be found to be quite highly significant. If the bivariate normality of (X, Y) is doubtful a t-test of significance of correlation coefficient is not valid.

Even when bivariate normality of data can not be justified, one may still be interested to test significance of association between X and Y. Rank correlation is one approach to obtain a measure of association and also to test its significance.

There are also situations when the one or both the variables can not be measured on ratio scale. One may find it more convenient or only suitable way to assign a measure of performance by assigning ranks or some way of assigning an order to the data. Such situations may often arise in biological studies as indicated by Sokal and Rohlf (1995). A non parametric method for testing significance of a measure of association between X and Y are, therefore, necessary.

In this chapter two measures of association are given and tests of their significance are provided. These measures are based on ranks of the sampled observations. One measure and its test (described in this section) has been suggested by Olmstead and Tukey (1947) and discussed in Steel and Torrie (1980), Mood (1950) and Sokal and Rohlf (1995). The other measure τ was suggested by Kendall (1938) and the critical values for test of significance have been provided by Box and Cox (1964) and by Burr (1960). The Kendall's τ is considered in the next section.

This test suggested by Olmstead and Tukey (1947) does not test equality of two distributions but is a *non-parametric test of association of two continuous variables* (cf. Steel and Torrie (1980)). This test is also called the *quadrant sum test.* Since extreme values affect the association between variables quite much, this test gives an importance to these.

This procedure differs from the classical test of correlation coefficient between the two continuous variables, since distribution of the variables are not assumed.

The steps for testing of association between two continuous variables using distribution free method are as follows.

Step 1: Plot the observed n pairs of observations (X_1, Y_1), (X_2, Y_2) ... (X_n, Y_n).
Step 2: Determine medians Me(X) and Me(Y) of X and Y observations, respectively.
Step 3: Draw vertical and horizontal lines through Me(X) and Me(Y). Mark the four quadrants with + and – sign. The two quadrants where X and Y of observations are greater than both the medians and where both are less than respective medians are given + sign. The other two quadrants are given – sign.
Step 4: Starting at the observation with highest Y-value move down in vertical direction to the observations until the vertical median line is crossed. Record the number of observations just before crossing.
Step 5: Starting at the observation with highest X-value i.e. from the extreme right move to left, until the horizontal median line is crossed. Record the number of observations just before crossing.
Step 6: Repeat steps 4 and 5 from the bottom and from left.
Step 7: Compute sum of the recorded numbers i.e. the quadrant sum.
Step 8: The quadrant sum QS is the test statistic for testing of association between the two variables. Large values indicate significant association.
The critical values are given in Table A.17. and the method is illustrated by an example given below.

Example 17.5: The data on 14 captive hens were obtained to study the use of ovulated follicles in determining eggs laid by ring necked pheasant.

Eggs laid:	39	29	46	28	31	25	49	57	51	21	42	38	34	47
Ovulated follicles:	37	34	52	26	32	25	55	65	44	25	45	26	29	30

Test if ovulated follicles (X) and eggs laid (Y) are associated (source: Steel and Torrie (1981), p. 277).

Solution:

1. The 14 pairs of values are plotted in the Fig. 17.1 given below.

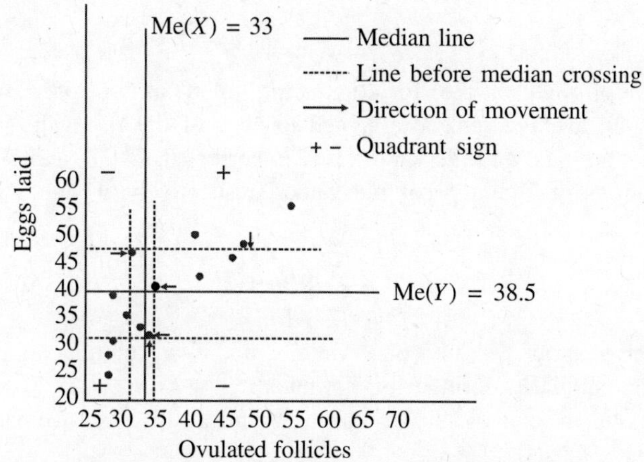

Fig. 17.1. Corner or quadrant sum test of association (source: Steel and Torrie (1981), p. 552).

2. The medians of X and Y are obtained next. Since there are 14 pairs of observations, the medians are 1/2 (7th value + 8th value). When the X values and Y values are arranged in ascending order of magnitude as:

Ordered Y :	21	25	28	29	31	34	38	39	42	46	47	49	51	57
Ordered X :	25	25	26	26	29	30	32	34	37	44	45	52	55	65

Me(X) = 1/2 (7th X value + 8th X value) = 1/2 (32 + 34) = 33
Me(Y) = 1/2 (7th Y value + 8th Y value) = 1/2 (38 + 39) = 38.5

3. Draw vertical line at Me(X) = 33 and a horizontal line at Me(Y) = 38.5. Put signs in the four quadrants.
4. Starting in the four directions the recorded numbers are

Down : 3
Left : 6
Up : 3
Right : 5

So the quadrant sum = 17

5. From the Table A.17, 5% significant value is 11.
Since quadrant sum, $QS > 11$, there is significant association between the two variables.

17.6 Coefficient of Concordance and Kendall's Tau Coefficient

A measure of association between the two variables is also called *coefficient of concordance*. Kendall (1938) suggested a measure τ which is also based on the ranks given to each of the X and Y observations. Several methods have been suggested which are all based on the ranks and the procedure of computation differ very slightly.

In this section two methods will be given.

Kendall's rank correlation coefficient obtained for sampled observations, denoted by τ is a measure of concordance between X and Y, and it is defined as

$$\tau = N \big/ n(n-1) \tag{17.26}$$

where n is the number of pairs of sampled observations (X_i, Y_i) and N is a function of ranks of the observed X's and Y's, which can be obtained in several ways. Two of these are explained below with the help of an actual data in Example 17.6. Some more details may be found in Sokal and Rohlf (1995).

Case of Ties

It may be noted that the *formula* (17.26) for computation of τ *holds if there are no ties*. If there are ties, as is the convention, average rank is assigned to each of the tied values. A correction term is to be computed, if there are ties, and the value for N is corrected. Let T_x and T_y denote the correction terms for tied observations in X and Y sample values, respectively. Then

$$T_x = \sum_{j=1}^{m} t_j(t_{j-1}) \tag{17.27}$$

where m is the number of groups of tied observations, and t_j are number of tied observations in j^{th} group of ties. T_y is also similarly defined for Y values.

The Kendall's measure of concordance τ, can be found in this case, from the following formula:

$$\tau = N \Big/ \left[\{n(n-1) - T_x\}\{n(n-1) - T_y\}\right]^{1/2} \tag{17.28}$$

To test significance of τ, i.e. to test the null hypothesis

H_0: Population value of $\tau = 0$ against

H_1: Population value of $\tau \neq 0$ i.e. there is some association in X and Y

the *statistic to be used, for large samples* ($n > 40$), is

$$U = \frac{\tau}{\sqrt{[2n(n+5)/9n(n-1)]}} \tag{17.29}$$

If H_0 is true $U \sim N(0, 1)$.

Hence the *test procedure* at 5% level is: Reject H_0 if $|U| > 1.96$.

When $n \leq 40$, this approximation is not good and a table for $n = 4$ to 40 may be used if there are no ties i.e. $m = 0$ for X as well as for Y. If ties are present, table given by Burr (1960) should be used.

Example 17.6: Again the data on 14 captive hens given in Example 17.5 is considered. Kendall's τ is used to find the coefficient of concordance and to test if the obtained τ is significant of association between X: the ovulated follicles and Y: the eggs laid.

The ordered X and Y are also taken from the solution of Example 17.5 and the computations are shown in the following table.

It is seen that there are two ties in X values, so $m = 2$. Both the ties are of length 2 i.e. value 25 occurs twice and value 26 also occurs twice. Hence $t_1 = 2 = t_2$.

So $$T_x = \sum_{j=1}^{2} t_j(t_j - 1) = t_1(t_1 - 1) + t_2(t_2 - 1) = 2(2-1) + 2(2-1) = 4$$

Also $n = 14$ and $T_y = 0$

So
$$\tau = \frac{N}{\left[\{14(14-1)-T_x\}\{14(14-1)\}\right]^{1/2}}$$

$$= \frac{N}{\sqrt{(14\times 13-4)(14\times 13)}} = \frac{N}{\sqrt{(182-4)182}} = N\Big/\sqrt{32396} = \frac{N}{179.99}$$

Computation of N

Since Y has no ties it is taken as first variable and R_1 denotes the ranks of Y and R_2 those of corresponding X. Now in the next table R_1 will be listed in ascending order and ranks of X's in the pair (X, Y) corresponding to ranked Y, will be written in R_2 column.

Procedure 1 for Computation of N

Starting with 1st value in R_2 column count the number of ranks larger than the 1st value of R_2. 1/2 is to be added for the first tie of a group. So the count for first value of $R_2 = 1.5$ is $13^1/_2 = C_1$. Do the same for all the R_2 values. Enter $C_1, C_2, \ldots, C_{14}$ in 3rd column. Find the total of these counts C_i. In other words find $\sum_{i=1}^{14} C_i = 78$

Now N is obtained from the relation

$$N = 4\left(\sum_{i=1}^{n} C_i\right) - n(n-1) = 4(78) - 14\times 13 = 312 - 182 = 130 \quad (17.29)$$

Computation of N for Example 17.6

R_1	R_2	C_i count of ranks > i^{th} rank in R_2	k_i
1	1.5 ×	13½	13
2	1.5 ×	12	12
3	3.5 ×	11½	11
4	8 ×	6	6
5	7 ×	6	6
6	5 ×	7	7
7	3.5 ×	7	7
8	9 ×	5	5
9	11 ×	3	3
10	12 ×	2	2
11	6 ×	3	3
12	13 ×	1	1
13	10 ×	1	1
14	14	0	0
	Total	78	77

Procedure 2 for Computation of N

If there are no tied rank in R_1's, this procedure can also be applied.

In this procedure (applicable for this data), R_2 values are read in ascending order, i.e. starting from 1, go to 2, 3, ..., 14. These are not read in the order they are present in R_2 column.

Start the first R_2 value = 1, and find k_1, the number of R_2 values (ranks) below the entry R_2 = 1 and cross out R_2 = 1. Next go to R_2 = 2 and count the number of entries below it, excluding the cross out ranks. Enter these counts k_i's in the 4th column of k_i. Find total of k_i i.e. $\Sigma k_i = 77$ for the above data.

Now N is obtained from the formula

$$N = 4\sum_{i=1}^{14} k_i - T_x - n(n-1) \tag{17.30}$$

or $$N = 4 \times 77 - 4 - 182 = 3084 - 182 = 122$$

Now $\tau = \frac{N}{179.99} = \frac{130}{179.99} = \tau_1$ by procedure 1 and $\tau = \frac{122}{179.99} = \tau_2$ by procedure 2. Now the test statistic by procedure 1 is

$$U_1 = \frac{\tau_1}{\sqrt{\{2(2n+5)/9n(n-1)\}}} = \frac{\tau_1}{\sqrt{2(28+5)/(126\times13)}} = \frac{\tau_1}{\sqrt{66/(126\times13)}}$$

and by procedure 2, the test statistic is

$$U_2 = \frac{\tau_2}{\sqrt{66/(126\times13)}}$$

Thus $$U_1 = .722/.201 = 3.59$$

and $$U_2 = .678/.201 = 3.37$$

Choose $\alpha = .05$ and normal table for critical point for two tail test.

Since $U_1 > 1.96$ and $U_2 > 1.96$, H_0 is rejected. Hence, it can be concluded that the eggs laid are associated with ovulated follicles.

18. Time Series Analysis

Some times numerical data on an individual, or on a population, or from an experiment in lab is collected chronologically. The time intervals are usually equally spaced, such as, yearly, monthly, ten yearly, weekly or hourly etc. Such series of observations which are ordered, are called *time series.* The observed characteristic changes with passage of time, so gives rise to series of observations which are comparable.

Time is not the only variable on which the observed characteristic may depend. Space variable may also be one, on which a numeric characteristic may depend. Thus pH value in soil may be different at various points along a strip of land. Even if the observed variable (value of the observed characteristic) Y does not depend on time, the data series may be handled using the *methods of time series analysis.*

Lab experiments on an animal or on a microbe under varying temperature or humidity conditions may also generate a data series, which also can be analyzed as a time series. Such types of statistical data are different from the data considered in the previous chapters, as time series are not an aggregate of values, but also have an order in which they appear.

Some examples of time series may be: the rainfall in a city in June for various years, the population of a country from census during a given period; import of petroleum in India in a series of months, weight of a new born baby every week for two years.

Normally the time intervals are considered of equal length and the observed random variable at time t is denoted by Y_t, where $t = 1, 2, 3, \ldots$. Thus the observation at time $t = 2$ will be denoted by Y_2. If past observations are also to be studied say, population size of India before 1900, one may use $t = -1, -2, \ldots,$ for population sizes in 1891, 1881, 1871 ... and $t = 0$ for denoting population in 1901, and $Y_0, Y_{-1}, Y_{-2}, Y_{-3}, \ldots$ will denote population sizes in 1901, 1891, 1881, 1871,..., respectively. For a better understanding of future, past is also to be studied.

Even though economists were largely involved in developing techniques for analysis of time series, but these are equally of interest to investigators in areas of demography, meteorology, sociology, geology, biology and public health.

Some real examples of time series are given as an illustration of their nature in practice.

Example 18.1: Table below gives the population of England and Wales in census years from 1811 onwards at 10 years interval (source: Yule and Kendall (1958), p. 611).

Figure 18.1 gives the typical time series of human population in England and Wales from 1811 to 1931.

Table 18.1 Population of England and Wales

Year	Population (millions) Y	Trend from fitted cubic parabola
1811	10.16	10.6
1821	12.00	11.8
1831	13.90	
1841	15.91	16.6
1851	17.93	
1861	20.07	20.5
1871	22.71	23.3
1881	25.97	26.2
1891	29.00	
1901	32.53	31.1
1911	36.07	
1921	37.89	37.9
1931	39.95	40.4
Total	314.09	

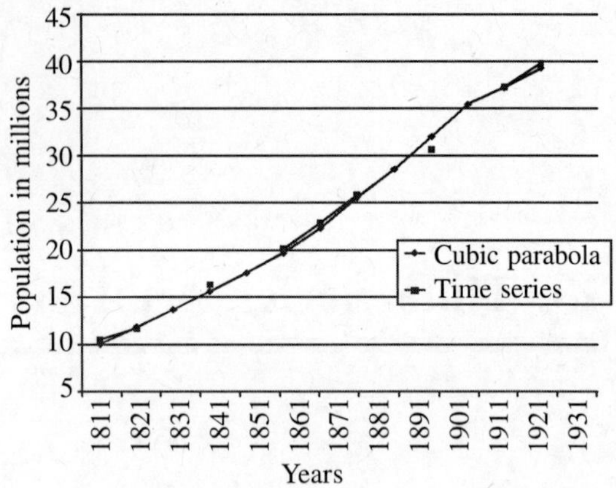

Fig. 18.1. Cubic parabola fitted to the population of England and Wales.

This time series is quite smooth in the sense that points lie very close to a smooth curve.

Example 18.2: The following table gives the data of newspaper advertising in the United States (US) during 1932-1960 (source: Croxton, Cowden and Klein (1975), p. 239).

Figure 18.2 shows the variation in the newspaper advertising in US during 1932 to 1960.

Table 18.2 Newspaper advertising in United States (1932-1960)

Year	Millions of lines	Year	Millions of lines
1932	1,164.8	1947	2,008.6
1933	1,065.5	1948	2,263.3
1934	1,178.9	1949	2,302.1
1935	1,246.0	1950	2,440.2
1936	1,380.0	1951	2,478.3
1937	1,409.0	1952	2,505.4
1938	1,225.4	1953	2,610.5
1939	1,243.6	1954	2,581.3
1940	1,268.6	1955	2,843.5
1941	1,313.2	1956	2,911.0
1942	1,241.8	1957	2,829.1
1943	1,396.4	1958	2,685.6
1944	1,361.3	1959	2,865.3
1945	1,391.6	1960	2,888.6
1946	1,729.6		

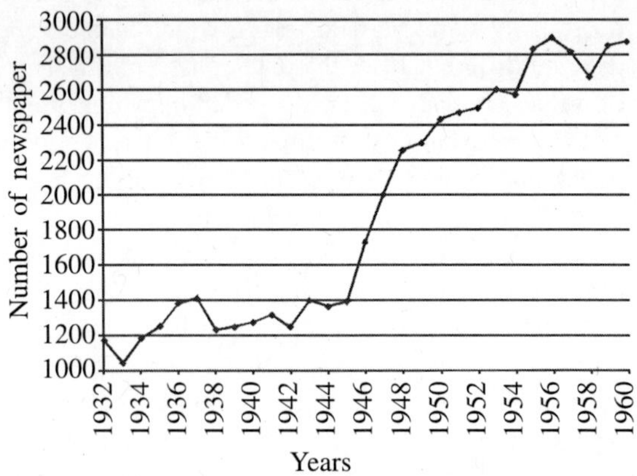

Fig. 18.2. Newspaper advertising in US during 1932-1960.

Example 18.3: Table below gives the sheep population of England and Wales for 1867-1939 (source: Yule and Kendall (1958), p. 612).

Table 18.3 Sheep population of England and Wales from 1867-1939

Year	Population (10,000)	Year	Population (10,000)	Year	Population (10,000)	Year	Population (10,000)
1867	2203	1886	1892	1905	1823	1924	1484
68	2360	87	1919	06	1843	25	1597
69	2254	88	1853	07	1880	26	1686
70	2165	89	1868	08	1968	27	1707
71	2024	90	1991	09	2029	28	1640
72	2078	91	2111	10	1996	29	1611
73	2214	92	2119	11	1933	30	1632
74	2292	93	1991	12	1805	31	1775
75	2207	94	1859	13	1713	32	1850
76	2119	95	1856	14	1726	33	1809
77	2119	96	1924	15	1752	34	1653
78	2137	97	1892	16	1795	35	1648
79	2132	98	1916	17	1717	36	1665
80	1955	99	1968	18	1648	37	1627
81	1785	1900	1928	19	1512	38	1791
82	1747	01	1898	20	1338	39	1797
83	1818	02	1850	21	1383		
84	1909	03	1841	22	1344		
85	1958	04	1824	23	1384		

Figure 18.3 shows the picture of yearly variation of sheep population given in Table 18.3.

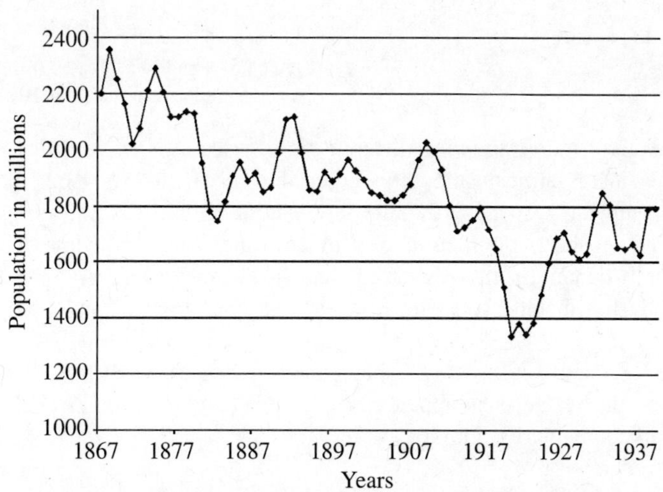

Fig. 18.3. Yearly sheep population of England and Wales.

Example 18.4: Table 18.4 and Fig. 18.4 show the percentage losses of ships per year in United Kingdom (UK) fleet (source: Yule and Kendall (1958), p. 613).

Table 18.4 Percentage UK ships lost per year during 1920-1938

Year	% of loss	Year	% of loss
1920	0.68	1930	0.49
1921	0.34	1931	0.17
1922	0.62	1932	0.30
1923	0.73	1933	0.39
1924	0.57	1934	0.42
1925	0.32	1935	0.41
1926	0.58	1936	0.24
1927	0.34	1937	0.46
1928	0.59	1938	0.50
1929	0.56		

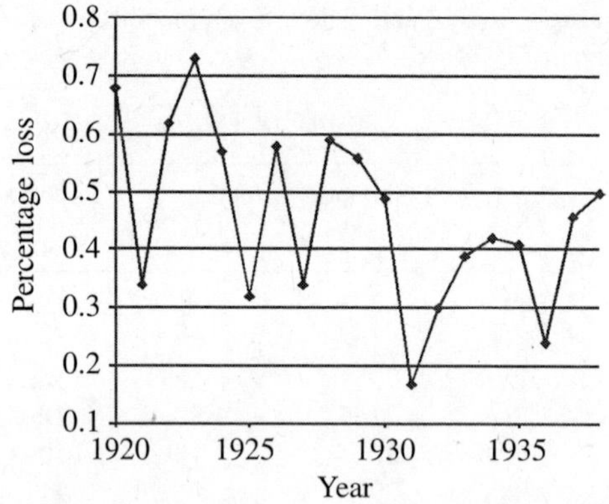

Fig. 18.4. UK vessels lost per year as a percentage of the total UK fleet.

Example 18.5: Crude birth rates (CBR) per 100 population of cattle in Great Britain are given in Table 18.5 and Fig. 18.5 (source: Yule and Kendall (1958), p. 614).

Table 18.5 CBR of cattle in great Britain during 1940-1945

Birth rate	1940	1941	1942	1943	1944	1945
Dec-Feb	33.2	35.2	35.2	34.8	37.6	36.0
March-May	45.2	44.0	46.4	44.8	41.2	42.0
June-Aug	33.2	38.8	35.6	32.0	32.8	30.0
Sept-Nov	40.0	32.8	34.4	38.4	36.8	35.2

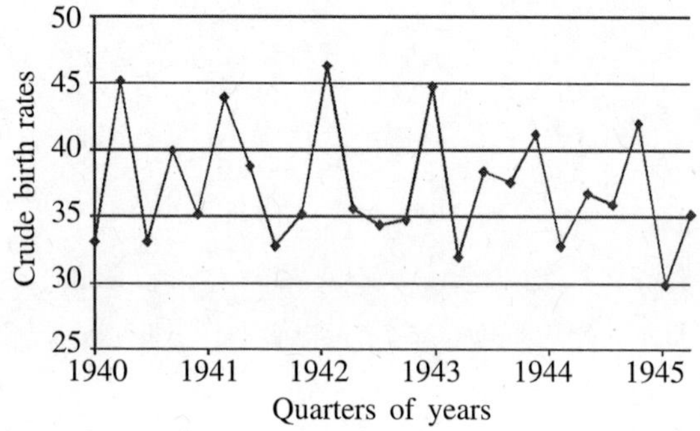

Fig. 18.5. Crude birth rates (number of births per 100) of cattle in Great Britain.

Looking at the above five figures showing different time series, it is evident that nature of time series are different. Figure 18.1 is almost smoothly increasing curve. In Fig. 18.2, though the newspaper advertising is moving upwards over the duration covered in figure, it is not a regular increase. Figure 18.3 shows a decline in sheep population of England and Wales of some what similar type but departure from smooth decline is more marked than in the increasing graph of Fig. 18.3.

Figures 18.2 and 18.3 include some local variations apart from a long term movement over the periods of graphs.

From the Table 18.4 and Fig. 18.4, the percentage of losses of British ships vary quite a large from year to year but the variation is apparently not very regular. In contrast the quaterly CBR's of cattles exhibit a marked regularity in Fig. 18.5.

18.1 Components of Time Series and their Determination

The above examples indicate that a time series may be considered to be consisting of a haphazard or random part and a systematic part. The part of a time series which can be explained or accounted for, is systematic part and the remaining part is the unsystematic or irregular or random.

The systematic part may be further assumed to be composed of three components: Trend (T), Seasonal (S) and Cyclic (C) component. A classical or traditional approach considers Y_t to be product of the said four components. In symbols

$$Y_t = T_t \times S_t \times C_t \times I_t \tag{18.1}$$
$$= (\text{Trend} \times \text{Seasonal} \times \text{Cyclic} \times \text{Irregular}) \text{ at time } t.$$

Another approach is to assume Y_t as sum of the four components. In symbols

$$Y_t = T_t + S_t + C_t + I_t \tag{18.2}$$
$$= (\text{Trend} + \text{Seasonal} + \text{Cyclic} + \text{Irregular}) \text{ at time } t.$$

Most of the time series data are appropriately represented as multiplicative model. So this chapter will consider the multiplicative model.

The analysis of time series includes isolation or determination of these components from it. Before giving methods of determining these components, a formal definition of these are given below.

Trend or secular trend is a long term, smooth and regular movement of the series, if observed for a sufficiently long period.

Trend may be of increasing type or of decreasing type. Some times in a time series, trend may change its direction. It may change from increasing direction for some period to a decreasing direction.

Seasonal fluctuations is a periodic movement in a time series where the period is less than or equal to one year. It is a movement which recurs or repeats at regular interval of time.

Seasonal fluctuations may be observed in daily passengers on roads in a week, monthly sales in departmental stores, bank operations per day in a month etc. The climatic conditions in different seasons, habits and customs which are followed by people at different times are main causes of such fluctuations.

Cyclical fluctuations are oscillatory movement in time series which has period of longer duration than a year. Moreover, cyclical fluctuations may not exhibit a regular periodicity. Business cycles are affected by previous activities and also by future prospects, so are not of exactly identical periods. Because of these interrelations, cyclic movements are not random.

Irregular or random fluctuations may be divided into two types: *episodic* and *accidental*. *Episodic* fluctuations can be identified to be linked with some specific events like, war, flood, earthquake, strike, severe storms, epidemics etc. *Accidental* movements are minor fluctuations which can not be attributed to some specific event.

All such movements which can not be accounted for by trend, or seasonal or cyclical fluctuations are irregular or random.

All the time series may not be composed of all the four components. One or more of the components may not be present in some. The time series on population of England and Wales given in Fig. 18.1 is dominated by only one component. Apart from trend, none of the other components are present in this time series.

The time series on CBR in Fig. 18.5 is nearly all seasonal fluctuation, whereas that of Fig. 18.4 contains a good amount of random fluctuation.

In the next subsections the methods of determining these components are considered.

18.1.1 Determination of Components of Time Series

In order to determine a component of any time series other components are to be removed. Normally trend is determined first. For determination of other components, trend is eliminated from the time series. Since a multiplicative model describes a time series in this chapter, the original series is divided by the determined trend values. In symbols, since

$$Y_t = T_t \times S_t \times C_t \times I_t$$

Trend is eliminated by computing

$$Y_t/T_t = S_t \times C_t \times I_t \tag{18.3}$$

For all t for which T_t can be determined.

A. Determination of Trend

Trend can be determined by eliminating the other three components viz. the seasonal, cyclical and irregular fluctuations from the series. Removal of seasonal can be easily done by taking average of values for the period which is also period of the seasonal variation. Removal of the other two components is also necessary. The following methods are, therefore, applied to obtain trend.

1. Fitting Trend by Inspection

The easiest and least laborious method is fitting trend by inspection of the graph of the time series and drawing the straight line or a polynomial which is closest to all the points and smoothly passes throughout the graph. A straight line trend may be drawn using a transparent ruler which appears to be a best fit. If a non-linear trend seems to be more appropriate it may be drawn free hand or using French Curves.

However, this method is subjective and the trend values may differ from one person to the other, for the same time-series.

2. Fitting a Polynomial by Least Square to Obtain Trend

The previous method being subjective, an objective method to give a unique value of trend may be used. This consists of fitting a polynomial of degree p to the observed values of time series.

As the trend is a broad or long term movement of the series, the observations may be represented as a polynomial of degree p:

$$Y_t = a_0 + a_1 t + a_2 t^2 + \dots + a_p t^p \tag{18.4}$$

The method of least square may be applied to determine the $p + 1$ coefficients $a_0, a_1, a_2, \dots, a_p$. For this define

$$S^2 = \sum \left[Y_t - (a_0 + a_1 t + a_2 t^2 + \dots + a_p t^p) \right]^2 \tag{18.5}$$

where summation is taken over all the values of Y_i and t_i in the time series ($i = 1, 2, \dots, n$.)

$$\frac{\partial S^2}{\partial a_j} = 0,\ j = 0, 1, \dots, p \tag{18.6}$$

gives the $p + 1$ equations called normal equations from which the $(p + 1)$ coefficients a_j, $j = 0, 1, 2, \dots, p$ can be determined.

It can be seen (cf. Yule and Kendall (1958), Croxton, Cowden & Klein (1975) and Goon et al. (2002)) that the normal equations are:

$$\left.\begin{aligned}
&\sum_i Y_{t_i} - a_0 n - a_1 \sum_i t_i - a_2 \sum_i t_i^2 \ldots - a_p \sum_i t_i^p = 0 \\
&\sum_i Y_{t_i} t_i - a_0 \sum_i t_i - a_1 \sum_i t_i^2 - a_2 \sum_i t_i^3 \ldots - a_i \sum_i t_i^{p+1} = 0 \\
&\sum_i Y_{t_i} t_i^2 - a_0 \sum_i t_i^2 - a_1 \sum_i t_i^3 - a_2 \sum_i t_i^4 \ldots - a_p \sum_i t_i^{p+2} = 0 \\
&\ldots\ldots\ldots\ldots\ldots\ldots\ldots\ldots\ldots\ldots\ldots\ldots\ldots\ldots \\
&\sum_i Y_{t_i} t_i^p - a_0 \sum_i t_i^p - a_i \sum_i t_i^{p+1} - a_2 \sum_i t_i^{p+2} \ldots - a_p \sum_i t_i^{2p} = 0
\end{aligned}\right\} \tag{18.7}$$

For ease of computations to solve the $(p + 1)$ normal equations (18.7) consider the first $2m + 1$ pairs of values $\left(Y_{t_1}, t_1\right), \left(Y_{t_2}, t_2\right), \ldots, \left(Y_{t_{2m+1}}, t_{2m+1}\right) \ldots$ If n is odd $2m + 1 = n$ and last pair $\left(Y_{t_n}, t_n\right)$ is deleted in normal equations, if n is even. Now the origin is shifted to t_{m+1}, so that the t_i's ($i = 1, 2, \ldots, 2m + 1$) become $-m, -(m - 1), \ldots -2, -1, 0, 1, 2, \ldots (m - 1), m$.

With this transformation the sum of odd powers of t_i's will become zero, i.e.

$$\sum_{i=-m}^{m} t_i^r = 0 \text{ if } r \text{ is odd} = 1, 3, 5, \ldots \tag{18.8}$$

With this change of origin of t to t_{m+1}, the normal equations become (for p an odd power)

$$\sum_i Y_{t_i} - a_0 n - a_2 \sum_i t_i^2 - a_4 \sum_i t_i^4 - \ldots = 0$$

$$\sum_i Y_{t_i} - a_i \sum_i t_i^2 - a_3 \sum_i t_i^4 - \ldots = 0$$

$$\ldots\ldots\ldots\ldots\ldots\ldots\ldots\ldots\ldots\ldots$$

$$\sum_i Y_{t_i} t_i^{p-1} - a_0 \sum_i t_i^{p-1} - a_2 \sum_i t_i^{p+1} - a_4 \sum_i t_i^{p+3} \ldots = 0$$

$$\sum_i Y_{t_i} t_i^p - a_i \sum_i t_i^{p+1} - a_3 \sum_i t_i^{p+3} - \ldots = 0$$

This reduces the number of terms in each equation.

Moreover, the equations of one set contain a_0, a_2, a_4 …. and of the second set contain a_1, a_3, a_5 ….. It may also be noted that sums of even powers of t_i is twice the sums of same even powers of m natural integers, where, if $n = 13$, $m = 6$ and

$$\sum_{i=1}^{13} t_i^2 = 2\left(\sum_{k=1}^{m} k^2\right) = 2\,\frac{m(m+1)(2m+1)}{6} = 2 \times \frac{6(7)(13)}{6} = 182$$

Using formula for sums of powers of natural numbers the coefficients of a_j's in normal equations can be written. One needs to compute the sums involving the products of powers of t_i's with Y_{t_i}.

Example 18.6: Considering data of Example 18.1, if a cubic degree polynomial $Y = a_0 + a_1t + a_2t^2 + a_3t^3$ is fitted, one obtains

$$Y = 23.299 + 2.895t + 0.06153t^2 - 0.01147t^3$$

When plotted, this gives a smooth increasing curve very close to the original time series data points. Table 18.1 contains the trend values from the above fitted cubic parabola and Fig. 18.1 shows this cubic parabola also.

3. Method of Moving Averages

The method of fitting a polynomial to whole of the time series data has one disadvantage. If after passage of time some new data come up and add up to the existing series, the determination of polynomial trend has to be done afresh.

Hence a simpler method of trend determination can be used. The method is known as the method of moving averages.

In this method $2m + 1$ is not equated to n, but m is chosen a small number at will. Polynomial of degree p is fitted to these first $(2m + 1)$ points or terms of the time series. By convention, origin is taken at the $(m + 1)^{\text{th}}$ term, i.e. the middle of the first $(2m + 1)$ terms.

The normal equations will then be

$$\sum_{t=-m}^{m} Y_t t^j - a_0 \sum_{t=-m}^{m} t^j - a_1 \sum_{t=-m}^{m} t^{j+1} - \ldots - a_p \sum_{t=-m}^{m} t^{(j+p)} = 0 \tag{18.9}$$

$$\text{for } j = 0, 1, \ldots, p$$

The coefficients $a_0, a_1, \ldots, a_p$ from these normal equations give the polynomial of best fit in the least square sense, to the first $2m + 1$ terms of the time series. This polynomial may be used to determine the trend value at the origin, which is at $(m + 1)^{\text{th}}$ point. Thus the trend value is obtained by putting $t = 0$ in the polynomial (18.4), giving the trend value as a_0.

Hence only a_0 needs to be determined from (18.9). It can be seen that except the first term $\sum_{t=-m}^{m} Y_t t^j$, all other sums are sums of powers of integers from $-m$ to m, so depend only on m and p.

Substituting values of these sums, and solving for a_0, one obtains a_0 a linear function of Y_t's, given by

$$a_0 = b_1 Y_{t_1} + b_2 Y_{t_2} + \ldots + b_{2m+1} Y_{t_{2m+1}} \tag{18.10}$$

b's depend only on m and p. (18.10) indicates that a_0 is a weighted average of first $(2m + 1)$ values of time series. The weights b's can be determined if m and p are known.

The process of determining trend value ($= a_0$) is repeated by starting from Y_{t_2} i.e. using terms $Y_{t_2}, Y_{t_3}, \ldots, Y_{t_{2m+2}}$ at the mid point of these and shifting one point till the last term is used for determination of the trend for mid points of the successive $2m + 1$ terms. Thus the trend value can not be determined corresponding to the first m and the last m terms. The weights (b's) to determine the weighted averages of successive $2m + 1$ terms (which are the a_0's, the trend values) are same. Hence the name of the process is moving averages.

The trend values smoothen the series since these values correspond to the best fit at mid point of the $2m + 1$ local values by a polynomial.

As a simplest illustration of the method of moving average the case when $p = 1$ and $m = 1$ is considered now.

It means it is desired to fit straight line to set of $2m + 1 = 3$ values. From (18.9) the normal equations to fit straight line to 3 values of time series will be

$$\left.\begin{aligned} &\sum_{t=-m}^{m} Y_t t^0 - a_0 \sum_{t=-m}^{m} t^0 - a_1 \sum_{t=-m}^{m} t^1 = 0 \quad \text{and} \\ &\sum_{t=-m}^{m} Y_t t - a_0 \sum_{t=-m}^{m} t^1 - a_1 \sum_{t=-m}^{m} t^2 = 0 \end{aligned}\right\} \tag{18.10}$$

Since $\sum_{t=-m}^{m} t^j = 0$ for odd values of j, the normal equations become

$$\left.\begin{aligned} &\sum_{t=-1}^{1} Y_t - a_0(3) = 0 \\ &\sum_{t=-1}^{1} Y_t t - a_1 \sum_{t=-1}^{1} t^2 = 0 \end{aligned}\right\} \tag{18.11}$$

Trend value is given by the value of a_0. From the first of (18.11), it is given by

$$a_0 = \frac{1}{3}\left(Y_{-1} + Y_0 + Y_1\right) \tag{18.12}$$

Thus the trend value is simple average of 3 values of time series and is centered at the middle value of the three.

This way one may continue by finding simple average of 2nd, 3rd and 4th values of original time series, centered at 3rd value.

These trend values can, therefore, be called 3-point moving averages. Only corresponding to the first and at the last term in the time series, there will not be any trend value.

In general if a straight line is fitted to $2m + 1$ points, i.e. $p = 1$, from (18.10) a_0 can be determined as

$$a_0 = \frac{1}{2m+1}\left[Y_{-m} + Y_{-(m+1)} + \ldots + Y_{-1} + Y_0 + Y_1 + \ldots + Y_{(m-1)} + Y_m\right] \tag{18.13}$$

In words one can say that this method gives the trend value at any point as the simple average of $(2m + 1)$ values centered at the middle of these values i.e. at $(m + 1)^{th}$ point.

Now a case when $p = 2$ and $m = 2$ is considered, i.e. fitting a quadratic function to $2m + 1 = 5$ points. Now from (18.9) the normal equations will be

$$\sum_{t=-2}^{2} Y_t t^0 - a_0 \sum_{t=-2}^{2} t^0 - a_1 \sum_{t=-m}^{m} t^1 - a_2 \sum_{t=-m}^{m} t^2 = 0$$

$$\sum_{t=-2}^{2} Y_t t^1 - a_0 \sum_{t=-2}^{2} t^1 - a_1 \sum_{t=-2}^{2} t^2 - a_2 \sum_{t=-2}^{2} t^3 = 0$$

$$\sum_{t=-2}^{2} Y_t t^2 - a_0 \sum_{t=-2}^{2} t^2 - a_1 \sum_{t=-2}^{2} t^3 - a_2 \sum_{t=-2}^{2} t^4 = 0$$

Now $\sum_{t=-2}^{2} t^0 = 5, \sum_{t=-2}^{2} t^2 = 10$ and $\sum_{t=-2}^{2} t^4 = 34$; so to determine a_0, 1st and 3rd equations are needed. These are given by

$$\sum_{t=-2}^{2} Y_t - 5a_0 - 10a_2 = 0 \text{ and}$$

$$\sum_{t=-2}^{2} t^2 Y - 10a_0 - 34a_2 = 0$$

Solving for a_0 by eliminating a_2 (i.e. multiply 1st equation by 17/5 and subtracting it from 2nd equation one gets

$$\sum_{t=-2}^{2} t^2 Y_t - 10a_0 - 34a_2 - \left(\frac{17}{5} \sum_{t=-2}^{2} Y_t - 17a_0 - 34a_2 \right) = 0$$

or

$$7a_0 = \frac{17}{5}\left[Y_{-2} + Y_{-1} + Y_0 + Y_1 + Y_2\right] - \left[4Y_{-2} + Y_{-1} + 0Y_0 + Y_1 + 4Y_2\right]$$

which on simplification gives

$$a_0 = \frac{1}{35}\left[-3Y_{-2} + 12Y_{-1} + 17Y_0 + 12Y_1 - 3Y_2\right] \tag{18.14}$$

Thus the trend value a_0 for a quadratic fit to 5 points is again a weighted average of 5 points on the time series. It may be noted that the weights satisfy the following conditions.

(a) The sum of the weights is unity.
(b) The weights are symmetrical about their middle value. Thus one should get the same trend values if one starts from first to last value of time series or from last to the first value.

The weights for the two cases can, therefore, be written as

Case 1 $p = 1$, $m = 1$ weights: $\frac{1}{3}$[1, 1, ...] and for

Case 2 $p = 2$, $m = 2$ weights: $\frac{1}{35}$[–3, 12, 17, ...]

Following the same techniques, as in Yule and Kendall (1958), the weights for the following values of p and m are given in the following Table 18.6.

It may be verified that the same weights result for $p = 2k + 1$ and for $p = 2k$, where $k = 1, 2, \ldots$.

As an illustration of the method of moving average, the time series of sheep population is considered in the Example 18.7 after the Table 18.6.

Table 18.6 Weights for moving averages by fitting polynomial

p	m	Weights
1 Straight line	Any integer ≥ 1	$\frac{1}{2m+1}$ [1, 1,... 1,...]
2 or 3 Quadratic or cubic	2	$\frac{1}{35}$ [–3, 12, 17,...]
	3	$\frac{1}{21}$ [–2, 3, 6, 7,...]
	4	$\frac{1}{231}$ [–21, 14, 39, 54, 59,...]
	5	$\frac{1}{429}$ [–36, 9, 44, 69, 84, 89,...]
4 or 5 Quartic or quintic	3	$\frac{1}{231}$ [5, –30, 75, 131,...]
	4	$\frac{1}{429}$ [15, –55, 30, 135, 179,...]
	5	$\frac{1}{429}$ [18, –45, –10, 60, 120, 143,...]

Example 18.7: Consider Example 18.3 giving the time series on sheep population of England and Wales. Trend values are to be determined for 1920 to 1939 using

(a) 3 point simple moving average (MA)
(b) 5 point quadratic moving average and
(c) 7 point simple moving average.

Solution: Table 18.7 below gives the three trend values desired. The weights are

(a) $\frac{1}{3}$ [1, 1, 1]

(b) $\frac{1}{35}$ [–3, 12, 17,...] and

(c) $\frac{1}{7}$ [1, 1, 1, 1,...]

The data for the years 1920 to 1939 are taken in column 3 from Table 18.3 and the trend values are given in subsequent columns.

Columns 5, 7 and 9 contain the computed values of the three desired trend values. Figure 18.6 shows the original series of population size every year and the three trend values.

Figure 18.6 indicates that 7-point simple moving average gives the best trend values, as it smoothens out the series better than the other two.

Table 18.7 Computation of trend for sheep population during 1920-1939

S.No.	Year	Population (10,000)	3 point simple MA		5 point quadratic MA		7 point simple MA	
			Weighted sum (WS)	MA = $\frac{1}{3}$WS	Weighted sum (WS)	MA = $\frac{1}{35}$WS	Weighted sum (WS)	MA = $\frac{1}{7}$WS
1	1920	1338	-	-	-	-	-	-
2	1921	1383	4065	1355	-	-	-	-
3	1922	1344	4111	1370	47586	1359.6	-	-
4	1923	1384	4212	1404	48524	1386.4	10216	1459.4
5	1924	1484	4465	1488.3	51910	1483.1	10588	1512.1
6	1925	1597	4767	1589	55916	1597.6	10842	1548.9
7	1926	1686	4990	1663.3	58938	1683.9	11109	1587.0
8	1927	1707	5033	1677.7	59307	1694.5	11357	162.4
9	1928	1640	4958	1652.7	57742	1649.8	11648	1664.0
10	1929	1611	4883	1627.7	61530	1758.0	10051	1435.9
11	1930	1632	5018	1672.7	57906	1654.5	10174	1453.4
12	1931	1775	5257	1752.3	61699	1762.8	10120	1445.7
13	1932	1850	5434	1811.3	74318.1	2123.4	10128	1446.9
14	1933	1809	5312	1770.7	62520	1786.3	10132	1447.4
15	1934	1653	5110	1703.3	59040	1686.9	10127	1446.7
16	1935	1648	4966	1655.3	57524	1643.5	10143	1449.0
17	1936	1665	4940	1646.7	57273	1636.4	10090	1441.4
18	1937	1627	5083	1694.3	58796	1679.9	-	-
19	1938	1791	5215	1738.3	-	-	-	-
20	1939	1797	-	-	-	-	-	-

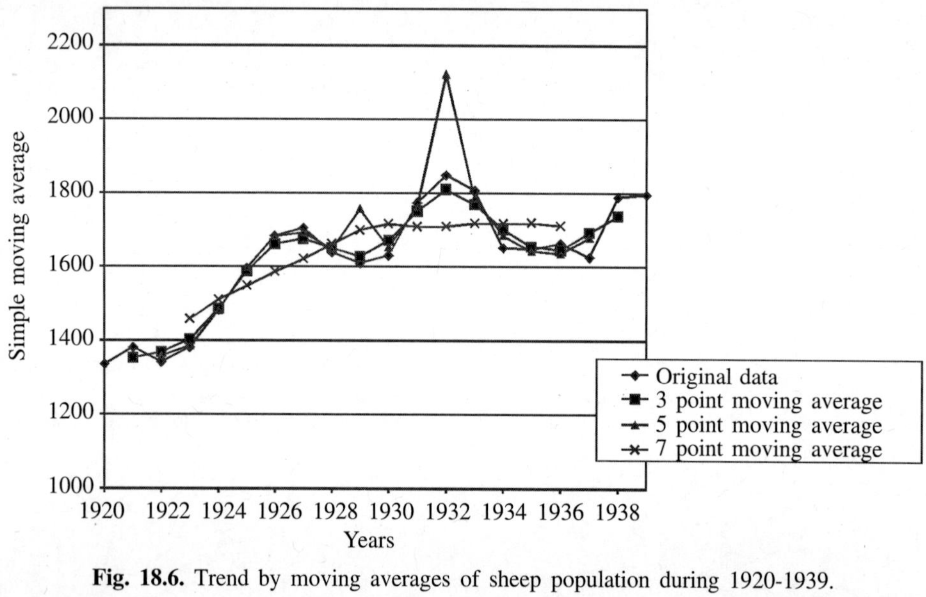

Fig. 18.6. Trend by moving averages of sheep population during 1920-1939.

4. Some Other Formulae

(a) It is quite clear that simple moving averages are computationally quite easy to apply as they have equal weights. However, one may repeat the simple moving averaging on the resulting series. Yule and Kendall (1958) observe that this process gives quite good substitute for accurate formulae without loss of efficiency of measurement of trend.

For example a simple 3-point moving average may be applied twice. First using $\frac{1}{3}(Y_1 + Y_2 + Y_3)$ on the whole series. Applying 3-point moving average again will require to compute

$$\frac{1}{3}\left[\frac{1}{3}(Y_1 + Y_2 + Y_3) + \frac{1}{3}(Y_2 + Y_3 + Y_4) + \frac{1}{3}(Y_3 + Y_4 + Y_5)\right]$$

$$= \frac{1}{9}\left[Y_1 + 2Y_2 + 3Y_3 + 2Y_4 + Y_5\right]$$

This moving average weights can be written as

$$\frac{1}{9}(1, 2, 3, ...) \qquad (18.15)$$

(b) There are two formulae which are commonly used in actuarial work viz. Spencer's 15-point and 21-point formula. Interested reader may consult the book by Yule and Kendall (1958) for details.

5. Moving Average with Even Number of Points

The number of points chosen for fitting polynomial to calculate the moving average were odd (=2m+1). If for some reason moving average is taken from even number of points, then the resulting value can not be related to a definite time point on the series. For example to eliminate seasonal effects, 4 month's average, from a monthly series, is to be calculated. Now this four month's average relates to middle of 2nd and 3rd month of a quarter; not to any of the four months. Similarly, in a monthly time series if 12 month moving average is obtained, the moving average value falls at the end of June and beginning of July. It will not relate to the middle of any month.

In such cases, a convention is to take the simple average of two consecutive moving averages (of even number of points) and thus center the resulting average as the trend value at a time point in the series. Thus if 12 month moving average is obtained, the 12 point averages are further averaged taking 2 points throughout the series of 12-point averages. Thus if the series begins at January, the first trend value (the first average of the 2 consecutive 12-month moving averages) will correspond to the trend value for July. Subsequent 2-point averages will be trend values for subsequent months in the time series.

(B) Measurement of Seasonal Fluctuations

Many time series, such as economic and meteorological series may be affected by seasons. Measurement of seasonal variation may be required for two purposes. One may like to get rid of seasonal variations and concentrate on the other components of the time series. Another purpose is to give as much importance in its measure as is given to measurement of trend. Planning of business and marketing can be done efficiently by understanding seasonal fluctuations.

Elimination of seasonal effects is a simple matter, since it can be achieved by taking observation time points at an interval larger than the period of seasonal fluctuations. Thus if observations are taken on a fixed date in a year, i.e. interval of observation is one year, the seasonal effect is eliminated. Similarly, by considering annual rainfall data (i.e. total rainfall in a year) effect of seasons is eliminated.

Methods of measurement of seasonal fluctuations are also often necessary. Hence some methods are considered now.

Attention is given to monthly movements in a year, not to the seasonal movements which may vary in pattern from year to year.

In general, these methods involve averaging, in some way, all the Januaries, all the Februaries, and so on. The methods differ in the way the time series is adjusted before averaging for 6 weekly working days or the four quarters of the year or the 12 months of the year. Various methods have been discussed by Goon et al. (2002), Croxton, Cowden and Klein (1975) and Yule and Kendal (1958). Three methods are given below.

The methods consist of finding averages and then applying appropriate correction factor to make total of the *seasonal indices* to 600 for weekly working days, to 400 for 4 quarters and to 1200 for the 12 months of the year.

1. Seasonal Indices for Unadjusted Data

This method should be applied if there is no appreciable cyclic movements or trend in the time series. Hence without adjusting the data, the averages are obtained for each day of week, for various weeks each quarter or each month for various years.

The averages over the weeks or years are obtained subsequently. The seasonal indices for 6 working days or 4 quarters, or 12 months are finally obtained by applying a minor correction to make the sums of indices 600 or 400 or 1200. Example 18.8 given below clarifies this method.

Example 18.8: The following table gives the number of books issued by a library on the 6 working days of nine weeks excluding the weeks with holidays. It is required to obtain seasonal indices from unadjusted data (source: Croxton, Cowden and Klein (1975), p. 286).

Table 18.8 Computation of seasoned index for 6 week days from data on number of issued library books

Week beginning	Monday	Tuesday	Wednesday	Thursday	Friday	Saturday
Feb 8	665	748	722	734	604	456
Feb 15	701	787	686	822	649	730
Feb 2	1,000	939	816	703	506	535
Mar 1	642	612	792	712	277	691
Mar 8	862	794	700	739	607	470
Mar 15	597	819	627	703	609	510
Apr 5	754	884	1,224	777	744	603
Apr 12	696	765	748	703	714	578
Apr 19	834	979	862	906	675	498
Total	6,751	7,327	7,127	6,799	5,385	5,071
Average = Total/9	750.1	814.1	797.4	755.4	598.3	563.4
SI	105.2	114.2	111.8	105.9	83.9	79.0

Solution: From the data without any adjustment from each of the nine weeks find the averages of number of issued books on each of the six days Mondays through Saturdays. For this find the totals of number of books on these days in nine weeks and write these totals in the first row below the table.

In the second row below the table find averages by dividing the totals by 9, the number of weeks. In the third row write the seasonal indices for each of the week days by correcting the obtained averages by multiplying each of the 6 averages by the correction factor.

$$CF = 6 \times 100/\text{Total of all the six averages}$$

Thus the seasonal index (SI) for Monday = Average for Monday × CF

or $$SI = \text{Average} \times 6 \times 100/\text{Total of averages}$$

$$= \text{Average} \times 600/(750.1 + 814.1 + ... + 563.4)$$

Thus $$\text{SI for Monday} = 750.1 \times 600/4278.7$$

Similarly the seasonal indices are obtained for the rest of the days in the last row.

2. Ratio-to Moving Average Method

The above method was an illustration of assessing intra-week movement of the time series. For intra-year movement of a series, seasonal indices are calculated from monthly data in the time series.

If daily data is available one can obtain seasonal indices giving idea of intra-week and intra-month movements. This and the next method are restricted to seasonal indices for intra-year movements, so monthly data are used.

Before using these methods, the investigator should be assured that the series has seasonal fluctuations and also contains trend and cyclic variations. The series is first adjusted by eliminating the trend and the cyclic variations by dividing it by the measure of trend. In this method centered 12-month moving average is used as measure of trend, since averaging will smooth intra-year fluctuations to a large extent.

In a *multiplicative model* division by moving-average retains in the series the seasonal and irregular movements. Seasonal indices are obtained by first averaging the 12 month's figures for each month, thus likely to eliminate the irregular or random fluctuations to a large extent, if not fully. The monthly averages are finally multiplied by CF = 1200/total of monthly averages to give the SI.

The example given below illustrates the process for the data on consumption of newsprint by US Publishers during Jan 1954 to June 1964.

Example 18.9 The following Table 18.9 gives the said data on monthly consumption of newsprint. It is desired to find seasonal indices for the 12 months using ratio to moving average method. (source: Croxton, Cowden and Klein (1975), p. 295-297)

Table 18.9 Consumption of newsprint by US Publishers*

Year and Month	Consumptions (in short tons) Y_t	12 month total T_{12}	Centered 12-month total CT_{12}	Ratio to MA in % $= (Y_t/CT_{12}) \times 1200$
1954				
January	363			
February	346			
March	400			
April	415			
May	422			
June	384			
		4683		
July	339		4693.5	86.64
		4704		
August	361		4713.5	91.92
		4723		

Contd...

Table 18.9 Continued

Year and Month	Consumptions (in short tons) Y_t	12 month total T_{12}	Centered 12-month total CT_{12}	Ratio to MA in % $= (Y_t/CT_{12}) \times 1200$
September	388		4742.5	98.16
		4762		
October	437		4770.5	109.92
		4779		
November	420		4795.5	105.12
		4812		
December	408		4831.0	101.40
		4850		
1955				
January	384		4869.5	94.68
		4889		
February	365		4901.0	89.40
		4913		
March	439		4911.5	107.28
		4910		
April	432		4904.0	105.72
		4898		
May	455		4900.5	111.48
		4903		
June	422		4932.5	102.72
		4962		
July	378		4979.5	91.08
		4997		
August	385			
September	425			
October	479			
November	462			
December	419			
⋮	⋮			
1964				
January	455			
February	452			
March	578			
April	528			
May	550			
June	490			

*Note: Table is given in part to show calculation of ratio to MA in last column.

Solution: The method consists of the following steps:

Step 1: To find 12-month moving average (centered) first find 12 month moving total and write it at middle of 12 months.

Thus the 1st 12 month moving total = Sum of first 12 month consumptions

$$= \text{Consumptions in Jan 54} + \text{Feb 54} + \ldots\ldots + \text{Dec 54}$$

$$= 363 + 346 + \ldots\ldots + 408 = 4683$$

and write it between June 54 and July 54.

In the subsequent moving totals subtract the first value i.e. Jan 54 value and add Jan 55 value.

In figures

2nd Moving Total = First moving total – Jan 54 value + Jan 55 value

$$= 4683 - 363 + 384 = 4704$$

Written between July 54 and Aug 54.

Continue to find the moving totals in this way moving down upto June 1964 to get the moving total between Dec 1963 and Jan 1964.

Step 2: Now find the centered moving total by averaging the moving totals between June-July 1954 and July-Aug 1954 and write it in line with July 1954. By averaging the two consecutive moving totals between lines, moving totals are centered starting from July 1954 and continue upto Dec 1963.

Thus the first centered moving total = (4683 + 4704)/2

= 4693.5 and is written in the line of July 1954.

Step 3: It is evident that centered moving totals are not available for the first six months: Jan 1954 to June 1954 and for the last six months of the series Jan 1964 to June 1964. The moving averages (12-months centered) can be determined for July 1954 to Dec 1963.

Find the ratio to MA by the formula

$$= (\text{Original value of series/Centered 12-month moving total}) \times 1200$$

Hence the ratio to moving average for July 1954

$$= (339/4693.5) \times 1200 = 86.64.$$

Subsequently moving down the table, ratio to MA for Aug 1954 to Dec 1963 can be found. But the above table illustrates the method for July 1954 to July 1955. The interested reader may consult the book by Croxton, Cowden and Klein (1975) for more details.

Step 4: To find the seasonal indices for each month Jan to Dec write month wise the ratio to MA values for the 9 years 1955 to 1963, since values for Jan to June 1954 are not available as in the Table 18.10 below.

Table 18.10 Ratio to MA in percentage

Year	Jan.	Feb.	Mar.	Apr.	May	Jun.	Jul.	Aug.	Sep.	Oct.	Nov.	Dec.
1955	94.6**	89.4	106.8	104.3	108.9	100.5	89.7	91.0	100.0	112.3	107.9	97.8
1956	93.7	92.5**	103.4	107.0	107.9	97.4	89.5	92.8	100.1	109.8	107.9	102.3
1957	94.0	89.4	107.2**	102.5	108.3	101.1	**87.4**	**90.5**	102.3	110.2	107.8	104.3
1958	92.7	87.7	104.5	**102.0**	**105.4**	98.7	88.4	93.7	99.3	112.2	110.1	**92.6**
1959	92.4	89.4	105.7	107.2	110.6	**97.2**	89.8	**94.3**	99.7	109.0	107.8	100.9
1960	94.5	90.8	102.6	103.8	110.7	100.2	91.2	91.5	99.1	112.8	108.7	100.3
1961	92.8	86.3	103.3	105.6	107.1	98.4	90.7	91.2	**98.4**	111.5	108.5	102.6
1962	94.0	89.6	103.3	104.3	106.9	98.0	91.5**	96.6**	105.7**	113.2**	112.4**	97.2
1963	**82.7**	**78.3**	**95.5**	107.3**	112.7**	104.5**	89.8	93.1	101.3	**108.3**	**106.6**	105.8**
Mean of 7 middle values	93.4	88.9	104.2	105.0	108.6	99.2	89.9	92.5	100.3	111.1	108.4	100.8
SI	93.22	88.73	104.00	104.80	108.39	99.01	89.73	92.32	100.11	110.89	108.19	100.61

Looking at the ratio to MA values in different months it is seen that the lowest ratio is found 78.3 in Feb 1963 and highest value is 113.2 in October 1962. Even within months there are some variations. Hence instead of using mean of nine years for each month, the two extreme values, i.e. the minimum and the maximum in each of 12 months are excluded before finding the arithmetic mean of the months over the 9 years. Excluding the minimum and maximum the mean for each of the 12 months is obtained from the remaining 7 values of ratio to MA for each month. In the Table 18.10, these extreme values are denoted by bold the minimum, and putting a ** over the maximum. These two values are not used in finding the means for the 12 months. Such modified means are written below the monthly values in Table 18.10. Now to get seasonal index each mean is multiplied by the correction factor (CF).

$$CF = 1200/(\text{Total of 12 means})$$

$$= 1200/1202.3 = .9981$$

This gives the seasonal indices (SI) after multiplication with CF. The calculated values are given in the last row. It can be verified that the total of modified seasonal indices now is 1200.

3. Ratio to Trend Method

In this method an appropriate equation is fitted to determine the trend values using least square method. The trend values are then determined for each of the months in the time series. The ratios of values in time series to corresponding trend values are determined as percentage. In symbols

$$\text{Ratio to trend percentage} = (\text{Value in series/Trend value}) \times 100$$

The different percentage ratios to trend for each of the 12 months are then averaged over the years. Finally the seasonal index for each month is obtained by multiplying each month average by correction factor defined in the last method.

Thus it may be noted that this method differs from the previous method of ratio to MA in that the trend value from a fitted equation is used in place of MA.

4. Link Relatives Method

In this method link relatives for each month are obtained from the formula

$$LR_t = (Y_t/Y_{t-1}) \times 100 \quad (18.16)$$

where LR_t is link relative for t^{th} month.

The link relatives are then averaged for each of the 12 months over the years as in the previous methods and adjusted to total to 1200.

This method was extensively used in the past, but not in current use because it can not eliminate other components of time series efficiently.

C. Determination of Cyclical Fluctuations

After measurement of two components of time series viz. the trend and the seasonal variation, one needs to measure the third component, the cyclical fluctuations.

In a multiplicative model, the time series values are

$$Y_t = T_t \times S_t \times C_t \times I_t$$

Hence to get $C_t \times I_t$, the original series is divided by trend values (T_t) and by seasonal indices (S_t). This can be accomplished in any of the following three ways:

(a) Y_t is first divided by trend value T_t and written as percentage of trend. In symbols first find

$$(Y_t/T_t) \times 100 \text{ to get } S_t \times C_t \times I_t \text{ for all the } t \quad (18.17)$$

Next divide the resulting series by S_t to de-seasonalize the series from which trend has been already removed to get $C_t \times I_t$.

(b) Y_t is first divided by S_t to get $T_t \times C_t \times I_t$ as residual and the resulting series is divided by T_t to get $C_t \times I_t$. At both the steps the resulting values are multiplied by 100 to change them to percentages. In symbols, first find

$$(Y_t/S_t) \times 100 \text{ to get } T_t \times C_t \times I_t$$

Next find $(T_t \times C_t \times I_t/T_t)$ 100 to get $C_t \times I_t$. (18.18)

(c) One may first multiply $T_t \times S_t$ in which the seasonal indices S_t are in decimals, not as percentages. Thus the trend and seasonal movements called *normal* values are obtained first. The original series Y_t is subsequently divided by these 'normal' values to get $C_t \times I_t$. In symbols, first obtain

$$T_t \times S_t \text{ and then find } Y_t/(T_t \times S_t) = C_t \times I_t \quad (18.19)$$

All the three methods give identical results except very minor differences (of the order of 0.1) due to rounding.

If an equation is fitted to yearly data to obtain the trend, slight modification of the equation may give monthly trend values centered at the month points. Thus for straight line trend if the equation of straight line is

$$Y = a + bX \quad (18.20)$$

For X in years, the modified straight line trend will become

$$Y = a^* + b^*X^*$$

where $$a^* = a/12 \text{ and } b^* = (b/144) \times 0.5 \quad (18.21)$$

Thus one gets monthly trend values and one can proceed for obtaining seasonal indices for the 12 months.

Now the first two methods to get $C_t \times I_t$ can be applied.

However, if yearly data forms the time series the seasonal fluctuation are already smoothed out. Hence the trend values obtained from yearly data may be considered to be the product $S_t \times T_t$. Division of yearly data Y_t by the yearly trend provides $C_t \times I_t$.

Now *smoothing* of $C_t \times I_t$ *is done to remove the irregular component* leaving the cyclical movements alone.

This is done by taking centered 2-point moving average of the $C_t \times I_t$ values in the residual series. Thus using 1/4 (1, 2, 1) moving average of residual ($C_t \times I_t$) series one obtains centered 2-month moving average. These are used to divide $C_t \times I_t$ values and multiplied by 100 to give the cyclical movements C_t in percentages.

D. Determination of Irregular Component

Once C_t is obtained, $C_t \times I_t$ residual series is divided by C_t to give the irregular components. Thus using residual method one can determine all the components of time series.

There are other sophisticated methods also available for determination of cyclical movements using periodogram and correlogram analysis. The interested reader may consult the books by Yule and Kendal (1958), Croxton et al. (1975) and Goon et al. (2002) for detailed discussion of the said two methods.

18.2 Autocorrelation in Time Series

In this section suppose that the time series has no trend or that the trend has been removed. The series will thus fluctuate more or less irregularly about some central value which might be the mean of the entire series.

If the variation about mean is entirely random, the successive values in the series are independent and the series may be a random sample from an unknown population. In practice, sometimes there may be departures from this entirely random arrangement. Two types of departures are normally observed: (a) peaks and troughs appear in the series and (b) neighboring terms of the series may be correlated.

Peaks and Troughs

If Y_t satisfies the condition that

$$Y_{t-1} < Y_t > Y_{t+1} \tag{18.22}$$

then Y_t is a peak.

However, if

$$Y_{t-1} > Y_t < Y_{t+1} \tag{18.23}$$

then Y_t is a trough.

In either case Y_t is a *turning point* and the interval between turning points is called a phase. If two or more values are equal and are greater than the neighboring values, they are regarded as single peak situated at the center of the range of equal values. Similar may be the case of troughs.

Let p be the number of turning points in a series of n terms. If the series is random from any population

$$E(p) = \frac{2}{3}(n-2) \text{ and} \tag{18.24}$$

$$V(p) = (16n - 29)/90 \tag{18.25}$$

In such a case, if n is large, according to Yule and Kendall (1958), p tends to a fairly normal distribution.

One can, therefore, test for randomness of the time-series by counting the number of turning points.

As an illustration consider the data on the Wolfer's sun spot numbers given in Table 18.11 and Fig. 18.7.

Table 18.11 Wolfer's sunspot numbers of the years (1853-1900)

Year	Number	Year	Number	Year	Number	Year	Number
1853	39.0	1865	30.5	1877	12.3	1889	6.3
1854	20.6	1866	16.3	1878	3.4	1890	7.1
1855	6.7	1867	7.3	1879	6.0	1891	35.6
1856	4.3	1868	37.3	1880	32.3	1892	73.0
1857	22.8	1869	73.9	1881	54.3	1893	84.9
1858	54.8	1870	139.1	1882	59.7	1894	78.0
1859	93.8	1871	111.2	1883	63.7	1895	64.0
1860	95.7	1872	101.7	1884	63.5	1896	41.8
1861	77.2	1873	66.3	1885	52.2	1897	26.2
1862	59.1	1874	44.7	1886	25.4	1898	26.7
1863	44.0	1875	17.1	1887	13.1	1899	12.1
1864	47.0	1876	11.3	1888	6.8	1900	9.5

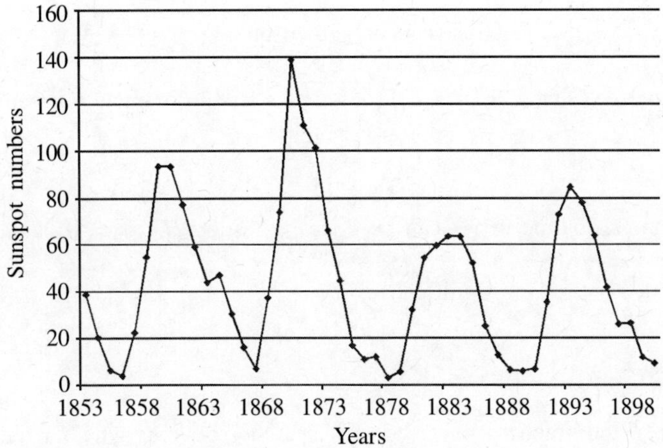

Fig. 18.7. Wolfer's sunspot numbers of the years (1853-1900).

In this series n = 48 and observed p = 14.

$$E(p) = \frac{2}{3}(48-2) = 30.67 \text{ and } V(p) = (16 \times 48 - 29)/90 = 8.21$$

$$\frac{p - E(p)}{\sqrt{V(p)}} = \frac{14 - 30.67}{2.87} = -5.81$$

Thus p departs significantly from normal distribution and the series can not be regarded as random. Another type of departure from randomness in a time series is presence of correlation between successive numbers in the series. Such type of correlation is called serial correlation or autocorrelation.

A formal definition of autocorrelation of order k or lag k (ρ_k) in a series is given by

$$\rho_k = \frac{\text{Cov}(Y_t, Y_{t+k})}{\sqrt{V(Y_t)V(Y_{t+k})}} \tag{18.26}$$

In a long time series $V(Y_t)$ and $V(Y_{t+k})$ are practically equal. For a short observed series variance of Y_{t+k} is also estimated from all the n terms and hence (18.26) becomes

$$\rho_k = \frac{\text{Cov}(Y_t, Y_{t+k})}{V(Y_t)} \tag{18.27}$$

Let y_t be deviations of the series members from the mean of the series, then the sample autocorrelation of lag k will be:

$$r_k = \left[\frac{\sum_{t=1}^{n-k} y_t y_{t+k}}{n-k}\right] \Bigg/ \left[\sum y_t^2/n\right] = \frac{n}{n-k}\left[\frac{\sum_{t=1}^{n-k} y_t y_{t+k}}{\sum_{t=1}^{n} y_t^2}\right] \tag{18.28}$$

If the time series is a random series or Y_t's are independent Cor $(Y_t, Y_{t+k}) = 0$. Hence theoretically for a random series

$$\rho_k = 0 \text{ and it can be shown that } V(r_k) \approx \frac{1}{n-k} \tag{18.29}$$

For large n, it may be seen that $Z = (r_k - \rho_k)\sqrt{n-k}$ has an approximate normal distribution. If series is random $\rho_k = 0$, so

$$Z = r_k\sqrt{n-k} \sim N(0, 1) \tag{18.30}$$

So a test of randomness of time series can be done by testing H_0: $\rho_k = 0$ against H_1: $\rho_k \neq 0$, using Z of (18.30).

H_0 is rejected at 5% level of significance, if

$$|Z| > 1.96 \tag{18.31}$$

A Remark

Often by inspection of the graph of time series, it becomes obvious whether the series is systematic to some extent or there are random fluctuations. The two tests mentioned above may be used to conclude if departure from randomness exists.

Oscillatory behaviour is studied by periodogram analysis and also by correlogram. Correlogram is a graph between k on the X-axis and r_k on the Y-axis.

If the time series is autoregressive (to be defined below) type, the theoretical nature of correlogram will have $r_k = 1$ for $k = 0$ and will decrease in magnitude to 0 for $k = \infty$. A departure of correlogram from this also indicates presence of oscillatory movements in the time series.

Autoregressive Time Series

As already mentioned earlier, and also observed by Yule and Kendall (1958) and Gujrati (2004), in general, movements in a time series are determined by two factors (a) a group of systematic properties and (b) a series of external shocks or error term which is a random variable.

Thus one can write the terms of a time series by

$$Y_t = f(t) + u_t \tag{18.32}$$

where $f(t)$ is a linear or a non-linear function of time and some explanatory variables in the system.

U_t is the disturbance or error term is a random variable, that has well defined probabilistic properties.

Further, instead of just mentioning that u_t's are generated such that $E(u_t u_{t+s}) \neq 0$ for $s \neq 0$, there are situations when the disturbance or error terms are generated with the following relationship:

$$u_t = \rho u_{t-1} + \in_t, \quad -1 < \rho < 1 \tag{18.33}$$

where ρ is a constant known as coefficient of autocovariance and $\in_t$ is stochastic disturbance term satisfying the following ordinary least square (OLS) assumptions:

$$E(\in_t) = 0,\ V(\in_t) = \sigma^2 \text{ and Cor } (\in_t, \in_{t+k}) = 0 \text{ for } k \neq 0 \tag{18.34}$$

(18.33) indicates that the value of disturbance at time $t(= u_t)$ is linearly related (ρ times) to its value at the previous time point plus a purely random error term.

The model given by equation (18.33) is known as first order autoregressive model or process and denoted as AR(1) model.

It is called autoregressive because it may be considered to be a regression of u_t on itself but lagged by one period of time i.e. on u_{t-1}. It is called first order AR model since u_t and its immediate past value u_{t-1} involved.

In general an autoregressive model of order q ($AR(q)$) is given by (cf. Wetherill et al. (1986) and Gujrati (2004))

$$u_t = \rho_1 u_{t-1} + \rho_2 u_{t-2} + \rho_3 u_{t-3} + \ldots + \rho_q u_{t-q} + \epsilon_t \tag{18.35}$$

In the $AR(1)$ model, ρ the coefficient of autocovariance is also called the first-order coefficient of autocorrelation or the coefficient of autocorrelation of lag 1.

The marginal variance of u_t is given by

$$V(u_t) = E(u_t^2) = \sigma_\epsilon^2 / (1 - \rho^2) \tag{18.36}$$

$$\text{Cov}\,(u_t, u_{t+k}) = E(u_t, u_{t+k}) = \rho^k \frac{\sigma_\epsilon^2}{1 - \rho^2} \quad \text{(18.37}$$

$$\text{Corr}\,(u_t, u_{t+k}) = \rho_k = \rho^k \tag{18.38}$$

Because of symmetry property of covariances and correlations

$$\text{Cov}\,(u_t, u_{t+k}) = \text{Cov}\,(u_t, u_{t-k}) \text{ and}$$

$$\text{Corr}\,(u_t, u_{t+k}) = \text{Corr}\,(u_t, u_{t-k}) = \rho^k \tag{18.39}$$

ρ lies between $-1 < \rho < 1$, so for large value of k, ρ_k $(= \rho^k)$ becomes very small, as mentioned in the discussion about correlogram.

Some other models of time series have also been suggested. These are (a) MA (q), the moving average of order q model (b) ARMA (p, q), the autoregressive of order p and moving average of order q model (c) ARIMA (p, d, q) the autoregressive integrated moving average model. A brief and symbolic description is given below.

For detailed discussions, the interested reader may consult advanced books on econometrics, time-series or regression analysis (e.g. Gujrati (2004), Wetherill et al. (1986)).

(a) MA(q), the moving average model of order q is represented as:

$$u_t = \epsilon_t + \alpha_1 u_{t-1} + \alpha_2 u_{t-2} + \ldots + \alpha_q u_{t-q} \tag{18.40}$$

(b) ARMA (p, q) is a combined p autoregressive and q moving average series, given by

$$u_t = \rho_1 u_{t-1} + \rho_2 u_{t-2} + \ldots + \rho_p u_{t-p} + u_t + \alpha_1 u_{t-1} + \ldots + \alpha_q u_{t-q} \tag{18.41}$$

(c) The ARIMA (p, d, q) is an autoregressive integrated moving average time series. It has p autoregressive terms, d is the number of times the series has to be differenced before it becomes stationary and q is the number of moving average terms.

An ARIMA (2, 1, 2) times series becomes stationary if differenced once i.e. $d = 1$. The differenced series then becomes ARMA (2, 2) series. If $d = 0$, the original series is stationary.

The stationarity in time series and its testing is discussed in the next section.

18.3 Stationarity in Time Series, Transformation and Tests of Stationarity

A time series is stationary, if its probabilistic characteristics e.g. mean, variance and covariance do not depend on time (or are time invariant). This implies that these characteristics of a time series do not change with time. Thus the mean and variance of, a stationary time series are constant over time and the covariance between the two time periods depends only on the distance or gap or lag between the two time periods not on the actual time (t) at which the covariance is computed.

In symbols

$$\text{Cor}\ (u_t, u_{t+k}) = \rho_k \text{ does not depend on } t. \text{ It depends on } k \text{ only.}$$

If a time series is not stationary, it will have a time varying mean or a time varying variance or both. If the time series is not stationary, its behavior can be studied only for the period, the series is available. The results can not be generalized for future time periods or for the past time periods. Therefore, for the purpose of forecasting or prediction, non-stationary time series may not be of practical value.

Hence one must be able to decide, whether a time series is stationary or not. Some tests are there for this purpose, which should be applied before prediction. Once a model has been decided i.e. the coefficients should be estimated for proper prediction. Choice of an appropriate model i.e. determination of p, d and q is also a problem, which must be solved before estimation of required coefficients and subsequently for prediction.

Softwares have been developed which can be carefully used to decide the model of the time series. An appropriate estimation procedure has to be opted for finally prediction for future.

18.3.1 Tests of Stationarity in Time Series

In this section, therefore, some .common methods of testing stationarity are discussed briefly.

(a) Graphical Inspection

Before applying a formal test, it is advisable to plot the time series data on a graph as has been done in Fig. 18.1-18.5. Such a plot gives a clue about the possible nature of time series. Thus the graph of sheep population in England and Wales in Fig. 18.2 shows a decline in population suggesting perhaps the average population over time has been changing. However, the graphs of percentage loss of ships and the crude birth rates, in Fig. 18.3 and 18.4, respectively, do not give a feel that the mean is changing over the period.

Such an intuitive feel is an starting point of more formal tests.

(b) The Correlogram

As already described a correlogram is a graph with lag k on the X-axis and the sample autocorrelation coefficients r_k on the Y-axis. If the series is stationary the autocorrelations will be closed to zero for various lags. Hence if correlogram is of this nature, one can say that the time series is likely to be stationary.

A test of hypothesis H_0: $\rho_k = 0$ against H_1: $\rho_k \neq 0$ may also be performed. For this the normal test given in (18.30) and (18.31) may be applied. A decision about the choice of k for testing H_0 is also needed. A rule of thumb is to take $k = n/3$ or $n/4$ (c.f. Gujrati (2004)).

(c) *Q*-Statistic

Box and Pierce (1970) developed a test of the joint hypothesis that all the ρ_k up to certain value of k are simultaneously equal to zero. In symbols to test H_0: $\rho_1 = \rho_2 = \ldots = \rho_m = 0$ against H_1: All the m ρ_k's are not equal to zero.

They suggest to use the statistic

$$Q = n\sum_{k=1}^{m}\hat{\rho}_k^2 = n\sum_{k=1}^{m} r_k^2 \tag{18.42}$$

where n: number of terms in time series and m: maximum lag length.

For large n, Q has approximately a χ^2-distribution with m d.f. if H_0 is true. So H_0 is rejected at 100 α% level if

$$Q > \chi^2_{m,\alpha} \tag{18.41}$$

(d) *LB* Statistic, suggested by Ljung-Box (LB) in 1978 is a variant of Q statistic. It is given by

$$LB = n(n+2)\sum_{k=1}^{m}\left(\frac{r_k^2}{n-k}\right) \tag{18.42}$$

LB statistic also follows χ^2-distribution with m d.f. if H_0 is true. So H_0 is rejected if

$$LB > \chi^2_{m,\alpha} \tag{18.43}$$

LB test is found to be more powerful than the Q-test if n is small.

(e) The Unit Root Test

This test is currently the most popular test of stationarity. Suppose that the time series is a unit root process given by

$$Y_t = \rho Y_{t-1} + u_t \quad -1 < \rho < 1 \tag{18.44}$$

Subtracting Y_{t-1} on both the sides

$$Y_t - Y_{t-1} = \rho Y_{t-1} - Y_{t-1} + u_t$$

or

$$\Delta Y_t = (\rho - 1)\, Y_{t-1} + u_t \tag{18.45}$$

This equation is regression equation of first difference of time series Y_t on Y_{t-1}. If $\rho = 1$ the equation (18.44) indicates that the series will be stationary. If this holds or root is unity the value of regression coefficient $(\rho - 1)$ in (18.45) will be zero.

So the method of testing for stationarity consists of the following steps:

1. Find the first difference $Y_t - Y_{t-1} = \Delta Y_t$ of the time series.
2. Find the regression coefficient (slope) of line of regression of ΔY_t on Y_{t-1}.
3. The estimated slope i.e. estimate of $\delta = \rho - 1$ follows the distribution whose critical points are given by Dickey and Fuller (1979) and are given below.

Table 18.12 1% and 5% critical points of Dickey and Fuller Unit Root Test (source: Gujrati (2004), p. 975)

n	25	50	100	250	500	∞
1% critical value	–2.66	–2.62	–2.60	–2.58	–2.58	–2.58
5% critical value	–1.95	–1.95	–1.95	–1.95	–1.95	–1.95

4. If H_0: $\delta = 0$ *is not rejected*, the *time series is non-stationary.*
   If H_0 *is rejected the time series is stationary.*

The test is named as DF unit root test, as Dickey and Fuller developed this test.

18.3.2 Transformation of Non-Stationary Time Series

Since a non-stationary time series leads to difficulty in prediction and drawing other conclusions, a transformation may be applied on the time series to make the series stationary. This requires to find if the time series is difference stationary series (DSS) or it is trend stationary series (TSS). The following transformations may be done to make the series stationary.

(a) If the series has a unit root, the first differences on the series are stationary.
(b) If the series is stationary around a trend as in Fig. 18.2, the time series Y_t is regressed over time. In other words a trend is fitted to all the time series points. The residuals from this regression i.e. ratio to trend values (for multiplicative model) and differences if additive model is presumed, become a stationary series.

Such a transformation is called detrending.

18.4 Prediction or Forecasting

Prediction or forecasting is an important part of time-series analysis. For some it may be the most important. One method is very common. It is based on ARIMA model and also known as Box-Jenkins (1978) methodology. The topic of time series forecasting or prediction is very vast and many books are available on this topic. The interested reader may consult any of the books by Box and Jenkins (1978), Makridakis et al. (1998), Pindyck and Rubinfield (1998), Gujrati (2004).

As observed by Gujrati (2004), there are five approaches to economic prediction based on time series data: (a) Exponential smoothing method (b) Single-equation regression models (c) Simultaneous-equation regression models (d) Autoregressive integrated moving average (ARIMA) model and (e) Vector auto regression (VAR).

The ARIMA or Box-Jenkins (BJ) methodology, which is most popular, does not construct single equation or simultaneous-equation models. It consists of analyzing the probabilistic or stochastic properties of time series under the philosophy: *Let the data speak for themselves*, unlike the regression models, the BJ method allows Y_t to be explained by past or lagged values of Y-itself and stochastic error terms. For a *univariae* ARIMA model i.e. pertaining to a single time series, the steps to be followed are given in brief. For details the interested reader may consult the above mentioned books and manuals of statistical software packages like SAS.

The BJ methodology determines the values of p, d, q of ARIMA model before forecasting. The steps are as follows:

Step 1 Identification and Checking for Stationarity from

(i) Autocorrelation function (ACF) and partial autocorrelation function (PACF) with tentative choice of p, d, q.
(ii) Autocorrelation check using Box-Pierce and LB statistics.
(iii) Dickey-Fuller (DF) or Augmented Dickey-Fuller (ADF) Unit Root tests.
(iv) Tentative order Selection Tests (SCAN: smallest eigen value of ARMA process correlation matrix).

Step 2 Estimation for Tentative Model

(i) Computation of Least Square Estimates of the parameters of tentative ARIMA model.

Step 3 Diagnostic Checking

(i) Residuals from the model are white noise i.e. have no autocorrelation (using LB test) indicates that whether another model fits better.
(ii) Akaike Information Criterion (AIC) and Schwartz Bayesian Criterion (SBC) are evaluated to see goodness of model.
(iii) Repeat the above three steps with changed (p, d, q) if LB test is rejected and/or if AIC and SBC can be improved.

Step 4 Forecasting or Prediction

Prediction or forecasting can be done after selection of proper model. Mostly, the forecasts obtained by this method are more reliable than those from traditional models. However, each case in the available data should also be checked.

19. Bioassay

The title of the chapter bioassay is a combination of two words: biological and assay. Bioassay may, therefore, be considered as a commonly used shorthand term. Biological assay is a type of scientific experiment for determination of nature and constitution of a stimulus which may be physical or chemical or biological or physiological or psychological. In its widest term, according to Finney (1949, 1964 and 1981), bioassay may be defined as measurement of potency of any stimulus by means of its reactions which it produces in living matter. It is a form of biological experiment to determine the potencies of treatments on a standard scale. It thus goes beyond comparing the magnitudes of effects of treatments (as in ANOVA and multiple comparisons).

Bioassays are essential in pharmaceutical, medical and agricultural sciences for development and launching of new drugs, vitamins, hormones, pesticides and fungicides. Potency of a natural product or herbal material can not be estimated without a bioassay.

Though the basic principle of biological assay was in use for a long time but it could not become an instrument for precise measurement until recently. The serious scientific use began at the close of 19th century when Ehrlich attempted to standardize diphtheria antitoxin. Standardization of drugs or toxicants became a practice before launching a new product by pharmaceutical industries and recommending marketing of a new pesticide and fungicide by agricultural scientists.

Thorough attention to this field was given by Dale (1939), Gautier (1945), Hartley (1935, 1945), Miles (1948). Irwin (1937) and Bliss & Cattell (1943) described many of the ideas and methods of this new science in their papers. Bliss and Cattell also gave a bibliography of the developments in the area of bioassay during the first half of the 20th century. The books by Finney (1952, 1964, 1945, and 1981) are perhaps the standard texts explaining very clearly detailed statistical techniques of bioassay. A more recent book by Govindarajulu (2001) includes the new developments in the theory and practice of bioassay during the later half of 20th century.

In this chapter a brief and complete account of some basic techniques of bioassay is given. The statistical method is included in all, but without giving a proof, as it may be too complicated and beyond the level of mathematics expected in the readers.

The next section gives the types of biological assays and considers one quantitative assay. In subsequent sections some more quantitative and qualitative response assays are considered, which are commonly used.

19.1 Types of Biological Assays, Direct Assays

Before describing the types of biological assays, it is in place that the basic structure of a bioassay is given first.

Structure of Bioassay

The typical bioassay consists of a stimulus applied to subject. A stimulus may be a vitamin, a hormone, a medicinal drug, a fungicide, a pesticide, etc, whereas, a subject may be an animal, a piece of tissue from an animal, a plant, a bacterial culture, a group of patients suffering from a given disease say diabetes or hypertension.

The intensity of the stimulus may be varied, as desired by the investigator. Thus varying the doses given to the subjects, which can be precisely measured as a weight, or volume or a concentration. Application of stimulus produces a biological response in the subject, which also can be precisely measured. The magnitude of the change will depend upon the dose.

Response of the subject may be (i) change in weight of whole animal or of some particular organ, (ii) an analytical value of blood sugar, (iii) blood cholesterol level, (iv) bone ash percentage, (v) recovery from symptoms of dietary deficiency, and (vi) occurrence or non-occurrence of a muscular contraction after application of stimulus.

The relationship between dose and response is not exact always, since it is obscured due to random variations between replicate subjects. Moreover, if the response is such that it can be produced more than once in the same subject, the responsiveness of the subject at different times may affect the relationship. However, still the relationship can be used to indicate the potency of a dose from knowledge of the response produced by it.

Biological assays estimate the potency of stimulus relative to its *standard preparation.* This standard preparation may be a convenient working standard adopted as a scale of reference by the experimenter or it may be of a wider use.

The Permanent Commission on Biological Standardization, established by the Health Organization of the League of Nations, has taken an initiative for defining, preparing and supplying the standard preparation of many therapeutic substances. If such substances are to be assayed, the standard preparation may be a sample of either the international standard or of a readily available working standard whose potency against the international standard has been always accurately evaluated.

Unit of the potency of standard is expressed either in the common units of weight or volume or in the international units adopted for insulin or penicillin.

A test preparation of stimulus of unknown potency is applied to biological subjects and the mean response to a *selected dose* is determined by bioassay. A dose of the standard preparation is also determined which produces the same average response. Experimentation with several doses of the test and the standard preparation is almost always done, before equating the doses of test and standard preparations.

The ratio of the two equally effective doses is an estimate of the potency of the test preparation relative to that of the standard preparation.

Some criteria of international standards Finney (1964) mentions the following criteria of 'the biological standard' for international adoption:

1. Standard is a dry preparation of an arbitrarily chosen but representative sample of the substance for which it is to serve as standard. When a pure substance is available, the samples must comply with accurately defined physical constants. Since the standard must satisfy world requirements for many years, the quantity of standard must be large.
2. The standard must be stable. So must be prepared in the absolutely dry condition and preserved constantly at or below freezing point, in sealed containers and protected from the action of light, moisture and oxygen.
3. The standard must be dispensed in such a form as to be readily accessible and capable of being brought into use by the experimenters with a minimum trouble, additional manipulation or delay.

Importance of Statistics in Biological Assay

In the light of available information the design of experiment should be constructed in such a way as to give the most useful and reliable results. The methods of biological assay are based on some of the general principles of statistics, and the experimenter should be aware of those. The method of analysis of the data should also be properly chosen so that the best use of the observed data is made, according to the adopted experimental design. Then only a proper estimate of potency of test preparation can be obtained.

While designing the experiment, the number and magnitude of the doses of each preparation to be tested, the number of subjects to be used at each dose, the system of allocating subjects to various doses, the order in which the subjects will be treated and the response measured and related characteristics of the experiment should be well defined. If the design of experiment is not proper, i.e. adequate thought has not been given to all the points of exploration, no analysis of experimental data can provide answers on such points. According to Fisher, even an imperfect analysis of a well designed experiment may give substantially correct conclusions.

Since the main object of every biological assay is to estimate certain parameters, especially the potency of the test preparation relative to standard and to assign a confidence limit to it, the theoretical and practical problems of statistical estimation play a very important role in bioassay. The validity of the assay needs to be checked by various significance tests, without which, the assay is worthless.

Before giving the types of biological assays some notations and definitions are given below.

Stimulus, subject, dose and response are already defined.

S will denote the standard preparation and T will denote test preparation

ρ denotes the true value of the potency

R denotes the estimate of ρ i.e. $\hat{\rho} = R$.

Apart from these very frequently used terms, others will be defined when used, if different from various terms already used in the previous chapters of this book.

Types of Biological Assay

Bioassays may be qualitative or quantitative.

Qualitative bioassays are used to assess the physical effects of substances on living organisms, which can not be measured as a quantitative variable. It may be an abnormal development or deformity.

Quantitative bioassays involve measurement of the response of stimulus on living organisms. Mainly three types of biological assays, apart from qualitative assays, are commonly used:

(i) Direct assays
(ii) Indirect assays based on quantitative response
(iii) Indirect assays based on quantal ('all or nothing' or 'presence or absence') response.

Statistical analyses required in the last two types of bioassays are very similar. Dose-response regression relationship is used in both of the last two. But the full calculations in the quantal responses are a little more complex because different weights on observations are needed.

Logical structure of the quantal response assays is more closely related to the direct assays than to the quantitative response assay listed as (ii) above. Hence an understanding of direct assays is needed first for proper understanding of the last two.

19.1.1 Direct Assays

In this section the principle and procedure of direct assays are given. It may also be found in the books by Finney (1964) and Govindarajulu (2001).

The doses of the standard and the test preparations sufficient to produce a pre-specified response are directly measured. The ratio between these doses is an estimate of the potency of the test preparation relative to the standard preparation.

This potency is the amount of standard preparation equivalent in effect to one unit of the test.

Such a technique is practicable only for certain stimuli and subjects. This method is applicable if the measurement of just the needed dose is possible. The response must be clear cut and easy to recognize. The dose must be capable of administration in such a manner that the exact amount needed to produce the specified response can be given and measured.

This critical dose generally varies from subject to subject and also from one application to other, for the same subject. The value of potency for direct assay is, therefore, only an estimate of the true potency ρ, and always an average of the critical doses of test and standard are used to calculate it.

For accuracy, it is required that applications of test and the standard preparations be done on the same subject so that differences between subjects do not creep in the estimation process.

However, it is not always possible, because the same subject may not be available for use again, since the response may be death or the application of stimulus once may bring some changes in the subject which can not be comparable to the condition of the subject before previous application.

Experimental Procedure of Direct Assay

Let A be the test preparation whose potency is to be determined relative to a preparation B, which is the standard.

Subjects are allotted randomly to the two preparations. The subjects are infused with the respective preparations till the prespecified response is produced. The just needed dose to produce the response is measured for each subject.

Suppose the *fatal dose* or *tolerance* of the k subjects under each of the two preparations, are determined as, under

Test preparation $\quad A : x_{A1}, x_{A2}, \ldots\ldots, x_{Ak}$

Standard preparation $\quad B : x_{B1}, x_{B2}, \ldots\ldots, x_{Bk}$

Now calculate average doses
$$\bar{X}_A = \frac{1}{k}\sum_{i=1}^{k} x_{A_i} \tag{19.1}$$

and
$$\bar{X}_B = \frac{1}{k}\sum_{j=1}^{k} x_{B_j} \tag{19.2}$$

The estimate of relative potency of A relative to the standard B is given by

$$R = \frac{\bar{X}_B}{\bar{X}_A} \tag{19.3}$$

Indicating that 1 unit of A is equivalent to R units of the standard preparation.

Obtaining an estimate of potency for individual observations, known to have inherent variation is not sufficient without some indication of the limits of error which might be present. Hence the precision and the confidence limits of R are considered.

Precision of the Estimate *R*

To obtain the standard error of R, the following result is used stating the coefficient of variation (C.V.) of the ratio of two independent estimates as function of C.V.'s of the two individual estimates a and b.

The Result

$$(\text{C.V. of } a/b)^2 = (\text{C.V. of } a)^2 + (\text{C.V. of } b)^2 \tag{19.4}$$

where a and b are two independent estimates and definition of

$$\text{C.V. of } a = (\text{Standard error of } a)/a \tag{19.5}$$

The result (19.4) holds if C.V. of b is small.

Applying this result on R

$$(\text{C.V. of } R)^2 = (\text{C.V. of } \bar{X}_B)^2 + (\text{C.V. of } \bar{X}_A)^2 \tag{19.6}$$

Since the true standard errors involving the variability in the population of test and standard, i.e. σ_A^2 and σ_B^2 are not known, the relation (19.6) will be changed into the estimate of coefficient of variation, in place of C.V. Thus

The equations (19.5) and (19.6), therefore, become

$$\frac{\hat{V}(R)}{R^2} = \frac{\hat{V}(\bar{X}_B)}{\bar{X}_B^2} + \frac{\hat{V}(\bar{X}_A)}{\bar{X}_A^2}$$

or

$$\hat{V}(R) = R^2\left[\frac{\hat{V}(\bar{X}_B)}{\bar{X}_B^2} + \frac{\hat{V}(\bar{X}_A)}{\bar{X}_A^2}\right] = \frac{1}{\bar{X}_A^2}\left[\hat{V}(\bar{X}_B) + R^2\hat{V}(\bar{X}_A)\right] \tag{19.7}$$

It is well known that

$$\hat{V}(\bar{X}_A) = \frac{s_A^2}{k} \text{ and } \hat{V}(\bar{X}_B) = \frac{s_B^2}{k} \tag{19.8}$$

where s_A^2 and s_B^2 are the unbiased estimators of σ_A^2 and σ_B^2, respectively, given by

$$s_A^2 = \frac{1}{(k-1)}\sum_{i=1}^{k}\left(X_{A_i} - \bar{X}_A\right)^2 \text{ and } s_B^2 = \frac{1}{(k-1)}\sum_{j=1}^{k}\left(X_{B_j} - \bar{X}_B\right)^2 \tag{19.9}$$

Substituting from (19.8) and (19.9) in (19.7)

$$\hat{V}(R) = \frac{1}{\bar{X}_A^2}\left[\frac{s_B^2}{k} + R^2\frac{s_A^2}{k}\right] = \frac{1}{k\bar{X}_A^2}\left[s_B^2 + R^2 s_A^2\right] \tag{19.10}$$

If $\sigma_A^2 = \sigma_B^2$, then one may obtain the pooled estimate of the common variance ($\sigma_A^2 = \sigma_B^2$) as

$$s^2 = \frac{(k-1)s_B^2 + (k-1)s_A^2}{2(k-1)} = \frac{\sum_{j=1}^{k}\left(X_{B_j} - \bar{X}_B\right)^2 + \sum_{i=1}^{k}\left(X_{A_i} - \bar{X}_A\right)^2}{2k-2} \tag{19.11}$$

In this case, the (19.7) and (19.10) becomes

$$\hat{V}(R) = \frac{1}{k\bar{X}_A^2}\left[s^2 + R^2 s^2\right] = \frac{s^2}{k\bar{X}_A^2}(1+R^2) \tag{19.12}$$

If the number of observations receiving the test and standard preparations are not same (= k), but are different, say are n_1 and n_2, respectively, then

$$\hat{V}(R) = \frac{s^2}{\bar{X}_A^2}\left(\frac{1}{n_2} + \frac{R^2}{n_1}\right) \tag{19.13}$$

The 100 (1 – α)% confidence limits for the unknown potency of test preparation are given by (R_L, R_U), where

$$R_L = R - \hat{\sigma}_R\, t_{\alpha/2, n_1+n_2-2}$$

and

$$R_U = R + \hat{\sigma}_R\, t_{\alpha/2, n_1+n_2-2} \tag{19.14}$$

where $\hat{\sigma}_R = \sqrt{\hat{V}(R)} = \frac{s}{\bar{X}_A}\left(\frac{1}{n_2} + \frac{R^2}{n_1}\right)^{1/2}$ and $t_{\alpha/2,(n_1+n_2-2)}$ is two tailed (100 α)% significant point of $t_{n_1+n_2-2}$ distribution.

In other words $t_{\alpha/2,(n_1+n_2-2)}$ is a point of Students' t-distribution with $n_1 + n_2 - 2$ d.f. such that

$$\rho\left[\left|t_{n_1+n_2-2}\right| > t_{\alpha/2, n_1+n_2-2}\right] = \alpha$$

The above procedure of estimating potency of test preparation by direct assay can be illustrated by the following example taken from Finney (1964, p. 22).

Example 19.1: The following table shows the fatal doses or tolerances of two groups of cats for two tinctures of *strophanthus* (*A* and *B*). The tincture *B* is to be regarded as standard preparation. The fatal doses are recorded as quantities per kg of body weight of cats. Seven cats were randomly assigned to the two preparations.

Table 19.1 Tolerances of cats for tinctures of *Strophanthus*

Serial No.	*Strophanthus A* (in 0.01 cc per kg)	*Strophanthus B* (in 0.01 cc per kg)
1	1.55	2.42
2	1.58	1.85
3	1.71	2.00
4	1.44	2.27
5	1.24	1.70
6	1.89	1.47
7	2.34	2.20
Total	11.75	13.91
Mean	1.68	1.99

Solution: Here $n_1 = n_2 = k = 7$. Also assume that variability of tolerance for tinctures are equal.

Here $\bar{X}_B = 0.0199$ and $\bar{X}_A = 0.0168$

So

$$R = \frac{\bar{X}_B}{\bar{X}_A} = \frac{0.0199}{0.0168} = 1.18$$

1 cc of tincture of *Strophanthus A* is equivalent to 1.18 cc of tincture of *Strophanthus B*.

Now to find the confidence limits, first the $\hat{V}(R)$ is determined

$$\hat{V}(R)=\frac{s^2}{k\bar{X}_A^2}(1+R^2)$$

where s^2 is the estimated pooled variance given by

$$s^2=\frac{\sum_i (X_{A_i}-\bar{X}_A)^2+\sum_j (X_{B_j}-\bar{X}_B)^2}{7+7-2}$$

$$=\frac{1}{12}\left[\left\{\sum_i X_{A_i}^2-\frac{\left(\sum X_{A_i}\right)^2}{7}\right\}+\left\{\sum_j X_{B_j}^2-\left(\sum X_{B_j}\right)^2\Big/7\right\}\right]$$

$$s^2=\frac{1}{12}\left[\left\{1.55^2+1.58^2+...+2.34^2-\frac{(11.75)^2}{7}\right\}\right.$$

$$\left.+\left\{2.42^2+1.85^2+...+2.20^2-\frac{(13.91)^2}{7}\right\}\right]$$

$$=\frac{1}{12}[\{20.4819-19.7232\}+\{28.3227-27.6412\}]$$

$$=\frac{1}{12}[.7587+.6815]=.1200$$

Now
$$\hat{V}(R)=\frac{s^2}{7.\bar{X}_A^2}[1+R^2\}=\frac{.1200}{7\times(1.68)^2}[1+1.18^2]$$

$$= 0.0145$$

$$\hat{\sigma}_R=\sqrt{\hat{V}(R)}=\sqrt{0.0145}=.120$$

The 95% confidence limits for potency of A is given by (R_L, R_U) where

$$R_L=R-\hat{\sigma}_R t_{.025,12} \text{ and } R_U=R+\hat{\sigma}_R t_{.025,12}$$

Now from Table A.4 of Student's t-distribution, $t_{.025,12} = 2.179$.

So $R_L = 1.18 - 0.12 \times 2.179 = 1.18 - .2615 = .92$

and $R_U = 1.18 + 0.12 \times 2.179 = 1.18 + .2615 = 1.44$

So 95% confidence interval for potency of A is (.92, 1.44).

It can be concluded with 95% confidence that 1 cc of test tincture A is not less potent than .92 cc of standard tinctures B, and can not be more potent than 1.44 cc of standard tincture B.

Fieller's Theorem for Confidence Limits of a Ratio

Estimate of potency is, many times a ratio of estimators of two parameters when obtained from a bioassay. Therefore, the theorem first given by Fieller (1940), can be used to obtain the confidence limits of the true potency of test preparation. The theorem:

Suppose that θ_1 and θ_2 are two parameters, and let

$$\rho = \theta_1/\theta_2 \tag{19.15}$$

Let $\hat{\theta}_1$ and $\hat{\theta}_2$ be two unbiased linear estimators of θ_1 and θ_2, respectively, from observations with normally distributed errors.

Thus $E(\hat{\theta}_1) = \theta_1$ and $E(\hat{\theta}_2) = \theta_2$. Moreover, let

$$V\left(\hat{\theta}_1\right) = s^2 v_{11},\ V\left(\hat{\theta}_2\right) = s^2 v_{22} \text{ and } \mathrm{Cov}\left(\hat{\theta}_1, \hat{\theta}_2\right) = s^2 v_{12} \tag{19.16}$$

where s^2 is an error mean square with v d.f.

Now $R = \dfrac{\hat{\theta}_1}{\hat{\theta}_2}$ will naturally be taken as an estimate of ρ.

According to *Fieller's theorem* the 100(1 – α)% confidence interval for ρ is given by

$$\frac{1}{1-g}\left[R - g\frac{v_{12}}{v_{22}} \pm \frac{st_{\alpha/2,\,v}}{\hat{\theta}_2}\left\{v_{11} - 2Rv_{12} + R^2 v_{22} - g\left(v_{11} - \frac{v_{12}^2}{v_{22}}\right)\right\}^{1/2}\right] \tag{19.17}$$

where $t_{\alpha/2,\,v}$ is upper tail 100 α/2% point of Student's *t*-distribution with v d.f. and

$$g = (t_{\alpha/2,\,v})^2 s^2 v_{22}/\hat{\theta}_2^2 \tag{19.18}$$

When $\hat{\theta}_2$ is large compared to the SE $(\hat{\theta}_2)$ i.e. $\hat{\theta}_2$ is much larger than $\hat{V}(\hat{\theta}_2) = s^2 v_{22}$, g will be quite small.

If g can be neglected the confidence interval of R will be

$$R \pm \frac{st_{\alpha/2,\,v}}{\hat{\theta}_2}(v_{11} - 2Rv_{12} + R^2 v_{22})^{1/2} \tag{19.19}$$

If $\hat{\theta}_1$ and $\hat{\theta}_2$ are obtained from two independent samples $\mathrm{Cov}(\hat{\theta}_1, \hat{\theta}_2) = s^2 v_{12} = 0$, and the confidence interval will be

$$R \pm \frac{st_{\alpha/2,\,v}}{\hat{\theta}_2}(v_{11} + R^2 v_{22})^{1/2} \tag{19.20}$$

This is same as that given in (19.14).

It may be noted that in statistical analysis of bioassay data, normality is often assumed. By virtue of central limit theorem, if the tolerance distribution is not markedly different from normal, the averages of a number of measured tolerances may be considered to be very close to normal. So the techniques which are developed assuming normality may not be giving misleading conclusions.

19.1.2 Dilution Assays

It is a form of direct assays, when the test preparation behaves as though it were simply a dilution of the standard preparation in a diluent which is completely inert on the response used for assay.

All the valid assays may not be of this character, but almost all the assays in common use are, either strictly or approximately dilution assays.

In some assays, there may be strong reasons for believing that all the constituents of the test preparation except one, are without any effect on the studied response of the subjects. An assay of the preparation against a standard preparation of the effective constituent, is then equivalent to an analysis for determining the content of the constituent. This may be described as an analytical dilution assay.

Some times the two preparations may behave as if they are qualitatively same in effective constituent, though they are known not to be the same. The statistical techniques applicable on analytical dilution assays may still be useful but such inferences are of limited value. Such an assay may be called a comparative dilution assay. As an example, two insecticides, of related but different chemical composition, when applied on one species of insect may behave as if one was a dilution of the other. However, if applied on other species or the experimental conditions change, the apparent dilution may not be possibly concluded.

Theoretical Background of Dilution Assay

Suppose that for two preparations A and B, the dose X_A of A and dose X_B of B, are equivalent. Moreover, B behaves as a dilution of A by a factor ρ, the relative potency of A.

This implies that

$$\rho X_A = X_B \tag{19.21}$$

for all possible pairs of doses X_A and X_B.

If (19.21) holds,

$$V(X_B) = \rho^2 V(X_A) \tag{19.22}$$

Variances of X_A and X_B, therefore, can not be equal unless $\rho = 1$.

So pooling of sample variances of X_A and X_B to obtain the estimate of common variance can not be done.

However, taking log on both sides of (19.21)

$$\log X_A + \log \rho = \log X_B \tag{19.23}$$

and then

$$V(\log X_A) = V(\log X_B) \tag{19.24}$$

It is evident that variances of log tolerances of the two preparations A and B are same. Moreover, since tolerance can never be negative, whereas log tolerance can be so, the assumption of normality is more appropriate for log tolerances.

All the values of doses must be positive, and an upper bound of dose need not always be given. So the distribution of doses is positively skewed and is not extended on negative side.

Distribution of log tolerance extends to negative as well as positive side and is likely to be symmetric. The assumption of normality of log tolerance is, therefore, more justifiable.

Estimation of Potency in Dilution Assay

Above discussions, especially (19.23) and (19.24) suggest that the log tolerances of the two preparations can be combined together for obtaining estimate of variance.

Let

$$M = \log \rho,\ Y_B = \log X_B \text{ and } Y_A = \log X_A \tag{19.25}$$

From (19.23) $M = Y_B - Y_A$ and hence one can estimate M by

$$\hat{M} = \bar{Y}_B - \bar{Y}_A \tag{19.26}$$

and

$$V(\hat{M}) = V\left(\bar{Y}_B - \bar{Y}_A\right) = s^2\left(\frac{1}{n_2} + \frac{1}{n_1}\right) \tag{19.27}$$

where s^2 is the common variance of log tolerances of A and B and n_1 and n_2 are the number of observations (tolerances) on the preparations A and B, respectively.

Since normality holds for distributions of log tolerances i.e. of Y_A and Y_B, 100 (1 – α)% confidence limits (M_L, M_U) for M are given by

$$M_L = \hat{M} - \left(t_{\alpha/2,\, n_1+n_2-2}\right) s\left(\frac{1}{n_1} + \frac{1}{n_2}\right)^{1/2} \tag{19.28}$$

and

$$M_U = \hat{M} + \left(t_{\alpha/2,\, n_1+n_2-2}\right) s\left(\frac{1}{n_1} + \frac{1}{n_2}\right)^{1/2} \tag{19.29}$$

$t_{\alpha/2, n_1+n_2-2}$ is 100 α/2% upper tail point of Student's t-distribution with $(n_1 + n_2 - 2)$ d.f. and

$$s^2 = \left[\sum_{i=1}^{n_1}\left(Y_{A_i} - \bar{Y}_A\right)^2 + \sum_{j=1}^{n_2}\left(Y_{B_j} - \bar{Y}_B\right)^2 + \right] \Big/ (n_1 + n_2 - 2) \tag{19.30}$$

Antilogs of $\hat{M}$, M_L and M_U give the estimates of relative potency of A and its confidence limits, respectively.

The primary object of a direct dilution assay is to obtain the most precise estimate of a difference between two mean log doses, which is possible for the available data.

To illustrate the method of estimation of potency in a dilution assay again consider Example 19.1 on the assay of *Strophanthus*.

Example 19.2: Determine the relative potency of *Strophanthus A* by using log tolerances of subjects considered in Example 19.1.

Solution: Relative potency of tincture *A* of *Strophanthus* against tincture *B* of *Strophanthus* is now determined using log tolerances. Table 19.2 gives the log tolerances of the two groups subjected to the two tinctures from the Table 19.1.

For obtaining $\bar{Y}_A$ and $\bar{Y}_B$ the totals of log tolerances are written in 8th row and the mean $\bar{Y}_A = \frac{1}{7}\sum_{i=1}^{7} Y_{A_i}$ and $\bar{Y}_B = \frac{1}{7}\sum_{j=1}^{7} Y_{B_j}$ are obtained as 0.217 and 0.292, respectively.

Table 19.2 Log tolerances from Table 19.1

Serial No.	*Strophanthus* $A(Y_A)$	*Strophanthus* $B(Y_B)$
1	0.190	0.384
2	0.199	0.267
3	0.233	0.301
4	0.158	0.356
5	0.093	0.230
6	0.276	0.167
7	0.369	0.342
Total of log tolerances	1.518	2.047
Mean	0.217	0.292
Sum of squares of log tolerances	.3760	.6339

$$\hat{\sigma}_A^2 = \frac{1}{7-1}\left[\sum_{i=1}^{7} Y_{A_i}^2 - 7\,\bar{Y}_A^2\right] = \frac{1}{6}\left[.3760 - \left(\sum_{i=1}^{7} Y_{A_i}\right)^2 \Big/ 7\right]$$

$$= \frac{1}{6}[.3760 - 1.518^2/7] = \frac{1}{6}[.3760 - .3292] = .0468/6 = .0078$$

and

$$\hat{\sigma}_B^2 = \frac{1}{7-1}\left[\sum_{j=1}^{7} Y_{B_j}^2 - 7\,\bar{Y}_B^2\right] = \frac{1}{6}\left[.6339 - \left(\sum_{j=1}^{7} Y_{B_i}\right)^2 \Big/ 7\right]$$

$$= \frac{1}{6}[.6339 - 2.047^2/7] = \frac{1}{6}[.6339 - .5986] = .0353/6 = .0059$$

Now a test of equality of variances of A and B is performed before obtaining the pooled estimate of common variance. Thus first to test $H_0 : \sigma_A^2 = \sigma_B^2$ against $H_1 : \sigma_A^2 > \sigma_B^2$ using F-statistic given by

$$F = \frac{\hat{\sigma}_A^2}{\hat{\sigma}_B^2} = \frac{.0468/6}{.0353/6} = 1.326$$

Now from Table A.6, $F_{6,\,6,\,.05} = 3.05$.

Since calculated $F = 1.326 < F_{6,\,6,\,.05} = 3.05$, H_0 is not rejected. Thus the pooled estimate of common variance $\sigma_A^2 = \sigma_B^2$ is obtained as

$$s^2 = \left\{\left[\sum_{i=1}^{7} Y_{A_i}^2 - 7\,\bar{Y}_A^2\right] + \left[\sum_{j=1}^{7} Y_{B_j}^2 - 7\,\bar{Y}_B^2\right]\right\} \Big/ (7+7-2)$$

$$= \frac{1}{12}[.0468 + .0353] = \frac{.0821}{12} = .0068$$

Now $\hat{M} = \bar{Y}_B - \bar{Y}_A = .292 - .217 = 0.075$.

Since upper tail 2.5% point of Student's t-distribution with 12 d.f. from Table A.4 is = 2.179

$$M_L = \hat{M} - (t_{.025,\,12})\, s\left(\frac{1}{7} + \frac{1}{7}\right)^{1/2} = .075 - 2.179 \times .082 \times \sqrt{2/7}$$

$$= 0.075 - 0.096 = -.021 = \bar{1}.979$$

and

$$M_U = \hat{M} + (t_{.025,\,12})\, s\left(\frac{1}{7} + \frac{1}{7}\right)^{1/2} = .075 + .096 = .171$$

Now taking antilogs of $\hat{M}$, M_L and M_U the relative potency of *Strophanthus A* relative to *Strophanthus B* is obtained as 1.189 cc per mg and the confidence limits of ρ are .953 cc per mg and 1.483 cc per mg.

Some Remarks on Direct Assays

Finney (1964) mentions that Miles and Perry (1950) noted some points about direct assays. To mention a few:

If two stimuli of unequal potency are applied at equal rates, the subjects receiving the less potent stimulus will have longer average times under treatment than those receiving the more potent stimulus. If there is any time lag in producing the specified response in the subject or any cumulative effect other than additive one, the comparison of two stimuli will be biased. If one has some idea of the relative potency of test preparation, one can overcome this difficulty by application of equipotent doses of the two stimuli per unit of time. However, it is almost impossible to say how much the validity of potency estimate is vitiated, even if the two stimuli differ only slightly.

Even when the bias produced by time lag or cumulative effect is absent, the technical difficulty may prevent experimenters from ensuring that subjects receive just the right dose to produce the characteristic response. For example to determine individual tolerances of aphids for an insecticide is impossible, and a different method of assay must be sought.

Ideally the trials of standard and test preparation should be done on the same subject, so that estimation is independent of the differences between subjects. But often, this is not possible, since the same subject can not be used again, because some times the response is death and many times the initial conditions of the subject do not remain comparable with other subjects and with the conditions before the application of drugs.

Now, therefore, indirect assays are considered which are based on the dose response relationship.

19.2 Indirect Assays and Dose Response Relationship

An indirect assay consists of giving specified doses of the test and the standard preparation to each of several subjects and a magnitude of some property of the subjects is measured. These properties may be weight of the animal, the weight of a particular organ, time of survival or recovery etc. All are, therefore, *quantitative responses*. If only presence or absence of a response is recorded, it is a *quantal response*.

In this section the indirect assays based on quantitative responses are considered, whereas the quantal response assays are considered in the next section.

This method suggested by Bliss (1940b) takes account of the relationship between dose and the magnitude of a response. If form of this relationship is known for the two preparations, equally effective doses may be estimated and by applying appropriate statistical procedure the precision of the estimate can also be found. This method is quite reasonable unless the variance of the response for a particular dose in direct assay, is very small. As already mentioned this method of indirect assay makes use of dose response relationship, so a discussion on dose response regression is given below.

19.2.1 The Dose Response Regression

Let u be the response of a subject receiving a dose z of a stimulus. For a given population of subjects or for repeated trials where independence of response is not affected, let $E(u) = U$.

A response u clearly depends on the administered dose, and so also the expected response U, depends on z. Let this relationship be given by:

$$U = F(z) \tag{19.31}$$

where $F(z)$ is a single valued function of z. Then this equation (19.31) is called the dose-response regression of u on z.

Now suppose for a standard preparation S and for the test preparation T, the dose-response regression equations are

$$U_S = F_S(z) \tag{19.32}$$

and

$$U_T = F_T(z) \tag{19.33}$$

For a given value of U, let Z_T and Z_S be the required doses of the test and the standard to produce average response U.

Z_T and Z_S will then be equally effective doses of the two preparations at the level of response U. Hence the relative potency of T relative to S at level of response U is given by Z_S/Z_T.

Now a diagram may be plotted to represent this potency Z_S/Z_T as a function of U. If the two stimuli T and S are completely unrelated, the curve obtained could be of any form and a single value for relative potency for all the U may not be available.

However, if the following assumptions of an analytical dilution assay are satisfied, one may have a single value of relative potency for a given U:

(i) Let the factor X be the effective constituent of the standard preparation S. Then the response produced by the known amount of factor X is actually due to the factor X itself, not due to some other substance associated with it say an impurity.
(ii) The response produced by the test preparation T, is only due to the presence of factor X, without any increase, decrease or modification, due to presence of any other substance in it.

In other words, it is assumed that S contains no substance, other than factor X, contributing to the response and that T may be regarded as a dilution of the S in a completely inert diluent.

As already mentioned, if the two preparations S and T contain the same effective constituent and all other constituents are without effect on U, so that one behaves as a dilution of the other in an inert diluent, the ratio Z_S/Z_T, the relative potency of T must be independent of U. In symbols

$$Z_S/Z_T = \rho \text{ for all } U \tag{19.34}$$

The graph of Z_S/Z_T (on Y-axis) and U (on X-axis) will be a line parallel to X-axis at $Y = \rho$.

In such a case, the regression functions F_S and F_T will be related as

$$F_T(z) = F_S(\rho z), \text{ for all } z \tag{19.35}$$

where ρ is the relative potency of T.

This relation (19.35) called condition of similarity, is a prerequisite of all the dilution assays. The fundamental assay problem is then as follows:

A standard preparation, whose content of effective constituent is either known or defined in arbitrary units has the dose response regression function

$$U = F(z) \tag{19.36}$$

under specified testing conditions.

A test preparation is known to have the same effective constituent, and the amount present in unit dose is to be estimated, making use of the fact that dose-response regression function for this preparation will be

$$U = F(\rho z) \tag{19.37}$$

where ρ is the relative potency of the preparation, which is the number of units of effective constituent contained in, or equivalent to unit dose of test preparation.

An estimate R of ρ must be formed from observations on dose response tests, usually after simultaneous estimation of parameters of regression function. Nature of the function F leads to some variation in the estimation of potency. If F could be written as a linear function of u and z or some functions of u and of z, the method becomes an application of linear regression. So the next section discusses some transformations linearizing the regression function (19.37).

Linearizing Transformations

Finney (1947a) observed that the general dose response regression, in some cases, can be adequately represented in the following form for standard preparation (SP)

$$U_s = f(\alpha + \beta z^{\lambda}) \tag{19.38}$$

in which α, β and λ are the parameters and f is a completely specified monotonic function of z.

Depending on whether λ is very close to 0 or $\lambda \to 0$ or it is non zero quantity, different linearizing transformations will be used and accordingly the estimate of potency, the experimental procedure and computational process may be easy or complex.

If $\lambda \neq 0$, regression function (19.38) holds. If $\lambda \to 0$, it can be mathematically proved that the regression function becomes

$$U_s = f(\alpha + \beta \log z) \tag{19.39}$$

where β is modified from that in (19.38).

If the transformations of dose Z is done in (19.38) and (19.39) amounting to choosing a new scale for dose, these regression functions become linear regression equations. The transformations are:

(a) If $\lambda \neq 0$, the new scale (or transformation) of Z is chosen as $X = Z^{\lambda}$ in (19.38)

(b) If $\lambda \to 0$, the new scale (or transformation) of Z is chosen as $X = \log Z$ in (19.38).

Using these, both the regression functions for SP will become

$$U_s = f(\alpha + \beta X) \tag{19.40}$$

These transformations of dose are also called dose metameters.

Now if a transformation of response, also called response metameter, is defined as

$$Y = f^{-1}(U_s) \tag{19.41}$$

The regression functions in both the situations i.e. when $\lambda \neq 0$ and $\lambda = 0$ become one as

$$Y = \alpha + \beta X \tag{19.42}$$

which is the simple linear regression equation.

Hence using the appropriate dose metameter X and the response metameter Y, the statistical technique of estimating the parameters α and β can be easily applied.

The similarity condition $F_T(z) = F_S(\rho z)$ implies that if the regression function for SP is (19.42), the regression function for TP will be given by

$$Y_T = \alpha + \beta\, \rho^{\lambda} X_T \qquad \text{if } \lambda \neq 0 \tag{19.43}$$

and

$$Y_T = \alpha + \beta \log \rho + \beta X_T \quad \text{if } \lambda = 0 \tag{19.44}$$

Thus using the two dose and response metameters given above, the regression functions of the SP and TP both are transformed to straight lines simultaneously.

If $\lambda \neq 0$ the straight lines for SP and TP are

$$Y_S = \alpha + \beta X_S \text{ and } Y_T = \alpha + \beta\, \rho^{\lambda} X_T \text{ and} \tag{19.45}$$

if $\lambda = 0$, these straight lines are

$$Y_S = \alpha + \beta X_S \text{ and } Y_T = \alpha + \beta \log \rho + \beta X_T \tag{19.46}$$

A. Looking at the straight lines in (19.45) it is evident that the slopes of SP and TP are not equal, so these lines intersect. Moreover

$$\frac{\text{Slope of test regression line}}{\text{Slope of standard regression line}} = \rho^{\lambda}$$

So $$R = \hat{\rho} = \left(\frac{\text{Estimated slope of test line of regression}}{\text{Estimated slope of standard line of regression}}\right)^{1/\lambda} \quad (19.47)$$

B. Regression equations (19.46) indicate that the lines of regression of SP and TP are parallel as shown in Fig. 19.1.

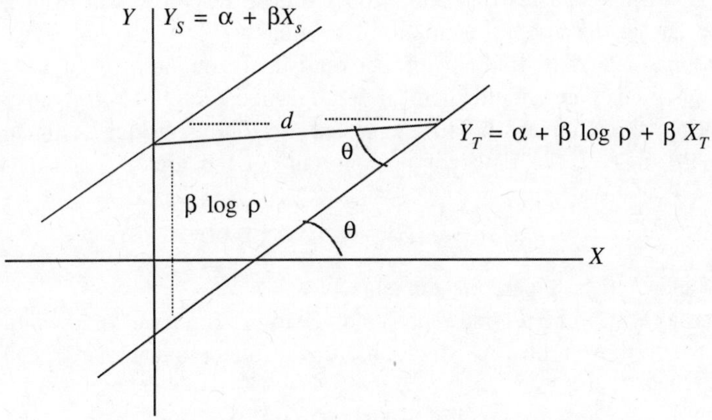

Fig. 19.1. Regression lines for SP and TP for $\lambda = 0$.

Hence

$$\beta = \tan\theta = \frac{\beta \log \rho}{d} = \frac{\text{Vertical distance between the two regression lines}}{\text{Horizontal distance between the two regression lines}} \quad (19.48)$$

or $$\log \rho = d \Rightarrow R = \hat{\rho} = \text{Anti log of } d \quad (19.49)$$

Thus the estimate of potency is antilog of the horizontal distance between the regression lines of TP and SP.

On the basis of these discussions, the following observations are evident.

1. If λ is known and f is also known by definition, a method of estimating potency of TP consists of the following steps:

   (a) Record the data on doses and responses of TP and SP.

   (b) Choose dose and response metameters and obtain these for the data on TP and SP so that the resulting transforms of TP and SP fall on intersecting lines or on parallel lines. The lines may be fitted empirically by eye observations of transformed data of SP and TP, or may be fitted by calculations of Least-Square Method. Now ρ can be estimated from (19.47) or (19.49), as the case may be.

2. A more general problem requires λ also to be estimated from the observed record of doses and responses of each assay. The computations then become quite complicated. However, since data from an assay are often not extensive enough to provide accurate values of λ. A better policy is, therefore, to draw graphs for the SP and TP for determining the form of f and the value of λ.

The most common values of λ are 0 and 1. Some other simple values are $\lambda = \frac{1}{2}, \frac{1}{3}, -1, -\frac{1}{2}$.

The response metameter is also usually taken a simple function such as

$$Y = U, Y = \log U, Y = \frac{1}{U} \text{ or } Y = U^{1/2}$$

Drawing graphs for various choices of f and λ on the dose response data from an assay, the values of λ and f can be chosen for proper estimation of potency.

With this attempt, one may find, by proper statistical goodness of fit tests, that straight line regression is not justified. One should then go for non-linear regression given below, very briefly.

Non-linear regression Emmens (1940) suggested the following logistic function to represent the quantitative response, in case the range of response is not too great:

$$U = \frac{1}{2} H[1 + \tanh(\alpha + \beta X)] \tag{19.50}$$

where $X = \log Z$ and tanh is hyperbolic tangent.

Emmens found (19.50) to be a good representation for a number of assay data. If H were known, or could be estimated from earlier investigations, the response metameter given by

$$Y = \tanh^{-1}\left(\frac{2U - H}{H}\right) \tag{19.51}$$

reduces the logistic regression (19.50) into a linear regression

$$Y = \alpha + \beta X \tag{19.52}$$

Now the methods mentioned for linear regression can be applied to estimate ρ.

However, if all these linearizing transforms are not suitable, the estimation of potency for the assay is much more complicated. One may think of quadratic regression given by

$$U_S = \alpha + \beta X_S + \gamma X_S^2 \text{ for SP} \tag{19.53}$$

and therefore, for TP

$$U_T = \alpha + \beta\rho^{\lambda} X_T + \gamma\rho^{2\lambda} X_T \qquad \text{for } \lambda \neq 0 \tag{19.54}$$

or

$$U_T = \alpha + \beta(\log \rho + X_T) + \gamma(\log \rho + X_T)^2 \qquad \text{for } \lambda = 0 \tag{19.55}$$

Assay with these regressions are not in common use, due to the troublesome estimation procedure of α, β, γ and ρ simultaneously.

Now once the regression equations for SP and TP are finalized, one of the methods of estimation should be used from those given below.

19.2.2 Methods of Estimation of Potency

After finalization of the dose response regression from previous experiments, it may be used for assay purpose in one of the following three ways:

(i) Standard curve estimation
(ii) Standard slope estimation and
(iii) Simultaneous trial method.

(i) Standard Curve Estimation

This method is applicable for other regression functions also than the straight line regression. The following steps should be followed:

(a) Determine the average response u for a specified dose z of TP.
(b) Find the metametric value of u (= y) and determine from the regression equation the value of x (= x_0) for this y.
(c) Compute the dose z_0 corresponding to the dose metameter x_0. So the dose z units of TP is equivalent to z_0 units of SP.
(d) The relative potency of TP = z_0/z (19.56)

This method would be a valid estimate of potency, if the sample of subjects for the assay were drawn from the same population and tested under the same conditions as that used in the earlier calculation of the regression line.

This method, however, has the following drawbacks:

(a) Even slight and possibly random variations in experimental conditions within a lab are liable to alter the true dose-response regression function from day to day in an unpredictable manner.
(b) This method is based on the assumption that a response curve once determined can be used in future assays, which is not admissible. However, there are few exceptions to this drawback.
(c) The curve estimated in one lab can not be used as a standard in other lab for comparison of TP's. It is because, the curve may be quite irrelevant because of differences in strains of animals and experimental technique.

Thus in general it may be suggested that the test of SP must be run simultaneously with the assay of any TP.

(ii) Standard Slope Estimation

For some types of responses, though the regression is likely to shift from day to day, the shift may be in position of regression only. The sensitivity of the subjects to variation of doses may be unaltered. In other words the slopes or rate of change of response per unit change of dose metameter may remain constant.

Thus the standard slope method is applicable only if response metameter and dose metameter regression is linear.

If experience has shown that the assumption of constant slope (i.e. linear regression on log dose) can be justified for SP, then a TP can be assayed relative to SP by simultaneous experimentation with one dose of SP and one dose of TP. The doses of SP and TP should be chosen well within the range of linearity and as equivalent as possible.

Let X_S and X_T be the log doses of SP and TP, respectively, and $\bar{Y}_S$ and $\bar{Y}_T$ be the corresponding mean response metameters. Then the log of relative potency is estimated as

$$\log\hat{\rho} = M = X_S - X_T - \frac{\bar{Y}_S - \bar{Y}_T}{b} \tag{19.57}$$

where b is the estimate of the common slope of linear regressions of SP and TP shown in Fig. 19.1 and solving for log $\hat{\rho}$ from these two estimated equations, the above expression in (19.57) follows. Now $\hat{\rho}$ is antilog of M.

Some Remarks

(a) When the data available are responses to one dose of SP and TP, the validity of standard slope method can not be tested. An experimenter may be confident that the regression line has not changed (i.e. b has not changed) and so uses this method. But he must be aware that the change in b even if unexpected, may risk the accuracy of potency estimate.

(b) In (19.57), if $\bar{Y}_S$ and $\bar{Y}_T$ are almost equal the estimate of potency M will not be affected significantly due to change in b. So a choice of equivalent dose metameters are very much desirable.

(c) Value of b measures the sensitivity of subjects to change of doses. It may vary with season, with different breeds of subjects, or erratically due to lack of complete control over the experimental conditions.

(iii) Simultaneous Trial Estimation

In view of the remarks given above and some draw backs of the standard curve estimation, this method overcomes the said objections to the above two methods. This method of estimation performs simultaneous trials of both the preparations, the SP and TP under strictly comparable conditions, using two or more doses of each.

In simultaneous trial, all the quantities required for computation of results of assay are obtainable from the responses in the current assay. If the design of experiment is well chosen, any validity tests, if needed, can be done using the same data. The statistical analysis and estimation are entirely self-contained. The preliminary investigation is used (a) to confirm that the regression is nearly linear and (b) to locate the suitable range of doses.

The position and slope of the regression line are estimated afresh for each assay. The apparent precision of the estimate is less than that from standard slope method.

This method consists of the following steps:

(a) Empirical responses for both, the SP and TP are plotted against the dose metameter X, and two provisional regression lines for each of them must be drawn, subject to constraint of either parallelism or intersection at $X = 0$.

(b) The estimate of ρ is found either from the ratio of slopes (i.e. $\lambda \neq 0$) or from horizontal distance between two parallel lines.

In this method each assay provides its own estimates of the parameters, though linear regression is assumed.

Widest application of this method is when regression lines are parallel. So the next subsection deals analysis of parallel line assays. Examples of computation of estimates of relative potency by the first two methods may be seen by interested reader in the book by Finney (1964).

19.2.3 Parallel Line Assay

As already seen, $\lambda = 0$ leads to parallel regression lines for SP and TP given by

$$Y_S = \alpha + \beta X_S \text{ and } Y_T = \alpha + \beta \log \rho + \beta X_T \tag{19.58}$$

The response metameter and the dose metameter giving these are $Y = f^{-1}(U)$ and $X = \log Z$, respectively.

If X_T and X_S are the equipotent values of dose metameter X, $\log \rho = \dfrac{Y_T - Y_S}{\beta}$ from (19.58)

Hence estimate of

$$\log \rho = M = \frac{Y_T - Y_S}{b} \tag{19.59}$$

where b is the estimate of the slopes of regression lines in (19.58).

If the data on several doses of SP and TP and their responses on subjects have been recorded, the estimated regression lines will be

$$Y_S - \bar{Y}_S = b\left(X_S - \bar{X}_S\right) \text{ and } Y_T - \bar{Y}_T = b\left(X_T - \bar{X}_T\right) \tag{19.60}$$

Subtracting 1st equation from 2nd of (19.60) and since X_T and X_S are equipotent, from (19.59), one gets

$$\log \hat{\rho} = M = \bar{X}_S - \bar{X}_T - \frac{\bar{Y}_S - \bar{Y}_T}{b} \tag{19.61}$$

and $\hat{\rho}$ can be determined from above.

Confidence Limits

Equation (19.61) can be written as

$$M - \left(\bar{X}_S - \bar{X}_T\right) = \frac{\bar{Y}_T - \bar{Y}_S}{b} \tag{19.62}$$

$\bar{X}_S - \bar{X}_T$ depends on the choice of doses in the experiment. So the confidence limits of M can be determined from the confidence limits of $\left(\bar{Y}_S - \bar{Y}_T\right)/b$ and adding $\bar{X}_S - \bar{X}_T$ to these limits.

Since b is independent of $\bar{X}_S - \bar{X}_T$, Fieller's theorem can be used to find the confidence limits of the ratio on RHS of (19.62).

The confidence limits for $M - \left(\bar{X}_S - \bar{X}_T\right)$ is given by Fieller's theorem as

$$\frac{1}{(1-g)}\left[M - \left(\bar{X}_S - \bar{X}_T\right) \pm t_{\alpha/2,\nu}\, \frac{s}{b}\left\{(1-g)\left(\frac{1}{N_S} + \frac{1}{N_T}\right) + \frac{\left(M - \bar{X}_S + \bar{X}_T\right)^2}{\sum S_{xx}}\right\}\right] \tag{19.63}$$

where $g = \dfrac{t^2 s^2}{b^2 \sum S_{xx}}$ and s^2 is mean square error from the ANOVA table for the two regression lines for SP and TP. The method of constructing ANOVA table will be now given below after explaining how the experimental data is collected and used for estimation and ANOVA.

Experimental Data and ANOVA

Symmetry in the number and spacing of doses and the number of subjects given the chosen doses has advantages in ease of computation and in precision of the estimators. However, (a) accidental loss of symmetry after the start of experiment, (b) storage of experimental material and (c) shortage of sufficient number of subjects, may lead to unsymmetrical assays unavoidably. So the structure and analysis of such experiment is discussed here.

In the *experimental set up* suppose that k doses of SP $s_1, s_2, \ldots, s_k$ and l doses of TP $t_1, t_2, \ldots, t_l$ are applied to subjects. i^{th} dose of SP is applied to n_i subjects with the mean response $\bar{Y}_{s_i}$ and the jth dose of TP is applied to m_j subjects with mean response $\bar{Y}_{t_j}$. Table below gives this arrangement:

Table 19.3 Experimental set up of parallel line assay

	SP						TP					
Doses	s_1	s_2	…	s_i	…	s_k	t_1	t_2	…	t_j	…	t_l
R	y_{11}^S	y_{21}^S	…	y_{i1}^S	…	y_{k1}^S	y_{11}^T	y_{21}^T	…	y_{j1}^T	…	y_{l1}^T
E	y_{12}^S	y_{22}^S	…	y_{i2}^S	…	y_{k2}^S	y_{12}^T	y_{22}^T	…	y_{j2}^T	…	y_{l2}^T
S	-	-		-		-	-	-		-		-
P	-	-		-		-	-	-		-		-
O	-	-		-		-	-	-		-		-
N	-	-		-		-	-	-		-		-
S	$y_{1n_1}^S$	$y_{1n_2}^S$	…	$y_{in_i}^S$	…	$y_{kn_k}^S$	$y_{1m_1}^T$	$y_{2m_2}^T$	…	$y_{jm_j}^T$	…	$y_{lm_l}^T$
E												
S												
Replicates	n_1	n_2	…	n_i	…	n_k	m_1	m_2	…	m_j	…	m_l
Mean response	$\bar{Y}_{s_1}$	$\bar{Y}_{s_2}$	…	$\bar{Y}_{s_i}$	…	$\bar{Y}_{s_k}$	$\bar{Y}_{t_1}$	$\bar{Y}_{t_2}$	…	$\bar{Y}_{t_j}$	…	$\bar{Y}_{t_l}$
Log dose	x_{s_1}	x_{s_2}	…	x_{s_i}	…	x_{s_k}	x_{t_1}	x_{t_2}	…	x_{t_j}	…	x_{t_l}

Analysis As a first step a diagram should be drawn to show mean responses at each level of dose and the log doses X's. This will enable rapid estimation of potency and is also a protection against gross error or misinterpretation of statistical analysis. It will also help in choosing proper dose and response metameters.

Before performing ANOVA, the assumption of equality of variances of responses at different doses should be tested using Bartlett's test of homogeneity of variances, given in Chapter 8.

Now various sum of squares (SS) for ANOVA table should be computed following the procedure given below:

For SP compute the following totals

$$n = \sum_{i=1}^{k} n_i = \text{total number of subjects to whom SP was applied} \tag{19.64}$$

$$\bar{X}_S = \left(\sum_{i=1}^{k} n_i x_{s_i} \right) \Big/ n \tag{19.65}$$

$$\bar{Y}_{s_i} = \left(\sum_{p=1}^{n_i} y_{ip}^s \right) \Big/ n_i = Y_{i\bullet}^s / n_i = \frac{\text{Total response for } i^{\text{th}} \text{ dose of SP}}{n_i} \tag{19.66}$$

$$\bar{Y}_S = \sum_{i=1}^{k} Y_{i\bullet}^s / n \tag{19.67}$$

$$S^S_{x^2} = \sum_{i=1}^{k} n_i x_{s_i}^2 - \left(\sum_{i=1}^{k} n_i x_{s_i} \right)^2 \Big/ n \tag{19.68}$$

$$S^S_{xy} = \sum_{i=1}^{k} x_{s_i} Y^S_{i.} - \left(\sum_{i=1}^{k} n_i x_{s_i} \right) \left(\sum_{i=1}^{k} Y^S_{i.} \right) \Big/ n \tag{19.69}$$

For TP compute, in the same way, the following totals

$$m = \sum_{j=1}^{l} m_j = \text{total number of subjects to whom TP was applied} \tag{19.70}$$

$$\bar{X}_T = \left(\sum_{j=1}^{l} m_j x_{t_j} \right) \Big/ m \tag{19.71}$$

$$\bar{Y}_{t_j} = \left(\sum_{q=1}^{m_j} y^T_{jq} \right) \Big/ m_j = Y^T_{j\bullet} / m_j = \frac{\text{Total response for } j^{\text{th}} \text{ dose of TP}}{m_j} \tag{19.72}$$

$$\bar{Y}_T = \sum_{j=1}^{l} Y^T_{j\bullet} / m \tag{19.73}$$

$$S^T_{x^2} = \sum_{j=1}^{l} m_j x_{t_j}^2 - \left(\sum_{j=1}^{l} m_j x_{t_j} \right)^2 \Big/ m \tag{19.74}$$

$$S^T_{xy} = \sum_{j=1}^{l} x_{t_j} Y^T_{j\bullet} - \left(\sum_{j=1}^{l} m_j x_{t_j} \right) \left(\sum_{j=1}^{l} Y^T_{j\bullet} \right) \Big/ m \tag{19.75}$$

$$N = n + m \text{ and } G = \sum_{i=1}^{k} Y^S_{i\bullet} + \sum_{j=1}^{l} Y^T_{j\bullet} = \text{TSP} + \text{TTP} \tag{19.76}$$

After these totals, *the SS are computed* as

$$\text{Correction factor} = \text{CF} = G^2/N \tag{19.77}$$

$$\text{SS for preparations} = \text{SSP} = (\text{TSP})^2/n + (\text{TTP})^2/m - \text{CF} \tag{19.78}$$

$$\text{SS for Pooled Regression} = \text{SSR} = \left(S^S_{xy} + S^T_{xy} \right)^2 \Big/ \left(S^S_{x^2} + S^T_{x^2} \right) \tag{19.79}$$

$$\text{Total SS} = \text{SST} = \sum_{i=1}^{k} \sum_{p=1}^{n_i} \left(y^S_{ip} \right)^2 + \sum_{j=1}^{l} \sum_{q=1}^{m_j} \left(y^T_{jq} \right)^2 - \text{CF}$$

$$= \text{Sum of squares of all the } n + m \text{ responses} - \text{CF} \tag{19.80}$$

$$\text{SS between doses} = \text{SSD} = \sum_{i=1}^{k}\left(Y_{i.}^{S}\right)^2 \Big/ n_i + \sum_{j=1}^{l}\left(Y_{j.}^{T}\right)^2 \Big/ m_j - \text{CF} \tag{19.81}$$

$$\text{SS within doses or SS due to Error} = \text{SSE} = \text{TSS} - \text{SSD} \tag{19.82}$$

SS due to Parallelism = Difference between fitting two independent regressions and one pooled regression

$$= \text{SS Par} = \left[\frac{\left(S_{xy}^{S}\right)^2}{S_{x^2}^{S}} + \frac{\left(S_{xy}^{T}\right)^2}{S_{x^2}^{T}}\right] - \text{SSR} \tag{19.83}$$

$$\text{SS due to linearity} = \text{SSL} = \text{SSD} - \text{SSP} - \text{SSR} - \text{SS Par} \tag{19.84}$$

Now the ANOVA table can be prepared as in the Table 19.4:

Table 19.4 ANOVA for parallel line assay

Source of variation	d.f.	SS	MSE = SS/d.f.	F
Preparation	1	SSP	SSP = MSP	MSP/MSE
Regression	1	SSR	SSR = MSR	MSR/MSE
Parallelism	1	SS Par	SS Par = MS Par	MS Par/MSE
Linearity	$k + l - 4$ (by subtr.)	SSL (by subtr.)	SSL/$(k + l - 4)$ = MSL	MSL/MSE
Doses	$k + l - 1$	SSD		
Error (within doses)	$N - (k + l)$ (by subtr.)	SSE (by subtr.)	SSE/$[N - (k + l)]$ = MSE	
Total	$N - 1$	TSS		

The significance of variation due to various sources can be tested by computing the F-statistics by taking ratio of mean squares with the MSE.

As already mentioned, the MSE is taken as the estimate of variance (= s^2) for computing the confidence limits.

Some Remarks

(a) Deviations from linearity, if significant, indicates that choice of dose of either SP or TP may be beyond the range of linearity of the regression of response. May be assumption of linearity is not justified, so method for non-linear regression may be more appropriate.

(b) If the two preparations i.e. SP and TP differ too much in response, assay results may not be satisfactory.

(c) If the tests for linearity and preparations do not give significantly high MSE's large MSE for parallelism will point towards violation of the basic assumption of parallel line assay. So the analysis will become invalid.

(d) The MSE for regression must be significantly large. Without a dose-response relationship supported by significant regression, estimates of potency can not be used.

If ANOVA gives logical results the estimate of potency given in (19.61) can be obtained after the estimate of common regression coefficient b is known. b is estimated as

$$b = \left(S_{xy}^S + S_{xy}^T\right) \Big/ \left(S_{x^2}^S + S_{x^2}^T\right) \tag{19.85}$$

Finney (1964) has given a numerical example of this type of parallel line assay of vitamin D_3 in cod-liver oil. 3 doses of SP and 4 doses of TP have been chosen. The doses are so chosen that the x values reduce to –2, 0, 2 for SP and are –3, – 1, 1, 3 for TP. The number of subjects are $n_1 = 9$, $n_2 = 10$ and $n_3 = 8$ for SP and $m_1 = 7$, $m_2 = 6$, $m_3 = 8$ and $m_4 = 7$ for TP.

The interested reader may refer to the computations given in the said book. The expressions given in equations (19.64) to (19.85) are computed for the example in the book. The expressions given in these equations (19.64) to (19.85) can be very well applied in any parallel line assay.

For ease of calculation generally doses in parallel line assay are chosen at equally logarithmic scaling i.e. x_{s_i} and x_{t_j} are at equal distance. Also for an efficient design $l = k$ and $n_i = m_j$ are chosen.

Such a design is also known as $2k$-point symmetrical design. The simplest symmetrical $2k$-point parallel line assay has $k = 2$ i.e. is 4-point. But it is inadequate for validity tests. For that k should be at least 3.

19.2.4 Slope Ratio Assay

The previous section discussed parallel line assay, which can be used if $\lambda = 0$. If $\lambda \neq 0$ the dose response regression is given by $U_S = f(\alpha + \beta Z^\lambda)$ for SP, which is transformed to $Y_S = \alpha + \beta X_S$ by use of metameters $X = Z^\lambda$ and $Y = f^{-1}(U)$. For TP, due to condition of similarity the regression is transformed to $Y_T = \alpha + \beta\rho^\lambda X_T$.

These are two straight lines intersecting at $X = 0$ and have slopes β and $\beta\rho^\lambda$. The straight lines intersect at $X = 0$, since it is reasonable that TP and SP will have same response if zero dose of each is administered. Differences in slopes is due to difference in potencies of TP and SP.

If b_T and b_S are estimated linear regression coefficients or slopes for TP and SP, respectively, it is clear that, the estimate of potency of TP

$$= \hat{\rho} = R = \left(\frac{b_T}{b_S}\right)^{1/\lambda} \tag{19.86}$$

Assays with these set up are called slope ratio assays.

Mostly in practice $\lambda = 1$ is assumed. In such a situation

$$\hat{\rho} = R = \left(b_T / b_S\right) \tag{19.87}$$

If λ is not known from previous experiments, analysis needs some complex computations. So is not given here.

The interested reader may again consult books by Finney (1964) and Govindarajulu (2001).

Slope ratio assays are chiefly used for microbiological field, where the test subject is no longer a single animal, but is an inoculum of a bacterial culture. It is added to a dose of the SP and TP and is incubated under standard conditions. The response is some measure of bacterial growth that takes place in a fixed time. Often response is a measure of turbidity or of the amount of alkali required to neutralize the acid formed during bacterial growth. The theoretical principle remains same as that for animals as subjects.

Moreover, the response actually measured seems to give homoscedastic linear regression without transformation. Hence $Y = U$ gives a suitable response metameter.

After describing the direct and indirect bioassays based on quantitative responses, the next section deals with quantal (all or nothing) responses.

19.3 Quantal Response Assays

As already mentioned in Section 19.2, indirect assays also deal with some stimulus-subject systems, for which quantitative measurement of a response due to action of stimulus is not possible or is impractical. In such cases, all that can be done is to record if a subject is or is not manifesting a certain reaction of the stimulus. Such type of responses are called quantal responses. Often the quantal response used is death. However, a quantal response may be any other easily recognizable change such as, onset of a skin infection, development of impotency to fertilize etc. Fungicides may be assayed by means of spore germination, insulin may be assayed by observing convulsions in mouse, vitamins by their success in curing the deficiency symptoms, etc.

Many quantal responses used for assay purpose are irreversible changes in the subjects which respond. So each subject can be used only once. The insect that has died and the spore that has germinated can not be used again. Some quantal responses, however, may not be of permanent nature. Such responding subject can, therefore, be used again after giving sufficient time gap to enable the subject returning to normalcy.

If the studied response is quantal, its occurrence or non-occurrence in the subject will depend on the intensity of the stimulus. For any subject there will be a specific level of intensity of the stimulus, below which the response will not appear. Such a value of intensity is called a *threshold or a limen or tolerance.* The term tolerance is very common in use.

The tolerance of any subject is, therefore, defined as that dose of stimulus, which would be just sufficient to produce the characteristic response. To any *lesser dose* the subject will *fail to respond* and to a *greater dose*, it will *always respond.*

The tolerance value may be fixed for a subject or it may vary at different testing occasions due to uncontrolled internal or external conditions. The experimenter can discuss and derive his conclusions on the basis of the value observed at the instant of the test conducted by him.

It is quite natural, that tolerance of one subject will vary from that of another subject in the population of subjects under study. Analysis of quantal response assay data, therefore, requires identification of the frequency distribution of the tolerance for the population studied.

Quantal Response Assay and Direct Assay

It was mentioned in Section 19.2, that there is a logical similarity between the quantal response assays and the direct assays. A brief discussion is appropriate here.

In a direct assay, the tolerance is measured on each subject. The estimation of potency is a comparison of average tolerances of the TP with SP.

In a quantal response assay a selected dose is applied to a subject and it is observed whether the response is produced or not and to record if the tolerance is less or more than the administered dose. The objective is still to infer about mean tolerance.

There are practical difficulties in measuring the dose just needed to produce a specified response i.e. to determine tolerance. But information whether the tolerances of each of a set of subjects is less or more than the administered dose, which is much simpler, can be used to estimate mean of the tolerance distribution.

In spite of the logical similarity of quantal response assay with direct assay, the statistical design and analysis of quantal response assay are more like the design and analysis of indirect assays with quantitative responses. The scheme is to choose a set of doses of SP and TP. A batch of subjects are tested at each of the selected doses. The estimate of potency is obtained, using the relationship between the mean response and the dose.

The number of subjects needed, is in general greater than that needed for direct assay using quantitative response.

Percentage response is not likely to have a linear regression on any simple dose metameter. An approximate result may be obtained by ignoring the actually observed non-linear part of the relationship. But it is advisable that the method should take account of the form of the tolerance distribution.

19.3.1 Probit Analysis

The assumption that the log tolerances follow normal distribution leads to a method using probit transformation which achieves linearization of the regression by metametric transformation of percentage responses.

Other probability models of the log tolerances are also applicable in some cases of responses. One such will be briefly discussed in the next subsection.

Let z denote the dose or intensity of stimulus, and let $f(z)$ be pdf of Z. Let P be the proportion of population whose tolerance is less than z_0. Then P is given by

$$P = \int_0^{z_0} f(z)dz \tag{19.88}$$

and

$$\int_0^{\infty} f(z)dz = 1 \tag{19.89}$$

since dose can not be negative and a very high dose (= ∞) will produce response in all the subjects in the population.

One can also write that

$$dP = f(z)dz \tag{19.90}$$

where $dP = P$ [tolerance of subjects in the population lies between z and $z + dz$] (19.91)

The tolerance distribution in a natural scale is generally highly positively skewed and its range is (0, ∞). A suitable transformation of dose scale often converts the tolerance distribution very close to the normal or Gaussian distribution. As already seen the logarithmic dose metameter $x = \log z$ is one such transformation. It gives negative values of x if z is very small. This transformation also advocated by Galton (1879), Clark (1983), Hemmingsen (1933), Bliss (1935) and others, is now-a-days a standard choice.

A general distribution of the log dose metameter x (henceforth will denote the log dose metameter in this chapter), containing two parameters, one of location and one of scale may be taken as

$$dP = \beta f(\alpha + \beta x) \tag{19.92}$$

If the tolerance distribution is normal, i.e. $N(\mu, \sigma^2)$

$$dP = \frac{1}{\sigma\sqrt{2\pi}} \exp\left[-\frac{1}{2\sigma^2}(x-\mu)^2\right] dx \tag{19.93}$$

Comparing (19.92) and (19.93) $\quad \mu = -\alpha/\beta$ and $\sigma = 1/\beta$ (19.94)

From (19.92), P the probability that tolerance of a subject chosen is less than x is given by

$$P = \int_0^x \beta f(\alpha + \beta u)du \tag{19.95}$$

For any specified tolerance density function $f(t)dt$, let Y be such that

$$P = \int_{-\infty}^{Y} f(t)dt \tag{19.96}$$

Then Y is called *the equivalent deviate of P*.

A value of P or Y determines the other uniquely. A table can be prepared giving values of Y for various P. From this table for various response rates P, the values of metametric transformation Y can be read and vice-versa.

Put $t = \alpha + \beta u$ in (19.95). It reduces (19.95) into

$$P = \int_{0}^{\alpha+\beta x} f(t)dt \tag{19.97}$$

From (19.96) and (19.97)

$$Y = \alpha + \beta x \tag{19.98}$$

If $f(\ )$ is a known function (19.98), the linear relationship between the response rate transform Y and dose metameter x, completely specifies the tolerance distribution. Hence by determining the estimates of α and β the tolerance distribution is specified.

Thus estimating the parameters of tolerance distribution becomes equivalent to determining the line of regression of Y-transform of the probability of response on x.

The probability of response is estimated as $\hat{P}$, the proportion of subjects who are observed to response.

The process of finding $\hat{\alpha} = a$ and $\hat{\beta} = b$ will be very similar to that followed in simple linear regression analysis after determining Y and x for SP and TP.

In an assay, each of the two preparations will have linear regression equation of the form (19.98). Because of condition of similarity for a dilution assay the values of β must be identical. So the computational procedure for parallel line assay should be followed for obtaining combined estimate of $\beta = b$. Estimate of log ρ is obtained from the horizontal distance between the two regression lines (as given in (19.62)). However, in these assays a weighting function w, apart from the number of subjects (n) receiving the chosen doses, is also used. In parallel line assay n was used as weight, whereas in quantal response assays an appropriate weighting function, multiplied by n ($= nw$) is used as weights. An example illustrates the procedure. If tolerance distribution (i.e. of x) is normal with mean μ and variance σ^2 i.e. is $N(\mu, \sigma^2)$

$$dP = \frac{1}{\sqrt{2\pi}\sigma}\exp\left[-\frac{1}{2\sigma^2}(x-\mu)^2\right]$$

Then the distribution of $t = (x - \mu)/\sigma$ will be $N(0, 1)$, and

$$dP = \frac{1}{\sqrt{2\pi}}e^{-t^2/2}$$

So normal equivalent deviate (N.E.D.) Y of P is given by

$$P = \int_{\infty}^{Y} \frac{1}{\sqrt{2\pi}} d^{-t^2/2} dt \tag{19.99}$$

Bliss (1934, a, b) introduced *probit of P* which is the normal equivalent deviate of P but increased by 5. Since almost all (99.73%) observations lie between ±3 of $N(0, 1)$ distribution, the chances of negative value of probit will be almost impossible, since it is increased by 5 over the $N(0, 1)$ value.

Thus the metametric transformation of response rate is

$$P = \int_{\infty}^{Y-5} \frac{1}{\sqrt{2\pi}} e^{-t^2/2} dt \tag{19.100}$$

Special tables of probits have been prepared from the tables of area under standard normal tables. A table given by Finney (1981) is reproduced in the Appendix as Table A.16.
From this table, for a given value of response rate, the value of probit can be read.

The weights to be attached for computation of estimated regression lines with x and y are

$$W = Z^2/PQ \tag{19.101}$$

where, Z is the ordinate of $N(0, 1)$ pdf at Y, given by

$$Z = \frac{1}{\sqrt{2\pi}} \exp\left\{-\frac{1}{2}Y^2\right\} \tag{19.102}$$

and $Q = 1 - P$.

Bliss (1935a) and Fisher and Yates (1964, Table XI) have tabulated weighting coefficient w as a function of Y. It is reproduced as Table A.18 in the Appendix.

For example (a) if response rate P is .63, from Table A.16 corresponding to row with response rate = .63, the expected probit Y = 5.3319.

(b) if response rate is .23, the NED will be negative. But since probit is increased by 5, its value can be seen as 4.2612 in the row with response rate value = .23 from the Table A.16.

The probit regression line can be thus obtained from the observed value of response rate for various values of dose metameter x.

19.3.2 Logit Analysis

Logistic function is also found to be applicable in a wide range of phenomena. Berkson (1944) discussed its use to represent the regression of quantal response rate on dose in place of the normal curve.

In this case the tolerance distribution is given by $f(t) = \text{sech}^2\, t$

and
$$P = \frac{1}{1 + e^{-(\alpha+\beta x)}} \tag{19.103}$$

Hence, the metametric transformation suggested by Berkson is

$$P = e^Y/(1 + e^Y) \tag{19.104}$$

or
$$Y = \log_e (P/Q) \tag{19.105}$$

where Y is called the logit of P.

Fisher and Yates (1963) and Wilson and Worcester (1943a, b, c) suggested almost similar transformation, given by

$$f(t) = \frac{1}{2}\text{sech}^2 t$$

$$P = \frac{1}{2}(1 + \tanh Y) = \frac{e^{2Y}}{1 + e^{2Y}} \tag{19.106}$$

so that $$Y = \frac{1}{2}\log_e(P/Q)$$

Now $Z = \frac{1}{2}\text{sech}^2 Y = 2PQ$, so the weighting coefficient is

$$w = Z^2/PQ = 4PQ \tag{19.107}$$

Worcester and Wilson (1943a, b, c) have given tables to assist computations for logit analysis when doses are equally spaced and equal number of subjects at each dose.

Finney (1964) has given the required tables for logit analysis. Interested reader may refer to this book.

19.4 Estimation of Potency

In the context of quantal responses of stimulus there are several ways to state its potency. Estimation of potency is done by assessing any one of the following:

(a) Minimal lethal dose or minimal effective dose.
(b) Median lethal dose (LD_{50}) or median effective dose (ED_{50}).
(c) LD_{90} or ED_{90}.

These are described and defined one by one.

(a) Minimal Lethal Dose or Minimum Effective Dose

In the beginning of quantal response assay technique, *minimal lethal or minimal effective* dose was used to assess potency of a stimulus. It was not clear, whether minimal lethal dose was the smallest dose sufficient to kill at least one subject or smallest dose sufficient to kill all the subjects.

Apparent meaning of this term involves the assumption that there is a dose, of a poison which is just sufficient to kill all or most of the living subjects. A dose, even if very little smaller than the said dose, will not kill any living subject of the same species. This is possible only if all the subjects have the same tolerance, which is not true. It is now very clear that tolerance is a random variable and there exists a probability distribution to describe its nature.

(b) Median Effective Dose

The concept of minimal lethal dose is not an unambiguous one and finding it, depends on the presence of a subject with least possible tolerance. If a larger group of subjects is taken, there may be a subject with even lower tolerance. It is very likely that the extreme values of tolerance may change considerably from one group of subjects to other of the same species. It was, therefore, suggested that a better assessment of potency is possible by the *median lethal dose* of a poison or in general the *median effective dose* of a stimulus. These are denoted by LD_{50} or ED_{50}. This is the dose that *produces the specified response in 50 per cent of the tested subjects.* If the tolerance distribution is symmetric, the median effective dose will be same as the mean of the tolerance distribution. It can be shown that estimating LD_{50} or ED_{50} is more satisfactory than estimating dose producing a small (5%) or large (95%) percentage of response.

If equal number of subjects are tested for their response to various doses, estimate of potency will be more precise.

(c) LD_{90} or ED_{90}

The dose *just sufficient to produce the specified critical response in 90% of the subjects* will be denoted by LD_{90} or ED_{90} for a poison or a stimulus, respectively. In general LD_q or ED_q ($0 < q < 100$) denotes the dose just sufficient to produce response in q percent of subjects.

The condition of similarity implies that the tolerance distributions for the SP and TP will be identical except for a factor ρ in the scales of their dose metameter. The *relative potency*, ρ therefore, *may be estimated* by the ratio of ED_{50} or mean doses for any form of tolerance distribution. In symbols:

$$\hat{\rho} = R = \frac{(LD_{50})_S}{(LD_{50})_T} \tag{19.108}$$

If logarithmic dose metameter is used, the difference between corresponding statistics for location of tolerance distributions such as ED_5, ED_{90}, median, mean, on the metametric scale will be estimate of log ρ.

In symbols

$$\text{Estimate of } \log \rho = M = \log (LD_{50})_S - \log (LD_{50})_T \tag{19.109}$$

In the next section the details of computational procedure for probit analysis are given.

19.5 Computational Procedure by Probit Analysis

The steps for probit analysis in a quantal response assay are listed and the expressions of various functions are given in a form suitable for computation.

As explained in the discussion of quantal response assay and probit analysis the steps are summarized below.

Step 1: Form a table to give in its columns

(i) x, the dose metameter normally log dose
(ii) n, the *number* of subjects, *tested* at each dose
(iii) r, the number of subjects, responding at each dose
(iv) $p = r/n$, the empirical response rate in percentage i.e. 100 p.
(v) From Table A.16 read the probit (called *empirical probit*) for each p or 100 p (= Y_E).

Step 2: Compare scatter diagrams of (x, Y_E) for SP and TP and draw two parallel lines by eye judgement which best fit the points from SP and TP.

(i) From the regression lines read the values of Y, the *expected probit* for both SP and TP for each x.
(ii) Expected probit Y is used for obtaining *weighting coefficients* w from Table 19.6. Tabulate w and nw for each x.
(iii) For each Y and 100 p determine the *working probit* y from Table A.16.
(iv) Tabulate nwx and nwy

For an easy access a shortened version of Table A.16 and that of weighting coefficients w are given below in Tables 19.5 and 19.6. An interested reader may consult the detailed table of weighting coefficients w in the book by Finney (1981).

Step 3: Determine the regression lines for SP and TP using nw, nwx and nwy by computing the following for each preparation.

Table 19.5 Transformation of percentage to probits

%	0	1	2	3	4	5	6	7	8	9
0	-	2.67	2.95	3.12	3.25	3.36	3.45	3.52	3.59	3.66
10	3.72	3.77	3.82	3.87	3.92	3.96	4.01	4.05	4.08	4.12
20	4.16	4.19	4.23	4.26	4.29	4.33	4.36	4.39	4.42	4.45
30	4.48	4.50	4.53	4.56	4.59	4.61	4.64	4.67	4.69	4.72
40	4.75	4.77	4.80	4.82	4.85	4.87	4.90	4.92	4.95	4.97
50	5.00	5.03	5.05	5.08	5.10	5.13	5.15	5.18	5.20	5.23
60	5.25	5.28	5.31	5.33	5.36	5.39	5.41	5.44	5.47	5.50
70	5.52	5.55	5.58	5.61	5.64	5.67	5.71	5.74	5.77	5.81
80	5.84	5.88	5.92	5.95	5.99	6.04	6.08	6.13	6.18	6.23
90	6.28	6.34	6.41	6.48	6.55	6.64	6.75	6.88	7.05	7.33
-	0.0	0.1	0.2	0.3	0.4	0.5	0.6	0.7	0.8	0.9
99	7.33	7.37	7.41	7.46	7.51	7.58	7.65	7.75	7.88	8.09

Table 19.6 The weighting coefficient, $w = Z^2/PQ$

Y	0.0	0.1	0.2	0.3	0.4	0.5	0.6	0.7	0.8	0.9
1	0.001	0.001	0.001	0.002	0.002	0.003	0.005	0.006	0.008	0.011
2	0.015	0.019	0.025	0.031	0.040	0.050	0.062	0.076	0.092	0.110
3	0.131	0.154	0.180	0.208	0.238	0.269	0.302	0.336	0.370	0.405
4	0.439	0.471	0.503	0.532	0.558	0.581	0.601	0.616	0.627	0.634
5	0.637	0.634	0.627	0.616	0.601	0.581	0.558	0.532	0.503	0.471
6	0.439	0.405	0.370	0.336	0.302	0.269	0.238	0.208	0.180	0.154
7	0.131	0.110	0.092	0.076	0.062	0.050	0.040	0.031	0.025	0.019
8	0.015	0.011	0.008	0.006	0.005	0.003	0.002	0.002	0.001	0.001

(i) $\bar{x} = \left(\sum nwx\right)/\sum nw$ and $\bar{y} = \left(\sum nwy\right)/\sum nw$ (19.110)

(ii) $$S_{x^2} = \sum nwx^2 - \frac{\left(\sum nwx\right)^2}{\sum nw}$$ (19.111)

(iii) $$S_{y^2} = \sum nwy^2 - \frac{\left(\sum nwy\right)^2}{\sum nw}$$ (19.112)

(iv) $$S_{xy} = \sum nwxy - \frac{\left(\sum nwx\right)\left(\sum nwy\right)}{\sum nw}$$ (19.113)

(v) $$b = \frac{S_{xy}}{Sx^2}$$ (19.114)

The lines of regression now can be written as

(a) for SP $Y^S = \bar{y}^S + b^S\left(X^S - \bar{x}^S\right)$ and (19.115)

(b) for TP $Y^T = \bar{y}^T + b^T\left(X^T - \bar{x}^T\right)$ (19.116)

Step 4: From these two regression lines find Y^S and Y^T for each x. Compare Y^S and Y^T with the expected probits found in the Step 2(i). If these do not match (can be tested by ANOVA for regression given in Table 19.4) repeat the Step 2 (ii) to Step 4 till a match is obtained using Y^T and Y^S in place of Y_E.

Step 5: Test for linearity and parallelism is also done as in Table 19.4. MSE of ANOVA will be used as s^2 for estimating the precision and confidence limits.

Step 6: If the tests validate the assumptions of linearity and parallelism, the estimate of log ρ (= M) is computed as

$$M = \bar{x}^S - \bar{x}^T - \frac{\bar{y}^S - \bar{y}^T}{b} \tag{19.117}$$

Step 7: The 95% confidence limits of log ρ (M_L, M_U) are given by

$$(M_L, M_U) = \left(\bar{x}^S - \bar{x}^T\right) + \left(\frac{1}{1-g}\right)\left[\left\{M - \left(\bar{x}^S - \bar{x}^T\right)\right\} \pm \frac{1.96}{b}\left\{(1-g)\left(\frac{1}{\sum_S nw} + \frac{1}{\sum_T nw}\right) + \frac{\left(M - \left(\bar{x}^S - \bar{x}^T\right)\right)^2}{\left(S^S_{x^2} + S^T_{x^2}\right)}\right\}^{2}\right] \tag{19.118}$$

where $g = \dfrac{1.96^2}{b^2\left(S^S_{x^2} + S^T_{x^2}\right)}$, $\Sigma_S nw$ and $\Sigma_T nw$ are totals of nw for SP and TP, respectively. Take negative sign on RHS of (19.118) while calculating M_L and positive sign, if obtaining M_U.

Step 8: *Median effective dose* ED_{50} for SP and TP is obtained by finding x corresponding to expected probit $Y = 5$ from the lines of regression for SP and TP, respectively.

Let x^S be the x value for $Y = 5$ from regression line for SP, then

$$\text{Estimate of log } ED_{50} = x^S = m^S$$

$$\text{Estimate for } ED_{50} = \text{antilog } x^S \text{ for SP}$$

Similarly, let x^T be the x value for $Y = 5$ from regression line for TP, then

$$\text{Estimate of } ED_{50} = \text{antilog } x^T \text{ for TP}$$

$$\text{and Estimate of log } ED_{50} = x^T = m^T.$$

Variances of Estimates of log ED_{50}

1. Variances of m (= estimated log ED_{50}) are estimated as

(a) if m is *close* to $\bar{x}$, $\hat{V}(m) = \dfrac{1}{b^2 \sum nw}$ (19.119)

(b) if m is *far from* $\bar{x}$

$$\hat{V}(m) = \frac{1}{b^2}\left[\frac{1}{\sum nw} + \frac{(m-\bar{x})^2}{S_{x^2}}\right] \tag{19.120}$$

The 95% confidence limits (m_L, m_U) for log ED_{50} is given by

$$m + \frac{g}{1-g}(m-\bar{x}) \pm \frac{1.96}{b(1-g)}\left[\frac{1-g}{\sum nw} + \frac{(m-\bar{x})^2}{S_{x^2}}\right]^{1/2} \tag{19.121}$$

where $g = 1.96^2/(bS_{x^2})$ and take negative sign for m_L and positive sign for m_U.

Estimates of log LD_{90} or log ED_{90} for SP and TP are also obtained as mentioned above by reading out the values of x for $Y = Y_{90} = 6.28$ the probit of $P = 90\%$.

The 95% confidence limits can also be found from (19.121) after replacing m by the estimates of log LD_{90} or log ED_{90}.

The logit analysis is also done in a similar manner with a difference that the appropriate tables meant for logit analysis should be used in place of the tables for probit analysis.

Books by Finney (1964 and 1981) contain several numerical examples. Interested readers may consult any one of these, if there is any doubt in any step.

20. Multivariate Analysis I

The title of this chapter "Multivariate Analysis" is the branch of statistics which deals with multivariate data, defined briefly in Chapter 9. Since the last quarter of 20th century, in almost all the fields of scientific enquiry, it is found necessary to collect information on several (say p) characteristics of, say n objects. This will give a set of p values for each of the n objects. Observations on an individual may be denoted by a random variable **X** which will have p components $(X_1, X_2, ..., X_p)'$. In this chapter the random variable X is a $p \times 1$ vector given by

$$X_{p\times 1} = \begin{pmatrix} X_1 \\ X_2 \\ . \\ . \\ . \\ X_p \end{pmatrix} = \mathbf{X_p} \tag{20.1}$$

If observations are taken on n objects, one would get n vectors $X_{(1)}, X_{(2)}, ..., X_{(n)}$, each of p-components, which can also be written in a matrix form. They may be written as

$$X_{p\times n} = \left[X_{(1)}\ X_{(2)}\ ......\ X_{(j)}\ ......\ X_{(n)}\right]$$

$$= \begin{bmatrix} X_{11} & X_{21} & ... & X_{j1} & ... & X_{n1} \\ X_{12} & X_{22} & ... & X_{j2} & ... & X_{n2} \\ \vdots & \vdots & \vdots & \vdots & \vdots & \vdots \\ X_{1p} & X_{2p} & ... & X_{jp} & ... & X_{np} \end{bmatrix} \tag{20.2}$$

where the matrix will be a $p \times n$ matrix, the n columns will be data on p characteristics of the n objects. Thus $X_{(j)}$ will be a $p \times 1$ vector containing the information about p characteristics of jth object, $(j = 1, 2, …, n)$.

The data can also be written as

$$X_{n \times p} = \begin{bmatrix} X'_{(1)} \\ X'_{(2)} \\ \cdot \\ \cdot \\ \cdot \\ X'_{(i)} \\ \cdot \\ \cdot \\ \cdot \\ X'_{(n)} \end{bmatrix} \tag{20.3}$$

Now the information of an object will be in a row and the p columns will be the components of the $p \times 1$ random variable written as row vector by transposing i.e. $X'_{p \times 1}$. The ith row ($i = 1, 2, \ldots, n$) will be containing the information on p characteristics of ith object and so will be having p values, denoted by transpose of $X_{(i)}$ denoted as $X'_{(i)}$ (See Appendix C for the notations and basic operations on matrices and vectors).

Since the p-characteristics are observed on the same individual, the components of the r.v. $X_{p \times 1}$ (from now onwards written as X_p in bold) i.e. $X_1, X_2, \ldots, X_p$ are correlated. In a scientific study, one may require to state probability of occurrence of some event of practical interest involving some or all of such correlated variables. For example, suppose a poverty alleviation program in a country has been implemented. From time to time, progress of the program is evaluated. Since several pre-specified different measures on a household will be necessary for assessment of the progress in poverty alleviation program, the appropriate multivariate analysis will be required for drawing valid conclusions.

The multivariate analysis, therefore, has two components. One is to develop the basic theory of the methods. The other component is to consider the practical problems involving multivariate data and to suggest the appropriate existing multivariate technique or if possible to develop new techniques or extend the existing ones to take care of new unsolved needs.

In this chapter some of the basic results will be given without proof, which are used for solving some of the practical problems, faced not only by statisticians but also by social, behavioral, biological and medical researchers.

The explosion of work in the last few decades and need for application of multivariate methods and very fast advancement of computer technology side by side, made it possible that without support of statistical theory, no scientific research is complete.

In this chapter problems related to comparison of means and classification of a multivariate observation, in two or more populations, will be considered. The next chapter will deal with some methods of dimension reduction of data.

As already mentioned, multivariate analysis deals with multivariate data. So the discussions and results are about the vector variable, and the variance covariance and correlations between the component variables are represented in a matrix. In earlier chapters mostly a univariate random variable has been considered, so its mean and variance were also known or unknown constants.

Some Basics

If the study variable is a p component r.v. its mean will be a vector of p-components and its variance covariance and the correlation matrix will be $p \times p$ matrices. The vector and the matrices will be denoted in bold type of capital letters. The dimension will be either indicated as suffix or will be mentioned in the text. Thus the p-component vector r.v. X will be denoted by $\mathbf{X_p}$ and the data matrix on n objects will be denoted by $\mathbf{X}_{p\times n}$ and $\mathbf{X}_{n\times p}$. If $\mathbf{X_p}$ denotes a p-component column vector, $\mathbf{X'_p}$ will denote a row vector.

The objective of applied multivariate analysis is to find answers to some scientific enquiry related to several characteristics, or to generate strong belief about natural phenomena, or to draw some conclusions which form the basis of future decisions.

Often some hypothesis is proposed, multivariate data is collected and some necessary computations are done on the collected data to reject or not to reject the hypothesis. The hypothesis may be modified if it is rejected, for further investigations. New data may be collected and used to examine the modified hypothesis.

About a decade ago application of results of multivariate analysis to real data and their interpretation for further actions required considerable effort, long hours to await results even from computers, and quite a lot of hand computations to check validity of computer results. These difficulties have been overcome now, because of enormously fast speed, enhanced memory of PC's, now easily available at each desk. Many of these have access to statistical packages such as SPSS, SAS, BMDP etc. which can handle now sophisticated and complex multivariate techniques. However, these must be used after fully understanding the assumptions which are required for the statistical tool to be used.

Application Areas

Multivariate analysis came into shape with F. Galton's use of correlation in the second half of nineteenth century. C. Spearman introduced factor analysis in education and psychology to explain human intelligence. R.A. Fisher introduced many notions to explain phenomena in genetics. Regression models formulated by Gauss for multidimensional applications in astronomy, are now further developed for use in economics (this branch is called econometrics) and as experimental design models in agriculture. Latent structure models and multidimensional scaling methods are used in education, psychology and sociology. Classification, discrimination and clustering methods are applied in anthropology, taxonomy and marketing.

Before coming to various multivariate methods, mentioned above, it is appropriate to summarize the main theoretical results used in various multivariate techniques. Efficient and clear statement of such results require vectors and matrices notations. For a novice in this field, a review of matrix algebra is given in Appendix C.

Theoretical Background of Multivariate Analysis

As already mentioned, a multivariate observation on an object can be represented as a vector of p-components. Symbolically, the r.v. representing these observations can be denoted by a column vector $\mathbf{X_p}$.

Any probabilistic inference about the r.v. $\mathbf{X_p}$ requires a knowledge of the probability distribution of the r.v. $\mathbf{X_p}$. This chapter considers all the components of $\mathbf{X}_p$ are continuous. If any of these components are not continuous, the common procedures can not be applied and some modifications are necessary. Such special cases will also be considered, mostly in the next chapter.

The concepts of joint, marginal and conditional distributions given in Chapter 5 may be refreshed before these are discussed for multivariate r.v. $\mathbf{X_p}$.

Cdf F(x)

Thus for a bivariate r.v. $\begin{pmatrix} X \\ Y \end{pmatrix}$, the joint cdf is defined as

$$F(x, y) = P[X \leq x, \quad \text{and} \quad Y \leq y]$$

$$= \int_{-\infty}^{x} \int_{-\infty}^{y} f(x, y)\, dy\, dx \tag{20.4}$$

where $f(x, y)$ is joint pdf of $\begin{pmatrix} X \\ Y \end{pmatrix}$.

Similarly if $\mathbf{X_p}$ is a p-component r.v., its joint cdf is

$$F(\mathbf{x}) = F(x_1, x_2, \ldots, x_p) = P[X_1 \leq x_1, X_2 \leq x_2, \ldots., X_p \leq x_p] \tag{20.5}$$

$$= \int_{-\infty}^{x_p} \int_{-\infty}^{x_{p-1}} \ldots\ldots\ldots. \int_{-\infty}^{x_2} \int_{-\infty}^{x_1} f(x_1, x_2, \ldots\ldots, x_p)\, dx_1\, dx_2 \ldots\ldots dx_p \tag{20.6}$$

where $f(x_1, x_2, \ldots., x_p)$ is joint pdf of $\mathbf{X_p}$ and is also denoted by $f(\mathbf{x})$.
As for univariate r.v. X, one can write the probability of an event related to $\mathbf{X_p}$ from the knowledge of $f(\mathbf{x})$. Thus

$$P(\mathbf{X} \in R) = \int_{R} \ldots. \int f(\mathbf{x})\, d\mathbf{x} \tag{20.7}$$

For any p-dimensional region R.
Joint pdf f(x) is related to cdf $F(\mathbf{x})$ as

$$\mathbf{f}(\mathbf{x}) = \frac{\partial^p (F(\mathbf{x}))}{\partial x_1 \partial x_2 \ldots \partial x_p} \tag{20.8}$$

Marginal densities Some times an experimenter finds that a subset of components of $\mathbf{X_p}$ i.e. a sub-vector of $\mathbf{X_p}$ is of interest. *Marginal distribution* of sub-vector will then be required for inferential purpose.

Let **Y** and **Z** be two sub-vectors of $\mathbf{X_p}$, given as

$$\mathbf{X}_p = \begin{pmatrix} \mathbf{Y}_q \\ \mathbf{Z}_{p-q} \end{pmatrix} \tag{20.9}$$

So **Y** has q components from the total p components of $\mathbf{X_p}$ and **Z** has the remaining $p - q$ components of $\mathbf{X_p}$. Then as in bivariate case, the marginal densities $g(\mathbf{y})$ and $h(\mathbf{z})$ of **Y** and **Z**, respectively, are given by

$$g(\mathbf{y}) = \int_{-\infty}^{\infty} f(\mathbf{x})\, d\mathbf{z} = \int_{-\infty}^{\infty} f(\mathbf{y}, \mathbf{z})\, d\mathbf{z} \tag{20.10}$$

and

$$h(\mathbf{z}) = \int_{-\infty}^{\infty} f(\mathbf{x})\, d\mathbf{y} = \int_{-\infty}^{\infty} f(\mathbf{y}, \mathbf{z})\, d\mathbf{y} \tag{20.11}$$

Conditional densities can also be written as for bivariate case. Thus the conditional density of **Y** given **Z** = **z** is defined as

$$\phi(\mathbf{y}/z) = \frac{f(\mathbf{y}, \mathbf{z})}{h(\mathbf{z})} = \frac{f(\mathbf{x})}{h(\mathbf{z})} \tag{20.12}$$

Similarly,

$$\psi(\mathbf{z}/\mathbf{y}) = \frac{f(\mathbf{y}, \mathbf{z})}{g(\mathbf{y})} = \frac{f(\mathbf{x})}{g(\mathbf{y})} \tag{20.13}$$

Independence of two random vectors **Y** and **Z** holds if any one of the following is satisfied:

(i) $f(\mathbf{y}, \mathbf{z}) = g(\mathbf{y}) \,.\, h(\mathbf{z})$ (20.14)

(ii) $F(\mathbf{x}) = G(\mathbf{y}) \,.\, H(\mathbf{z})$ (20.15)

where $G(\mathbf{y})$ and $H(\mathbf{z})$ are cdf's of **Y** and **Z**, respectively.

(iii) $\phi(\mathbf{y}/\mathbf{z}) = g(\mathbf{y})$ (20.16)

Transformation of Variables

Let the density of $\mathbf{X_p}$ be $f(x_1, x_2, ..., x_p) = f(\mathbf{x})$. Consider the p real valued functions

$$y_i = h_i\,(\mathbf{x}), \quad i = 1, 2, ..., p \tag{20.17}$$

If the transformation from **x** to **y** is such that, one can write the inverse transformation for the **x** values, where $f(\mathbf{x}) > 0$, as:

$$x_i = g_i\,(\mathbf{y}), \quad i = 1, 2, ..., p \tag{20.18}$$

Let the random variables $Y_1, Y_2, ..., Y_p$ be defined as

$$Y_i = h_i\,(\mathbf{X_p}), \quad i = 1, 2, ..., p \tag{20.19}$$

Then the joint density of $\mathbf{Y}_p$ can be written as

$$Q(\mathbf{y}) = f\,\{g_1(\mathbf{y}), g_2(\mathbf{y}), ..., g_p(\mathbf{y})\}\; J(\mathbf{y}) \tag{20.20}$$

where $J(\mathbf{y})$ is the Jacobian of the transformation given by

$$\begin{aligned} J(\mathbf{y}) &= \text{mod } |D_{p\times p}| \\ &= \text{Absolute value of determinant of } D_{p\times p} \end{aligned} \tag{20.21}$$

where

$$D = \begin{bmatrix} \frac{\partial g_1}{\partial y_1} & \frac{\partial g_1}{\partial y_2} & \cdots & \frac{\partial g_1}{\partial y_p} \\ \frac{\partial g_2}{\partial y_1} & \frac{\partial g_2}{\partial y_2} & \cdots & \frac{\partial g_2}{\partial y_p} \\ \cdot & \cdot & & \cdot \\ \cdot & \cdot & & \cdot \\ \cdot & \cdot & & \cdot \\ \frac{\partial g_p}{\partial y_1} & \frac{\partial g_p}{\partial y_2} & \cdots & \frac{\partial g_p}{\partial y_p} \end{bmatrix}, \text{ assuming that all the derivatives exist} \qquad (20.22)$$

$$P[\mathbf{Y} \in S] = \int_S \cdots\cdots \int Q(\mathbf{y})\, d\mathbf{y} \qquad (20.23)$$

Mean and Variance – Covariance of $\mathbf{X}_p$

Mean or expected value of the vector X_p is given by

$$E(\mathbf{X}_p) = \begin{bmatrix} E(X_1) \\ E(X_2) \\ \cdot \\ \cdot \\ \cdot \\ E(X_p) \end{bmatrix} = \begin{bmatrix} \mu_1 \\ \mu_2 \\ \cdot \\ \cdot \\ \cdot \\ \mu_p \end{bmatrix} = \mu \qquad (20.24)$$

Thus μ is the mean vector of $\mathbf{X_p}$.

Let σ_{ij} = Cov (X_i, X_j) (defined in Chapter 9, Section 9.1)

$= E\,[(X_i - EX_i)\,(X_j - EX_j)];\ i, j = 1, 2, ..., p$ (20.25)

$= \sigma_{ji}$

and $\sigma_{ii} = \sigma_i^2 = V(X_i),\ i = 1, 2, ..., p$

Hence the matrix $\mathbf{\Sigma}_{p\times p} = ((\sigma_{ij}))$ is a symmetric matrix.

The diagonal elements σ_{ii} (= σ_i^2) are the variances of X_i's and the off-diagonal elements are the covariances of the components of X_p.

The matrix $\mathbf{\Sigma}_{p\times p}$ is, therefore, called the **variance-covariance matrix** of $\mathbf{X}_p$ or simply **covariance matrix** of $\mathbf{X}_p$.

Correlation Matrix of X_p

Let $\boldsymbol{\rho}_{p\times p} = ((\rho_{ij}))$ denote the correlation matrix of $\mathbf{X}_p$, then as defined in Chapter 9,

$$\rho_{ij} = \text{Correlation between } X_i \text{ and } X_j = \frac{\text{Cov}\,(X_i, X_j)}{\sqrt{V(X_i)V(X_j)}} = \rho_{ji}\ \ i, j = 1, 2, ..., p \qquad (20.26)$$

Now $$\rho_{ii} = 1 \text{ for } i = 1, 2, \ldots, p \tag{20.27}$$

Hence all the diagonal elements of correlation matrix $\boldsymbol{\rho}$ are 1 and the off-diagonal elements are correlations between the components of $\mathbf{X}_p$. Since $\rho_{ij} = \rho_{ji}$, the correlation matrix is also a symmetric matrix.

Let $$\mathbf{D}_{p\times p} = \text{diag}\left(\sigma_1^2, \sigma_2^2, \ldots, \sigma_p^2\right)$$

So $\mathbf{D}^{1/2} = \text{diag}\ (\sigma_1, \sigma_2, \ldots, \sigma_p)$ and one can write

$$\mathbf{\Sigma} = \mathbf{D}^{1/2}\ \boldsymbol{\rho}\ \mathbf{D}^{1/2} \tag{20.28}$$

Covariance Matrix of $Y = AX + b$

A linear transformation of $\mathbf{X}_\text{p}$ can be written as given above such that $\mathbf{Y}$ is k-component vector, $\mathbf{A}$ is a $k \times p$ matrix of constants, $\mathbf{b}$ is a k-component vector of constants. Then the covariance matrix of $\mathbf{Y}$, $\mathbf{\Sigma}_y$ is given by

$$\mathbf{\Sigma}_y = \boldsymbol{A}\mathbf{\Sigma}_x\ \boldsymbol{A}' \tag{20.29}$$

Multivariate Normal Distribution

It is already seen that if a univariate r.v. is studied, the assumption of normal distribution is often made and this simplifies the inferential procedure very much. Due to the central limit theorem, even if the distribution is not exactly normal but is uni-modal and not highly skewed, for large samples, the procedures for normal distribution remain valid.

For multivariate data, many statistical procedures are based on the properties of multivariate normal (MVN) distribution. MVN distribution plays an important role in multivariate data analysis, as normal distribution in univariate data. The generalization of univariate central limit theorem applies for multivariate data also, if the sample size is large. Moreover, the standardized sums of large number of independent data vectors tend to follow MVN distribution, is also a reason for application of this distribution in many real life problems.

Hence the nature of this distribution and its important properties are given here and an analogy of the pdf of MVN r.v. with that of univariate normal distribution is explained.

The pdf of a univariate normal r.v. X can be written as

$$f(x) = \frac{1}{\sqrt{2\pi}\,\sigma}\exp\left[-\frac{1}{2\sigma^2}(x-\mu^2)\right]$$

$$= k\exp\left[-\frac{1}{2}(x-\mu)\frac{1}{\sigma^2}(x-\mu)\right] \tag{20.30}$$

Now if X is replaced by a p-component vector $\mathbf{X}_p$ so x by $\mathbf{x}_p$, μ by p-component vector $\boldsymbol{\mu}$ and $\frac{1}{\sigma^2}$ by $\mathbf{\Sigma}^{-1}$ (the inverse of the covariance matrix of $\mathbf{X}_p$), the square term $\frac{(x-\mu)^2}{\sigma^2}$ by a quadratic form $(\mathbf{x} - \boldsymbol{\mu})'\ \Sigma^{-1}\ (\mathbf{x} - \boldsymbol{\mu})$, one gets

$$f(\mathbf{x}) = K\exp\left[-\frac{1}{2}(\mathbf{x}-\mu)'\,\Sigma^{-1}(\mathbf{x}-\mu)\right] \tag{20.31}$$

The value of K, when evaluated from the result that $\int_{-\infty}^{\infty} \ldots\ldots \int_{-\infty}^{\infty} f(\mathbf{x})\, dx_1\, dx_2 \ldots. dx_p = 1$, is obtained as

$$K = (2\pi)^{-p/2} \, |\,\boldsymbol{\Sigma}\,|^{-1/2}$$

The multivariate normal density, therefore, can be written as

$$f(\mathbf{x}) = (2\pi)^{-p/2} |\Sigma|^{-1/2} \exp\left[-\frac{1}{2}(\mathbf{x}-\mu)' \Sigma^{-1} (\mathbf{x}-\mu)\right] \tag{20.32}$$

This density exists only if $|\boldsymbol{\Sigma}| > 0$ or that $\boldsymbol{\Sigma}$ is a non-singular matrix. Symbolically it can be written that

$$\mathbf{X}_{\mathbf{p}} \sim N(\mu, \boldsymbol{\Sigma})$$

Properties of $N(\mu, \Sigma)$

It can be proved that

(a) $E(\mathbf{X}_p) = \mu$ i.e. the **mean vector** of $\mathbf{X}_p$ is μ (20.33)

(b) **Covariance matrix** of $\mathbf{X}_p = \boldsymbol{\Sigma}$ (20.34)

(c) If $\mathbf{X}_{\mathbf{p}} \sim N(\mu, \boldsymbol{\Sigma})$, then if correlation matrix $\rho = \mathbf{I}$
i.e. if $\rho_{ij} = 0$ for $i \neq j = 1, 2, ..., p$; the components of $\mathbf{X}_p$ are **uncorrelated** and **independent**.

(d) If $\mathbf{X}_p \sim N(\mu, \boldsymbol{\Sigma})$, a transformation $\mathbf{Y} = \boldsymbol{\Sigma}^{-1/2}(\mathbf{x} - \mu)$ standardizes $\mathbf{X}_p$. This is because the distribution of $\mathbf{Y}$ is $N(\mathbf{O}, I_p)$, where I_p is identity matrix of order p, given by

$$I_p = \begin{pmatrix} 1\,0\,0 \ldots\ldots 0 \\ 0\,1\,0 \ldots\ldots 0 \\ \ldots\ldots\ldots\ldots\ldots \\ 0\,0\,0 \ldots\ldots 1 \end{pmatrix} \text{is a } p \times p \text{ identity matrix}$$

(e) If $\mathbf{X}_p \sim N(\mu, \boldsymbol{\Sigma})$, the marginal distribution of any sub-vector of $\mathbf{X}_p$ is also normally distributed with a correspondingly partitioned mean vector and covariance matrix. In symbols, let

$$\mathbf{X} = \begin{pmatrix} \mathbf{Y} \\ \mathbf{Z} \end{pmatrix} \mu = \begin{pmatrix} \mu_y \\ \mu_z \end{pmatrix} \text{ and } \Sigma = \begin{pmatrix} \Sigma_{11} & \Sigma_{12} \\ \Sigma_{21} & \Sigma_{22} \end{pmatrix}$$

where $\mathbf{Y}$ is $q \times 1$ sub-vector of $\mathbf{X}_p$, $\mathbf{Z}$ is $(p - q) \times 1$ sub-vector of remaining components of $\mathbf{X}_p$, μ_y is $q \times 1$ vector of mean of components in $\mathbf{Y}$ and μ_z is $(p - q) \times 1$ vector of means of remaining components in $\mathbf{Z}$ and $\boldsymbol{\Sigma}_{11}$ is $q \times q$ covariance matrix of $\mathbf{Y}$ and $\boldsymbol{\Sigma}_{12}$ is $q \times (p - q)$ matrix $\boldsymbol{\Sigma}_{21}$ is $(p - q) \times q$ matrix and $\boldsymbol{\Sigma}_{22}$ is $(p - q) \times (p - q)$ covariance matrix of $\mathbf{Z}$ These will result if $\boldsymbol{\Sigma}$ is partitioned into 4 parts by drawing a horizontal line after the q rows and another vertical line after the q columns of $\boldsymbol{\Sigma}$.

Then if $\boldsymbol{\Sigma}_{11}$ and $\boldsymbol{\Sigma}_{22}$ are non-singular, then

$$\mathbf{Y} \sim N_p(\mu_y, \boldsymbol{\Sigma}_{11}) \text{ and } \mathbf{Z} \sim N_{p-q}(\mu_z, \boldsymbol{\Sigma}_{22}) \tag{20.35}$$

(f) Let $\mathbf{X}_p$ with distribution $N(\mu, \boldsymbol{\Sigma})$ be partitioned into sets $\mathbf{X}^{(1)}$ and $\mathbf{X}^{(2)}$ given by

$$\mathbf{X}^{(1)} = \begin{pmatrix} X_1 \\ X_2 \\ \cdot \\ \cdot \\ \cdot \\ X_q \end{pmatrix} \text{ and } \mathbf{X}^{(2)} = \begin{pmatrix} X_{q+1} \\ X_{q+2} \\ \cdot \\ \cdot \\ \cdot \\ X_p \end{pmatrix}$$

So $\mathbf{X}^{(1)}$ is q-component vector and $\mathbf{X}^{(2)}$ is $p - q$ component vector. Then $\mathbf{X} = \begin{pmatrix} \mathbf{X}^{(1)} \\ \mathbf{X}^{(2)} \end{pmatrix}$ and

$$E(\mathbf{X}) = \mu = \begin{pmatrix} \mu^{(1)} \\ \mu^{(2)} \end{pmatrix}, \text{ where } \mu^{(j)} = E\left(\mathbf{X}^{(j)}\right), j = 1, 2 \tag{20.36}$$

Further let Cov $(\mathbf{X}^{(1)}) = \mathbf{\Sigma}_{11}$, Cov $(\mathbf{X}^{(2)}) = \mathbf{\Sigma}_{22}$ and

$$\text{Cov}\ (\mathbf{X}^{(1)}, \mathbf{X}^{(2)}) = \text{Cov}\ (\mathrm{X}^{(2)}, \mathrm{X}^{(1)}) = \mathbf{\Sigma}_{12} = \mathbf{\Sigma}_{21} = \mathbf{0} \tag{20.37}$$

Under this condition, the covariance matrix of $\mathbf{X}_p$ can be written as

$$\Sigma = \begin{pmatrix} \Sigma_{11} & \Sigma_{12} \\ \Sigma_{21} & \Sigma_{22} \end{pmatrix} = \begin{pmatrix} \Sigma_{11} & \mathbf{O} \\ \mathbf{O} & \Sigma_{22} \end{pmatrix} \tag{20.38}$$

and

$$\Sigma^{-1} = \begin{pmatrix} \Sigma_{11}^{-1} & \mathrm{O} \\ \mathrm{O} & \Sigma_{22}^{-1} \end{pmatrix} \tag{20.39}$$

and

$$|\mathbf{\Sigma}| = |\mathbf{\Sigma}_{11}| \cdot |\mathbf{\Sigma}_{22}| \tag{20.40}$$

Under the conditions (20.33) to (20.38), it can be proved that marginal distributions of $\mathbf{X}^{(1)}$ is $N\ (\mu^{(1)}, \mathbf{\Sigma}_{11})$ and of $\mathbf{X}^{(2)}$ is $N\ (\mu^{(2)}, \mathbf{\Sigma}_{22})$.

These **results** can be **stated** as : If $\mathbf{X}_p$ has MVN distribution $N(\mu, \mathbf{\Sigma})$ is partitioned into two sub-vectors $\mathbf{X}_q^{(1)}$ and $\mathbf{X}_{p-q}^{(2)}$ then, the necessary and sufficient condition that $\mathbf{X}^{(1)}$ and $\mathbf{X}^{(2)}$ are independent is that $\mathbf{\Sigma}_{12} = 0$ i.e. covariance of a variable in one set with a variable of the second set is 0.

(g) If $\mathbf{X}_p$, μ and $\mathbf{\Sigma}$ are partitioned as in (f), then the **linear transformations** $\mathbf{Y}^{(1)} = \mathbf{X}^{(1)} - \mathbf{\Sigma}_{12}\ \mathbf{\Sigma}_{22}^{-1}\ \mathbf{\Sigma}^{(2)}\ \mathbf{X}^{(2)}$ and $\mathbf{Y}^{(2)} = \mathbf{X}^{(2)}$ are independently normally distributed as

$$\mathbf{Y}^{(1)} \sim N\left(\nu^{(1)}, \Sigma_{11} - \Sigma_{12}\Sigma_{22}^{-1}\Sigma_{21}\right) \text{ and } \mathbf{Y}^{(2)} \sim N\left(\nu^{(2)}, \Sigma_{22}\right) \tag{20.41}$$

where $\mathbf{v}^{(1)} = \mu^{(1)} - \mathbf{\Sigma}_{12}\mathbf{\Sigma}_{22}^{-1}\ \mu^{(2)}$ and $\mathbf{v}^{(2)} = \mu^{(2)}$.

Let $\mathbf{\Sigma}_{11} - \mathbf{\Sigma}_{12}\mathbf{\Sigma}_{22}^{-1}\mathbf{\Sigma}_{21} = \mathbf{\Sigma}_{11.2}$, then one can write that

$$\mathbf{Y}^{(1)} \sim N\left(\nu^{(1)}, \Sigma_{11.2}\right) \text{ and } \mathbf{Y}^{(2)} \sim N\left(\nu^{(2)}, \Sigma_{22}\right)$$

(h) Let the components of $\mathbf{X} = \begin{pmatrix} \mathbf{X}^{(1)} \\ \mathbf{X}^{(2)} \end{pmatrix}$ be divided into two groups as above.

Moreover, let $\boldsymbol{\mu}$ and $\boldsymbol{\Sigma}$ be also similarly partitioned so that $\boldsymbol{\Sigma}_{12}$ is covariance matrix of $\mathbf{X}^{(1)}$ and $\mathbf{X}^{(2)}$. Then if $\mathbf{X} \sim N(\boldsymbol{\mu}, \boldsymbol{\Sigma})$, the **conditional** distribution of $\mathbf{X}^{(1)}$ given $\mathbf{X}^{(2)} = \mathbf{x}^{(2)}$ is normal with mean $\boldsymbol{\mu}^{(1)} + \boldsymbol{\Sigma}_{12}\boldsymbol{\Sigma}_{22}^{-1}(\mathbf{x}^{(2)} - \boldsymbol{\mu}^{(2)})$ with covariance matrix $\boldsymbol{\Sigma}_{11} - \boldsymbol{\Sigma}_{12}\boldsymbol{\Sigma}_{22}^{-1}\boldsymbol{\Sigma}_{21}$. In symbols.

If
$$\mathbf{X} = \begin{pmatrix} \mathbf{X}^{(1)} \\ \mathbf{X}^{(2)} \end{pmatrix} \sim N(\mu, \Sigma) \text{ and } \mu = \begin{pmatrix} \mu^{(1)} \\ \mu^{(2)} \end{pmatrix}, \Sigma = \begin{pmatrix} \Sigma_{11} & \Sigma_{12} \\ \Sigma_{21} & \Sigma_{22} \end{pmatrix}$$

Then
$$\left(\mathbf{X}^{(1)}/\mathbf{X}^{(2)} = \mathbf{x}^{(2)}\right) \sim N\left(\mu^{(1)} + \Sigma_{12}\Sigma_{22}^{-1}(\mathbf{x}^{(2)} - \mu^{(2)}), \Sigma_{11.2}\right) \tag{20.42}$$

where
$$\Sigma_{11.2} = \Sigma_{11} - \Sigma_{12}\Sigma_{22}^{-1}\Sigma_{21} \tag{20.43}$$

(i) Let $\mathbf{x}_1, \mathbf{x}_2, ..., \mathbf{x}_N$ be a random sample of size N from $\mathbf{X} \sim N(\boldsymbol{\mu}, \boldsymbol{\Sigma})$ $(N > p)$. Also let

$$\bar{\mathbf{x}} = \frac{1}{N}\sum_{\alpha=1}^{N} \mathbf{x}_\alpha,\ A = \sum_{\alpha=1}^{N} (\mathbf{x}_\alpha - \bar{\mathbf{x}})(\mathbf{x}_\alpha - \bar{\mathbf{x}})' = (A_{ij});\ i, j = 1, 2, \ldots, p$$

$$A_{ij} = \sum_{\alpha=1}^{N} \left(\mathbf{x}_{i\alpha} - \mathbf{x}_i\right)\left(\mathbf{x}_{j\alpha} - \bar{\mathbf{x}}_j\right)',\ \bar{\mathbf{x}}_i = \frac{1}{N}\sum_{\alpha=1}^{N} \mathbf{x}_{i\alpha} \tag{20.44}$$

and $\boldsymbol{\psi} = \boldsymbol{\Sigma}^{-1}$.
Then A is a non-singular matrix.
It can be shown that the m.l.e.'s of $\boldsymbol{\mu}$ and $\boldsymbol{\psi}$ are given by

$$\hat{\mu} = \bar{x},\ \hat{\psi} = NA^{-1} \text{ and } \hat{\Sigma} = A/N \tag{20.45}$$

(j) **Sampling Distribution of the Sample Mean $\bar{\mathbf{x}}$**
If $\mathbf{X} \sim N(\boldsymbol{\mu}, \boldsymbol{\Sigma})$, the mean $\bar{\mathbf{x}}$ of a sample of size N from $\mathbf{X}$ is distributed as $N(\boldsymbol{\mu}, \boldsymbol{\Sigma}/N)$ and is independent of $\hat{\boldsymbol{\Sigma}}$, the m.l.e. of $\boldsymbol{\Sigma}$.

(k) **Sampling Distribution of $N\,\hat{\boldsymbol{\Sigma}}$**

$$\mathbf{N}\,\hat{\boldsymbol{\Sigma}} \text{ is distributed as } \sum_{\alpha=1}^{N-1} \mathbf{Z}_\alpha \mathbf{Z}'_\alpha, \text{ where } \mathbf{Z}_\alpha \sim N(0, \boldsymbol{\Sigma}) \text{ is independent of } \mathbf{Z}_\beta\ (\alpha \neq \beta) \tag{20.46}$$

(l)
$$E(\hat{\Sigma}) = \frac{1}{N} E\left[\sum_{\alpha=1}^{N-1} \mathbf{Z}_\alpha \mathbf{Z}'_\alpha\right] = \frac{N-1}{N}\Sigma \tag{20.47}$$

$\hat{\boldsymbol{\Sigma}}$ is, therefore, not an unbiased estimator of $\boldsymbol{\Sigma}$

but
$$S = \frac{1}{N-1} A = \frac{1}{N-1}\sum_{\alpha=1}^{N} (x_\alpha - \bar{x})(x_\alpha - \bar{x})' \tag{20.48}$$

is an unbiased estimator of $\boldsymbol{\Sigma}$.

A reader interested in the proofs of these results, may consult books on multivariate analysis such as Anderson (1972), Press (1972) and Dillon and Goldstein (1984) etc.

After giving a summary of these results, the first section starts with testing of hypothesis concerning means of one or two multivariate normal populations.

20.1 Hoteling's T^2 and Mahalanobis D^2

Suppose that an experimenter has multivariate data, which can be assumed to have MVN distribution. He might be interested to infer whether the sample data has come from a MVN population with a specified mean vector $\boldsymbol{\mu}_0$. For univariate data the inferential procedures have been given in Chapter 8. It may be recalled that if the variance σ^2 of the univariate r.v. X is not known and sample size is small, Student's t-distribution of the test-statistic is used to decide whether to reject or not to reject $H_0 : \mu_X = \mu_0$.

If $\mathbf{X}$ is a p-component column vector and a test of $H_0 : \boldsymbol{\mu}_\mathbf{X} = \boldsymbol{\mu}_0$ is to be conducted when covariance matrix of $\mathbf{X}$, $\boldsymbol{\Sigma}$ is not known, a multivariate analogue of the univariate t-distribution is used.

For a univariate X, the Student's t-statistic for testing H_0 is

$$t = \frac{\sqrt{n}\left(\bar{X} - \mu_0\right)}{s} \tag{20.49}$$

Square of t is given by

$$t^2 = \frac{n\left(\bar{X} - \mu_0\right)^2}{s^2}$$

$$= n\left(\bar{X} - \mu_0\right)\left(s^2\right)^{-1}\left(\bar{X} - \mu_0\right) \tag{20.50}$$

A multivariate analogue of this may be written as

$$T^2 = N\left(\bar{\mathbf{X}} - \mu_0\right)S^{-1}\left(\bar{\mathbf{X}} - \mu_0\right) \tag{20.51}$$

The distribution of T^2-statistic was derived by Hotelling (1931). For an arbitrary value of μ_X, the distribution was derived by Bose and Roy (1938) and by Hsu (1938). This T^2-statistic is called the **Hotelling's T^2**.

Result

Let $U = \left(\dfrac{T^2}{N-1}\right)\left(\dfrac{N-p}{p}\right)$, where T^2 is given in (20.51) then

$$U \sim F_{p.N-p} \text{ if } H_0 \text{ is true} \tag{20.52}$$

This result is utilized in testing of population mean, if a random sample $\mathbf{x}_1, \mathbf{x}_2, \ldots, \mathbf{x}_N$ is available on a p-component random variable $\mathbf{X}_p \sim N(\boldsymbol{\mu}, \boldsymbol{\Sigma})$. In this **one sample** problem it is desired to test $H_0 : \boldsymbol{\mu} = \boldsymbol{\mu}_0$ against $H_1 : \boldsymbol{\mu} \neq \boldsymbol{\mu}_0$.

There arise two cases: (a) $\boldsymbol{\Sigma}$ is known (b) $\boldsymbol{\Sigma}$ is unknown.

(a) $\boldsymbol{\Sigma}$ is known: Define

$$U = N\left(\bar{\mathbf{x}} - \mu_0\right)'\Sigma^{-1}\left(\bar{\mathbf{x}} - \mu_0\right) \tag{20.53}$$

It can be seen that $\boldsymbol{\Sigma}$ being known, U is a statistic and it is used to test H_0. If H_0 is true, U follows χ^2-distribution with p d.f. In symbols

$$U \sim \chi_p^2 \text{ if } H_0 \text{ is true} \tag{20.54}$$

So the *test procedure* is

$$\text{Reject } H_0 \text{ if } U > \chi^2_{p,\alpha} \tag{20.55}$$

(b) **Σ is unknown:** In this case Hotelling's T^2-statistic is used.
Thus calculate

$$T^2 = N(\bar{\mathbf{x}} - \mu_0)' \mathbf{S}^{-1} (\bar{\mathbf{x}} - \mu_0)$$

and
$$U = \left(\frac{T^2}{N-1}\right)\left(\frac{N-p}{p}\right) \tag{20.56}$$

which has F-distribution with p and $N - p$ d.f. if H_0 is true. In symbols

$$U \sim F_{p,\, N-p} \text{ if } H_0 \text{ is true} \tag{20.57}$$

where U is defined in (20.56).
So the *test procedure* to test H_0 against H_1 at 100 α% level of significance is

$$\text{Reject } H_0 \text{ if } F > F_{p,\, N-p,\, \alpha} \tag{20.58}$$

$\chi^2_{p,\,\alpha}$ and $F_{p,N\text{-}p,\,\alpha}$ are the 100 α% upper tail significant points of χ^2-distribution with p d.f. and of F-distribution with p and N-p d.f., respectively.

Mahalanobis D²

Define
$$D^2 = (\bar{\mathbf{x}} - \mu_0)' \mathbf{S}^{-1} (\bar{\mathbf{x}} - \mu_0), \text{ then} \tag{20.59}$$

it measures the distance between the sample mean $\bar{x}$ and the population mean μ_0. This is called **Mahalanobis D²**. Looking at (20.51) and (20.59) it is clear that

$$T^2 = ND^2 \tag{20.60}$$

and from (20.56)

$$\left(\frac{ND^2}{N-1}\right)\left(\frac{N-p}{p}\right) \sim F_{p,N-p} \text{ if } H_0 \text{ is true} \tag{20.61}$$

Two Sample problem. Suppose $\mathbf{X}_p \sim N(\mu_1, \Sigma)$ and $\mathbf{Y}_p \sim N(\mu_2, \Sigma)$.
Random samples $\mathbf{x}_1, \mathbf{x}_2, ..., \mathbf{x}_{N_1}$ and $\mathbf{y}_1, \mathbf{y}_2, ..., \mathbf{y}_{N_2}$ have been selected independently on **X** and **Y**, respectively. It is desired to test the hypothesis that the two samples are drawn from the populations with same mean vector. In symbols, it is desired to test

$$H_0 : \mu_1 = \mu_2 \text{ against } H_1 : \mu_1 \neq \mu_2 \tag{20.62}$$

To test H_0 against H_1, the following procedure is adopted:

(i) Find
$$\bar{\mathbf{x}} = \frac{1}{N_1}\sum_{i=1}^{N_1} \mathbf{x}_i \text{ and } \bar{\mathbf{y}} = \frac{1}{N_2}\sum_{j=1}^{N_2} \mathbf{y}_j \tag{20.63}$$

(ii) Calculate
$$S = \frac{1}{N_1 + N_2 - 2}\left[\sum_{i=1}^{N_1}(\mathbf{x}_i - \bar{\mathbf{x}})(\mathbf{x}_i - \bar{\mathbf{x}})' + \sum_{j=1}^{N_2}(\mathbf{y}_j - \bar{\mathbf{y}})(\mathbf{y}_j - \bar{\mathbf{y}})'\right] \tag{20.64}$$

(iii) Calculate
$$T^2 = \frac{N_1 N_2}{N_1 + N_2} (\bar{\mathbf{x}} - \bar{\mathbf{y}})' \mathbf{S}^{-1} (\bar{\mathbf{x}} - \bar{\mathbf{y}}) \tag{20.65}$$

(iv) $V = \dfrac{N_1 + N_2 - p - 1}{(N_1 + N_2 - 2)p}$ T^2 follows F-distribution with p and $N_1 + N_2 - p - 1$ d.f. if H_0 is true. In symbols.

$$\frac{N_1 + N_2 - p - 1}{(N_1 + N_2 - 2)p} T^2 \sim F_{p, N_1 + N_2 - p - 1} \text{ if } H_0 \text{ is true} \tag{20.66}$$

(v) So the *test procedure* is

$$\text{Reject } H_0 \text{ if } V > F_{p, N_1 + N_2 - p - 1, \alpha} \tag{20.67}$$

where $F_{p,N_1+N_2-p-1,\alpha}$ is upper tail 100 α% point of F_{p,N_1+N_2-p-1}.

Mahalanobis Distance between two populations of **X** and **Y** is

$$(\mu_1 - \mu_2)' \Sigma^{-1} (\mu_1 - \mu_2) \tag{20.68}$$

in terms of parameters.

$$D^2 = (\bar{\mathbf{x}} - \bar{\mathbf{y}})' \mathbf{S}^{-1} (\bar{\mathbf{x}} - \bar{\mathbf{y}}) \tag{20.69}$$

is the distance between two samples from MVN populations of X and Y, where **S** is defined in (20.64).

This is also called **Mahalanobis D²**, the distance between two samples from MVN populations. Comparing (20.65) with (20.69), one can write

$$T^2 = \frac{N_1 N_2}{N_1 + N_2} D^2 \tag{20.70}$$

T^2 is also called sometimes, the **Studentized Mahalanobis D²**.

The distribution of

$$\frac{N_1 N_2 \, D^2}{(N_1 + N_2)(N_1 + N_2 - p)} \; \frac{N_1 + N_2 - p - 1}{p} \text{ is } F_{(p, N_1 + N_2 - p - 1)} \text{ if } H_0 : \mu_1 = \mu_2 \text{ is true}$$

Multivariate Techniques an Overview

Apart from the problem of concluding about the mean of the p-component MVN vector from one sample or from two samples, an experimenter may be interested in exploring various aspects of the structure of data.

When the interest is on the association between the two sets of variables in the p-component multivariate observations, **dependence** methods should be used. Thus $\mathbf{X}_p$ is partitioned into two sub vectors. One sub-vector may be **dependent** or **criterion** or **response** variable set, and the rest is the **independent** or **explanatory** or **predictor** variable set.

When interest centers on mutual association across all the p-variables in the data vector, with no distinction among the components of the variable in the data, the methods used are **interdependence** methods.

Multivariate methods commonly used in the said two classes are listed below in Table 20.1.

Table 20.1 Multivariate methods

Dependence Methods	Interdependence Methods
(a) Multiple correlation and Regression (only one criterion, rest predictor)	(d) Discriminant Analysis
	(e) Principal Component Analysis
(b) Canonical Correlation (criterion ≥ 2 rest predictor)	(f) Factor Analysis
	(g) Cluster Analysis
(c) MANOVA (multiple criterion)	(h) Multidimensional Scaling

The next section of this Chapter considers Discriminant Analysis and the last one deals with MANOVA. As already mentioned a description of the problem and its solution are given without giving proof of any result used, therein. For the proofs, the interested reader may consult the books on Multivariate Analysis such as by Anderson (1972), Press (1972), Dillon and Goldstein (1984), Rao (1973), Harris (1975), Morrison (1967) etc.

20.2 Discriminant Analysis: Classification in Two or More than Two Populations

Discriminant analysis or classification problem does not divide the components of the observation vector. On the basis of the measurements on *p*-characteristics of an object or individual, one wants to classify it as belonging to exactly one of the k mutually exclusive populations $\pi_1, \pi_2, \ldots, \pi_k$. The experimenter does not know from which population, the object has come in his sample. Only on the basis of the measurements on *p*-characteristics, the object is to be classified as coming from one population or the other.

There may be three types of situations, when one attempts to classify an object with a *p*-component observation vector $\mathbf{X}_p$:

(a) Observation $\mathbf{X}_p$ may belong to one of two MVN populations when the parameters are (i) known or (ii) unknown.
(b) Observation $\mathbf{X}_p$ may belong to one of k mutually exclusive MVN populations with same covariance matrix $\mathbf{\Sigma}$ but different mean vectors and when the parameter $\mathbf{\Sigma}$ is (i) known or is (ii) unknown.
(c) Observation vector $\mathbf{X}_p$ may not have a MVN distribution.

General Criteria of Good Classification

For explaining the ideas behind the procedure of classification, consider the problem of classification into one of two populations. The generalization to one of k populations follows easily.

So given a data vector $\mathbf{X}_p$ on an object, it is to be classified as belonging to π_1 or π_2. the *p*-dimensional data space will be divided into two regions R_1 and R_2, such that if $\mathbf{X}_p \in R_1$ it is classified as coming from π_1 and to π_2 otherwise.

In this procedure, there are possibilities of two types of error

(a) Object actually belongs to π_1, but is classified as from π_2.
(b) Object actually belongs to π_2, but is classified as from π_1.

To obtain a good procedure, the loss or cost of misclassification must be minimized. For this define the costs of misclassification as

$$C\ (2/1) = \text{Cost of misclassifying in } \pi_2 \text{ when it belongs to } \pi_1 \text{ and}$$
$$C\ (1/2) = \text{Cost of misclassifying in } \pi_1 \text{ when it belongs to } \pi_2 \qquad (20.71)$$

Table 20.2 Costs of misclassification

True Situation	Decision Taken	
	π_1	π_2
π_1	0	$C(2/1)$
π_2	$C(1/2)$	0

To obtain a good classification procedure, the loss or cost of misclassification is minimized. Two cases arise to determine the good classification procedure

(i) a priori probabilities q_1 and q_2, that an observation chosen at random belongs to π_1 or π_2 are known.
(ii) q_1 and q_2 are not known.

In the first case since $p_1(\mathbf{x})$ and $p_2(\mathbf{x})$ the joint pdf's of $\mathbf{X}$ in π_1 and π_2, respectively are known, the average losses can be found. A **good procedure** is one for which the **average loss is minimum.** For the derivation of expression of the average loss reader may refer to books by Anderson (1958) and Press (1972).

It is now required to determine the regions R_1 and R_2 for which average loss is minimum. It can be seen from the said books, that

$$R_1 : \frac{p_1(\mathbf{x})}{p_2(\mathbf{x})} > k, \quad \text{where } k = \frac{q_2\ C(1/2)}{q_1\ C(2/1)} \text{ and}$$

$$R_2 : \frac{p_1(\mathbf{x})}{p_2(\mathbf{x})} < k \qquad (20.72)$$

The procedure given in (20.72) forms a base for a good classification. Thus the procedures can be given as the following for various cases:
(a) (i) To classify $\mathbf{X}$ (p-component data vector) in one of two MVN populations $N(\boldsymbol{\mu}^{(1)}, \boldsymbol{\Sigma})$ or $N(\boldsymbol{\mu}^{(2)}, \boldsymbol{\Sigma})$, when $\boldsymbol{\mu}^{(1)}$, $\boldsymbol{\mu}^{(2)}$ and $\boldsymbol{\Sigma}$ are known. R_1 and R_2 can be obtained as:

$$R_1 : \mathbf{X}'\ \boldsymbol{\Sigma}^{-1}\ (\boldsymbol{\mu}^{(1)} - \boldsymbol{\mu}^{(2)}) > 1/2\ (\boldsymbol{\mu}^{(1)} + \boldsymbol{\mu}^{(2)})'\ \boldsymbol{\Sigma}^{-1}\ (\boldsymbol{\mu}^{(1)} - \boldsymbol{\mu}^{(2)}) + C \text{ and}$$
$$R_2 : \mathbf{X}'\ \boldsymbol{\Sigma}^{-1}\ (\boldsymbol{\mu}^{(1)} - \boldsymbol{\mu}^{(2)}) < 1/2\ (\boldsymbol{\mu}^{(1)} + \boldsymbol{\mu}^{(2)})'\ \boldsymbol{\Sigma}^{-1}\ (\boldsymbol{\mu}^{(1)} - \boldsymbol{\mu}^{(2)}) + C \qquad (20.73)$$

where $$C = \log \frac{q_2 C(1/2)}{q_1 C(2/1)}$$ **which becomes 0**, if $q_2\ C(1/2) = q_1\ C(2/1)$ (20.74)

The L.H.S. of (20.73) viz. $\mathbf{X}'\ \boldsymbol{\Sigma}^{-1}\ (\boldsymbol{\mu}^{(1)} - \boldsymbol{\mu}^{(2)})$ is a linear function of components of $\mathbf{X}$, since $\boldsymbol{\Sigma}^{-1}\ (\boldsymbol{\mu}^{(1)} - \boldsymbol{\mu}^{(2)}) = \mathbf{d}$ (say), is a $p \times 1$ vector. This linear function is known as **linear discriminant function.** $\mathbf{X}'$d can be determined if $\mathbf{d} = \boldsymbol{\Sigma}^{-1}\ (\boldsymbol{\mu}^{(1)} - \boldsymbol{\mu}^{(2)})$ is known. So $\mathbf{d}$ can be determined either by computing

Σ^{-1} ($\mu^{(1)} - \mu^{(2)}$) or by solving Σ **d** = $\mu^{(1)} - \mu^{(2)}$ for **d**, as a system of p linear equations for p components of **d**.

If **q_1, q_2, C(1/2) and C(2/1) are known** then C can be found and R_1 and R_2 can be determined.
(ii) If q_1 and q_2 are not known but q_2 $C(1/2) = q_1$ $C(2/1)$ holds (20.73) can be used with $C = 0$.

If **d** has been obtained either by computing Σ^{-1} ($\mu^{(1)} - \mu^{(2)}$) or by solving Σ **d** = ($\mu^{(1)} - \mu^{(2)}$), classification of an object can be done as

$$\text{If } \mathbf{X}' d > 1/2\ (\mu^{(1)} + \mu^{(2)})' \mathbf{d}, \text{ classify the object in } \pi_1 \text{ otherwise to } \pi_2 \tag{20.75}$$

(a) (ii) If $k = 2$ and the parameters of the two MVN populations are not known, the classification procedure is

$$\text{if } \mathbf{X}'\mathbf{S}^{-1}\left(\bar{\mathbf{X}}^{(1)} - \bar{\mathbf{X}}^{(2)}\right) \geq \frac{1}{2}\left(\bar{\mathbf{X}}^{(1)} + \bar{\mathbf{X}}^{(2)}\right)' \mathbf{S}^{-1}\left(\bar{\mathbf{X}}^{(1)} - \bar{\mathbf{X}}^{(2)}\right) \tag{20.76}$$

the **object belongs** to π_1, otherwise to π_2,
where **S** is the pooled covariance matrix defined as

$$\mathbf{S} = \frac{1}{N_1 + N_2 - 2}\left[\sum_{i=1}^{N_1}\left(\mathbf{X}_i^{(1)} - \bar{\mathbf{X}}^{(1)}\right)\left(\mathbf{X}_i^{(1)} - \bar{\mathbf{X}}^{(1)}\right)' + \sum_{j=1}^{N_2}\left(\mathbf{X}_j^{(2)} - \bar{\mathbf{X}}^{(2)}\right)\left(\mathbf{X}_j^{(2)} - \bar{\mathbf{X}}^{(2)}\right)'\right] \tag{20.77}$$

Thus there should be random samples of sizes N_1 from π_1 and N_2 from π_2.
(b) Suppose $\mathbf{X}_p$ is to be classified into one of the k MVN populations, i^{th} population is N ($\mu^{(i)}$, Σ), $i = 1, 2, ..., k$. If $\mu^{(i)}$, $i = 1, 2, \ldots, k$ and Σ are known, the classification rules are as given below.
(i) If costs of misclassification **C(i/j) (j ≠ i) are equal** and prior probabilities q_i ($i = 1, 2, \ldots, k$) are known, then according to Anderson (1958) classify the object in π_j if

$$\left[\mathbf{x} - \frac{1}{2}\left(\mu^{(i)} + \mu^{(i)}\right)\right]' \Sigma^{-1}\left(\mu^{(j)} - \mu^{(i)}\right) > \log\frac{q_i}{q_j} \text{ for } i = 1, 2, ..., k; i \neq j \tag{20.78}$$

L.H.S. of (20.78) is $\log\dfrac{p_j(\mathbf{x})}{p_i(\mathbf{x})}$.

(ii) If costs of misclassification $C(i/j)$ and q_i are equal, according to Press (1972), classify the object with observed data vector **x** in π_j if

$$\left(\mathbf{x} - \mu^{(i)}\right)' \Sigma_i^{-1}\left(\mathbf{x} - \mu^{(i)}\right) - \left(\mathbf{x} - \mu^{(j)}\right)' \Sigma_j^{-1}\left(\mathbf{x} - \mu^{(j)}\right) > \log\frac{|\Sigma_j|}{|\Sigma_i|} \tag{20.79}$$

for $i = 1, 2, ..., k$ ($i \neq j$).

It may be noted that the discriminant function in this case is not a linear one, but is a quadratic form of **x**.
(iii) Suppose that Σ is not known, determine the Mahalanobis distance D_i^2 of **x** from i^{th} population π_i. This is given by:

$$D_i^2 = \left(\mathbf{x} - \mu^{(i)}\right)' S^{-1}\left(\mathbf{x} - \mu^{(i)}\right); \quad i = 1, 2, ..., k \tag{20.80}$$

where S is the pooled estimate of the covariance matrix based on N_i random observations from π_i, given by

$$\mathbf{S} = \frac{1}{\left(\sum_{i=1}^{k} N_i - k\right)} \left[\sum_{i=1}^{k}\sum_{j=1}^{N_i}\left(\mathbf{x}_{ij} - \overline{\mathbf{x}}^{(i)}\right)\left(\mathbf{x}_{ij} - \overline{\mathbf{x}}^{(i)}\right)'\right]$$

$$\overline{\mathbf{x}}^{(i)} = \frac{1}{N_i}\sum_{j=1}^{N_i}\mathbf{x}_{ij};\ i = 1, 2, ..., k \tag{20.81}$$

Classify the object with data vector **x** in the jth population π_j, if

$$D_j^2 = \min\left(D_1^2, D_2^2, ..., D_k^2\right) \tag{20.82}$$

(c) Now suppose the observed data vector **x** is known to have come from one of several populations, which are not MVN.

The distance between the observation vectors **x** and **y** (both of p-components belonging to j^{th} group (j = 1, 2, ..., k) is given by

$$d_j^2\,(\mathbf{x}, \mathbf{y}) = (\mathbf{x} - \mathbf{y})'\, V_j^{-1}\,(\mathbf{x} - \mathbf{y}) \tag{20.83}$$

where V_j is of one of the following forms

$$V_j = \begin{cases} S_p, \text{ the pooled covariance matrix given in (20.81)} \\ \text{diag}\,(S_p) \\ S_j, \text{ the estimated covariance matrix of } j^{\text{th}} \text{ group} \\ \text{diag}\,(S_j) \\ I \end{cases} \tag{20.84}$$

Now with the group-specific densities p_j (**x**), and the associated prior probabilities q_j's, the posterior probabilities ϕ (j/**x**) that **x** belongs to π_j (j = 1, 2, ..., k) are evaluated.

The object is classified in population π_j if ϕ (j/**x**) is maximum among all the posterior probabilities. In symbols, classify in π_j if

$$\phi(j/\mathbf{x}) = \max_i\,(\phi(i/\mathbf{x})) \tag{20.85}$$

If this largest probability is less than a specified threshold, the object with observed data vector is not included in any of the k groups but is classified in other than the said k populations.

Estimation of Classification Error-Rate

A test data set is classified using the classification procedure and the number of misclassified observations is counted. Proportion of misclassification in the group gives the estimate of group-specific error-rate.

One can carry out **stepwise discriminant analysis** to find out a small subset of variables which may be sufficient for a satisfactory classification.

If a set of observations ($\mathbf{x}_p$) have been grouped in k classes on the basis of p-component data vector $\mathbf{x}_p$ on N objects, only a smaller subset of p variables is determined using the above mentioned technique.

N × (p + 1) matrix is prepared for N objects, including also the group number as first variable in the $N \times p$ observation matrix, for performing stepwise discriminant analysis. As indicated by Affifi and Clark (1984) and Dixon (1985), at each step a variable is entered which discriminates most between k groups. The selected variable is one with highest F-statistic for one-way analysis of covariance (ANCOVA), where the covariates are previously entered variables. For ANCOVA, the interested readers may consult books on Design of Experiments such as Cochran and Cox (1957), Steel and Torrie (1981), Goon et al. (2002), etc. The stepwise discriminant analysis is included in various software packages like BMDP.

Classification functions, which are linear combinations of entered variables, are obtained for each of the k groups. Values of classification functions are obtained for a subject and it is classified into a group for which value of the classification function is highest.

For detailed discussions and for using this procedure the interested reader may consult manual of BMDP statistical software, module 7M (Tabachnik and Fidel (1983)). Dixon (1985) and SAS Online Doc: Version 8. An application of the above procedure of Step-wise discriminant analysis is given in the next chapter in the Section 21.3 on cluster analysis.

20.3 MANOVA

Multivariate analysis of variance (MANOVA) may be considered to be a generalization of the ANOVA. Univariate one-way ANOVA tests if means of measured values of some observable characteristic in several (say k) groups or populations are equal. For a univariate case, random samples of n_i observations are available and Y_{ij}, i = 1, 2, ..., k, j = 1, 2, ..., n_i are the measured observations from i^{th} population. This technique has been described in Section 8.6 of Chapter 8 and the model and the layout of the experiment is also given in Chapter 14, Section 14.2.

MANOVA is a straightforward generalization of univariate ANOVA. Special care is required for testing related hypotheses. In this section only one-way layout of MANOVA is considered. For two-way lay-out and for higher order of MANOVA, the reader is referred to Press (1972), Harris (1975), etc.

MANOVA – One Way Layout

Let there be $N_i(i = 1, 2, ..., k)$ independent, random p-variate observation vectors, from k MVN populations $N(\boldsymbol{\mu}^{(i)}, \boldsymbol{\Sigma})$. It is desired to test if these $N_i(i = 1, 2, ..., k)$ observation vectors are drawn from MVN populations with equal mean vectors, when it is assumed that the covariance matrices of the various k groups or populations are same (= $\boldsymbol{\Sigma}$).

From the ith population, the vectors of observation are

$$\mathbf{y}_{i1}, \mathbf{y}_{i2}, \mathbf{y}_{i3}, ..., \mathbf{y}_{i\,Ni} \; (i = 1, 2, ..., k) \tag{20.86}$$

and each

$$\mathbf{y}_{ij} \sim N(\boldsymbol{\mu}^{(i)}, \boldsymbol{\Sigma}) \text{ for } (i = 1, 2, ..., k \text{ and } j = 1, 2, ..., N_i) \tag{20.87}$$

On the basis of these observations it is desired to test the hypothesis

$$H_0 : \boldsymbol{\mu}^{(1)} = \boldsymbol{\mu}^{(2)} = ... = \boldsymbol{\mu}^{(k)} \text{ against } H_1 : \text{all the } \boldsymbol{\mu}^{(i)} \text{'s are not equal} \tag{20.88}$$

As an example of real world problem, $\mathbf{y}_{ij}$, the p-component vector may be sales in a multi-product firm. The lth component of $\mathbf{y}_{ij}$ corresponds to sale of lth (l = 1, 2, ..., p) product using i^{th} production technique by the firm on j^{th} vector observation in sample of product using i^{th} technique (j = 1, 2, ..., N_i). The firm wants to identify if sale can be improved by an appropriate choice of technique out of k techniques.

Let $N = \sum_{i=1}^{k} N_i$. The observation matrix can be written as

$$\mathbf{Y}_{N\times p} = \left[\mathbf{Y}_{11}\ldots\mathbf{Y}_{1N_1}\,\mathbf{Y}_{21}\ldots\ldots\mathbf{Y}_{2N_2}\ldots\ldots\mathbf{Y}_{k1}\ldots.\mathbf{Y}_{kN_k}\right]' \tag{20.89}$$

Analogous to univariate one way ANOVA model, the model for MANOVA (one-way lay out) can be written as

$$\mathbf{y}_{ij} = \boldsymbol{\mu}^{(i)} + \mathbf{e}_{ij} \tag{20.90}$$

According to Press (1972) and Anderson (1972), to test H_0 against H_1 the likelihood ratio test-statistic used is

$$U = \frac{\left|N\hat{\boldsymbol{\Sigma}}_1\right|}{\left|N\hat{\boldsymbol{\Sigma}}_0\right|} \tag{20.91}$$

where $\hat{\boldsymbol{\Sigma}}_1$ and $\hat{\boldsymbol{\Sigma}}_0$ are the estimates of the common covariance matrix $\boldsymbol{\Sigma}$, when H_0 is not true and when H_0 is true, respectively.

If H_0 is true, U is distributed as $U_{p,\,k-1,\,N-k}$. In symbols

$$U = \frac{\left|N\hat{\boldsymbol{\Sigma}}_1\right|}{\left|N\hat{\boldsymbol{\Sigma}}_0\right|} \sim U_{p,k-1,N-k} \quad \text{if } H_0 \text{ is true} \tag{20.92}$$

For large N, Box (1949) gave an approximation that

$$-\,m \log U \sim \chi^2_{p(p-1)} \text{ if } H_0 \text{ is true} \tag{20.93}$$

where

$$m = (N - 1 - p) \tag{20.94}$$

The estimates of $\boldsymbol{\Sigma}$ under H_0 and H_1 are given by

$$N\hat{\boldsymbol{\Sigma}}_1 = \sum_{i=1}^{k}\sum_{j=1}^{N_i}\left(\mathbf{Y}_{ij} - \bar{\mathbf{Y}}^{(i)}\right)\left(\mathbf{Y}_{ij} - \bar{\mathbf{Y}}^{(j)}\right)' \tag{20.95}$$

where $\bar{\mathbf{Y}}^{(i)}$ = mean vector of sample from i^{th} population, and is

$$= \frac{1}{N_i}\sum_{j=1}^{N_i}\mathbf{Y}_{ij}$$

$$N\hat{\boldsymbol{\Sigma}}_0 = \sum_{i=1}^{k}\sum_{j=1}^{N_i}\left(\mathbf{Y}_{ij} - \bar{\mathbf{Y}}\right)\left(\mathbf{Y}_{ij} - \bar{\mathbf{Y}}\right)' \tag{20.96}$$

where $\bar{\mathbf{Y}} = \frac{1}{N}\sum_{i=1}^{k}\sum_{j=1}^{N_i}\mathbf{Y}_{ij} = \frac{1}{N}\sum_{i=1}^{k}N_i\bar{\mathbf{Y}}_i$ = mean vector of all the sample vectors (20.97)

It may be noted that $\hat{\boldsymbol{\Sigma}}_0$ is the estimate using mean of all the sample vectors i.e. assuming that $\boldsymbol{\mu}^{(1)} = \boldsymbol{\mu}^{(2)} = \ldots. = \boldsymbol{\mu}^{(k)}$; and $\hat{\boldsymbol{\Sigma}}_1$ is the weighted average of the estimates of $\boldsymbol{\Sigma}$ from samples from each of the k populations.

For any positive definite matrix P, $|P|$ = determinant of P (see Appendix C).

The test procedure is to reject H_0 against H_1 at level α

$$\text{if } U \le U_{p,\,k-1,\,N-k,\,\alpha} \tag{20.98}$$

and if N is large, using Box approximation, reject H_0 against H_1 at level α, if

$$-m \log U > \chi^2_{p(p-1),\alpha} \tag{20.99}$$

Anderson (1972) gave an example of Barnards' study of Egyptian skulls (1935) to explain the test procedure given above.

The data has been collected from 4 populations: (i) Late Predynastic (ii) Sixth to twelfth (iii) Twelfth to thirteenth and (iv) Ptolemaic Dynasties. Four measurements (i.e. 4 components of $\mathbf{Y}_{ij}$) are maximum breadth, basiaveolar length, nasal height and basibregmatic height. The numbers of observations are 91, 162, 70 and 75 from these populations, respectively. The data have been summarized to give $\bar{\mathbf{y}}^{(1)}$, $\bar{\mathbf{y}}^{(2)}$, $\bar{\mathbf{y}}^{(3)}$ and $\bar{\mathbf{y}}^{(4)}$, the $\hat{\mathbf{\Sigma}}_1$ and $\hat{\mathbf{\Sigma}}_0$. It is desired to test $H_0 : \boldsymbol{\mu}^{(1)} = \boldsymbol{\mu}^{(2)} = \boldsymbol{\mu}^{(3)} = \boldsymbol{\mu}^{(4)}$ against H_1 : all the four $\boldsymbol{\mu}^{(i)}$'s are not equal.

Solution: In this problem $k = 4$, $p = 4$, $N_1 = 91$, $N_2 = 162$, $N_3 = 70$ and $N_4 = 75$

$$\left(\bar{\mathbf{Y}}^{(1)}\bar{\mathbf{Y}}^{(2)}\bar{\mathbf{Y}}^{(3)}\bar{\mathbf{Y}}^{(4)}\right) = \begin{pmatrix} 133.582418 & 134.582418 & 134.371429 & 135.306667 \\ 98.307692 & 96.462963 & 95.857143 & 95.040000 \\ 50.835165 & 51.148148 & 50.100000 & 52.093333 \\ 133.000000 & 134.882716 & 133.642857 & 131.466667 \end{pmatrix}$$

$$N\hat{\Sigma}_1 = \begin{pmatrix} 9661.997470 & 445.573301 & 1130.623900 & 2148.584210 \\ 445.573301 & 9073.115027 & 1239.211990 & 2255.812722 \\ 1130.623900 & 1239.211990 & 3938.320351 & 1271.054662 \\ 2148.584210 & 2255.812722 & 1271.054662 & 8741.508 \end{pmatrix}$$

$$N\hat{\Sigma}_0 = \begin{pmatrix} 9785.178098 & 214.197666 & 1217.929248 & 2019.820216 \\ 214.197666 & 9559.460890 & 1131.716372 & 2381.126040 \\ 1217.929248 & 1131.716372 & 4088.731856 & 1133.473898 \\ 2019.820216 & 2381.126040 & 1133.473898 & 9382.242 \end{pmatrix}$$

The test statistic

$$U = \frac{\left|N\hat{\Sigma}_1\right|}{\left|N\hat{\Sigma}_0\right|} = \frac{2.4269054 \times 10^5}{2.9544775 \times 10^5} = 0.8214344$$

$N = 398$, $N - k = 398 - 4 = 394$, $k = 4$, $p = 4$, $m = 398 - 1 - 4 = 393$
Since N is large

$$-m \log U_{4,3,394} \sim \chi^2_{(4\times 3)} = \chi^2_{12} \text{ if } H_0 \text{ is true}$$

$$-m \log U = 77.30 \text{ and } \chi^2_{12,.01} = 26.2$$

So H_0 is rejected at 1% level of significance.

21. Multivariate Analysis II

In the last chapter multivariate techniques of comparing of mean vectors of two or more MVN populations have been considered. The Hotelling's T^2, Mahalanobis-D^2 and MANOVA – one way layout have been considered and test procedures have been described.

As an exploration of an interdependence structure of MVN data, the problem of classification of an observed data vector into one of two or into one of several MVN populations have been considered. In this chapter some more interdependence methods will be considered. The first section considers the method of Principal Component Analysis (PCA). The subsequent two sections discuss the Factor Analysis and Cluster Analysis. The background, their applications, their procedures and interpretation of results are explained with the help of some real numerical examples.

The dependence method of multiple regression and multiple correlation, has only one criterion or response or dependent variable, whereas canonical correlation methods are applied if criterion, or response or dependent variables are more than one. The rest of the components of the p-component vector $\mathbf{X}_p$ are explanatory or predictor or independent variables. Basic concepts of these two methods will be briefly discussed before the section on PCA.

Multiple Correlation

The idea of partial and multiple correlations was developed first, for three variables, by Karl Pearson (1896). Yule (1897 a and b) further extended the theory. From the property (h) of MVN distribution about the conditional density of $\mathbf{X}^{(1)}$ given $\mathbf{X}^{(2)} = \mathbf{x}^{(2)}$ given in Chapter 20, it may be recalled that

$$E(\mathbf{X}^{(1)}/\mathbf{X}^{(2)} = \boldsymbol{x}^{(2)}) = \mu^{(1)} + \sum_{12}\sum_{22}^{-1}(\mathbf{x}^{(2)} - \mu^{(2)}) \tag{21.1}$$

It is, therefore, evident that (21.1) also represents the regression of $X^{(1)}$ on $x^{(2)}$ and the matrix $\sum_{12}\sum_{22}^{-1}$ is the matrix of regression coefficient. RHS of (21.1) is called the regression function.

$$\text{Covariance matrix of } (X^{(1)}/X^{(2)} = \boldsymbol{x}^{(2)}) = \sum_{11..2} = \sum_{11} - \sum_{12}\sum_{22}^{-1}\sum_{21} \tag{21.2}$$

It may be written as $((\sigma_{ij.q+1,\, q+2,\, ...,\, p}))_{q\times q}$; where the elements $\sigma_{ij.q+1,\, ...,\, p}$ are the partial covariances of ith and j^{th} components of $X^{(1)}$ given $X^{(2)} = \mathbf{x}^{(2)}$.

One may, therefore, define the partial correlation between X_i and X_j, $(i, j = 1, 2, ..., q)$ when $X^{(2)}$ is fixed (i.e. $X_{q+1}, X_{q+2}, ..., X_p$ are fixed) as

$$\rho_{ij.q+1,q+2,\ldots,p} = \frac{\sigma_{ij.q+1,\ldots,p}}{\sqrt{(\sigma_{ii.q+1,\ldots,p})}\sqrt{\sigma_{jj.q+1,\ldots,p}}} \tag{21.3}$$

Since variance, covariance and correlation coefficients are independent of change of origin, for these discussions it is assumed that $\boldsymbol{\mu} = \mathbf{0}$.

Let $\boldsymbol{\beta}$ be the ith row of $\sum_{12}\sum_{22}^{-1}(i = 1, 2, \ldots, q)$. The p-q elements of this row vector $\boldsymbol{\beta}$ are the regression coefficients of X_i on $\mathbf{X}^{(2)}$.

If $\boldsymbol{\sigma}_{(i)}$ is i^{th} row of $\boldsymbol{\Sigma}_{12}$, $\qquad \boldsymbol{\beta} = \boldsymbol{\sigma}_{(i)}\,\boldsymbol{\Sigma}_{22}^{-1}$ (21.4)

Results

Anderson (1972) has shown that if $\boldsymbol{\alpha}$ is row vector of p-q constants

(a) $V(X_i - \boldsymbol{\alpha}\mathbf{X}^{(2)})$ is minimum if $\boldsymbol{\alpha} = \boldsymbol{\beta}$ and (21.5)
(b) Correlation between X_i and $\boldsymbol{\alpha}\mathbf{X}^{(2)}$ is maximum if $\boldsymbol{\alpha} = \boldsymbol{\beta}$ (21.6)

Therefore, the multiple correlation between X_i and $\mathbf{X}^{(2)}$ is defined as the maximum correlation between X_i and $\boldsymbol{\alpha}\mathbf{X}^{(2)}$. Because of (b) part of the result above in equation (21.5), the multiple correlation between X_i and $\mathbf{X}^{(2)}$ (i = 1, 2, ..., q) is given by

$$\bar{R}_{i.q+1,q+2,\ldots,p} = \frac{\text{Cov}\,(\boldsymbol{\beta}\mathbf{X}^{(2)}, X_i)}{\sqrt{V(\boldsymbol{\beta}\mathbf{X}^{(2)})}\sqrt{V(X_i)}} = \frac{\sigma_{(i)}\sum_{22}^{-1}\sigma'_{(i)}}{\sqrt{\sigma_{(i)}\sum_{22}^{-1}\sigma'_{(i)}}\sqrt{\sigma_{ii}}}$$

$$= \frac{\sqrt{\sigma_{(i)}\sum_{22}^{-1}\sigma'_{(i)}}}{\sqrt{\sigma_{ii}}} \tag{21.7}$$

$$1 - \bar{R}^2_{i.q+1,q+2,\ldots,p} = \frac{\sigma_{ii} - \sigma_{(i)}\sum_{22}^{-1}\sigma'_{(i)}}{\sigma_{ii}} \tag{21.8}$$

From (21.2) the numerator of (21.8) is $\sigma_{ii.q+1,\ldots,p}$. Hence, one can write that

$$\sigma_{ii.q+1,\ldots,p} = \sigma_{ii}\left(1 - \bar{R}^2_{i.q+1,\ldots,p}\right) \tag{21.9}$$

(21.9) implies that

(a) Conditional variance $\sigma_{ii.q+1,\ldots,p} \le \sigma_{ii}$ (21.10)
(b) Larger the value of $\bar{R}_{i.q+1,\ldots,p}$, greater is the reduction in conditional variance of X_i.
(c) $\bar{R}^2_{i.q+1,\ldots,p} = 1 - \dfrac{\sigma_{ii.q+1,\ldots,p}}{\sigma_{ii}}$ (21.11)

represents the proportion of total variability of X_i accounted for by the regression of X_i on $\mathbf{X}^{(2)}$.

The value of $\bar{R}^2_{i.q+1,\ldots,p}$ close to 1 implies that the regression equation is a good fit to the data and its value close to 0 indicates a poor fit.

The value of $\bar{R}^2_{i.q+1, \ldots, p}$ lies between 0 to 1. In symbols

$$0 \le \bar{R}^2_{i.q+1, \ldots, p} \le 1 \tag{21.12}$$

Moreover, since $\mathbf{\Sigma}_{22}$ is a positive definite matrix, the numerator of RHS of equation (21.7) is positive. Hence the multiple correlation is always positive. In symbols

$$\bar{R}_{i.q+1, \ldots, p} \ge 0 \tag{21.13}$$

$\bar{R}_{i.q+1, \ldots, p}$ is the multiple correlation of X_i on $\mathbf{X}^{(2)}$ in the population. If a random sample $\mathbf{X}_1, \mathbf{X}_2, \ldots, \mathbf{X}_N (N > p)$ is available, the sample multiple correlation of X_i on $\mathbf{X}^{(2)}$ is given by

$$R_{i.q+1, \ldots, p} = \sqrt{\frac{\mathbf{a}_{(i)} \mathbf{A}_{22}^{-1} \mathbf{a}_{(i)}}{a_{ii}}} \tag{21.14}$$

where $\mathbf{a}_{(i)}$ is i^{th} row of $\mathbf{A}_{12}$, a_{ii} is i^{th} diagonal element of $\mathbf{A}_{11}$ and $\mathbf{A}_{22}$ is from a partition of matrix $\mathbf{A}$ (defined in 20.44) similar to that of $\mathbf{A}$. The matrix $\mathbf{A}$ is again given as

$$\mathbf{A} = \sum_{\alpha=1}^{N} (\mathbf{X}_\alpha - \bar{\mathbf{X}})(\mathbf{X}_\alpha - \bar{\mathbf{X}})' \tag{21.15}$$

where $\bar{\mathbf{X}} = \frac{1}{N} \sum_{\alpha=1}^{N} \mathbf{X}_\alpha$ and the partition of $\mathbf{A}$ is given as

$$\mathbf{A} = \begin{pmatrix} \mathbf{A}_{11} & \mathbf{A}_{12} \\ \mathbf{A}_{21} & \mathbf{A}_{22} \end{pmatrix} \tag{21.16}$$

$\mathbf{A}_{11}$ is a $q \times q$ matrix and $\mathbf{A}_{22}$ is $(p - q) \times (p - q)$ matrix of the elements of $\mathbf{A}$.

From (21.14), it can be seen that

$$U = \frac{R^2_{i.q+1, \ldots, p}}{1 - R^2_{i.q+1, \ldots, p}} = \frac{\mathbf{a}_{(i)} A_{22}^{-1} \mathbf{a}'_{(i)}}{a_{ii.q+1, \ldots, p}}$$

Moreover, $F = U\left(\frac{N-p}{p-1}\right)$ has F-distribution with $(p - 1)$ and $(N - p)$ d.f. if multiple correlation in population $(= \bar{R}_{i.q+1, \ldots, p})$ is equal to 0.

Thus to test $H_0 : \bar{R}_{i.q+1, \ldots, p} = 0$ against $H_1 : \bar{R}_{i.q+1, \ldots, p} \ne 0$ the test statistic used is

$$F = \left(\frac{N-p}{p-1}\right) U \tag{21.17}$$

and, H_0 is rejected at $100\alpha\%$ level of significance if

$$F > F_{p-1, N-p, \alpha} \tag{21.18}$$

It may be summarized that multiple correlation is correlation of the criterion variable (X_i) with its estimated (or predicted) value from multiple linear regression equation.

Example 21.1: Sundaram (2002) conducted a hospital based study in Varanasi city. The aim of multiple regression analysis was to find the quantitative relationship between the Water Quality Parameters of River Ganga at Varanasi and the Incidence of waterborne enteric (WBE) diseases in Varanasi, since river Ganga is its major source of drinking water (though well treated by Water Works before distribution). Hence the 21 water quality parameters as listed below are being used as regressor variables.
1. Dissolved oxygen 2. Biochemical oxygen demand 3. Temperature 4. pH 5. Conductivity 6. Sodium 7. Chloride 8. Fluoride 9. Nitrate 10. Potassium 11. Magnesium 12. Carbonate 13. Bi-carbonate 14. Iron 15. Calcium 16. Sulphate 17. Nitrite 18. Boron 19. Silicate 20. Total coliform 21. Faecal coliform

The following five WBE diseases: (i) Diarrhoea, (ii) Dysentery, (iii) Hepatitis, (iv) Typhoid and (v) Gastro-enteritis; have been considered and number of indoor patients of these diseases in the five hospitals during January 1996 to June 1999 have been recorded.

Incidence of WBE diseases (Y variable = I_m) is defined below for this analysis, because the available data is not adequate to enable applying standard definition of incidence.

The Monthly Incidence (I_m)

$I_m = (N_m/D_m) \times 100$, where

N_m is the total no. of patients of the 5 selected Hospitals (SSL, SSPG, VSM, HSS and IDH) of any age or sex residing in Varanasi city who newly contacted one or more of the said diseases (noted from the appropriate hospital registers in m^{th} month

D_m is the total no. of patients of all the 5 hospitals residing in Varanasi city admitted for general medical treatment including those in pediatrics in m^{th} month
where, m = Jan. 96, Feb. 96, ..., June 99

The analysis resulted in multiple $R = 0.903$ Intercept = $\beta_0 = -204.98$

Value of multiple $R^2 = 0.815$ means that more than 81% of the variance in incidence rate has been explained by the model. It is found that only three parameters pH($p = 0.021$) Nitrate ($p = 0.003$) and Nitrite ($p = 0.021$) are significant in the full multiple regression model; while except temperature ($p = 0.09$) all other variables are too insignificant in this model.

In the forward stepwise regression method, while considering the 75% of the total variation, eight parameters (temperature, iron, nitrate, pH, nitrite, total coliforms, boron and sulphate) are sufficient to be entered. It is also to be noted that while in the full model the three parameters (pH, nitrate and nitrite) were found significant, but in forward stepwise regression method, temperature and iron entered before these parameters.

Canonical Correlation

The theory of canonical correlations was proposed by Hotelling (1933, 1936) for studying the linear relationships between two sets of variates. He suggested to study only a few uncorrelated linear combinations of the variables in each of the two sets instead of all the components in the two sets. This is especially useful if the sets of variates consist of very large say 100, 1000 or more variables. For studying interrelations between these two large sets, the explorer would like to find only a few linear combinations which are most highly correlated and consider only those few instead of the complete large set.

Let the MVN variable $\mathbf{X}_p \sim N(\boldsymbol{\mu}, \boldsymbol{\Sigma})$ be partitioned in two subsets $\mathbf{X} = \begin{pmatrix} \mathbf{X}^{(1)} \\ \mathbf{X}^{(2)} \end{pmatrix}$, $\mathbf{X}^{(1)}$ with p_1 components and $\mathbf{X}^{(2)}$ with the rest p_2 (= $p - p_1$) components.

Accordingly
$$\Sigma = \begin{pmatrix} \Sigma_{11} & \Sigma_{12} \\ \Sigma_{21} & \Sigma_{22} \end{pmatrix}$$

Canonical correlation analysis tries to find linear combinations

$$U = \mathbf{a}'\mathbf{X}^{(1)} \text{ and } V = \mathbf{b}'\mathbf{X}^{(2)} \text{ such that } \mathbf{a}\boldsymbol{\Sigma}_{11}\mathbf{a} = \mathbf{b}\boldsymbol{\Sigma}_{22}\mathbf{b} = 1 \tag{21.19}$$

Since only variances, covariances and correlations are considered, it is assumed that $\boldsymbol{\mu} = \mathbf{0}$. Hence $E(\mathbf{X}) = \mathbf{0}$, so $E(U) = 0 = E(V)$, and correlation between

$$U \text{ and } V = \mathbf{a}'\ \boldsymbol{\Sigma}_{12}\mathbf{b} \tag{21.20}$$

The problem of determining **a** and **b** which maximize (21.10) and satisfy (20.19) is obtained by solving the matrix equations

$$\begin{aligned} -\lambda\boldsymbol{\Sigma}_{11}\mathbf{a} + \boldsymbol{\Sigma}_{12}\ \mathbf{b} = \mathbf{0} \\ \boldsymbol{\Sigma}_{21}\mathbf{a} - \lambda\boldsymbol{\Sigma}_{22}\ \mathbf{b} = \mathbf{0} \end{aligned} \tag{21.21}$$

or

$$\begin{pmatrix} -\lambda\Sigma_{11} & \Sigma_{12} \\ \Sigma_{21} & -\lambda\Sigma_{22} \end{pmatrix}\begin{pmatrix} \mathbf{a} \\ \mathbf{b} \end{pmatrix} = \mathbf{0} \tag{21.22}$$

For a non-zero solution for **a** and **b**, the following must be satisfied:

$$\begin{vmatrix} -\lambda\Sigma_{11} & \Sigma_{12} \\ \Sigma_{21} & -\lambda\Sigma_{22} \end{vmatrix} = 0 \tag{21.23}$$

For details and proofs the interested reader may consult the books by Anderson (1972), Press (1972), Dillon and Goldstein (1984).

(21.23) is a polynomial of λ of degree p and will have p roots, which can be arranged in order of magnitude as

$$\lambda_1 \geq \lambda_2 \geq \ldots \geq \lambda_p \tag{21.24}$$

Anderson (1972) and Press (1972) have shown that out of these p roots $p_2 - p_1$ are 0 and $2p_1$ [$= p - (p_2 - p_1)$] are of the form $\pm\lambda_i$ for $i = 1, 2, \ldots, p_1$. The set of roots $(\lambda_1, \lambda_2, \ldots, \lambda_p) = (\lambda_1, \lambda_2, \ldots, \lambda_{p_1}, 0, 0, \ldots, 0, -\lambda_{p_1}, \ldots, -\lambda_1)$

Two equivalent forms of equation (21.23) are given by

$$\left|\Sigma_{12}\Sigma_{22}^{-1}\Sigma_{21} - \lambda^2\Sigma_{11}\right| = 0 \text{ and } \left(\Sigma_{12}\Sigma_{22}^{-1}\Sigma_{21} - \lambda^2\Sigma_{11}\right)\mathbf{a} = \mathbf{0} \tag{21.25}$$

$$\left|\Sigma_{21}\Sigma_{11}^{-1}\Sigma_{12} - \lambda^2\Sigma_{22}\right| = 0 \text{ and } \left(\Sigma_{21}\Sigma_{11}^{-1}\Sigma_{12} - \lambda^2\Sigma_{22}\right)\mathbf{b} = \mathbf{0} \tag{21.26}$$

It can be seen that out of total p_1 solutions of λ^2 from (21.25) and total p_2 solutions of λ^2 from (21.26), the non-zero solutions of λ^2 from both (21.25) and (21.26) are identical.

Solving equation (21.22) or (21.25) and (21.26) for $\lambda = \lambda_1$ gives $\mathbf{a}_{(1)}$ and $\mathbf{b}_{(1)}$ and the linear combinations $U_1 = \mathbf{a}'_{(1)}\mathbf{X}^{(1)}$ and $V_1 = \mathbf{b}'_{(1)}\mathbf{X}^{(2)}$ give the first canonical variates. λ_1, the maximum correlation between U_1 and V_1 is called the first canonical correlation.

In the same way solving (21.22) or (21.25) and (21.26) for $\lambda = \lambda_2$ one gets the solutions $\mathbf{a}_{(2)}$ and $\mathbf{b}_{(2)}$. The linear combinations $U_2 = \mathbf{a}'_{(2)}\mathbf{X}^{(1)}$ and $V_2 = \mathbf{b}'_{(2)}\mathbf{X}^{(2)}$ are the second canonical variables. U_2

and V_2 are uncorrelated to U_1 and V_1, respectively, $V(U_2) = V(V_2) = 1$ and corr $(U_2, V_2) = \lambda_2$ the second largest root of (21.23). λ_2 is called the second canonical correlation. The number of positively correlated canonical variates is at most equal to p_1. One proceeds in this way with the subsequent roots, to get the pair of canonical variates. The rth of canonical variates pair are obtained by solving (21.22) or (21.25) and (21.26) for $\lambda = \lambda_r$ as $U_r = \mathbf{a}'_{(r)}\mathbf{X}^{(1)}$ $V_r = \mathbf{b}'_{(r)}\mathbf{X}^{(2)}$. U_r and V_r are both of unit variance and are uncorrelated with the first $(r - 1)$ canonical variates and have maximum correlation λ_r, called the rth canonical correlation.

A Result

Let

$$\mathbf{U} = \begin{pmatrix} U_1 \\ U_2 \\ \cdot \\ \cdot \\ \cdot \\ U_{p_1} \end{pmatrix} = \mathbf{AX}^{(1)},\ \mathbf{V}^{(1)} = \begin{pmatrix} V_1 \\ V_2 \\ \cdot \\ \cdot \\ \cdot \\ V_{p_1} \end{pmatrix} = \mathbf{B}_1\mathbf{X}^{(2)} \text{ and } \mathbf{V}^{(2)} = \begin{pmatrix} V_{p_1+1} \\ \cdot \\ \cdot \\ \cdot \\ V_{p_2} \end{pmatrix} = \mathbf{B}_2\mathbf{X}^{(2)} \tag{21.27}$$

and

$$\mathbf{Var}\begin{pmatrix} \mathbf{U} \\ \mathbf{V} \end{pmatrix} = \mathbf{Var}\begin{pmatrix} \mathbf{U} \\ \mathbf{V}^{(1)} \\ \mathbf{V}^{(2)} \end{pmatrix} = \begin{pmatrix} \mathbf{I} & \mathbf{D}_\lambda & \mathbf{O} \\ \mathbf{D}_\lambda & \mathbf{I} & \mathbf{O} \\ \mathbf{O} & \mathbf{O} & \mathbf{I} \end{pmatrix} \tag{21.28}$$

where $\mathbf{D}_\lambda$ = diag $(\lambda_1, \lambda_2, ..., \lambda_{p_1})$ and $\lambda_1 \geq \lambda_2 \geq \lambda_3 \ldots \geq \lambda_{p_1}$.

These are the population canonical correlations and canonical variates. Estimates of these are obtained by using the corresponding partitions of the sample estimate of covariance matrix $\mathbf{\Sigma} = \hat{\mathbf{\Sigma}} = \mathbf{A}/N$ (see equations (21.15) and (21.16)). For a solved numerical example giving computational details, the reader may consult book by Anderson (1972). Now software packages are available for the routine procedure as well as new refinements.

21.1 Principal Component Analysis (PCA)

As already mentioned, in this section the first interdependence technique is discussed. Whenever, a complex problem in any study area is explored, often data on a large number of relevant variables are collected. In the beginning of a study a basic theory can not be relied for choice of variables to be included and thus mostly a large number of variables are measured. It results in a large number of relationships to be explained, making it beyond comprehension. The researcher wants to reduce the dimension of the data by choosing only those variables which may be most relevant. He, therefore, seeks the ways of rearranging or summarizing the data, so that with as little loss of information as possible, the dimension of the study problem is reduced.

In such situations some times the interest is to identify the variables with highest variability, so that the variables with low variability, i.e. behaving as approximately a constant may be discarded. By focusing attention only on a few high variance variables, the study is conveniently possible on subspace of dimension less than p.

The PCA finds the new set of orthogonal and uncorrelated normalized linear combinations of original variables (called principal components (PC)). The first PC has largest variance, second PC, the next highest and so on. In general the rth PC has maximum variance among all normalized linear combinations which are uncorrelated with the first $(r - 1)$ PC's. The first PC has a large proportion of the total variance of the original variables, the second one also a large proportion but less than the first. Thus only the first few ($\leq p$) PC's are sufficient to explain almost all the variability in the data, thus reducing the dimension of the problem.

This approach was introduced by Karl Pearson (1901) who considered the case of non-stochastic variables. It was generalized by Hotelling (1933) to the case of stochastic variables.

Principal Components in the Population

Let $\mathbf{X}_p$ be a p-component random vector with covariance matrix. Since interest is only in variances and covariances, it is assumed that the mean vector of $\mathbf{X}_p$ is $\mathbf{O}$. In symbols

$$E(\mathbf{X}) = \mathbf{O} \text{ and } \mathrm{Cov}(\mathbf{X}) = \mathbf{\Sigma}$$

Further assume that the elements of $\mathbf{X}_p$ are measured in the same units, otherwise the requirement of principal component $U = \boldsymbol{\beta}'\mathbf{X}$ being a normalized linear combination is not sensible. If it is not reasonable to assume that all the components of $\mathbf{X}_p$ are measured in the same units, a large sample size is feasibly taken, and the standardized variables may be used. The sample covariance matrix $\mathbf{A}/N$ of $\mathbf{X}_p$ then changes to the sample correlation matrix.

Anderson (1972) gave the details of matrix algebra technique for determining the PC's. In short the steps are given below.

To determine first principal component (FPC) the normalized linear combination, $\boldsymbol{\beta}'\mathbf{X}(= \beta_1 X_1 + \beta_2 X_2 + \ldots + \beta_p X_p)$, which maximizes

$$V(\boldsymbol{\beta}'\mathbf{X}) = \boldsymbol{\beta}'\mathbf{\Sigma}\boldsymbol{\beta} \tag{21.29}$$

is found. In other words, it is required to find a p-component vector $\boldsymbol{\beta}$, which maximizes (21.29) and satisfies

$$\boldsymbol{\beta}'\boldsymbol{\beta} = 1 \tag{21.30}$$

Using Lagrange multiplier λ, $\boldsymbol{\beta}$ is solution of the matrix equation

$$(\mathbf{\Sigma} - \lambda\mathbf{I})\,\boldsymbol{\beta} = \mathbf{O} \tag{21.31}$$

In order to get a solution of (21.31) satisfying (21.30), λ must satisfy

$$|\mathbf{\Sigma} - \lambda I| = 0 \tag{21.32}$$

LHS of (21.32) is a polynomial in λ of degree p, so has p roots. Let these roots be ordered as $\lambda_1 \geq \lambda_2 \geq \ldots \geq \lambda_p$.

From (21.31)
$$V(\boldsymbol{\beta}'\mathbf{X}) = 1 \tag{21.33}$$

So for determining FPC having maximum variance, choose the largest root λ_1 to solve equation (21.31) and obtain the normalized solution $\boldsymbol{\beta}^{(1)}$. Then

$$U_1 = \boldsymbol{\beta}'^{(1)}\mathbf{X} \text{ is the FPC} \tag{21.34}$$

It can be seen that (cf. Appendix C) λ_1 is the largest characteristic root of $\mathbf{\Sigma}$ and $\boldsymbol{\beta}^{(1)}$ is normalized characteristic vector of $\mathbf{\Sigma}$ corresponding to λ_1.

Now a linear combination $\boldsymbol{\beta}'\mathbf{X}$ is to be found, that has maximum variance among all normalized linear combinations uncorrelated to U_1. It is found that second PC is found as the characteristic vector $\boldsymbol{\beta}^{(2)}$ of $\mathbf{\Sigma}$ corresponding to second largest characteristic root λ_2 (obtained from (21.31)).

Thus $$U_2 = \boldsymbol{\beta}^{(2)'}\mathbf{X} \tag{21.35}$$

is second PC with variance λ_2.

This procedure is continued. At the rth step, the normalized linear combination $U_r = \boldsymbol{\beta}^{(r)'}\mathbf{X}$ is determined by solving equation (21.31) for $\lambda = \lambda_r$. Thus $\boldsymbol{\beta}^{(r)}$ is r^{th} characteristic vector of $\boldsymbol{\Sigma}$ corresponding to λ_r.

Some Results

Thus for $r = 1, 2, ..., p$, the p principal components $\mathbf{U} = [U_1, U_2, ..., U_p] = [\boldsymbol{\beta}^{(1)'}\mathbf{X}, \boldsymbol{\beta}^{(2)'}\mathbf{X}, ..., \boldsymbol{\beta}^{(p)'}\mathbf{X}]$ give the transformation of the original variable $\mathbf{X}$ into $\mathbf{U}$ such that covariance matrix of $\mathbf{U}$ = diag $(\lambda_1, \lambda_2, ..., \lambda_p)$

1. If $\boldsymbol{\Sigma}$ is positive definite $\lambda_1, \lambda_2, ..., \lambda_p > 0$.
2. If $\boldsymbol{\Sigma}$ is positive semi definite some $\lambda_i = 0$.
3. The proportion of total variation explained by first r PC's of

$$\mathbf{X} = \left(\sum_{i=1}^{r} \lambda_i\right) \Bigg/ \left(\sum_{i=1}^{p} \lambda_i\right) \tag{21.36}$$

Thus using an appropriate first few number of PC's, the desired proportion of total variance of $\mathbf{X}$ can be explained.

A note: Readers not conversant with matrix algebra should read the section on eigen values and eigen vectors in Appedix C.

Principal Components from Sample

Let $\mathbf{x}_1, \mathbf{x}_2, ..., \mathbf{x}_N$ $(N > p)$ be N observations from $\mathbf{X}_p \sim N(\boldsymbol{\mu}, \boldsymbol{\Sigma})$, and $\mathbf{S} = \mathbf{A}/N$. Then $\mathbf{S}$ is MLE of $\boldsymbol{\Sigma}$. λ_i and $\boldsymbol{\beta}^{(i)}$ are unique functions of $\boldsymbol{\Sigma}$, if first non-zero components of $\boldsymbol{\beta}^{(i)}$ are positive. Hence replacing $\boldsymbol{\Sigma}$ by $\mathbf{S}$ in (21.31) one obtains the MLE of λ_i and $\boldsymbol{\beta}^{(i)}$ as l_i and $\mathbf{b}^{(i)}$, $(i = 1, 2, ..., p)$ (cf. Girshick (1936) and Anderson (1958)).

Sampling Properties of Principal Components

All the above discussions hold without any assumption about distribution of $\mathbf{X}_p$. The sampling properties are derived by Girshick (1936), and Anderson (1958, 1963). Press (1972) has given many sampling properties which may be used for drawing inference about PC's. Some of these are given below, under the assumption that $\mathbf{X}_p \sim N(\mathbf{O}, \boldsymbol{\Sigma})$, where $\boldsymbol{\Sigma}$ is positive definite with distinct characteristic roots $\lambda_1, \lambda_2, ..., \lambda_p > 0$; with corresponding characteristic vectors $\boldsymbol{\beta}^{(1)}, \boldsymbol{\beta}^{(2)}, ..., \boldsymbol{\beta}^{(p)}$. $\hat{\boldsymbol{\Sigma}} = \mathbf{A}/N$ is MLE of $\boldsymbol{\Sigma}$. Let $l_1 > l_2 > ... > l_p > 0$ are the characteristic roots of $\hat{\boldsymbol{\Sigma}} = \mathbf{S}$ and $\mathbf{b}^{(1)}, \mathbf{b}^{(2)}, ..., \mathbf{b}^{(p)}$ are the corresponding characteristic vectors.

(a) l_i $(i = 1, 2, ..., p)$ is distributed independently of the elements of $\mathbf{b}^{(i)}$.

(b) $\sqrt{N-1}(l_i - \lambda_i) \sim N(0, 2\lambda_i^2)$ and it is distributed independently of other characteristic roots.

(c) For positive semi-definite $\boldsymbol{\Sigma}$, the upper bound of largest root λ_1 is given by

$$\lambda_1 \le \max_i \left(\sum_{j=1}^{p} |\sigma_{ij}|\right), i = 1, 2, ..., p \tag{21.37}$$

In words
Variance of FPC ≤ largest of the row sums of absolute values of elements of **Σ**.

Some Tests Useful before Applying PCA

There are situations when use of PCA does not advantageously reduce the dimension of the study data. So it is advisable that some tests are applied to know if the PCA will be useful, before starting the PCA. Two such situations and the two tests appropriate for these situations are given below.

(A) If one has $\lambda_1 = \lambda_2 = ... = \lambda_p$, it implies that the contribution of each PC in total variance is same. So λ_1 is a characteristic root of $\mathbf{\Sigma}$ with multiplicity p. So $(\mathbf{\Sigma} - \lambda I)$ has rank $p - p = 0$. Thus the covariance matrix of $\mathbf{X}_p \sim N(\boldsymbol{\mu}, \mathbf{\Sigma})$ is $\mathbf{\Sigma} = \lambda_1\mathbf{I}$.

In other words the components of **X** are independently distributed with variance λ_1, which is same as that of PC's. So there is no advantage of using PCA. One can work with **X**.

So before using PCA, it may be required to test if $\lambda_1 = \lambda_2 = \ldots = \lambda_p$. This is equivalent to the hypothesis that the components of $\mathbf{X}_p$ are independent.
To test $H_0 : \lambda_1 = \lambda_2 = \ldots = \lambda_p \equiv H_0 : X_i$'s $(i = 1, 2, ..., p)$ are independent

The statistic to be used is

$$V = -N\left(1 - \frac{2p+11}{6N}\right)\log|\mathbf{R}| \tag{21.38}$$

where **R** is the sample correlation matrix defined as

$$\mathbf{R} = ((r_{ij})) \text{ and } r_{ij} = \frac{\sum_{\alpha=1}^{N}(X_{i\alpha} - \bar{X}_i)(X_{j\alpha} - \bar{X}_j)}{\left[\left\{\sum_{\alpha=1}^{N}(X_{i\alpha} - \bar{X}_i)^2\right\}\left\{\sum_{\alpha=1}^{N}(X_{j\alpha} - \bar{X}_j)^2\right\}\right]^{1/2}}$$

V has χ^2-distribution with $\nu = p(p - 1)/2$ d.f. for large N, if H_0 is true. So H_0 is rejected at 100α% level of significance if

$$V \geq X^2_{\nu,\, \alpha} \tag{21.39}$$

(B) Let $\lambda_1, \lambda_2, \ldots, \lambda_p$ be the characteristic roots of **Σ**. Now $\lambda_1 > \lambda_2 > \ldots > \lambda_k > \lambda_{k+1} > \ldots > \lambda_p$, so λ_p is very small. If one wants to work with only first k PC's, he should first test the hypothesis $H_0 : \lambda_{k+1} = \lambda_{k+2} = \ldots = \lambda_p$ against H_1 : All the equalities in H_0 do not hold.

Press (1972) gives the test statistic Q for this test, if N is large, defined as

$$Q = (p - k)(N - 1)(\log l - 1) \tag{21.40}$$

where $l = \frac{1}{p-k}(l_{k+1} + l_{k+2} + ... + l_p)$.

If H_0 is true Q has χ^2-distributions with f d.f.

The test procedure is to reject H_0 at 100α% level of significance if,

$$Q > \chi^2_{f,\alpha} \tag{21.41}$$

where $f = \frac{(p-k)(p-k+1)}{2} - 1$.

As an illustration of PCA technique the computational details are not given, since the determination of PC's and their variances are same as the determination of characteristic roots and characteristic vectors of $\mathbf{\Sigma}$ or $\hat{\mathbf{\Sigma}}$. For this various mathematical and statistical softwares exist, which must be used for $p > 3$.

If a reader is interested in the various steps to be followed in these computations, he may refer to the examples given in Anderson (1972), Press (1972) and Dillon and Goldstein (1984). The following table gives an idea of reduction in dimension of the problem with smaller number of PC's explaining the high percentage of total variation.

Table 21.1 Percentage of total variation explained by PC's

Data	Value of p	Percentage of total variation explained by PC's	Number of PC's and percentage of total variation accounted by them
1. Four measurements on varieties of Iris (Anderson (1972))	4	$PC^{(1)}$ about 78% $PC^{(2)}$ about 12%	First 2 PC's account about 90%
2. 9 optical densities in 3 colours in pictures (Press (1972))	9	$PC^{(1)}$-61% $PC^{(2)}$-14% $PC^{(3)}$-9%	First 2 PC's account about 75%
3. 93 semantic differential scales of Jazz (Dillon and Goldstein (1984))	93		First ten PC's account about 62.3%.

21.2 Factor Analysis

Factor analysis is the second technique that deals with interdependence of the components of **X**. It attempts to account for the correlation pattern in a set of observable random variables in terms of a *minimal number of unobservable or latent* random variables called factors which are linear combinations of the original variables.

It is a branch of statistical science, but extensively used in psychology, so it is mistaken as a concept of psychological theory. Approach of factor analysis (FA) might be used as a first step in a sequence of investigations which are aimed at developing insight into the relationships among variables. A second step might involve more direct input response analyses of the data (such as the regression model analyses).

A matrix of N observations on a p-component correlated random variables, $\mathbf{X}_{N \times p}$ is examined to see if the data could have been generated by a linear model involving m ($< p$) unobservable random variables (called factors) that are in a certain sense fundamental to the data generating process. These fundamental variables (factors) or their linear combinations are used to explain the observed data. The main solution of FA is choice of weights (i.e. the factor loading matrix) which should be used in linear combinations of factors. Thus a regrouping of original variables emerges which can be interpreted meaningfully most of the times. A satisfactory solution results in a small numbers (m) of factors which convey all the essential information in the original set of variables of p ($> m$) components. The *chief aim of FA* is to *attain scientific parsimony or economy of description.*

A *brief history of factor analysis* clarifies how the theory of FA developed according to its areas of application.

In principle, the concept of *latent factors* was first propounded by Galton in 1988. The actual mathematical model of factor structure has originated in (1904). Charles Spearman delivered monumental work while developing his *two-factor theory*. It suggests that the correlations among a set of intelligence test scores can be generated by (a) the latent factor of *general intellectual ability* and (b) a second set of factors reflecting the *unique qualities* of the individual tests. Though Spearman is considered as father of the concepts of FA, who devoted about forty years of his work in this area, the statistical aspects were first emphasized by Karl Pearson (1901) in the name of 'principal axes'. The first two decades of the twentieth century saw considerable amount of work on the psychological theories and mathematical formation of FA.

The concept of *multiple factor theory* took over the two-factor theory of Spearman though the earlier method of two factor theory was used to extract other factors also. The concept of multiple factor arose through work by Garnett (1919) and Thurstone (1947).

An account of development of FA may be found by interested readers in work by Burt (1966) and Armstrong (1967).

Exploratory and Confirmatory Factor Analysis

If the experimenter has no prior knowledge of underlying structure of the data, hence of the type and number of factors involved in data generating process, he uses the technique of FA to explore the common structure underlying the data. FA with this purpose is called *exploratory*.

On the other hand, when the data analyst has prior knowledge of the common underlying structure of the data, he wants to confirm or change the previously known structure in the follow up data. This use of FA is called *confirmatory*.

21.2.1 Mathematical Formulation of Factor Analysis Model

Let $\mathbf{X}_p$ be an observable random vector of response. Since only the correlation structure is of interest, without loss of generality one can assume that $E(X_p) = O$. Further, it is assumed that $\mathbf{X}_p$ is generated by a linear combination of orthogonal, unobservable m factors common to all $\mathbf{X}'_p$**s,** and upon which some error or disturbance is present. In symbols:

$$\begin{aligned}
X_1 &= \lambda_{11}F_1 + \lambda_{12}F_2 + \dots + \lambda_{1m}F_m + e_1 \\
X_2 &= \lambda_{21}F_1 + \lambda_{22}F_2 + \dots + \lambda_{2m}F_m + e_2 \\
&\dots\dots\dots\dots\dots\dots\dots\dots\dots\dots \\
X_i &= \lambda_{i1}F_1 + \lambda_{i2}F_2 + \dots + \lambda_{im}F_m + e_i \\
&\dots\dots\dots\dots\dots\dots\dots\dots\dots\dots \\
X_p &= \lambda_{p1}F_1 + \lambda_{p2}F_2 + \dots + \lambda_{pm}F_m + e_p
\end{aligned} \tag{A}$$

where F_j : j^{th} common factor variate, $j = 1, 2, \dots, m$. m is much less than p.

λ_{ij} : parameter (or weight) reflecting the importance of jth factor in composition of ith response, called the loading of ith response on j^{th} factor

$i = 1, 2, \dots, p$ and $j = 1, 2, \dots, m$

e_i : i^{th} specific factor variate and error or residual variate.

Let $\mathbf{X}_p = \begin{pmatrix} X_1 \\ X_2 \\ \cdot \\ \cdot \\ \cdot \\ X_i \\ \cdot \\ \cdot \\ \cdot \\ X_p \end{pmatrix}, \mathbf{F}_m = \begin{pmatrix} F_1 \\ F_2 \\ \cdot \\ \cdot \\ \cdot \\ F_j \\ \cdot \\ \cdot \\ \cdot \\ F_m \end{pmatrix}, \mathbf{e}_p = \begin{pmatrix} e_1 \\ e_2 \\ \cdot \\ \cdot \\ \cdot \\ e_i \\ \cdot \\ \cdot \\ \cdot \\ e_p \end{pmatrix}$ and

$$\wedge_{p\times m} = \begin{pmatrix} \lambda_{11}\lambda_{12}\ldots\lambda_{1j}\ldots\lambda_{1m} \\ \lambda_{21}\lambda_{22}\ldots\lambda_{2j}\ldots\lambda_{2m} \\ \cdot \\ \cdot \\ \cdot \\ \lambda_{i1}\lambda_{i2}\ldots\lambda_{ij}\ldots\lambda_{im} \\ \cdot \\ \cdot \\ \cdot \\ \lambda_{p1}\lambda_{p2}\ldots\lambda_{pj}\ldots\lambda_{pm} \end{pmatrix}$$

Then the factor model can be written in matrix form as

$$\mathbf{X}_p = \wedge\mathbf{F} + \mathbf{e} \tag{21.42}$$

Here $\mathbf{X}_p$: observable, $\mathbf{F}_m$: Non-observable common factors

$\mathbf{e}_p$: non-observable errors + specific factors

$\wedge_{p\times m}$: matrix of factor loadings, non-observable.

It is assumed that:

(a) $\mathbf{X}_p \sim N_p(\mathbf{O}, \mathbf{\Sigma})$ (b) $\mathbf{F}_m \sim N(\mathbf{O}, \mathbf{I}_m)$, $m \le p$

In words F_j's are iid. $N(0, 1)$ random variables

(c) $\mathbf{e}_p \sim N(\mathbf{O}, \boldsymbol{\psi})$, where $\boldsymbol{\psi} = \text{diag}(\psi_1, \psi_2, \ldots, \psi_p)$

Thus components of $\boldsymbol{e}$ are also mutually independent.

ψ_i are called *specificity* of ith response, or specific variance of ith response $i = 1, 2, \ldots, p$.

(d) $\mathbf{F}$ and $\boldsymbol{e}$ are independent so $E(\mathbf{Fe}') = 0$

(e) $\wedge' \boldsymbol{\psi}^{-1} \wedge = \mathbf{D} = \text{diag}(J_1, J_2, \ldots, J_m)$, where $J_1 > J_2 > \ldots > J_m$.

This is imposed for a unique solution of the parameter estimation equation.

Under the above assumptions, from (21.42) one can find that

$$\mathbf{\Sigma} = \wedge\wedge' + \boldsymbol{\psi} \tag{21.43}$$

Equation (21.43) implies that

$$\sigma_{ij} = \lambda_{i1}\lambda_{j1} + \lambda_{i2}\lambda_{j2} + \dots + \lambda_{im}\lambda_{jm} \quad \text{for } i \neq j \text{ and} \tag{21.44}$$

$$\sigma_{ii} = \lambda_{i1}^2 + \lambda_{i2}^2 + \dots + \lambda_{im}^2 + \psi_i \tag{21.45}$$

So

$$\sum_{k=1}^{m} \lambda_{ik}^2 = \sigma_{ii} - \psi_i \tag{21.46}$$

are the diagonal elements of $\wedge\wedge'$ and are called communalities of the response.

It can be seen that the covariance matrix of the observable and common factor variates is given by

$$\text{Cov}\left(\mathbf{X}_p, \mathbf{F}'_m\right) = E\left(\mathbf{XF}'\right) = \wedge \tag{21.47}$$

Thus λ_{ij} = Covariance of ith response with j^{th} common factor.

Purpose of the factor analysis is to determine the elements of the factor loading matrix $\wedge$ and ψ on the basis of N observations on the response variable $\mathbf{X}_p$ (where $N > p$).

The procedure customarily is to **fix m**, the number of common factors before hand, and then estimate $\wedge$ and ψ. Estimation is followed by interpretation of various factors on the basis of factor loadings.

Some Terms

(a) The set of equations given in (A) is also called factor pattern

(b) The $p \times m$ matrix of factor loadings $\wedge$ (columns associated to factors) is referred as pattern matrix.

(c) A matrix of the correlation between the response variables X_i's and the common factors is also of interest and is called factor structure matrix or structure matrix.
In general the elements of a *structure matrix* are different from the elements of a *pattern matrix*. If X's are standardized, so that $\mathbf{\Sigma}$ is reduced to correlation matrix, then λ_{ij} are the correlations of the ith response X_i with j^{th} factor F_j. In such a case the elements of *structure matrix* and *pattern matrix* are same.

(d) Communality of a variable (or response) X_i, is also denoted by

$$h_i^2 \left(= \sigma_{ii} - \psi_i\right) = \sum_{k=1}^{p} \lambda_{ik}^2 \text{ from (21.46)}$$

is the *common variance* of X_i. It is also the portion of a variable's total variance accounted for by the common factors.

(e) The unique or specific variance of a variable ψ_i is also called specificity of variable X_i . ψ_i is the extent or portion left unexplained by the common factors.

(f) Though the factor analysis model given above, deals with variance covariance matrix $\mathbf{\Sigma}$, often the original variables are standardized so $\mathbf{\Sigma}$ in equation (21.43) may be replaced by $\mathbf{R}$, the correlation matrix to give

$$\mathbf{R} = \wedge\wedge' + \psi \tag{21.48}$$

The reduced correlation matrix $\mathbf{R}^*$ is obtained when the diagonal elements of $\mathbf{R}$ (each = 1) are replaced by communalities h_i^2 i.e. $\mathbf{R}^* = \mathbf{R} - \psi$.

For obtaining the solution of factor analysis problem, the estimates of ψ_i' s are subtracted from the observed (or sample) correlation matrix.

For detailed discussions of the above ideas the books by Press (1972), Dillon and Goldstein (1984), Harris (1995), Morrison (1967) etc. may be consulted.

Details of computational procedure are available in Harman (1967) and Mulaik (1972).

In this chapter some methods are listed and two methods are briefly discussed.

21.2.2 Factor Analysis Procedures

Broadly speaking the procedures of obtaining solution of factor analysis are of two types: (a) Methods requiring estimates of communality and (b) Methods requiring estimate of number of factors. Before going to methods requiring estimates of communality, some commonly used estimates of communality are given as mentioned in Dillon and Goldstein (1984) and Harman (1967).

1. The simplest estimate of communality of ith response is the maximum correlation between X_i and other $(p - 1)$ response variables without regard to its sign. Thus the maximum of absolute values of off-diagonal elements of ith row of **R** is used as estimate of communality of X_i and i^{th} diagonal element of **R** is replaced by this value.
2. Second estimate is given by

$$h_i^2 = \frac{r_{ij}r_{ik}}{r_{jk}} \tag{21.49}$$

   where r_{ij} and r_{ik} are two highest off-diagonal elements of ith row of **R**.
3. Third estimate is the average correlation of X_i. In symbols

$$h_i^2 = \left(\sum_{i \neq j=1}^{p} r_{ij}\right) \Big/ (p-1) \tag{21.49a}$$

   Range of h_i^2 is (0.0, 1.0). Value 0 indicates that X_i has no correlation with any other variable and 1 indicates that variation of X_i is totally accounted for by the factors.
   Some methods of communality estimation give the estimate > 1, e.g. if $r_{ij}r_{jk} > r_{jk}$, $h_i^2 > 1$. Such cases are called *Heywood cases* and the communality estimates in such cases are set as .99 or 1.0.
4. Squared multiple correlation (SMC) of X_i on the other $(p - 1)$ variables is another estimator of communality. This finds a good application since almost all the statistical software packages provide its value.
5. Iterative procedure are also nowadays used often, due to access to high-speed computers. The method consists in substituting initial estimates in the diagonal correlation matrix for factor analysis with pre-determined number of factors. The row sums of squares of resulting factor loading matrix are used as subsequent estimates. The process is repeated till stable estimates of communalities are obtained by giving a proper stopping rule.

(a) Methods Requiring Communality Estimates are:
1. Principal factor method 2. Triangular decomposition method, 3. Centroid method.

(b) Methods Requiring Estimate of Number of Factors are:
1. Maximum likelihood method, 2. Minres method, 3. Canonical factor, Image factor, Alpha factor and Multiple group method.

A vivid description of these methods are given in the book by Harman (1976). Harris (1975), Dillon and Goldstein (1984) and Press (1972) describe some of these with complete mathematical base. In this chapter, due to paucity of space only a differentiating description of principal factor (PF), centroid and maximum likelihood (ML) methods will be given.

(a) 1. Principal Factor Method

Karl Pearson (1901) laid the foundation for the method of principal axes which was subsequently modified by Hotelling (1933) to give the basis of principal factor method. This method is most widely used method of factor analysis. This method requires considerable calculations feasible only on high speed computer. Kelley (1935) suggested the computational procedure, which was first applied on computer by Wrigley and Neuhaus (1952, 1955).

The PF method extracts factors such that each factor makes the maximum possible contribution to the total communality. This is achieved by applying the method of PCA to the reduced correlation matrix **R***. It has been shown by Harman (1976) that the solution is equivalent to determination of eigen values and eigen vectors of **R***. The largest eigen value and the corresponding normalized eigen vector of **R*** give the maximum contribution to communality of the first factor and this is taken as the first factor.

This procedure continues on the adjusted reduced correlation matrix by removing the contribution of first factor. Thus the first adjusted reduced correlation matrix is

$$\mathbf{R}_1^* = \mathbf{R}^* - \boldsymbol{\lambda}_1\boldsymbol{\lambda}_1' \tag{21.50}$$

where $\boldsymbol{\lambda}_j$ is j^{th} column of $\boldsymbol{\Lambda}$.

Now the largest eigen value of $\mathbf{R}_1^*$ and corresponding eigen vector are determined to get the second factor and its contribution to communality. In a similar manner, at the mth stage the adjusted reduced correlation matrix is

$$\mathbf{R}_{m-1}^* = \mathbf{R}^* - \sum_{j=1}^{m-1} \lambda_j \lambda_j' \tag{21.51}$$

If elements of adjusted reduced correlation matrix $\mathbf{R}_m^*$ become small the process is terminated. Since the communalities in the diagonal elements of **R*** are fractional, **R*** may not remain positive definite. So some eigen values may turn out to negative. Such eigen values and corresponding eigen vectors are discarded.

A rule of thumb (cf. Dillon & Goldstein (1984)), therefore, is to extract factors until the sum of eigen values is close to the total communality in **R***.

(a) 2. Triangular Decomposition (TD) Method

This procedure yields the general factor as the first factor, which has significant loading for all the observable variables X_i's . The second factor has zero correlation with X_1. The third factor has zero correlations with X_1 and X_2, and so on. Thus the factor structure/loading matrix is truncated lower triangular, because the loadings above diagonal are zero.

For details of computation with numerical example, the interested reader may consult Harris (1975).

(a) 3. Centroid Method

This method was developed by Thurstone (1931). Before the access to computer was readily available, this method could be applied. This is of historical interest now.

The name of the method is due to its close relationship to the concept of center of gravity also called centroid.

Harris (1975), Bennett and Bowers (1976) and Harman (1976) give the method of determining the first centroid factor by

$$\left(\frac{1}{p}\sum_{k} a_{k1}, 0, 0, ..., 0\right) \tag{21.52}$$

where
$$a_{j1} = \frac{\sum_{k=1}^{p} r_{jk}}{\sqrt{\sum_{j=1}^{p}\sum_{k=1}^{p} r_{jk}}} = \frac{S_j}{\sqrt{T}};\ j = 1, 2, ..., p \tag{21.53}$$

S_j is the sum of all the correlations in j^{th} column of correlation matrix **R** and T is total of all the p^2 elements of **R**, including the diagonal terms in both S_j and T. The diagonal elements of **R** are replaced by h_i^2, the commmunalities. Bennett and Bowers (1976) have used the largest correlation in the row as h_i^2.

The first factor loadings are given by $a_{j1} (j = 1, 2, ..., p)$.

Now the first factor residuals are obtained and the second centroid factor is obtained as the first, but from the adjusted residual matrix.

The elements of the first residual matrix are given by

$$_1r_{ij} = r_{ij} - {_1e_{ij}} \quad i, j = 1, 2, ..., p \tag{21.54}$$

where r_{ij} is the $(i, j)^{th}$ element of **R** and

$$_1e_{ij} = a_{i1}a_{j1} \quad i, j = 1, 2, ..., p \tag{21.55}$$

obtained from first factor loadings calculated from (21.53).

If the elements of residual matrix become zero it implies that there are no more factors and the computation stops.

This process is repeated to obtain m centroid factors. Numerical examples may also be seen in the above books and in the book by Frutcher (1954).

(b) 1. Maximum Likelihood (ML) Method

The ML method was given by Lawley (1940). The number of factors m is fixed before hand and then Λ and ψ are estimated. Estimation is followed by testing hypothesis on number of factors. ML method finds the factor structure that maximizes the likelihood of obtaining the observed correlation matrix given that the proposed structure is the true structure in population.

Lawley and Maxwell (1963) and Anderson and Rubin (1956) derived the maximum likelihood estimators of Λ and ψ which are obtained by solving the following two matrix equations given by Press (1972).

$$\text{diag}\left(\hat{\psi} + \hat{\Lambda}\hat{\Lambda}'\right) = \text{diag}\,\mathbf{S} \text{ and} \tag{21.56}$$

$$\mathbf{S}\hat{\psi}^{-1}\hat{\Lambda} = \hat{\Lambda}\left(\mathbf{I} + \hat{\Lambda}'\hat{\psi}^{-1}\hat{\Lambda}'\right) \tag{21.57}$$

where $\mathbf{S} = \mathbf{A}/N$, the m.l.e. of $\mathbf{\Sigma}$.

(21.56) and (21.57) do not give unique solutions unless assumption (e) of the factor model holds. Jöreskog (1967) gave rapidly converging algorithm for solution of these equations on computer.

Numerical examples may be seen in the books by Harman (1976) and Dillon and Goldstein (1984).

(b) 2. Minres Method

Factor analysis seeks to describe the inter correlation matrix of the correlation of observed variables as accurately as possible from a small number of unobservable factors. Minimum residuals or *minres* method minimizes the sum of the squared discrepancies between the observed and reproduced correlations.

Getting a solution to factor analysis conceptually is to minimize the residual correlation as the least square method in regression analysis. But for factor model, the solution using this concept could not be obtained, though several researchers worked in this direction. Harman and Jones (1966) suggested the practical solution which is possible only on computer with an efficient algorithm giving fast results developed by Harman and Fukuda (1966).

The book by Harman (1976) contains the detailed theory and computational procedure with a numerical example.

(b) 3. Canonical Factor, Image Factor, Alpha Factor and Multiple Group Methods.

All the above methods use correlation matrix for determination of factor solutions. The variables are thus standardized. After the solution is obtained the standardized variables may be rescaled to attain certain desired objectives.

Canonical factor analysis rescales the metric of the unique part. The idea of maximizing the correlation between linear combinations of two sets of variables has been used to develop this method of factor analysis.

It is desired to find that factor which is maximally related to the observed variables. Solution depends on a canonical correlation analysis of the factors to be obtained (the common factors) and the observed variables (cf. Rao, 1955).

The procedure finds the factor variable which is predictable from the observed data with maximum precision. Next a second factor variable, which is orthogonal to the first and has maximum relation with observed variables, is determined. This is continued till the prespecified number m of factors is obtained.

The theory, computational procedure and numerical example may be found in a lucid way in Harman (1976).

Image factor analysis employs the error of prediction in regression of each observed variable on the remaining $(p - 1)$ variables.

Alpha factor analysis rescales the common parts.

The multiple group method factorizes the correlation matrix into several multiple factors simultaneously, which are usually oblique instead of being orthogonal.

These methods are not used quite often so is not discussed in this chapter. The interested reader may consult Harman (1976) for more details.

21.2.3 Test of Number of Factors

As already mentioned some FA procedures pre-specify the number of factors. So a formal test of the appropriate number of factors should be done to justify the choice of factors.

The maximum likelihood nature of the estimates of factor loadings leads to a formal test of adequacy of m-factor model for reproducing the observed covariances or correlations.

One wants to test whether the model $\mathbf{X}_p = \Lambda\mathbf{F} + \mathbf{e}$ fits the observed data or not. An equivalent hypothesis is $H_0 : \mathbf{\Sigma} = \Lambda\Lambda' + \psi$ against $H_1 : \mathbf{\Sigma}$ is any $p \times p$ positive definite symmetric matrix. Λ is $p \times m$ factor loading matrix.

Likelihood ratio test is employed for this as given in Press (1972), Harman (1976) and in Morrison (1967) and Overall and Klett (1972). To test H_0 against H_1 the likelihood ratio test statistic is given by

$$U_m = -2\log\lambda = (N-1)\log\frac{\left|\hat{\Sigma}_0\right|}{\left|\hat{\Sigma}_1\right|} \tag{21.58}$$

If N is large and H_0 is true $U_m \sim \chi_r^2$ (21.59)

where r = Number of unconstrained parameters under H_1 – Number of unconstrained parameters under H_0

$$= \frac{1}{2}[(p-m)^2 - (p+m)] \tag{21.60}$$

Even if $n = (N - 1)$ is large, the convergence of the distribution is very slow so Bartlett (1954) suggested to replace $n = (N - 1)$ in (21.58) by n', given by

$$n' = n - \frac{2p+5}{6} - \frac{2m}{3} \tag{21.61}$$

The choice of n' makes a fast convergence of the distribution of U_m to χ_r^2.

Under H_0, $\mathbf{\Sigma}$ is estimated by $\hat{\mathbf{\Sigma}}_0 = \hat{\psi} + \Lambda\Lambda'$ where $\hat{\psi}$ and $\hat{\Lambda}$ are obtained by solving (21.56) and (21.57) $\hat{\mathbf{\Sigma}}_1 = \mathbf{A}/N$.

Thus while performing the FA using ML method, the test statistic can also be evaluated. The software packages for ML method contain the routine for testing H_0.

21.2.4 Interpretation of Factors

Most important but a difficult part of the factor analysis is interpretation of results, especially identifying or naming the factors. The factor loadings λ_{ij} are the correlations between the ith component X_i of $\mathbf{X}_p$ and j^{th} factor F_j. Values of these give a clue of the variables which are highly related (i.e. have high correlation) with various factors. Examining the factor loading matrix, one identifies the variables most highly related to various factors and thus one can name a factor. The following steps may be followed to do this in a systematic way:

1. For a given sample size N find the smallest value of λ_{ij} which is significant of the relationship of X_i with F_j. Since λ_{ij} are the correlations, the t-test of correlations give the following bound for $100\alpha\%$ significant value of λ_{ij}^2

$$\lambda_{ij}^2 > \frac{t_{N-2,\alpha/2}^2}{N-2+t_{N-2,\alpha/2}^2} \tag{21.62}$$

For $N = 100$ and $\alpha = .05$, it gives approximately that

$$|\lambda_{ij}| > 0.2 \tag{21.63}$$

For sample sizes < 100, the commonly used lower bound for $|\lambda_{ij}|$ is taken as 0.3. So any loading larger in magnitude than 0.3, is considered significant.

2. For each factor i.e. looking column wise in the factor loading matrix mark the significant loadings, thus identifying the variables significantly related to the factors.
3. If there are some variables which do not have significant loading with any factor, its communality may also be examined. One may look into the importance of this variable in the study very carefully and critically, and may decide to delete this variable in future.

If a variable does not have an overlap of significant loadings with more than one factor, i.e. a row of Λ has more than one significant loading, the identification or naming of factor becomes difficult. Otherwise, the variables (X_i's) which have significant loadings with a factor are used to identify the factor and there is clear parsimony.

If a variable or some variables have significant factor loadings with more than one factor, rotation is one technique to overcome the problem of factor interpretation. The idea of rotation is given below.

21.2.5 Factor Rotation

As mentioned above, some times the factor solution does not give a simple structure and identification of factors is difficult. Factor rotation is one technique which is commonly used to achieve a simple structure and hence a parsimony in interpretation.

There are two methods of rotating factor axes. The first is orthogonal rotation which preserves the 90° angle between two axes as before. The second method is called oblique rotation, which has no such restriction.

In the orthogonal rotation only one angle say θ is to be specified and each of the axes makes an angle θ with the corresponding original axes.

In oblique rotation various axes may have different angles with the corresponding original axes and these angles are to be specified, to perform the rotation.

There are three readily available orthogonal rotation algorithms: *varimax*, *quartimax* and *equimax*. The five popular methods of oblique rotation are: the *oblimax*, *quartimin*, *covarimin*, *biquatimin* and *oblimin*.

Varimax method of orthogonal rotation is the most popular and most of the statistical software packages, such as SPSS, BMDP and SAS, use it either with raw factor loadings or with normalized loadings. For getting the normalized loadings the variables' (X_i's) loadings are divided by h_i, the communality.

Details of the computational form of the methods may be seen by the interested reader in the books by Harman (1976) and Rummel (1970). A brief idea of the principles of the methods and their relative merits and demerits are also given by Dillon and Goldstein (1984).

Dillon and Goldstein (1984) also give a summary comparison table of various FA solution techniques adapted from Gorsuch (1974).

21.2.6 Factor Scores

The methods of PCA and FA reduce the dimension of the data and the researcher may like to obtain the location of each subject ($\mathbf{X}_\alpha$) on each of the m factors. This is achieved by computing the *factor scores* of each observation.

The FA is conducted on the reduced correlation matrix **R***, not on the original correlation matrix **R** on which PCA is performed. Since PCA is performed on the original correlation or covariance matrix, principal component scores could be calculated by substituting $\mathbf{X} = \mathbf{X}_\alpha$ in $\boldsymbol{\beta}^{(i)\prime}\mathbf{X}$ or $\mathbf{b}^{(i)\prime}\mathbf{X}$.

The solution of the FA model being non-unique also prevents direct evaluation of factor scores. These are to be estimated. One of the successful method of estimation is multiple regression analysis. For the jth factor F_j; $j = 1, 2, ..., m$, the extracted factor score of αth respondent (or observation), $\alpha = 1, 2, ..., N$ will be given by

$$\hat{F}_{\alpha j} = b_1 X_{\alpha 1} + b_2 X_{\alpha 2} + ... + b_p X_{\alpha p} \tag{21.64}$$

where b_i, $i = 1, 2, ..., p$ are the regression coefficients of multiple linear regression equation of $F_{\alpha j}$ on $X_{\alpha 1}, X_{\alpha 2}, ..., X_{\alpha p}$. The regression equation can be written in matrix form as

$$\hat{\mathbf{F}}_{N \times m} = \mathbf{X}_{N \times p} \mathbf{B}_{p \times m} \tag{21.65}$$

Here $\hat{\mathbf{F}}_{N\times m}$ is matrix of estimated factor scores on each of m factors. The jth column of $\hat{\mathbf{F}}$ is score of respondents on j^{th} factor. The jth value in α^{th} row, $\alpha = 1, 2, ..., N$ is estimated score of αth respondent on j^{th} factor ($j = 1, 2, ..., m$).

$\mathbf{X}_{N\times p}$ is the original observation matrix α^{th} row is response of αth respondent. $\mathbf{B}_{p\times m}$ is matrix of regression coefficients, such that j^{th} column gives regression coefficients of jth factor to be used for estimating j^{th} factor score of the N respondents by multiplying with it the αth row of $\mathbf{X}_{N\times p}$.

If $\mathbf{Z}_{N\times p}$ is standardized observation matrix then $\hat{\mathbf{F}}_{N\times m}$ can be obtained as

$$\hat{\mathbf{F}}_{N \times m} = \mathbf{Z}_{N \times p} \mathbf{R}^{-1}_{p \times p} \wedge_{p \times m} \tag{21.66}$$

Dillon and Goldstein (1984) have given detailed proof of these results and an illustration of the factor scores is given by a numerical example of 14 nation varimax rotation factor pattern matrix.

Two examples of application of FA technique on a hospital survey data (Kar (2003)) and a socio-economic survey data on rural population of Eastern UP are given to visualize the results of FA using software packages.

Example 21.2: Factor analysis (FA) of hospital survey data in Varanasi city was done by Kar (2003) to identify the major factors responsible for episodes of the WBE diseases. The data was collected from OPD of four hospitals in city.

The following five WBE diseases, also mentioned in Example 21.1, have been considered: (i) Diarrhoea (ii) Dysentery (iii) Hepatitis (iv) Typhoid (v) Gastro-enteritis.

Out door patients of these diseases were interviewed with the structured questionnaire, from July 1998 to June 1999, once in a month. Table 21.2 gives the frequency and number of patients surveyed in the said five hospitals.

Table 21.3 gives the list of the studied variables.

For conducting FA each qualitative or nominal variable was quantified by giving a high score to such answers of questions generating the said variables, which may cause or are indicators of higher pollution or un-cleanliness.

Factor analysis was performed *using SPSS version 10.2* to identify the factors responsible for the said enteric diseases. Following Table 21.4 gives the factor loadings. As some variables had significant factor loadings for more than one factor, varimax rotation was performed. Table 21.5 gives the rotated factor loadings. This rendered the interpretation and naming of factors possible.

Table 21.2 Source of data on episode of WBE diseases from OPD

Sl. No.	Name of hospitals	Type of hospital	No. of patients	Freq./month	Source of data
1	SSL BHU	Attached to institute of medical science in a central university.	10	1	Outdoor Gastro and Pediatrics
2	SSPG. Kabir Chowra	State government hospital attached to state university Ayurvedic college	10	1	Outdoor Gen. Med. and Pediatrics
3	HSS. Chowk	Private	10	1	Outdoor Gen. Med. and Pediatrics
4	VSM. Ramghat	Private	10	1	Outdoor Gen. Med. and Pediatrics

Table 21.3 Variables studied

Sl. No.	Variables studied and their names	Sl. No.	Variables studied and their names	S l. No.	Variables studied and their names
1	**Age**	9	Distance from Ganga **DistGang**	17	Scores of food habit: taken outside-**FhOut**
2	**Sex**	10	Source of drinking water-**WatSoDr**	18	Scores of food habit: Uncovered-**FhUncov**
3	Education of guardian **PrEd**	11	Source of water for cooking-**WatSoFoo**	19	Scores of food habit Children on breast feeding-**FhBf**
4	Education of patient **PtEd**	12	Source of water for cleaning utensils-**WatSoUt**	20	Personal hygiene: taking bath-**PhyBth**
5	Occupation of guardian **PrOc**	13	Treatment of water at home-**WatTreat**	21	Personal hygiene: Hand washing after defecation **PhyWh**
6	Occupation of patient **PtOc**	14	Use of refrigerator **Refr**	22	Type of latrine :for adult **LatAdu**
7	Income of patient **PtInc**	15	Scores of food habit: Cooked and eaten fresh **FhCf**	23	Type of latrine:for children-**LatChild**
8	Income of guardian **PrInc**	16	Scores of food habit: Cooked and eaten stale **FhCl**	24	Source of water for cleaning clothes-**WatSoClo**

Table 21.4 Factor loadings

Variables	FAC 1	FAC 2	FAC 3	FAC 4	FAC 5	FAC 6	FAC 7	FAC 8	FAC 9	FAC 10	FAC 11	FAC 12	FAC 13
Age	0.042	**0.463**	0.019	**0.391**	–0.07	–0.057	0.03	–0.08	0.042	–0.104	–0.035	0.056	–0.051
Sex	0.127	0.144	0.236	0.044	0.094	–0.103	0.077	–0.15	0.007	0.269	0.056	0.06	–0.021
PrEd	0.016	–0.27	**0.326**	–0.276	–0.16	0.187	–0.16	0.118	–0.02	0.138	0.008	–0.01	–0.02
PtEd	0.076	0.059	**0.364**	0.037	–0.09	0.049	–0.15	0.188	–0.08	–0.101	–0.12	0.038	0.008
PrOc	0.013	**–0.67**	0.078	0.248	0.154	–0.185	–0.07	0.007	0.012	0.061	–0.041	0.018	–0.013
PtOc	0.076	**0.574**	0.141	–0.24	0.053	–0.154	–0.04	–0.05	0.119	0.017	–0.006	–0.01	0.01
PtInc	–0.001	**–0.59**	0.084	**0.305**	0.121	–0.166	–0.09	–0.06	0.018	–0.014	0.043	–0.03	0.035
PrInc	0.222	**0.423**	0.189	–0.101	0.024	–0.216	–0.12	0.023	–0.04	0.051	0.056	–0	0.053
DistGang	0.033	0.08	**0.364**	0.106	–0.03	–0.19	–0.03	–0.05	0.047	–0.01	–0.103	–0	0.048
WatSoDr	**0.683**	–0.11	–0.15	–0.12	0.026	0.024	–0.15	0.013	0.024	–0.1	0.077	0.075	0.011
WatSoFoo	**0.924**	–0.07	–0.09	–0.03	–0.05	0.003	–0.12	–0.06	0.047	–0.058	0.041	0.029	0.004
WatSoClo	**0.974**	–0.05	–0.02	0.01	–0.05	–0.046	0.105	0.038	–0.02	0.072	–0.062	–0.09	–0.03
WatSoUt	**0.98**	–0.07	–0.04	0.015	–0.06	–0.044	0.103	0.024	–0.01	0.029	–0.059	–0.02	–0.006
WatTreat	0.113	–0.07	0.191	0.033	0.001	0.103	0.096	0.01	0.213	–0.08	0.096	–0.05	0.034
Refr	–0.014	–0.17	–0.22	–0.175	–0.04	–0.089	0.225	–0.04	–0.1	0	–0.055	0.039	0.091
FhCf	0.085	–0.08	0.205	–0.143	0.261	–0.091	0.039	–0.02	–0.22	–0.087	0.095	0.003	–0.037
FhCl	–0.008	–0.14	0.113	–0.174	**0.347**	–0.009	0.115	0.006	0.17	–0.063	–0.067	0.003	–0.037
FhOut	0.006	–0.22	0.209	–0.093	0.082	–0.051	0.144	0.227	0.066	–0.045	0.055	0.051	0.008
FhUncov	0.095	0.062	0.196	0.152	–0.12	0.06	0.32	0.148	–0.06	0.012	0.041	0.019	–0.002
FhBf	–0.017	–0.07	–0.1	**–0.311**	**0.333**	0.052	0.005	–0.05	0.001	0.012	–0.099	0.021	–0.005
PhyBth	0.097	–0.03	0.328	–0.06	0.007	0.066	0.049	–0.25	–0.15	–0.145	0.005	–0.04	–0.005
PhyWh	0.106	–0.23	0.235	–0.018	–0.12	**0.345**	0.048	–0.23	0.069	0.009	–0.025	0.043	0.023
LatAdu	0.173	0.161	–0.05	0.257	0.272	**0.329**	–0.06	0.035	–0.06	0.101	–0.004	0.002	0.035
LatChild	0.233	**0.389**	0.039	0.206	**0.379**	0.157	–0	0.09	–0.03	–0.012	0.002	–0.02	0.029

Table 21.5 Rotated factor loadings

Variables	FAC 1	FAC 2	FAC 3	FAC 4	FAC 5	FAC 6	FAC 7	FAC 8	FAC 9	FAC 10	FAC 11	FAC 12	FAC 13
Age	0.003	0.22	0.222	0.208	-0.27	0.016	0.068	–0.41	0.058	0.06	0.013	0.013	–0.107
Sex	0.093	0.052	0.096	0.084	0.011	0.052	0.053	–0.02	–0.06	0.431	–0.01	–0.01	–0.006
PrEd	0.018	–0.072	0.072	–0.12	0.01	0.144	0.019	0.572	–0.02	0.014	0.013	0.004	–0.028
PtEd	0.044	0.029	0.378	0.089	-0.11	0.031	0.107	0.224	–0.07	–0.09	–0.05	0.035	–0.051
PrOc	0.047	**–0.75**	0.023	–0.08	0.078	-0.019	–0	0.061	0.002	0.039	–0.02	0	–0.015
PtOc	0.03	**0.543**	0.26	0.021	0.109	–0.108	–0.08	–0.08	–0.04	0.215	0.094	–0	0.003
PtInc	0.023	**–0.703**	0.039	–0.04	-0.01	0.025	–0.03	–0.01	–0.04	0.016	0.074	0.005	0.025
PrInc	0.178	0.341	0.287	0.059	-0.05	–0.186	–0.06	–0.01	–0.13	0.223	0.054	0.031	0.049
DistGang	0.006	–0.035	**0.4**	–0.02	-0.03	0.054	0.075	–0.02	–0.05	0.172	0.022	–0.04	0.016
WatSoDr	**0.69**	–0.013	–0.065	0.012	0.053	–0.03	–0.17	0.058	–0.03	–0.11	0.054	0.168	–0.007
WatSoFoo	**0.926**	–0.003	0.014	0.024	-0.03	0.045	–0.14	0.006	–0.01	–0.03	0.065	0.093	–0.013
WatSoClo	**0.98**	–0.001	0.015	0.035	0.005	–0.011	0.094	0.004	0.001	0.058	–0.03	–0.08	0
WatSoUt	**0.989**	–0.008	0.015	0.016	-0	0.009	0.085	–0.02	0.002	0.024	–0.03	–0.04	0.014
WatTreat	0.098	–0.045	0.075	0.03	0.064	0.177	0.141	0.046	0.032	–0	0.25	0.021	0
Refr	0.001	–0.044	–0.226	–0.24	0.135	–0.028	0.062	–0.06	–0.03	–0.06	–0.16	–0	0.152
FhCf	0.062	–0.069	0.052	0.045	0.191	–0.045	0.033	0.08	–0.39	0.07	–0.01	0.032	0.003
FhCi	–0.018	–0.065	0.038	0.043	**0.455**	0.038	0.036	0.031	–0.05	0.024	0.082	0.005	–0.036
FhOut	0.008	–0.158	0.07	–0.06	0.194	–0.05	0.26	0.175	–0.05	–0.02	0.096	0.091	–0.003
FhUncov	0.083	0.042	0.04	0.047	-0.11	0.07	0.436	–0.01	–0.03	0.042	0.006	–0.02	–0.001
FhBf	–0.017	0.036	–0.127	0.045	**0.416**	–0.024	–0.16	0.075	–0.07	–0.01	–0.09	0.003	0.028
PhyBth	0.07	0.001	0.152	–0.02	0.017	0.302	0.019	0.047	–0.33	0.049	0.001	–0.05	0.009
PhyWh	0.098	–0.108	–0.021	–0	-0.01	**0.507**	0.033	0.193	0.004	0.012	0.017	0.01	–0.007
LatAdu	0.124	0.025	–0.084	**0.542**	-0.04	0.062	–0.02	–0.03	0.044	0.004	–0.04	–0	0.006
LatChild	0.162	0.195	0.093	**0.56**	0.071	–0.072	0.037	–0.16	–0.06	0.041	0.037	0.001	–0.001

Note: Varimax rotation has been applied from SPSS and the factors are now identifiable. No variable now has significant loadings with more than one factor.

Table 21.6 gives the percentage of variance explained by each factor.

Table 21.6 Variance explained by factors

Sl. No.	Factors	Positive eigen values	Cumulated eigen values	% Variance explained by each factor	Cumulated variance
1	Factor 01	3.443	3.443	36.17	36.17
2	Factor 02	1.991	5.434	20.92	57.09
3	Factor 03	0.926	6.36	9.73	66.82
4	Factor 04	0.804	7.164	8.45	75.26
5	Factor 05	0.654	7.818	6.87	82.13
6	Factor 06	0.516	8.334	5.42	87.55
7	Factor 07	0.349	8.683	3.67	91.22
8	Factor 08	0.297	8.98	3.12	94.34
9	Factor 09	0.2	9.18	2.1	96.44
10	Factor 10	0.19	9.37	2	98.43
11	Factor 11	0.092	9.461	0.97	99.39
12	Factor 12	0.032	9.493	0.34	99.73
13	Factor 13	0.026	9.519	0.27	100

Results of Factor Analysis Indicate that

- Out of 13, only 6 factors explain more than 87% of variation in episodes of water borne enteric diseases.
- Factor loading matrix in Table 21.4 indicates that the three variables: Age, Ptlnc and Fhbf have absolute value of factor loadings > .3 for two factors. So identification of factors is difficult. Hence varimax rotation was applied to get Table 21.5.
- Using the factor loading matrix (obtained by orthogonal rotation) naming of these factors are possible, as given below.

Factors Explaining more than 87% of Variance

1. Water source
2. Social status
3. Affinity to Ganga
4. Latrine types
5. Food habits
6. Personal hygiene

Example 21.3: To measure the impact of UP's top position to eradicate poverty, reduce income inequalities, remove social and economic disparities and improve quality of life and living standard of rural eastern UP 103 variables related to 9 areas of socio-economic conditions were considered on 156 villages of four Eastern UP districts. FA was performed for identifying factors responsible for socio-economic disparities of weaker section. Data were collected in a UGC funded project under the author in 1989-91.

Statistical software CSS has limit of 50 variables, and BMDP4M of 60 variables for factor analysis. So 54 variables were chosen for principal factor analysis by BMDP4M to give following results:

Table 21.6 Variances explained by various factors (Carmine theta = 0.87)

Factor	Eigen values	Cumulative proportion of variance		VP[1]
		In data space	In factor space	
1	6.8427	0.1267	0.1688	4.168
2	4.6885	0.2135	0.2845	4.048
3	3.5311	0.2789	0.3717	3.427
4	3.1812	0.3378	0.4502	3.032
5	2.8479	0.3906	0.5204	2.673
6	2.5096	0.4371	0.5824	2.396
7	2.1236	0.4764	0.6348	2.180
8	2.0863	0.5150	0.6863	1.985
9	1.8820	0.5499	0.7327	1.941
10	1.7096	0.5815	0.7749	1.922
11	1.6292	0.6117	0.8151	1.710
12	1.4633	0.6388	0.8512	1.672
13	1.4150	0.6650	0.8861	1.670
14	1.2826	0.6888	0.9178	1.629
15	1.1900	0.7108	0.9471	1.580
16	1.1266	0.7317	0.9749	1.530
17	1.01162	0.7505	1.0000	1.482

Note: 1. VP[1] for each factor is sum of the squares of the elements in the column of the factor pattern matrix corresponding to that factor.
2. Cumulative proportion of variance up to 30 factors is 92%.
3. Eigen values > 1 upto 17 factors only.

Table 21.7 Identified factors for socio-economic disparities in weaker section of rural population of eastern UP

Factor number	Name of factors	Percentage of variance explained
1.	Wages in farm	12.67
2.	Unemployment	8.68
3.	Awareness of current important areas	6.54
4.	Household amenities and assets	5.89
5.	Working pool	5.28
6.	Cash wage in non-farm areas	4.65
7.	Working children	3.93
8.	Sound health	3.86
9.	Non-veg and toxic food	3.49
10.	Connectivity and health care	3.16
11.	Availing loan	3.02
12.	Rich food	2.71
13.	Old male and women welfare	2.62
14.	Education and information propagation for weaker section	2.38
15.	Basic sufficiency	2.20
16.	Health and home conscience	2.09
17.	Up to date health care and looks	1.88

17 factors account for 75% and first 8 account for more than 50% of total variability.

Computation of Factor Scores

Factor scores were obtained for the sampled villages by the said package, to assess the position of observed units on factor scales. Following method was used to obtain the factor scores for blocks and tehsils of the four districts:

Factor score for block = Sum of the factor scores for 6 villages in the block

Factor score for tehsils = Average of the factor scores for the blocks

Factor score for districts = Average of the factor scores for the tehsils

Table 21.8 gives these for all the tehsils of the four districts.

Composite factor score for district or tehsil = Weighted average of district or tehsil factor scores on first **8** factors using corresponding eigen values as weights.

In Table 21.9 Figures in ( ) are district ranks and in [ ] are tehsil ranks on composite factor scores.

The composite factor scores are the resultant of interactions of the eight factors, which together account for more than 50% of total variation, and may be an appropriate scale to rank the districts and the tehsils in order of their socio-economic development.

1. On this scale Jaunpur is at the top Varanasi is at the second position, Mirzapur at the third and Gorakhpur the lowest. Their combined scores are 0.2288, 0.0409, –0.0424 and –0.5276, respectively.
2. Out of the thirteen tehsils under study, Kerakat tehsil of Jaunpur ranks highest with score 1.0438, Mirzapur tehsil of the same district is 2nd with score 0.9728, Chakia tehsil of Varanasi ranks 3rd with score 0.9201. All of these have very close high ranks.
3. Shahganj tehsil of Jaunpur ranks 4th with composite score 0.7313, followed by Chandauli tehsil with composite score 0.5131, which appear satisfactory, being at the average.

Table 21.8 Average factor scores for tehsils and districts

	Factors							
	1	2	3	4	5	6	7	8
Varanasi district	1.651	–.653	1.022	–1.862	–1.242	1.091	–1.191	–.697
1. Chakia	2.229	2.545	–3.022	1.500	1.375	–1.555	0.00	2.056
2. Chandauli	4.667	3.107	–0.203	–1.289	–3.295	–.906	–3.875	–3.608
3. Gyanpur	2.896	–6.187	1.642	–4.459	2.500	3.463	–.422	.604
4. Varanasi	–3.190	–2.078	5.672	–.203	2.343	3.362	–.468	–1.838
Jaunpur district	–.934	.752	–.162	2.044	.834	.195	.640	–.443
1. Kerakat	.079	–.722	4.253	3.529	.052	–1.496	3.194	1.205
2. Jaunpur	–5.180	5.030	–1.414	1.938	.826	1.264	3.072	.409
3. Machhalishahar	–1.117	.949	–7.489	–.301	–1.816	2.912	0.635	–2.609
4. Mariahu	–1.017	–1.084	3.479	0.224	1.896	–2.129	–2.764	.810
5. Shahganj	0.535	–.416	.362	4.832	3.211	.425	–.940	–2.031
Gorakhpur district	1.886	.393	–3.437	.198	–1.113	–4.850	.625	–1.067
1. Bansgaon	2.418	0.268	–2.475	–2.445	3.303	–6.858	.174	–3.219
2. Gorakhpur	1.353	–0.197	–3.589	2.840	–5.528	–2.843	1.076	1.084
Mirzapur district	–2.852	–.769	2.691	–1.395	2.113	2.939	.543	3.393
1. Chunar	–3.141	2.956	–2.726	–.744	1.337	6.840	–0.262	2.336
2. Mirzapur	–2.563	–4.493	5.418	–2.046	2.890	–.962	1.347	4.449

Table 21.9 Composite factor scores for tehsils and districts

Tehsils	Districts	Composite factor scores	
	Varanasi	0.0409	(2)
1. Chakia		0.9201	[3]
2. Chandauli		0.5131	[5]
3. Gyanpur		–0.0505	[8]
4. Varanasi		–0.328	[7]
	Jaunpur	0.2288	(1)
1. Kerakat		1.0438	[1]
2. Jaunpur		0.0795	[6]
3. Machhalishahar		–0.1139	[9]
4. Mariahu		–1.1704	[13]
5. Shahganj		0.7313	[4]
	Gorakhpur	–0.5276	(4)
1. Bansgaon		–0.4507	[11]
2. Gorakhpur		–0.4902	[12]
	Mirzapur	–0.0424	(3)
1. Chunar		–0.2037	[10]
2. Mirzapur		0.9728	[2]

21.3 Cluster Analysis

Again suppose a set of N objects are observed for their p-characteristics. These may be measured in numeric quantities or counts, may be qualitative (nominal or ordinal). The objects may be human beings, a quadrat, a pollen spectrum or cities. The variables may be weight, length, area, population size, the species of pollen, colour of eyes, hair or flowers.

The earlier multivariate techniques assumed that the measurements had a common distribution or that the objects were *homogeneous*. In this section the interest is to search if the objects could be coming from a certain smaller number ($< N$) of natural groups or classes or clusters. The class of procedures that tries to form groups or clusters of the p-component data sets of N objects, such that objects within a group (or cluster) are similar to one another and those of different groups or cluster differ, is called cluster analysis. How close the p-component data vector should be for being grouped in one cluster i.e. degree of homogeneity desired and the number of clusters there ought to be, which depends on the degree of heterogeneity desired, are specified before applying clustering procedure.

This type of clustering is called clustering by objects. Another type of grouping may be of the p-variables on the data set to display relationships within and between sets of variables. This class of procedure is called *clustering of variables or attributes.*

Researchers have been engaged in the process of assigning objects into initially undefined classes in such a way that objects of a class, are in some sense, similar to one another. A historical development of classification theory is presented by Glimour (1951), Cain (1962) and Blackith and Reyment (1971).

Clustering problem is how to partition N objects into $K(K < N)$ groups so that in some sense, individuals within a group are homogeneous and those in different groups are heterogeneous? To seek a solution to such a loosely structured problem at least following questions should be answered:

(a) How many clusters are inherent in the data?
(b) How should cluster boundaries be established?
(c) To decide a measure of inter-object similarity and the precise meaning of similarity.
(d) A procedure for forming the clusters should be specified based on the chosen measure of similarity.
(e) Should all possible pairs of N objects in p-dimensional space (or p points in N-space) be scrutinized for similarities? This will be of tremendous magnitude for enumerative techniques. According to King (1967) the total number of ways of partitioning 25 objects in 5 groups is between 2 to 3 quadrillion.

With the fast development of computers and ready accessibility to it, the application of cluster analysis has penetrated in diverse areas of taxonomy, ecology, medicine, biology, economics, psychology, sociology, political science, archaeology, pattern recognition, linguistics, market research and even micro array data analysis.

Definition of Cluster

N objects with p-characteristics may be represented as N points in p-dimensional space, each of the p variables are represented in one of the areas in this space.

Clusters may be described as continuous regions in this space having a high density of points separated by regions with relatively low density.

As mentioned above, a measure of similarity or distance is an essential criterion for choosing a clustering technique among so many. So first this measure is defined in the next subsection. Various clustering techniques or methods are listed in the sub-section 21.3.2 and a few will be given in some detail. For a vivid discussion, the reader may refer to books by Gordon (1981), Press (1972), Dillon and Goldstein (1984) and Chapter 23 of SAS online DOC Version 8.

21.3.1 Distance and Similarity Matrices

Any clustering procedure is based on the measure of similarity or distance between objects. A matrix of $N \times N$ elements can be formed giving the similarity values (s_{ij}) or distances (d_{ij}) between object i and object j ($i, j = 1, 2, ..., N$). Distances between i^{th} and j^{th} object d_{ij} can be readily transformed into similarity measure s_{ij} and vice-versa. Some such transformations (or relationships) are given by

1. $s_{ij} = 1/(1 + d_{ij}) \Rightarrow d_{ij} = \dfrac{1}{s_{ij}} - 1$ (21.67)

2. $s_{ij} = c - d_{ij} \Rightarrow d_{ij} = c - s_{ij}$ (21.68)

for some constant $c(i, j = 1, 2, ..., N)$.

If $s_{ii}(i = 1, 2, ..., N)$ do not all have same value and dissimilarity (or distance) measure possessing metric properties is desired, complications occur in the transformations from s_{ij} to d_{ij}.

Some distance measures (such as Euclidean distance) are not scale invariant and usually, therefore, some scaling method is undertaken to ensure that $s_{ii} = 1(i = 1, 2, ..., N)$. This may be achieved by standardizing the raw data.

Gower (1966) observed that

$$d_{ij} = \sqrt{2(1 - s_{ij})} \tag{21.69}$$

can represent the Euclidean distance if $\mathbf{S} = ((s_{ij}))$ is non-negative definite.

If $\mathcal{S}$ denotes the set of N objects in hand the dissimilarity (or distance) between a pair of objects d is a function of $\mathcal{S} \times \mathcal{S}$ of real numbers satisfying

(a) $d_{ij} \geq 0$ for all $i, j \in \mathcal{S}$ (21.70)
(b) $d_{ii} = 0$ for all $i \in \mathcal{S}$ and (21.71)
(c) $d_{ij} = d_{ji}$ for $i, j \in \mathcal{S}$ (21.72)
(d) $d_{ij} + d_{ik} \geq d_{jk}$ for any $i, j, k \in \mathcal{S}$ (21.73)

Because of these properties of d_{ij} one *can completely specify the dissimilarities of N objects by a lower triangular matrix of N(N – 1)/2 d_{ij}'s.*

There are different types of similarity or distance measures. Only some of these will be defined in this sub-section. A more comprehensive list may be found in Cormack (1971), Anderberg (1973) and Diday and Simon (1976).

A particular similarity or distance measure is appropriate for a particular type of variable. Hence the measures are given for quantitative variables, qualitative (nominal and ordinal) variables and for binary variables.

It is easy to compute the matrix **S** or **D** (= $((d_{ij}))$) if all the p-components of variables on the objects are of same type. But if the p-variables on the objects are of mixed type (i.e. some quantitative and some ordinal and/or nominal and/or binary) the method is different.

(a) Quantitative Variables

If the variables are quantitative, distance-type of measures are used.

Let $\mathbf{X}_i' = (X_{i1}, X_{i2}, ..., X_{ip})$, $i = 1, 2, ..., N$ denote the measurements on the ith object or individual. The most general distance measure is Minkowski metric defined as

$$d_{ij} = \left\{ \sum_{k=1}^{p} \left| X_{ik} - X_{jk} \right|^{r} \right\}^{1/r} \tag{21.74}$$

If **r = 2**, it reduces to the familiar Euclidean distance given by

$$d_{ij} = \left\{ \sum_{k=1}^{p} \left| X_{ik} - X_{jk} \right|^{2} \right\}^{1/2} \tag{21.75}$$

If **r = 1**, the resulting distance given below is called the absolute or city-block metric

$$d_{ij} = \sum_{k=1}^{p} \left| X_{ik} - X_{jk} \right| \tag{21.76}$$

Most commonly used distance measure is Euclidean, which is not scale invariant. So one should be careful while using raw data since simple change or variation of scale badly distorts the distance between objects.

Another distance measure is Mahalanobis D^2 given by

$$D^2 = (\mathbf{X}_i - \mathbf{X}_j)' \, \mathbf{S}^{-1}(\mathbf{X}_i - \mathbf{X}_j) \tag{21.77}$$

where this $\mathbf{S}$ is unbiased estimate of $\mathbf{\Sigma}$ defined in (20.48).

(b) Binary Variables

Suppose that each of the p binary variables can be divided unambiguously into one of two states only, as '+', indicating presence and '-', indicating absence of an attribute. Then for calculating similarity or distance between two objects the relationship between i^{th} and j^{th} objects is summarized as the following 2 × 2 table:

Table 21.10 2 × 2 contingency table for binary variables

		Object j	
		Number of variables '+'	Number of variables '–'
Object i	Number of variables '+'	a	b
	Number of variables '–'	c	d

Many similarity measures can be defined depending on a, b, c, d. The following give some of the more popular ones:

(i) Simple matching coefficient

$$s_{ij} = \frac{a+d}{a+b+c+d} \text{ and } d_{ij} = \frac{b+c}{a+b+c+d} \tag{21.78}$$

(ii) Jaccard's coefficient

$$s_{ij} = \frac{a}{a+b+c} \text{ and } d_{ij} = \frac{b+c}{a+b+c} \quad (21.79)$$

(iii) $s_{ij} = \dfrac{2a}{2a+b+c}$ and $d_{ij} = \dfrac{b+c}{2a+b+c}$ (21.80)

(iv) $s_{ij} = \dfrac{2(a+d)}{2(a+d)+b+c}$ and $d_{ij} = \dfrac{b+c}{2(a+d)+b+c}$ (21.81)

(v) $s_{ij} = \dfrac{a}{a+2(b+c)}$ and $d_{ij} = \dfrac{2(b+c)}{a+2(b+c)}$ (21.82)

(vi) $s_{ij} = \dfrac{a}{a+b+c+d}$ and $d_{ij} = \dfrac{b+c+d}{c+b+c+d}$ (21.83)

It can be seen that the major difference between these measures is due to incorporation of number of (–, –) i.e. d, and weights given to unmatched pairs of variables i.e. b and c and number of (+, +) i.e. a.

(c) Nominal Variables

A nominal variable with s states is divided into s binary variables. If an observation is of mth state of the nominal variable, then m^{th} of these divided binary variable is assigned '+' sign and the rest of $(s - 1)$ variables are assigned –. If there are p nominal variables in the observation vector, similar division is done for each of the p. For two observations i and j a table of 2 rows is prepared and from the counts of (+, +) = a, of (+, –) = b, of (–, +) = c and (– , –) = d, the 2 × 2 contingency table is prepared as described in (b) above. The similarity or dissimilarity or distance measure is computed using one of the six formulae.

(d) Ordinal Variables

For an ordinal variable, easiest way is to give a numeric value to various states. These numeric values may be at equal difference, if the states possess equally more or less of the attribute. If it is not so, numeric values may be at unequal difference. For example complexion of individuals may be recorded as black, wheatish and fair. In such case the numeric value assigned may be 0, 1, 3 instead of 0, 1, 2, if fair complexion differs from wheatish more than wheatish differs from black.

Distance measure may subsequently be computed for these ordinal variables.

(e) Variables of Mixed Type

Often the studied objects are described in terms of variables of mixed type such as length or weight (numeric), hair colour (ordinal), mental status (nominal), availability of four wheeler (binary). In order to classify such objects it is necessary to apply special method of computing similarity or distance between two objects. One such method has been suggested by Gower (1971).

Let s_{ijk} is the similarity index between i^{th} and j^{th} object (i, j = 1, 2, ..., N) for k^{th} (k = 1, 2, ..., p) variable. These variables may be of different types and indices are obtained using the appropriate method listed above.

Now the overall similarity coefficient for i^{th} and j^{th} object is obtained by the following formula given by Gower (1971) as:

$$s_{ij} = \left(\sum_{k=1}^{p} w_{ijk}\, s_{ijk}\right) \Bigg/ \left(\sum_{k=1}^{p} w_{ijk}\right) \tag{21.84}$$

where w_{ijk} is 1, if comparison is valid for k^{th} variables; otherwise $w_{ijk} = 0$.
Other methods are as given below.

(i) The first is to convert some of the variables of other type into the type of majority present. Any variable in the sequence: numeric, ordinal, nominal, binary can be converted into that appearing later in the sequence. As already mentioned, numeric and ordinal variables can be converted into numeric type and nominal ones into binary form. After the transformation into one type, apply the appropriate method for the reduced type of variable.

(ii) Another method is to carry out separate analyses over all the N objects, considering set of variables of one type in an analysis. Synthesize the results of these analyses to form final clusters (or groups).

21.3.2 Clustering Methods

The two popular kinds of clustering methods are hierarchical method and partitioning method.

Hierarchical Techniques

The hierarchical techniques perform successive fusion or divisions of the data. In these, the allocation of an object is irrevocable. In other words, if an object is classified in a cluster it never joins any other cluster.

There are two ways of hierarchical clustering: *agglomerative* and *divisive*. Agglomerative methods proceed by forming a series of fusions of N objects into groups, eventually result in all the N objects falling in one cluster. Divisive methods partition the set of N objects into finer and finer subdivisions so that finally each object becomes a cluster of its own.

Hence the primary question is when to stop. Output of both the methods of clustering can be graphically represented by a *dendogram*, which is a two-dimensional tree like diagram indicating the fusion or partition at each successive stages.

Following gives a list of some methods of agglomerative hierarchically clustering p-component data on N objects:

1. Single linkage or nearest neighbour. 2. Average linkage, 3. Complete linkage, 4. Density linkage (Wong's hybrid), 5. k^{th} nearest-neighbour, 6. Two-stage density linkage, 7. McQuitty's similarity analysis, 8. Median method, 9. Ward's minimum-variance.

In general, each observation begins in a cluster by itself. The two closest clusters are merged to form a new cluster replacing the two old clusters. Various clustering methods differ in how the distance between two clusters is computed.

For numerical data it is advisable to scale or standardize the data, since large variability may lead to overlapping with another group or cluster.

Books by Gordon (1981), Hartigan (1975), Sneath and Sokal (1973), Everitt (1980) give a detailed discussion on clustering methods. A brief description is available in Press (1972), Dillon and Goldstein (1984) and SAS Online Doc™ Version.

In this subsection a few methods are described, which are in common use.

For clarity of discussion the following **notations** are defined, which are used in the following pages.

1. G : The number of clusters at any given level of hierarchy
2. $\mathbf{X}_i$: i^{th} observation vector of p-components
3. C_K : K^{th} cluster which is a subset of N objects
4. $\mathbf{N}_K$: Number of observations in C_K
5. $\bar{\mathbf{X}}$: Sample mean vector $= \frac{1}{N}\sum_{i=1}^{N} X_i$
6. $\bar{\mathbf{X}}_K$: Mean vector for cluster $C_K = \frac{1}{N_K}\sum_{i\in C_K} X_i$
7. $\|\mathbf{X}\|$: Euclidean length of vector $X = \sqrt{\sum_{i=1}^{p} X_i^2}$
8. T : $\sum_{i=1}^{N}\|X_i - \bar{X}\|^2 = \sum_{i=1}^{N}\sum_{j=1}^{p}(X_{ij} - \bar{X}_j)^2$ = Total sum of squared deviations of N objects from $\bar{X}$
9. W_K : $\sum_{i\in C_K}\|\mathbf{X}_i - \bar{\mathbf{X}}_k\|^2$ = Sum of squared deviations of objects in C_K from the mean vector of C_K.
10. P_G : $\sum_J W_J$, where summation is over the G clusters at the Gth level of hierarchy.
11. β_{KL} : $W_M - W_K - W_L$ if $C_M = C_K \cup C_L$
12. d_{ij} : Distance or dissimilarity measures between i^{th} and j^{th} objects.
13. D_{KL} : A distance or dissimilarity measure between C_K and C_L.

In the following it is assumed that C_K and C_L are merged to form new cluster C_M. The distances of new cluster C_M from other clusters C_j are also computed using a formula, which may differ with the similarity measure used in the method.

1. Single Linkage or the Nearest Neighbour Method

This method was proposed in 1951 and reinvented by Sneath (1957).

(i) At the first stage two objects A and B are joined to form a cluster if

$$d_{AB} = \min_{i,j} d_{ij} \tag{21.85}$$

(ii) Now the distances between this cluster C_M and other (clusters) unclustered observations C_J are obtained by finding

$$D_{MJ} = \min_{i\in C_M}\min_{j\in C_J}(d_{ij})$$

If
$$D_{JM} = \min_J (D_{MJ}), \text{ then} \tag{21.86}$$

at the second stage C_J is joined with the first stage cluster C_M.

The steps (i) and (ii) are repeated till the two clusters are fused to form a single cluster of all the N elements.

For a numerical example the reader is referred to the Euclidean distance matrix for 5 objects A, B, C, D, E. Among the elements of 5×5 distance matrix $D^{(1)}$, suppose $d_{AB} = \min (d_{ij})$ for $i \neq j = A, B, C, D, E$.

So A and B form a cluster. Now the distances of C, D, E from cluster AB is recalculated. Distances between C, D, E remain same as before. Hence the resulting distance matrix $D^{(2)}$ will be 4×4 and except 1st row and 1st column, 2nd to 4th row and column elements are same as of $D^{(1)}$. The first row and column elements are computed by finding distances from mean vector of cluster AB and elements C, D and E.

If it is found that in $D^{(2)}$, $d_{DE} = d_{ED}$ is minimum, D and E form a second cluster. C remains unclustered. No other element joins the first cluster AB.

Now a 3×3 distance matrix $D^{(3)}$ is computed for distance between AB, C and DE. In $D^{(3)}$, suppose distance of C and AB is smaller than $d_{C(DE)}$ or $d_{(AB)(DE)}$. Hence C joins AB. Lastly the two clusters ABC and DE join together.

The dendogram summarizes the clustering procedure. In this diagram the distance is shown on Y-axis and the objects are shown on X-axis. It is a tree type of diagram and is shown below in Fig. 21.1.

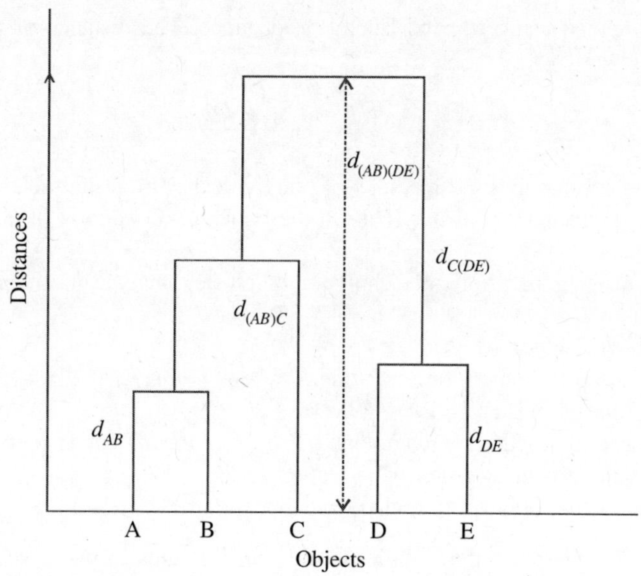

Fig. 21.1. Dendogram for clustering 5 objects by single linkage.

2. The Average Linkage Method

In this method the distance between two clusters is defined as the average of distances between all pairs of objects, one from one cluster and the other from the second cluster. This average distance between cluster K and L is given by

$$D_{KL} = \frac{1}{N_K N_L} \sum_{i \in C_K} \sum_{j \in C_L} d_{ij} \tag{21.87}$$

3. Complete Linkage or the Farthest Neighbour Method

This method is based on the farthest element. The distance between two clusters is given by

$$D_{KL} = \max_{i \in C_K} \max_{j \in C_L} (d_{ij}) \tag{21.88}$$

The clusters which join should satisfy

$$D_{JM} = \max (D_{JK}, D_{JL}) \tag{21.89}$$

4. The Centroid Method

In the centroid method, the distance between two clusters is defined as the squared Euclidean distance between the centroids or means of two clusters. In symbols

$$D_{KL} = \left\| \bar{\mathbf{X}}_K - \bar{\mathbf{X}}_L \right\|^2 \tag{21.90}$$

The clusters with smallest distance are joined. The distance between the new amalgamated cluster C_M and other clusters C_J is given by

$$D_{JM} = \frac{N_K D_{JK} + N_L D_{JL}}{N_M} - \frac{N_K N_L D_{KL}}{N_M^2} \tag{21.91}$$

5. The Median Method

The distances are computed using squared Euclidean distance. The distance of new cluster C_M from other clusters C_J is given by

$$D_{JM} = \frac{D_{JK} + D_{JL}}{2} - \frac{D_{KL}}{4}$$

These are some of the commonly used methods of hierarchical cluster analysis. There are many more methods also and the interested reader may consult the books by Gordon (1981), Hartigan (1975) and Dillon and Goldstein (1984).

Now one method using partitioning techniques of cluster analysis is given below.

6. K – Means Clustering Method

Let X_{ij} denote the value of ith object on j^{th} variable, where $i = 1, 2, \ldots, N$ and $j = 1, 2, \ldots, p$.
The procedure for *K*-means clustering is as follows:
Search for a partition with small error component *E* by moving object from one cluster to another until no transfer of an individual results in reduction in *E*.
There are some methods to form K starting points. Four of these are:

(a) Choose the first *K* objects in the sample as the initial *K* cluster mean vectors.
(b) Choose the first *K* objects that are mutually furthest apart.
(c) Start with a trial value for *K*, that is larger than is thought necessary, and proceed to form cluster centers spaced at intervals of one standard deviation (*sd*) on each variables, and
(d) Choose *K* and the initial cluster configurations based on prior knowledge.

The steps given below summarize the method of forming clusters on the basis of the *K*-means algorithm.

Step 1: Denote the sum of row elements by *Sum(i)*, then an initial cluster can be formed by considering case i as part of the jth cluster, where j is the integral part of K [*Sum(i)* – *Min*]/(*Max* – *Min*) + 1, in which Max and Min are the maximum and minimum values of Sum(i), respectively.

Step 2: Now calculate $B(l, j)$, which is the mean of the jth variable overall cases in the lth cluster.

Step 3: The distance between the ith case and l^{th} cluster is given by

$$D(i, l) = \left[\sum_{j=1}^{p} (X_{ij} - B(l, j))^2 \right]^{1/2}$$

The error of any particular partition is defined as

$$E[P(n, K)] = \sum_{i=1}^{N} D[i, 1(i)]^2$$

where $l(i)$ is the cluster containing the ith case.

Step 4: We now check to see whether any movement of any case from one cluster to another results in a reduction in E. Compute for every case the following value:

$$R_{l(i),l} = \frac{n(l)\{D(i, l)\}^2}{n(l)+1} - \frac{n(l(i))\{D(i, l(i))\}^2}{n(l(i))-1}$$

where

$n(l)$ = number of cases in the lth cluster, and

$l(i)$ = cluster containing the ith case.

Step 5: Carrying out the calculations in Step 4, if all the $R_{1(i), 1}$'s are positive and hence no shifting will result in decreased error, the final result is reached. Otherwise repeat the whole procedure starting from Step 2.

Two examples of cluster analysis are given.

Example 21.4: Consider again the data on water borne enteric diseases from OPD of five hospitals, considered in Example 21.2 for factor analysis. Applying single linkage analysis for clustering by statistical software package SPSS 10.0, the following clusters were obtained. There were 343 patients from 12 municipal wards of Varanasi city. The variables are same as those for factor analysis.

Table 21.10 below gives the number of patients in each of the 5 clusters.

Table 21.10 Number of WBE disease cases in 5 clusters

Sl. No.		Cluster	No. of cases in each cluster
1	Cluster	1	195
		2	1
		3	48
		4	47
		5	55
2	Valid		343
3	Missing		0

The Table 21.11 gives the distances between the final cluster centers. Table 21.12 gives the ward wise number of WBE disease patients in various clusters.

Table 21.11 Distance between final cluster centers

Cluster	1	2	3	4	5
1		15.738	7.152	7.768	11.893
2	15.738		12.880	12.887	18.395
3	7.152	12.880		8.600	12.827
4	7.768	12.887	8.600		11.457
5	11.893	18.395	12.827	11.457	

Table 21.12 Number of cases in clusters: ward wise

Ward	Cluster 1	Cluster 2	Cluster 3	Cluster 4	Cluster 5
1	26	-	6	9	9
2	32	-	6	3	0
3	42	-	7	10	10
4	28	-	4	5	4
5	23	-	4	10	14
6	12	-	5	2	2
7	9	-	5	4	5
8	2	-	0	0	1
9	9	-	3	1	2
10	3	-	3	0	0
11	4	-	0	0	3
12	5	1	2	3	5
Total	195	1	45	47	55

From Table 21.12 it may be concluded that:

(i) Ward 3 (Chowk) has highest number of cases in clusters 1, 3 and 4. Being most highly densely populated area, seems also to be most prone to disease.

(ii) Overall looking at the table, the Wards 1, 2, 3, 4 (Kotwali, Chetganj, Chowk, Dashaswamedh), the most densely populated central part of Varanasi city, appear to be more prone to considered water borne diseases.

Example 21.5: Figure 21.2 and Table 21.13 give the fifteen locations of India from where sixteen ecotypes of *Hydra* were collected by Prasad and Mookerjee (1985).

Table 21.13 Locations of 16 ecotypes of *hydra*

Sl. No.	Distant locations	Sl. No.	Distant locations
1.	Srinagar (SRN)	9.	Bolpur (BLP)
2.	Jammu (JMU)	10.	Chandan Nagar (CNR)
3.	Imphal (IMP)	11.	Santiniketan (SNT)
4.	Shillong (SHL)	12.	Green Hydra from Hyderabd (GHYD)
5.	Delhi (DLH)	13.	White Hydra from Hyderabad (WHYD)
6.	Lucknow (LKH)	14.	Madurai (MDR)
7.	Pune (PNA)	15.	Tirupati (TRP)
8.	Calcutta (CAL)	16.	Trivendrum (TRV)

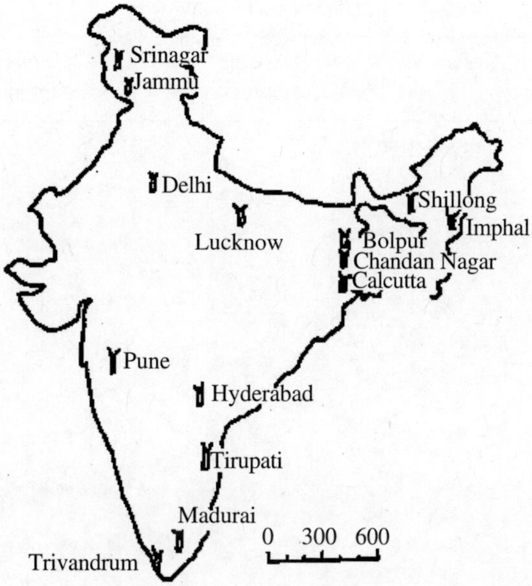

Fig. 21.2. Map of India showing the sites of collection of the sixteen ecotype of *hydra*.

Table 21.14 Variables studied for cluster formation of the 16 ecotypes of *Hydra*

Sl. No.	Morphology	Sl. No.	Regeneration	Sl. No.	Asexual reproduction	Sl. No.	Sexual reproduction	Sl. No.	Cellular aspects
1.	Normal length of body column		Time taken for hypotome regeneration at		Population doubling time at	27.	Pres./absence of circannual rhythm of repro		Density of nematocyst at different parts of tentacle:
2.	Length of body column when fully stretched	7.	29°C	13.	29°C	28.	Gonad induction by low tempera	31.	Basal
	Relative length of the tentacle as compared to the body length	8.	23°C	14.	23°C	29.	Reproductive system: Gonads absent/ monocious/ dioceous condition	32.	Middle
3.	In normal conditions	9.	15°C		Time taken for bud development at	30.	Shape of testes: Testes absent/conical/ round/almost rectangular	33.	Tip region
4.	In fully stretched conditions		Time taken for basal disc	15.	29°C			34.	Shape of holotrichous

Contd...

Table 21.14 Continued

Sl. No.	Morphology	Sl. No.	Regeneration	Sl. No.	Asexual reproduction	Sl. No.	Sexual reproduction	Sl. No.	Cellular aspects
			regeneration at						isorhiza slipper shaped/ cylindrical/ pyriform/ bean shaped
5.	Distance between the basal disc and budding region	10.	29°C	16.	23°C			35.	Presence/ absence of symbiotical gae I endodermal cells
6.	Order of emergence of tentacles on buds	11.	23°C	17.	15°C				
		12.	15°C		Time taken for bud maturation at				
				18.	29°C				
				19.	23°C				
				20.	15°C				
					Time taken for successive bud initiation at				
				21.	29°C				
				22.	23°C				
				23.	15°C				
					Max. No. of buds per hydra observed at				
				24.	29°C				
				25.	23°C				
				26.	15°C				

35 variables at various levels of organization – morphological, physiological and cellular levels have been considered (see Table 21.14).

Out of 35 variables obtained on the *Hydra*, variable numbers 27, 28 and 35 are binary. Variables 29, 30 and 34 are nominal variables and have been assigned numeric values depending on the rating of the attribute present. For example variable 30: Shape of testes – absent, conical, round and almost rectangular have been assigned numeric values 0, 1, 2, 3, respectively. The variables 31 to 33 are quantitative.

For binary variables the dissimilarity measures are obtained using simple matching coefficient given in Gordon (1981) and Dillon and Goldstein (1984). Cluster analysis was performed on each of the five groups of the variables and also on all the 35 variables, using CSS software package. The results of 6 morphological variables and for all the 35 variables are given in Table 21.15. Dendograms for these have been obtained by UPGMA method (Johnson (1967), Spencer (1984) and Singh (1989)). Computation of distance matrices has been done at each step by a computer program written by author of this book. Figures 21.3 and 21.4 show the dendograms for the 6 morphological variables and for all the 35 variables.

Table 21.15 gives the number of clusters considering only first six variables on morphology and from all the 35 variables. The cases are also listed in the table.

Table 21.15 Results of cluster analysis: on six and on all the variables

Variables	Cluster numbers						
	1	2	3	4	5	6	7
Morphology	DLH	IMP	TRV	GHYD	CAL		
	SHL	PNA	SRN				
		BLP	LKN				
		JMU	TRP				
		WHYD	SNT				
			CNR				
All			MDR				
Variables	TRV	PNA	DLH	IMP	CNR	GHYD	CAL
(1-35)	SRN	JMU	SHL	BLP	WHYD		
	LKN						
	MDR						
	SNT						
	TRP						

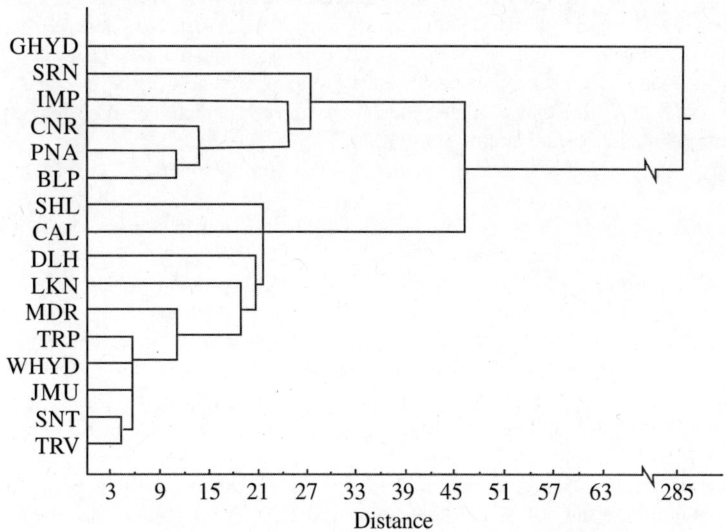

Fig. 21.3. Dendrogram for morphogenetic variable.

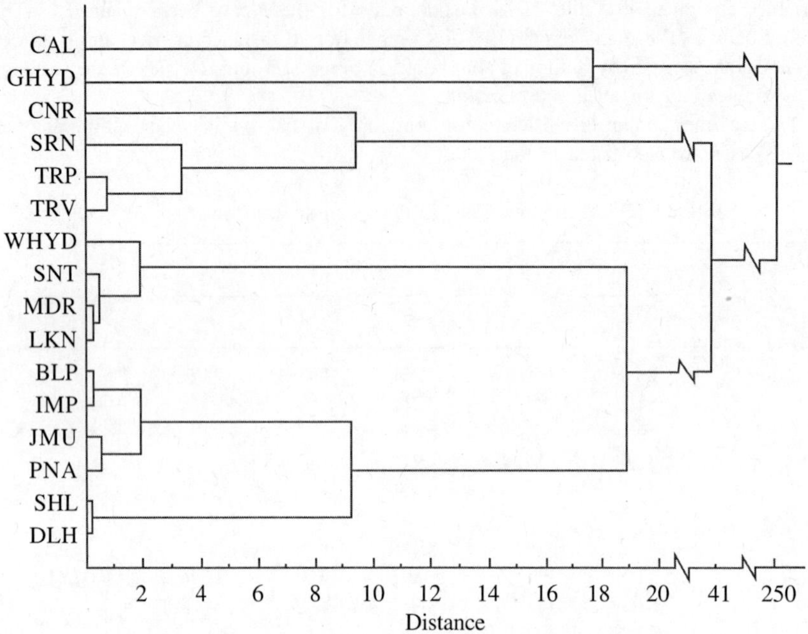

Fig. 21.4. Dendrogram for all variables considered together.

Example 21.6: Now stepwise discriminant analysis has been applied on the raw data matrix used in above cluster analysis (Rastogi et al. (1992)). The cluster number shown in Table 21.15 are used as grouping code for each ecotype, and entered as the first variable for the said discriminant analysis on the input data matrix of order 16 × 36. Module 7M of BMDP package (Tabachnick and Fidel (1983), Afifi and Clark (1984) Dixon (1985)) has been used with stepwise forward inclusion of variables (cf. Rastogi and Pandey (1995)).

Classification functions have been obtained for each group. The values of these for an ecotype are obtained by substituting the observed X_{36} for that ecotype. The *Hydra* ecotype is then classified into a group for which the evaluated classification function is maximum.

Table 21.16 gives the coefficients of classification function for six selected variables and the corresponding constants.

It is observed that only 6 variables are defining the 7 classification functions. The rest of the variables are insignificant for discriminating between *Hydra* clusters.

Only the first CV accounts for 88.36%, whereas first three CV's account for 99.62% of total variation. The 4th, 5th and 6th contribute only 0.38%, which is very small. Figure 21.5 shows, therefore, only three-dimensional representation of discriminant analysis with the first 3 CV's as the axes *X*, *Y*, *Z*. The values of these CV's for 16 ecotypes, calculated for observed values of variables for each ecotype, are plotted in this figure. It appears that there is an overlapping of groups 3 and 4. It may be noted that this overlapping is only an impression due to 3 dimensional figure drawn on a 2 dimensional paper. In fact groups 3 and 4 are different as evident from the calculated values of *X*, *Y*, *Z* for clusters 3 and 4.

The effectiveness of proposed classification procedure is evaluated by finding the percentage of correct classification. Results of cluster analysis is considered as correct class of ecotypes. The classification matrix given in Table 21.17 shows that classification of all the 16 ecotypes in 7 clusters is 100% correct.

Table 21.16 Coefficients for canonical variables

Sl. No.	Variables	CV1	CV2	CV3	CV4	CV5	CV6
1.	Length of body column in fully stretched condition (X_2)	23.53	10.68	–2.62	–3.95	–0.45	–0.40
2.	Relative length of tentacles as compared to the body length in fully stretched condition (X_4)	0.43	–0.01	0.02	–0.00	0.01	0.00
3.	Time taken for bud development at 22±1°C (X_{15})	0.46	0.29	0.07	0.06	–0.04	–0.02
4.	Time taken for bud maturation at 22±1°C (X_{118})	0.09	–0.54	–0.11	–0.02	–0.02	–0.05
5.	Time taken for successive bud initiation at 15 22±1°C (X_{21})	–0.54	–0.13	–0.09	0.23	0.09	0.04
6.	Maximum no. of buds per hydra observed at 22±1°C (X_{126})	10.38	6.096	0.86	–3.05	0.46	–1.65
	Constants	–230.51	–51.80	18.23	16.60	–2.44	7.03

Cumulative proportion of total dispersion accounted for by CV's
0.88355 0.98636 0.99623 0.99985 0.99996 1.0000

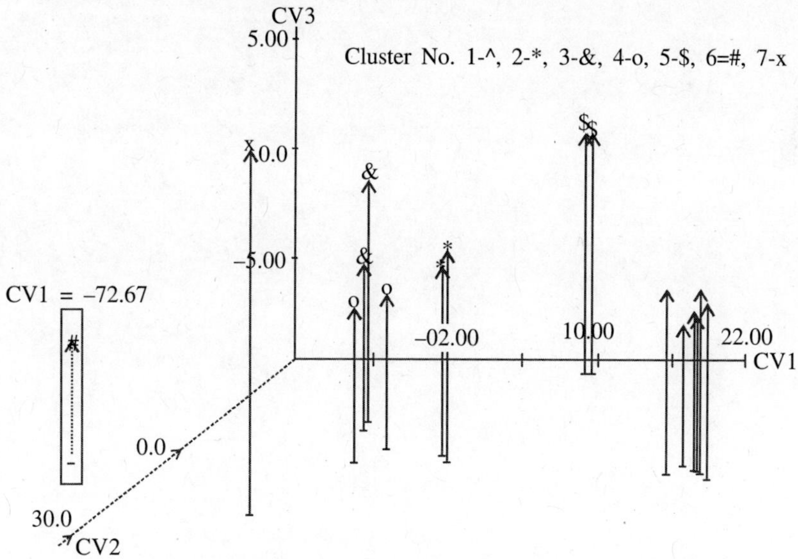

Fig. 21.5. Discrimination of clusters.

Table 21.17 Classification matrix

Cluster numbers	Percent correct	Number of cases classified into groups						
		1	2	3	4	5	6	7
1.	100.0	6	0	0	0	0	0	0
2.	100.0	0	2	0	0	0	0	0
3.	100.0	0	0	2	0	0	0	0
4.	100.0	0	0	0	2	0	0	0
5.	100.0	0	0	0	0	2	0	0
6.	100.0	0	0	0	0	0	1	0
7.	100.0	0	0	0	0	0	0	1
Total	100.0	6	2	2	2	2	1	1

22. Bioinformatics and Computational Biology

Bioinformatics is the branch of science dealing with biology, information technology and statistics. It is a discipline to understand, organize and interpret the information associated with biological molecules.

According to NCBI, bioinformatics is the field of science in which biology, computer science, and information technology merge into a single discipline. The ultimate goal of the field is to enable the discovery of new biological insights and to create a global perspective from which unifying principles in biology can be discerned.

NIH gave a working definition of bioinformatics as – research, development, or application of computational tools and approaches for expanding the use of biological, medical, behavioral or health data, including those to acquire, store, organize, archive, analyze or visualize such data. It helps to describe, analyze, simulate and predict the biological processes and deals with

1. Creation of *databases* for storage and handling large biological data.
2. Development of *algorithms* and application of *statistics* to determine relationships among various large biological data sets.
3. Use of these tools for analysis and interpretation of biological data on DNA, RNA and protein sequences, protein structures, protein functions, functions of nucleic acid and protein complexes.

22.1 Concepts of Bioinformatics: A Digital Laboratory

Basically *biology* has been an *observational science* rather than a *deductive science*. Although recent developments have not altered this basic orientation but the nature of data has radically changed. In the last few decades, the *biological data* have become not only much more *quantitative and precise* but also *discrete*. It made possible the determination of the genome sequence of an individual organism or clone not only completely but in principle, exactly also.

It is well known that *experimental error* can never be avoided fully but for modern genome sequencing it is *extremely low*. This knowledge has not been able to convert biology into a deductive science, though life does obey principles of physics and chemistry. This is because life is too complex and too dependent on past contingencies, to enable deducing its detailed properties from basic principles.

A *second characteristic* of bioinformatics data is its *huge volume*. Currently nucleotide sequence data banks contain 16×10^9 bases (or 16 Gbp, where Gbp means Geiga byte base pairs). Approximate size of human genome is 3.2×10^9 letters or 2 huges (HUman Genome EquivalentS), where 1 huge is comparable to the number of characters appearing in six years' complete issues of New York Times (cf. Lesk (2003)).

Bioinformatics may thus be *defined* as the science of using information related to molecular biology to understand biological phenomena. It is a part of the larger science of computational biology, which is an application of quantitative analytical techniques in modeling and solving problems in biological systems. Bioinformatics has application of statistical methods, pattern recognition and some of the computational methods.

Rastogi et al. (2004) mention that according to Fredj Tekaia, bioinformatics is "the mathematical, statistical and computing methods that aim to solve biological problems using DNA and amino acid sequences and related information."

The fundamental objectives of bioinformatics are:

(a) It allows a researcher to access existing databases and also to submit new entries, if found not already existing.
(b) It develops tools and resources that help in analysis of biological data to determine similarities and dis-similarity if any, with the previously existing data. It requires extensive knowledge of computational procedure and also thorough understanding of biological phenomena.
(c) After analyzing the biological data, the results are to be interpreted in a biologically meaningful manner. One can also conduct global analyses of the available data to determine the common principles which apply on many biological systems and to demarcate the features which are unique to some.

Some of the applications of bioinformatics are listed below:

1. Information related to biomolecules can be mapped.
2. Sequences can be compared, usually by aligning corresponding segments and looking for matches and mismatches in their sequences. Genes and proteins that are sufficiently similar are likely to be related (homologous, orthologous or paralogous).
3. Is used in primer design.
4. Used to predict function of gene products.
5. Similarity and relatedness of proteins is used to trace family trees.
6. Molecular modelers and structure biologists use for prediction of the secondary structure of proteins and to handle vast and complex data from X-ray crystallography, nuclear magnetic resonance (NMR) and electron microscopy for creating 3-D models of molecules.
7. Supports applications to areas of medicine, agriculture and other scientific fields of medical imaging/image analysis.

There are plenty of books on bioinformatics, some short and precise, some detailed ones, explaining the purpose and techniques. Some of the good books are by Westhead et al. (2003), Lesk (2003) and Rastogi et al. (2004). A fundamental book encompassing all the basics, details of tools, computational aspects and giving the mathematical approach behind them is by Mount (2001). A glimpse in brief is presented in this chapter. For all the deep insight and full details, the reader must look into one of the books mentioned above.

Various types of data are analyzed in bioinformatics. In the next section some databases and tools will be mentioned.

22.2 Databases and Tools of Bioinformatics

All the books on bioinformatics give a description of the types of data and the major databases. Rastogi et al. (2004) have given one table giving the types of data and the topics which are explored by analysis of those data. These types of data include the DNA sequence, protein sequence, 3-D structure, genomes and gene expression. In another table they give various data bases and their website address (≡ URL: uniform resource locater). The table below lists some such site addresses and their details.

Table 22.1 Some sites of databases and their particulars

URL	Database	Particulars
	Nucleotide sequences	
http://www.ncbi.nlm.nih.gov/genbank	Gen Bank	*National Centre for Biotechnology Information* (*NCBI*) home page. Resource for public database, bioinformatics tools and applications. Links to many useful sites and resources for bioinformatics softwares
http://www.ebi.ac.uk/emble	EMBL	*European Molecular Biology Laboratory.* European Bioinformatics Institute Outstation. A resource for biological databases and softwares, with excellent tutorial support.
http://www.ddbj.nig.ac.jp/	DDBJ	*DNA Data Bank of Japan.* New sequences can be deposited in any of the above databases since they exchange data on a daily basis.
	Protein sequence (primary)	
www.expasy.ch/sprot/sprot-top.html	SWISS-PROT	Swiss Institute of Bioinformatics (SIB) annotate the sequences and describe protein functions, domain structure and post-translational modifications.
www.mips.biochem.mpg.de/proj/protseqdb	PIR-International	*Protein Information Resource*, a division of National Biomedical Research Foundation (NBRF), US. Has collaboration with Munich Information Centre for Protein Sequences (MIPS) and Japanese International Protein Sequence Database (JIPSD) Nature same as described above.
	Protein sequence (Secondary)	
www.expasy.ch/prosite	PROSITE	Part of SWISS-PROT. Data base of short protein sequence patterns and profiles, which characterize significant sites. ExPASy (Expert Protein Analysis System).
www.bioinf.man.ac.uk/dbbrowser/PRINTS/PRINTS.html	PRINTS	Provides a compendium of protein fingerprints – groups of conserved motifs that characterize a protein family.

Contd...

Table 22.1 Continued

URL	Database	Particulars
www.sanger.ac.uk/pfam	Pfam	Database of protein families defined as domains (contiguous segments of entire protein sequences).
	Genome sequences	
www.ncbi.nlm.nih.gov/entrez/query.fcgi/db=genome	ENTREZ	NCBI maintains databases and links of access to them.
www.bioinfo.mbb.yale.edu/	Gene Census	Provides an entry point for genome analysis with an interactive whole genome comparison from evolutionary perspective. Allows building of phylogenetic trees on different criteria, analysis of single genome and retrieval of information for individual genes.
	Macromolecular structures	
www.rcsb.org/pdb	PDB (Protein Data Bank)	Main primary database for 3D structures of macromolecules (obtained from *X*-ray crystallography and NMR) such as proteins, RNA, DNA and various complexes.
www.biochem.ucl.ac.uk/bsm/cath	CATH (Class Architecture, Topology, Homologous super-family)	Is a hierarchical classification of protein domain structures which clusters them in four major structural levels.
www.scop.mrc-lmb.cam.ac.uk/scop	SCOP (Structural Classification of Proteins)	Database for classifying hierarchi-cally the 3D structures of proteins. It is updated as new structures are deposited in PDB.
	Integrated databases	
www.ebi.ac.uk/interpro www.expasy.ch/srs5	Inter Pro and SRS (Sequence Retrieval System)	EBI runs as SRS service for anyone. Indexes a large number of database, provides a well-defined web-interface, allows programs on websites to create links for SRS at EBI.
www.ncbi.nlm.nih.gov/entrez	ENTREZ	Provides gateways to DNA and protein sequences, genome mapping data, 3D macromolecular structures and the PubMed bibliographic database. SRS provides search facilities for a large number of databases and provides links between them. Provides well defined web-interface to define links to query ENTREZ.

The above table does not give the exhaustive list of the websites of various databases. For more details an internet search may be adequate. It is also useful to consult some of the books on bioinformatics like Mount (2001), Rastogi et al. (2004), Westhead et al. (2003), Lesk (2003) etc. for more details on account of possible data analysis problem.

In order to understand biological processes properly the building blocks of life, the basic macromolecules must be well understood. Before starting the next section, it is appropriate that some definitions and a description of life's base molecules and functions be given briefly. For more details the reader may refer to Sen et.al (2006), Harper's illustrated Biochemistry (latest edition) etc.

Some Basics

1. **DNA** (Deoxyribo Nucleic Acid)/RNA (Ribo Nucleic Acid) is a polymer of nucleotides, that contains genetic information (genetic codes), to control biological processes. DNA consists of two strands of poly-nucleotides each consisting of deoxyribose residue, a phosphate group and nitrogenous base.
2. **Nitrogenous bases** are four in number: A – Adenine, T – Thymine, C – Cytosine and G – Guanine. Thymine (T) is usually substituted with U – Uracil in RNA.
3. **Codons** Triplets of the first 4 bases are called codons and there are $4^3 = 64$ codons.
4. **Amino acids:** Codons map into 20 amino acids (AA's) known as

   (i) Arginine, (ii) Histidine, (iii) Lysine, (iv) Aspartate, (v) Glutamate, (vi) Glycine, (vii) Serine, (viii) Threonine (ix) Tyrosine, (x) Cysteine, (xi) Asparagine, (xii) Glutamine, (xiii) Alanine, (xiv) Valine, (xv) Leucine, (xvi) Isoleucine, (xvii) Proline, (xviii) Tryptophan, (xix) Phenylamine, (xx) Methionine

   As evident 64 codons form 20 AA's. Thus more than one codons are used to form almost each of the AA. The details may be seen in Jain (2008).
5. **Natural Classifications of 20 amino acids (AA's)** are based on their following four properties:

   (A) Chemical properties

   (a) **Acidic** (Asp, Glu) (b) **Aliphatic** (Ala, Gly, Ile, Leu, Val)
   (c) **Amide** (Asn, Gln) (d) **Aromatic** (Phe, Trp, Tyr)
   (e) **Basic** (Arg, His, Lys), (f) **Hydroxyle** (Ser, Thr)
   (g) **Imino** (Pro), (h) **Sulfur** (Cys, Met)

   (B) Functional Properties

   (a) **Acidic** (Asp, Glu) (b) **Basic** (Arg, His, Lys)
   (c) **Hydrophobic non-polar** (Ala, Ile, Leu, Met, Phe, Pro, Trp, Val)
   (d) **Polar uncharged** (Asn, Cys, Glu, Ser, Thr, Tyr, Gly)

   (C) Charge Properties

   (a) **Acidic** (Asp, Glu) (b) **Basic** (Arg, His, Lys)
   (c) **Neutral** (all other AA's)

(D) Structural Properties

(a) **Ambivalent** (Ala, Cys, Gly, Pro, Ser, Thr, Trp, Tyr)
(b) **External** (Arg, Asn, Asp, Gln, Glu, His, Lys)
(c) **Internal** (Ile, Leu, Met, Phe, Val)

For more details, the reader may consult Harper's Biochemistry, Jain (2008) and Sen et al. (2006). It is conventional to write nucleotides in lower case. The AA's are abbreviated by their first three letters e.g. Gly for Glycine, except the following AA's, which have other specific abbreviations:

(i) Isoleucine – Ile (ii) Glutamine – Gln
(iii) Asparagine – Asn (iv) Tryptophan – Trp

How hereditary information is stored, passed on and implemented is perhaps the fundamental problem of biology. Three types of maps have been essentially used for this purpose: 1. Linkage map of genes, 2. Banding patterns of chromosomes, 3. DNA sequences.

DNA sequences of the four nitrogeneous bases are dealt directly with stored hereditary information in its physical form.

DNA molecules are long, *linear*, chain molecules containing a message in a four letter alphabet (a, t, c, g) (cf. Lesk (2003). Even for microorganisms, the message is long, say of the order of 10^6 characters.

In DNA, the molecules comprising the alphabets are chemically similar and the structure of DNA is, to a first approximation uniform.

6. Proteins are made up of strings of AA's. They show a great variety of 3-D conformations, necessary *to support their diverse structural and functional roles*. The *AA sequence* of a protein *dictates its 3-D structure*.

For each natural AA sequence under normal physiological environment, there is a *unique stable, native state. If normal physiological conditions are disturbed, the protein unfolds to biologically inactive structure*. It will re-adopt the native structure indistinguishable from the original state. This *spontaneous folding to native state* is a *giant leap of nature* from *one dimensional world* of genetic and protein sequences to *3-dimensional world* inhabited by humans.

7. The translation of DNA sequences to AA sequences is very simple to describe logically, as it is specified by the genetic code.

Folding of polypeptide chain into precise 3-D structure is very difficult to describe logically. The translation requires immensely complicated machinery of ribosome, tRNA's and associated molecules. But protein folding occurs spontaneously. After these basics now the section on sequence analysis is considered.

22.3 Sequence Analysis

Let the term sequences (in biology) be defined first. According to Rastogi et al. (2004) these are the strings of characters, for convenience of understanding and performing various mathematical functions.

A general rule is that sequence determines shape, and shape determines function. So *by a study of sequence similarity*, eventually *one hopes to discover or validate similarity in shape and function*. Since there exists the paradigm:

- DNA sequence determines protein sequence,

- Protein sequence determines protein structure,
- Protein structure determines protein function,

most of the organized activities of bioinformatics are focused on analysis of data related to these processes.

Motivation of Sequence Analysis

A large variety of biological problems involve sequence or string of nucleotide or aminoacids. *Sequence alignment* can be useful for discovering information related to functions, structure and evolution. Thus *it may be used* for:

1. Restructuring long sequences of DNA from overlapping fragments.
2. Determining physical and genetic maps from probe data under various experiments' protocol.
3. Storing, retrieving and comparing DNA strings.
4. Comparing two or more strings for similarities to find related proteins.
5. Exploring frequently occurring patterns of nucleotides.
6. Finding informative elements in protein and DNA sequence.
7. Identifying an unknown sequence.
8. Finding other members of multistage families.

The basic idea is that similar sequences produce similar proteins. Thus to predict the characteristic of a protein using only its sequence data, the structure function information on known protein and similar sequences available in database can be used.

Example 22.1: When considering protein folding, it usually suffices that two protein sequences are identical at 25% of their positions for 3-D structures to be almost identical. Classical example is the establishment of an association between cancer and uncontrolled cell growth. This was established by comparing the sequence of a cancer associated gene against sequence of protein known to be responsible for cell growth. The high correlation between these two sequences proved the connection between cancer and cellular growth (Source: Westhead et al. (2003), p. 60).

Again some *terms related to sequence analysis are defined* before discussing the procedure of analysis.

Identical: When a corresponding character is shared between two species or populations, that character is called identical.

Similarity is the degree to which two species or populations share identities. In sequences, similarity has both, a quantitative and a qualitative aspect.

Similarity measure gives a quantitative answer, indicating the degree of similarity between two sequences.

Homologous: When characters are similar due to common ancestry, they are said to be homologous.

Analogous: When characters are similar due to convergent evolution they are analogous.

Homology is, therefore, not synonymous with similarity. Homology is a judgment and similarity is a measurement.

An *alignment* is a mutual arrangement of two sequences, which is a sort of qualitative answer. It exhibits where the two sequences are similar and where they differ.

An *optimal alignment* is one that exhibits the most correspondences and the least differences.

It has been noted that DNA sequences are constructed from only four nucleic acids denoted by a, t, g, c. So any sequence that consists of a mixture of those four alphabets will show some similarity with another similarly constructed sequence. It is desired to *distinguish between* this type of '*chance*' *similarity* and the one that is the result of a *real evolutionary and/or functional relationship.*

Therefore, a proper *way of quantifying similarity is needed.* The following points must be considered before doing so.

The *resemblance* of two DNA sequences may be because *all the contemporary genetic material* has one *ancestral ancient* DNA.

According to this theory, during course of evolution mutations occur, resulting in differences between contemporary species. Most of these changes are due to local mutations, which modify DNA sequence in a specific manner. Local modifications between nucleotide sequence or more generally, between strings over an arbitrary alphabet can be either (a) an *insertion* of a letter or several letters to the sequence, (b) a *deletion* of a letter or more from the sequence. (c) *substitution* of an existing sequence by another sequence.

Insertion and deletion are the reverse of each other or are reciprocal actions. Since insertion and deletion of the same character one after the other results in the original sequence. These two, in short, are called *Indel.*

Distance between two given strings is the *minimal sum of weights* for a set of mutations transforming one into the other.

Similarity derives its definition from the concept of one ancestral ancient DNA: by *assigning weights* corresponding to *resemblance.* Given *two strings* the *similarity* between them is *the maximal sum of such weights.*

Quantifying similarity begins with an alignment.

Example 22.2: Consider the sequences given by

S_1: AATTGATTGCGCATTTAAAGGG

S_2: AACTGA- - -CGCATCTTAAGGG

(Source: Westhead et al. (2003), p. 60).

In this the two sequences are aligned or equivalenced by writing the DNA bases of S_1 above those of S_2. AA, TGA, CGCAT, T and AAGGG of S1 align with identical bases in S_2. After AA in S_1, T of S_1 has C in S_2. After CGCAT, T of S_1 has again C in S_2 and A of S_1 has T of S_2. Thus there are 3 mismatches, 16 letters match and there is a gap of 3 characters in S_2 against TTG of S_1.

The gap of 3 in S_2 may be due to insertion of 3 characters TTG in S_1 or may be due to deletion of TTG from S_2. Thus this gap is an *indel.* Unless the ancestral sequence is available it can not be distinguished whether there is insertion in S_1 or a deletion in S_2.

If the two sequences are simple as S_1 and S_2, it is easy to find the best alignment. However, if the two sequences are long and unrelated it is much more difficult.

The special computational methods are required to find the best alignment.

Sequence Matching Methods

There are following major methods of sequence matching:

1. Global alignment
2. Local alignment and
3. Gap penalty.

1. Global alignment is applicable when there are two strings S and T of roughly same length and question is to find what is similarity (or difference) between the two.

Global alignment is done across the entire length to include as many matches as possible upto and including the sequence end.

2. Local alignment is applicable when there are two strings S and T, and question is to find the maximum similarity (or minimum difference) between a substring of S and a substring of T. Also to determine what are these most similar substrings.

Sometimes the two strings may not be highly similar as a whole, but may contain regions that are highly similar.

Thus the task is to find and extract a pair of regions, one from each of S and T, that exhibit high similarity. This is called local alignment or local similarity.

Example 22.3: Consider two strings S: a b c x d e x and T : x x x c d e. (Source: Rastogi et al. (2004), p. 140).
Let the two be aligned as T : x x x c d e
Let a **score 2** be given to each match and **-1** to each mismatch.
Then the substring α : c x d e and β : c d e of S and T, respectively, have optimal alignment.

In many biological applications, *local similarity* is far more meaningful than global similarity. This is true for comparing long stretches of non-coding DNA, since only small regions within those strings may be related.

When comparing protein sequences, local alignment is also critical because proteins from very different families often share the same structural or functional subunits. Local alignment is an appropriate tool for searching such modules.

3. Gap penalty can be applied on strings of different lengths also. Let there be two strings S and T. Question is to define a gap as any maximal, consecutive run of spaces in a single string of a given alignment, and the length of the gap as the number of *indel* operations on it. It is desired to determine the similarity between the two strings, for a given weight function for gaps (depending on their lengths).

To measure value of an alignment, scores are given for matches, mismatches and spaces.

Gaps help create alignments that better conform to underlying biological models and more closely fit patterns that one expects to find in meaningful alignment.

The number of continuous gaps and number of spaces are to be taken into account, while calculating an alignment mark.

Example 22.4: Consider the following two strings S and T aligned as

S : a t c - - g c – t g c a c c
T : a – c g t g c t t - - - c c.

This alignment has 4 gaps with total 7 spaces. It can be further described as having 7 matches, no mismatch, 4 gaps with 7 spaces.

Methods (or Algorithms) of Alignment

As has been already mentioned similarity of two sequences is measured by determining quantitatively the measure of sequence similarity and difference. For this first step is to align the two sequences. It has been already noted that for short sequences and closely related ones, aligning is fairly straight forward. But for long sequences and if they are not closely related, it is much more difficult. In this case, consider the following methods of alignment.

(A) Dot matrix method is useful for simple alignment, does not show sequences or produce optimal alignment.
(B) Dynamic programming is a computational method, which produces optimal alignment by starting an alignment from one end (as in dot matrix), then keeping track of all possible best alignments to that point.

(C) Heuristic methods are fast computational machine based methods. These may not be as accurate as dynamic programming methods.

There are several books which give in detail various methods. Some of the good books are by Rastogi et al. (2004), Mount (2001), Lesk (2003) etc. In this chapter some methods are described briefly as follows:

1. One method is to write the two sequences and slide one over the other. At each position note the number of matches. The alignment with the largest number of matches will be taken as optimal.

Example 22.5: Consider the following two sequences

S_1: g g c t a a c g a c c g and S_2 = g c t t a c c c g c c g

which are first put S_2 after S_1. S_1 is above S_2.

By sliding the S_1 towards right, suppose at the following positions the number of matches are counted.

```
ggctaacgaccg →         |
                       | Number of matches = 1
          gct tacccgccg|

  ggctaacgaccg →|
                | Number of matches = 7
  gcttacccgccg  |

    ggctaacgaccg →|
                  | Number of matches = 3
gcttacccgccg      |
```

It appears that the alignment with 7 matches is best of the above three alignments.

2. The same can be done by using a matrix also. The sequences are put on columns and rows i.e. horizontally and vertically on the matrix. On the positions of elements i.e. at (i, j)th position in the matrix put a character if there is a match and leave blank all the other positions.

   Instead of writing a character at the positions of matches a 'dot' may be placed at respective positions. A diagonal stretch of dots will indicate the regions where the two sequences are similar.

Example 22.6: For the sequences S_1 and S_2 of Example 22.5, the matrix using 'A' for a match, is obtained as:

S_1

		g	*g*	*c*	*t*	*a*	*a*	*c*	*g*	*a*	*c*	*c*	*g*
	g	A	A						A				A
	c			A				A			A	A	
	t				A								
	t				A								
	a					A	A			A			
S_2	*c*			A				A			A	A	
	c			A				A			A	A	
	c			A				A			A	A	
	g	A	A						A				A
	c			A				A			A	A	
	c			A				A			A	A	
	g	A	A						A				A

For detecting more distant similarities it may be better to use a large sequence of say 20, 30 or 50 bases and specify the percentage of matching (say 55%).

Similarity measure associates a numeric value with the pair of sequences, such that higher value indicates greater similarity.

A distance measure also associates a numeric value with a pair of sequence such that higher value implies smaller similarity.

Two ways of measuring distance between two character strings are: (a) Hamming distance and (b) Levenshtein distance.

Hamming distance defined between two sequences of equal length is the number of positions with mismatching characters.

Example 22.7: Consider

(a) Seq. 1 : a t c g and Seq. 2 : c t a g

(b) Seq. 1 : a a t a t g c t a g c t a

Seq. 2 : t a c t a g c t t a a t c

Counting the number of mismatches, the Hamming distance (Hd) for

(a) Hd = 2

(b) Hd = 8.

This measure is very useful in some cases but it is not suitable for sequences of unequal lengths.

Levenshtein distance or edit distance between two sequences, not necessarily of equal lengths, considers the simple one character edit operations. The distance is the minimal number of edit operations required to change one string into another.

An edit operation is a deletion, an insertion or alteration of a single character in either sequence.

Example 22.8: (a) Consider two sequences

S_1 : c g – t a c and S_2 : a g t t c c

In these, alteration of first c to a and a to c in S_1 and insertion of t in the blank space, makes the two sequences identical. Hence Levenshtein distance (Ld) = 3.

(b) Consider S_1 : a a t t g a t t g t t g c g c a t t t a a a g g g

S_2 : a a c t g a - - - c g c a t c t t a a g g g

Ld = 15.

The gap in S_2 could result either from insertion of 3 characters in S_1 or from deletion of 3 characters from S_2. Unless the ancestral sequence is available, these possibilities can not be distinguished. So the gap is also called *indel*.

Scoring schemes: All the sequence analyses, in general depend on scoring matrices. The scoring system must account for residue substitution and insertion or deletion. The gaps in a sequence involve gap penalties.

One scoring scheme may be to assign positive score (say 1) for matching, negative score (say – 1) for mismatching and gap penalty to be subtracted for each gap. Gaps and gap penalties are necessary to obtain best alignment.

Three policies are in common use:

(a) *Constant penalty* say 1 or A for each gap.

(b) *Proportional penalty* is proportional to the length (x) of the gap (= Bx)

(c) *Affine penalty* is of the form $A + Bx$, where A is *gap opening* penalty and B is *gap extension* penalty for each space in gap.

Sometimes the penalty is taken as $A + B (x - 1)$. So unless $x \geq 2$, the gap extension penalty is not added to gap opening penalty.
Scoring Matrix (SM) is important because (i) choice of SM can strongly influence outcome of sequence analysis. (ii) Biological equivalence of SM is implicitly related to principles of evolution. (iii) Elements of SM specify the weight to assign a given comparison

(a) by cost of replacing one residue with another
(b) by a measure of similarity for replacement.

Similarity is used for database searching and *distance* is more applicable for reconstruction of phylogenetic tree.

Maximization of similarity is same as *minimizing a distance.* Similarity matrices and distance matrices are convertible one into another by using some mathematical transformation described in Chapter 21, Section 21.3.

Hd and Ld given above measure the *dissimilarity* of sequences. *Similar sequences* give *small distance* and *dissimilar* sequences give *large distance.* It is common to define *scores* as measure of sequence similarity. So *similar sequences* give *high score* and *dissimilar sequences* give *low scores.*

Computational Methods of Alignment

As already mentioned these methods are required if the sequences are long and not closely related. Two methods based on dynamic programming are commonly used. These methods are:

(a) Needleman-Wunsch (NW) method and
(b) Smith-Waterman (SW) method.

Both the methods are closely related but the *main difference* is that *NW algorithm finds global similarity* and *SW method finds local similarity.*

NW alignment tries to cover as much of the sequence as possible starting at leftmost end of one of the two sequences and ends at rightmost end of the other sequence.

SW method is mostly used because real biological sequences are often not similar over their entire lengths, but only in local positions.

e.g. (i) genes of different organisms with similar exons but dissimilar introns.
(ii) proteins which share only certain domains.

Computational methods of dynamic programming result in an alignment score that reflects degree of relatedness of two sequences.

Another method is *percentage of identically aligned residues* (PIAR). For the two sequences of Example 22.8 (b) PIAR = 7/22 × 100 = 32%.

Algorithm for Pairwise Alignment

Though based on dynamic programming, there are *three variants* of the alignment problem:

(i) entire sequence against entire sequence
(ii) region of one sequence against entire other sequence
(iii) region of one against region of other sequence.

Global Alignment or NW Alignment

The global alignment was first applied to biological sequences by Needleman and Wunsch (1970). Their logic was based on the following points:

(a) Over the course of evolution, some positions undergo base and amino acid substitutions, insertion or deletion. Because insertion and deletion are rare compared to base substitution, it requires to penalize gaps more heavily than mismatches. So using scores of +1 for match, –1 for mismatch and –2 for each gap introduced, the similarity measure (Sim) is calculated by

$$\text{Sim} = +1\ (n_m) - 1\ (n_{mm}) - 2\ (n_{gaps}) \tag{22.1}$$

where n_m: number of matches, n_{mm}: number of mismatches and n_{gaps}: number of gaps

The alignment with highest value of Sim is the optimal alignment.
NW used the concept that the solution to the large problem is simplified if solution to a smaller subset of larger problem is known. The method is based on the following principles:

1. Sequence S of length m is written at 0th column (or as row elements vertically), and sequence T of length n is written as 0th row (or as column elements on the top horizontally).
2. All possible alignments are contained in the score matrix

$$A_{(m+1)(n+1)} = ((a_{ij})),\ i = 0, 1, 2, ..., m;\ j = 0, 1, 2, ..., n \tag{22.2}$$

3. The optimal alignment will be the path through array that has the highest score i.e. largest matches and fewest mismatches and gaps.
4. Alignment problem requires to enter score at every position (i, j) in A i.e. to find $a(i, j)$. For this one has to look at three adjacent cells in A, viz. $a(i, j-1)$, $a(i-1, j-1)$ and $a(i-1, j)$

Steps given below are followed for calculation of $a(i, j)$ in A.
Step 1: One or more gaps are added to the beginning of either sequence by running across the top of the array and progressively adding a gap penalty as score in each 0th row and 0th column elements.
Step 2: Applying three scoring rules i.e. cost of three edit operations calculate

$$a(i, j) = \max\begin{cases} a(i, j-1) - 2 \\ a(i-1, j-1) + p(i, j) \\ a(i-1, j) - 2 \end{cases} \tag{22.3}$$

where $i = 1, 2, ..., m$; $j = 1, 2, ..., n$ and $p(i, j)$ is score for a match.

Also put an arrow pointing back to one of the adjacent cell that gave higher score or to more than one adjacent cells, if there is a tie.
Correspondence between individual edit operations and steps between adjacent cells in A are

$(i-1, j-1) \rightarrow (i, j)$ corresponds to substitution of $s_i \rightarrow t_j$

$(i-1, j) \rightarrow (i, j)$ corresponds to deletion of s_i from S

$(i, j-1) \rightarrow (i, j)$ corresponds to insertion of t_j in S at position j.

Calculations of $a(i, j)$ starts from $a(1, 1)$ and goes up to $a(m, n)$.
Step 3 is *construction of optimal alignment*. When calculations are complete $a(m, n)$ is the optimal similarity measure of the sequences S and T.

Start at the last position $a(m, n)$ and work backwards to $a(1, 1)$ using the arrows (or pointers). The path gives the optimal alignment.

The path giving the optimal alignment may not be unique. Some times one may get more than one optimal alignment. NW method or algorithm is applied for global alignment. For local alignment the SW alignment is used, as described below.

Local Alignment or SW Alignment

Smith-Waterman's dynamic programming algorithm is used for local alignment. This method gives highest scoring local match between two sequences. (cf. Smith and Waterman (1981 a, b))

Instead of using NW algorithm the SW algorithm can be used for matching two sequences, which have matched regions (i) that is only a fraction of their lengths (ii) that have different lengths and (iii) where one sequence is a fragment of the other.

Scoring system of SW method is different from NW method as given below:

(i) Scoring system must have negative scores for mismatches.
(ii) When a scoring matrix element (i.e. $a(i, j)$) becomes negative, it is set to zero. This effectively terminates the alignment upto that point.

The optimal alignments are obtained by starting at highest scoring positions in the completed score matrix and following a track back path upto a zero score.

For more details and computational examples, the books by Rastogi et al. (2004), Mount (2001) and Lesk (2003) may be consulted.

Merits and Demerits of Computational Methods

1. Merit is that the method is guaranteed to give global optimum. It will find the best alignment score with no approximation, given the choice of substitution matrix $P = ((p(i, j))$ and gap penalty.
2. Demerit is that many alignments may give the same optimal score and none of them may correspond to biologically correct alignment.
   For example, W. Fitch and T. Smith while comparing α and β-chains of chicken haemoglobin, found 17 alignments giving the same optimal score. Only one of these 17 is correct on the basis of the structures. There are 1317 alignments with scores within 5% of optimal score.
3. Another technical demerit is, time required to align two sequences of lengths n and m is proportional to $n \times m$. So dynamic programming method is not suitable for searching an entire database of sequence for a match to a probe sequence.

It is even less convenient for 'all-against-all' alignment than it is for matching of a probe sequence with length of entire database.

22.3.1 Protein Sequences

It is well known that evolutionary mechanism of gene duplication allows organisms to acquire redundant copies of genes. The redundant copies are then free to evolve new functions and become *homologous genes* with different functions.

Protein sequences can be aligned to maximize AA (amino acid) identities, but this will not reveal distant evolutionary relationships. Proteins (as already discussed) are formed from the naturally occurring 20 AA's. The changes in nucleic acid occur due to evolution but protein coding sequences evolve under strong functional constraints. *Changes in these sequences generally survive only if they do not have deleterious effect on the structure and function of the protein.*

Usually, *protein coding genes evolve much more slowly* than most other parts of genome, because of the need to maintain protein structure and function.

Evolutionary changes in protein sequences tend to involve *substitutions between AA's with similar properties*, so that structure and function of protein is not affected. A great deal of bioinformatics is, therefore, concerned with the detection of evolutionary relationships between protein sequences. For analyses of protein sequences variety of *substitution scoring schemes* have been proposed.

1. The simplest scheme is to assign +1 to identical AA pairs and 0 to non-identical pairs.
2. Assign score +1 for a match within a class of similar physicochemical type of AA's and –1 for residues in different classes.
3. One may desire a more precise substitution score from a combination of properties of AA's. Alternatively, one may try to *let the proteins teach an appropriate scoring scheme.*
   M.O. Dayhoff (1978) did this first, by collecting substitution frequency statistics in the protein sequences then known. Her results were used for many years but now due to availability of much larger set of sequences, newer matrices have been constructed.
   A likely change should score higher than a rare one: is the basic principle. But if there have been multiple substitutions at certain sites, this will bias the statistics. To avoid this sufficiently similar sequences should be used for estimating relative frequency of changes. In this case, one can assume that no position has changed more than once.
4. Scoring matrices have been constructed by $p(i, j)$ elements. A value from scoring matrix is to be added for AA substitution, instead of adding +1 for a match and –1 for a mismatch.

PAM and BLOSUM Scoring Matrices

The two popular scoring methods for protein sequences are PAM (Percent Accepted Mutation) and BLOSUM (BLOcks SUbstitution Matrix).
As suggested by MO Dayhoff it is

(a) based on explicit evolutionary model
(b) represents a specific evolutionary distance and
(c) ranges from identical to completely random.

The 'accepted' in PAM means fixed population and is, therefore, a more complex process than simply mutation.

PAM is a measure of sequence divergence. PAM1 implies 1% accepted mutation (cf. Lesk (2003)). Thus two sequences PAM 1 apart have 99% identical residues AA's. For sequences within PAM1 level of divergence, it is likely that there have been no more than one change at any position.

PAM 250 represents 250% mutation, i.e. an average of 2.5 accepted mutation per residue – a very distant relationship.

PAM approach can be summarized as

- Carefully examine the kinds of mutations that occur in closely related protein sequences i.e. at short evolutionary times.
- Extrapolate these differences to greater mutational distance/longer times.
- Tabulate actual mutations by looking at proteins that are sufficiently closely related that there is no ambiguity in alignment.
- Sequences no more than 15% different may be chosen so that changes can be thought of as a single evolutionary step.
- Consider a tree to correctly count changes.

The relation between PAM score and percentage sequence identity is given by the following table.

Table 22.2 PAM score and percentage sequences identity (Source: Lesk (2003), p. 174)

PAM score	0	30	80	110	200	250
% identity	100	75	50	60	25	20

The PAM_{250} (of MO Dayhoff) expresses scores as log odds values given by

$$\text{Score of mutation } i \leftrightarrow j = \left(\log_{10} \frac{\text{Observed } i \leftrightarrow j \text{ mutation rate}}{\text{Mutation rate expected from AA frequencies}} \right) \times 10 \tag{22.4}$$

Multiplication by 10 on RHS of (22.4) avoids decimal point. If value of PAM score is 2 for a mutation $C \leftrightarrow S$, then

$$.2 = \log_{10} (\text{relative expectation value of mutation}) \tag{22.5}$$

Hence

$$\text{Relative expectation frequency of mutation} = 10^{.2} = 1.6 \tag{22.6}$$

Equation (22.6) implies that in related sequences, the mutation would be expected to occur 1.6 times more frequently than occurring randomly from AA frequencies.

Probability of two independent mutations = Product of the probabilities of the two mutations (22.7)

But by using logarithm, addition of the probabilities (or PAM scores) of mutations will be on RHS of (22.7). Thus will lead to computational ease.

BLOSUM was developed by S. Henikoff and J.G. Henikoff (1992, 1994a, b) for scoring substitutions in AA sequence comparison. Their goal was to replace Dayhoff PAM matrix with another that would perform best to identify distant relationships, making use of much larger data than Dayhoff. BLOSUM matrices are based on the BLOCKS database of aligned protein sequences.
BLOSUM is (a) based on empirical frequencies.
(b) with narrower range than PAM matrix.
PROSITE is a set of motifs.

In proteins short and localized patterns, sometimes identify molecules that share a common function, even if there is no obvious overall relationship between their sequences. PROSITE is a collection of these 'signature' patterns.

BLOSUM approach has been summarized in Rastogi et al. (2004) as follows:

- Locally align each feature to get blocks (which are locally conserved regions i.e. more constrained regions likely to be related to structure/function). Blocks contain sequences at all different evolutionary distances and may be highly biased.
- Dealing with bias and distance.

(i) Cluster all sequences with less than $X\%$ identities.
(ii) Count as 1 sequence all the clustered sequences.
(iii) if X is 100, it removes the identical sequences.
(iv) if $X < 100$, it reduces the weight on closely related sequences.

- Calculate substitution frequencies and log-odds matrix. This gives a BLOSUM X table
- (i) $X = 62$ gives BLOSUM 62. It clusters more than 62% identical sequences.
  (ii) $X = 80$ gives BLOSUM 80. It clusters more than 80% identical sequences.

Calculation of scores are done in following steps:
Step 1: Calculate the ratio

$$R = \frac{\text{Number of observed pairs of AA's at any position}}{\text{Number of pairs expected from the overall AA frequencies}}$$

Step 2: As in PAM, scores are log odds i.e. $(\log_{10} R) \times 10$. (22.8)
Step 3: To avoid overweighting closely-related sequences, Henikoffs replaced groups of proteins, that have sequence identities higher than a threshold by either a single representative or a weighted average.

The threshold 62% produces the commonly used BLOSUM 62 substitution matrix. This is offered by all the programs as an option and is the default in most. BLOSUM matrices which have replaced Dayhoff matrix for most applications.

Scoring of insertions/deletions or 'gap weighting'

To form a *complete scoring scheme* for alignments, *in addition to the substitution matrix*, a way of scoring gaps is also needed.

For alignment of DNA sequences, CLUSTAL-W recommends use of identity matrix for substitution and gap. *10 is for initiation of gap as* gap penalty and *0.1 for each gap extension.*

For *aligning protein sequences*, the recommendations are to use *BLOSUM 62 matrix for substitutions* and *gap penalties 11 for gap initiation and 1 for gap extension by 1 residue.*

Lesk (2003) has given the PAM 250 and BLOSUM 62 matrices. For more detailed discussions, the reader may consult it and books by Rastogi et al. (2004) and Westhead et al. (2003). In the next subsection a brief discussion on the software tools for sequences alignment is given.

22.3.2 FASTA and BLAST

Sequence similarity searches are required to extract sequences from a database that are similar to a query sequence. Dynamic programming algorithms (the NW and SW) are guaranteed to find the best alignment of two sequences with a given scoring scheme. The process is often quite slow and takes hours for search of a large database.

A better and faster method focuses on short regions of 'local' similarity. FASTA and BLAST (Basic Local Alignment Search Tool) are used for this purpose, because of their high speed than SW-algo for local alignment.

FASTA, developed by Lipman and Pearson in 1985, considers exact matches between short substrings. If a significant number of such exact matches are found, FASTA uses dynamic programming (DP) algo to complete optimal alignment.

This approach allows to *trade off speed for precision: the larger the number of parameters chosen, smaller the exact matches.*

This method is faster but looses precision i.e. it becomes less likely that the optimal alignment contains enough exact matching words of a length, which is under the control of user. The default value is commonly 2 for protein sequences and 6 for DNA sequences.

Experience shows that with sensibly chosen parameters, FASTA misses very few cases of significant homology.

BLAST, developed by Altschul et al. in 1990, is another method based on similar idea. It focuses on no gap alignment of a certain fixed length.

Instead of requiring exact matches, BLAST uses a scoring function to measure similarity (instead of distance). For proteins, one can argue that segment pairs with no gaps and high similarity measure indicate regions of functional similarity.

For a given threshold parameter BLAST reports to the user all database entries which have a segment pair matching with query sequence that scores higher than the threshold parameter.

If the scoring function used has a probabilistic interpretation, BLAST can also give an assessment of the statistical significance of the matches it reports.

Multiple Sequence Alignment (MSA)

1. MSA illustrates relationships between two or more sequences. When the sequences involved are diverse, the conserved residues are often key residues associated with maintenance of structural stability or biological function.
2. Multiple alignments can reveal many clues about protein structure and function.
3. It is a tool to determine levels of homology, hence relatedness between members of a series of globally related sequences.
4. It is important for finding similar domains in a set of sequences for doing phylogenetic analysis.

The goals of MSA are

(a) high sensitivity – picking up even very distant relationships.
(b) high selectivity – minimizing number of sequences reported that are not truly homologous.
(c) If one has to find the evolution and the sequence conservation in a group of sequences, they are to be compared simultaneously by MSA.

How is MSA Done?

The basic process is same as pair wise alignment. There are conventions regarding scoring of multiple alignment. Pair wise alignments can be merged when they align one sequence to all others, when a linear order of the given sequence is maintained or when sequences pairs with pair wise alignments form a tree.

Global and Local Alignment problems are also tackled.

Manual processes can not work for large problems. So some of the computer analysis methods used are: (1) Two-step progressive alignment method (Clustal and Pileup approaches) Clustal W (for Global) Block Maker (for Local) and (2) Automated tools (MACAW, MEME).

The details of these and some more procedures may be looked in the books on bioinformatics such as Rastogi et al. (2004).

22.4 Application of Hidden Markov Model (HMM)

Apart from the two tools mentioned above for a fast and efficient search of sequences in databases, especially the protein sequences, the newly developed Hidden Markov Model is also found to be very useful.

It is a class of probabilistic graphical model, used for about the past three decades, to model problems of speech recognition, ion channel recording, optical character recognition, modeling of coding and non-coding regions of DNA, of protein binding and to model, align and analyze entire protein families.

HMM is a stochastic model which retains the statistical properties of the observed real world data, and accurately models its source. HMM has also the ability to simulate the source.

Gene finding or gene prediction in DNA is also one of the quite important problem of computational biology. Completely sequenced genomes have become readily available and it is desired to extract biological knowledge from these data. One may explain molecular interactions that occur in cells and may define important cellular pathways, using HMM.

Theory behind the HMM, various approaches and applications may be found in books by Baldi and Brunak (2003), Jones and Pevzner (2005) and Ewens and Grant (2006). A point wise tutorial on this topic is also available on website www.cfar.umd.edn/~kanungo given by Tapas Kanungo. In this section a brief theoretical background on the lines of T. Kanungo is given and an application on gene finding is given.

Theory of HMM

As the name suggests, HMM is related to Markov Chain. So a Markov chain is described first.

Markov Chain: Consider a system which can be described at time t as being in any one of a set of N distinct states 1, 2, ..., N. At various discrete time points say t = 1, 2, ..., T (the time points may not necessarily be equally spaced on time scale), the state of the system may change, with a specific probability.

Let X_t denote the family of random variables denoting the state of the system at time t. Then $[X_t = i]$ is the event that the system is in state i at time t.

Such a system is said to be a Markov chain if

$$\text{Prob}\left[X_{t_n} = j/X_{t_{n-1}} = i,\ X_{t_{n-2}} = i_{n-2} = ...;\ X_{t_1} = i_1\right]$$

$$= \text{Prob}\left[X_{t_n} = j/X_{t_{n-1}} = i\right] = p_{ij}(n) \tag{22.9}$$

(22.9) implies that the probability of the system being in a state j at time t_n depends only on the state of the system just preceding the t_n, not on the earlier past states.

Without loss of generality, the state of system may be changing with change in some other scale, say geographical area, on length etc. So t may be used as a default index for each index set. One can use the subscript n of t as n^{th} time point and X_n may be used to denote the state of the system at discrete observation time point t_n. The equation (22.9) can be, therefore, simplified as

$$\text{Prob}\left[X_n = j/X_{n-1} = i, X_{n-2} = i_{n-2}, ..., X_i = i_1\right]$$

$$= \text{Prob}\left[X_n = j/X_{n-1} = i\right] = p_{ij}(n) \tag{22.10}$$

$p_{ij}(n)$ = Probability of transition from state i to j in one step at time n.

If the one step transition probability does not depend on the time point n, the Markov chain is said to have property of stationarity. In symbols, if

$$p_{ij}(n) = p_{ij} \tag{22.11}$$

the Markov chain is stationary.

The one step transition probability matrix (TPM) is given by

$$P = ((p_{ij})) \tag{22.12}$$

where p_{ij} satisfies the following constraints

$$p_{ij} \geq 0 \text{ and } \sum_j p_{ij} = 1 \tag{22.13}$$

In words, (i) the elements of P are non-negative and
(ii) the row sums of P are 1.

Some Results

1. *Probability of system being in a set of states i.e.*

$$\text{Prob}\left[X_1, X_2, ..., X_T\right] = \text{Prob}\left(X_T/X_{T-1}\right).\text{Prob}\left(X_{T-1}/X_{T-2}\right)......\text{Prob}\left(X_2/X_1\right)\text{Prob}\left(X_1\right)$$

$$= \text{Prob}\left[X_1\right] \text{Prob}\left[X_2/X_1\right]...... \text{Prob}\left(X_T/X_{T-1}\right) \quad (22.14)$$

$$= a_1\, p_{12}\, p_{23} \cdots\cdots p_{T-1,T} \quad (22.15)$$

2. *The Probability distribution is given by*

$$a_i = \text{Prob}\ [X_t = i], \quad i = 1, 2, ..., N \quad (22.16)$$

Thus $A = \begin{pmatrix} a_1 \\ a_2 \\ \cdot \\ \cdot \\ \cdot \\ a_N \end{pmatrix}$ gives the probability distribution of the system being in various states.

3. *Probability that the system remains in state i for exactly d time units is given by*

$$p_i(d) = a_i p_{ii}^{d-1}(1 - p_{ii}) \quad (22.17)$$

4. *Expected value of the duration d in state i is given by*

$$E(d) = \sum_{d=1}^{\infty} d\, p_i(d) = 1/(1 - p_{ii}) \quad (22.18)$$

HMM

The HMM is more general and more flexible than Markov chain. HMM also has a set of N states 1, 2, ..., N which are not observable. The observations are probabilistic function of states. There exists a transition probability with which there occurs a transition of one state to another. States being unobservable or hidden, the name Hidden Markov Model is given.

Definition

(i) N is the number of hidden states

(ii) The sequence of states visited by system is denoted by

$$Q: \{q_1, q_2, ..., q_t ...\}$$

State of the system at time t is q_t which may be one of 1, 2, ..., N.

(iii) The state emits an observable symbol from a set V: {1, 2, ..., M} O_t is the emitted symbol at time t.

(iv) The state transition probability matrix (TPM) is given by $P = ((p_{ij}))$, where

$$p_{ij} = \text{Prob}\ (q_{t+1} = j/q_t = i)\ 1 \le i, j \le N \quad (22.19)$$

(v) The observation emission probabilities, for each state j and symbol k in V is given by $B = ((b_j(k)))$, where

$$b_j(k) = \text{Prob}\left(O_t = k/q_t = j\right);\ k = 1, 2, \ldots\ldots, M;\ j = 1, 2, \ldots\ldots, N \tag{22.20}$$

(vi) The initial state distribution is given by

$$A = \begin{pmatrix} a_1 \\ a_2 \\ \cdot \\ \cdot \\ \cdot \\ a_N \end{pmatrix} \text{ where } a_i = \text{Prob } (q_t = i);\ 1 \le i \le N \tag{22.21}$$

(vii) The HMM may be denoted by $\lambda = (P, B, A)$ (22.22)

HMM is commonly used for solving the following three basic problems:

1. Given observation $O = (o_1, o_2, \ldots., o_T)$ and model $\lambda = (P, B, A)$, efficiently compute Prob (O/λ).
   - Hidden states complicate the evaluation of the above probability.
   - Given two models λ_1 and λ_2, this can be used to choose the better one.

2. Given observation $O = (o_1, o_2, \ldots, o_T)$ and model λ find the optimal state sequence $q = (q_1, q_2, \ldots, q_T)$.
   - Optimality criterion has to be decided (e.g. whether it should be maximum likelihood).
   - 'Explanation' for the data is to be given.

3. Given $O = (o_1, o_2, \ldots, o_T)$, estimate model parameters $\lambda = (P, B, A)$ that maximize Prob (O/λ).

For computing Prob $(o_1, o_2, \ldots, o_T/\lambda)$ the following algorithm (algo) is used:

Algorithm
Let $q = (q_1, q_2, \ldots, q_T)$ be a state sequence.
Assuming that the observations $o_1, o_2, \ldots, o_T$ are independent

$$\text{Prob }(O/q, \lambda) = \prod_{t=1}^{T} \text{Prob }(o_t/q_t, \lambda) = P(O/q, \lambda)$$

$$= b_{q_1}(o_1)\, b_{q_2}(o_2) \ldots b_{q_T}(o_T) \tag{22.23}$$

Probability of a particular state sequence is:

$$\text{Prob }(q/\lambda) = a_{q_1} p_{q_1 q_2}\ p_{q_2 q_3} \ldots\ldots\ p_{q_{T-1} q_T} = P(q/\lambda) \tag{22.24}$$

Also

$$\text{Prob }(O, q/\lambda) = \text{Prob }(O, \lambda)\ \text{Prob }(q/\lambda) \tag{22.25}$$

Enumerate paths and sum probabilities:

$$\text{Prob}\ (O/\lambda) = \sum_q P(O/q, \lambda)\ P(q/\lambda) \tag{22.26}$$

The Forward Algorithm to evaluate the probability in (22.26) is as follows:
Step 1: Define forward variable α_t (i) as:

$$\alpha_t(i) = \text{Prob}\ (o_1, o_2, \ldots\ldots, o_t, q_t = i/\lambda) \tag{22.27}$$

Step 2: $\alpha_t(i)$ is the probability of observing the partial sequence $(o_1, o_2, \ldots, o_t)$ such that the state q_t is i.

(i) Initialize $\alpha_1(i) = a_i\ b_i\ (o_1)$; $i = 1, 2, \ldots, N$
(ii) Use the following to subsequent α's

$$\alpha_{t+1}(j) = \left[\sum_{i=1}^{N} \alpha_t(i)/o_{ij}\right] b_j(o_{t+1}) \tag{22.28}$$

$$j = 1, 2, \ldots, N$$

Step 3: Hence compute $\text{Prob}\ (O/\lambda) = \sum_{i=1}^{N} \alpha_T(i)$ See Fig. 22.1.

Backward algorithm has the following steps:
Step 1: Define backward variable $\beta_t(i)$ as:

$$\beta_t(i) = \text{Prob}\ (o_{t+1}, o_{t+2}, \ldots, o_T/q_t = i, \lambda) \tag{22.29}$$

$$i = 1, 2, \ldots, N$$

$\beta_t(i)$ is the probability of observing the partial sequence $(o_{t+1}, o_{t+2} \ldots, o_T)$ such that the state q_t is i.
Step 2:

(i) Initialize $\beta_T(i) = 1$
(ii) Now, it can be seen that

$$\beta_t(i) = \sum_{j=1}^{N} p_{ij} b_j(o_{t+1}) \beta_{t+1}(j),\quad 1 \le i \le N, t = (T-1), \ldots\ldots, 1 \tag{22.30}$$

See Fig. 22.2.

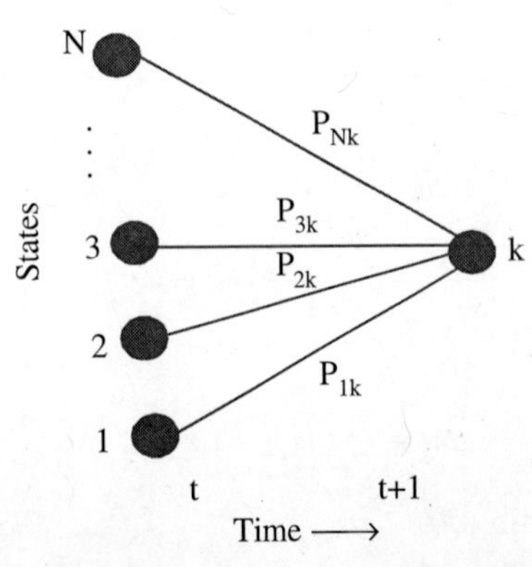

Fig. 22.1. Forward procedure: intuition.

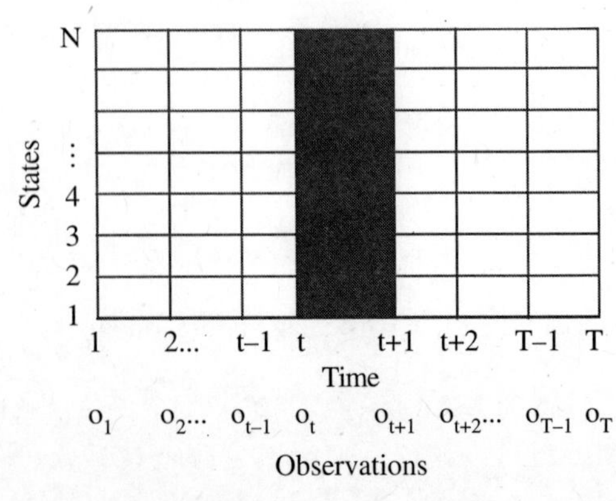

Fig. 22.2. Backward algorithm.

Solution to Problem 2 using HMM is obtained from the following steps:
Step 1: To choose the most likely path, find the path $(q_1, q_t, \ldots, q_T)$ that maximizes the likelihood:

$$\text{Prob}\ (q_1, q_2, \ldots, q_T/O, \lambda)$$

Step 2: Solution is obtained by *Dynamic Programming*. For that define:

$$\delta_t(i) = \max_{q_1, q_2, \ldots, q_{t-1}} \text{Prob}\ (q_1, q_2, \ldots, q_t = i, o_1, o_2, \ldots, o_t/\lambda) \tag{22.31}$$

$$i = 1, 2, \ldots, N$$

Then $\delta_t(i)$ is the highest probability path ending in state i
By induction we have:

$$\delta_{t+1}(j) = \max_i [\delta_t(i)\ p_{ij}].b_j(o_{t+1});\ \ j = 1, 2, \ldots, N \tag{22.32}$$

Viterbi Algorithm for finding highest probability path is as follows.
Step 1: Initialization:

$$\delta_1(i) = a_i b_i(o_1), \text{ and } \psi_1(i) = 0$$

$$1 \le i \le N \tag{22.33}$$

Step 2: Recursion:

$$\delta_t(j) = \max_{1 \le i \le N} \left[\delta_{t-1}(i)\ p_{ij}\right] b_j(o_t) \tag{22.34}$$

$$\psi_t(j) = \arg\max_{1 \le i \le N} \left[\delta_{t-1}(i)\ p_{ij}\right]$$

$$2 \le t \le T,\ 1 \le j \le N \tag{22.35}$$

Step 3: Termination:

$$P^* = \max_{1 \le i \le N} [\delta_T(i)] \tag{22.36}$$

$$q_T^* = \arg\max_{1 \le i \le N} [\delta_T(i)] \tag{22.37}$$

Step 4: Maximum probability path (state sequence) is obtained by backtracking as

$$q_t^* = \psi_{t+1}(q_{t+1}^*),\quad t = T-1, T-2, \ldots, 1 \tag{22.38}$$

Solution to Problem 3 is obtained as follows:
It is desired to estimate $\lambda = (P, B, A)$ to maximize Prob (O/λ)

No analytical method exists because of complexity. Only iterative solution is given using Baum-Welch (BW) Algorithm.

The steps for BW Algorithm are:
Step 1: Let initial model be λ_0.
Step 2: Compute new λ based on λ_0 and observation O.
Step 3: If log Prob (O/λ) – log Prob (O/λ_0) < DELTA, stop.
Else set λ as λ_0 and go to step 2.

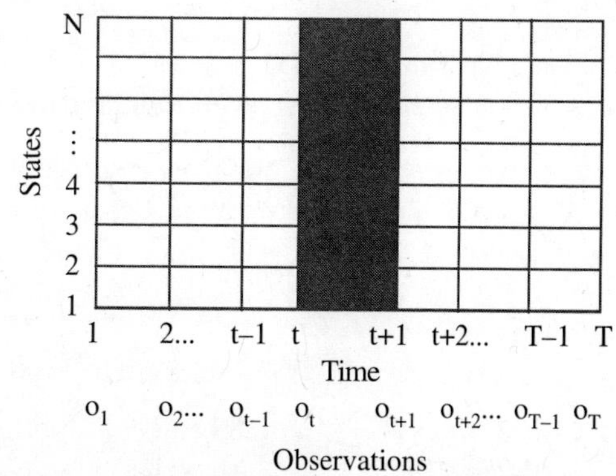

Fig. 22.3. Viterbi algorithm.

Prior to BW algo, Baum-Welch preliminaries are to be followed as:

1. Define $\xi(i, j)$ as the probability of being in state i at time t and in state j at time $t + 1$.

$$\xi_t(i, j) = \frac{\alpha_t(i) p_{ij} b_j(o_{t+1}) \beta_{t+1}(j)}{P(O/\lambda)}$$

$$= \frac{\alpha_t(i) p_{ij} b_j(o_{t+1}) \beta_{t+1}(j)}{\sum_{i=1}^{N} \sum_{j=1}^{N} \alpha_t(i) a_{ij} b_j(o_{t+1}) \beta_{t+1}(j)} \quad (22.39)$$

   where $\beta_t(i)$ is given by (22.29).
2. Define $\gamma_t(i)$ as probability of being in state i at time t, given the observation sequence.

$$\gamma_t(i) = \sum_{j=1}^{N} \xi_t(i, j) \quad (22.40)$$

3. Then $\sum_{t=1}^{T} \gamma_t(i)$ is the expected number of times state i is visited, and
4. $\sum_{t=1}^{T-1} \xi_t(i, j)$ is the expected number of transitions from state i to state j.

Baum-Welch: Update Rules are:

1. π_i = expected frequency in state i at time $t = 1$ is $\gamma_1(i)$. (22.41)
2. $\bar{p}_{ij}$ = (expected number of transition from state i to state j)/(expected number of transitions from state i):

$$\bar{p}_{ij} = \frac{\Sigma \xi_t(i, j)}{\Sigma \gamma_t(i)} \quad (22.42)$$

3. $\bar{b}_j(k)$ = (expected number of times in state j and observing symbol k)/(expected number of times in state j)

$$\bar{b}_j(k) = \frac{\Sigma_{t, o_t = k} \gamma_t(j)}{\Sigma_t \gamma_t(j)} \quad (22.43)$$

   Following these, solution to problem 3 can be obtained.

An Application of HMM

Das (2005) applied HMM on gene finding problem with following steps:

- Developed 3 HMM for three regions of DNA viz. EXON, INTRON and Inter-genic Region
- Trained these with multiple observation sequences using Baum-Welch Algorithm
- Tested each model with test data sets by finding the score

Dimensions

	Initial State Prob.	Symbol Emmision Prob.	TPM
EXON	89 × 1	89 × 4	89 × 89
INTRON	10 × 1	10 × 4	10 × 10
Intergenic region	10 × 1	10 × 4	10 × 10

It appears from Table 22.4 that for various regions gene identification is done quite correctly using HMM.

Table 22.4 Performance of models during training

Model	Seq. length	No. of seq.	Correct %
EXON	40-250	250	99.9
INTRON	40-200	150	99.0
Intergenic	80-400	250	99.9

22.5 Microarray Data

Microarrays are simply ordered sets of DNA molecules of known sequence fixed to solid surfaces. A 2D array, typically on a glass, filter, or silicon wafer, upon which genes or gene fragments are deposited or synthesized in a predetermined spatial order allowing them to be made available as probes in a high-throughput, parallel manner, is named as microarray.

Usually rectangular, they can consist of a few hundred to hundreds of thousands of sets. Each individual feature goes on the array at precisely defined location on the substrate. The identity of the DNA molecule fixed to each feature never changes. Scientists use that fact in calculating their experimental results.

Microarray analysis permits scientists to detect thousands of genes in a small sample simultaneously and to analyze the expression of those genes. As a result, it promises to enable biotechnology and pharmaceutical companies to identify drug targets – the proteins with which drugs actually interact. Since it can also help identify individuals with similar biological patterns, microarray analysis can assist drug companies in choosing the most appropriate candidates for participating in clinical trials of new drugs. In future, this emerging technology has the potential to help medical professionals select the most effective drugs, or those with the fewest side effects, for individual patients.

Working of Microarray

Each microarray holds a DNA sequence that is a reverse complement to the target RNA sequence. Although the specific protocols differ, the microarray approach first involves isolating RNA or messenger RNA from appropriate biological samples, making the RNA (or a copy of it) fluorescent, hybridizing it to the microarray, washing off the excess and scanning the microarray under laser light. This can be explained as follows:

The labelled RNA is hybridized to a microarray for a period of time, after which the excess is washed off and the microarray is scanned under laser light. With oligonucleotide microarrays, for which all probes have been designed to be theoretically similar with regard to hybridization temperature and binding affinity, each microarray measures a single sample and provides an absolute measurement level for each RNA molecule, although this absolute measurement might not correlate exactly with concentration in terms of micrograms per unit volume. With cDNA microarrays, for which each probe has its own hybridization characteristic, each microarray measures two samples, and provides a relative measurement level for each RNA molecule. Regardless of the technique, the end result is 4,000-50,000 measurements of gene expression per biological sample. As a complete experiment might involve anywhere up to hundreds of microarrays, the resultant RNA-expression data sets can vary greatly in size.

Use of Microarrays

Microarray usage in drug discovery is expanding, and its applications include basic research and target discovery, biomarker determination, pharmacology, toxicogenomics, target selectivity, development of prognostic tests and disease-subclass determination. As the cost of microarrays continues to drop, it is clear that microarrays are becoming more integral to the drug discovery process. In addition to the obvious use of functional genomics in basic research and target discovery, such as finding genes expressed in significantly different patterns across samples, there are many other specific uses in this domain. These include:

1. Biomarker determination: to find genes that correlate with and presage disease progression, but are easier to measure and follow in clinical trials;
2. Pharmacology: to determine differences in gene expression in tissues exposed to various doses of compounds;
3. Geno-toxicity, to find gene-expression patterns in a model tissue or organism exposed to a compound and their use as early predictors of adverse events in humans;
4. Target selectivity, to define a compound by the gene-expression pattern it provokes in a target tissue and then compare it with other compounds using these patterns;
5. Prognostic tests, to find a set of genes that accurately distinguishes one disease from another; and
6. Disease-subclass determination, to find multiple subcategories of tumors in a single clinical diagnosis.

Many free and commercial software packages are now available to analyze microarray data sets, although it is still difficult to find a single off-the-shelf software package that answers all functional-genomics questions. As the field is still young, when developing a bioinformatics analysis pipeline, it is more important to have a good understanding of both, the biology involved and the analytical techniques rather than having the right software.

Raw Data from Microarrays

Microarray data comprises images from hybridized arrays representing hybridization signal intensities for individual spots. These may be generated by single fluorescent, dual fluorescent, radioactive or colorometric labels and the recording methods differ in each case.

Most array experiments nowadays are carried out using glass spotted microarrays or high-density oligonucleotide chips. In both cases, the substrate has minimal autofluorescence as a fluorescent probe can be used. The data are acquired by confocal laser scanning of the hybridized array at the appropriate excitation wavelength and recording at the appropriate emission wavelength (or channel). A single label is used for oligonucleotide chips, so different arrays. However, in the case of spotted arrays, two probes can be labeled with different fluorophores and hybridized simultaneously to the same array allowing different fluorophores and hybridized simultaneously to the same array allowing differential gene expression to be monitored directly.

Precautions in Data Acquisition

It is essential to record signal intensities from individual spots accurately as errors in data recording cannot be detected or corrected at a later stage. The signal must be corrected for background hybridization. Controls must be used when comparing results across different arrays.

Need for Softwares in Analysis

Software for reading microarrays is generally provided with the recording equipment (scanner or phosphorimager) but manual adjustment is necessary to compensate for variations in array manufacture. DNA arrays may contain many thousands of features. Therefore, data acquisition and analysis must be automated.

The software for initial image processing is normally provided with the scanner (or phosphorimager). This allows the boundaries of individual spots to be determined and the total signal intensity to be measured over the whole spot (this is termed the signal volume).

Locating spots precisely can be a problem, particularly if there is distortion on the array surface. Therefore, it is often necessary to align and grid manually. This is very important because signal intensities can vary across individual spots and the shape and size of different spots may not be uniform.

Most importantly, the signal intensity has to be corrected for background which may be generated by nonspecific hybridization, autofluorescence, dust and other contaminants or poor hybridization technique (e.g. partial dehydration). The background can vary over the array surface, so signal intensities must be normalized for local background values. Correction for background is difficult when the signal intensity for a particular spot is itself very low.

Control features should be included on the array to measure nonspecific hybridization and variable hybridization across arrays. For example, Affymetrix GeneChips incorporate a set of mismatching oligonucleotides for each perfect match set to determine nonspecific hybridization. Controls are particularly important where duplicate arrays are being manufactured or experimental protocol can influence the signal intensities on different arrays. The bottom line is that errors and artifacts introduced before or during data acquisition cannot be detected or corrected at a later stage.

Gene Expression Matrices

The raw data from microarray experiments are converted into tables known as gene expression matrices. The rows represent genes and the columns represent experimental conditions. The data in the table are signal intensities, representing relative levels of gene expression.

Whichever platform is used, the aim of data processing is to convert the hybridization signals into numbers, which can be used to build a gene expression matrix. The hybridization signals are intensity of colour spots in a microarray image that is obtained from microarray chip as initial output of microarray experiment. This image is analyzed by special software that measures the intensity of each spot in terms of a ratio of red and green colours which forms that particular spot. So now hybridization signals are converted into numbers. Gene expression matrix is made of these numbers.This step is known as quantitative analysis. Essentially this matrix can be regarded as a table in which the rows represent genes (the different features on the array) and the columns represent treatments or conditions used in the experiment.

22.6 Probabilistic Modelling and Clustering of Microarray Data

In this section the role of statistics in the field of bioinformatics is considered. In the subsection 22.6.1 the significance of sequence similarity search is considered in brief. For details the books already referred for sequence analysis may be consulted. In the subsection 22.6.2 applications of statistical techniques in microarray data analysis are briefly described. An example of cluster analysis of microarray data has been given as an illustration.

22.6.1 Statistical Significance of Search (or Alignment)

It is desired to know when an identified sequence similarity is significant i.e. when it represents a real biological or evolutionary relationship, not a chance similarity between unrelated sequences.

Hence to know what values of similarity measure (= S) indicate real biological relationship, one requires to know $P[S \geq s]$ (= p) for unrelated sequences.

If p is very small, it implies that match is significant of real biological relationship. The similarity search methods report the calculated p-value.

The experimenter has to choose the threshold value of p for significance.

A related quantity reported by most of the search softwares is E (Expected) Value.

$$E = \text{expected frequency of the event } [S \geq s]$$
$$= \text{number of scores of values} \geq s \text{ that would be expected to occur by chance}$$

p and E values can be calculated for two matched sequences or for the entire database search.

The *relation of S and p* depends on the distribution of scores from the control population (non-normal).

A rough guide to interpret p-values is

$p \leq 10^{-100}$	Exact match
$10^{-100} < p \leq 10^{-50}$	Very nearly identical
$10^{-50} < p \leq 10^{-10}$	Closely related, homology certain
$10^{-5} < p \leq 10^{-1}$	Usually distant relatives
$p > 10^{-1}$	Match probably insignificant

E is found by $N\,p$, where N : size of database probed.
p does not depend on N. $0 < p < 1$ but $0 < E < N$

A rough guide to interpret E is

$E \leq .02$	sequences probably homologous
$.02 < E \leq 1$	homology can not be ruled out
$E > 1$	match may be expected to be by chance.

These are based on the assumption of extreme value distribution

$$F(x) = e(-e^{-ax-b}) \qquad \text{in FASTA.}$$

Statistics are useful guide, but not a substitute for careful thinking over the results.

Sensitivity and Specificity

Evaluation of results of a database search is usually done in terms of two complementary measures, *sensitivity* and *specificity*.

Suppose threshold values of E or p are chosen. Sequences with significant similarity are called *hits*.

Table 22.5 Sensitivity and specificity

Real state	Search partitions	
	Hits (+ve)	Non-hits (–ve)
True	n_{tp} (real relation)	n_{tn} real relation no hit
False	n_{fp} no relation but hit	n_{fn} no relation no hit

$$\text{Sensitivity} = S_n = n_{tp}/(n_{tp} + n_{fn}) \tag{22.44}$$

$$\text{Specificity} = S_p = n_{tp}/(n_{tp} + n_{fp}) \tag{22.45}$$

Sensitivity attempts to report all the real hits. Lower the false negatives n_{fn}, higher the sensitivity. Specificity attempts to report only the real hits. Lower the false positive n_{fp}, higher the specificity.

$$S_n = 1 \text{ if } n_{fn} = 0,\ S_p = 1 \text{ if } n_{fp} = 0$$

Thus high values of S_n and S_p indicate that the procedure is working well.

22.6.2 Cluster Analysis of Microarray Data

In this subsection an application of statistical technique of cluster analysis on microarray data is given. The *K*-means clustering method has been applied by Rai, Maurya and Singh (2006) on a micro-array data on yeast. Data file has 99 genes. $K = 10$ has been chosen to form 10 clusters of genes of yeast cell. The computer program has been developed in C. The result is given below:

DATA FILE NAME:-yeast1.txt

```
*    analyzing file format                              *
*    step 0:-file has 101 row and 10 columns            *
*    step 1:-gene name is present                       *
*    step 2:-gene weight is given                       *
*    step 3:-Experiment weight is given                 *
*    Process of analyzing file format is complete       *
*    file has 99 Genes                                  *
*    file has 7 samples
```

Cluster	No. of genes	Genes in cluster		
cluster 1	4	YBR005W	YBR072W	YCR012W
		YCR013C		
cluster 2	1	YCR055C		
cluster 3	1	YCR057C		
cluster 4	30	YAL003W	YAL025C	YAL035W
		YAL038W	YAL040C	YAR015W
		YBL027W	YBL039C	YBL054W
		YBL072C	YBL099W	YBR029C
		YBR031W	YBR032W	YBR066C
		YBR084C-A	YBR101C	YBR154C
		YBR181C	YBR287W	YCL042W
		YCL059C	YCL064C	YCR007C
		YCR021C	YCR031C	YCR056W
		YBR030W	YBR079C	YBR177C
cluster 5	21	YAL004W	YAL005C	YAL010C
		YAL012W	YAL015C	YAL034C
		YAR027W	YBL067C	YBL108W
		YBR059C	YBR078W	YBR082C
		YBR086C	YBR132C	YBR145W
		YBR178W	YBR196C	YBR214W

Contd...

Continued

Cluster	No. of genes	Genes in cluster		
		YBR222C	YCL040W	YCR048W
cluster 6	1	YCLX03C		
cluster 7	20	YAL018C	YAL055W	YBL009W
		YBL010C	YBL078C	YBL084C
		YBR045C	YBR063C	YBR064W
		YBR148W	YBR168W	YBR184W
		YBR232C	YBR233W	YBR250W
		YBR257W	YBR268W	YBR285W
		YCL048W	YCR002C	
cluster 8	14	YAL054C	YAL062W	YAL067C
		YAR003W	YAR007C	YBL015W
		YBL043W	YBR001C	YBR088C
		YBR186W	YBR231C	YCR005C
		YCR010C	YCR062W	
cluster 9	1	YCR061W		
cluster 10	6	YAL036C	YBL042C	YBR025C
		YBR172C	YBR247C	YCLX02C

Thus cluster analysis has grouped together the genes with similar properties. The biologist can interpret these results.

23. Computer Techniques

Chapter 11 of this book gave an introduction to computers, its brief history, ways of man-machine communication, a brief note on low and high level languages and various operating systems normally available on desktops and laptops.

After describing various statistical techniques, simple as well as complicated, it is appropriate that an overview of problem solving on computer be also given. Some researchers may be interested to develop their own way of solving a computational problem, as there may be large number of repetitions of computing a statistic for different sets of collected data. For these, some idea of programming will be given in the first section of this chapter.

For complex univariate statistical problems and for multidimensional statistical techniques, it has been mentioned that solution to some examples have been obtained by statistical packages. Hence the second section of this chapter contains a brief description on using some statistical packages.

23.1 Programming in FORTRAN and C++

Whenever some problem is to be solved, the researcher should analyze it and decide the sequence of tasks which should be performed. Before starting to write the program, the first and compulsory step is to write various tasks in proper sequence so that solution to the problem is achieved.

This list of tasks in a proper sequence may be called the *algorithm,* which must terminate in a finite number of steps, to solve a problem. Thus one can say that while solving a problem, it should be broken into sub-problems. If any of the sub-problems are also not very simple and easy, it also should be further subdivided so that one knows how to accomplish those. When such *stepwise refinement* of the problem under consideration has been done, the next step is to write the instructions for computer to perform the listed tasks.

Writing the set of instructions for computer to solve a given problem is called programming. With the emergence of so many high level languages, program is written in one of these. Choice of a high level language is done depending on the type of problem to be solved. COBOL was used mostly for business and commercial purposes. A high level language C has become very popular because of its good properties such as:

(i) it is flexible and structured
(ii) it includes certain low level features also
(iii) it can be compiled into small object programs which execute efficiently
(iv) its compilers are available for most of PC's and mainframes
(v) it is largely machine independent. So programs in C can be easily ported from one computer to the other.

For numerical computating problems in science and engineering, FORTRAN, PASCAL and ALGOL are the convenient languages. Of these three, FORTRAN continued enhancement as versions II, III, IV, 77, 90/95. FORTRAN 90, is perhaps the most highly used language for statistical computations. In the next subsection a brief overview of program writing in FORTRAN is given, followed by programming in C^{++}.

23.1.1 Programming in FORTRAN

While starting to write a program for solving a problem, it has been mentioned in the previous section, that an algorithm must be prepared giving the proper sequence in which various tasks are to be undertaken to solve a problem.

Another point which requires proper thinking is, to specify very clearly the elements (or data) on which the tasks are to be performed. These set of elements or data must be provided to the computer through the program. There are different types of data and it must be declared in the program for computer, to store and work on it accordingly.

In fact any program is composed of at least three independent sections (a) Initialization, opening files etc. in the main program (b) Sub-programs (consisting of procedures as functions or subroutines) and (c) Closing of files and end statement, instructing the computer to terminate the execution of program.

If a child has to learn a language for communication, he must learn various characters used in the language. He should know how to combine characters to form meaning full words. Subsequently correct words are combined in a grammatically correct form to convey the thought with correct meaning.

Similarly before writing a program in a high level language the characters which are used, must be known to the programmer. FORTRAN allows the following characters.

(i) Uppercase and lower case English alphabets are treated as equivalent in FORTRAN except when they appear in character constant (52 in all).
(ii) 0-9 decimal digits (10 in all)
(iii) Special characters Δ = + - * / ( ) ' . : $; ! ? " < > _ & % (21 in all).

Thus a total of 83 characters are now supported by FORTRAN. The last 9 special characters are new entrants.

The decimal digits and the 52 alphabets are called alphanumeric characters and the rest 21 are called special characters.

Whatever be the computer language used for programming, some operations on data are always needed to achieve the desired results.

Data handled by FORTRAN are of various types and FORTRAN program must declare its type, to enable FORTRAN give suitable location for its storage, retrieval and performing various operations on it.

As in algebra, in FORTRAN also there are two types of quantities. One which do not change their value throughout the program run, called constants, and the other which are stored in memory, keep changing their values as the program run takes place. These are called variables and are given a name.

Constants may be integer or real. They may be signed + or –. If no sign is given, as in algebra, it is taken as positive.

Variables in FORTRAN have a specific storage location and variable name is the name tag for the memory location containing the value of variable.

Variable is *defined* if a number is stored in its named memory location. A variable is defined either by an input statement (i.e. reading a value of the variable) or by a statement of the type $x = 5.0$.

Variables in FORTRAN may be of the following types: 1. Integer, 2. Real, 3. Double Precision, 4. Complex, 5. Logical, 6. Character.

Integers are signed or unsigned whole numbers, e.g. 34, –7, 1, 0

Real variables contain decimal digits also. Either these appear with a decimal point or in exponent form, e.g. –1.23, 0.2348E07, 54.879, 0.45612E-14.

Integer variable names start with any one of the following 6 characters: i, j, k, l, m, n. Real variable names start with a character other than the said 6. Thus variable names starting from A to H and from O to Z are real. These are also called *floating point* variables. Normally a string of one to six letters or numbers, (first being always a letter) are used to denote a variable name.

Higher accuracy to real numbers i.e. more decimal digits can be achieved by declaring such variables as *Double Precision.*

These types of statements are called type declaration statements. Use of these may change a variable with name starting with I, J, K, L, M, N as real or a normally real variable name to be used in the FORTRAN program as an integer variable.

The syntax of type declaration statement is

Type Variable names (separated by commas).

Type may be one of the following: INTEGER, REAL, LOGICAL, CHARACTER, DOUBLE PRECISION, COMPLEX.

Values of single precision number may be printed in the exponent form as say 0.34 E5 or 0.12 E-14 but a double precision variable is printed as 0.34 D5 or 0.12 D-14. LOGICAL variables can take only two values: .TRUE. or .FALSE. The periods before after the words are must.

Complex variables take the value of the form a+ib and they are denoted in FORTRAN by (a, b) i.e. real and imaginary parts separated by comma and enclosed in parentheses.

This type declaration statements may be more than one in a program and should be used before the first use of the variable.

After defining FORTRAN variables and their types, the programmer proceeds to write FORTRAN expressions, for computation of arithmetic expressions.

A FORTRAN expression is a series of variables and constants connected by arithmetic operators. Following are the valid arithmetic operators in FORTRAN: +, – , *, / and **. These are used for addition, subtraction, multiplication, division and exponentiation (e.g. a^x = a**x), respectively.

In a FORTRAN expression all the variables and constants must be of the same mode.

Integer expressions or real expressions included in parentheses are integer expressions and real expressions, respectively.

Some Remarks

1. Real variable or constant may have an integer exponent but an integer variable or constant can not have real exponent. Thus I**A, K**2.5 are invalid.
2. Result of an integer expression is integer, e.g. if J = 5, K = 2, J/K = 2 not 2.5. The fraction part will be chopped off. Thus if J = 2, K = 3, J/K = 0.

Hierarchy of Arithmetic Operators in Expression

The order of arithmetic operations are: Exponentiation first (**), multiplication (*) and division (/) second, and addition (+) and subtraction (–) last.

All the expressions are examined from left to right and the above hierarchy is followed. e.g. in A + D** F/C + D, D^F is evaluated first, then D^F/C is computed and lastly $A + \frac{D^F}{C} + D$ is calculated.

Hierarchy can be altered by use of parenthesis.

If an expression has a single parenthesis, the expression inside will be evaluated first following the hierarchy of arithmetical operators inside the parentheses. Then the left to right evaluation of the expression will be done following hierarchy rules.

Next step in writing a FORTRAN program is to write a FORTRAN arithmetic statement. It has the following form.

$$\text{Variable name} = \text{Expression} \tag{23.1}$$

Remarks

1. LHS of arithmetic statement can not be an expression.
2. RHS of arithmetic statement, i.e. the expression after '=' sign is evaluated and the value is stored in the variable name on LHS of '=' sign.
3. A real variable may be set equal to integer expression and an integer variable name set equal to a real expression. Thus if

(a) X = J/K and J = 5 and K = 2 then X = 2.0
(b) I = X/Y and X = 7.0 and Y = 3.0 then I = 2

4. A variable name can not be used in an expression unless it is defined. Apart from an input statement (Read), it can also be defined by an arithmetic statement.

Variable name = Value of variable

e.g. X = 2.4, I = 2

5. If all the variables on RHS of arithmetic statement are defined variables, the variable on the LHS is also defined.

Relational Operators and Relational Expressions

There are following *relational operators*: .LT. (less than), .LE. (less than equal to), .GT. (greater than), .GE. (greater than equal to), .EQ. (equal to) and .NE. (not equal to).
Variables joined by relational operators are *relational expressions.*
Result of relational expression is .TRUE. or .FALSE.

Logical Operators and Logical Expressions

The logical operators are:
.NOT. (negating the value of operand), .AND. (gives a result .TRUE. only if both the operands are .TRUE. Otherwise the result is .FALSE.), .OR. (gives a result .TRUE. unless both the operands are .FALSE.)
.NOT. operator precedes the single operand and .AND. and .OR. are used between two operands.
Logical constants or variables, and relational expressions joined by logical operators are called *logical expressions* and give values .TRUE. or .FALSE. as stated above.

Precedence or Hierarchy of Relational and Logical Operators

There is no precedence of relational operators. Left to right rule is followed and if there are parentheses, the inner most parenthesis is executed first.

In logical operators .NOT. is evaluated first, .AND. is the second and .OR. is last to be evaluated. As other operators, the parentheses are evaluated first.

The FORTRAN statements of the form

Variable name = Expression (23.2)

described above are also known as assignment statements.

Input-Output (I/O) statements are also essential part of a computer program. These provide a means of instructing a computer to read data from an input unit for solving a problem and to transfer the results of computation from the computer memory to one of output units, after the execution of the program is over. The statement for the first part is called the input statement and that for the second part is called the output statement. Both are needed in almost every program so are called I/O statements.

I/O statements consist of two parts: an executable part and another a declarative part. The *executable part* commands certain items to be read into the memory or transfer out of it.

The *declarative part* gives information about the exact form in which the data will be found in input file and the form it is to be printed out. The declarative part is called format statement. Format statement is not essential. One may give format free I/O statements also.

Input data is then presented in a flexible standard form and the output is also printed in a standard form.

Format Free or List Directed Read Statement

The general forms are:

(i) READ, List
(ii) READ *, List
(iii) Read (*,*) List (23.3)

The first statement instructs data to be read from the input device. The values of different variables in the list entered on the card are separated by a space. Computer reads and assigns the values sequentially to the listed variable names.

The second statement reads data from the default source, usually the keyboard and assigns values to various variables in the list.

The first '*' in the third statement indicates that data has to be read from keyboard and the second '*' means that FORTRAN is free to decide format of the data.

Types of various variables in the list is determined from the type declaration statement, if any, or the usual convention of variable name used.

Character variables are generally enclosed in pair of apostrophes or quote marks. The complex variable is given by two real variables separated by a comma and are enclosed in a pair of parentheses.

Each READ statement causes a new record of data to be read.

Lists of two READ statements should not be put in the same line. Value of any one variable should not be split in two lines.

Format Free or List Directed Output Statement

The general form of format free output statements are

(i) PRINT, list
(ii) PRINT *, list and
(iii) WRITE (*,*) list (23.4)

The first and second PRINT statements write the listed variables on the default device, usually now the monitor screen, in the same sequence as the variables in the list.

Values of the listed variables are printed in the order of the list. If a value can not be printed in the remaining space in the line, it is carried to the next line, leaving the insufficient spaces blank in previous line. Character data can be split in more than one line, if necessary.

Numerical data are printed with one blank between two values. The character data are not separated in this way from one another.

Formatted I/O Statements

The formatted I/O statements are in general of the form

READ (i, j) list and WRITE (i, j) list (23.5)

i is an integer code giving the input unit for READ or WRITE. j is the label of FORMAT statement which specifies the form in which the data is to be read or written.

Statement label is a 1 to 5 digit number used before a FORTRAN statement to refer to the statement. It should be an unsigned nonzero number. It is written (was punched earlier on cards) in the column 1 to 5 before a statement.

General form of a formatted READ statement is

READ (i, n) list

n FORMAT (s_1, s_2, ...) (23.6)

Here i is I/O unit code and n is format statement number. s_1, s_2 .... are the format specifications for variables in the list.

FORMAT statement is a non-executable statement. Its function is to supply information to the compiler about how the data will be found for reading and how results are to be presented.

The following terms are defined before giving method of writing format specifications.

A Record: This is contents of one 80 column card (nowadays of one line of data file having 80 columns per line). A record for printing is normally a 132 column line.

A Field: A record consists of a number of fields. A field is a set of columns on card or on a line of data file, reserved for one variable. The number of columns in a field is called width of field. e.g. numbers 128 and 2.231 have fields' widths of 3 an 5. Decimal point is also counted in field width. General forms of format specifications are given in the following Table 23.1

Table 23.1 Format specifications for various variable types

Variable types	Format specification	Rule
Integer	In	n is an integer ≥ width of field
Real (floating point)	Fw.d	w and d are integers, w is field width. w ≥ d + 4; d is number of decimal digits (4 = 1 for decimal point +1 for sign of number +1 for a space before number + numeric value before decimal)
Real (exponent form)	Ew.d	w ≥ d + 6, w and d are integers as above. Extra 6 for E, sign of number and of exponent; decimal point, two for exponent
Real (double precision)	Dw.d	w ≥ d + 6, w and d as above.
Alpha numeric or character	Aw	w: width of variable field = number of characters in the variable.
Logical	Lw	L1 or L5 (L1 for T or F, w > 1 for some blanks before the value).

Note: 1. In I and F some extra width should be provided for sign of out put

2. Better to use w = 15 in E and D specifications.

Apart from the format specifications *control edit descriptors* are also required. The format specification do not determine the location of the field within a record where a variable in the list is printed. They do not determine explicitly the gaps between two output variables. It is, therefore, necessary that format statement also contains the carriage control (start of printing location) information also.

Table 23.2 Control edit descriptors

nX	To include n blank spaces
Tn	To start printing from n^{th} column.
TRn	To start printing n columns at the right of last print
'String'	To print the string of characters enclosed in ' '
"String"	To print the string of characters enclosed in " "
/	To leave the current line and begin printing remaining items in the next line.
$r(s_1, s_2 ...)$	To repeat r times the format specifications in the parentheses.

An example of formatted READ may be

```
      READ (5, 10) A, C, X1, NUM, SUM
   10 FORMAT (2X, 2(F 7.2, 2X), E 15.2, I6, 2X, E 15.6)
```

With this format the variables should be entered in the following way:

(i) Two columns should be left blank (due to first 2X).
(ii) Values of A and C should be entered in fields for each as next 7 and 2 spaces. 2 digits are for decimal digits, one place is for decimal point and there are 4 spaces for mantissa of A and C including place for sign if negative. After the 7 spaces width for A two blanks should be left before entering value of C. C also has same field width and 2 blanks are left after value of C.
(iii) Value of X1 is in exponent format with field width = 15 with 2 decimal digits only. If X1 = .13 E – 10, this requires 7 columns (right most) in field of 15 spaces. So left 7 spaces should be left blank and then enter 0.13 E – 10 in the remaining 8 spaces in the field width of X1.
(iv) Value of NUM is to be entered in next 6 columns. The digits in the value of NUM should be right aligned, so that the left spaces are filled with zeros. If a space is left at right one extra zero will be read by computer. So if NUM = 130 it should be entered as ΔΔΔ130 (Δ ⇒ a blank space).
(v) After 2 more blanks for 2X, the value of SUM is entered as real in exponent form, as explained for X1, but there are 6 spaces for decimal digits.

A general form of a formatted WRITE statement is WRITE (i, n) list

$$\text{n FORMAT } (s_1, s_2, \ldots..) \tag{23.7}$$

where i is an output unit code, n is label of FORMAT statement giving format specifications s_1, s_2, ... mentioned in Tables 23.1 and 23.2. An example of formatted WRITE statement is given as

```
      WRITE (6, 15) A, B, X, ITER, SUM
   15 FORMAT (5X, 3 (F 7.2, 2X)/10X, 'ITERATION =',I6, 2X, 'SUM =', E 15.6)
```

With this format statement, the output will be printed as:

(i) 5 columns will be left blank, then A, B and X will be written in floating point mode containing 2 decimal digits and field width of 7. So mantissa will be of 4 digits. 2 blank spaces will be left after writing A and also after writing B and X in the next 7 spaces. If mantissa of A, B and X are of less than 4 digits including sign, the left spaces will be left unused or blank.

(ii) After writing X the line (or card) ends, due to /. ITER and SUM are written in the next line. Due to 10 X after / , 10 blank spaces are left and from the 11th column in 2nd line the character string enclosed in 1st pair of apostrophes will be written and next the value of ITER in field width of 6. So the value will be written such that the extra space, if any, than the number of digits in value of ITER will be left blank on L.H.S. of the printed digits.

Again 2 blank spaces will be left for 2X, then the character string enclosed in ' ' will be printed before the value of SUM in exponent form with 6 decimal digits.
Suppose SUM is obtained in 121 iterations as 12.361453. Thus in the second line the output will be
ΔΔΔΔΔΔΔΔΔΔITERATION = ΔΔ121ΔΔSUM = ΔΔΔ0.123614 E02
It will look like

ITERATION = 121 SUM = 0.123614 E02

Structure of FORTRAN Program

As already mentioned a FORTRAN program has basically 3 parts (a) Main Program (b) Sub programs and (c) Closing of files and end of program.

A brief description of FORTRAN statements has been given. One must also know that how it should be printed.
The following points should be kept in mind to prepare a FORTRAN program file.

1. FORTRAN statements are to be written between 7th to 72nd columns.
2. If one line is not sufficient and a FORTRAN statement needs more space than 7th column to 72nd column of a line, it may be continued in the next line(s) (or card(s)) by entering some character in 6th column of subsequent line(s) of a statement. Some users prefer to give natural numbers 1, 2, ... for the continuation lines for ease of clarity.
3. Columns 1 to 5 are used for label or statement number. These statement numbers may be of 1 to 5 digits integer. Two statements can not have same label in a program segment i.e. in the main or in a subprogram. There should be a unique statement number for each of the labeled statements in the main program or in the subprograms. Same label may appear in two or more program segments.
4. Comment or program description may be given by entering C in the first column. It is advisable to give comment lines to describe the purpose or logic of the program or program segment. There may be more than one comment line within a program to describe what is being done in the following part of a program.
5. Each program has an END statement. The main program has also at least one STOP statement and functions subprogram and subroutine subprograms have at least one RETURN also before END.

Consider now an example of a simple FORTRAN program.

Example 23.1: Suppose it is desired to solve the equation

$$cx - d = e$$

To write a FORTRAN program for solving this problem, first the steps for obtaining the solution should be written as the following algorithm:

1. Read c, d, and e
2. Find solution x = (d + e)/c
3. Write c, d, e and solution for x.

The FORTRAN program for solving this problem is

```
C   SOLUTION OF LINEAR EQUATION
    READ (*,*) C, D, E
    X = (D + E)/C
    WRITE (*, *) C, D, E, X
    STOP
    END
```

Subscripted Variables in FORTRAN

Variables discussed so far were single valued. But there are situations when another type of variables called subscripted variables are also prevalent.

For example an investigator may go to a high school and measure the weights of class 10 students. If W is used as the name of the variable 'weight' one may give the name W for the weights w_1, w_2, ..., w_{30} of 30 class 10 students of the said school.

Thus W is a subscripted variable having 30 values. It is called a subscripted variable.

W_1 is weight of first student, W_2 that of second, ..., W_{30} that of 30th student of class 10. In FORTRAN the subscript is enclosed in parentheses as: W(1),W(2), ..., W(30).

These can also be referred as W(I), I = 1, 2, ..., 30. The variable W in this case takes 30 values.

If heights and weights of the class 10 students are measured and are given the variable name S, then S will be a 30 × 2 matrix. 1st column of S gives heights and 2nd column the weights of 30 students. When subscripted variables are used in FORTRAN, a **DIMENSION** statement is given in the program before the first use of such variable.

One DIMENSION statement may include more than one subscripted variables. This statement gives information to the FORTRAN compiler to provide sufficient number of memory locations for each subscripted variable in the program.

General form of the DIMENSION statement is

$$\text{DIMENSION } V_1\ (n_1),\ V_2\ (n_2,\ n_3),\ V_3\ (n_1,\ n_3,\ n_4)\ \ldots.\ V(n) \tag{23.8}$$

V_1, V_2, V_3 .... V are each subscripted FORTRAN variable names.

Each of the values in parentheses are positive integer constants specifying the maximum value of the subscripts which could be taken by various variables.

An example of dimension statement is

DIMENSION S(30, 2),W(30), X(4, 5)

Control Statements

Normally the statements in the FORTRAN program are executed in the order they are written. However, for some problems it is desired that the order of execution be possible to change than the normal flow.

Such statements which are used to change the order of execution or in other words, move the control of execution as required by programmer, are called Control Statements.

The FORTRAN Control Statements are:

1. *GOTO statement :* Its general form is

$$\text{GOTO n} \tag{23.9}$$

where n is the label of an executable FORTRAN statement.

2. *Logical IF Statement* has the general form

$$\text{IF (e) } s_1 \tag{23.10}$$

$$s_2$$

where e is a logical expression and s_1 and s_2 are FORTRAN statements.
s_1 is any valid executable statement except another logical IF or DO statement.
If e is true s_1 is executed and s_2 is executed next, otherwise only s_2 is executed.

Sometimes it is desired that if e is true, only s_1 be executed not s_2. For this one may use

```
IF (e) GOTO n
s2
.
.
.
n s1
```

3. *The Arithmetic IF Statement* has the general form

$$\text{IF (a.e) } n_1, n_2, n_3 \tag{23.11}$$

a.e. is an arithmetic expression, and n_1, n_2, n_3 are the statement numbers or labels.
If a.e. is negative, control goes to statement number n_1,
If a.e. is zero, control goes to statement number n_2 and
If a.e. is positive, control goes to statement number n_3.

3. *The Computed GOTO Statement* has the general form

$$\text{GO TO } (n_1, n_2, \ldots, n_m), \text{ i} \tag{23.12}$$

where n_1, n_2, …, n_m are statement numbers, and i is a non-subscripted integer variable name.

The control is passed to n_1 or n_2 or …. n_m depending on whether the computed value of i is 1 or 2 or …. or m. The value of i must never exceed m.

Example 23.2: Suppose 30 values of heights of students upto 1 decimal point in cm, are given. It is desired to find mean height using logical IF statement.
The FORTRAN program is written as:

```
C    EXAMPLE 23.2 CALCULATING MEAN HEIGHT OF 30 STUDENTS
     XN = 30.0
C    H is height
     N = 0
     SUM = 0
10   READ (5,20) H
     N = N + 1
     SUM = SUM + H
     IF (N.LT.30) GO TO 10
     AVE = SUM / XN
     WRITE (6, 25) AVE
20   FORMAT (5X, F 10.3)
```

```
25  FORMAT (5X, 'AVERAGE HEIGHT = ', F 10.3)
    STOP
    END
```

REPEATED Calculations

Sometimes problem solving on computer requires that some computations are repeated a specified number of times, values of some of the variables may differ each time. One such problem may be summation of a finite series. In such a case one has to set up a loop, where the control moves back and forth several times and branches out when the specified number of repetitions are over. This is accomplished in FORTRAN by a **DO statement**.

The general form of a DO statement is

```
        DO n I = m1, m2, m3
        s1
        s2
        .
        .
n       sk
```
(23.13)

I is a valid non-subscripted integer variable. m_1, m_2, m_3 are non-subscripted positive integer variables or positive integer constants.

m_1 is the starting value taken by I; m_2 is the largest value of the index I, which after each flow upto statement n is incremented by m_3. If m_3 is not present it is taken as 1.

Hence another valid form of DO statement is

```
        DO n i = m1, m2
        s1
        s2
        .
        .
        .
n       sk
```
(23.14)

As an illustration of DO statement consider calculation of variance of heights of 30 class ten students using DO loop and height as subscripted variable in the following Example 23.3.

Example 23.3: FORTRAN program for calculation of variance of heights.

```
C   CALCULATION OF VARIANCE OF HEIGHTS USING DO LOOP
C   HEIGHT H AS SUBSCRIPTED VARIABLE
    DIMENSION H(30)
    DO 10 I = 1, 30
C   READING HEIGHTS FROM 30 LINES
    READ (5, 11) H (I)
11  FORMAT (2X, F 6.2)
10  CONTINUE
C   CALCULATION OF SUM OF HEIGHT: SUMH
C   SUM OF SQUARE OF HEIGHT: SUMHSQ
    SUMH = 0.0
    SUMHSQ = 0.0
    DO 20 N = 1,30
```

```
    SUMH = SUMH + H(N)
    SUMHSQ = SUMHSQ + H(N) *H(N)
20  CONTINUE
    HMEAN = SUMH / 30.0
    HVAR = SUMHSQ/30.0 - HMEAN*HMEAN
    WRITE (6, 30) HMEAN, HVAR
30  FORMAT (5X, 'MEAN = ', F8.2, 4X, 'VAR (H) ≡', F 10.2)
    STOP
    END
```

A Remark

Reading values of H using DO loop as given in Example 23.3 can be modified by DO type statement given as

```
       READ (5, 10) (H(I), I = 1, 30)
10     FORMAT (10 F 8.2)                                  (23.15)
```

In this 10 values of H are printed in 80 columns of a line. Each height has field width of 8 spaces. 3 lines will be used for creating the input data file. Reading will continue till all the 30 values are read.

A matrix A(10, 5) can also be read using a nested DO statement

```
       Read (5, 30) (A(I, J), I = 1, 10), J = 1, 5)
30     FORMAT (10 F 8.2)                                  (23.16)
```

For J = 1, A(I, J) is read for I = 1, 10, then J is incremented by 1 i.e. for J = 2 A(I, J) is read for I = 1, 2, ... 10. This way J is incremented from 1 to 5. Thus in 5 lines, the 10 elements of A giving columns 1, 2, 3, 4, 5 of A are to be input.

FUNCTIONS and SUBROUTINES

Another important set of programs, called subprograms, are used often in a FORTRAN program.

FUNCTIONS

Some mathematical functions which are extensively used, are included in the compiler's library. Such functions are called in built functions. These are various trigonometric functions, e^x, $\log_e x$, $\log_{10} x$, $|x|$, $|N|$, square root, min $(x_1, x_2, \ldots, x_n)$, max $(x, x_2, \ldots x_n)$, real to integer conversion and integer to real conversion etc.

The list and the specific name of these *built in functions* may be seen in the compiler used or various books on FORTRAN programming may be consulted (cf. Rajaraman (1982), Kumar (1988), Chapman (1999) and Joshi (2003).

A general way of using the function in the program is to write

function name (argument). (23.17)

in any FORTRAN statement.

Apart from the built in functions, sometimes need arises for user defined function(s). In FORTRAN there are two ways of defining functions. One is through FUNCTION statement and another through FUNCTION subprogram.

The *general form of function statement* is

$$\text{function name } (X_1, X_2, \ldots, X_K) = s_1 \quad (23.18)$$

$X_1, X_2, ..., X_K$ are dummy variables and s_1 is a valid FORTRAN expression involving $X_1, X_2, ...,$ X_K.

Within the program the function name with actual arguments is used and the computer computes its value by evaluating s_1 for the actual values of the variables replacing the dummy variables.

While using a function statement it should be declared by giving statement

FUNCTION function name(s). (23.19)

Various function names (may be more than one) which are used as statement function may be listed to inform the compiler. The function statements are generally given after the declaration statements, and before the first program statement.

The function subprogram is a set of program statements which are included outside the main program. The general form of function subprogram is

```
FUNCTION name (a1, a2, ...., an)
..........................
..........................
name = e (a1, a2, ...., an)
RETURN
..........................
name = e (a1, a2, ......, an)
RETURN
END                                              (23.20)
```

The first statement indicates the beginning of FUNCTION subprogram. The comment lines if necessary may be put after the first line. The function name must appear at least once on LHS and an expression of arguments on RHS assigns a value to the function name. The RETURN statement passes the calculated value of function to the calling program.

Calling program may be the main program or any other subprogram where the function name with arguments appears in a statement.

There are certain specific rules which must be followed in writing the function subprogram and giving a calling statement.

Another subprogram is SUBROUTINE subprogram. A function subprogram returns only **one value** to the calling program. The subroutine subprogram can return more than one values also.

Sometimes a value is not returned but some operations are performed. Then also a subroutine may be used. For example a matrix may be needed to be transposed.

The values which are to be returned to calling program are also included as list of arguments. The name of the subroutine does not appear any where in the subroutine subprogram.

The *general form of a subroutine subprogram* is

```
SUBROUTINE  name (list of arguments)
.
.
.
RETURN
.
.
.
END
```

List of arguments are dummy variables in subprogram.

Computation or operation of various tasks in subroutine is evoked by calling program by giving a 'call' statement as

Call name (list of arguments)

The list of arguments includes some input parameters for which the subroutine will do computations or operations and also the variables through which the subroutine will pass the values to calling program.

The position of arguments in calling and subroutine subprogram should match in variable types and also in whether they are input variables or variables giving returned values.

There are definite rules for writing a subroutine sub-program which must be followed. These may be seen in the books on FORTRAN programming already referred.

Following these discussions, at least simple programs can be written by a user. An example of using subroutine subprogram is given below. Examples of function statement and function subprogram may be looked into the above mentioned and other books on FORTRAN programming.

Example 23.4: Subroutine subprogram to calculate mean and standard deviation (sd) of marks in English of class 6 to class 10 students of a school in the September class test with full marks 10.
Let the marks in English be denoted by ES. Suppose there are 20 students in each class and all the students were present. Thus ES may be considered as a matrix of 5 rows and 20 columns.
So ES (5, 20) is English score matrix in which 20 values in rows 1 to 5 are the marks of students of class 6 to 10. 1/2 mark in total is not rounded. So marks are upto 1 decimal place. Subroutine for computing mean and sd will be used in the FORTRAN program.

```
C   MAIN PROGRAM
C   MEAN AND SD OF ENGLISH SCORES OF
C   20 STUDENTS OF CLASS 6 T0 10 OF A SCHOOL
C   ES ENGLISH SCORES, CSE (J) SCORE OF JTH
C   STUDENT OF A CLASS
    DIMENSION ES(5, 20), XMN(5), SD(5), CSE(20)
    READ (5, 10) ((ES(I, J), J = 1, 20), I = 1, 5)
C   READING MARKS CLASSWISE (OR ROWS), 20 SCORES
C   WITH ONE GAP
10  FORMAT (20 (1X, F 3.1))
    DO 20 IC = 1, 5
C   DEFINING CSE(J) : SCORE OF CLASS IC (=1 ⇒ class 6)
    DO 30 J = 1, 20
30  CSE(J) = ES(IC, J)
    CALL MNSD (CSE, N, XMN (IC), SD (IC))
    ICLASS = IC + 5
20  WRITE (6, 40) ICLASS, XMN(IC), SD(IC)
40  FORMAT (5X, 'CLASS=', I3, 2X 'MEAN=', F5.2, 2X,
   1 'STANDARD DEVIATION = ', F8.3)
    STOP
    END

    SUBROUTINE MNSD (X, N, XMEAN, STD)
C   SUBROUTINE FOR CALCULATION OF MEAN & SD OF X
    DIMENSION X(20)
C   SX = SUM OF X SX2 = SUM OF X SQUARE
```

```
      SX = 0.0
      SX2 = 0.0
      DO 40 J = 1, 20
      SX = SX + X (J)
      SX2 = SX + (X(J)) * (X(J))
40    CONTINUE
      XN = N
      XMEAN = SX / XN
      VX = SX2 / XN - XMEAN * XMEAN
      STD = SQRT (VX)
      RETURN
      END
```

As already described till now, analysis of the problem and writing of algorithm is the first step in problem solving on computer. The second step is to write the steps in an appropriate order in a high level language (FORTRAN in this subsection), to solve the problem.

These two steps do not involve a computer anywhere. These are to be done by the user. The third step i.e. running or executing the program involves computer.

A program written in a high level language is not understood by computer, since it understands only machine codes in binary form. This work i.e. translation of program in high level to machine language is done by what is called *compiler*. These are written by professional programmers. The process of changing the high level language program into machine codes is called compilation which gives, what is called an *object program*.

The steps involving computer for solving a problem may be listed as

1. Preparing source program file
2. Compilation of source program
3. Execution of object program

Depending on the operating system (OS) loaded on the computer and the compiler available in the computer system, some command lines are to be given to compile the source program file. The name of the source code file should also be according to the prescribed rule with appropriate extension like .for or .f.

The compiler indicates the syntax errors, which should be carefully corrected for successful compilation. A specific command (depending on the OS on the system) for executing and giving the execution errors listing, is also added. Removing carefully all the execution errors and the compilation errors, called debugging, is the most important and sometimes cumbersome task. Debugging must be completed before results of the problem may be obtained. A user should have a test data and self computed result for testing the computer program or debugging.

Then only the problem solving on computer can be completed. If necessary, for different data sets, the same program may be used with different data files, provided the variables and dimensions match and there is no overflow or underflow.

The compilation errors and execution errors list is provided in the compiler's manual. These must be carefully read or the professionals operating the computer systems may be consulted for debugging.

Readers interested in writing programs can also use FORTRAN libraries for statistical computations. A few such libraries are given below:

(i) *Statlib* site address – www.lib.stat.cmu.edu/apstat/
(ii) *Fortran tools and library* site address – www.fortran.com/tools.html
(iii) *NAG Fortran library* site address – www.nag.co.uk
(iv) Apophenia: a library of statistical functions
These are open source of statistical functions which can be easily accessed.

The given list of FORTRAN program statements is not the exhaustive one. It is a developing language so will have new additions as the needs arise.

So an interested reader must supplement the above subsection with other current books on FORTRAN 90/95 by Adams et al. (1994, 1997), Chapman (1998a, b), Rajaraman (1997) and Joshi (2003) etc.

In the next subsection a brief description of programming in C and C^{++} is considered. The programming technique is basically same whether the high level language is FORTRAN or C or C^{++}. The steps and the components are almost same. However, the syntax and formation of the components such as the main and the sub-program, are slightly and sometimes widely different in form.

Some of these differences are given in the next subsection briefly. An idea of difference in the nature of 'C' program is evident by an example of very simple 'C' program.

C^{++} is also briefly mentioned after C.

23.1.2 Programming in C and C^{++}

Every C program consists of one or more *functions,* one of which must be called main. Execution of program begins by executing the instructions in it. Instructions in C consist of seeming algebraic expressions, certain English keywords such as: *if, else, for, do* and *while* etc. and operators (*unary, arithmetic, relational* and *logical, conditional*, and assignment)
Each *function* must contain

(i) A function *heading*, which is the function name, followed by optional list of *arguments* enclosed in parentheses.
(ii) A list of argument *declarations*, if arguments are included in the heading.
(iii) A *compound statement*, which comprises the remainder of the function.

Arguments are referred as *parameters* and are symbols containing information which is passed between function and other parts of the program.

Each *compound statement* is enclosed within a pair of braces { }. Braces may contain elementary statements (called *expression statements*) and other compound statements which may be nested. Each expression statement must end with ; (semicolon).

Comments (remarks), which serve as an aid to the human readers of the program, are a form of programming etiquette, may appear any where in the program. However, these must be placed within the delimiters /* and */.

Example 23.5: Consider a very simple C program that reads the radius of a circle, calculates its area and writes the calculated area, as given below (source: Gottfried (1991), p. 10)

```
    * include < stdio.h >                          /* Library file access * /
    /* program to calculate area of a circle */    /* Title giving purpose * /
    main ( )                                       /* Function heading * /
{
    float radius, area                             /* variables declaration */
    printf ("Radius = ?");                          /* Output statement (prompt) */
```

```
        scanf ("%f", & radius);                         /* Input statement */
        area = 3.14159 * radius * radius                /* assignment statement */
        printf ("Area = % f", area)                     /* output statement */
    }
```

This program may be written in more than one ways. One may include one more procedure function for calculating area of circle for a given radius passed from main function.

However, the program in C as given above normally does not have comments on RHS of every line. However, the second comment line is normally provided in each program. This program has almost all the *features* of a C program. These are *listed below* for further emphasis for a beginner.

1. It is customary to type program instructions in lower case. Most comments are also typed in lower case, but for distinguishing from instruction statements, may be typed in upper case. Upper case and lower case characters are not equivalent in C.
2. The first line contains a reference to a special file, called *stdio.h*, which contains information that must be included in the program when it is compiled. The required information is automatically handled by the computer.
3. Second line is the comment, which specifies the purpose of the program.
4. Third line is a heading for function *main*, which does not include any arguments in the ( ) parentheses.
5. The remaining five lines of the program are enclosed in { } are the compound statement within *main*.
   (i) The first line in the compound statement is *variable type declaration* as floating point type for radius and area.
   (ii) Remaining four lines are expression statements. The second line (*printf*) generates a request for information, viz. a value for radius. This is entered in the computer via the third line (*scanf*)
   (iii) The fourth line is also an expression statement called an *assignment statement*. This statement causes the area to be calculated from the given value of radius. This *assignment statement* contains an algebraic expression containing arithmetic operator *, which is multiplication operator.
   (iv) The last line (*printf*) causes the calculated value of area to be displayed. The numerical value will be preceded by a brief label: 'Area ='.
6. Each expression statement in the compound statement ends with a semicolon, which is a requirement for all the expression statements.
7. Liberal use of spacing and blank lines separate different parts of the program into logically identifiable components, also indicating subordinate relationships among various instructions. These are encouraged as good programming practices.

When the C program is run the program results in an interactive dialogue. The user has to respond to them if the program so requires.

The program given in Example 23.5 results in

radius = ? **3** 3 is entered by user if the area is required for circle of radius 3.

Area = 28.274309

is printed as the output, which is computed value of area of circle.

Fundamentals of C

There are many popular dialects in C. The standard specification is the ANSI C. As in any language, C also has its *set* of *characters*. There are some *rules to write a statement*. There are *identifiers and key words, allowed data types* (*constants and variables*) single *valued or arrays*. The statements will *involve expressions.*

All these basic elements combine to form a comprehensive program.

The characters set used in C consists of (i) upper case English alphabets (A – Z) (ii) the lower case English letters (a to z) (iii) the numerical digits 0 to 9 and (iv) certain special characters given in Table 23.3.

Table 23.3 Special characters in C

!	*	+	\	"	<
#	(	=	¦	{	>
%	)	~	;	}	/
^	-	[	:	,	?
&	_		'	.	Δ
	(Underscore)	]	(Apostrophe)		(Blank)

@ and $, which are included within strings and comments are also allowed in most versions of C. *Some combinations of characters* called *escape sequences*, are used to indicate special meanings such as backspace, new line and horizontal tap by \b, \n and \t, respectively.

Identifiers in C are names given to various program elements, such as variables, functions and arrays. *Identifiers* consist of letters and digits in any order, but first character must be a letter, preferably a lower case. Uppercase and lowercase letters are not interchangeable in C. Underscore may also be included, as it is also considered a letter. It is often used in the middle of an identifier but rarely used in that way. The following are some *identifiers.*

TABLE x y12 sum_1 –temperature names area tax-rate

An identifier should contain enough characters so that its meaning is clear. There are also certain reserved words in C, which have standard predefined meaning, called *key words.*

These can be used only for their standard purpose in C. The following Table 23.4 gives the key words of C as given by Gottfried (1991).

Table 23.4 Standard key words in C

auto	extern	sizeof	break
float	static	case	for
struct	char	goto	switch
const	if	typedef	continue
int	union	default	long
unsigned	do	register	void
double	return	volatile	else
short	while	enum	signed
Additional Keywords Available in some Compilers			
ada	far	near	asm
fortran	pascal	entry	huge

This list is not an exhaustive one. The reader should consult an appropriate manual of the C compiler loaded on the computer system to be used, for other key words.

Another very important element in a C program is *data* on which processing is to be done by computer.

Data types Several types of data are supported by C. The basic data types are given in the following Table 23.5 and also the memory requirement for each type.

Some basic data types can be augmented with data type *qualifiers* such as *short, long, signed, unsigned.* Use of qualifiers alters the range of data, which also depends on whether the computer word length is 32 bits or 64 or 16 bits. So a user should confirm from manual of the compiler.

Table 23.5 Data types in C

Data type	Notation	Memory requirements (normally)
Integer quantity	Int	2 bytes or 1 words (may vary with compilers)
Single character	Char	1 byte
Floating point number (a number with decimal point and/or an exponent)	Float	1 Word (4 bytes)
Double-precision floating point number (i.e. more significant figures, and an exponent which may be larger in magnitude)	Double	2 Words (8 bytes)

There are four basic types of constants in C, namely: *integer constants*, *floating-point constants, character constants* and *string constants*. Integer and floating-point constants may be denoted by *numeric constants.* Numeric constants in C should satisfy the following conditions.

1. There should not be comma or blank in it.
2. Constant may be preceded by – (negative) sign.
3. Value of constant should be within the range for the compiler, which may vary from one compiler to the other.

If w is word length on a computer, an integer may fall within -2^{w-1} to $+2^{w-1} -1$, whereas an unsigned integer may vary from 0 to $2^w - 1$.

Variables and Arrays are also one of the most important elements of C. A *variable* is used to represent some specific type of information within a designated position. A variable represents a numerical quantity or a character constant, which should be assigned before the variable name is used in the program. A given variable may be assigned different values at various places in the program. An *array* is another kind of variable, that is extensively used in C. It refers to a collection of data items, which all have the same name (the array in C is similar to arrays in FORTRAN). The elements of an array are distinguished from one another by the subscripts associated to the array name. In *C the subscripts* are enclosed in [ ]. An array of n elements has subscripts from 0 to n – 1. The first element is referred to as x[0], the second by x[1] and the nth by x[n – 1]. Arrays in C may be categorized in the following ways:

Integer arrays, character arrays, one-dimensional arrays and multidimensional arrays. For example y[1] [2] is second element of 1st row of a two dimensional array y.

A *general form of a one-dimensional array definition* may be given as

storage-class data-type array[expression] (23.21)

For example some definitions are

(i) int x[1000] (ii) char caste[30] (iii) float ny[12]

Initial values to arrays *may be assigned* in the following general form

Data-type array [expression] = {value 1, value 2, .., value n} (23.22)

value 1, value 2, …, value n are the values of 1st, 2nd, … nth elements of the array, respectively. Some array definitions and assignment of initial values are as given below:

int class[10] = {1, 2, 3, 4, 5, 6, 7, 8, 9, 10}

Static float x[6] = {–3.0, 0., 2.5}

It may be noted that if all the elements of the array are not initialized, the remaining elements are assigned value 0.

Thus the last three elements of array x viz. x[4] = x[5] = x[6] = 0.

A multidimensional array can be defined in a general form as:

$$\text{Storage-class} \quad \text{data-type} \quad \text{array}[e_1]\,[e_2]\,\ldots.\,[e_k] \tag{23.23}$$

$e_1, e_2, \ldots, e_k$ are k positive valued integer expressions giving the maximum number of array elements. In other words $e_i - 1$, i = 1, 2, …, k is the maximum value of ith subscript.

Some examples of array definitions are as given below

float marks[30][5]

static double rec[50][10]

If a multidimensional array definition also assigns initial values, it should be taken care that the order of values listed should be such that the desired value is assigned to various array elements. The general rule is that the last (or right most) subscript increases most rapidly and the first (or left most) the least rapidly. Thus the elements of a two-dimensional array are assigned values by rows. Thus elements of 1st row will be assigned values in the list first, then to the elements of 2nd row and so on.

Thus one can define and initialize a two-dimensional array as

(i) int rec[3][4] = {10, 20, 30, 40, 50, 60, 70, 80, 90, 100, 110, 120}

(ii) Also a group may be formed by enclosing in pair of { } and the above array rec may also be defined as:

```
int rec[3][4] = {
                {10, 20, 30, 40},
                {50, 60, 70, 80}
                {90, 100, 110, 120}
                };
```

As earlier if fewer elements of a row are assigned values in the list, the rest of the row elements are assigned value 0. For example suppose definition of a two-dimensional array is given as

```
int   x[2][5]
= {
   {11, 12, 13, 14},
   {17, 18, 19}
            };
```

Then the 5th element in 1st row i.e. x[0][5] is assigned value 0 and the 4th and 5th elements 2nd row i.e. x[1][4] = x[1][5] = 0.

Declaration in C

A declaration in C associates a specific data-type to one or more variables.
A general form of a declaration is

$$\text{Data-type } v_1, v_2, \ldots, v_k \tag{23.24}$$

where v_1, v_2, ... v_k are the variable names.

For example

```
int    x, rec, class
float  age, income, height
```

Expressions in C

C also uses expressions as other high level languages like FORTRAN. The expressions in C may be of the following form:

(a) it may be a single data item, such as number or a character
(b) it may consist of a single entry such as a constant, a variable, an array element or a reference to function
(c) it may also be a combination of such entities interconnected with one or more *operators*.
(d) it may represent logical conditions which may be either true or false, which are represented in C by 1 and 0, respectively.

Some expressions in C are as given below:
(i) a + b (ii) x = y (iii) c = a + b (iv) r ≤ s (v) u = = v (vi)+ + i
2nd and 3rd are assigning value of y to x, and of a+b to c, respectively.
4th expression takes value 1 if r is less than or equal to s, otherwise 0.
5th expression also takes value 1 if u and v are found equal by comparison otherwise takes value 0.
6th expression increments value of i by 1.

Operators in C

Thus it is very clear that knowing about operators is very essential for programming in C. There are following types of operators in C.

(a) **Arithmetic Operators** are following five

+ - * / and %

In words addition, subtraction, multiplication, division and remainder after integer division (also called *modulus operator*). There is no exponentiation operator (as in FORTRAN). A library function (pow) carries out exponentiation.

Division of one integer by another integer truncates the decimal digits but if at least one is floating point number the result is floating point number.

(b) Unary Operators

C has the following three unary operators which operate on a single operand. So these operators are used before a numerical constant, a variable or an expression.

- , + + and - - (23.25)

The first operator makes a constant or variable or an expression negative.
The second is incremental operator and the third is decremental operator.
Thus + + k increases k by 1 and - - k decreases k by 1.
If the incremental or decremental operator follows the operand, the value of operand is altered after it is utilized. If the operator precedes, the value of operand will be altered before it is utilized for the intended purpose in the program.

(c) Relational and Logical Operators

C has four *relational operators*

(i) < (ii) <= (iii) > (iv) > = (23.26)

which are less than, less than or equal to, greater than, greater than or equal to.

There are also the following two equality operators

(v) = = implies equal to and (vi) ! = implies not equal to (23.27)

Logical operators or logical connectives in C are only two

(i) & & (*logical and*)
(ii) ¦ ¦ (*logical or*)
(iii) ! implies negation of a logical expression (*logical not*)

(d) Conditional Operator

In C there is one conditional operator ? :. An expression making use of conditional operator is called conditional expression.

A conditional expression is written in the form

$$e_1 \; ? \; e_2 \; : \; e_3 \qquad (23.28)$$

e_1, e_2, e_3 are three expressions. In a conditional expression (23.28) e_1 is evaluated first. If e_1 is true i.e. its value is non zero, then e_2 is evaluated and its value becomes the value of the conditional expression (23.28).

If e_1 is false i.e. its value is zero, then e_3 is evaluated and this becomes the value of conditional expression (23.28).

It may be noted that value of the conditional expression is obtained by evaluating either e_2 or e_3.

(e) Assignment Operators

There are six assignment operators in C. They are

= += -= *= /= %=

(i) Most commonly used assignment operator is = , written in the form

v = e (23.29)

where v is a variable and e an expression. This operator assigns the value of the expression e to the variable v.

Multiple assignments of the form

$$v_1 = v_2 = ... = v_k = e \tag{23.30}$$

is also permissible in C. It operates from right to left, i.e. value of the expression e is assigned to v_k. Value of v_k to v_{k-1}, v_{k-1} to v_{k-2} .... and so on lastly of v_2 to v_1.

(ii) + = operator is used in general as

$$e_1 \mathrel{+}= e_2 \tag{23.31}$$

It is equivalent to $e_1 = e_1 + e_2$

(iii) - = is written in general as

$$e_1 \mathrel{-}= e_2 \tag{23.32}$$

which is equivalent to $e_1 = e_1 - e_2$.

(iv) * = is in general written as

$$e_1 \mathrel{*}= e_2 \tag{23.33}$$

which is equivalent to $e_1 = e_1 * e_2$.

(v) /= is in general used as

$$e_1 \mathrel{/}= e_2 \tag{23.34}$$

which is equivalent to $e_1 = e_1 / e_2$

(vi) %= is in general used as

$$e_1 \mathrel{\%}= e_2 \tag{23.35}$$

which is equivalent to $e_1 = e_1 \% e_2$.

In all these e_1 is usually a variable or an array element.

The operator precedence are given below from highest to lowest in Table 23.6 (cf. Gottfried (1991)).

The *unary, conditional and assignment* operators *associate* the operands from *right to left,* whereas other operators associate from left to right.

After knowing about characters, variables, declaration, operators and expressions one becomes capable of writing statements to form a simple complete program in C. A formal brief description of *statement in C* is given now.

Table 23.6 Precedence of operators in C

Operators category	Operators
Unary	- ++ -- ! sizeof (type)
Arithmetic	* / %
Arithmetic	+ -
Relational	< <= > >=
Equality	= !=
Logical *and*	&&
Logical *or*	¦¦
Conditional	? :
Assignment	= += -= *= /= %=

Statements in C

A *statement* instructs the computer to carry out some action. There are three different classes of statements in C:

(a) Expression statements (b) Compound statements and (c) Control statements.

An *expression statement* consists of an expression followed by a semicolon. This causes the expression to be evaluated.

A *compound statement* consists of several individual statements enclosed in a pair of braces { }. The individual statements may be expression statements, compound statements or control statements. A compound statement does not end with a semicolon.

Control statements are used to create special program effects such as logical tests, loops and branches. Mostly control statements require that other statements are contained within them.

For commonly used operations or calculations C is accompanied by a number of library functions. They include functions to carry out input/output operations, open and close files, test for end-of-file, carrying out various mathematical calculations, trigonometric, logarithmic and exponential function, determining absolute values, square roots, etc.

A library function (or built in function is accessed by writing the function name, followed by a list of arguments separated by commas and enclosed in parentheses ( ) as in FORTRAN. A list of library functions commonly appearing in C programs is provided in any book on C programming (cf. Gottfried (1991), Kerninghan and Ritchie (1991), Plauger (1992)).

For a complete list, the user should refer to a reference manual provided with C compiler.

Any program needs I/O functions. Library of C has several I/O functions such as: *getchar, putchar, scanf, printf, gets and puts*, which are commonly used for transferring information to and from computer via standard I/O devices, such as key board and monitor.

getchar and *putchar transfer single character to and from computer*

gets and *puts* facilitate input and output of strings.

There are specific syntax or rule which must be followed for proper transfer to and from computer.

A potential programmer must consult a book referred above or the user reference before using the appropriate I/O function.

As in FORTRAN, C-program also requires repeated computations which was performed by control or DO statements in FORTRAN. These facilities are available in C also, but these are not exactly of the same form. A C-programmer must consult the said books and reference manual for being able to write correct loops.

Apart from the user defined functions and function subprograms as mentioned in FORTRAN, the C language has more than these facilities, clearly explained with examples and exercises in the referred books. The interested reader must consult at least one of these thoroughly before going to run the C-program. Some extra facilities in C are 1. *pointers*, 2. *while*, 3. *do-while*, 4. *for*, 5 *if-else* and 6. *go to* statements, which are often used.

Running a C program

As already mentioned after an overall program strategy is finalized and following the syntax rules, the C-program has been written, the *program has to be entered into computer*. Only after a compiler has *translated this high level* language program into *machine code*, i.e. changed into an object program, the results from computer can not be output.

Program is entered into computer using an *editor*

Next step is to *compile* it. It is usually accomplished by giving a single command

compile file name (23.36)

Now the compiled object program is linked with one or more library routines to be able to execute the program. It is also accomplished by a single command such as

link file name (23.37)

In UNIX operating system

$$\text{cc file name.c} \tag{23.38}$$

results in an object program *a.out*.

By giving the command *'a.out'* execution may start giving output file.

As in FORTRAN, the user may find that compilation errors and execution errors are hindering the computer run and giving results in output file.

Debugging must be done, using reference manual, C-programming books or taking help from computer operators, if run in a computer center.

During 1980's a new variant of C, called C^{++} was developed at Bell Laboratories, USA. Release 1.1 came out in 1986 and developments continued and last release came as C^{++} in 2008.

Knowledge of C language is not necessary for learning to write programs in C^{++}. Some part of vocabulary of C is common with C^{++}. But learning to use C^{++} does not require learning only new set of syntax and symantics. New ways of programming gives the power of C^{++}.

C^{++} is an *object-oriented – programming language.* It has become possible through the mechanism of *inheritance*.

A book entitled "C^{++} Beginners Guide" gives step by step approach even for first time programmers.

Margaret Ellis and Bjarre Stroustrup have developed and brought out "C^{++} Language Reference" which contains C^{++} key words and C^{++} operators.

Interested readers may contact professionals of C^{++} for any complex problems, if a difficulty is faced due to non-availability of the required program to solve a new problem not matching the conditions mentioned in statistical packages.

Large libraries of C/C^{++} functions are available for interested programmers e.g.

(i) Apophenia: A library of statistical functions.

(ii) Scythe Lib: an open source C^{++} class library of statistical functions.

23.2 Use of Statistical Packages

Writing a computer program for a problem by every user is not an easy job. In scientific studies, a researcher has to analyze bulk of data and some of the statistical techniques are used by researchers of different fields all of whom may not be proficient in programming. So some professional programmers in collaboration with statisticians or some statisticians who are highly refined expert in statistical theory as well as good programmers have developed software packages for statistical analysis of data.

There exist, therefore, hundreds of software packages in the market. Some of these deal with some areas of statistics not all, whereas there are some which elaborately deal almost all the areas of applications of statistical theory. Some of these have included even slightly different statistical techniques developed gradually with passage of time. Some of these techniques are applicable when the data satisfy some conditions.

There are three statistical software packages which are most widely used and have included vast variants of techniques in almost all the areas of statistical theory. These are SAS (Statistical Analysis System), SPSS (Statistical Package for Social Sciences) and BMDP (Bio MeDical Programs).

The BMDP was developed in 1961 at University of California at Los Angeles, USA. The SPSS was developed in 1968 by NH Nie at Stanford University, USA. The SAS was developed by Antony J. Barr in 1966 at the University of North Carolina, USA.

All of these are very comprehensive packages and contain programs for simplest statistics like mean, median etc. to complex techniques like MANOVA and ARIMA models. Each of these have a base module and other specific modules. The base module contains almost all the most commonly used basic statistical analysis programs and have specific modules for specialized statistical analysis techniques, such as modules for regression analysis, time series analysis, design of experiments, survival analysis, multivariate analysis, nonparametric methods, etc.

Each statistical package gives a method of creating data file, which is to be used for various statistical analysis. Many times the researchers have their data file on an Excel spreadsheet or an ASCII file created for use in a FORTRAN program. The packages mostly include instructions to be followed, for proper way of importing the already created said data files.

In every statistical package for this purpose, there is a 'data manager'. There is also a graphic option, through which graphs (such as scatter diagram, pie chart, bar diagram, histogram, frequency curve, regression lines, dendogram etc. can be plotted nicely with labels and headings. The numerical results are printed in a specific format with headings and the user must know what these results signify. Only then a proper interpretation of the results is possible leading to valid conclusions, which the data want to say.

Now in the following subsections a brief outline of these three statistical packages is given.

23.2.1 SPSS

SPSS has been developing into newer versions by including more and more enhanced techniques in it. It is a very comprehensive system, as it can analyse data from almost any type of files in the computer system. The required statistical analysis can be performed quite easily by an experienced user.

Data files can be created by Data Editor of SPSS even by a new entrant in computer world, for analysis by SPSS. If SPSS icon is clicked the spreadsheet like that in EXCEL, opens up and data can be entered in various columns after giving the variable names in the first row of the SPSS spreadsheet.

After entering the data, go to 'file' from the top row of menu. Opt 'save', give the data file the name of the created file and click 'save' in the dialog box opening after opting 'save' in 'file' menu. The file will be saved with extension .sav in SPSS format.

The statistical analyses available in SPSS package are listed in Table 23.7.

Table 23.7 Statistical analyses in SPSS

1. Reports
2. Descriptive statistics
3. Tables
4. Compare means
5. General linear model
6. Mixed models
7. Correlation
8. Regression
9. Log linear
10. Classify
11. Data reduction
12. Scale
13. Non parametric tests
14. Time series
15. Survival
16. Multiple response, missing value analysis
17. Complex samples

If any of these is to be applied, say 'Regression', click Analyze in the top line of Menu. The whole list of analyses will open up. Choose 'Regression' and click it and further choose 'Linear ….' SPSS analysis window will open. Type the data file name with extension .sav, if it is created and saved by the user. If the file already exists in some other format (i.e. with extension .xls or any other format) it should be first imported in SPSS before starting analysis.

For importing an existing data file, go to File in top menu. Click 'Open' and 'Data….'. A dialog box for importing will open. Choose the file name after opting the existing extension, by browsing to specific drive and folder, click the required data file.

File will be imported in SPSS format with extension .sav. Give a file name to this data file and click save option besides the file name in the dialog box opened after 'File', 'OPEN' and 'Data …..'. The existing data file may also be imported if a proper path (directory, folder and file name with extension) is typed in the space for data file name and click save.

For performing desired analysis say 'Regression', choose it by clicking over 'Regression'. A dialog box for entering 'dependent' and 'independent variables' opens. For simple linear regression choose 'linear', and only one independent variable and one dependent variable is choosen from the list of variables displayed on the LHS of newly opened analysis dialog box.

Click 'Ok' and results of regression analysis are displayed. Save the output file by opting 'File' and 'save as'. If a scatter plot is also to be taken, a choice should be made of graph from the top menu of Data Editor. Click at scatter plot. Choose the simple design of scatter plot, click define in scatter plot design box. Give the variables for Y-axis and for X-axis from the variables in the imported (or created) data file.

Now choose Title and click it. Give the appropriate title of the graph. Click continue and OK in the window prior to that with Continue. The scatter plot is ready and displayed on the monitor.

Click 'File' from Menu line and click 'save as'. Give file name and click **save**. Scatter plot file will be saved with .spo extension.

To see the saved files choose 'File', 'OPEN' and 'output….'. the output files will be listed in a dialogue box. Select one and click 'Open'. The selected output file will be displayed. It can be printed by giving 'Print' command.

'Regression' in the 'Analyse' has several options of regression analysis viz. multiple, stepwise (forward/backward), non-linear etc. Choosing various options, different types of regression analysis can be performed. The selection of 'dependent' and 'independent' variables for analysis are chosen subsequently as in simple linear regression. For multiple regression, two or more independent variables from the listed data file variables are chosen. There are also options for printing estimated statistics and results of testing the significance of regression coefficients. The ANOVA table for goodness of fit and multiple correlation coefficient are also printed in a default way.

SPSS is a menu driven package. So can be easily handled by any user. User's Guide of SPSS may be consulted to have an idea of various options giving various enhancements or uncommon advanced techniques. The interpretation of results in output file may also be found in brief or reference to the employed statistical techniques are also provided.

A Remark

The multivariate techniques of *Factor analysis* and *Principal component analysis* are contained in item no. 11: *Data Reduction in 'Analyze'* Menu on top row.

The *Cluster analysis* and *Discriminant analysis* are given in item no. 10: *Classify* of *'Analyze'* Menu.

Details of the available techniques can be viewed by clicking an item listed with Analyze,icon of SPSS menu bar on top line. An appropriate choice can be made for a specific purpose.

SPSS user's guide such as SPSS.inc.(1999) may be used for step by step procedure and to know the theoretical background behind the techniques given in SPSS 10.2. Now even SPSS 17.0 is available.

23.2.2 BMDP

The BMDP (BioMeDical Programs) developed in 1961 is also a good comprehensive and convenient statistical software package. A list of statistical features available in BMDP are given in Table 23.8.

There are several programs available with each of the above listed features performing different forms of analysis and resulting in different ways of presentation in result output. In all there are 44 programs for different purposes. The program name must be opted for a specific variant of a particular statistical analysis. The Statistical Solutions of the website www.statsol.ie/index.php gives the list of BMDP program names and the specific analysis performed, reproduced in Table 23.8.

New versions of BMDP run the programs in either interactive or batch mode. The prompt driven method is designed for easy use.

The desired program name has to be given for the required analysis. Various options for different variants or enhancements or refined analysis should be chosen by user.

In interactive mode, the instructions can be edited after the program starts using a built in full screen editor. Repeated running of program is also possible.

For certain programs special interactive options are available, through which some predictor variables may be entered or deleted and effect of this can be seen instantly in the output files. For some, one may add a paragraph and can see the altered output.

BMDP Manual, Users Digest or Data Manager Manual may be seen for detailed guidelines.

Table 23.8 Statistical features in BMDP

1.	Data manipulation
2.	Data description
3.	Group comparison
4.	Plots and histograms
5.	Frequency tables
6.	Correspondence analysis
7.	Regression
8.	Maximum likelihood estimation
9.	Non-linear regression
10.	Analysis of variance
11.	Multivariate analysis
12.	Nonprametric analysis
13.	Cluster analysis
14.	Missing values
15.	Survival analysis
16.	Time series

23.2.3 SAS

SAS (Statistical Analysis System) is another very widely used statistical package. It was originally developed for mainframe computer, so can be used for analysis of very small to very large data sets.

SAS has been designed to encompass the following statistical analyses with almost all possible variants, enhancements and refinements. Table 23.9 gives the statistical analyses possible through SAS.

It appears, it is designed to suit every researchers' data in hand and the way he intends to analyze it. There are different PROCEDURES given as PROC name. The SAS manuals such as Introductory Guide and Language Guides (cf. SAS (1985a,b, 1987a,b)) and on line literature gives a complete description of what SAS does and how it is done.

The user must choose a specific PROC and its options have to be written as a small set of commands. Milton and Tsokos (1983) and Forthofer and Lee (2006) have given several examples of simple (descriptive statistics) to the complex non-parametric tests being done using SAS. Even ARIMA (auto regressive integrated moving average) modeling for prediction in time series problems are also dealt in SAS.

Detailed chapter wise description of the techniques have been given in User Guide. A little practice makes a researcher frequent user of SAS for various research problems he needs to solve

Table 23.9 Statistical analyses in SAS

	Statistical analyses		Statistical analyses
1.	Analysis of variance	8.	Psychometric analysis
2.	Mixed models	9.	Cluster analysis
3.	Regression	10.	Non-parametric analysis
4.	Categorical data analysis	11.	Survey data analysis
5.	Bayesian analysis	12.	Multiple imputation for missing values
6.	Multivariate analysis	13.	Study planning
7.	Survival analysis		

However, a brief and simple description of problem solving in SAS has been provided in Milton and Tsokos (1983) from the step O (zero) to enable SAS read an existing data or to provide data directly to SAS using appropriate DATA statement.

The interested reader must go through the said book and the book by Forthofer and Lee (2006).

It may be noted that after giving CARDS statement data can be entered through keyboard itself.

The output can be displayed graphically also using the SAS graphical procedures GPLOT and GCHART. In the absence of SAS Graphics, CHART and PLOT can also be used.

From these two books an interested reader can find examples of methods of using the following SAS procedures:

Milton and Tsokos (1983)

PROC CHART	PROC CORR	PROC MEANS	PROC FREQ
PROC GLM	PROC PLOT	PROC PRINT	PROC SORT
PROC TTEST			

Forthofer and Lee (2006)

PROC FREQ	PROC GPLOT	PROC GCHART	PROC UNIVARIATE DATA
PROC PRINT	PROC SORT	PROC CORR	PROC UNIVARIATE PLOT
PROC NPAR1WAY WILCOXON		PROC FREQ ORDER = DATA	
PROC LIFETEST PLOTS		PROC TTEST	PROC ANOVA
PROC REG	PROC LOGISTIC		

Appendices

List of Tables

Appendix A: Statistical and Mathematical Tables

Table A.1 The binomial distribution: individual terms [n = 5(1)15; selected values of p upto p = 1/2]

n	x	p = 0.01	p = 0.02	p = 0.05	p = 1/16	p = 0.10	p = 1/9	p = 3/16	p = 0.20
5	0	.95099	.90392	.77378	.72420	.59049	.55493	.35409	.32768
	1	.04803	.09224	.20363	.24140	.32805	.34683	.40857	.40960
	2	.00097	.00376	.02143	.03218+	307290	.08671	.18857	.20480
	3	.00001	.00008	.00113	.00215	.00810	.01084	.04352	.05120
	4	–	–	.00003	.00007	.00045	.00067+	.00502	.00640
	5	–	–	–	–	.00001	.00002	.00023	.0032
6	0	.94148	.88584	.73509	.67893	.53144	.49327	.28770	.26214
	1	.05706	.10847	.23214	.27158	.35429	.36995	.39835	.39322
	2	.00144	.00554	.03054	.04526	.09842	.11561	.22982	.24576
	3	.00002	.00015	.0214	.00402	.01458	.01927	.07072–	.08192
	4	–	–	.00009	.00020	.00121+	.00181	.01224	.01536
	5	–	–	–	.00001	.00006–	.00009	.00113	.00154
	6	–	–	–	–	–	–	.00004	.00006
7	0	.93207	.86813	.69834	.63650	.47830	.43846	.23376	.20972
	1	.06590	.12401+	.25728	.29703	.37201	.38366–	.37760+	.36700
	2	.00200	.00760	.04062	.05941	.12400	.14387	.26142	.27525
	3	.00003	.00025+	.00357–	.00660	.02296	.02997	.10055	.11469
	4	–	.00001	.00018+	.00044	.00255	.00375	.02320	.02867
	5	–	–	.00001	.00002	.00017	.00028	.00321	.00430
	6	–	–	–	–	.00001	.00001	.00025	.00036
	7	–	–	–	–	–	–	.00001	.00001
8	0	.92274	.85076	.66342	.59672	.43047	.38974	.18993	.16777
	1	.07457	.13890	.27934–	.31825	.38263+	.38975–	.35063	.33555–
	2	.00264	.00992	.05145+	.07426	.14881–	.17051	.28321–	.29360
	3	.00005	.00041	.00542	.00990	.03307	.04263	.13071	.14680
	4	–	.00001	.00035+	.00082+	.00459	.00666	.03770	.04587+

Contd...

Source: Rao, Mitra, Mathai and Ramamurthy (1975) Formula and Tables for Statistical Work

Table A.1 Contd.

n	x	$p = 0.01$	$p = 0.02$	$p = 0.05$	$p = 1/16$	$p = 0.10$	$p = 1/9$	$p = 3/16$	$p = 0.20$
	5	–	–	.00002	.00005–	.00041	.00067	.00696	.00918
	6	–	–	–	–	.00002	.00004	.00081–	.00115
	7	–	–	–	–	–	–	.00005	.00008
9	0	.91352	.83375	.63025	.55942	.38742	.34644	.15432	.13422
	1	.08304+	.15314	.29854	.33566–	.38742	.38974	.32050	.30199
	2	.00336	.01250	.06285	.08951	.17219	.19488–	.29585	.30199
	3	.00008	.00059+	.00772	.01392	.04464	.05683+	.15930	.17616
	4	–	.00002	.00061	.00139	.00744	.01066	.05514	.06606
	5	–	–	.00003	.00010–	.00083	.00133	.01273	.01651
	6	–	–	–	–	.00006	.00011	.0195+	.00276–
	7	–	–	–	–	–	.00001	.00020–	.00029
	8	–	–	–	–	–	–	.00001	.00002
10	0	.90438	.81707	.59874	.52446	.34868	.30795	.12538	.10737
	1	.09135	.16675	.31512	.34964	.38742	.38493	.28934	.26844
	2	.00416–	.01532–	.07464–	.10489	.19371	.21652	.30047	.30199
	3	.00011	.00083	.01047+	.01865	.05739+	.07218–	.18491	.20133
	4	–	.00003	.00097–	.00218	.01117–	.01579	.07467	.08808
	5	–	–	.00006	.00017	.00148+	.00236+	.02068	.02642
	6	–	–	–	.00001	.00014	.00025	.00398	.00551
	7	–	–	–	–	.00001	.00002	.00052	.00078+
	8	–	–	–	–	–	–	.00005	.00008–
11	0	.89534	.80073	.56880	.49168	.31381	.27373	.10187	.08590
	1	.09948	.17946	.32931	.36057	.38355	.37638	.25860	.23622
	2	.00502	.01834	.08665+	.12019	.21308	.23524	.29839–	.29528
	3	.00016–	.00112	.01369–	.02403+	.07103	.08821	.20657	.22148
	4	–	.00005	.00144	.00321	.01578	.02205	.09534	.11073
	5	–	–	.00010+	.00030	.00245+	.00380	.03080	.03876
	6	–	–	.00001	.00002	.00028–	.00048	.00711	.00968+
	7	–	–	–	–	.00002	.00005–	.0117	.00173
	8	–	–	–	–	–	–	.00014	.00022
	9	–	–	–	–	–	–	.00001	.00002
12	0	.88638	.78472	.54036	.46095	.28243	.24332	.08277	.06872
	1	.10745–	.19217+	.34128	.36876	.37657	.36497	.22921	.20616
	2	.00596+	.02157	.09879	.13633–	.23013	.25092	.29093–	.28347
	3	.00021–	.00147	.01733	.03004+	.08523	.10455	.22379	.23622
	4	–	.00007	.00206–	.00451	.02131	.02940	.11619+	.13287+
	5	–	–	.00017	.00048	.00379	.00588	.04291–	.05315
	6	–	–	.00001	.00004	.00049	.00086	.01155	.01551–
	7	–	–	–	–	.00005	.00009	.00228	.00332
	8	–	–	–	–	–	.00001	.00033	.00052
	9	–	–	–	–	–	–	.00004–	.00006

Contd...

Table A.1 Contd.

n	x	p = 0.01	p = 0.02	p = 0.05	p = 1/16	p = 0.10	p = 1/9	p = 3/16	p = 0.20
13	0	.87752	.76902	.51334	.43214	.25419	.21628	.06725	.05498
	1	.11523	.20403	.35124–	.37453–	.36715+	.35146	.20176	.17867
	2	.00698	.02498	.11091+	.14980+	.24478–	.26359	.27935	.26800+
	3	.00026	.00187	.02141–	.03662	.09972	.12081	.23638	.24567
	4	.00001	.00010	.00281+	.00611–	.02770	.03775	.13637	.15355
	5	–	–	.00027	.00073	.00554	.00850–	.05664+	.06909+
	6	–	–	.00002	.00007	.00082	.00142	.01743	.02304–
	7	–	–	–	–	.00009	.00017+	.00403–	00575+
	8	–	–	–	–	.00001	.00002	.00069+	.00108
	9	–	–	–	–	–	–	.00009	.00015
	10	–	–	–	–	–	–	.00001	.00002
14	0	.86875	.75364	.48767+	.40513	.22877	.19225	.05464	.04398
	1	.12285	.21533	.35934	.37813–	.35586	.33644	.17654	.15393
	2	.00806+	.02856	.12294–	.16385	.25701	.27335	.26480	.25014
	3	.00033	.00233	.02588	.04370–	.11423	.13668	.24444–	.25014
	4	.00001	.00013	.00374+	.00801	.03490	.04698	.15512	.17197
	5	–	.00001	.00040–	.00106+	.00776	.01175	.07159	.08599
	6	–	–	.0000.3	.00011	.00129	.00220	.02479–	.03224
	7	–	–	–	.00001	.00016	.00031	.00653+	.00921
	8	–	–	–	–	.00002	.00004–	.00132	.00202
	9	–	–	–	–	–	–	.00020	.00033+
	10	–	–	–	–	–	–	.00003–	.00005–
15	0	.86006	.73857	.46329	.37981	.20589	.17089	.04440	.03518
	1	.13031	.22609	.36576	.37981	.34315	.32041+	.15368	.13195–
	2	.00921	.03230	.13475	.17725	.26690	.28037–	.24825	.23089+
	3	.00041–	.00286	.03073	.05120	.12850+	.15186	.24825	.25014
	4	.00001	.00017	.00486–	.01025–	.04284	.05695	.17187	.18761–
	5	–	.00001	.00056	.00150	.01047	.01566	.08726	.10318
	6	–	–	.00005	.00016+	.00194	.00326	.03356	.04299
	7	–	–	–	.00002–	.00028	.00053–	.00996	.01382
	8	–	–	–	–	.00003	.00006+	.00229+	.00346–
	9	–	–	–	–	–	.00001	.00042–	.00067
	10	–	–	–	–	–	–	.00005+	.00010
	11	–	–	–	–	–	–	.00001	.00001

Note: To obtain 5 decimal accuracy for individual terms add (subtract) 1 in the last place if there is +(–) sign against an entry.

Table A.1 Contd.

n	x	p = 1/4	p = 0.30	p = 1/3	p = 0.40	p = 7/16	p = 4/9	p = 1/2
5	0	.23730	.16807	.13169	.07779	.05631	.05292	.03125
	1	.39551	.36015	.32922	.25920	.21900	.21169	.15625
	2	.26367	.30870	.32921+	.34560	.34066	.33870	.31250
	3	.08789	.13230	.16461	.23040	.26496	.27096	.31250
	4	.01465	.02835	.04115	.07680	.10304	.10839–	.15625
	5	.00098	.00243	.00412	.01024	.01603	.01734	.03125
6	0	.17798	.11765	.08779	.04666	.03168	.02940	.01562
	1	.35596	.30252+	.26338–	.18662	.14782	.14113	.09375
	2	.29663	.32414	.32921+	.31104	.28743	.28225	.23438
	3	.13183+	.18522	.21948	.27648	.29808	.30107	.31250
	4	.02396	.05953	.08231–	.13824	.17388	.18064	.23437
	5	.00440–	.01021	.01646	.3686	.05410	.05780+	.09375
	6	.00024	.00073	.00137	.00410	.00701	.00771	.01563
7	0	.13348	.08235	.05883	.02799	.01782	.01633	.00781
	1	.31147–	.24707–	.20484+	.13064	.09701	.09147	.05469
	2	.31146	.31765	.30727	.26127	.22635	.21953	.16406
	3	.17303	.22689	.25606	.29031–	.29342	.29271	.27344
	4	.05768	.09724	.12803	.19353+	.22822	.23416	.27344
	5	.01154	.02501–	.03841	.07742–	.10650	.11240	.16406
	6	.00128	.00357	.00640	.01720	.02761	.02997	.05469
	7	.00006	.00022	.00046	.00164	.00307	.00343	.00781
8	0	.10011	.05765	.03902	.01680	.01002	.00907	.00391
	1	.26697	.19765	.15607	.08958	.06237–	.05808	.03125
	2	.31146	.29647+	.27313	.20901+	.16976+	.16261	.10937
	3	.20764	.25413–	.27313	.27870–	.26408	.26019–	.21876
	4	.08652	.13613+	.17071	.23224	.25674	.26018	.27344
	5	.02307	.04668	.06828	.12386	.15976–	.16652	.21875
	6	.00385	.01000	.01707	.04129	.06212+	.06660+	.10937
	7	.00036+	.00122	.00244	.00786	.01381	.01522	.03125
	8	.00002	.00007	.00015	.00066	.00134	.00152	.00391
9	0	.07508	.04035	.02601	.01008	.00564	.00504	.00195
	1	.22526–	.15565	.11706	.06046+	.03946	.03630	.01758
	2	.30034	.26683	.23411	.16125–	.12278	.11615	.07031
	3	.23359+	.26683	.27313	.25082	.22282	.21682	.16407–
	4	.11680	.17153	.20484+	.25082	.25995	.26018	.24609
	5	.03894–	.07352–	.10243–	.16722	.20218+	.20815	.24609
	6	.00865	.02100	.03414	.07432	.10484	.11101	.16407–
	7	.00123+	.00386	.0731+	.02123	.03495	.03806	.07031
	8	.00011–	.00041	.00092–	.00354	.00679+	.00761	.01758
	9	–	.00002	.00005	.00026	.00059	.00068	.00195
10	0	.05631	.02825	.01734	.00605	.00317	.00280	.00098
	1	.18772–	.12106	.08671	.04031	.02467–	.02241	.00976+

Contd...

Table A.1 Contd.

n	x	p = 1/4	p = 0.30	p = 1/3	p = 0.40	p = 7/16	p = 4/9	p = 1/2
	2	.28156+	.23347	.19509	.12093	.08632+	.08066	.04395
	3	.25029	.26683	.26012	.21499	.17905	.17208	.11718+
	4	.14599+	.20012	.22761	.25082	.24371	.24091	.20508
	5	.05840	.10292	.13657–	.20066	.22746	.23127	.24610–
	6	.01622	.03676	.05690	.11148	.14743	.15418	.20507+
	7	.00309	.00900	.01626	.04247	.06552	.07049–	.11719
	8	.00039	.00145	.00304+	.01061+	.01911	.02114	.04395
	9	.00003	.00013+	.00034	.00158–	.00330	.00376	.00976+
	10	–	.00001	.00001	.00010	.00026	.00030	.00098
11	0	.04224	.01977	.01158	.00363	.00178	.00156	.00049
	1	.15486	.09322	.06359	.02660+	.01527–	.01369	.00537
	2	.25810	.19975	.15896	.08869–	.05935	.05477	.02685+
	3	.25810	.25682	.23845	.17736+	.13848	.13145	.08057
	4	.17207	.22014–	.23844+	.23649	.21542	.21032	.16113
	5	.08030	.13208	.16691	.22073–	.23457	.23555+	.22559
	6	.02677	.05660+	.08346	.14715	.18244	.18845–	.22559
	7	.00637	.01733	.02981	.07007	.10135+	.10768	.16113
	8	.00106	.00371	.00745	.02336	.03942	.04307	.08057
	9	.00012	.00053	.00124	.00519	.01022	.01149	.02685+
	10	.00001	.00005	.00012	.00069	.00159	.00184	.00537
	11	–	–	.00001	.00004	.00011	.00013	.00049
12	0	303168	.01384	.00771	.00218	.00100	.00086	.00024
	1	312670+	.07119–	.04624	.01741	.00937–	.00830	.00293
	2	.23230–	.16779	.12717	.06385	.04006	.03652–	.01612–
	3	.25810	.23970	.21195	.14190–	.10386	.09737	.05371
	4	.19358	.23114	.23845	.21284	.18176	.17526	.12085
	5	.10324	.15849+	.19076	.22703	.22619	.22434	.19336
	6	.04015	.07925	.11127	.17658	.20525	.20938	.22558+
	7	.01147	.02911	.04769	.10090	.13683	.14358	.19336
	8	.00239	.00780	.01490	.04204	.06651+	.07179	.12085
	9	.00035	.00148+	.00332–	.01246	.02300–	.02552	.05371
	10	.00004	.00019	.00049+	.00249	.00536+	.00613	.01612–
	11	–	.00002–	.00005	.00030	.00076	.00089	.00293
	12	–	–	–	.00002	.00005	.00006	.00024
13	0	.02376	.00969	.00514	.00131	.00056	.00048	.00012
	1	.10295	.05398	.03340	.01132	.00571	.00499	.00159
	2	.20589+	.13881	.10019+	.04527+	.02663	.02398–	.00952
	3	.25165	.21813	.18369	.11068	.07595	.07032	.03491
	4	.20971	.23370+	.22962–	.18446	.14768	.14065–	.08728
	5	.12583	.18029	.20665	.22136–	.20675	.20252+	.15711–
	6	.05592	.10302	.13777	.19676	.21441	.21603	.20947
	7	.01864	.04416–	.06889–	.13117	.16677–	.17283–	.20947
	8	.00466	.01419	.02583	.06559	.09727+	.10369	.15711–
	9	.00086	.00338	.00717+	.02429	.04204-	.04609	.08728

Contd...

Table A.1 Contd.

n	x	p = 1/4	p = 0.30	p = 1/3	p = 0.40	p = 7/16	p = 4/9	p = 1/2
	10	.00012	.00058	.00144	.00647+	.01307+	.01475	.03491
	11	.00001	.00007	.00019+	.00118	.00278–	.00321+	.00952
	12	–	–	.00002	.00013	.00036	.00043	.00159
	13	–	–	–	.00001	.00002	.00003	.00012
14	0	.01782	.00678	.00343	.00078	.00032	.00027	.00006
	1	.08315	.04070–	.02397+	.00732–	.00345+	.00298+	.00086–
	2	.18016	.11336	.07793	.03169	.01748	.01554	.00555
	3	.24021	.19433	.15586	.08452	.05437	.04973–	.02222
	4	.22019	.22903	.21431	.15495	.11630	.10939	.06109+
	5	.14680	.19632–	.21431	.20659+	.18091	.17502	.12220–
	6	.07340	.12620	.16073+	.20660	.21106	.21003	.18328+
	7	.02796	.06181	.09184+	.15741	.18761	.19203	.20948-
	8	.00816	.02318	.04019–	.09182	.12767+	.13442	.18328+
	9	.00181	.00662	.01339	.04081	.06621–	.07169	.12220–
	10	.00030	.00142	.00335	.01360	.02574+	.02867+	.06109+
	11	.00004	.00022	.00061	.00330	.00729–	.00834	.02222
	12	–	.00003–	.00007+	.00055	.00141+	.00167	.00555
	13	–	–	.00001	.00006	.00017	.00021	.00086–
	14	–	–	–	–	.00001	.00001	.00006
15	0	.01336	.00475	.00228	.00047	.00018	.00015	.00003
	1	.06682	.03052	.01713	.00470	.00208	.00178	.00046
	2	.15591	09156	.05995	.02194	.01135–	.00996	.00320
	3	.22520	.17004	.12988	.06339	.03823	.03453	.01389
	4	.22520	.21862	.19482	.12678	.08920+	.08287	.04165+
	5	.16514+	.20613	.21431–	.18594	.15264	.14585	.09165–
	6	.09175	.14724	.17859	.20659+	.19787	.19447	.15274
	7	.03932	.08113	.11481	.17709-	.19787	.20003	.19638
	8	.01311	.03477	.05740	.11805+	.15390	.16002	.19638
	9	.00340	.01159	.02233–	.06122–	.09309+	.09957	.15274
	10	.00067+	.00298	.00669+	.02448+	.04345	.04780–	.09165–
	11	.00011–	.00058	.00152	.00742	.01536	.01738	.04165+
	12	.00001	.00008	.00026–	.00165	.00398	.00463	.01389
	13	–	.00001	.00003	.00025	.00072–	.00086	.00320
	14	–	–	–	.00003–	.00008	.00009+	.00046
	15	–	–	–	–	–	.00001	.00003

Table A.2 The Poisson distribution (entries in body of table give the probability of x or less when the expected number λ is that given in the left margin of the table i.e. cdf $F(x)$)

$\lambda \backslash x \rightarrow$ $\downarrow$	0	1	2	3	4	5	6	7	8	9
0.02	.980	1.000								
0.04	.961	.999	1.000							
0.06	.942	.998	1.000							
0.08	.923	.997	1.000							
0.10	.905	.995	1.000							
0.15	.861	.990	.999	1.000						
0.20	.819	.982	.999	1.000						
0.25	.779	.974	.998	1.000						
0.30	.741	.963	.996	1.000						
0.35	.705	.951	.994	1.000						
0.40	.670	.938	.992	.999	1.000					
0.45	.938	.925	.989	.999	1.000					
0.50	.607	.916	.986	.998	1.000					
0.55	.577	.894	.982	.998	1.000					
0.60	.549	.878	.977	.997	1.000					
0.65	.522	.861	.972	.996	.999	1.000				
0.70	.497	.844	.966	.994	.999	1.000				
0.75	.472	.827	.959	.993	.999	1.000				
0.80	.449	.809	.953	.991	.999	1.000				
0.85	.427	.791	.945	.989	.998	1.000				
0.90	.407	.772	.937	.987	.998	1.000				
0.95	.387	.754	.929	.984	.997	1.000				
1.00	.368	.736	.920	.981	.996	.999	1.000			
1.1	.333	.699	.900	.974	.995	.999	1.000			
1.2	.301	.663	.879	.966	.992	.998	1.000			
1.3	.273	.627	.857	.957	.989	.998	1.000			
1.4	.247	.592	.833	.946	.986	.997	.999	1.000		
1.5	.223	.558	.809	.934	.981	.996	.999	1.000		
1.6	.202	.525	.783	.921	.976	.994	.999	1.000		
1.7	.183	.493	.757	.907	.970	.992	.998	1.000		
1.8	.165	.463	.731	.891	.964	.990	.997	.999	1.000	
1.9	.150	.434	.704	.875	.956	.987	.997	.999	1.000	
2.0	.135	.406	.677	.857	.947	.983	.995	.999	1.000	
2.2	.111	.355	.623	.819	.928	.975	.993	.998	1.000	
2.4	.091	.308	.570	.779	.904	.964	.988	.997	.999	1.000
2.6	.074	.267	.518	.736	.877	.951	.983	.995	.999	1.000
2.8	.061	.231	.469	.692	.848	.935	.976	.992	.998	.999
3.0	.050	.199	.423	.647	.815	.916	.966	.988	.996	.999
3.2	.041	.171	.380	.603	.781	.895	.955	.983	.994	.998
3.4	.033	.147	.340	.558	.744	.871	.942	.977	.992	.997
3.6	.027	.126	.303	.515	.706	.844	.927	.969	.988	.996
3.8	.022	.107	.269	.473	.668	.816	.909	.960	.984	.994
4.0	.018	.092	.238	.433	.629	.785	.889	.949	.979	.992
4.2	.015	.078	.210	.395	.590	.753	.867	.936	.972	.989

Contd...

Source: Rao, Mitra, Mathai and Ramamurthy (1975) Formula and Tables for Statistical Work

Table A.2 Contd.

λ\x → ↓	0	1	2	3	4	5	6	7	8	9
4.4	.012	.066	.185	.359	.551	.720	.844	.921	.964	.985
4.6	.010	.056	.163	.326	.513	.686	.818	.905	.955	.980
4.8	.008	.048	.143	.294	.476	.651	.791	.887	.944	.975
5.0	.007	.040	.125	.265	.440	.616	.762	.867	.932	.968
5.2	.006	.034	.109	.238	.406	.581	.732	.845	.918	.960
5.4	.005	.029	.095	.213	.373	.546	.702	.822	.903	.951
5.6	.004	.024	.082	.191	.342	.512	.670	.797	.886	.941
5.8	.003	.021	.072	.170	.313	.478	.638	.771	.867	.929
6.0	.002	.017	.062	.151	.285	.446	.606	.744	.847	.916
2.8	1.000									
3.0	1.000									
3.2	1.000									
3.4	.999	1.000								
3.6	.999	1.000								
3.8	.998	.999	1.000							
4.0	.997	.999	1.000							
4.2	.996	.999	1.000							
4.4	.994	.998	.999	1.000						
4.6	.992	.997	.999	1.000						
4.8	.990	.996	.999	1.000						
5.0	.986	.995	.998	.999	1.000					
5.2	.982	.993	.997	.999	1.000					
5.4	.977	.990	.996	.999	1.000					
5.6	.972	.988	.995	.998	.999	1.000				
5.8	.965	.984	.993	.997	.999	1.000				
6.0	.957	.980	.991	.996	.999	.999	1.000			
6.2	.002	.015	.054	.134	.259	.414	.574	.716	.826	.902
6.4	.002	.012	.046	.119	.235	.384	.542	.687	.803	.886
6.6	.001	.010	.040	.105	.213	.355	.511	.658	.780	.869
6.8	.001	.009	.034	.093	.192	.327	.480	.628	.755	.850
7.0	.001	.007	.030	.082	.173	.301	.450	.599	.729	.830
7.2	.001	.006	.025	.072	.156	.276	.420	.569	.703	.810
7.4	.001	.005	.022	.063	.140	.253	.392	.539	.676	.788
7.6	.001	.004	.019	.055	.125	.231	.365	.510	.648	.765
7.8	.000	.004	.016	.048	.112	.210	.338	.481	.620	.741
8.0	.000	.003	.014	.042	.100	.191	.313	.453	.593	.717
8.5	.000	.002	.009	.030	.074	.150	.256	.386	.523	.653
9.0	.000	.001	.006	.021	.055	.116	.207	.324	.456	.587
9.5	.000	.001	.004	.015	.040	.089	.165	.269	.392	.522
10.0	.000	.000	.003	.010	.029	.067	.130	.220	.333	.458

λ\x → ↓	10	11	12	13	14	15	16	17	18	19
6.2	.949	.975	.989	.995	.998	.999	1.000			
6.4	.939	.969	.986	.994	.997	.999	1.000			
6.6	.927	.963	.982	.992	.997	.999	.999	1.000		
6.8	.915	.955	.978	.990	.996	.998	.999	1.000		

Contd...

Table A.2 Contd.

λ\x → ↓	10	11	12	13	14	15	16	17	18	19
7.0	.901	.947	.973	.987	.994	.998	.999	1.000		
7.2	.887	.937	.967	.984	.993	.997	.999	.999	1.000	
7.4	.871	.926	.961	.980	.991	.996	.998	.999	1.000	
7.6	.854	.915	.954	.976	.989	.995	.998	.999	1.000	
7.8	.835	.902	.945	.971	.986	.993	.997	.999	1.000	
8.0	.816	.888	.936	.966	.983	.992	.996	.998	.999	1.000
8.5	.763	.849	.909	.949	.973	.986	.993	.997	.999	.999
9.0	.706	.803	.876	.926	.959	.978	.989	.995	.998	.999
9.5	.645	.752	.836	.898	.940	.967	.982	.991	.996	.998
10.0	.583	.697	.792	.864	.917	.951	.973	.986	.993	.997

λ\x → ↓	20	21	22
8.5	1.000		
9.0	1.000		
9.5	.999	1.000	
10.0	.998	.999	1.000

λ\x → ↓	0	1	2	3	4	5	6	7	8	9
10.5	.000	.000	.002	.007	.021	.050	.102	.179	.279	.397
11.0	.000	.000	.001	.005	.015	.038	.079	.143	.232	.341
11.5	.000	.000	.001	.003	.011	.028	.060	.114	.191	.289
12.0	.000	.000	.001	.002	.008	.020	.046	.090	.155	.242
12.5	.000	.000	.000	.002	.005	.015	.035	.070	.125	.201
13.0	.000	.000	.000	.001	.004	.011	.026	.054	.100	.166
13.5	.000	.000	.000	.001	.003	.008	.019	.041	.079	.135
14.0	.000	.000	.000	.000	.002	.006	.014	.032	.062	.109
14.5	.000	.000	.000	.000	.001	.004	.010	.024	.048	.088
15.0	.000	.000	.000	.000	.001	.003	.008	.018	.037	.070

λ\x → ↓	10	11	12	13	14	15	16	17	18	19
10.5	.521	.639	.742	.825	.888	.932	.960	.978	.988	.994
11.0	.460	.579	.689	.781	.854	.907	.944	.968	.982	.991
11.5	.402	.520	.633	.733	.815	.878	.924	.954	.974	.986
12.0	.347	.462	.576	.682	.772	.844	.899	.937	.963	.979
12.5	.297	.406	.519	.628	.725	.806	.869	.916	.948	.969
13.0	.252	.353	.463	.573	.675	.764	.835	.890	.930	.957
13.5	.211	.304	.409	.518	.623	.718	.798	.861	.908	.942
14.0	.176	.260	.358	.464	.570	.669	.756	.827	.883	.923
14.5	.145	.220	.311	.413	.518	.619	.711	.790	.853	.901
15.0	.118	.185	.268	.363	.466	.568	.664	.749	.819	.875

λ\x → ↓	20	21	22	23	24	25	26	27	28	29
10.5	.997	.999	.999	1.000						
11.0	.995	.998	.999	1.000						

Contd...

Table A.2 Contd.

λ/x → ↓	20	21	22	23	24	25	26	27	28	29
11.5	.992	.996	.998	.999	1.000					
12.0	.988	.994	.997	.999	.999	1.000				
12.5	.983	.991	.995	.998	.999	.999	1.000			
13.0	.975	.986	.992	.996	.998	.999	1.000			
13.5	.965	.980	.989	.994	.997	.998	.999	1.000		
14.0	.952	.971	.983	.991	.995	.997	.999	.999	1.000	
14.5	.936	.960	.976	.986	.992	.996	.998	.999	.999	1.000
15.0	.917	.947	.967	.981	.989	.994	.997	.998	.999	1.000

λ\x→ ↓	4	5	6	7	8	9	10	11	12	13
16	.000	.001	.004	.010	.022	.043	.077	.127	.193	.275
17	.000	.001	.002	.005	.013	.026	.049	.085	.135	.201
18	.000	.000	.001	.003	.007	.015	.030	.055	.092	.143
19	.000	.000	.001	.002	.004	.009	.018	.035	.061	.098
20	.000	.000	.000	.001	.002	.005	.011	.021	.039	.066
21	.000	.000	.000	.001	.001	.003	.006	.013	.025	.043
22	.000	.000	.000	.001	.001	.002	.00	.008	.015	.028
23	.000	.000	.000	.000	.000	.001	.002	.004	.009	.017
24	.000	.000	.000	.000	.000	.000	.001	.003	.005	.011
25	.000	.000	.000	.000	.000	.000	.001	.001	.003	.006

λ\x→ ↓	14	15	16	17	18	19	20	21	22	23
16	.368	.467	.566	.659	.742	.812	.868	.911	.942	.963
17	.281	.371	.468	.564	.655	.736	.805	.861	.905	.937
18	.208	.287	.375	.469	.562	.651	.731	.799	.855	.899
19	.150	.215	.292	.378	.469	.561	.647	.725	.793	.849
20	.105	.157	.221	.297	.381	.470	.559	.644	.721	.787
21	.072	.111	.163	.227	.302	.384	.471	.558	.640	.816
22	.048	.077	.117	.169	.232	.306	.387	.472	.556	.637
23	.031	.052	.082	.123	.175	.238	.310	.389	.472	.555
24	.020	.034	.056	.087	.128	.180	.243	.314	.392	.473
25	.012	.022	.038	.060	.092	.134	.185	.247	.318	.394

λ\x→ ↓	24	25	26	27	28	29	30	31	32	33
16	.978	.987	.993	.996	.998	.999	.999	1.000		
17	.959	.975	.985	.991	.995	.997	.999	.999	1.000	
18	.932	.955	.972	.983	.990	.994	.997	.998	.999	1.000
19	.893	.927	.951	.969	.980	.988	.993	.996	.998	.999
20	.843	.888	.922	.948	.966	.978	.987	.992	.995	.997
21	.782	.838	.883	.917	.944	.963	.976	.985	.991	.994
22	.812	.777	.832	.877	.913	.940	.959	.973	.983	.989
23	.635	.708	.772	.827	.873	.908	.936	.956	.971	.981
24	.554	.632	.704	.768	.823	.868	.904	.932	.953	.969
25	.473	.553	.629	.700	.763	.818	.863	.900	.929	.950

Contd...

Table A.2 Contd.

λ\x→ ↓	34	35	36	37	38	39	40	41	42	43
19	.999	1.000								
20	.999	.999	1.000							
21	.997	.998	.999	.999	1.000					
22	.994	.996	.998	.999	.999	1.000				
23	.988	.993	.996	.997	.999	.999	1.000			
24	.979	.987	.992	.995	.997	.998	.999	.999	1.000	
25	.966	.978	.985	.99	.994	.997	.998	.999	.999	1.000

Source: Rao, Mitra, Mathai and Ramamurthy (1975) Formula and Tables for Statistical Work

Table A.3 The standard normal distribution: ordinates at z and area from 0 to z [z = 0.000 (0.001) 0.299]

Ordinate ϕ (z)	z	Probability integral $A(z)$ from 0 to z									
		0	1	2	3	4	5	6	7	8	9
.398942	0.00	.000000	.000399	.000798	.001197	.001596	.001995	.002394	.002793	.003192	.003590
.398922	0.01	.003989	.004388	.004787	.005186	.005585	.005984	.006383	.006782	.007181	.007579
.398862	0.02	.007973	.008377	.008776	.009175	.009574	.009973	.010371	.010770	.011169	.011568
.398796	0.03	.11966	.012365	.012764	.013163	.013561	.013960	.014359	.014757	.015156	.015555
.398623	0.04	.15953	.016352	.016751	.017149	.017548	.017946	.018345	.018743	.019142	.019540
.398444	0.05	.019939	.020337	.020736	.021134	.021532	.021931	.022329	.022727	.023126	.023524
.398225	0.06	.023922	.024320	.024719	.025117	.025515	.025913	.026311	.026709	.027407	.027505
.397966	0.07	.027903	.028301	.028699	.029097	.029495	.029893	.030290	.030688	.031086	.031484
.397668	0.08	.031881	.032279	.032677	.033074	.033472	.033869	.034267	.034664	.035062	.035459
.397330	0.09	.035856	.036254	.036651	.037048	.037445	.037843	.038240	.038637	.039034	.039431
.396953	0.10	.039828	.040225	.040622	.041019	.041415	.041812	.042209	.042606	.043002	.043399
.396536	0.11	.043795	.044192	.044588	.044985	.045381	.045777	.046174	.046570	.046966	.047362
.396080	0.12	.047758	.048154	.048550	.048946	.049342	.049738	.050134	.050530	.050926	.051321
.395585	0.13	.051717	.052112	.052508	.052903	.053299	.053694	.054089	.054485	.054880	.055275
.395052	0.14	.055670	.056065	.056460	.056855	.057250	.057645	.058039	.058434	.058829	.059223
.394479	0.15	.059618	.060012	.060407	.060801	.061195	.061589	.061983	.062378	.062772	.063166
.393863	0.16	.063559	.063953	.064347	.064741	.065134	.065528	.065922	.066315	.066708	.067102
.393219	0.17	.067495	.067888	.068281	.068674	.069067	.069460	.069853	.070546	.070639	.071031
.392531	0.18	.071424	.071816	.072209	.072601	.072993	.073385	.073778	.074170	.074562	.074954
.391806	0.19	.075345	.075737	.076129	.076521	.076912	.077304	.077695	.078086	.078477	.078869
.391043	0.20	.079260	.079651	.080042	.080432	.080823	.081214	.081605	.081995	.082386	.082776
.390242	0.21	.083166	.083556	.083946	.084337	.084726	.085116	.085506	.085896	.086285	.086675
.389404	0.22	.087064	.087454	.087843	.088232	.088621	.089010	.089399	.089788	.090177	.090566
.388529	0.23	.090954	.091343	.091731	.092119	.092508	.092896	.093284	.093672	.094059	.094447
.387617	0.24	.094835	.095222	.095610	.095997	.096385	.096772	.097159	.097546	.097933	.098320
.386668	0.25	.098706	.099093	.099479	.099866	.100252	.100638	.101025	.101411	.101797	.102182
.385683	0.26	.102568	.102954	.103339	.103725	.104110	.104495	.104880	.105265	.105650	.106035
.384663	0.27	.106420	.106804	.107189	.107573	.107958	.108342	.108726	.109110	.109494	.109878
.383606	0.28	.110261	.110645	.111028	.111412	.111795	.112178	.112561	.112944	.113327	.113709
.382515	0.29	.114092	.114474	.114857	.115239	.115621	.116003	.116385	.116767	.116148	.117530

Contd...

Entry in the table after the column of values of z is the proportion of area under the entire standard normal curve, which is between $x = 0$ and a positive value of z. Ordinate values ϕ (z) are given in first column. Areas and ordinates for negative values of z are obtained by symmetry.

Source: Rao, Mitra, Mathai and Ramamurthy (1975) Formula and Tables for Statistical Work

Table A.3 Contd.

[z = 0.300 (0.001) 0.599]

Ordinate ϕ (z)	z	Probability integral $A(z)$ from 0 to z									
		0	1	2	3	4	5	6	7	8	9
.381388	0.30	.117911	.118293	.118674	.119055	.119436	.119817	.120198	.120578	.120958	.121339
.380226	0.31	.121720	.122100	.122480	.122860	.123239	.123619	.123999	.124378	.124758	.125137
.379031	0.32	.125516	.125895	.126274	.126652	.127031	.127409	.127788	.128166	.128544	.128922
.377801	0.33	.129300	.129678	.130055	.130433	.130810	.131187	.131565	.131942	.132318	.132695
.376537	0.34	.133072	.133448	.133825	.134201	.134577	.134953	.135329	.135704	.136080	.136455
.375240	0.35	.136831	.137206	.137581	.137956	.138331	.138705	.139080	.139454	.139826	.140202
.373911	0.36	.140576	.140950	.141324	.141698	.142071	.142444	.142817	.143190	.143563	.143936
.372548	0.37	.144309	.144681	.145054	.145426	.145798	.146170	.146542	.146913	.147285	.147656
.371154	0.38	.148027	.148398	.148769	.149140	.149511	.149881	.150252	.150622	.150992	.151362
.369728	0.39	.151732	.152101	.152471	.152840	.153209	.153579	.153947	.154316	.154685	.155053
.368270	0.40	.155422	.155790	.156158	.156526	.156894	.157261	.157629	.157996	.158363	.158730
.366782	0.41	.159097	.159464	.159830	.160197	.160563	.160929	.161295	.161661	.162026	.162392
.365263	0.42	.162757	.163122	.163487	.163852	.164217	.164582	.164946	.165310	.165674	.166038
.363714	0.43	.166402	.166766	.167129	.167893	.167856	.168219	.168582	.168944	.169307	.169669
.362135	0.44	.170031	.170394	.170755	.171117	.171479	.171840	.172201	.172562	.172923	.173254
.360527	0.45	.173645	.174005	.174366	.174726	.175086	.175445	.175805	.176164	.176524	.176883
.358890	0.46	.177242	.177601	.177959	.178318	.178676	.179034	.179392	.179750	.180108	.180465
.357225	0.47	.180822	.181180	.181537	.181893	.182250	.182607	.182963	.183319	.183675	.184031
.355533	0.48	.184386	.184742	.185097	.185452	.185807	.186162	.186516	.186871	.187225	.187579
.353812	0.49	.187933	.188287	.188640	.188994	.189347	.189700	.190053	.190405	.190758	.191110
.352065	0.50	.191462	.191814	.192166	.192518	.192869	.193221	.193572	.193923	.194273	.194624
.350292	0.51	.194974	.195324	.195674	.196024	.196374	.196723	.197073	.197422	.197771	.198120
.348493	0.52	.198468	.198817	.199165	.199513	.199861	.200208	.200556	.200903	.201250	.201597
.346668	0.53	.201944	.202291	.202637	.202983	.203329	.203675	.204021	.204366	.204711	.205057
.344818	0.54	.205401	.205746	.206091	.206435	.206779	.207123	.207467	.207811	.208154	.208497
.342944	0.55	.208840	.209183	.209526	.209868	.210211	.210553	.210895	.211236	.211578	.211919
.341046	0.56	.212260	.212601	.212942	.213283	.213623	.213963	.214303	.214643	.214983	.215322
.339124	0.57	.215661	.216000	.216339	.216678	.217016	.217354	.217692	.218030	.218368	.218705
.337180	0.58	.219043	.219380	.219717	.220083	.220390	.220726	.221062	.221398	.221734	.222069
.335213	0.59	.222405	.222740	.223075	.223409	.223744	.224078	.224412	.224746	.225080	.225414

Contd...

Table A.3 Contd.

[z = 0.600 (0.001) 0.899]

Ordinate ϕ (z)	z	Probability integral $A(z)$ from 0 to z									
		0	1	2	3	4	5	6	7	8	9
.333225	0.60	.225747	.226080	.226413	.226746	.227078	.227411	.227743	.228075	.228406	.228738
.331215	0.61	.229069	.229400	.229731	.230062	.230392	.230723	.231053	.231383	.231712	.232042
.329184	0.62	.232371	.232700	.233029	.233358	.233686	.234014	.234343	.234670	.234998	.235325
.327133	0.63	.235653	.235980	.236307	.236633	.236960	.237286	.237612	.237938	.238263	.238589
.325062	0.64	.238914	.239239	.239563	.039888	.240212	.240536	.240860	.241184	.241508	.241831
.322972	0.65	.242154	.242477	.242799	.243122	.243444	.243766	.244088	.244410	.244731	.245052
.320864	0.66	.245373	.245697	.246014	.246335	.246655	.246975	.247294	.247614	.247933	.248252
.318737	0.67	.248571	.248890	.249208	.249526	.249844	.250162	.250480	.250797	.251114	.251431
.316593	0.68	.251748	.252064	.252381	.252697	.253012	.253328	.253643	.253959	.254274	.254588
.314432	0.69	.254903	.255217	.255531	.255845	.256159	.256472	.256786	.257099	.257411	.257724
.312254	0.70	.258036	.258348	.258660	.258972	.259284	.259595	.259906	.260217	.260527	.260838
.310060	0.71	.261148	.261458	.261768	.262077	.262386	.262695	.263004	.263313	.263621	.263930
.307851	0.72	.264238	.264545	.264853	.265160	.265467	.265774	.266081	.266387	.266693	.266999
.305627	0.73	.267305	.267610	.267916	.268221	.268526	.268830	.269135	.269439	.269743	.270047
.303389	0.74	.270350	.270653	.270956	.271259	.271562	.271864	.272166	.272468	.272770	.273071
.301137	0.75	.273373	.273674	.273974	.274275	.274575	.274876	.275175	.275475	.275775	.276074
.298872	0.76	.276373	.276671	.276970	.277268	.277566	.277864	.278162	.278459	.278756	.279053
.296595	0.77	.279350	.276947	.279943	.280239	.280535	.280830	.281126	.281421	.281715	.282010
.294305	0.78	.282305	.282599	.282893	.283186	.283480	.283773	.284066	.248359	.284652	.284944
.292004	0.79	.285236	.285528	.285820	.286111	.286402	.286693	.286984	.287274	.287565	.287855
.289692	0.80	.288145	.288434	.288724	.289013	.289302	.289590	.289879	.290167	.290455	.290742
.287369	0.81	.291030	.291317	.291604	.291891	.292178	.292464	.292750	.293036	.293321	.293607
.285036	0.82	.293892	.294177	.294462	.294746	.295030	.295314	.295598	.295881	.296165	.296448
.282694	0.83	.296731	.297013	.297296	.297578	.297860	.298141	.298423	.298704	.298985	.299265
.280344	0.84	.299546	.299826	.300106	.300386	.300665	.300945	.301224	.301502	.301781	.302059
.277985	0.85	.302337	.302615	.302893	.303170	.303448	.303724	.304001	.304278	.304554	.304830
.275618	0.86	.305105	.305381	.305656	.305931	.306206	.306481	.306755	.307029	.307303	.307576
.273244	0.87	.307850	.308123	.308396	.308668	.308941	.309213	.309485	.309757	.310028	.310299
.270864	0.88	.310570	.310841	.311112	.311382	.311652	.311922	.312191	.312461	.312730	.312998
.268477	0.89	.313267	.313535	.313804	.314071	.314339	.314606	.314874	.315141	.315407	.315674

Contd...

Table A.3 Contd.

[z = 0.900 (0.001) 1.199]

Ordinate ϕ (z)	z	Probability integral $A(z)$ from 0 to z									
		0	1	2	3	4	5	6	7	8	9
.266085	0.90	.315940	.316206	.316472	.316737	.317002	.317267	.317532	.317797	.318061	.318325
.263688	0.91	.318589	.318852	.319116	.319379	.319642	.319904	.320167	.320429	.320691	.320952
.261286	0.92	.321214	.321475	.321736	.321996	.322257	.322517	.322777	.323037	.623296	.323555
.258881	0.93	.323814	.324073	.324332	.324590	.324848	.325106	.325363	.325621	.328878	.326135
.256471	0.94	.326391	.326647	.326904	.327160	.327415	.327671	.327926	.328181	.328435	.328690
.254059	0.95	.328944	.329198	.329452	.329705	.329958	.330211	.330464	.330716	.330969	.331221
.251644	0.96	.331472	.331724	.331975	.332226	.332477	.332728	.332978	.333228	.333478	.333727
.249228	0.97	.333977	.334226	.334475	.334723	.334972	.335220	.335468	.335715	.335963	.336210
.246809	0.98	.336457	.336704	.336950	.337196	.337442	.337688	.337933	.338179	.338424	.338668
.244390	0.99	.338913	.339157	.339401	.339645	.339889	.340132	.340375	.340618	.340860	.341103
.241971	1.00	.341345	.341587	.341828	.342070	.342311	.342552	.342792	.343038	.343273	.343513
.239551	1.01	.343752	.343992	.344231	.344470	.344709	.344947	.345185	.345423	.345661	.345899
.237132	1.02	.346136	.346373	.346610	.346846	.347082	.347318	.347554	.347790	.348025	.348260
.234714	1.03	.348495	.348730	.348964	.349198	.349432	.349666	.349899	.350132	.350365	.350598
.232297	1.04	.350830	.351062	.351294	.351526	.351757	.351989	.352219	.352450	.352681	.352911
.229882	1.05	.353141	.353371	.353600	.353830	.354059	.354287	.354516	.354744	.354972	.355200
.227470	1.06	.355428	.355655	.355882	.356109	.356336	.356562	.356788	.357014	.357240	.357465
.225060	1.07	.357690	.357915	.358140	.358364	.358589	.358813	.359036	.359560	.359488	.359706
.222653	1.08	.359929	.360151	.360374	.360596	.360818	.361039	.361261	.361482	.361702	.361923
.220251	1.09	.362143	.362364	.362583	.362803	.363023	.363242	.363461	.363679	.363898	.364116
.217852	1.10	.364334	.364552	.364769	.364986	.365203	.365420	.365637	.365853	.366069	.366285
.215458	1.11	.366500	.366716	.366931	.367146	.367360	.367575	.367789	.368003	.368217	.368430
.213069	1.12	.368643	.368856	.369069	.369281	.369493	.369705	.369917	.370129	.370340	.370551
.210686	1.13	.370762	.370972	.371183	.371393	.371603	.371812	.372022	.372231	.372440	.372648
.208308	1.14	.372857	.373065	.373273	.373481	.373688	.373895	.374102	.374309	.374516	.374722
.205936	1.15	.374928	.375134	.375339	.375545	.375750	.375955	.376159	.376364	.376568	.376772
.203571	1.16	.376976	.377179	.377382	.377585	.377788	.377991	.378193	.378395	.378597	.378798
.201214	1.17	.379000	.379201	.379401	.379602	.379802	.380003	.380203	.380402	.380602	.380801
.198863	1.18	.381000	.381199	.381397	.381595	.381793	.381991	.382189	.382386	.382583	.382780
.196520	1.19	.382977	.383173	.383369	.383565	.383761	.383956	.384152	.384347	.384541	.384736

Contd...

Table A.3 Contd.

[z = 1.200 (0.001) 1.499]

Ordinate ϕ (z)	z	Probability integral $A(z)$ from 0 to z									
		0	1	2	3	4	5	6	7	8	9
.194186	1.20	.384930	.385124	.385318	.385512	.385705	.385898	.386091	.386284	.386476	.386669
.191860	1.21	.386861	.387052	.387244	.387435	.387626	.387817	.388008	.388198	.388388	.388578
.189543	1.22	.388768	.388957	.389146	.389335	.389524	.389712	.389901	.390089	.390277	.390464
.187235	1.23	.390651	.390839	.391025	.391212	.391399	.391585	.391771	.391956	.392142	.392327
.184937	1.24	.392512	.392697	.392882	.393066	.393250	.393434	.393618	.393801	.393984	.394167
.182649	1.25	.394350	.394533	.394715	.394897	.395079	.395261	.395442	.395623	.395804	.395985
.180371	1.26	.396165	.396346	.396526	.396705	.396885	.397064	.397243	.397422	.397601	.397779
.178104	1.27	.397958	.398136	.398313	.398491	.398668	.398845	.399022	.399199	.399375	.399551
.175847	1.28	.399727	.399903	.400079	.400254	.400429	.400604	.400778	.400953	.401127	.401301
.173602	1.29	.401475	.401648	.401821	.401994	.402167	.402340	.402512	.402684	.402856	.403028
.171369	1.30	.403200	.403371	.403542	.403713	.403883	.404054	.404224	.404394	.404563	.404733
.169147	1.31	.404902	.405071	.405240	.405409	.405577	.405745	.405913	.406081	.406248	.406415
.166937	1.32	.406582	.406749	.406916	.407082	.407248	.407414	.407580	.407746	.407911	.408076
.164740	1.33	.408241	.408405	.408570	.408734	.408898	.409062	.409225	.409389	.409552	.409715
.162555	1.34	.409877	.410040	.410202	.410364	.410526	.410687	.410849	.411010	.411171	.411332
.160383	1.35	.411492	.411652	.411812	.411972	.412132	.412291	.412450	.412609	.412768	.412927
.158225	1.36	.413085	.413243	.413401	.413559	.413716	.413873	.414031	.414187	.414344	.414500
.156080	1.37	.414657	.414813	.414968	.415124	.415279	.415434	.415589	.415744	.415898	.416053
.153948	1.38	.416207	.416361	.416514	.41668	.416821	.416974	.417127	.417279	.417437	.417584
.151831	1.39	.417736	.417887	.418039	.418190	.418341	.418492	.418643	.418793	.418973	.419094
.149727	1.40	.419243	.419393	.419542	.149692	.419841	.419989	.420138	.420286	.420434	.420582
.147639	1.41	.420730	.420878	.421025	.421172	.421319	.421466	.421612	.421759	.421905	.422050
.145564	1.42	.422196	.422342	.422487	.422632	.422777	.422921	.423066	.423210	.423354	.423498
.143505	1.43	.423641	.423785	.423928	.424071	.424214	.424356	.424499	.424641	.424783	.424925
.141460	1.44	.425066	.425208	.425349	.425490	.425631	.425771	.425911	.426052	.426191	.426331
.139431	1.45	.426471	.426610	.426749	.426888	.427027	.427165	.427304	.427442	.427580	.427717
.137417	1.46	.427855	.427992	.428129	.428266	.428403	.428540	.428676	.428812	.428948	.429084
.135418	1.47	.729219	.429354	.429490	.429624	.429759	.429894	.430028	.430162	.430296	.430430
.133435	1.48	.430563	.430697	.430830	.430963	.431096	.431228	.431360	.431493	.431625	.431756
.131468	1.49	.431888	.432019	.432150	.432281	.432412	.432543	.432673	.432803	.432933	.433063

Contd...

Table A.3 Contd.

Ordinate $\phi(z)$	z	Probability integral $A(z)$ from 0 to z [z = 1.500 (0.001) 1.799]									
		0	1	2	3	4	5	6	7	8	9
.129518	1.50	.433193	.433322	.433451	.433580	.433709	.433838	.433966	.434095	.434223	.434351
.127583	1.51	.434478	.434606	.434733	.434860	.434987	.435114	.435240	.435367	.435493	.435619
.125665	1.52	.435745	.435870	.435995	.436121	.436246	.436370	.436495	.436619	.436744	.436868
.123763	1.53	.436992	.437115	.437239	.437362	.437485	.437608	.437731	.437853	.437976	.438798
.121878	1.54	.438220	.438342	.438463	.438585	.438706	.438827	.438948	.439068	.439189	.439309
.120009	1.55	.439429	.439549	.439669	.439788	.439908	.440027	.440146	.440265	.440383	.440502
.118157	1.56	.440620	.440738	.440856	.440974	.441091	.441209	.441326	.441443	.441559	.441676
.116323	1.57	.441792	.441909	.442025	.442141	.442256	.442372	.442487	.442602	.442717	.442832
.114505	1.58	.442947	.443061	.443175	.443289	.443403	.443517	.443630	.443744	.443857	.443970
.112704	1.59	.444083	.444195	.444308	.444420	.444532	.444644	.444756	.444867	.444979	.445090
.110921	1.60	.445201	.445312	.445422	.445533	.445643	.445753	.445863	.445973	.446082	.446192
.109155	1.61	.446301	.446410	.446519	.446628	.446736	.446845	.446953	.447061	.447169	.447276
.107406	1.62	.447384	.447491	.447598	.447705	.447812	.447919	.448025	.448131	.448238	.448343
.105675	1.63	.448449	.448555	.448660	.448766	.448871	.448975	.449080	.449185	.449289	.449393
.103961	1.64	.449497	.449601	.449705	.449809	.449912	.450015	.450118	.450221	.450324	.450426
.102265	1.65	.450529	.450631	.450733	.450835	.450936	.451038	.451139	.451240	.451341	.451442
.100586	1.66	.451543	.451643	.451744	.451844	.451944	.452044	.452143	.452243	.452342	.452441
.098925	1.67	.452540	.452639	.452738	.452836	.452935	.453033	.453131	.453229	.453326	.453424
.097282	1.68	.453521	.453619	.453716	.453812	.453909	.454006	.454102	.454198	.454294	.454390
.095657	1.69	.454486	.454582	.45467	.454772	.454867	.454962	.455057	.455152	.455246	.455340
.094049	1.70	.455435	.455529	.455622	.455716	.455809	.455903	.455996	.456089	.456182	.456275
.092459	1.71	.456367	.456459	.456552	.456644	.456736	.456827	.456919	.457010	.457102	.457193
.090887	1.72	.457284	.457375	.457465	.457556	.457646	.45736	.457826	.457916	.458006	.458095
.089333	1.73	.458185	.458274	.458363	.458452	.458541	.458630	.458718	.458806	.458895	.458983
.087796	1.74	.459070	.459158	.459246	.458933	.459420	.459508	.459595	.459681	.459768	.459854
.086277	1.75	.459941	.460027	.460113	.460199	.460285	.460370	.460456	.460541	.460626	.460711
.084776	1.76	.460796	.460881	.460965	.461050	.461134	.461218	.461302	.461386	.461470	.461553
.083293	1.77	.461636	.461720	.461803	.461886	.461968	.462051	.462134	.462216	.462298	.462380
.081828	1.78	.462462	.462544	.462625	.462707	.462788	.462869	.462950	.463031	.463112	.463193
.080380	1.79	.463273	.463353	.463434	.463514	.463593	.463673	.463753	.463832	.463911	.463991

Contd...

Table A.3 Contd.

[z = 1.800 (0.001) 2.099]

Ordinate ϕ (z)	z	Probability integral $A(z)$ from 0 to z									
		0	1	2	3	4	5	6	7	8	9
.078950	1.80	.464070	.464149	.464227	.464306	.464384	.464463	.464541	.464619	.464697	.464774
.077538	1.81	.464852	.464930	.465007	.465084	.465161	.465238	.465315	.465391	.465468	.465544
.076413	1.82	.465620	.465697	.465773	.465848	.465924	.465999	.466075	.466150	.466225	.466300
.074766	1.83	.466375	.466450	.466524	.466599	.466773	.466747	.466821	.466895	.466969	.467042
.073407	1.84	.467116	.467189	.467262	.467335	.467408	.467481	.467554	.467626	.467699	.467771
.072065	1.85	.467843	.467915	.467987	.468059	.468130	.468202	.468273	.468344	.468415	.468486
.070740	1.86	.468557	.468628	.468698	.468769	.468839	.468909	.468979	.469049	.469119	.469189
.069433	1.87	.469258	.469327	.469397	.469466	.469535	.469604	.469672	.469741	.469809	.469878
.068144	1.88	.469946	.470014	.470082	.470150	.470218	.470285	.470353	.470420	.470487	.470554
.066871	1.89	.470621	.470688	.470755	.470821	.470887	.470954	.471020	.471086	.471152	.471218
.065616	1.90	.471283	.471349	.471414	.471480	.471545	.471610	.471675	.471740	.471804	.471869
.064378	1.91	.471933	.471998	.472062	.472126	.472190	.472254	.472317	.472381	.472444	.472508
.063157	1.92	.472571	.472634	.472697	.472760	.472823	.472885	.472948	.473010	.473072	.473135
.061952	1.93	.473197	.473258	.473320	.473382	.473443	.473505	.473566	.473627	.473688	.473749
.060765	1.94	.473810	.473871	.473931	.473992	.474052	.474113	.474173	.474233	.474293	.474352
.059595	1.95	.474412	.474471	.474531	.474590	.474649	.474708	.474767	.474826	.474885	.474944
.058441	1.96	.475002	.475060	.475119	.475177	.475235	.475293	.475351	.475408	.475466	.475523
.057304	1.97	.475581	.475638	.475695	.475752	.475809	.475866	.475923	.475979	.476036	.476092
.056183	1.98	.476148	.476204	.476260	.476316	.476372	.476428	.476483	.476539	.476594	.476649
.055079	1.99	.476705	.476760	.476814	.476869	.476924	.476979	.477033	.477087	.477142	.477196
.053991	2.00	.477250	.477304	.477358	.477411	.477465	.477518	.477572	.477625	.477678	.477731
.052919	2.01	.477784	.477837	.477890	.477943	.477995	.478048	.478100	.478152	.478204	.478256
.051864	2.02	.478308	.478360	.478412	.478563	.478515	.478566	.478618	.478669	.478720	.478771
.050824	2.03	.478822	.478873	.478923	.478974	.479024	.479075	.479125	.479175	.479225	.479275
.049800	2.04	.479325	.479375	.479424	.479474	.479523	.479573	.479622	.479671	.479720	.479769
.048792	2.05	.479818	.479867	.479915	.479964	.480012	.480060	.480109	.480157	.480205	.480253
.047800	2.06	.480301	.480348	.480396	.480444	.480491	.480538	.480586	.480633	.480680	.480727
.046823	2.07	.480774	.480821	.480867	.480914	.480960	.481007	.481053	.481099	.481145	.481191
.045861	2.08	.481237	.481283	.481329	.481374	.481420	.481465	.481511	.481556	.481601	.481646
.044915	2.09	.481691	.481736	.481781	.481825	.481870	.481915	.481959	.482003	.482047	.482092

Contd...

Table A.3 Contd.

[z = 2.100 (0.001) 2.399]

Ordinate ϕ (z)	z	Probability integral $A(z)$ from 0 to z									
		0	1	2	3	4	5	6	7	8	9
.043984	2.10	.482136	.482180	.482223	.482267	.482311	.482354	.482398	.482441	.482485	.482528
.043067	2.11	.482571	.482614	.482657	.482700	.482742	.482785	.482828	.482870	.482912	.482955
.042166	2.12	.482997	.483039	.483081	.483123	.483165	.483207	.483248	.483290	.483331	.483373
.041280	2.13	.483414	.483455	.483497	.483538	.483579	.483619	.483660	.483701	.483742	.483782
.040408	2.14	.483826	.483863	.483903	.483943	.483984	.484024	.484064	.484103	.484143	.484183
.039550	2.15	.484222	.484262	.484301	.484341	.484380	.484419	.484458	.484497	.484536	.484575
.038707	2.16	.484614	.484652	.484691	.484729	.484768	.484806	.484844	.484883	.484921	.484959
.037878	2.17	.484997	.485034	.495072	.485110	.485147	.485185	.485222	.485260	.485297	.485334
.037063	2.18	.485371	.485408	.485445	.485482	.485519	.485556	.485992	.485629	.485665	.485702
.036262	2.19	.485738	.485774	.485810	.485846	.485882	.485918	.485954	.485990	.486025	.586061
.035475	2.20	.486097	.486132	.486167	.486203	.486238	.486273	.486308	.486343	.486378	.486413
.034701	2.21	.486447	.486482	.486517	.486551	.486586	.486620	.486654	.486688	.486723	.486757
.033941	2.22	.486791	.486825	.486858	.486892	.486926	.486959	.486993	.487026	.487060	.487093
.033194	2.23	.487126	.487159	.487193	.487226	.487258	.487291	.487324	.487357	.487389	.487422
.032460	2.24	.487455	.487487	.487519	.487552	.487584	.487616	.487648	.487680	.487712	.487744
.031740	2.25	.487776	.487807	.487839	.487870	.487902	.487933	.487965	.487996	.488027	.488058
.031032	2.26	.488089	.488120	.488151	.488182	.488213	.488244	.488274	.488305	.488335	.488366
.030337	2.27	.488396	.488427	.488457	.488487	.488517	.488547	.488577	.488607	.488637	.488666
.029655	2.28	.488696	.488726	.488755	.488785	.488814	.488844	.488873	.488902	.488931	.488960
.028985	2.29	.488989	.489018	.489047	.489076	.489105	.489133	.489162	.489191	.489219	.489248
.028327	2.30	.489276	.489304	.489332	.489361	.489389	.489417	.489445	.489473	.489500	.489528
.027682	2.31	.489556	.489584	.489611	.489639	.489666	.489694	.489721	.489748	.489775	.489802
.027048	2.32	.489830	.489857	.489884	.489910	.489937	.489964	.489991	.490017	.490044	.490070
.026426	2.33	.490097	.490123	.490150	.490176	.490202	.490228	.490254	.490280	.490306	.490332
.025817	2.34	.490358	.490384	.490410	.490435	.490461	.490486	.490512	.490537	.490563	.490588
.025218	2.35	.490613	.490938	.490664	.490689	.490714	.490739	.490764	.490788	.490813	.490838
.024631	2.36	.490863	.490887	.490912	.490936	.490961	.490985	.491009	.491034	.491058	.491082
.024056	2.37	.491106	.491130	.491154	.491178	.491202	.491226	.491249	.491273	.491297	.491320
.023491	2.38	.491344	.491367	.491391	.491414	.491437	.491460	.491484	.491507	.491530	.491553
.022937	2.39	.491576	.491599	.491622	.491644	.491667	.491690	.491712	.491735	.491758	.491780

Contd...

Table A.3 Contd.

[z = 2.400 (0.001) 2.699]

Ordinate ϕ (z)	z	Probability integral $A(z)$ from 0 to z									
		0	1	2	3	4	5	6	7	8	9
.022395	2.40	.491802	.491825	.491847	.491869	.491892	.491914	.491936	.491958	.491980	.492002
.021862	2.41	.492024	.492046	.492067	.492089	.492111	.492132	.492154	.492175	.492197	.492218
.021341	2.42	.492240	.492261	.492282	.492304	.492325	.492346	.492367	.492388	.492409	.492430
.020829	2.43	.492451	.492471	.492492	.492513	.492534	.492554	.492575	.492595	.492616	.492636
.020328	2.44	.492656	.492677	.492697	.492717	.492737	.492757	.492777	.492797	.492817	.492837
.019837	2.45	.492857	.492877	.492897	.492916	.492936	.492956	.492975	.492995	.493014	.493034
.019356	2.46	.493053	.493072	.493092	.493111	.493130	.493149	.493168	.493187	.493206	.493225
.018885	2.47	.493244	.493263	.493282	.493301	.493320	.493338	.493357	.493375	.493394	.493412
.018423	2.48	.493431	.493449	.493468	.493486	.493504	.493522	.493541	.493559	.493577	.493595
.017971	2.49	.493613	.493631	.493649	.493667	.493684	.493702	.493720	.493738	.493755	.493773
.017528	2.50	.493790	.493808	.493825	.493843	.493860	.493877	.493859	.493912	.493929	.493946
.017095	2.51	.493963	.493981	.493998	.494015	.494031	.494048	.494065	.494082	.494099	.494115
.016670	2.52	.494132	.494149	.494166	.494182	.494199	.494215	.494232	.494248	.494264	.494281
.016254	2.53	.494297	.494313	.494329	.494345	.494362	.494378	.494394	.494410	.494426	.494442
.015843	2.54	.494457	.494473	.494489	.494505	.494520	.494536	.494552	.494567	.494583	.494598
.015449	2.55	.494614	.494629	.494645	.494660	.494675	.494691	.494706	.494721	.494736	.494751
.015060	2.56	.494766	.494781	.494796	.494811	.494826	.494841	.494856	.494871	.494886	.494900
.014678	2.57	.494915	.494930	.494944	.494959	.494973	.494988	.495002	.495017	.495031	.495046
.014305	2.78	.495060	.495074	.495089	.495103	.495117	.495131	.495145	.495159	.495173	.495187
.013940	2.79	.495201	.495215	.495229	.495243	.495257	.495270	.495284	.495298	.495312	.495325
.013583	2.60	.495339	.495352	.495366	.495379	.495393	.495406	.495420	.495433	.495446	.495460
.013234	2.61	.495473	.495486	.495499	.495512	.495526	.495539	.495552	.495565	.495578	.495591
.012892	2.62	.495604	.495616	.405629	.495642	.495655	.495668	.495680	.495693	.495706	.495718
.012558	2.63	.495731	.495743	.495756	.495768	.495781	.495793	.495806	.495818	.495830	.495842
.012232	2.64	.495855	.495867	.495879	.495891	.495903	.495915	.495928	.495940	.495952	.495963
.011912	2.65	.495975	.495987	.495999	.496011	.496023	.496035	.496046	.496058	.496070	.496081
.011600	2.66	.496093	.496105	.496116	.496128	.496139	.496151	.496162	.496173	.496185	.496196
.011295	2.67	.496207	.496219	.496230	.496241	.496252	.496264	.496275	.496286	.496297	.496308
.010997	2.68	.496319	.496330	.496341	.496352	.496363	.496374	.496384	.496395	.496406	.406417
.010706	2.69	.496427	.496438	.496449	.496459	.496470	.496481	.496491	.496502	.496512	.496523

Contd...

Table A.3 Contd.

[z = 2.700 (0.001) 2.999]

Ordinate ϕ (z)	z	Probability integral $A(z)$ from 0 to z									
		0	1	2	3	4	5	6	7	8	9
.010421	2.70	.496533	.496543	.496554	.496564	.496574	.496585	.496595	.496605	.496615	.496626
.010143	2.71	.496636	.496646	.496656	.496666	.496676	.496686	.496696	.496706	.496716	.496726
.009871	2.72	.496736	.496746	.496756	.496765	.496775	.496785	.496795	.496804	.496814	.496824
.009606	2.73	.496833	.496843	.496852	.496862	.496871	.496881	.496890	.496900	.496909	.496919
.009347	2.74	.496928	.4966937	.496947	.496956	.496965	.496974	.496984	.496993	.497002	.497011
.009094	2.75	.497020	.497029	.497038	.497047	.497056	.497065	.497074	.497083	.497092	.497101
.008846	2.76	.497110	.497119	.497128	.497136	.497145	.497154	.497163	.497171	.497180	.497189
.088605	2.77	.497197	.497206	.497214	.497223	.497231	.497240	.497248	.497257	.497265	.497274
.008370	2.78	.497282	.497290	.497299	.497307	.497315	.497324	.497332	.497340	.497348	.497356
.008140	2.79	.497365	.497373	.497381	.497389	.497397	.497405	.497413	.497421	.497429	.497437
.007915	2.80	.497445	.497453	.497461	.497469	.497476	.497484	.497492	.497500	.497507	.497515
.007697	2.81	.497523	.497531	.497538	.497546	.497554	.497561	.497569	.497576	.497584	.497591
.007483	2.82	.497599	.497606	.497614	.497621	.497629	.497636	.497643	.497651	.497658	.497665
.007274	2.83	.497673	.497680	.497687	.497694	.497702	.497709	.497716	.497723	.497730	.497737
.007071	2.84	.497744	.497751	.497758	.497765	.497772	.497779	.497786	.497793	.497800	.497807
.006873	2.85	.497814	.497821	.497828	.497835	.497841	.497848	.497855	.497862	.497868	.497875
.006679	2.86	.497882	.497888	.497895	.497902	.497908	.497915	.497922	.497928	.497935	.497941
.006491	2.87	.497948	.497954	.497961	.497967	.497973	.497980	.497986	.497993	.497999	.498005
.006307	2.88	.498012	.498018	.498024	.498030	.498037	.498043	.498049	.498055	.498062	.498068
.006127	2.89	.498074	.498080	.498086	.498092	.498098	.498104	.498110	.498116	.498122	.498128
.005953	2.90	.498134	.498140	.498146	.498152	.498158	.498164	.498170	.498175	.498181	.498187
.005782	2.91	.498193	.498199	.498204	.498210	.498216	.498222	.498227	.498233	.498239	.498244
.005616	2.92	.498250	.498255	.498261	.498267	.498272	.498278	.498283	.498289	.498294	.498300
.005454	2.93	.498305	.498311	.498316	.498321	.498327	.498332	.498338	.498343	.498348	.498354
.005296	2.94	.498359	.498634	.498370	.498375	.498380	.498385	.498390	.498396	.498401	.498406
.005143	2.95	.498411	.498416	.498421	.498426	.498432	.498437	.498442	.498447	.498452	.498457
.004993	2.96	.498462	.498467	.498472	.498477	.498482	.498487	.498491	.498496	.498501	.498506
.004847	2.97	.498511	.498516	.498521	.498525	.498530	.498535	.498540	.498545	.498549	.498554
.004705	2.98	.498559	.498563	.498568	.498573	.498577	.498582	.498587	.498591	.498596	.498601
.004567	2.99	.498605	.498610	.498614	.498619	.498623	.498628	.498632	.498637	.498641	.498646

Contd...

Table A.3 Contd.

[z = 3.000 (0.01) 4.09]

Ordinate ϕ (z)	z	Probability integral $A(z)$ from 0 to z									
		0	1	2	3	4	5	6	7	8	9
.004432	3.0*	.498650	.498694	.498736	.498777	.498817	.498856	.498893	.498930	.498965	.498999
.003267	3.1	.499032	.499065	.499096	.499126	.499155	.499184	.499211	.499238	.499264	.499289
.002384	3.2	.499313	.499336	.499359	.499381	.499402	.499423	.499443	.499462	.499481	.499499
.001723	3.3	.499517	.499534	.499550	.499566	.499581	.499596	.499610	.499624	.499638	.499651
.001232	3.4	.499663	.499675	.499687	.499698	.499709	.499720	.499730	.499740	.499749	.499758
.000873	3.5	.499767	.499776	.499784	.499792	.499800	.499807	.499815	.499822	.499828	.499835
.000612	3.6	.499841	.499847	.499853	.499858	.499864	.499869	.499874	.499879	.499883	.499888
.000425	3.7	.499892	.499896	.499900	.499904	.499908	.499912	.499915	.499918	.499922	.499925
.000292	3.8	.499928	.499931	.499933	.499936	.499938	.499941	.499943	.499946	.499948	.499950
.000199	3.9	.499952	.499954	.499956	.499958	.499959	.499961	.499963	.499964	.499966	.499967
.000134	4.0*	.499968	.499979	.499987	.499991	.499995	.499997	.499998	.499999	.499999	.500000

*Note the change in the interval of tabulation.

Table A.4 Critical values of the *t* distribution

ν	α(2): 0.50 α(1): 0.25	0.20 0.10	0.10 0.05	0.05 0.025	0.02 0.01	0.01 0.005	0.005 0.0025	0.002 0.001	0.001 0.0005
1	1.000	3.078	6.314	12.706	31.821	63.657	127.321	318.309	636.619
2	0.816	1.886	2.920	4.303	6.965	9.925	14.089	22.327	31.599
3	0.765	1.638	2.353	3.182	4.541	5.841	7.453	10.215	12.924
4	0.741	1.533	2.132	2.776	3.747	4.604	5.598	7.173	8.610
5	0.727	1.476	2.015	2.571	3.365	4.032	4.773	5.893	6.869
6	0.718	1.440	1.943	2.447	3.143	3.707	4.317	5.208	5.959
7	0.711	1.415	1.895	2.365	2.998	3.499	4.029	4.785	5.408
8	0.706	1.397	1.860	2.306	2.896	3.355	3.833	4.501	5.041
9	0.703	1.383	1.833	2.262	2.821	3.250	3.690	4.297	4.781
10	0.700	1.372	1.812	2.228	2.764	3.169	3.581	4.144	4.587
11	0.697	1.363	1.796	2.201	2.718	3.106	3.497	4.025	4.437
12	0.695	1.356	1.782	2.179	2.681	3.055	3.428	3.930	4.318
13	0.694	1.350	1.771	2.160	2.650	3.012	3.372	3.852	4.221
14	0.692	1.345	1.761	2.145	2.624	2.977	3.326	3.787	4.140
15	0.691	1.341	1.753	2.131	2.602	2.947	.286	3.733	4.073
16	0.690	1.337	1.746	2.120	2.583	2.921	3.252	3.686	4.015
17	0.689	1.333	1.740	2.110	2.567	2.898	3.222	3.646	3.965
18	0.688	1.330	1.734	2.101	2.552	2.878	3.197	3.610	3.922
19	0.688	1.328	1.729	2.093	2.539	2.861	3.174	3.579	3.883
20	0.687	1.325	1.725	2.086	2.528	2.845	3.153	3.552	3.850
21	0.686	1.323	1.721	2.080	2.518	2.831	3.135	3.527	3.819
22	0.686	1.321	1.717	2.074	2.508	2.819	3.119	3.505	3.792
23	0.685	1.319	1.714	2.069	2.500	2.807	3.104	.485	3.768
24	0.685	1.318	1.711	2.064	2.492	2.797	3.091	3.467	3.745
25	0.684	1.316	1.708	2.060	2.485	2.787	3.078	3.450	3.725
26	0.684	1.315	1.706	2.056	2.479	2.779	3.067	3.435	3.707
27	0.684	1.314	1.703	2.052	2.473	2.771	3.057	3.421	3.690
28	0.683	1.313	1.701	2.048	2.467	2.763	3.047	3.408	3.674
29	0.683	1.311	1.699	2.045	2.462	2.756	3.038	3.396	3.659
30	0.683	1.310	1.697	2.042	2.457	2.750	3.030	3.385	3.646
31	6.682	1.309	1.696	2.040	2.453	2.744	3.022	3.375	3.633
32	0.682	1.309	1.694	2.037	2.449	2.738	3.015	3.365	3.622
33	0.682	1.308	1.692	2.035	2.445	2.733	3.008	3.356	3.611
34	0.682	1.307	1.691	2.032	2.441	2.728	3.002	3.348	3.601
35	0.682	1.306	1.690	2.030	2.438	2.724	2.996	3.340	3.591
36	0.681	1.306	1.688	2.028	2.434	2.719	2.990	3.333	3.582
37	0.681	1.305	1.687	2.026	2.431	2.715	2.985	3.326	3.574
38	0.681	1.304	1.686	2.024	2.429	2.712	2.980	3.319	3.566
39	0.681	1.304	1.685	2.023	2.426	2.708	2.976	3.313	3.558
40	0.681	1.303	1.684	2.021	2.423	2.704	2.971	3.307	3.551

Contd...

Note: Entries are critical values with α = α(2) for two sided tests and α = α(1) for upper tail test. Put negative sign before entries for lower tail test with α = α(1).

Source: Zar (2004), Biostatistical Analysis

Table A.4 Contd.

ν	α(2): 0.50 α(1): 0.25	0.20 0.10	0.10 0.05	0.05 0.025	0.02 0.01	0.01 0.005	0.005 0.0025	0.002 0.001	0.001 0.0005
41	0.681	1.303	1.683	2.020	2.421	2.701	2.967	3.301	3.544
42	0.680	1.302	1.682	2.018	2.418	2.698	2.963	3.296	3.538
43	0.680	1.302	1.681	2.017	2.416	2.695	2.959	3.291	3.532
44	0.680	1.301	1.680	2.015	2.414	2.692	2.956	3.286	3.526
45	0.680	1.301	1.679	2.014	2.412	2.690	2.952	3.281	3.520
46	0.680	1.300	1.679	2.013	2.410	2.687	2.949	3.277	3.515
47	0.680	1.300	1.678	2.012	2.408	2.685	2.946	3.273	3.510
48	0.680	1.299	1.677	2.011	2.407	2.682	2.943	2.269	3.505
49	0.680	1.299	1.677	2.010	2.405	2.680	2.940	3.265	3.500
50	0.679	1.299	1.676	2.009	2.403	2.678	2.937	3.261	3.496
52	0.679	1.298	1.675	2.007	2.400	2.674	2.932	3.255	3.488
54	0.679	1.297	1.674	2.005	2.397	2.670	2.927	3.248	3.480
56	0.679	1.297	1.673	2.003	2.395	2.667	2.923	3.242	3.473
58	0.679	1.296	1.672	2.002	2.392	2.663	2.918	3.237	3.466
60	0.679	1.296	1.671	2.000	2.390	2.660	2.915	3.232	3.460
62	0.678	1.295	1.670	1.999	2.388	2.657	2.911	3.227	3.454
64	0.678	1.295	1.669	1.998	2.386	2.655	2.908	3.223	3.449
66	0.678	1.295	1.668	1.907	2.384	2.652	2.904	3.218	3.444
68	0.678	1.294	1.668	1.995	2.382	2.650	2.902	3.214	3.439
70	0.678	1.294	1.667	1.994	2.381	2.648	2.899	3.211	3.435
72	0.678	1.293	1.666	1.993	2.379	2.646	2.896	3.207	3.431
74	0.678	1.293	1.666	1.993	2.378	2.644	2.894	3.204	3.427
76	0.678	1.293	1.665	1.992	2.376	2.642	2.891	3.201	3.423
78	0.678	1.292	1.665	1.991	2.375	2.640	2.889	3.198	3.420
80	0.678	1.292	1.664	1.990	2.374	2.639	2.887	3.195	3.416
82	0.677	1.297	1.664	1.989	2.373	2.637	2.885	3.193	3.413
84	0.677	1.292	1.663	1.989	2.372	2.636	2.883	3.190	3.410
86	0.677	1.291	1.663	1.988	2.370	2.634	2.881	3.188	3.407
88	0.677	1.291	1.662	1.987	2.369	2.633	2.880	3.185	3.405
90	0.677	1.291	1.662	1.987	2.368	2.632	2.878	3.183	3.402
92	0.677	1.291	1.662	1.986	2.368	2.630	2.876	3.181	3.399
94	0.677	1.291	1.661	1.986	2.367	2.269	2.875	3.179	3.397
96	0.677	1.290	1.661	1.985	2.366	2.628	2.873	3.177	3.395
98	0.677	1.290	1.661	1.984	2.365	2.627	2.872	3.175	3.393
100	0.677	1.290	1.660	1.984	2.364	2.626	2.871	3.174	3.390
105	0.677	1.290	1.659	1.983	2.362	2.623	2.868	3.170	3.386
110	0.677	1.289	1.659	1.982	2.361	2.621	2.865	3.166	3.381
115	0.677	1.289	1.658	1.981	2.359	2.619	2.862	3.163	3.377
120	0.677	1.289	1.658	1.980	2.358	2.617	2.860	3.160	3.373
125	0.676	1.288	1.657	1.979	2.357	2.616	2.858	3.157	3.370
130	0.676	1.288	1.657	1.978	2.355	2.614	2.856	3.154	3.367
135	0.676	1.288	1.656	1.978	2.354	2.613	2.854	3.152	3.364
140	0.676	1.288	1.656	1.977	2.353	2.611	2.852	3.149	3.361
145	0.676	1.287	1.655	1.976	2.352	2.610	2.851	3.147	3.359
150	0.676	1.287	1.655	1.976	2.351	2.609	2.849	3.145	3.357

Contd...

Table A.4 Contd.

ν	α(2): 0.50 α(1): 0.25	0.20 0.10	0.10 0.05	0.05 0.025	0.02 0.01	0.01 0.005	0.005 0.0025	0.002 0.001	0.001 0.0005
160	0.676	1.287	1.654	1.975	2.350	2.607	2.846	3.142	3.352
170	0.676	1.287	1.654	1.974	2.348	2.605	2.844	3.139	3.349
180	0.676	1.286	1.653	1.973	2.347	2.603	2.842	3.136	3.345
190	0.676	1.286	1.6533	1.973	2.346	2.602	2.840	3.134	3.342
200	0.676	1.286	1.653	1.972	2.345	2.601	2.839	3.131	3.340
250	0.675	1.285	1.651	1.969	2.341	2.596	2.832	3.123	3.330
300	0.675	1.284	1.650	1.968	2.339	2.592	2.828	3.118	3.323
350	0.675	1.284	1.649	1.967	2.337	2.500	2.825	3.114	3.319
400	0.675	1.284	1.649	1.966	2.336	2.588	2.823	3.111	3.315
450	0.675	1.283	1.648	1.965	2.335	2.587	2.821	3.108	3.312
500	0.675	1.283	1.648	1.965	2.334	2.586	2.820	3.107	3.310
600	0.675	1.283	1.647	1.964	2.333	2.584	2.817	3.104	3.307
700	0.675	1.283	1.647	1.963	2.332	2.583	2.816	3.102	3.304
800	0.675	1.283	1.647	1.963	2.331	2.582	2.815	3.100	3.303
900	0.675	1.282	1.647	1.963	2.330	2.581	2.814	3.099	3.301
1000	0.675	1.282	1.646	1.962	2.330	2.581	2.813	3.098	3.300
∞	0.6745	1.2816	1.6449	1.9600	2.3263	2.5758	2.8070	3.0902	3.2905

Table A.5 Percentage points of the χ^2-distribution

ν	Q						
	0.995	0.990	0.975	0.950	0.900	0.750	0.500
1	392704.10^{-10}	157088.10^{-9}	982069.10^{-9}	393214.10^{-8}	0.0157908	0.1015308	0.454937
2	0.0100251	0.0201007	0.0506356	0.102587	0.210720	0.575364	1.38629
3	0.0717212	0.114832	0.215795	0.351846	0.584375	1.212534	2.36597
4	0.206990	0.297110	0.484419	0.710721	1.063623	1.92255	3.35670
5	0.411740	0.554300	0.831211	1.145476	1.61031	2.67460	4.35146
6	0.675727	0.872085	1.237347	1.63539	2.20413	3.45460	5.34812
7	0.989265	1.239043	1.68987	2.16735	2.83311	4.25485	6.34581
8	1.344419	1.646482	2.17973	2.73264	3.48954	5.07064	7.34412
9	1.734926	2.087912	2.70039	3.32511	4.16816	5.89883	8.34283
10	2.15585	2.55821	3.24697	3.94030	3.86518	6.73720	9.34182
11	2.60321	3.05347	3.81575	4.57481	5.57779	7.58412	10.3410
12	3.07382	3.57056	4.40379	5.22603	6.30380	8.43842	11.3403
13	3.56503	4.10691	5.00874	5.89186	7.04150	9.29906	12.3398
14	4.07468	4.66043	5.62872	6.57063	7.78953	10.1653	13.3393
15	4.60094	5.22935	6.26214	7.26094	8.54675	11.0365	14.3389
16	5.14224	5.81221	6.90766	7.96164	9.31223	11.9122	15.3385
17	5.69724	6.40776	7.56418	8.67176	10.0852	12.7919	16.3381
18	6.26481	7.01491	8.23075	9.39046	10.8649	13.6753	17.3379
19	6.84398	7.63273	8.90655	10.1170	11.6509	14.5620	18.3376
20	7.43386	8.26040	9.59083	10.8508	12.4426	15.4518	19.3374
21	8.03366	8.89720	10.28293	11.5913	13.2396	16.3444	20.3372
22	8.64272	9.54249	10.9823	12.3380	14.0415	17.2396	21.3370
23	9.26042	10.19567	11.6885	13.0905	14.8479	18.1373	22.3369
24	9.88623	10.8564	12.4011	13.8484	15.6587	19.0372	23.3367
25	10.5197	11.5240	13.1197	14.6114	16.4734	19.9393	24.3366
26	11.1603	12.1981	13.8439	15.3791	17.2919	20.8434	25.3364
27	11.8076	12.8786	14.5733	16.1513	18.1138	21.7494	26.3363
28	12.4613	13.5648	15.3079	16.9279	18.9392	22.6572	27.3363
29	13.1211	14.2565	16.0471	17.7083	19.7677	23.5666	28.3362
30	13.7867	14.9535	16.7908	18.4926	20.5992	24.4776	29.3360
40	20.7065	22.1643	24.4331	26.5093	29.0505	33.6603	39.3354
50	27.9907	29.7067	32.3574	34.7642	37.6886	42.9421	49.3349
60	35.5346	37.4848	40.4817	43.1879	46.4589	52.2938	59.3347
70	43.2752	45.4418	48.7576	51.7393	55.3290	61.6983	69.3344
80	51.1720	53.5400	57.1532	60.3915	64.2778	71.1445	79.3343
90	59.1963	61.7541	61.6466	69.1260	73.2912	80.6247	89.3342
100	67.3276	70.0648	74.2219	77.9295	82.3581	90.1332	99.41
X	–2.5758	–2.3263	–1.9600	–1.6449	–1.2616	–0.6745	0.0000

Contd...

Note: $Q = Q\left(\chi_v^2\right) = 1 - P\left(\chi_v^2\right) = 2^{-v/2}\left\{\Gamma\left(\frac{1}{2}V\right)\right\}^{-1} \int_{\chi^2}^{\infty} e^{-x/2}\, X^{v/2-1}\, dx.$

The entries are 100 α% upper tail values of χ^2

Source: Rao, Mitra, Mathai and Ramamurthy (1975) Formula and Tables for Statistical Work

Table A.5 Contd.

ν	Q						
	0.250	0.100	0.050	0.025	0.010	0.005	0.001
1	1.32330	2.70554	3.84146	5.02389	6.63490	7.87944	10.828
2	2.77259	4.60517	5.99147	7.37776	9.21034	10.5966	13.816
3	4.10835	6.25139	7.81473	9.34840	11.3449	12.8381	16.266
4	5.38527	7.77944	9.48773	11.1433	13.2767	14.8602	18.467
5	6.62568	9.23635	11.0705	12.8325	15.0863	16.7496	20.515
6	7.84080	10.6446	12.5916	14.4494	16.8119	18.5476	22.458
7	9.03715	12.0170	14.0671	16.0128	18.4753	20.2777	24.322
8	10.2188	13.3616	15.5073	17.5346	20.0902	21.9550	26.125
9.	11.3887	14.6837	16.9190	19.0228	21.6660	23.5893	27.877
10.	12.5489	15.9871	18.3070	20.4831	23.2093	25.1882	29.588
11.	13.7007	17.2750	19.6751	21.9200	24.7250	26.7569	31.264
12.	14.8454	18.5494	21.0261	23.3367	26.2170	28.2995	32.909
13.	15.9839	19.8119	22.3621	24.7356	27.6883	29.8194	34.528
14.	17.1170	21.0642	23.6848	26.1190	29.1413	31.3193	36.123
15.	18.2451	22.3072	24.9958	27.4884	30.5779	32.8013	37.697
16.	19.3688	23.5418	26.2962	28.8454	31.9999	34.2672	39.252
17.	20.4887	24.7690	27.5871	30.1910	33.4087	35.7185	40.790
18.	21.6049	25.9894	28.8693	31.5264	34.8053	37.1564	42.312
19.	22.7178	27.2036	30.1435	32.8523	36.1908	38.5822	43.820
20.	23.8277	28.4120	31.4104	34.1696	37.5662	39.9968	45.315
21.	24.9348	29.6151	32.6705	35.4789	38.9321	41.4010	46.797
22.	26.0393	30.8133	33.9244	36.7807	40.2894	42.7956	48.268
23.	27.1413	32.0069	35.1725	38.0757	41.6384	44.1813	49.728
24.	28.2412	33.1963	36.4151	39.3641	42.9798	45.5585	51.179
25.	29.3389	34.3816	37.6525	40.6465	44.3141	46.9278	52.620
26.	30.4345	35.5631	38.8852	41.9232	45.6417	48.2899	54.052
27.	31.5284	36.7412	40.1133	43.1944	46.9630	49.6449	55.476
28.	32.6205	37.9159	41.3372	44.4607	48.2782	50.9933	56.892
29.	33.7109	39.0875	42.5569	45.7222	49.5879	52.3356	58.302
30	34.7998	40.2560	43.7729	46.9792	50.8922	53.6720	59.703
40.	45.6160	51.8050	55.7585	59.3417	63.6907	66.7659	73.402
50.	56.3336	63.1671	67.5048	71.4202	76.1539	79.4900	86.661
60.	66.9814	74.3970	79.0819	83.2976	88.3794	91.9517	99.607
70.	77.5766	85.5271	90.5312	95.0231	100.425	104.215	112.317
80.	88.1303	96.5782	101.879	106.629	112.329	116.321	124.839
90.	98.6499	107.565	113.145	118.136	124.116	128.299	137.208
100.	109.141	118.498	124.342	129.561	135.807	140.169	149.449
X	+0.6745	+1.2816	+1.6449	+1.9600	+2.3263	+2.5758	+3.0902

For ν > 100 take

$$\chi^2 = \nu\left\{1 - \frac{2}{9\nu} + X\sqrt{\frac{2}{9\nu}}\right\}^3 \quad \text{or } \chi^2 = \frac{1}{2}\left\{X + \sqrt{(2\nu - 1)}\right\}$$

according to the degree of accuracy required. *X* is the standardized normal deviate corresponding to *P* = 1 – *Q*, and is shown in the bottom line of the table.

Table A.6 Upper tail values of *F*

Denominator df	Probability of a larger *F*	Numerator df								
		1	2	3	4	5	6	7	8	9
1.	.100	39.86	49.50	53.59	55.83	57.24	58.20	58.91	59.44	59.86
	.050	161.4	199.5	215.7	224.6	230.2	234.0	236.8	238.9	240.5
	.025	647.8	799.5	864.2	899.6	921.8	937.1	948.2	956.7	963.3
	.010	4052.0	4999.5	5403.0	5625.0	5764.0	5859.0	5928.0	5982.0	6022.0
	.005	16211.0	20000	21615.0	22500.0	23056.0	23437.0	23715.0	23925.0	24091.0
2.	.100	8.53	9.00	9.16	9.24	9.29	9.33	9.35	9.37	9.38
	.050	18.51	19.00	19.16	19.25	19.30	19.33	19.35	19.37	19.38
	.025	38.51	39.00	39.17	39.25	39.30	39.33	39.36	39.37	39.39
	0.010	98.50	99.00	99.17	99.25	99.30	99.33	99.36	99.37	99.39
	0.005	198.5	199.0	199.2	199.2	199.3	199.3	199.4	199.4	199.4
3.	.100	5.54	5.46	5.39	5.34	5.31	5.28	5.27	5.25	5.24
	0.050	10.13	9.55	9.28	9.12	9.01	8.94	8.89	8.85	8.81
	0.025	17.44	16.04	15.44	15.10	14.88	14.73	14.62	14.54	14.47
	.010	34.12	30.82	29.46	28.71	28.24	27.91	27.67	27.49	27.35
	.005	55.55	49.80	47.47	46.19	45.39	44.84	44.43	44.13	43.88
4.	.100	4.54	4.32	4.19	4.11	4.05	4.01	3.98	3.95	3.94
	.050	7.71	6.94	6.59	6.39	6.26	6.16	6.09	6.04	6.00
	.025	12.22	10.65	9.98	9.60	9.36	9.20	9.07	8.98	8.90
	.010	21.20	18.00	16.69	15.98	15.52	15.21	14.98	14.80	14.66
	.005	31.33	26.28	24.26	23.15	22.46	21.97	21.62	21.35	21.14
5.	.100	4.06	3.78	3.62	3.52	3.45	3.40	3.37	3.34	3.32
	.050	6.61	5.79	5.41	5.19	5.05	4.95	4.88	4.82	4.77
	.025	10.01	8.43	7.76	7.39	7.15	6.98	6.85	6.76	6.68
	.010	16.26	13.27	12.06	11.39	10.97	10.67	10.46	10.29	10.16
	.005	22.78	18.31	16.53	15.56	14.94	14.51	14.20	13.96	13.77
6.	.100	3.78	3.46	3.29	3.18	3.11	3.05	3.01	2.98	2.96
	.050	5.99	5.14	4.76	4.53	4.39	4.28	4.21	4.15	4.10
	.025	8.81	7.26	6.60	6.23	5.99	5.82	5.70	5.60	5.52
	.010	13.75	10.92	9.78	9.15	8.75	8.47	8.26	8.10	7.98
	.005	18.63	14.54	12.92	12.03	11.46	11.07	10.79	10.57	10.39
7.	.100	3.59	3.26	3.07	2.96	2.88	2.83	2.78	2.75	2.72
	.050	5.29	4.74	4.35	4.12	3.97	3.87	3.79	3.73	3.68
	.025	8.07	6.54	5.89	5.52	5.29	5.12	4.99	4.90	4.82
	.010	12.25	9.55	8.45	7.85	7.46	7.19	6.99	6.84	6.72
	.005	16.24	12.40	10.88	10.05	9.52	9.16	8.89	8.68	8.51
8.	.100	3.46	3.11	2.92	2.81	2.73	2.67	2.62	2.59	2.56
	.050	5.32	4.46	4.07	3.84	3.69	3.58	3.50	3.44	3.39
	.025	7.57	6.06	5.42	5.05	4.82	4.65	4.53	4.43	4.36
	.010	11.26	8.65	7.59	7.01	6.63	6.37	6.18	6.03	5.91
	.005	14.69	11.04	9.60	8.81	8.30	7.95	7.69	7.50	7.34

Contd...

Source: Rao, Mitra, Mathai and Ramamurthy (1975), Formula and Tables for Statistical Work, which reproduced a portion of "Tables of percentage points of the inverted beta (F) distribution," Biometrika, vol. 33 (1943) by M. Merrington and C.M. Thompson and from Table 18 of Biometrika Tables for Statisticians, vol. 1, Cambridge University Press, 1954, edited by E.S. Pearson and H.O. Hartley, Reproduced with permission of the authors, editors, and Biometrika trustees.

Table A.6 Contd.

Denominator df	Probability of a larger *F*	Numerator df								
		1	2	3	4	5	6	7	8	9
9.	.100	3.36	3.01	2.81	2.69	2.61	2.55	2.51	2.47	2.44
	.050	5.12	4.26	3.86	3.63	3.48	3.37	3.29	3.23	3.18
	.025	7.21	5.71	5.08	4.72	4.48	4.32	4.20	4.10	4.03
	.010	10.56	8.02	6.99	6.42	6.06	5.80	5.61	5.47	5.35
	.005	13.61	10.11	8.72	7.96	7.47	7.13	6.88	6.69	6.54
10.	.100	3.29	2.92	2.73	2.61	2.52	2.46	2.41	2.38	2.35
	.050	4.96	4.10	3.71	3.48	3.33	3.22	3.14	3.07	3.02
	.025	6.94	5.46	4.83	4.47	4.24	4.07	3.95	3.85	3.78
	.010	10.04	7.56	6.55	5.99	5.64	5.39	5.20	5.06	4.94
	.005	12.83	9.43	8.08	7.34	6.87	6.54	6.30	6.12	5.97
11.	.100	3.23	2.86	2.66	2.54	2.45	2.39	2.34	2.30	2.27
	.050	4.84	3.98	3.59	3.36	3.20	3.09	3.01	2.95	2.90
	.025	6.72	5.26	4.63	4.28	4.04	3.88	3.76	3.66	3.59
	.010	9.65	7.21	6.22	5.67	5.32	5.07	4.89	4.74	4.63
	.005	12.23	8.91	7.60	6.88	6.42	6.10	5.86	5.68	5.34
12.	.100	3.18	2.81	2.61	2.48	2.39	2.33	2.28	2.24	2.21
	.050	4.75	3.89	3.49	3.26	3.11	3.09	2.91	2.85	2.80
	.025	6.55	5.10	4.47	4.12	3.89	3.73	3.61	3.51	3.44
	.010	9.33	6.93	5.95	5.41	5.06	4.82	4.64	4.50	4.39
	.005	11.75	8.51	7.23	6.52	6.07	5.76	5.52	5.35	5.20
13.	.100	3.14	2.76	2.56	2.43	2.35	2.28	2.23	2.20	2.16
	.050	4.67	3.81	3.41	3.18	3.03	2.92	2.83	2.77	2.71
	.025	6.41	4.97	4.35	4.00	3.77	3.60	3.48	3.39	3.31
	.010	9.07	6.70	5.74	5.21	4.86	4.62	4.44	4.30	4.19
	.005	11.37	8.19	6.93	6.23	5.79	5.48	5.25	5.08	4.94
14.	.100	3.10	2.73	2.52	2.39	2.31	2.24	2.19	2.15	2.12
	.050	4.60	3.74	3.34	3.11	2.96	2.85	2.76	2.70	2.65
	.025	6.30	4.86	4.24	3.89	3.66	3.50	3.38	3.29	3.21
	.010	8.86	6.51	5.56	5.04	4.69	4.46	4.28	4.14	4.03
	.005	11.06	7.92	6.68	6.00	5.56	5.26	5.03	4.86	4.72
15.	.100	3.07	2.70	2.49	2.36	2.27	2.21	2.16	2.12	2.09
	.050	4.54	3.68	3.29	3.06	2.90	2.79	2.71	2.64	2.59
	.025	6.20	4.77	4.15	3.80	3.58	3.41	3.29	3.20	3.12
	.010	8.68	6.36	5.42	4.89	4.56	4.32	4.14	4.00	3.89
	.005	10.80	7.70	6.48	5.80	5.37	5.07	4.85	4.67	4.54
16.	.100	3.05	2.67	2.46	2.33	2.24	2.18	2.13	2.09	2.06
	.050	4.49	3.63	3.24	3.01	2.85	2.74	2.66	2.59	2.54
	.025	6.12	4.69	4.08	3.73	3.50	3.34	3.22	3.12	3.05
	.010	8.53	6.23	5.29	4.77	4.44	4.20	4.03	3.89	3.78
	.005	10.58	7.51	6.30	5.64	5.21	4.91	4.69	4.52	4.38
17.	.100	3.03	2.64	2.44	2.31	2.22	2.15	2.10	2.06	2.03
	.050	4.45	3.59	3.20	2.96	2.81	2.70	2.61	2.55	2.49
	.025	6.04	4.62	4.01	3.66	3.44	3.28	3.16	3.06	2.98
	.010	8.40	6.11	5.18	4.67	4.34	4.10	3.93	3.79	3.68

Contd...

Table A.6 Contd.

Denominator df	Probability of a larger F	Numerator df								
		1	2	3	4	5	6	7	8	9
	.005	10.38	7.35	6.16	5.50	5.07	4.78	4.56	4.39	4.25
18.	.100	3.01	2.62	2.42	2.29	2.20	2.13	2.08	2.04	2.00
	.050	4.41	3.55	3.16	2.93	2.77	2.66	2.58	2.51	2.46
	.025	5.98	4.56	3.95	3.61	3.38	3.22	3.10	3.01	2.93
	.010	8.29	6.01	5.09	4.58	4.25	4.01	3.84	3.71	3.60
	.005	10.22	7.21	6.03	5.37	4.96	4.66	4.44	4.28	4.14
19.	.100	2.99	2.61	2.40	2.27	2.18	2.11	2.06	2.02	1.98
	.050	4.38	3.52	3.13	2.90	2.74	2.63	2.54	2.48	2.42
	.025	5.92	4.51	3.90	3.56	3.33	3.17	3.05	2.96	2.88
	.010	8.18	5.93	5.01	4.50	4.17	3.94	3.77	3.63	3.52
	.005	10.07	7.09	5.92	5.27	4.85	4.56	4.34	4.18	4.04
20.	.100	2.97	2.59	2.38	2.25	2.16	2.09	2.04	2.00	1.96
	.050	4.35	3.49	3.10	2.87	2.71	2.60	2.51	2.45	2.39
	.025	5.87	4.46	3.86	3.51	3.29	3.13	3.01	2.91	2.84
	.010	8.10	5.85	4.94	4.43	4.10	3.87	3.70	3.56	3.46
	.005	9.94	6.99	5.82	5.17	4.76	4.47	4.26	4.09	3.96
21.	.100	2.96	2.57	2.36	2.23	2.14	2.08	2.02	1.98	1.95
	.050	4.32	3.47	3.07	2.84	2.68	2.57	2.49	2.42	2.37
	.025	5.83	4.42	3.82	3.48	3.25	3.09	2.97	2.87	2.80
	.010	8.02	5.78	4.87	4.37	4.04	3.81	3.64	3.51	3.40
	.005	9.83	6.89	5.73	5.09	4.68	4.39	4.18	4.01	3.88
22.	.100	2.95	2.56	2.35	2.22	2.13	2.06	2.01	1.97	1.93
	.050	4.30	3.44	3.05	2.82	2.66	2.55	2.46	2.40	2.34
	.025	5.79	4.38	3.78	3.44	3.22	3.05	2.93	2.84	2.76
	.010	7.95	5.72	4.82	4.31	3.99	3.76	3.59	3.45	3.35
	.005	9.73	6.81	5.65	5.02	4.61	4.32	4.11	3.94	3.81
23.	.100	2.94	2.55	2.34	2.21	2.11	2.05	1.99	1.95	1.92
	.050	4.28	3.42	3.03	2.80	2.64	2.53	2.44	2.37	2.32
	.025	5.75	4.35	3.75	3.41	3.18	3.02	2.90	2.81	2.73
	.010	7.88	5.68	4.76	4.26	3.94	3.71	3.54	3.41	3.30
	.005	9.63	6.73	5.58	4.95	4.54	4.26	4.05	3.88	3.75
24.	.100	2.93	2.54	2.33	2.19	2.10	2.04	1.98	1.94	1.91
	.050	4.26	3.40	3.01	2.78	2.62	2.51	2.42	2.36	2.30
	.025	5.72	4.32	3.72	3.38	3.15	2.99	2.87	2.78	2.70
	.010	7.82	5.61	4.72	4.22	3.90	3.67	3.50	3.36	3.26
	.005	9.55	6.66	5.52	4.89	4.49	4.20	3.99	3.83	3.69
25.	.100	2.92	2.53	2.32	2.18	2.09	2.02	1.97	1.93	1.89
	.050	4.24	3.39	2.99	2.76	2.60	2.49	2.40	2.34	2.28
	.025	5.69	4.29	3.69	3.35	3.13	2.97	2.85	2.75	2.68
	.010	7.77	5.57	4.68	4.18	3.85	3.63	3.46	3.32	3.22
	.005	9.48	6.60	5.46	4.84	4.43	4.15	3.94	3.78	3.64
26.	.100	2.91	2.52	2.31	2.17	2.08	2.01	1.96	1.92	1.88
	.050	4.23	3.37	2.98	2.74	2.59	2.47	2.39	2.32	2.27
	.025	5.66	4.27	3.67	3.33	3.10	2.94	2.82	2.73	2.65

Contd...

Table A.6 Contd.

Denominator df	Probability of a larger F	Numerator df 1	2	3	4	5	6	7	8	9
	.010	7.72	5.53	4.64	4.14	3.82	3.59	3.42	3.29	3.18
	.005	9.41	6.54	5.41	4.79	4.38	4.10	3.89	3.73	3.60
26.	.100	2.91	2.52	2.31	2.17	2.08	2.01	1.96	1.92	1.88
	.050	4.23	3.37	2.98	2.74	2.59	2.47	2.39	2.32	2.27
	.025	5.66	4.27	3.67	3.33	3.10	2.94	2.82	2.73	2.65
	.010	7.72	5.53	4.64	4.14	3.82	3.59	3.42	3.29	3.18
	.005	9.41	6.54	5.41	4.79	4.38	4.10	3.89	3.73	3.60
27.	.100	2.90	2.51	2.30	2.17	2.07	2.00	1.95	1.91	1.87
	.050	4.21	3.35	2.96	2.73	2.57	2.46	2.37	2.31	2.25
	.025	5.63	4.24	3.65	3.31	3.08	2.92	2.80	2.71	2.63
	.010	7.68	5.49	4.60	4.11	3.78	3.56	3.39	3.26	3.15
	.005	9.34	6.49	5.36	4.74	4.34	4.06	3.85	3.69	3.56
28.	.100	2.89	2.50	2.29	2.16	2.06	2.00	1.94	1.90	1.87
	.050	4.20	3.34	2.95	2.71	2.56	2.45	2.36	2.29	2.24
	.025	5.61	4.22	3.63	3.29	3.06	2.90	2.78	2.69	2.61
	.010	7.64	5.45	4.57	4.07	3.75	3.53	3.36	3.23	3.12
	.005	9.28	6.44	5.32	4.70	4.30	4.02	3.81	3.65	3.52
29.	.100	2.89	2.50	2.28	2.15	2.06	1.99	1.93	1.89	1.86
	.050	4.18	3.33	2.93	2.70	2.55	2.43	2.35	2.28	2.22
	.025	5.59	4.20	3.61	3.27	3.04	2.88	2.76	2.67	2.59
	.010	7.60	5.42	4.54	4.04	3.73	3.50	3.33	3.20	3.09
	.005	9.23	6.40	5.28	4.66	4.26	3.98	3.77	3.61	3.48
30.	.100	2.88	2.49	2.28	2.14	2.05	1.98	1.93	1.88	1.85
	.050	4.17	3.32	2.92	2.69	2.53	2.42	2.33	2.27	2.21
	.025	5.57	4.18	3.59	3.25	3.03	2.87	2.75	2.65	2.57
	.010	7.56	5.39	4.51	4.02	3.70	3.47	3.30	3.17	3.07
	.005	9.18	6.35	5.24	4.62	4.23	3.95	3.74	3.58	3.45
40.	.100	2.84	2.44	2.23	2.09	2.00	1.93	1.87	1.83	1.79
	.050	4.08	3.23	2.84	2.61	2.45	2.34	2.25	2.18	2.12
	.025	5.42	4.05	3.46	3.13	2.90	2.74	2.62	2.53	2.45
	.010	7.31	5.18	4.31	3.83	3.51	3.29	3.12	2.99	2.89
	.005	8.83	6.07	4.98	4.37	3.99	3.71	3.51	3.35	3.22
60.	.100	2.79	2.39	2.18	2.04	1.95	1.87	1.82	1.77	1.74
	.050	4.00	3.15	2.76	2.53	2.37	2.25	2.17	2.10	2.04
	.025	5.29	3.93	3.34	3.01	2.79	2.63	2.51	2.41	2.33
	.010	7.08	4.98	4.13	3.65	3.34	3.12	2.95	2.82	2.72
	.005	8.49	5.79	4.73	4.14	3.76	3.49	3.20	3.13	3.01
120.	.100	2.75	2.35	2.13	1.99	1.90	1.82	1.77	1.72	1.68
	.050	3.92	3.07	2.68	2.45	2.29	2.17	2.09	2.02	1.96
	.025	5.15	3.80	3.23	2.89	2.67	2.52	2.39	2.30	2.22
	.010	6.85	4.79	3.95	3.48	3.17	2.96	2.79	2.66	2.56
	.005	8.18	5.54	4.50	3.92	3.55	3.28	3.09	2.93	2.81
∞	.100	2.71	2.30	2.08	1.94	1.85	1.77	1.72	1.67	1.63
	.050	3.84	3.00	2.60	2.37	2.21	2.10	2.01	1.94	1.88
	.025	5.02	3.69	3.12	2.79	2.57	2.41	2.29	2.19	2.11
	.010	6.63	4.61	3.78	3.32	3.02	2.80	2.64	2.51	2.41
	.005	7.88	5.30	4.28	3.72	3.35	3.09	2.90	2.74	2.62

Table A.6 Contd.

Numerator df											
10	12	15	20	24	30	40	60	120	∞	P	Denom. df
60.19	60.71	61.22	61.74	62.00	62.26	62.53	62.79	63.06	63.33	.100	1
241.9	243.9	245.9	248.0	249.1	250.1	251.1	252.3	253.3	254.3	.050	
968.6	976.7	984.9	993.1	997.2	1001	1006	1010	1014	1018	.025	
6056	6106	6157	6209	6235	6261	6287	6313	6339	6366	.010	
24224	24426	24630	24836	24940	25044	25148	25253	25359	25465	.005	
9.39	9.41	9.42	9.44	9.45	9.46	9.47	9.47	9.48	9.49	.100	2
19.40	19.41	19.43	19.45	19.45	19.46	19.47	19.48	19.49	19.50	.050	
39.40	39.41	39.43	39.45	39.46	39.46	39.47	39.48	39.49	39.50	.025	
99.40	99.42	99.43	99.45	99.46	99.47	99.47	99.48	99.49	99.50	.010	
199.4	199.4	199.4	199.4	199.5	199.5	199.5	199.5	199.5	199.5	.005	
5.23	5.22	5.20	5.18	5.18	5.17	5.16	5.15	5.14	5.13	.100	3
8.78	8.74	8.70	8.66	8.64	8.62	8.59	8.57	8.55	8.53	.050	
14.42	14.34	14.25	14.17	14.12	14.08	14.04	13.99	13.95	13.90	.025	
27.23	27.05	26.87	26.69	26.60	26.50	26.41	26.32	26.22	26.13	.010	
43.69	43.39	43.08	42.78	42.62	42.47	42.31	42.15	41.99	41.83	.005	
3.92	3.90	3.87	3.84	3.83	3.82	3.80	3.79	3.78	3.76	.100	4
5.96	5.91	5.86	5.80	5.77	5.75	5.72	5.69	5.66	5.63	.050	
8.84	8.75	8.66	8.56	8.51	8.46	8.41	8.36	8.31	8.26	.025	
14.55	14.37	14.20	14.02	13.93	13.84	13.75	13.65	13.56	13.46	.010	
20.97	20.70	20.44	20.17	20.03	19.89	19.75	19.61	19.47	19.32	.005	
3.30	3.27	3.24	3.21	3.19	3.17	3.16	3.14	3.12	3.10	.100	5
4.74	4.68	4.62	4.56	4.53	4.50	4.46	4.43	4.40	4.36	.050	
6.62	6.52	6.43	6.33	6.28	6.23	6.18	6.12	6.07	6.02	.025	
10.05	9.89	9.72	9.55	9.47	9.38	9.29	9.20	9.11	9.02	.010	
13.62	13.38	13.15	12.90	12.78	12.66	12.53	12.40	12.27	12.14	.005	
2.94	2.90	2.87	2.84	2.82	2.80	2.78	2.76	2.74	2.72	.100	6
4.06	4.00	3.94	3.87	3.84	3.81	3.77	3.74	3.70	3.67	.050	
5.46	5.37	5.27	5.17	5.12	5.07	5.01	4.96	4.90	4.85	.025	
7.87	7.72	7.56	7.40	7.31	7.23	7.14	7.06	7.97	6.88	.010	
10.25	10.03	9.81	9.59	9.47	9.36	9.24	9.12	9.00	8.88	.005	
2.70	2.67	2.63	2.59	2.58	2.56	2.54	2.51	2.49	2.47	.100	7
3.64	3.57	3.51	3.44	3.41	3.38	3.34	3.30	3.27	3.23	.050	
4.76	4.67	4.57	4.47	4.42	4.36	4.31	4.25	4.20	4.14	.025	
6.62	6.47	6.31	6.16	6.07	5.99	5.91	5.82	5.74	5.65	.010	
8.38	8.18	7.97	7.75	7.65	7.53	7.42	7.31	7.19	7.08	.005	
2.54	2.50	2.46	2.42	2.40	2.38	2.36	2.34	2.32	2.29	.100	8
3.35	3.28	3.22	3.15	3.12	3.08	3.04	3.01	2.97	2.93	.050	
4.30	4.20	4.10	4.00	3.95	3.89	3.84	3.78	3.73	3.67	.025	
5.81	5.67	5.52	5.36	5.28	5.20	5.12	5.03	4.95	4.86	.010	
7.21	7.01	6.81	6.61	6.50	6.40	6.29	6.18	6.06	5.95	.005	
2.42	2.38	2.34	2.30	2.28	2.25	2.23	2.21	2.18	2.16	.100	9
3.14	3.07	3.01	2.94	2.90	2.86	2.83	2.79	2.75	2.71	.050	
3.96	3.87	3.77	3.67	3.61	3.56	3.51	3.45	3.39	3.33	.025	
5.26	5.11	4.96	4.81	4.73	4.65	4.57	4.48	4.40	4.31	.010	
6.42	6.23	6.03	5.83	5.73	5.62	5.52	5.41	5.30	5.19	.005	

Contd...

Table A.6 Contd.

Numerator df											
10	12	15	20	24	30	40	60	120	∞	P	Denom. df
2.32	2.28	2.24	2.20	2.18	2.16	2.13	2.11	2.08	2.06	.100	10
2.98	2.91	2.85	2.77	2.74	2.70	2.66	2.62	2.58	2.54	.050	
3.72	3.62	3.52	3.42	3.37	3.31	3.26	3.20	3.14	3.08	.025	
4.85	4.71	4.56	4.41	4.33	4.25	4.17	4.08	4.00	3.91	.010	
5.85	5.66	5.47	5.27	5.17	5.07	4.97	4.86	4.75	4.64	.005	
2.25	2.21	2.17	2.12	2.10	2.08	2.05	2.03	2.00	1.97	.100	11
2.85	2.79	2.72	2.65	2.61	2.57	2.53	2.49	2.45	2.40	.050	
3.53	3.43	3.33	3.23	3.17	3.12	3.06	3.00	2.94	2.88	.025	
4.54	4.40	4.25	4.10	4.02	3.94	3.86	3.78	3.69	3.60	.010	
5.42	5.24	5.05	4.86	4.76	4.65	4.55	4.44	4.34	4.23	.005	
2.19	2.15	2.10	2.06	2.04	2.01	1.99	1.96	1.93	1.90	.100	12
2.75	2.69	2.62	2.54	2.51	2.47	2.43	2.38	2.34	2.30	.050	
3.37	3.28	3.18	3.07	3.02	2.96	2.91	2.85	2.79	2.72	.025	
4.30	4.16	4.01	3.86	3.78	3.70	3.62	3.54	3.45	3.36	.010	
5.09	4.91	4.72	4.53	4.43	4.33	4.23	4.12	4.01	3.90	.005	
2.14	2.10	2.05	2.01	1.98	1.96	1.93	1.90	1.88	1.85	.100	13
2.67	2.60	2.53	2.46	2.42	2.38	2.34	2.30	2.25	2.21	.050	
3.25	3.15	3.05	2.95	2.89	2.84	2.78	2.72	2.66	2.60	.025	
4.10	3.96	3.82	3.66	3.59	3.51	3.43	3.34	3.25	3.17	.010	
4.82	4.64	4.46	4.27	4.17	4.07	3.97	3.87	3.76	3.65	.005	
2.10	2.05	2.01	1.96	1.94	1.91	1.89	1.86	1.83	1.80	.100	14
2.60	2.53	2.46	2.39	2.35	2.31	2.27	2.22	2.18	2.13	.050	
3.15	3.05	2.95	2.84	2.79	2.73	2.67	2.61	2.55	2.49	.025	
3.94	3.80	3.66	3.51	3.43	3.35	3.27	3.18	3.09	3.00	.010	
4.60	4.43	4.25	4.06	3.96	3.86	3.76	3.66	3.55	3.44	.005	
2.06	2.02	1.97	1.92	1.90	1.87	1.85	1.82	1.79	1.76	.100	15
2.54	2.48	2.40	2.33	2.29	2.25	2.20	2.16	2.11	2.07	.050	
3.06	2.96	2.86	2.76	2.70	2.64	2.59	2.52	2.46	2.40	.025	
3.80	3.67	3.52	3.37	3.29	3.21	3.13	3.05	2.96	2.87	.010	
4.42	4.25	4.07	3.88	3.79	3.69	3.58	3.48	3.37	3.26	.005	
2.03	1.99	1.94	1.89	1.87	1.34	1.81	1.78	1.75	1.72	.100	16
2.49	2.42	2.35	2.28	2.24	2.19	2.15	2.11	2.06	2.01	.050	
2.90	2.89	2.79	2.68	2.63	2.57	2.51	2.45	2.38	2.32	.025	
3.69	3.55	3.41	3.26	3.18	3.10	3.02	2.93	2.84	2.75	.010	
4.27	4.10	3.92	3.73	3.64	3.54	3.44	3.33	3.22	3.11	.005	
2.00	1.96	1.91	1.86	1.84	1.81	1.78	1.75	1.72	1.69	.100	17
2.45	2.38	2.31	2.23	2.19	2.15	2.10	2.06	2.01	1.96	.050	
2.92	2.82	2.72	2.62	2.56	2.50	2.44	2.38	2.32	2.25	.025	
2.59	3.46	3.31	3.16	3.08	3.00	2.92	2.83	2.75	2.65	.010	
4.14	3.97	3.79	3.61	3.51	3.41	3.31	3.21	3.10	2.98	.005	
1.98	1.93	1.89	1.84	1.81	1.78	1.75	1.72	1.69	1.66	.100	18
2.41	2.34	2.27	2.19	2.15	2.11	2.06	2.02	1.97	1.92	.050	
2.87	2.77	2.67	2.56	2.50	2.44	2.38	2.32	2.26	2.19	.025	

Contd...

Table A.6 Contd.

Numerator df											
10	12	15	20	24	30	40	60	120	∞	P	Denom. df
3.51	3.37	3.23	3.08	3.00	2.92	2.84	2.75	2.66	2.57	.010	
4.03	3.86	3.68	3.50	3.40	3.30	3.20	3.10	2.99	2.87	.005	
1.96	1.91	1.86	1.81	1.79	1.76	1.73	1.70	1.67	1.63	.100	19
2.38	2.31	2.23	2.16	2.11	2.07	2.03	1.98	1.93	1.88	.050	
2.82	2.72	2.62	2.51	2.45	2.39	2.33	2.27	2.20	2.13	.025	
3.43	3.30	3.15	3.00	2.92	2.84	2.76	2.67	2.58	2.49	.010	
3.93	3.76	3.59	3.40	3.31	3.21	3.11	3.00	2.89	2.78	.005	
1.94	1.89	1.84	1.79	1.77	1.74	1.71	1.68	1.64	1.61	.100	20
2.35	2.28	2.20	2.12	2.08	2.04	1.99	1.95	1.90	1.84	.050	
2.77	2.68	2.57	2.46	2.41	2.35	2.29	2.22	2.16	2.09	.025	
3.37	3.23	3.09	2.94	2.86	2.78	2.69	2.61	2.52	2.42	.010	
3.85	3.68	3.50	3.32	3.22	3.12	3.02	2.92	2.81	2.69	.005	
1.92	1.87	1.83	1.78	1.75	1.72	1.69	1.66	1.62	1.59	.100	21
2.32	2.25	2.18	2.10	2.05	2.01	1.96	1.92	1.87	1.81	.050	
2.73	2.64	2.53	2.42	2.37	2.31	2.25	2.18	2.11	2.04	.025	
3.31	3.17	3.03	2.88	2.80	2.72	2.64	2.55	2.46	2.36	.010	
3.77	3.60	3.43	3.24	3.15	3.05	2.95	2.84	2.73	2.61	.005	
1.90	1.86	1.81	1.76	1.73	1.70	1.67	1.64	1.60	1.57	.100	22
2.30	2.23	2.15	2.07	2.03	1.98	1.94	1.89	1.84	1.78	.050	
2.70	2.60	2.50	2.39	2.33	2.27	2.21	2.14	2.08	2.00	.025	
3.26	3.12	2.98	2.83	2.75	2.67	2.58	2.50	2.40	2.31	.010	
3.70	3.54	3.36	3.18	3.08	2.98	2.88	2.77	2.66	2.55	.005	
1.89	1.84	1.80	1.74	1.72	1.69	1.66	1.62	1.59	1.55	.100	23
2.27	2.20	2.13	2.05	2.01	1.96	1.91	1.86	1.81	1.76	.050	
2.67	2.57	2.47	2.36	2.30	2.24	2.18	2.11	2.04	1.97	.025	
3.21	3.07	2.93	2.78	2.70	2.62	2.54	2.45	2.35	2.26	.010	
3.64	3.47	3.30	3.12	3.02	2.92	2.82	2.71	2.60	2.48	.005	
1.88	1.83	1.78	1.73	1.70	1.67	1.64	1.61	1.57	1.53	.100	24
2.25	2.18	2.11	2.03	1.98	1.94	1.89	1.84	1.79	1.73	.050	
2.64	2.54	2.44	2.33	2.27	2.21	2.15	2.08	2.01	1.94	.025	
3.18	3.03	2.89	2.74	2.66	2.58	2.49	2.40	2.31	2.21	.010	
3.59	3.42	3.25	3.06	2.97	2.87	2.77	2.66	2.55	2.43	.005	
1.87	1.82	1.77	1.72	1.69	1.66	1.63	1.59	1.56	1.52	.100	25
2.24	2.16	2.09	2.01	1.96	1.92	1.87	1.82	1.77	1.71	.050	
2.61	2.51	2.41	2.30	2.24	2.18	2.12	2.05	1.98	1.91	.025	
3.13	2.99	2.85	2.70	2.62	2.54	2.45	2.36	2.27	2.17	.010	
3.54	3.37	3.20	3.01	2.92	2.82	2.72	2.61	2.50	2.38	.005	
1.86	1.81	1.76	1.71	1.68	1.65	1.61	1.58	1.54	1.50	.100	26
2.22	2.15	2.07	1.99	1.95	1.90	1.85	1.80	1.75	1.69	.050	
2.59	2.49	2.39	2.28	2.22	2.16	2.09	2.03	1.95	1.88	.025	
3.09	2.96	2.81	2.66	2.58	2.50	2.42	2.33	2.23	2.13	.010	
3.49	3.33	3.15	2.97	2.87	2.77	2.67	2.56	2.45	2.33	.005	

Contd...

Table A.6 Contd.

Numerator df											
10	12	15	20	24	30	40	60	120	∞	P	Denom. df
1.85	1.80	1.75	1.70	1.67	1.64	1.60	1.57	1.53	1.49	.100	27
2.20	2.13	2.06	1.97	1.93	1.88	1.84	1.79	1.73	1.67	.050	
2.57	2.47	2.36	2.25	2.19	2.13	2.07	2.00	1.93	1.85	.025	
3.06	2.93	2.78	2.63	2.55	2.47	2.38	2.29	2.20	2.10	.010	
3.45	3.28	3.11	2.93	2.83	2.73	2.63	2.52	2.41	2.29	.005	
1.84	1.79	1.74	1.69	1.66	1.63	1.59	1.56	1.52	1.48	.100	28
2.19	2.12	2.04	1.96	1.91	1.87	1.82	1.77	1.71	1.65	.050	
2.55	2.45	2.34	2.23	2.17	2.11	2.05	1.98	1.91	1.83	.025	
3.03	2.90	2.75	2.60	2.52	2.44	2.35	2.26	2.17	2.06	.010	
3.41	3.25	3.07	2.89	2.79	2.69	2.59	2.48	2.37	2.25	.005	
1.83	1.78	1.73	1.68	1.65	1.62	1.58	1.55	1.51	1.47	.100	29
2.18	2.10	2.03	1.94	1.90	1.85	1.81	1.75	1.70	1.64	.050	
2.53	2.43	2.32	2.21	2.15	2.09	2.03	1.96	1.89	1.81	.025	
3.00	2.87	2.73	2.57	2.49	2.41	2.33	2.23	2.14	2.03	.010	
3.38	3.21	3.04	2.86	2.76	2.66	2.56	2.45	2.33	2.21	.005	
1.82	1.77	1.72	1.67	1.64	1.61	1.57	1.54	1.50	1.46	.100	30
2.16	2.09	2.01	1.93	1.89	1.84	1.79	1.74	1.68	1.62	.050	
2.51	2.41	2.31	2.20	2.14	2.07	2.01	1.94	1.87	1.79	.025	
2.98	2.84	2.70	2.55	2.47	2.39	2.30	2.21	2.11	2.01	.010	
3.34	3.18	3.01	2.82	2.73	2.63	2.52	2.42	2.30	2.18	.005	
1.76	1.71	1.66	1.61	1.57	1.54	1.51	1.47	1.42	1.38	.100	40
2.08	2.00	1.92	1.84	1.79	1.74	1.69	1.64	1.58	1.51	.050	
2.39	2.29	2.18	2.07	2.01	1.94	1.88	1.80	1.72	1.64	.025	
2.80	2.66	2.52	2.37	2.29	2.20	2.11	2.02	1.92	1.80	.010	
3.12	2.95	2.78	2.60	2.50	2.40	2.30	2.18	2.06	1.93	.005	
1.71	1.66	1.60	1.54	1.51	1.48	1.44	1.40	1.35	1.20	.100	60
1.99	1.92	1.84	1.75	1.70	1.65	1.59	1.53	1.47	1.39	.050	
2.27	2.17	2.06	1.94	1.88	1.82	1.74	1.67	1.58	1.48	.025	
2.63	2.50	2.35	2.20	2.12	2.03	1.94	1.84	1.73	1.60	.010	
2.90	2.74	2.57	2.39	2.29	2.19	2.08	1.96	1.83	1.69	.005	
1.65	1.60	1.55	1.48	1.45	1.41	1.37	1.32	1.26	1.19	.100	120
1.91	1.83	1.75	1.66	1.61	1.55	1.50	1.43	1.35	1.25	.050	
2.16	2.05	1.94	1.82	1.76	1.69	1.61	1.53	1.43	1.31	.025	
2.47	2.34	2.19	2.03	1.95	1.86	1.76	1.66	1.53	1.38	.010	
2.71	2.54	2.37	2.19	2.09	1.98	1.87	1.75	1.61	1.43	.005	
1.60	1.55	1.49	1.42	1.38	1.34	1.30	1.24	1.17	1.00	.100	∞
1.83	1.75	1.67	1.57	1.52	1.46	1.39	1.32	1.22	1.00	.050	
2.05	1.94	1.83	1.71	1.64	1.57	1.48	1.39	1.27	1.00	.025	
2.32	2.18	2.04	1.88	1.79	1.70	1.59	1.47	1.32	1.00	.010	
2.52	2.36	2.19	2.00	1.90	1.79	1.67	1.53	1.36	1.00	.005	

Table A.7 Upper percentage points of the studentized range $q_\alpha = \left(\bar{Y}_{max} - \bar{Y}_{min}\right)/S_{\bar{y}}$

Error df	α	*p* = number of treatment means									
		2	3	4	5	6	7	8	9	10	11
5	.05	3.64	4.60	5.22	5.67	6.03	6.33	6.58	6.80	6.99	7.17
	.01	5.70	6.97	7.80	8.42	8.91	9.32	9.67	9.97	10.24	10.48
6	.05	3.46	3.34	4.90	5.31	5.63	5.89	6.12	6.32	6.49	6.65
	.01	5.24	6.33	7.03	7.56	7.97	8.32	8.61	8.87	9.10	9.30
7	.05	3.34	4.16	4.68	5.06	5.36	5.61	5.82	6.00	6.16	6.30
	.01	4.95	5.92	6.54	7.01	7.37	7.68	7.94	8.17	8.37	8.55
8	.05	3.26	4.04	4.53	4.89	5.17	5.40	5.60	5.77	5.92	6.05
	.01	4.74	5.63	6.20	6.63	6.96	7.24	7.47	7.68	7.87	8.03
9	.05	3.20	3.95	4.42	4.76	5.02	5.24	5.43	5.60	5.74	5.87
	.01	4.60	5.43	5.96	6.35	6.66	6.91	7.13	7.32	7.49	7.65
10	.05	3.15	3.88	4.33	4.65	4.91	5.12	5.30	5.46	5.60	5.72
	.01	4.48	5.27	5.77	6.14	6.43	6.67	6.87	7.05	7.21	7.36
11	.05	3.11	3.82	4.26	4.57	4.82	5.03	5.20	5.35	5.49	5.61
	.01	4.39	5.14	5.62	5.97	6.25	6.48	6.67	6.84	6.99	7.13
12	.05	3.08	3.77	4.20	4.51	4.75	4.95	5.12	5.27	5.40	5.51
	.01	4.32	5.04	5.50	5.84	6.10	6.32	6.51	6.67	6.81	6.94
13	.05	3.06	3.73	4.15	4.45	4.69	4.88	5.05	5.19	5.32	5.43
	.01	4.26	4.96	5.40	5.73	5.98	6.19	6.37	6.53	6.67	6.79
14	.05	3.03	3.70	4.11	4.41	4.64	4.83	4.99	5.13	5.25	5.36
	.01	4.21	4.89	5.32	5.63	5.88	6.08	6.26	6.41	6.54	6.66
15	.05	3.01	3.67	4.08	4.37	4.60	4.78	4.94	5.08	5.20	5.31
	.01	4.17	4.83	5.25	5.56	5.80	5.99	6.16	6.31	6.44	6.55
16	.05	3.00	3.65	4.05	4.33	4.56	4.74	4.90	5.03	5.15	5.26
	.01	4.13	4.78	5.19	5.49	5.72	5.92	6.08	6.22	6.35	6.46
17	.05	2.98	3.63	4.02	4.30	4.52	4.71	4.86	4.99	5.11	5.21
	.01	4.10	4.74	5.14	5.43	5.66	5.85	6.01	6.15	6.27	6.38
18	.05	2.97	3.61	4.00	4.28	4.49	4.67	4.82	4.96	5.07	5.17
	.01	4.07	4.70	5.09	5.38	5.60	5.79	5.94	6.08	6.20	6.31
19	.05	2.96	3.59	3.98	4.25	4.47	4.65	4.79	4.92	5.04	5.14
	.01	4.05	4.67	5.05	5.33	5.55	5.73	5.89	6.02	6.14	6.25
20	.05	2.95	3.58	3.96	4.23	4.45	4.62	4.77	4.90	5.01	5.11
	.01	4.02	4.64	5.02	5.29	5.51	5.69	5.84	5.97	6.09	6.19
24	.05	2.92	3.53	3.90	4.17	4.37	4.54	4.68	4.81	4.92	5.01
	.01	3.96	4.54	4.91	5.17	5.37	5.54	5.69	5.81	5.92	6.02
30	.05	2.89	3.49	3.84	4.10	4.30	4.46	4.60	4.72	4.83	4.92
	.01	3.89	4.45	4.80	5.05	5.24	5.40	5.54	5.65	5.76	5.85
40	.05	2.86	3.44	3.79	4.04	4.23	4.39	4.52	4.63	4.74	4.82
	.01	3.82	4.37	4.70	4.93	5.11	5.27	5.39	5.50	5.60	5.69
60	.05	2.83	3.40	3.74	3.98	4.16	4.31	4.44	4.55	4.65	4.73
	.01	3.76	4.28	4.60	4.82	4.99	5.13	5.25	5.36	5.45	5.33
120	.05	2.80	3.36	3.69	3.92	4.10	4.24	4.36	4.48	4.56	4.64
	.01	3.70	4.20	4.50	4.71	4.87	5.01	5.12	5.21	5.30	5.38
∞	.05	2.77	3.31	3.63	3.86	4.03	4.17	4.29	4.39	4.47	4.55
	.01	3.63	4.12	4.40	4.60	4.76	4.88	4.99	5.08	5.16	5.23

Source: Steel and Torrie (1981), Principles and Procedures of Statistics: A Biometrical Approach who abridged from Biometrica Tables for Statisticians, vol. 1, 1954.

Table A.7 Contd.

p = number of treatment means									α	Error df
12	13	14	15	16	17	18	19	20		
7.32	7.47	7.60	7.72	7.83	7.93	8.03	8.12	8.21	.05	5
10.70	10.89	11.08	11.24	11.40	11.55	11.68	11.81	11.93	.01	
6.79	6.92	7.03	7.14	7.24	7.34	7.43	7.51	7.59	.05	6
9.49	9.65	9.81	9.95	10.08	10.21	10.32	10.43	10.54	.01	
6.43	6.55	6.66	6.76	6.85	6.94	7.02	7.09	7.17	.05	7
8.71	8.86	9.00	9.12	9.24	9.35	9.46	9.55	9.65	.01	
6.18	6.29	6.39	6.48	6.57	6.65	6.73	6.80	6.87	.05	8
8.18	8.31	8.44	8.65	8.66	8.76	8.85	8.94	9.03	.01	
5.98	6.09	6.19	6.28	6.36	6.44	6.51	6.58	6.64	.05	9
7.78	7.91	8.03	8.13	8.23	8.32	8.41	8.49	8.57	.01	
5.83	5.93	6.03	6.11	6.20	6.27	6.34	6.40	6.47	.05	10
7.48	7.60	7.71	7.81	7.91	7.99	8.07	8.15	8.22	.01	
5.71	5.81	5.90	5.99	6.06	6.14	6.20	6.26	6.33	.05	11
7.25	7.36	7.46	7.56	7.65	7.73	7.81	7.88	7.95	.01	
5.62	5.71	5.80	5.88	5.95	6.03	6.09	6.15	6.21	.05	12
7.06	7.17	7.26	7.36	7.44	7.52	7.59	7.66	7.73	.01	
5.53	5.63	5.71	5.79	5.86	5.93	6.00	6.05	6.11	.05	13
6.90	7.01	7.10	7.19	7.27	7.34	7.42	7.48	7.55	.01	
5.46	5.55	5.64	5.72	5.79	5.85	5.92	5.97	6.03	.05	14
6.77	6.87	6.96	7.05	7.12	7.20	7.27	7.33	7.39	.01	
5.40	5.49	5.58	5.65	5.72	5.79	5.85	5.90	5.96	.05	15
6.66	6.76	6.84	6.93	7.00	7.07	7.14	7.20	7.20	7.26	
5.35	5.44	5.52	5.59	5.66	5.72	5.79	5.84	5.90	.05	16
6.56	6.66	6.74	6.82	6.90	6.97	7.03	7.09	7.15	.01	
5.31	5.39	5.47	5.55	5.61	5.68	5.74	5.79	5.84	.05	17
6.48	6.57	6.66	6.73	6.80	6.87	6.94	7.00	7.05	.01	
5.27	5.35	5.43	5.50	5.57	5.63	5.69	5.74	5.79	.05	18
6.41	6.50	6.58	6.65	6.72	6.79	6.85	6.91	6.96	.01	
5.23	5.32	5.39	5.46	5.53	5.59	5.65	5.70	5.75	.05	19
6.34	6.43	6.51	6.58	6.65	6.72	6.78	6.84	6.89	.01	
5.20	5.28	5.36	5.43	5.49	5.55	5.61	5.66	5.71	.05	20
6.29	6.37	6.45	6.52	6.59	6.65	6.71	6.76	6.82	.01	
5.10	5.18	5.25	5.32	5.38	5.44	5.50	5.54	5.59	.05	24
6.11	6.19	6.26	6.33	6.39	6.45	6.51	6.56	6.61	.01	
5.00	5.08	5.15	5.21	5.27	5.33	5.38	5.43	5.48	.05	30
5.93	6.01	6.08	6.14	6.20	6.26	6.31	6.36	6.41	.01	
4.91	4.98	5.05	5.11	5.16	5.22	5.27	5.31	5.36	.05	40
5.77	5.84	5.90	5.96	6.02	6.07	6.12	6.17	6.21	.01	
4.81	4.88	4.94	5.00	5.06	5.11	5.16	5.20	5.24	.05	60
5.60	5.67	5.73	5.79	5.84	5.89	5.93	5.98	6.02	.01	
4.72	4.78	4.84	4.90	4.95	5.00	5.05	5.09	5.13	.05	120
5.44	5.51	5.56	5.61	5.66	5.71	5.75	5.79	5.83	.01	
4.62	4.68	4.74	4.80	4.85	4.89	4.93	4.97	5.01	.05	∞
5.29	5.35	5.40	5.45	5.49	5.54	5.57	5.61	5.65	.01	

Table A.8 Significant studentized ranges for 5% and 1% level new multiple-range test

Error df	Significance level	p = number of means for range being tested													
		2	3	4	5	6	7	8	9	10	12	14	16	18	20
1	.05	18.0	18.0	18.0	18.0	18.0	18.0	18.0	18.0	18.0	18.0	18.0	18.0	18.0	18.0
	.01	90.0	90.0	90.0	90.0	90.0	90.0	90.0	90.0	90.0	90.0	90.0	90.0	90.0	90.0
2	.05	6.09	6.09	6.09	6.09	6.09	6.09	6.09	6.09	6.09	6.09	6.09	6.09	6.09	6.09
	.01	14.0	14.0	14.0	14.0	14.0	14.0	14.0	14.0	14.0	14.0	14.0	14.0	14.0	14.0
3	.05	4.50	4.50	4.50	4.50	4.50	4.50	4.50	4.50	4.50	4.50	4.50	4.50	4.50	4.50
	.01	8.26	8.5	8.6	8.7	8.8	8.9	8.9	9.0	9.0	9.0	9.1	9.2	9.3	9.3
4	.05	3.93	4.01	4.02	4.02	4.02	4.02	4.02	4.02	4.02	4.02	4.02	4.02	4.02	4.02
	.01	6.51	6.8	6.9	7.0	7.1	7.1	7.2	7.2	7.3	7.3	7.4	7.4	7.5	7.5
5	.05	3.64	3.74	3.79	3.83	3.83	3.83	3.83	3.83	3.83	3.83	3.83	3.83	3.83	3.83
	.01	5.70	5.96	6.11	6.18	6.26	6.33	6.40	6.44	6.5	6.6	6.6	6.7	6.7	6.8
6	.05	3.46	3.58	3.64	3.68	3.68	3.68	3.68	3.68	3.68	3.68	3.68	3.68	3.68	3.68
	.01	5.24	5.51	5.65	5.73	5.81	5.88	5.95	6.00	6.0	6.1	6.2	6.2	6.3	6.3
7	.05	3.35	3.47	3.54	3.58	3.60	3.61	3.61	3.61	3.61	3.61	3.61	3.61	3.61	3.61
	.01	4.95	5.22	5.37	5.45	5.53	5.61	5.69	5.73	5.8	5.8	5.9	5.9	6.0	6.0
8	.05	3.26	3.39	3.47	3.52	3.55	3.56	3.56	3.56	3.56	3.56	3.56	3.56	3.56	3.56
	.01	4.74	5.00	5.14	5.23	5.32	5.40	5.47	5.51	5.5	5.6	5.7	5.7	5.8	5.8
9	.05	3.20	3.34	3.41	3.47	3.50	3.52	3.52	3.52	3.52	3.52	3.52	3.52	3.52	3.52
	.01	4.60	4.86	4.99	5.08	5.17	5.25	5.32	5.36	5.4	5.5	5.5	5.6	5.7	5.7
10	.05	3.15	3.30	3.37	3.43	3.46	3.47	3.47	3.47	3.47	3.47	3.47	3.47	3.47	3.48
	.01	4.48	4.73	4.88	4.96	5.06	5.13	5.20	5.24	5.28	5.36	5.42	5.48	5.54	5.55
11	.05	3.11	3.27	3.35	3.39	3.43	3.44	3.45	3.46	4.46	4.46	4.46	4.46	4.47	4.48
	.01	4.39	4.63	4.77	4.86	4.94	5.01	5.06	5.12	5.15	5.24	5.28	5.34	5.38	5.39
12	.05	3.08	3.23	3.33	3.36	3.40	3.42	3.44	3.44	3.46	3.46	3.46	3.46	3.47	3.48
	.01	4.32	4.55	4.68	4.76	4.84	4.92	4.96	5.02	5.07	5.13	5.17	5.22	5.24	5.26
13	.05	3.06	3.21	3.30	3.35	3.38	3.41	3.42	3.44	3.45	3.45	3.46	3.46	3.47	3.47
	.01	4.26	4.48	4.62	4.69	4.74	4.84	4.88	4.94	4.98	5.04	5.08	5.13	5.14	5.15
14	.05	3.03	3.18	3.27	3.33	3.37	3.39	3.41	3.42	3.44	3.45	3.46	3.46	3.47	3.47
	.01	4.21	4.42	4.55	4.63	4.70	4.78	4.83	4.87	4.91	4.96	5.00	5.04	5.06	5.07
15	.05	3.01	3.16	3.25	3.31	3.36	3.38	3.40	3.42	3.43	3.44	3.45	3.46	3.47	3.47
	.01	4.17	4.37	4.50	4.58	4.64	4.72	4.77	4.81	4.84	4.90	4.94	4.97	4.99	5.00
16	.05	3.00	3.15	3.23	3.30	3.34	3.37	3.39	3.41	3.43	3.44	3.45	3.46	3.47	3.47
	.01	4.13	4.34	4.45	4.54	4.60	4.67	4.72	4.76	4.79	4.84	4.88	4.91	4.93	4.94
17	.05	2.98	3.13	3.22	3.28	3.33	3.36	3.38	3.40	3.42	3.44	3.45	3.46	3.47	3.47
	.01	4.10	4.30	4.41	4.50	4.56	4.63	4.68	4.72	4.75	4.80	4.83	4.86	4.88	4.89
18	.05	2.97	3.12	3.21	3.27	3.32	3.35	3.37	3.39	3.41	3.43	3.45	3.46	3.47	3.47
	.01	4.07	4.27	4.38	4.46	4.53	4.59	4.64	4.68	4.71	4.76	4.79	4.82	4.84	4.85
19	.05	2.96	3.11	3.19	3.26	3.31	3.35	3.37	3.39	3.41	3.43	3.44	3.46	3.47	3.47
	.01	4.05	4.24	4.35	4.43	4.50	4.56	4.61	4.64	4.67	4.72	4.76	4.79	4.81	4.82
20	.05	2.95	3.10	3.18	3.25	3.30	3.34	3.36	3.38	3.40	3.43	3.44	3.46	3.46	3.47
	.01	4.02	4.22	4.33	4.40	4.47	4.53	4.58	4.61	4.65	4.69	4.73	4.76	4.78	4.79
22	.05	2.93	3.08	3.17	3.24	3.29	3.32	3.35	3.37	3.39	3.42	3.44	3.45	3.46	3.47
	.01	3.99	4.17	4.28	4.36	4.42	4.48	4.53	4.57	4.60	4.65	4.68	4.71	4.74	4.75
24	.05	2.92	3.07	3.15	3.22	3.25	3.31	3.34	3.37	3.38	3.41	3.44	3.45	3.46	3.47
	.01	3.96	4.14	4.24	4.33	4.39	4.44	4.49	4.53	4.57	4.62	4.64	4.67	4.70	4.72

Contd...

Table A.8 Contd.

Error df	Significance level	p = number of means for range being tested													
		2	3	4	5	6	7	8	9	10	12	14	16	18	20
26	.05	2.91	3.06	3.14	3.21	3.27	3.30	3.34	3.36	3.38	3.41	3.43	3.45	3.46	3.47
	.01	3.93	4.11	4.21	4.30	4.36	4.41	4.46	4.50	4.53	4.58	4.62	4.65	4.67	4.69
28	.05	2.90	3.04	3.13	3.20	3.26	3.30	3.33	3.35	3.37	3.40	3.43	3.45	3.46	3.47
	.01	3.91	4.08	4.18	4.28	4.34	4.39	4.43	4.47	4.51	4.56	4.60	4.62	4.65	4.67
30	.05	2.89	3.04	3.12	3.20	3.25	3.29	3.32	3.35	3.37	3.40	3.43	3.45	3.46	3.47
	.01	3.89	4.06	4.16	4.22	4.32	4.36	4.41	4.45	4.48	4.54	4.58	4.61	4.63	4.65
40	.05	2.86	3.01	3.10	3.17	3.22	3.27	3.30	3.33	3.35	3.39	3.42	3.44	3.46	3.47
	.01	3.82	3.99	4.10	4.17	4.24	4.30	4.34	4.37	4.41	4.46	4.51	4.54	4.57	4.59
60	.05	2.83	2.98	3.08	3.14	3.20	3.24	3.28	3.31	3.33	3.37	3.40	3.43	3.45	3.47
	.01	3.76	3.92	4.03	4.12	4.17	4.23	4.27	4.31	4.34	4.39	4.44	4.47	4.50	4.53
100	.05	2.80	2.95	3.05	3.12	3.18	3.22	3.26	3.29	3.32	3.36	3.40	3.42	3.45	3.47
	.01	3.71	3.86	3.98	4.06	4.11	4.17	4.21	4.25	4.29	4.35	4.38	4.42	4.45	4.48
∞	.05	2.77	2.92	3.02	3.09	3.15	3.19	3.23	3.26	3.29	3.34	3.38	3.41	3.44	3.47
	.01	3.64	3.80	3.90	3.98	4.04	4.09	4.14	4.17	4.20	4.26	4.31	4.34	4.38	4.41

Source: Steel and Torrie (1981) Principles and Procedures of Statistics: A Biometrical Approach, who abridged from D.B. Duncan, "Multiple range and multiple *F* tests," *Biometrics* 11, 1-42 (1955).

Table A.9A Table for *t* for one-sided comparisons between *p* treatment means and a control for a joint confidence coefficient of $P = .95$ and $P = .99$

Error df	*P*	*p* = number of treatment means, excluding control								
		1	2	3	4	5	6	7	8	9
5	.95	2.02	2.44	2.68	2.85	2.98	3.08	3.16	3.24	3.30
	.99	3.37	3.90	4.21	4.43	4.60	4.73	4.85	4.94	5.03
6	.95	1.94	2.34	2.56	2.71	2.83	2.92	3.00	3.07	3.12
	.99	3.14	3.61	3.88	4.07	4.21	4.33	4.43	4.51	4.59
7	.95	1.89	2.27	2.48	2.62	2.73	2.82	2.89	2.95	3.01
	.99	3.00	3.42	3.66	3.83	3.96	4.07	4.15	4.23	4.30
8	.95	1.86	2.22	2.42	2.55	2.66	2.74	2.81	2.87	2.92
	.99	2.90	3.29	3.51	3.67	3.79	3.88	3.96	4.03	4.09
9	.95	1.83	2.18	2.37	2.50	2.60	2.68	2.75	2.81	2.86
	.99	2.82	3.19	3.40	3.55	3.66	3.75	3.82	3.89	3.94
10	.95	1.81	2.15	2.34	2.47	2.56	2.64	2.70	2.76	2.81
	.99	2.76	3.11	3.31	3.45	3.56	3.64	3.71	3.78	3.83
11	.95	1.80	2.13	2.31	2.44	2.53	2.60	2.67	2.72	2.77
	.99	2.72	3.06	3.25	3.38	3.48	3.56	3.63	3.69	3.74
12	.95	1.78	2.11	2.29	2.41	2.50	2.58	2.64	2.69	2.74
	.99	2.68	3.01	3.19	3.32	3.42	3.50	3.56	3.62	3.67
13	.95	1.77	2.09	2.27	2.39	2.48	2.55	2.61	2.66	2.71
	.99	2.65	2.97	3.15	3.27	3.37	3.44	3.51	3.56	3.61
14	.95	1.76	2.08	2.25	2.37	2.46	2.53	2.59	2.64	2.69
	.99	2.62	2.94	3.11	3.23	3.32	3.40	3.46	3.51	3.56
15	.95	1.75	2.07	2.24	2.36	2.44	2.51	2.57	2.62	2.67
	.99	2.60	2.91	3.08	3.20	3.29	3.36	3.42	3.47	3.52
16	.95	1.75	2.06	2.23	2.34	2.43	2.50	2.56	2.61	2.65
	.99	2.58	2.88	3.05	3.17	3.26	3.33	3.39	3.44	3.48
17	.95	1.74	2.05	2.22	2.33	2.42	2.49	2.54	2.59	2.64
	.99	2.57	2.86	3.03	3.14	3.23	3.30	3.36	3.41	3.45
18	.95	1.73	2.04	2.21	2.32	2.41	2.48	2.53	2.58	2.62
	.99	2.55	2.84	3.01	3.12	3.21	3.27	3.33	3.38	3.42
19	.95	1.73	2.03	2.20	2.31	2.40	2.47	2.52	2.57	2.61
	.99	2.54	2.83	2.99	3.10	3.18	3.25	3.31	3.36	3.40
20	.95	1.72	2.03	2.19	2.30	2.39	2.46	2.51	2.56	2.60
	.99	2.53	2.81	2.97	3.08	3.17	3.23	3.29	3.34	3.38
24	.95	1.71	2.02	2.17	2.28	2.36	2.43	2.48	2.53	2.57
	.99	2.49	2.77	2.92	3.03	3.11	3.17	3.22	3.27	3.31
30	.95	1.70	1.99	2.15	2.25	2.33	2.40	2.45	2.50	2.54
	.99	2.46	2.72	2.87	2.97	3.05	3.11	3.16	3.21	3.24
40	.95	1.68	1.97	2.13	2.23	2.31	2.37	2.42	2.47	2.51
	.99	2.42	2.68	2.82	2.92	2.99	3.05	3.10	3.14	3.18
60	.95	1.67	1.95	2.10	2.21	2.28	2.35	2.39	2.44	2.48
	.99	2.39	2.64	2.78	2.87	2.94	3.00	3.04	3.08	3.12
120	.95	1.66	1.93	2.08	2.18	2.26	2.32	2.37	2.37	2.41
2.45	.99	2.36	2.60	2.73	2.82	2.89	2.94	2.99	3.03	3.06
∞	.95	1.64	1.92	2.06	2.16	2.23	2.29	2.34	2.38	2.42
	.99	2.33	2.56	2.68	2.77	2.84	2.89	2.93	2.97	3.00

Source: Steel and Torrie (1981), Principles and Procedures of Statistics: A Biometrical Approach, who reproduced it from CW Dunnett (1955) " A multiple comparison procedure for comparing several treatments with a control," Jour. Amer. Statist. Asso, 50, pp. 1096-1121

Table A.9B Table for t for two-sided comparisons between p treatment means and a control for a joint confidence coefficient of $P = .95$ and $P = .99$

Error df	P	p = number of treatment means, excluding control								
		1	2	3	4	5	6	7	8	9
5	.95	2.57	3.03	3.39	3.66	3.88	4.06	4.22	4.36	4.49
	.99	4.03	4.63	5.09	5.44	5.73	5.97	6.18	6.36	6.53
6	.95	2.45	2.86	3.18	3.41	3.60	3.75	3.88	4.00	4.11
	.99	3.71	4.22	4.60	4.88	5.11	5.30	5.47	5.61	5.74
7	.95	2.36	2.75	3.04	3.24	3.41	3.54	3.66	3.76	3.86
	.99	3.50	3.95	4.28	4.52	4.71	4.87	5.01	5.13	5.24
8	.95	2.31	2.67	2.94	3.13	3.28	3.40	3.51	3.60	3.68
	.99	3.36	3.77	4.06	4.27	4.44	4.58	4.70	4.81	4.90
9	.95	2.26	2.61	2.86	3.04	3.18	3.29	3.39	3.48	3.55
	.99	3.25	3.63	3.90	4.09	4.24	4.37	4.48	4.57	4.65
10	.95	2.23	2.57	2.81	2.97	3.11	3.21	3.31	3.39	3.46
	.99	3.17	3.53	3.78	3.95	4.10	4.21	4.31	4.40	4.47
11	.95	2.20	2.53	2.76	2.92	3.05	3.15	3.24	3.31	3.38
	.99	3.11	3.45	3.68	3.85	3.98	4.09	4.18	4.26	4.33
12	.95	2.18	2.50	2.72	2.88	3.00	3.10	3.18	3.25	3.32
	.99	3.05	3.39	3.61	3.76	3.89	3.99	4.08	4.15	4.22
13	.95	2.16	2.48	2.69	2.84	2.96	3.06	3.14	3.21	3.27
	.99	3.01	3.33	3.54	3.69	3.81	3.91	3.99	4.06	4.13
14	.95	2.14	2.46	2.67	2.81	2.93	3.02	3.10	3.17	3.23
	.99	2.98	3.29	3.49	3.64	3.75	3.84	3.92	3.99	4.05
15	.95	2.13	2.44	2.64	2.79	2.90	2.99	3.07	3.13	3.19
	.99	2.95	3.25	3.45	3.59	3.70	3.79	3.86	3.93	3.99
16	.95	2.12	2.42	2.63	2.77	2.88	2.96	3.04	3.10	3.16
	.99	2.92	3.22	3.41	3.55	3.65	3.74	3.82	3.88	3.93
17	.95	2.11	2.41	2.61	2.75	2.85	2.94	3.01	3.08	3.13
	.99	2.90	3.19	3.38	3.51	3.62	3.70	3.77	3.83	3.89
18	.95	2.10	2.40	2.59	2.73	2.84	2.92	2.99	3.05	3.11
	.99	2.88	3.17	3.35	3.48	3.58	3.67	3.74	3.80	3.85
19	.95	2.09	2.39	2.58	2.72	2.82	2.90	2.97	3.04	3.09
	.99	2.86	3.15	3.33	3.46	3.55	3.64	3.70	3.76	3.81
20	.95	2.09	2.38	2.57	2.70	2.81	2.89	2.96	3.02	3.07
	.99	2.85	3.13	3.31	3.43	3.53	3.61	3.67	3.73	3.78
24	.95	2.06	2.35	2.53	2.66	2.76	2.84	2.91	2.96	3.01
	.99	2.80	3.07	3.24	3.36	3.45	3.52	3.58	3.64	3.69
30	.95	2.04	2.32	2.50	2.62	2.72	2.79	2.86	2.91	2.96
	.99	2.75	3.01	3.17	3.28	3.37	3.44	3.50	3.55	3.59
40	.95	2.02	2.29	2.47	2.58	2.67	2.75	2.81	2.86	2.90
	.99	2.70	2.95	3.10	3.21	3.29	3.36	3.41	3.46	3.50
60	.95	2.00	2.27	2.43	2.55	2.63	2.70	2.76	2.81	2.85
	.99	2.66	2.90	3.04	3.14	3.22	3.28	3.33	3.38	3.42
120	.95	1.98	2.24	2.40	2.51	2.59	2.66	2.71	2.76	2.80
	.99	2.62	2.84	2.98	3.08	3.15	3.21	3.25	3.30	3.33
∞	.95	1.96	2.21	2.37	2.47	2.55	2.62	2.67	2.71	2.75
	.99	2.58	2.79	2.92	3.01	3.08	3.14	3.18	3.22	3.25

Source: Steel and Torrie (1981), Principles and Procedures of Statistics: A Biometrical Approach, who reproduced it from CW Dunnett (1955) "A multiple comparison procedure for comparing several treatments with a control," Jour. Amer. Statist. Asso. 50, pp. 1096-1121

Table A.10A Critical values for the Kolmogorov-Smirnov one-sample test (several α)

One-sided test, α =	.10	.05	.025	.01	.005
Two-sided test, α =	.20	.10	.05	.02	.01
n					
1	.900	.950	.975	.990	.995
2	.684	.776	.842	.900	.929
3	.565	.636	.708	.785	.829
4	.493	.565	.624	.689	.734
5	.447	.509	.563	.627	.669
6	.410	.468	.519	.577	.617
7	.381	.436	.483	.538	.576
8	.358	.410	.454	.507	.542
9	.339	.387	.430	.480	.513
10	.323	.369	.409	.457	.489
11	.308	.352	.391	.437	.468
12	.296	.338	.375	.419	.449
13	.285	.325	.361	.404	.432
14	.275	.314	.349	.390	.418
15	.266	.304	.338	.377	.404
16	.258	.295	.327	.366	.392
17	.250	.286	.318	.355	.381
18	.244	.279	.309	.346	.371
19	.237	.271	.301	.337	.361
20	.232	.265	.294	.329	.352
21	.226	.259	.287	.321	.344
22	.221	.253	.281	.314	.337
23	.216	.247	.275	.307	.330
24	.212	.242	.269	.301	.323
25	.208	.238	.264	.295	.317
26	.204	.233	.259	.290	.311
27	.200	.229	.254	.284	.305
28	.197	.225	.250	.279	.300
29	.193	.221	.246	.275	.295
30	.190	.218	.242	.270	.290
31	.187	.214	.238	.266	.285
32	.184	.211	.234	.262	.281
33	.182	.208	.231	.258	.277
34	.179	.205	.227	.254	.273
35	.177	.202	.224	.251	.269
36	.174	.199	.221	.247	.265
37	.172	.196	.218	.244	.262
38	.170	.194	.215	.241	.258
39	.168	.191	.213	.238	.255
40	.165	.189	.210	.235	.252
Approximation for $n > 40$:	$\frac{1.0730}{\sqrt{n}}$	$\frac{1.2239}{\sqrt{n}}$	$\frac{1.3581}{\sqrt{n}}$	$\frac{1.5174}{\sqrt{n}}$	$\frac{1.6276}{\sqrt{n}}$

Source: Steel and Torrie (1981), Principles and Procedures of Statistics: A Biometrical Approach, who extracted it from "Table of percentage points of Kolmogorov statistics," *J. Amer. Statist. Assoc.*, **51**: 111-112 (1956), with permission of the author, L.H. Miller, and the editor.

Table A.10B Critical values for the one sample Kolmogorov-Smirnov test (5% and 1% critical value for one-side and two-side tests)

	One-sided test D_n^+ or D_n^-		Two-sided D_n	
n	1%	5%	1%	5%
1	.990	.950	.995	.975
2	.900	.776	.929	.842
3	.785	.636	.829	.708
4	.689	.565	.734	.624
5	.627	.509	.669	.563
6	.577	.468	.617	.519
7	.538	.436	.576	.483
8	.507	.410	.542	.454
9	.480	.387	.513	.430
10	.457	.369	.489	.409
11	.437	.352	.468	.391
12	.419	.338	.449	.375
13	.404	.325	.432	.361
14	.390	.314	.418	.349
15	.377	.304	.404	.338
16	.366	.295	.392	.327
17	.355	.286	.381	.318
18	.346	.279	.371	.309
19	.337	.271	.361	.301
20	.329	.265	.352	.294
25	.295	.238	.317	.264
30	.270	.218	.290	.242
35	.251	.202	.269	.224
Over 35	$\frac{1.52}{\sqrt{n}}$	$\frac{1.22}{\sqrt{n}}$	$\frac{1.63}{\sqrt{n}}$	$\frac{1.36}{\sqrt{n}}$

Source: Rao, Mitra, Mathai and Ramamurthy (1975); Formula and Tables for Statistical Work.

Table A.11 Critical points of Wilcoxon Signed-Rank test

One-sided	Two-sided	$n = 5$	$n = 6$	$n = 7$	$n = 8$	$n = 9$	$n = 10$
$P = .05$	$P = .10$	1	2	4	6	8	11
$P = .025$	$P = .05$		1	2	4	6	8
$P = .01$	$P = .02$			0	2	3	5
$P = .005$	$P = .01$				0	2	3
One-sided	Two-sided	$n = 11$	$n = 12$	$n = 13$	$n = 14$	$n = 15$	$n = 16$
$P = .05$	$P = .10$	14	17	21	26	30	36
$P = .025$	$P = .05$	11	14	17	21	25	30
$P = .01$	$P = .02$	7	10	13	16	20	24
$P = .005$	$P = .01$	5	7	10	13	16	19
One-sided	Two-sided	$n = 17$	$n = 18$	$n = 19$	$n = 20$	$n = 21$	$n = 22$
$P = .05$	$P = .10$	41	47	54	60	68	75
$P = .025$	$P = .05$	35	40	46	52	59	66
$P = .01$	$P = .02$	28	33	38	43	49	56
$P = .005$	$P = .01$	23	28	32	37	43	49
One-sided	Two-sided	$n = 23$	$n = 24$	$n = 25$	$n = 26$	$n = 27$	$n = 28$
$P = .05$	$P = .10$	83	92	101	110	120	130
$P = .025$	$P = .05$	73	81	90	98	107	117
$P = .01$	$P = .02$	62	69	77	85	93	102
$P = .005$	$P = .01$	55	61	68	76	84	92
One-sided	Two-sided	$n = 29$	$n = 30$	$n = 31$	$n = 32$	$n = 33$	$n = 84$
$P = .05$	$P = .10$	141	152	163	175	188	201
$P = .025$	$P = .05$	127	137	148	159	171	183
$P = .01$	$P = .02$	111	120	130	141	151	162
$P = .005$	$P = .01$	100	109	118	128	138	149
One-sided	Two-sided	$n = 35$	$n = 36$	$n = 37$	$n = 38$	$n = 39$	
$P = .05$	$P = .10$	214	228	242	256	271	
$P = .025$	$P = .05$	195	208	222	235	250	
$P = .01$	$P = .02$	174	186	198	211	224	
$P = .005$	$P = .01$	160	171	183	195	208	
One-sided	Two-sided	$n = 40$	$n = 41$	$n = 42$	$n = 43$	$n = 44$	$n = 45$
$P = .05$	$P = .10$	287	303	319	336	353	371
$P = .025$	$P = .05$	264	279	295	311	327	344
$P = .01$	$P = .02$	238	252	267	281	297	313
$P = .005$	$P = .01$	221	234	248	262	277	292
One-sided	Two-sided	$n = 46$	$n = 47$	$n = 48$	$n = 49$	$n = 50$	
$P = .05$	$P = .10$	389	408	427	446	466	
$P = .025$	$P = .05$	361	379	397	415	434	
$P = .01$	$P = .02$	329	345	362	380	398	
$P = .005$	$P = .01$	307	323	339	356	373	

Source: Milton and Tsokos (1983) Statistical Methods in the Biological and Health Sciences, who reprinted it from W.H. Beyer (ed.), CRC Handbook of Tables for Probability and Statistics, 2nd ed., 1968, p. 400. Copyright CRC Press Inc., Boca Raton, Florida.

Table A.12 The Wald-Wolfowitz run test (1% critical values)

n_1	n_2																		n_1
	3	4	5	6	7	8	9	10	11	12	13	14	15	16	17	18	19	20	
3	-	-	-	-	-	-	-	-	-	2	2	2	2	2	2	2	2	2	3
4	-	-	-	-	-	2	2	2	2	2	2	2	3	3	3	3	3	3	4
5	-	-	-	2	2	2	2	3	3	3	3	3	3	3	3	4	4	4	5
6	-	-	2	2	2	3	3	3	3	3	3	4	4	4	4	4	4	4	6
7	-	-	2	2	3	3	3	3	4	4	4	4	4	5	5	5	5	5	7
8	-	2	2	3	3	3	3	4	4	4	5	5	5	5	5	6	6	6	8
9	-	2	2	3	3	3	4	4	5	5	5	5	6	6	6	6	6	7	9
10	-	2	3	3	3	4	4	5	5	5	5	6	6	6	7	7	7	7	10
11	-	2	3	3	4	4	5	5	5	6	6	6	7	7	7	7	8	8	11
12	2	2	3	3	4	4	5	5	6	6	6	7	7	7	8	8	8	8	12
13	2	2	3	3	4	5	5	5	6	6	7	7	7	8	8	8	9	9	13
14	2	2	3	4	4	5	5	6	6	7	7	7	8	8	8	9	9	9	14
15	2	3	3	4	4	5	6	6	7	7	7	8	8	9	9	9	10	10	15
16	2	3	3	4	5	5	6	6	7	7	8	8	9	9	9	10	10	10	16
17	2	3	3	4	5	5	6	7	7	8	8	8	9	9	10	10	10	11	17
18	2	3	4	4	5	6	6	7	7	8	8	9	9	10	10	11	11	11	18
19	2	3	4	4	5	6	6	7	8	8	9	9	10	10	10	11	11	12	19
20	2	3	4	4	5	6	7	7	8	8	9	9	10	10	11	11	12	12	20
$n_2 \rightarrow$	3	4	5	6	7	8	9	10	11	12	13	14	15	16	17	18	19	20	

(5% critical values)

n_1	n_2																			n_1
	2	3	4	5	6	7	8	9	10	11	12	13	14	15	16	17	18	19	20	
2	-	-	-	-	-	-	-	-	-	-	2	2	2	2	2	2	2	2	2	2
3	-	-	-	-	2	2	2	2	2	2	2	2	2	3	3	3	3	3	3	3
4	-	-	-	2	2	2	3	3	3	3	3	3	3	3	4	4	4	4	4	4
5	-	-	2	2	3	3	3	3	3	4	4	4	4	4	4	4	5	5	5	5
6	-	2	2	3	3	3	3	4	4	4	4	5	5	5	5	5	5	6	6	6
7	-	2	2	3	3	3	4	4	5	5	5	5	5	6	6	6	6	6	6	7
8	-	2	3	3	3	4	4	5	5	5	6	6	6	6	6	7	7	7	7	8
9	-	2	3	3	4	4	5	5	5	6	6	6	7	7	7	7	8	8	8	9
10	-	2	3	3	4	5	5	5	6	6	7	7	7	7	8	8	8	8	9	10
11	-	2	3	4	4	5	5	6	6	7	7	7	8	8	8	9	9	9	9	11
12	2	2	3	4	4	5	6	6	7	7	7	8	8	8	9	9	9	10	10	12
13	2	2	3	4	5	5	6	6	7	7	8	8	9	9	9	10	10	10	10	13
14	2	2	3	4	5	5	6	7	7	8	8	9	9	9	10	10	10	11	11	14
15	2	3	3	4	5	6	6	7	7	8	8	9	9	10	10	11	11	11	12	15
16	2	3	4	4	5	6	6	7	8	8	9	9	10	10	11	11	11	12	12	16
17	2	3	4	4	5	6	7	7	8	9	9	10	10	11	11	11	12	12	13	17
18	2	3	4	5	5	6	7	8	8	9	9	10	10	11	11	12	12	13	13	18
19	2	3	4	5	6	6	7	8	8	9	10	10	11	11	12	12	13	13	13	19
20	2	3	4	5	6	6	7	8	9	9	10	10	11	12	12	13	13	13	14	20
$n_2 \rightarrow$	2	3	4	5	6	7	8	9	10	11	12	13	14	15	16	17	18	19	20	

Source: Rao, Mitra, Mathai and Ramamurthy (1975); Formula and Tables for Statistical Work.

Table A.13 The Wilcoxon Matched pair signed rank test (5% and 1% critical values of W_-, W_+ and W)

n	W_- or W_+ (one-sided)		W (two-sided)	
6	-	2	-	0
7	0	3	-	2
8	2	5	0	4
9	3	8	2	6
10	5	10	3	8
11	7	13	5	11
12	10	17	7	14
13	13	21	10	17
14	16	25	13	21
15	20	30	16	25
16	24	35	20	30
17	28	41	23	35
18	33	47	28	40
19	38	53	32	46
20	43	60	38	52
21	49	67	43	59
22	58	75	49	66
23	62	83	55	73
24	69	91	61	81
25	77	100	68	89

Source: Rao, Mitra, Mathai and Ramamurthy (1975); Formula and Tables for Statistical Work.

Table A.14A The Wilcoxon-Mann-Whitney test (1% and 5% critical values of U_{12} or U_{21} for one-sided test)

(1% critical values of U_{12} or U_{21} for one-sided test)

n_1	n_2																				n_1
	1	2	3	4	5	6	7	8	9	10	11	12	13	14	15	16	17	18	19	20	
1	-	-	-	-	-	-	-	-	-	-	-	-	-	-	-	-	-	-	-	-	1
2	-	-	-	-	-	-	-	-	-	-	-	-	0	0	0	0	0	0	1	1	2
3	-	-	-	-	-	-	0	0	1	1	1	2	2	2	3	3	4	4	4	5	3
4	-	-	-	-	0	1	1	2	3	3	4	5	5	6	7	7	8	9	9	10	4
5	-	-	-	0	1	2	3	4	5	6	7	8	9	10	11	12	13	14	15	16	5
6	-	-	-	1	2	3	4	6	7	8	9	11	12	13	15	16	18	19	20	22	6
7	-	-	0	1	3	4	6	7	9	11	12	14	16	17	19	21	23	24	26	28	7
8	-	-	0	2	4	6	7	9	11	13	15	17	20	22	24	26	28	30	32	34	8
9	-	-	1	3	5	7	9	11	14	16	18	21	23	26	28	31	33	36	38	40	9
10	-	-	1	3	6	8	11	13	16	19	22	24	27	30	33	36	38	41	44	47	10
11	-	-	1	4	7	9	12	15	18	22	25	28	31	34	37	41	44	47	50	53	11
12	-	-	2	5	8	11	14	17	21	24	28	31	35	38	42	46	49	53	56	60	12
13	-	0	2	5	9	12	16	20	23	27	31	35	39	43	47	51	55	59	63	67	13
14	-	0	2	6	10	13	17	22	26	30	34	38	43	47	51	56	60	65	69	73	14
15	-	0	3	7	11	15	19	24	28	33	37	42	47	51	56	61	66	70	75	80	15

Contd...

Table A.14A Contd.

(1% critical values of U_{12} or U_{21} for one-sided test)

n_1	n_2: 1	2	3	4	5	6	7	8	9	10	11	12	13	14	15	16	17	18	19	20	n_1
16	-	0	3	7	12	16	21	26	31	36	41	46	51	56	61	66					16
17	-	0	4	8	13	18	23	28	33	38	44	49	55	60	66	71					17
18	-	0	4	9	14	19	24	30	36	41	47	53	59	65	70	76					18
19	-	1	4	9	15	20	26	32	38	44	50	56	63	69	75	82					19
20	-	1	5	10	16	22	28	34	40	47	53	60	67	73	80	87					20
n_2	1	2	3	4	5	6	7	8	9	10	11	12	13	14	15	16					

(5% critical values of U_{12} or U_{21} for one-sided test)

n_1	n_2: 1	2	3	4	5	6	7	8	9	10	11	12	13	14	15	16	17	18	19	20	n_1
1	-	-	-	-	-	-	-	-	-	-	-	-	-	-	-	-	-	-	0	0	1
2	-	-	-	-	0	0	0	1	1	1	1	2	2	2	3	3	3	4	4	4	2
3	-	-	0	0	1	2	2	3	3	4	5	5	6	7	7	8	9	9	10	11	3
4	-	-	0	1	2	3	4	5	6	7	8	9	10	11	12	14	15	16	17	18	4
5	-	0	1	2	4	5	6	8	9	11	12	13	15	16	18	19	20	22	23	25	5
6	-	0	2	3	5	7	8	10	12	14	16	17	19	21	23	25	26	28	30	32	6
7	-	0	2	4	6	8	11	13	15	17	19	21	24	26	28	30	33	35	37	39	7
8	-	1	3	5	8	10	13	15	18	20	23	26	28	31	33	36	39	41	44	47	8
9	-	1	3	6	9	12	15	18	21	24	27	30	33	36	39	42	45	48	51	54	9
10	-	1	4	7	11	14	17	20	24	27	31	34	37	41	44	48	51	55	58	62	10
11	-	1	5	8	12	16	19	23	27	31	34	38	42	46	50	54	57	61	65	69	11
12	-	2	5	9	13	17	21	26	30	34	38	42	47	51	55	60	64	68	72	77	12
13	-	2	6	10	15	19	24	28	33	37	42	47	51	56	61	65	70	75	80	84	13
14	-	2	7	11	16	21	26	31	36	41	46	51	56	61	66	71	77	82	87	92	14
15	-	3	7	12	18	23	28	33	39	44	50	55	61	66	72	77	83	88	94	100	15
16	-	3	8	14	19	25	30	36	42	48	54	60	65	71	77	83	89	95	101	107	16
17	-	3	9	15	20	26	33	39	45	51	57	64	70	77	83	89	96	102	109	115	17
18	-	4	9	16	22	28	35	41	48	55	61	68	75	82	88	95	102	109	116	123	18
19	0	4	10	17	23	30	37	44	51	58	65	72	80	87	94	101	109	116	123	130	19
20	0	4	11	18	25	32	39	47	54	62	69	77	84	92	100	107	115	123	130	138	20
n_2	1	2	3	4	5	6	7	8	9	10	11	12	13	14	15	16	17	18	19	20	

Source: Rao, Mitra, Mathai and Ramamurthy (1975); Formula and Tables for Statistical Work.

Table A.14B The Wilcoxon-Mann-Whitney test (1% and 5% critical values of U for two-sided test)

(1% critical values of U for two-sided test)

n_1	n_2																				n_1
	1	2	3	4	5	6	7	8	9	10	11	12	13	14	15	16	17	18	19	20	
1	-	-	-	-	-	-	-	-	-	-	-	-	-	-	-	-	-	-	-	-	1
2	-	-	-	-	-	-	-	-	-	-	-	-	-	-	-	-	3	-	0	0	2
3	-	-	-	-	-	-	-	-	0	0	0	1	1	1	2	2	2	2	3	3	3
4	-	-	-	-	-	0	0	1	1	2	2	3	3	4	5	5	6	6	7	8	4
5	-	-	-	-	0	1	1	2	3	4	5	6	7	7	8	9	10	11	12	13	5
6	-	-	-	0	1	2	3	4	5	6	7	9	10	11	12	13	15	16	17	18	6
7	-	-	-	0	1	3	4	6	7	9	10	12	13	15	16	18	19	21	22	24	7
8	-	-	-	1	2	4	6	7	9	11	13	15	17	18	20	22	24	26	28	30	8
9	-	-	0	1	3	5	7	9	11	13	16	18	20	22	24	27	29	31	33	36	9
10	-	-	0	2	4	6	9	11	13	16	18	21	24	26	29	31	34	37	39	42	10
11	-	-	0	2	5	7	10	13	16	18	21	24	27	30	33	36	39	42	45	48	11
12	-	-	1	3	6	9	12	15	18	21	24	27	31	34	37	41	44	47	51	54	12
13	-	-	1	3	7	10	13	17	20	24	27	31	34	38	42	45	49	53	57	60	13
14	-	-	1	4	7	11	15	18	22	26	30	34	38	42	46	50	54	58	63	67	14
15	-	-	2	5	8	12	16	20	24	29	33	37	42	46	51	55	60	64	69	73	15
16	-	-	2	5	9	13	18	22	27	31	36	41	45	50	55	60	65	70	74	79	16
17	-	-	2	6	10	15	19	24	29	34	39	44	49	54	60	65	70	75	81	86	17
18	-	-	2	6	11	16	21	26	31	37	42	47	53	58	64	70	75	81	87	92	18
19	-	0	3	7	12	17	22	28	33	39	45	51	57	63	69	74	81	87	93	99	19
20	-	0	3	8	13	18	24	30	36	42	48	54	60	67	73	79	86	92	99	105	20

(5% critical values of U for two-sided test)

n_1	n_2																				n_1
	1	2	3	4	5	6	7	8	9	10	11	12	13	14	15	16	17	18	19	20	
1	-	-	-	-	-	-	-	-	-	-	-	-	-	-	-	-	-	-	-	-	1
2	-	-	-	-	0	0	0	0	0	0	0	1	1	1	1	1	2	2	2	2	2
3	-	-	-	-	0	1	1	2	2	3	3	4	4	5	5	6	6	7	7	8	3
4	-	-	-	0	1	2	3	4	4	5	6	7	8	9	10	11	11	12	13	13	4
5	-	-	0	1	2	3	5	6	7	8	9	11	12	13	14	15	17	18	19	20	5
6	-	-	1	2	3	5	6	8	10	11	13	14	16	17	19	21	22	24	25	27	6
7	-	-	1	3	5	6	8	10	12	14	16	18	20	22	24	26	28	30	32	34	7
8	-	0	2	4	6	8	10	13	15	17	19	22	24	26	20	31	34	36	38	41	8
9	-	0	2	4	7	10	12	15	17	20	23	26	28	31	34	37	39	42	45	48	9
10	-	0	3	5	8	11	14	17	20	23	26	29	33	36	39	42	45	48	52	55	10
11	-	0	3	6	9	13	16	19	23	26	30	33	37	40	44	47	51	55	58	62	11
12	-	1	4	7	11	14	18	22	26	29	33	37	41	45	49	53	57	61	65	69	12
13	-	1	4	8	12	16	20	24	28	33	37	41	45	50	54	59	63	67	72	76	13
14	-	1	5	9	13	17	22	26	31	33	40	45	50	55	59	64	67	74	78	83	14
15	-	1	5	10	14	19	24	29	34	39	44	49	54	59	64	70	75	80	85	90	15
16	-	1	6	11	15	21	26	31	37	42	47	53	59	64	70	75	81	86	92	98	16
17	-	2	6	11	17	22	28	34	39	45	51	57	63	67	75	81	87	93	99	105	17
18	-	2	7	12	18	24	30	36	42	48	55	61	67	74	80	86	93	99	106	112	18
19	-	2	7	13	19	25	32	38	45	52	58	65	72	78	85	92	99	106	113	119	19
20	-	2	8	13	20	27	34	41	48	55	62	69	76	83	90	98	105	112	119	127	20

Source: Rao, Mitra, Mathai and Ramamurthy (1975); Formula and Tables for Statistical Work.

Table A.15A Critical values for the Kolmogorov-Smirnov two-sample test $n_1 = n_2$

One-sided test $1 - \alpha =$	0.90	0.95	0.975	0.99	0.995
Two-sided test $1 - \alpha =$	0.80	0.90	.95	0.98	0.99
n					
3	2/3	2/3			
4	3/4	3/4	3/4		
5	3/5	3/5	4/5	4/5	4/5
6	3/6	4/6	4/6	5/6	5/6
7	4/7	4/7	5/7	5/7	5/7
8	4/8	4/8	5/8	5/8	6/8
9	4/9	5/9	5/9	6/9	6/9
10	4/10	5/10	6/10	6/10	7/10
11	5/11	5/11	6/11	7/11	7/11
12	5/12	5/12	6/12	7/12	7/12
13	5/13	6/13	6/13	7/13	8/13
14	5/14	6/14	7/14	7/14	8/14
15	5/15	6/15	7/15	8/15	8/15
16	6/16	6/16	7/16	8/16	9/16
17	6/17	7/17	7/17	8/17	9/17
18	6/18	7/18	8/18	9/18	9/18
19	6/19	7/19	8/19	9/19	9/19
20	6/20	7/20	8/20	9/20	10/20
21	6/21	7/21	8/21	9/21	10/21
22	7/22	8/22	8/22	10/22	10/22
23	7/23	8/23	9/23	10/23	10/23
24	7/24	8/24	9/24	10/24	11/24
25	7/25	8/25	9/25	10/25	11/25
26	7/26	8/26	9/26	10/26	11/26
27	7/27	8/27	9/27	11/27	11/27
28	8/28	9/28	10/28	11/28	12/28
29	8/29	9/29	10/29	11/29	12/29
30	8/30	9/30	10/30	11/30	12/30
31	8/31	9/31	10/31	11/31	12/31
32	8/32	9/32	10/32	12/32	12/32
33	8/33	9/33	11/33	12/33	13/33
34	8/34	10/34	11/34	12/34	13/34
35	8/35	10/35	11/35	12/35	13/35
36	9/36	10/36	11/36	12/36	13/36
37	9/37	10/37	11/37	13/37	13/37
38	9/38	10/38	11/38	13/38	14/38
39	9/39	10/39	11/39	13/39	14/39
40	9/40	10/40	12/40	13/40	14/40
Approximation for $n > 40$:	$\frac{1.5174}{\sqrt{n}}$	$\frac{1.7308}{\sqrt{n}}$	$\frac{1.9206}{\sqrt{n}}$	$\frac{1.1460}{\sqrt{n}}$	$\frac{1.3018}{\sqrt{n}}$

Source: Steel and Torrie (1981); Principles and Procedures of Statistics: A Biometrical Approach, who extracted it from "Small-sample distribution for multi-sample statistics of the Smirnov type," *Ann. Math. Statist.*, **31**: 710-720 (1960), with permission of the authors, Z.W. Birnbaum and R.A. Hall, and the editor.

Table A.15B Critical values for the Kolmogorov-Smirnov two-sample test $n_1 \neq n_2$

One-side test $1 - \alpha$ =		0.90	0.95	0.975	0.99	0.995
Two-side test $1 - \alpha$ =		0.80	0.90	0.95	0.98	0.99
$n_1 = 3$	$n_2 = 4$	3/4	3/4			
	5	2/3	4/5	4/5		
	6	2/3	2/3	5/6		
	7	2/3	5/7	6/7	6/7	
	8	5/8	3/4	3/4	7/8	
	9	2/3	2/3	7/9	8/9	8/9
	10	3/5	7/10	4/5	9/10	9/10
	12	7/12	2/3	3/4	5/6	11/12
$n_1 = 4$	$n_2 = 5$	3/5	3/4	4/5	4/5	
	6	7/12	2/3	3/4	5/6	5/6
	7	17/28	5/7	3/4	6/7	6/7
	8	5/8	5/8	3/4	7/8	7/8
	9	5/9	2/3	3/4	7/9	8/9
	10	11/20	13/20	7/10	4/5	4/5
	12	7/12	2/3	2/3	3/4	5/6
	16	9/16	5/8	11/16	3/4	13/16
$n_1 = 5$	$n_2 = 6$	3/5	2/3	2/3	5/6	5/6
	7	4/7	23/35	5/7	29/35	6/7
	8	11/20	5/8	27/40	4/5	4/5
	9	5/9	3/5	31/45	7/9	4/5
	10	1/2	3/5	7/10	7/10	4/5
	15	8/15	3/5	2/3	11/15	11/15
	20	1/2	11/20	3/5	7/10	3/4
$n_1 = 6$	$n_2 = 7$	23/24	4/7	29/42	5/7	5/6
	8	1/2	7/12	2/3	3/4	3/4
	9	1/2	5/9	2/3	13/18	7/9
	10	1/2	17/30	19/30	7/10	11/15
	12	1/2	7/12	7/12	2/3	3/4
	18	4/9	5/9	11/18	2/3	13/18
	24	11/24	1/2	7/12	5/8	2/3
$n_1 = 7$	$n_2 = 8$	1/2	33/56	5/8	41/56	3/4
	9	31/63	5/9	40/63	5/7	47/63
	10	33/70	39/70	43/70	7/10	53/70
	14	3/7	1/2	4/7	9/14	5/7
	28	3/7	13/28	15/28	17/28	9/14
$n_1 = 8$	$n_2 = 9$	4/9	13/24	5/8	2/3	3/4
	10	19/40	21/40	23/40	27/40	7/10
	12	11/24	1/2	7/12	5/8	2/3
	16	7/16	1/2	9/16	5/8	5/8
	32	13/32	7/16	1/2	9/16	19/32
$n_1 = 9$	$n_2 = 10$	7/15	1/2	26/45	2/3	31/45
	12	4/9	1/2	5/9	11/18	2/3
	15	19/45	22/45	8/15	3/5	29/45
	18	7/18	4/9	1/2	5/9	11/18
	36	13/36	5/12	17/36	19/36	5/9

Contd...

Table A.15B Contd.

One-side test $1 - \alpha =$	0.90	0.95	0.975	0.99	0.995
Two-side test $1 - \alpha =$	0.80	0.90	0.95	0.98	0.99
$n_1 = 10$ $n_2 = 15$	2/5	7/15	1/2	17/30	19/30
20	2/5	9/20	1/2	11/20	3/5
40	7/20	2/5	9/20	1/2	
$n_1 = 12$ $n_2 = 15$	23/60	9/20	1/2	11/20	7/12
16	3/8	7/16	23/48	13/24	7/12
18	13/36	5/12	17/36	13/36	5/9
20	11/30	5/12	7/15	31/60	17/30
$n_1 = 15$ $n_2 = 20$	7/20	2/5	13/30	29/60	31/60
$n_1 = 16$ $n_2 = 20$	27/80	31/80	17/40	19/40	41/80
Approximation for large sample:					
$\sqrt{\frac{n_1 + n_2}{n_1 n_2}} \times$	1.0730	1.2239	1.3581	1.5174	1.6276

Source: Steel and Torrie (1981); Principles and Procedures of Statistics: A Biometrical Approach, who extracted it from "Distribution table for the deviation between two sample cumulative", *Ann. Math. Statist.,* **23**: 435-441 (1952), with permission of the author, F.J. Massey, Jr., and the editor.

Table A.16 Transformation of percentage to probit

%	0.0	0.1	0.2	0.3	0.4	0.5	0.6	0.7	0.8	0.9	1	2	3	4	5
0	–	1.9098	2.1218	2.2522	2.3479	2.4242	2.4879	2.5427	2.5911	2.6344	For more detail see values for 95-100				
1	2.6737	2.7096	2.7429	2.7738	2.8027	2.8299	2.8556	2.8799	2.9031	2.9251					
2	2.9463	2.9665	2.9859	3.0046	3.0226	3.0400	3.0569	3.0732	3.0890	3.1043					
3	3.1192	3.1337	3.1478	3.1616	3.1750	3.1881	3.2009	3.2134	3.2256	3.2376					
4	3.2493	3.2608	3.2721	3.2831	3.2940	3.3046	3.3151	3.3253	3.3354	3.3454					
5	3.3551	3.3648	3.3742	3.3836	3.3928	3.4018	3.4107	3.4195	3.4282	3.4368	9	18	27	36	45
6	3.4452	3.4536	3.4618	3.4699	3.4780	3.4859	3.4937	3.5015	3.5091	3.5167	8	16	24	32	40
7	3.5242	3.5316	3.5389	3.5462	3.5534	3.5605	3.5675	3.5745	3.5813	3.5882	7	14	21	28	36
8	3.5949	3.6016	3.6083	3.6148	3.6213	3.6278	3.6342	3.6405	3.6468	3.6531	6	13	19	26	32
9	3.6592	3.6654	3.6715	3.6775	3.6835	3.6894	3.6953	3.7012	3.7070	3.7127	6	12	18	24	30
10	3.7184	3.7241	3.7298	3.7354	3.7409	3.7464	3.7519	3.7574	3.7628	3.7681	6	11	17	22	28
11	3.7735	3.7788	3.7840	3.7893	3.7945	3.7996	3.8048	3.8099	3.8150	3.8200	5	10	16	21	26
12	3.8250	3.8300	3.8350	3.8399	3.8448	3.8497	3.8545	3.8593	3.8641	3.8689	5	10	15	20	24
13	3.8736	3.8783	3.8830	3.8877	3.8923	3.8969	3.9015	3.9061	3.9107	3.9152	5	9	14	18	23
14	3.9197	3.9242	3.9286	3.9331	3.9375	3.9419	3.9463	3.9506	3.9550	3.9593	4	9	13	18	22
15	3.9636	3.9678	3.9721	3.9763	3.9806	3.9848	3.9890	3.9931	3.9973	4.0014	4	8	13	17	21
16	4.0055	4.0096	4.0137	4.0178	4.0218	4.0259	4.0299	4.0339	4.0379	4.0419	4	8	12	16	20
17	4.0458	4.0498	4.0537	4.0576	4.0615	4.0654	4.0693	4.0731	4.0770	4.0808	4	8	12	16	19
18	4.0846	4.0884	4.0922	4.0960	4.0998	4.1035	4.1073	4.1110	4.1147	4.1184	4	8	11	15	19
19	4.1221	4.1258	4.1295	4.1331	4.1367	4.1404	4.1440	4.1476	4.1512	4.1548	4	7	11	15	18
20	4.1584	4.1619	4.1655	4.1690	4.1726	4.1761	4.1796	4.1831	4.1866	4.1901	4	7	11	14	18
21	4.1936	4.1970	4.2005	4.2039	4.2074	4.2108	4.2142	4.2176	4.2210	4.2244	3	7	10	14	17
22	4.2278	4.2312	4.2345	4.2379	4.2412	4.2446	4.2479	4.2512	4.2546	4.2579	3	7	10	13	17
23	4.2612	4.2644	4.2677	4.2710	4.2743	4.2775	4.2808	4.2840	4.2872	4.2905	3	7	10	13	16
24	4.2937	4.2969	4.3001	4.3033	4.3065	4.3097	4.3129	4.3160	4.3192	4.3224	3	6	10	13	16
25	4.3255	4.3287	4.3318	4.3349	4.3380	4.3412	4.3443	4.3474	4.3505	4.3536	3	6	9	12	16
26	4.3567	4.3597	4.3628	4.3659	4.3689	4.3720	4.3750	4.3781	4.3811	4.3842	3	6	9	12	15
27	4.3872	4.3902	4.3932	4.3962	4.3992	4.4022	4.4052	4.4082	4.4112	4.4142	3	6	9	12	15
28	4.4172	4.4201	4.4231	4.4260	4.4290	4.4319	4.4349	4.4378	4.4408	4.4437	3	6	9	12	15
29	4.4466	4.4495	4.4524	4.4554	4.4583	4.4612	4.4641	4.4670	4.4698	4.4727	3	6	9	12	14
30	4.4756	4.4785	4.4813	4.4842	4.4871	4.4899	4.4928	4.4956	4.4985	4.5013	3	6	9	11	14
31	4.5041	4.5070	4.5098	4.5126	4.5155	4.5183	4.5211	4.5239	4.5267	4.5295	3	6	8	11	14
32	4.5323	4.5351	4.5379	4.5407	4.5435	4.5462	4.5490	4.5518	4.5546	4.5573	3	6	8	11	14
33	4.5601	4.5628	4.5656	4.5684	4.5711	4.5739	4.5766	4.5793	4.5821	4.5848	3	5	8	11	14
34	4.5875	4.5903	4.5930	4.5957	4.5984	4.6011	4.6039	4.6066	4.6093	4.6120	3	5	8	11	14
35	4.6147	4.6174	4.6201	4.6228	4.6255	4.6281	4.6308	4.6335	4.6362	4.6389	3	5	8	11	13
36	4.6415	4.6442	4.6469	4.6495	4.6522	4.6549	4.6575	4.6602	4.6628	4.6655	3	5	8	11	13
37	4.6681	4.6708	4.6734	4.6761	4.6787	4.6814	4.6840	4.6866	4.6893	4.6919	3	5	8	11	13
38	4.6945	4.6971	4.6998	4.7024	4.7050	4.7076	4.7102	4.7129	4.7155	4.7181	3	5	8	10	13
39	4.7207	4.7233	4.7259	4.7285	4.7311	4.7337	4.7363	4.7389	4.7415	4.7441	3	5	8	10	13
40	4.7467	4.7492	4.7518	4.7544	4.7570	4.7596	4.7622	4.7647	4.7673	4.7699	3	5	8	10	13
41	4.7725	4.7750	4.7776	4.7802	4.7827	4.7853	4.7879	4.7904	4.7930	4.7955	3	5	8	10	13
42	4.7981	4.8007	4.8032	4.8058	4.8083	4.8109	4.8134	4.8160	4.8185	4.8211	3	5	8	10	13
43	4.8236	4.8262	4.8287	4.8313	4.8338	4.8363	4.8389	4.8414	4.8440	4.8465	3	5	8	10	13
44	4.8490	4.8516	4.8541	4.8566	4.8592	4.8617	4.8642	4.8668	4.8693	4.8718	3	5	8	10	13

Contd...

Table A.16 Contd.

%	0.0	0.1	0.2	0.3	0.4	0.5	0.6	0.7	0.8	0.9	1	2	3	4	5
45	4.8743	4.8769	4.8794	4.8819	4.8844	4.8870	4.8895	4.8920	4.8945	4.8970	3	5	8	10	13
46	4.8996	4.9021	4.9046	4.9071	4.9096	4.9122	4.9147	4.9172	4.9197	4.9222	3	5	8	10	13
47	4.9247	4.9272	4.9298	4.9323	4.9348	4.9373	4.9398	4.9423	4.9448	4.9473	3	5	8	10	13
48	4.9498	4.9524	4.9549	4.9574	4.9599	4.9624	4.9649	4.9674	4.9699	4.9724	3	5	8	10	13
49	4.9749	4.9774	4.9799	4.9825	4.9850	4.9875	4.9900	4.9925	4.9950	4.9975	3	5	8	10	13
50	5.0000	5.0025	5.0050	5.0075	5.0100	5.0125	5.0150	5.0175	5.0201	5.0226	3	5	8	10	13
51	5.0251	5.0276	5.0301	5.0326	5.0351	5.0376	5.0401	5.0426	5.0451	5.0476	3	5	8	10	13
52	5.0502	5.0527	5.0552	5.0577	5.0602	5.0627	5.0652	5.0677	5.0702	5.0728	3	5	8	10	13
53	5.0753	5.0778	5.0803	5.0828	5.0853	5.0878	5.0904	5.0929	5.0954	5.0979	3	5	8	10	13
54	5.1004	5.1030	5.1055	5.1080	5.1105	5.1130	5.1156	5.1181	5.1206	5.1231	3	5	8	10	13
55	5.1257	5.1282	5.1307	5.1332	5.1358	5.1383	5.1408	5.1434	5.1459	5.1484	3	5	8	10	13
56	5.1510	5.1535	5.1560	5.1586	5.1611	5.1637	5.1662	5.1687	5.1713	5.1738	3	5	8	10	13
57	5.1764	5.1789	5.1815	5.1840	5.1866	5.1891	5.1917	5.1942	5.1968	5.1993	3	5	8	10	13
58	5.2019	5.2045	5.2070	5.2096	5.2121	5.2147	5.2173	5.2198	5.2224	5.2250	3	5	8	10	13
59	5.2275	5.2301	5.2327	5.2353	5.2378	5.2404	5.2430	5.2456	5.2482	5.2808	3	5	8	10	13
60	5.2533	5.2559	5.2585	5.2611	5.2637	5.2663	5.2689	5.2715	5.2741	5.2767	3	5	8	10	13
61	5.2793	5.2819	5.2845	5.2871	5.2898	5.2924	5.2950	5.2976	5.3002	5.3029	3	5	8	10	13
62	5.3055	5.3081	5.3107	5.3134	5.3160	5.3186	5.3213	5.3239	5.3266	5.3292	3	5	8	11	13
63	5.3319	5.3345	5.3372	5.3398	5.3425	5.3451	5.3478	5.3505	5.3531	5.3558	3	5	8	11	13
64	5.3585	5.3611	5.3638	5.3665	5.3692	5.3719	5.3745	5.3772	5.3799	5.3826	3	5	8	11	13
65	5.3853	5.3880	5.3907	5.3934	5.3961	5.3989	5.4016	4.4043	5.4070	5.4097	3	5	8	11	14
66	5.4125	5.4152	5.4179	5.4207	5.4234	5.4261	5.4289	5.4316	5.4344	5.4372	3	5	8	11	14
67	5.4399	5.4427	5.4454	5.4482	5.4510	5.4538	5.4565	5.4593	5.4621	5.4649	3	6	8	11	14
68	5.4677	5.4705	5.4733	5.4761	5.4789	5.4817	5.4845	5.4874	5.4902	5.4930	3	6	8	11	14
69	5.4959	5.4987	5.5015	5.5044	5.8072	5.5101	5.5129	5.5158	5.5187	5.5215	3	6	9	11	14
70	5.5244	5.5273	5.5302	5.5330	5.5359	5.5388	5.5417	5.5446	5.5476	5.5505	3	6	9	12	14
71	5.5534	5.5563	5.5592	5.5622	5.5651	5.5681	5.5710	5.5740	5.5769	5.5799	3	6	9	12	15
72	5.5828	5.5858	5.5888	5.5918	5.5948	5.5978	5.6008	5.6038	5.6068	5.6098	3	6	9	12	15
73	5.6128	5.6158	5.6189	5.6219	5.6250	5.6280	5.6311	5.6341	5.6372	5.6403	3	6	9	12	15
74	5.6433	5.6464	5.6495	5.6526	5.6557	5.6588	5.6620	5.6651	5.6682	5.6713	3	6	9	12	16
75	5.6745	5.6776	5.6808	5.6840	5.6871	5.6903	5.6935	5.6967	5.6999	5.7031	3	6	10	13	16
76	5.7063	5.7095	5.7128	5.7160	5.7192	5.7225	5.7257	5.7290	5.7323	5.7356	3	7	10	13	16
77	5.7388	5.7421	5.7454	5.7488	5.7521	5.7554	5.7588	5.7621	5.7655	5.7688	3	7	10	13	17
78	5.7722	5.7756	5.7790	5.7824	5.7858	5.7892	5.7926	5.7961	5.7995	5.8030	3	7	10	14	17
79	5.8064	5.8099	5.8134	5.8169	5.8204	5.8239	5.8274	5.8310	5.8345	5.8381	4	7	11	14	18
80	5.8416	5.8452	5.8488	5.8524	5.8560	5.8596	5.8633	5.8669	5.8705	5.8742	4	7	11	14	18
81	5.8779	5.8816	5.8853	5.8890	5.8927	5.8965	5.9002	5.9040	5.9078	5.9116	4	7	11	15	19
82	5.9154	5.9192	5.9230	5.9269	5.9307	5.9346	5.9385	5.9424	5.9463	5.9502	4	8	12	15	19
83	5.9542	5.9581	5.9621	5.9661	5.9701	5.9741	5.9782	5.9822	5.9863	5.9904	4	8	12	16	20
84	5.9945	5.9986	6.0027	6.0069	6.0110	6.0152	6.0194	6.0237	6.0279	6.0322	4	8	13	17	21
85	6.0364	6.0407	6.0450	6.0494	6.0537	6.0581	6.0625	6.0669	6.0714	6.0758	4	9	13	18	22
86	6.0803	6.0848	6.0893	6.0939	6.0985	6.1031	6.1077	6.1123	6.1170	6.1217	5	9	14	18	23
87	6.1264	6.1311	6.1359	6.1407	6.1455	6.1503	6.1552	6.1601	6.1650	6.1700	5	10	15	19	24
88	6.1750	6.1800	6.1850	6.1901	6.1952	6.2004	6.2055	6.2107	6.2160	6.2212	5	10	15	21	26
89	6.2265	6.2319	6.2372	6.2426	6.2481	6.2536	6.2591	6.2646	6.2702	6.2759	5	11	16	22	27
90	6.2816	6.2873	6.2930	6.2988	6.3047	6.3106	6.3165	6.3225	6.3285	6.3346	6	12	18	24	29

Contd...

Table A.16 Contd.

%	0.0	0.1	0.2	0.3	0.4	0.5	0.6	0.7	0.8	0.9	1	2	3	4	5
91	6.3408	6.3469	6.3532	6.3595	6.3658	6.3722	6.3787	6.3852	6.3917	6.3984	6	13	19	26	32
92	6.4051	6.4118	6.4187	6.4255	6.4325	6.4395	6.4466	6.4538	6.4611	6.4684	7	14	21	28	35
93	6.4758	6.4833	6.4909	6.4985	6.5063	6.5141	6.5220	6.5301	6.5382	6.5464	8	16	24	31	39
94	6.5548	6.5632	6.5718	6.5805	6.5893	6.5982	6.6072	6.6164	6.6258	6.6352	9	18	27	36	45
95	6.6449	6.6546	6.6646	6.6747	6.6849	6.6954	6.7060	6.7169	6.7279	6.7392					
	97	100	101	102	105	106	109	110	113	115					
96	6.7507	6.7624	6.7744	6.7866	6.7991	6.8119	6.8250	6.8884	6.8522	6.8663					
	117	120	122	125	128	131	134	138	141	145					
97	6.8808	6.8957	6.9110	6.9268	6.9431	6.9600	6.9774	6.9954	7.0141	7.0335					
	149	153	158	163	169	174	180	187	194	202					
98.0	7.0537	7.0558	7.0579	7.0600	7.0621	7.0642	7.0663	7.0684	7.0706	7.0727	2	4	6	8	11
98.1	7.0749	7.0770	7.0792	7.0814	7.0836	7.0858	7.0880	7.0902	7.0924	7.0947	2	4	7	9	11
98.2	7.0969	7.0992	7.1015	7.1038	7.1061	7.1084	7.1107	7.1130	7.1154	7.1177	2	5	7	9	12
98.3	7.1201	7.1224	7.1248	7.1272	7.1297	7.1321	7.1345	7.1370	7.1394	7.1419	2	5	7	10	12
98.4	7.1444	7.1469	7.1494	7.1520	7.1545	7.1571	7.1596	7.1622	7.1648	7.1675	3	5	8	10	13
98.5	7.1701	7.1727	7.1754	7.1781	7.1808	7.1835	7.1862	7.1890	7.1917	7.1945	3	5	8	11	14
98.6	7.1973	7.2001	7.2029	7.2058	7.2086	7.2115	7.2144	7.2173	7.2203	7.2232	3	6	9	12	14
98.7	7.2262	7.2292	7.2322	7.2353	7.2383	7.2414	7.2445	7.2476	7.2508	7.2539	3	6	9	12	15
98.8	7.2571	7.2603	7.2636	7.2668	7.2701	7.2734	7.2768	7.2801	7.2835	7.2869	3	7	10	13	17
98.9	7.2904	7.2938	7.2973	7.3009	7.3044	7.3080	7.3116	7.3152	7.3189	7.3226	4	7	11	14	18
99.0	7.3263	7.3301	7.3339	7.3378	7.3416	7.3455	7.3495	7.3535	7.3575	7.3615	4	8	12	16	20
99.1	7.3656	7.3698	7.3739	7.3781	7.3824	7.3867	7.3911	7.3954	7.3999	7.4044	4	9	13	17	22
99.2	7.4089	7.4135	7.4181	7.4228	7.4276	7.4324	7.4372	7.4422	7.4471	7.4522	5	10	14	19	24
99.3	7.4573	7.4624	7.4677	7.4730	7.4783	7.4838	7.4893	7.4949	7.5006	7.5063	5	11	16	22	27
99.4	7.5121	7.5181	7.5241	7.5302	7.5364	7.5427	7.5491	7.5556	7.5622	7.5690	6	13	19	25	32
99.5	7.5758	7.5828	7.5899	7.5972	7.6045	7.6121	7.6197	7.6276	7.6356	7.6437					
99.6	7.6521	7.6606	7.6693	7.6783	7.6874	7.6968	7.7065	7.7164	7.7266	7.7370					
99.7	7.7478	7.7589	7.7703	7.7822	7.7944	7.8070	7.8202	7.8338	7.8480	7.8627					
99.8	7.8782	7.8943	7.9112	7.9290	7.9478	7.9677	7.9889	8.0115	8.0357	8.0618					
99.9	8.0902	8.1214	8.1559	8.1947	8.2389	8.2905	8.3528	8.4316	8.5401	8.7190					

Source: Finney (1981); Probit Analysis is indebted to Professor R.A. Fisher and Dr. F. Yates, and also to Messrs Oliver and Boyd, Ltd. of Edinburgh, for permission to reprint Table I from Table IX of their book *Statistical Tables for Biological, Agricultural and Medical Research.*

Table A.17 Working probits

Response (%)	Expected probit, *Y* Y = 2.0-2.9; 0-50% response									
	2.0	2.1	2.2	2.3	2.4	2.5	2.6	2.7	2.8	2.9
0	1.695	1.787	1.877	1.967	2.057	2.146	2.234	2.321	2.408	2.494
1	3.951	3.467	3.141	2.927	2.793	2.716	2.681	2.674	2.690	2.721
2	6.207	5.147	4.404	3.886	3.529	3.287	3.127	3.027	.972	.949
3	8.463	6.827	5.667	4.846	4.265	.857	.574	.380	3.254	3.176
4	–	8.507	6.931	5.806	5.002	4.428	4.020	.733	.536	.403
5	–	–	8.194	6.765	.938	.998	.467	4.086	.818	.631
6	–	–	9.458	7.725	6.474	5.569	4.913	4.440	4.099	3.858
7	–	–	–	8.684	7.210	6.139	5.360	.793	.381	4.085
8	–	–	–	9.644	.946	.710	.806	5.146	.663	.313
9	–	–	–	–	8.683	7.280	6.253	.499	.945	.540
10	–	–	–	–	9.419	.851	.699	.852	5.227	.767
11	–	–	–	–	–	8.421	7.146	6.205	5.509	4.995
12	–	–	–	–	–	.992	.592	.558	.791	5.222
13	–	–	–	–	–	9.562	8.039	.911	6.073	.449
14	–	–	–	–	–	–	.486	7.264	.355	.677
15	–	–	–	–	–	–	.932	.617	.636	.904
16	–	–	–	–	–	–	9.379	7.970	6.918	6.132
17	–	–	–	–	–	–	.825	8.323	7.200	.359
18	–	–	–	–	–	–	–	.676	.482	.586
19	–	–	–	–	–	–	–	9.029	.764	.814
20	–	–	–	–	–	–	–	.382	8.046	7.041
21	–	–	–	–	–	–	–	9.735	8.328	7.268
22	–	–	–	–	–	–	–	–	.610	.496
23	–	–	–	–	–	–	–	–	.892	.723
24	–	–	–	–	–	–	–	–	9.173	.950
25	–	–	–	–	–	–	–	–	.455	8.178
26	–	–	–	–	–	–	–	–	9.737	8.405
27	–	–	–	–	–	–	–	–	–	.633
28	–	–	–	–	–	–	–	–	–	.860
29	–	–	–	–	–	–	–	–	–	9.087
30	–	–	–	–	–	–	–	–	–	.315
31	–	–	–	–	–	–	–	–	–	9.542
32	–	–	–	–	–	–	–	–	–	.769
33	–	–	–	–	–	–	–	–	–	.997
34	–	–	–	–	–	–	–	–	–	–
35	–	–	–	–	–	–	–	–	–	–

Contd...

Source: Finney (1981); Probit Analysis.

Table A.17 Contd.

Response (%)	Expected probit, Y									
	Y = 3.0-3.9; 0-50% response									
	3.0	3.1	3.2	3.3	3.4	3.5	3.6	3.7	3.8	3.9
0	2.579	2.662	2.745	2.826	2.906	2.984	3.061	3.135	3.207	3.277
1	2.764	2.815	2.872	2.932	2.996	3.061	3.127	3.193	3.259	3.323
2	.949	.967	.998	3.039	3.086	.139	.194	.252	.310	.369
3	3.134	3.120	3.125	.145	.176	.216	.261	.310	.362	.415
4	.319	.272	.252	.251	.267	.293	.328	.369	.413	.461
5	.505	.424	.378	.358	.357	.370	.395	.427	.465	.507
6	3.690	3.577	3.505	3.464	3.447	3.447	3.461	3.485	3.516	3.553
7	.875	.729	.632	.570	.537	.525	.528	.544	.568	.599
8	4.060	.882	.758	.677	.627	.602	.595	.602	.619	.645
9	.246	4.034	.885	.783	.717	.679	.662	.660	.671	.690
10	.431	.186	4.012	.889	.808	.756	.728	.719	.722	.736
11	4.616	4.339	4.138	3.996	3.898	3.834	3.795	3.777	3.774	3.782
12	.801	.491	.265	4.102	.988	.911	.862	.835	.825	.828
13	.986	.644	.391	.208	4.078	.988	.929	.894	.877	.874
14	5.172	.796	.518	.315	.168	4.065	.996	.952	.928	.920
15	.357	.948	.645	.421	.258	.142	4.062	4.010	.980	.966
16	5.542	5.101	4.771	4.527	4.348	4.220	4.129	4.069	4.031	4.012
17	.727	.253	.898	.634	.439	.297	.196	.127	.083	.058
18	.913	.406	5.025	.740	.529	.374	.263	.185	.134	.104
19	6.098	.558	.151	.846	.619	.451	.330	.244	.186	.149
20	.283	.710	.278	.953	.709	.528	.396	.302	.237	.195
21	6.468	5.863	5.405	5.059	4.799	4.606	4.463	4.361	4.289	4.241
22	.653	6.015	.531	.165	.889	.683	.530	.419	.340	.287
23	.839	.168	.658	.272	.979	.760	.597	.477	.392	.333
24	7.024	.320	.785	.378	5.070	.837	.664	.536	.443	.379
25	.209	.472	.911	.484	.160	.914	.730	.594	.495	.425
26	7.394	6.625	6.038	5.591	5.250	4.992	4.797	4.652	4.546	4.471
27	.580	.777	.165	.697	.340	5.069	.864	.711	.598	.517
28	.765	.930	.291	.803	.430	.146	.931	.769	.649	.563
29	.950	7.082	.418	.910	.520	.223	.997	.827	.701	.608
30	8.135	.234	.545	6.016	.610	.300	5.064	.886	.752	.654
31	8.320	7.387	6.671	6.122	5.701	5.378	5.131	4.944	4.804	4.700
32	.506	.539	.798	.229	.791	.455	.198	5.002	.855	.746
33	.691	.692	.925	.335	.881	.532	.265	.061	.907	.792
34	.876	.844	7.051	.441	.971	.609	.331	.119	.958	.838
35	9.061	.996	.178	.548	6.061	.687	.398	.177	5.010	.884
36	9.247	8.149	7.305	6.654	6.151	5.764	5.465	5.236	5.061	4.930
37	.432	.301	.431	.760	.242	.841	.532	.294	.113	.976
38	.617	.454	.558	.867	.332	.918	.599	.353	.164	5.022
39	.802	.606	.685	.973	.422	.995	.665	.411	.216	.068
40	.987	.758	.811	7.079	.512	6.073	.732	.469	.267	.113
41	–	8.911	7.938	7.186	6.602	6.150	5.799	5.528	5.319	5.159

Contd...

Table A.17 Contd.

Response (%)	Y = 3.0-3.9; 0-50% response. Expected probit, Y									
	3.0	3.1	3.2	3.3	3.4	3.5	3.6	3.7	3.8	3.9
42	–	9.063	8.065	.292	.692	.227	.866	.586	.370	.205
43	–	.216	.191	.398	.782	.304	.932	.644	.422	.251
44	–	.368	.318	.505	.873	.381	.999	.703	.473	.297
45	–	.520	.445	.611	.963	.459	6.066	.761	.525	.343
46	–	9.673	8.571	7.717	7.053	6.536	6.133	5.819	5.576	5.389
47	–	.825	.698	.824	.143	.613	.200	.878	.628	.435
48	–	.978	.825	.930	.233	.690	.266	.936	.679	.481
49	–	–	.951	8.036	.323	.767	.333	.994	.731	.527
50	–	–	9.078	.143	.414	.845	.400	6.053	.782	.572

Response (%)	Y = 4.0-4.9; 0-50% response. Expected probit, Y									
	4.0	4.1	4.2	4.3	4.4	4.5	4.6	4.7	4.8	4.9
0	3.344	3.408	3.469	3.525	3.577	3.624	3.664	3.698	3.724	3.741
1	3.386	3.446	3.503	3.557	3.607	3.652	3.691	3.724	3.750	3.766
2	.427	.487	.538	.589	.637	.680	.719	.751	.775	.791
3	.468	.521	.572	.621	.667	.709	.746	.777	.801	.816
4	.510	.559	.607	.653	.697	.737	.773	.803	.826	.841
5	.551	.596	.641	.685	.727	.766	.800	.829	.852	.867
6	3.592	3.634	3.676	3.717	3.757	3.794	3.827	3.858	3.878	3.892
7	.634	.671	.710	.749	.787	.822	.854	.882	.903	.917
8	.675	.709	.745	.781	.817	.851	.882	.908	.929	.942
9	.716	.747	.779	.813	.847	.879	.909	.934	.954	.967
10	.758	.784	.814	.845	.877	.908	.936	.960	.980	.993
11	3.799	3.822	3.848	3.877	3.907	3.936	3.963	3.987	4.005	4.018
12	.840	.859	.883	.909	.937	.964	.990	4.013	.031	.043
13	.882	.897	.917	.941	.967	.993	4.017	.039	.057	.068
14	.923	.934	.952	.973	.997	4.021	.044	.065	.082	.093
15	.964	.972	.986	4.005	4.027	.050	.072	.092	.108	.119
16	4.006	4.010	4.021	4.038	4.057	4.078	4.099	4.118	4.133	4.144
17	.047	.047	.056	.070	.087	.106	.126	.144	.159	.169
18	.088	.085	.090	.102	.117	.135	.153	.170	.184	.194
19	.130	.122	.125	.134	.147	.163	.180	.196	.210	.219
20	.171	.160	.159	.166	.177	.192	.207	.223	.236	.245
21	4.212	4.198	4.194	4.198	4.207	4.220	4.235	4.249	4.261	4.270
22	.253	.235	.228	.230	.237	.248	.262	.275	.287	.295
23	.295	.273	.263	.262	.267	.277	.289	.301	.312	.320
24	.336	.310	.297	.294	.297	.305	.316	.327	.338	.345
25	.377	.348	.332	.326	.327	.334	.343	.354	.363	.370

Contd...

Table A.17 Contd.

Y = 4.0-4.9; 0-50% response

Response (%)	Expected probit, *Y*									
	4.0	4.1	4.2	4.3	4.4	4.5	4.6	4.7	4.8	4.9
26	4.419	4.385	4.366	4.358	4.357	4.362	4.370	4.380	4.389	4.396
27	.460	.423	.401	.390	.387	.391	.397	.406	.415	.421
28	.501	.461	.435	.422	.417	.419	.425	.432	.440	.446
29	.543	.498	.470	.454	.447	.447	.452	.459	.466	.471
30	.584	.536	.504	.486	.477	.476	.479	.485	.491	.496
31	4.625	4.573	4.539	4.518	4.507	4.504	4.506	4.511	4.517	4.522
32	.667	.611	.573	.550	.537	.533	.533	.537	.542	.547
33	.708	.649	.608	.582	.567	.561	.560	.563	.568	.572
34	.749	.686	.642	.614	.597	.589	.588	.590	.594	.597
35	.791	.724	.677	.646	.627	.618	.615	.616	.619	.622
36	4.832	4.761	4.711	4.678	4.657	4.646	4.642	4.642	4.645	4.648
37	.873	.799	.746	.710	.687	.675	.669	.668	.670	.673
38	.915	.836	.780	.742	.717	.703	.696	.695	.696	.698
39	.956	.874	.815	.774	.747	.731	.723	.721	.721	.723
40	.997	.912	.849	.806	.777	.760	.750	.747	.747	.748
41	5.039	4.949	4.884	4.838	4.807	4.788	4.778	4.773	4.773	4.774
42	.080	.987	.918	.870	.837	.817	.805	.799	.798	.799
43	.121	5.024	.953	.902	.867	.845	.832	.826	.824	.824
44	.163	.062	.988	.934	.897	.873	.859	.852	.849	.849
45	.204	.099	5.022	.966	.927	.902	.886	.878	.875	.874
46	5.245	5.137	5.057	4.998	4.957	4.930	4.913	4.904	4.900	4.900
47	.287	.175	.091	5.030	.987	.959	.941	.931	.926	.925
48	.328	.212	.126	.062	5.017	.987	.968	.957	.952	.950
49	.369	.250	.160	.094	.047	5.015	.995	.983	.977	.975
50	.411	.287	.195	.126	.078	.044	5.022	5.009	5.003	5.000

Y = 5.0-5.9; 0-50% response

Response (%)	Expected probit, *Y*									
	5.0	5.1	5.2	5.3	5.4	5.5	5.6	5.7	5.8	5.9
0	3.747	3.740	3.719	3.680	3.620	3.536	3.422	3.272	3.079	2.834
1	3.772	3.765	3.744	3.706	3.647	3.564	3.452	3.304	3.114	2.871
2	.797	.790	.770	.732	.675	.593	.482	.336	.148	.909
3	.822	.816	.795	.758	.702	.621	.512	.368	.183	.946
4	.847	.841	.821	.785	.729	.650	.542	.400	.217	.984
5	.872	.866	.846	.811	.756	.678	.572	.433	.252	3.021
6	3.897	3.891	3.872	3.837	3.783	3.706	3.602	3.465	3.287	3.059
7	.922	.916	.898	.863	.810	.735	.632	.497	.321	.097
8	.947	.942	.923	.890	.838	.763	.662	.529	.356	.134
9	.972	.967	.949	.916	.865	.792	.692	.561	.390	.172
10	.997	.992	.974	.942	.892	.820	.722	.593	.425	.209

Contd...

Table A.17 Contd.

Response (%)	Expected probit, Y									
	5.0	5.1	5.2	5.3	5.4	5.5	5.6	5.7	5.8	5.9
	Y = 5.0-5.9; 0-50% response									
11	4.022	4.017	4.000	3.968	3.919	3.848	3.752	3.625	3.459	3.247
12	.047	.042	.025	.994	.946	.877	.782	.657	.494	.284
13	.073	.068	.051	4.021	.973	.905	.812	.689	.528	.322
14	.098	.093	.077	.047	4.000	.934	.842	.721	.563	.360
15	.123	.118	.102	.073	.028	.962	.872	.753	.597	.397
16	4.148	4.143	4.128	4.099	4.055	3.990	3.902	3.785	3.632	3.435
17	.173	.168	.153	.126	.082	4.019	.932	.817	.666	.472
18	.198	.194	.179	.152	.109	.047	.962	.849	.701	.510
19	.223	.219	.204	.178	.136	.076	.992	.881	.735	.548
20	.248	.244	.230	.204	.163	.104	4.022	.913	.770	.585
21	4.273	4.269	4.256	4.230	4.191	4.132	4.052	3.945	3.804	3.623
22	.298	.294	.281	.257	.218	.161	.082	.977	.839	.660
23	.323	.320	.307	.283	.245	.189	.112	4.009	.873	.698
24	.348	.345	.332	.309	.272	.218	.142	.041	.908	.735
25	.373	.370	.358	.335	.299	.246	.172	.073	.942	.773
26	4.398	4.395	4.383	4.362	4.326	4.275	4.202	4.105	3.977	3.811
27	.423	.420	.409	.388	.353	.303	.232	.137	4.011	.848
28	.449	.445	.435	.414	.381	.331	.262	.169	.046	.886
29	.474	.471	.460	.440	.408	.360	.292	.201	.080	.923
30	.499	.496	.486	.466	.435	.388	.322	.233	.115	.961
31	4.524	4.521	4.511	4.493	4.462	4.417	4.352	4.265	4.149	3.999
32	.549	.546	.537	.519	.489	.445	.382	.297	.184	4.036
33	.574	.571	.563	.545	.516	.473	.412	.329	.219	.074
34	.599	.597	.588	.571	.544	.502	.442	.361	.253	.111
35	.624	.622	.614	.598	.571	.530	.472	.393	.288	.149
36	4.649	4.647	4.639	4.624	4.598	4.559	4.502	4.425	4.322	4.186
37	.674	.672	.665	.650	.625	.587	.532	.457	.357	.224
38	.699	.697	.690	.676	.652	.615	.562	.489	.391	.262
39	.724	.723	.716	.702	.679	.644	.592	.521	.426	.299
40	.749	.748	.742	.729	.706	.672	.622	.553	.460	.337
41	4.774	4.773	4.767	4.755	4.734	4.701	4.652	4.585	4.495	4.374
42	.799	.798	.793	.781	.761	.729	.682	.617	.529	.412
43	.825	.823	.818	.807	.788	.757	.712	.649	.564	.450
44	.850	.849	.844	.833	.815	.786	.742	.682	.598	.487
45	.875	.874	.869	.860	.842	.814	.772	.714	.633	.525
46	4.900	4.899	4.895	4.886	4.869	4.843	4.802	4.746	4.667	4.562
47	.925	.924	.921	.912	.897	.871	.832	.778	.702	.600
48	.950	.949	.946	.938	.924	.899	.862	.810	.736	.637
49	.975	.975	.972	.965	.951	.928	.892	.842	.771	.675
50	5.000	5.000	.997	.991	.978	.956	.922	.874	.805	.713

Contd...

Table A.17 Contd.

Response (%)	Expected probit, Y									
	6.0	6.1	6.2	6.3	6.4	6.5	6.6	6.7	6.8	6.9
	Y = 6.0-6.9; 0-50% response									
0	2.523	2.132	1.643	1.030	0.261	–	–	–	–	–
1	2.564	2.178	1.694	1.088	0.327	–	–	–	–	–
2	.606	.224	.746	.146	.394	–	–	–	–	–
3	.647	.270	.797	.205	.461	–	–	–	–	–
4	.688	.316	.849	.263	.528	–	–	–	–	–
5	.730	.362	.900	.321	.595	–	–	–	–	–
6	2.771	2.408	1.952	1.380	0.661	–	–	–	–	–
7	.812	.454	2.003	.438	.728	–	–	–	–	–
8	.854	.500	.055	.496	.795	–	–	–	–	–
9	.895	.546	.106	.555	.862	–	–	–	–	–
10	.936	.591	.158	.613	.928	0.067	–	–	–	–
11	2.978	2.637	2.209	1.671	0.995	0.144	–	–	–	–
12	3.019	.683	.261	.730	1.062	.221	–	–	–	–
13	.060	.729	.312	.788	.129	.299	–	–	–	–
14	.102	.775	.364	.846	.196	.376	–	–	–	–
15	.143	.821	.415	.905	.262	.453	–	–	–	–
16	3.184	2.867	2.467	1.963	1.320	0.530	–	–	–	–
17	.226	.913	.518	2.022	.396	.607	–	–	–	–
18	.267	.959	.570	.080	.463	.685	–	–	–	–
19	.308	3.005	.621	.138	.530	.762	–	–	–	–
20	.350	.050	.673	.197	.596	.839	–	–	–	–
21	3.391	3.096	2.724	2.255	1.663	0.916	–	–	–	–
22	.432	.142	.776	.313	.730	.993	0.062	–	–	–
23	.474	.188	.827	.372	.797	1.071	.152	–	–	–
24	.515	.234	.879	.430	.864	.148	.243	–	–	–
25	.556	.280	.930	.488	.930	.225	.333	–	–	–
26	3.598	3.326	2.982	2.547	1.997	1.302	0.423	–	–	–
27	.639	.372	3.033	.605	2.064	.379	.513	–	–	–
28	.680	.418	.085	.663	.131	.457	.603	–	–	–
29	.721	.464	.136	.722	.197	.534	.693	–	–	–
30	.763	.509	.188	.780	.264	.611	.784	–	–	–
31	3.804	3.555	3.239	2.838	2.331	1.688	0.874	–	–	–
32	.845	.601	.291	.897	.398	.766	.964	–	–	–
33	.887	.647	.342	.955	.465	.843	1.054	0.050	–	–
34	.928	.693	.394	3.014	.531	.920	.144	.156	–	–
35	.969	.739	.445	.072	.598	.997	.234	.262	–	–
36	4.011	3.785	3.497	3.130	2.665	2.074	1.324	0.369	–	–
37	.052	.831	.548	.189	.732	.152	.415	.475	–	–
38	.093	.877	.600	.247	.799	.229	.505	.581	–	–
39	.135	.923	.651	.305	.865	.306	.595	.688	–	–
40	.176	.969	.703	.364	.932	.383	.685	.794	–	–
41	4.217	4.014	3.754	3.422	2.999	2.460	1.775	0.900	–	–
42	.259	.060	.806	.480	3.066	.538	.865	1.007	–	–

Contd...

Table A.17 Contd.

Response (%)	Expected probit, Y									
	6.0	6.1	6.2	6.3	6.4	6.5	6.6	6.7	6.8	6.9
	Y = 6.0-6.9; 0-50% response									
43	.300	.106	.857	.539	.132	.615	.955	.133	0.035	–
44	.341	.152	.909	.597	.199	.692	2.046	.219	.162	–
45	.383	.198	.960	.655	.266	.769	.136	.326	.280	–
46	4.424	4.244	4.012	3.714	3.333	2.846	2.226	1.432	0.415	–
47	.465	.290	.063	.772	.400	.924	.316	.538	.542	–
48	.507	.336	.115	.830	.466	3.001	.406	.645	.669	–
49	.548	.382	.166	.880	.533	.078	.496	.751	.795	–
50	.589	.428	.218	.947	.600	.155	.586	.857	.922	–

Y = 3.0-3.9; 51-100% response

Response (%)	Expected probit, Y									
	3.0	3.1	3.2	3.3	3.4	3.5	3.6	3.7	3.8	3.9
51	–	–	9.205	8.249	7.504	6.922	6.467	6.111	5.834	5.618
52	–	–	.331	.355	.594	.999	.534	.170	.885	.664
53	–	–	.458	.462	.684	7.076	.600	.228	.937	.710
54	–	–	.585	.568	.774	.154	.667	.286	.988	.756
55	–	–	.711	.674	.864	.231	.734	.345	6.040	.802
56	–	–	9.838	8.781	7.954	7.308	6.801	6.403	6.091	5.848
57	–	–	.965	.887	8.045	.385	.868	.461	.143	.894
58	–	–	–	.993	.135	.462	.934	.520	.194	.940
59	–	–	–	9.100	.225	.540	7.001	.578	.246	.986
60	–	–	–	.206	.315	.617	.068	.636	.297	6.031
61	–	–	–	9.312	8.405	7.694	7.135	6.695	6.349	6.077
62	–	–	–	.419	.495	.771	.201	.753	.400	.123
63	–	–	–	.525	.585	.848	.268	.811	.452	.169
64	–	–	–	.631	.676	.926	.335	.870	.503	.215
65	–	–	–	.738	.766	8.003	.402	.928	.555	.261
66	–	–	–	9.844	8.856	8.080	7.469	6.986	6.606	6.307
67	–	–	–	.950	.946	.157	.535	7.045	.658	.353
68	–	–	–	–	9.036	.234	.602	.103	.709	.399
69	–	–	–	–	.126	.312	.669	.162	.761	.445
70	–	–	–	–	.216	.389	.736	.220	.812	.491
71	–	–	–	–	9.307	8.466	7.803	7.278	6.864	6.536
72	–	–	–	–	.397	.543	.869	.337	.915	.582
73	–	–	–	–	.487	.621	.936	.395	.967	.628
74	–	–	–	–	.577	.698	8.003	.453	7.018	.674
75	–	–	–	–	.667	.775	.070	.512	.070	.720
76	–	–	–	–	9.757	8.852	8.136	7.570	7.121	6.766
77	–	–	–	–	.848	.929	.203	.628	.173	.812

Contd...

Table A.17 Contd.

Y = 3.0-3.9; 51-100% response

Response (%)	Expected probit, *Y*									
	3.0	3.1	3.2	3.3	3.4	3.5	3.6	3.7	3.8	3.9
78	–	–	–	–	.938	9.007	.270	.687	.224	.858
79	–	–	–	–	–	.084	.337	.745	.276	.904
80	–	–	–	–	–	.161	.404	.803	.327	.950
81	–	–	–	–	–	9.238	8.470	7.862	7.379	6.995
82	–	–	–	–	–	.315	.537	.920	.430	7.041
83	–	–	–	–	–	.393	.604	.978	.482	.087
84	–	–	–	–	–	.470	.671	8.037	.533	.133
85	–	–	–	–	–	.547	.738	.095	.585	.179
86	–	–	–	–	–	9.624	8.804	8.154	7.636	7.225
87	–	–	–	–	–	.701	.871	.212	.688	.271
88	–	–	–	–	–	.779	.938	.270	.739	.317
89	–	–	–	–	–	.856	9.005	.329	.791	.363
90	–	–	–	–	–	.933	.072	.387	.842	.409
91	–	–	–	–	–	–	9.138	8.445	7.894	7.454
92	–	–	–	–	–	–	.205	.504	.945	.500
93	–	–	–	–	–	–	.272	.562	.997	.546
94	–	–	–	–	–	–	.339	.620	8.048	.592
95	–	–	–	–	–	–	.405	.679	.100	.638
96	–	–	–	–	–	–	9.472	8.737	8.151	7.684
97	–	–	–	–	–	–	.539	.795	.203	.730
98	–	–	–	–	–	–	.606	.854	.254	.776
99	–	–	–	–	–	–	.673	.912	.306	.822
100	–	–	–	–	–	–	.739	.970	.357	.868

Y = 4.0-4.9; 51-100% response

Response (%)	Expected probit, *Y*									
	4.0	4.1	4.2	4.3	4.4	4.5	4.6	4.7	4.8	4.9
51	5.452	5.325	5.229	5.158	5.108	5.072	5.049	5.035	5.028	5.025
52	.493	.363	.264	.190	.138	.101	.076	.062	.054	.051
53	.535	.400	.298	.222	.168	.129	.103	.088	.079	.076
54	.576	.438	.333	.254	.198	.157	.131	.114	.105	.101
55	.617	.475	.367	.286	.228	.186	.158	.140	.131	.126
56	5.659	5.513	5.402	5.318	5.258	5.214	5.185	5.167	5.156	5.151
57	.700	.550	.436	.351	.288	.243	.212	.193	.182	.177
58	.741	.588	.471	.383	.318	.271	.239	.219	.207	.202
59	.783	.626	.505	.415	.348	.299	.266	.245	.233	.227
60	.824	.663	.540	.447	.378	.328	.294	.271	.258	.252
61	5.865	5.701	5.574	5.479	5.408	5.356	5.321	5.298	5.284	5.277
62	.907	.738	.609	.511	.438	.385	.348	.324	.310	.303

Contd...

Table A.17 Contd.

Response (%)	Expected probit, *Y*									
	Y = 4.0-4.9; 51-100% response									
	4.0	4.1	4.2	4.3	4.4	4.5	4.6	4.7	4.8	4.9
63	.948	.776	.643	.543	.468	.413	.375	.350	.335	.328
64	.989	.814	.678	.575	.498	.441	.402	.376	.361	.353
65	6.031	.851	.712	.607	.528	.470	.429	.402	.386	.378
66	6.072	5.889	5.747	5.639	5.558	5.498	5.456	5.429	5.412	5.403
67	.113	.926	.781	.671	.588	.527	.484	.455	.437	.429
68	.155	.964	.816	.703	.618	.555	.511	.481	.463	.454
69	.196	6.001	.851	.735	.648	.583	.538	.507	.489	.479
70	.237	.039	.885	.767	.678	.612	.565	.534	.514	.504
71	6.279	6.077	5.920	5.799	5.708	5.640	5.592	5.560	5.540	5.529
72	.320	.114	.954	.831	.738	.669	.619	.586	.565	.555
73	.361	.152	.989	.863	.768	.697	.647	.612	.591	.580
74	.402	.189	6.023	.895	.798	.725	.674	.638	.617	.605
75	.444	.227	.058	.927	.828	.754	.701	.665	.642	.630
76	6.485	6.265	6.092	5.959	5.858	5.782	5.728	5.691	5.668	5.655
77	.526	.302	.127	.991	.888	.811	.755	.717	.693	.680
78	.568	.340	.161	6.023	.918	.839	.782	.743	.719	.706
79	.609	.377	.196	.055	.948	.868	.809	.770	.744	.731
80	.650	.415	.230	.087	.978	.896	.837	.796	.770	.756
81	6.692	6.452	6.265	6.119	6.008	5.924	5.864	5.822	5.796	5.781
82	.733	.490	.299	.151	.038	.953	.891	.848	.821	.806
83	.774	.528	.334	.183	.068	.981	.918	.874	.847	.832
84	.816	.565	.368	.215	.098	6.010	.945	.901	.872	.857
85	.857	.603	.403	.247	.128	.038	.972	.927	.898	.882
86	6.898	6.640	6.437	6.279	6.158	6.066	6.000	5.953	5.923	5.907
87	.940	.678	.472	.311	.188	.095	.027	.979	.949	.932
88	.981	.716	.506	.343	.218	.123	.054	6.006	.975	.958
89	7.022	.753	.541	.375	.248	.152	.081	.032	6.000	.983
90	.064	.791	.575	.407	.278	.180	.108	.058	.026	6.008
91	7.105	6.828	6.610	6.439	6.308	6.208	6.135	6.084	6.051	6.033
92	.146	.866	.644	.471	.338	.237	.162	.110	.077	.058
93	.188	.903	.679	.503	.368	.265	.190	.137	.102	.084
94	.229	.941	.713	.535	.398	.294	.217	.163	.128	.109
95	.270	.979	.748	.567	.428	.322	.244	.189	.154	.134
96	7.312	7.016	6.783	6.600	6.458	6.350	6.271	6.215	6.179	6.159
97	.353	.054	.817	.632	.488	.379	.298	.242	.205	.184
98	.394	.091	.852	.664	.518	.407	.325	.268	.230	.210
99	.436	.129	.886	.696	.548	.436	.353	.294	.256	.235
100	.477	.166	.921	.728	.578	.464	.380	.320	.281	.260

Contd...

Table A.17 Contd.

Response (%)	Expected probit, Y									
	Y = 5.0-5.9; 51-100% response									
	5.0	5.1	5.2	5.3	5.4	5.5	5.6	5.7	5.8	5.9
51	5.025	5.025	5.023	5.017	5.005	4.985	4.953	4.906	4.840	4.750
52	.050	.050	.048	.043	.032	5.013	.983	.938	.874	.788
53	.075	.075	.074	.069	.059	.041	5.013	.970	.909	.825
54	.100	.100	.100	.096	.087	.070	.043	5.002	.943	.863
55	.125	.126	.125	.122	.114	.098	.073	.034	.978	.901
56	5.150	5.151	5.151	5.148	5.141	5.127	5.103	5.066	5.012	4.938
57	.175	.176	.176	.174	.168	.155	.133	.098	.047	.976
58	.201	.201	.202	.201	.195	.183	.163	.130	.082	5.013
59	.226	.226	.227	.227	.222	.212	.193	.162	.116	.051
60	.251	.252	.253	.253	.250	.240	.223	.194	.151	.088
61	5.276	5.277	5.279	5.279	5.277	5.269	5.253	5.226	5.185	5.126
62	.301	.302	.304	.305	.304	.297	.283	.258	.220	.164
63	.326	.327	.330	.332	.331	.325	.313	.290	.254	.201
64	.351	.352	.355	.358	.358	.354	.343	.322	.289	.239
65	.376	.378	.381	.384	.385	.382	.373	.354	.323	.276
66	5.401	5.403	5.406	5.410	5.412	5.411	5.403	5.386	5.358	5.314
67	.426	.428	.432	.437	.440	.439	.433	.418	.392	.351
68	.451	.453	.458	.463	.467	.467	.463	.450	.427	.389
69	.476	.478	.483	.489	.494	.496	.493	.482	.461	.427
70	.501	.504	.509	.515	.521	.524	.523	.514	.496	.464
71	5.526	5.529	5.534	5.541	5.548	5.553	5.553	5.546	5.530	5.502
72	.551	.554	.560	.568	.575	.581	.583	.578	.565	.539
73	.577	.579	.585	.594	.603	.609	.613	.610	.599	.577
74	.602	.604	.611	.620	.630	.638	.643	.642	.634	.615
75	.627	.630	.637	.646	.657	.666	.673	.674	.668	.652
76	5.652	5.655	5.662	5.673	5.684	5.695	5.703	5.706	5.703	5.690
77	.677	.680	.688	.699	.711	.723	.733	.738	.737	.727
78	.702	.705	.713	.725	.738	.752	.763	.770	.772	.765
79	.727	.730	.739	.751	.765	.780	.793	.802	.806	.802
80	.752	.755	.764	.777	.793	.808	.823	.834	.841	.840
81	5.777	5.781	5.790	5.804	5.820	5.837	5.853	5.866	5.875	5.878
82	.802	.806	.816	.830	.847	.865	.883	.898	.910	.915
83	.827	.831	.841	.856	.874	.894	.913	.930	.944	.953
84	.852	.856	.867	.882	.901	.922	.943	.962	.979	.990
85	.877	.881	.892	.908	.928	.950	.973	.995	6.014	6.028
86	5.902	5.907	5.918	5.935	5.956	5.979	6.003	6.027	6.048	6.066
87	.927	.932	.943	.961	.983	6.007	.033	.059	.083	.103
88	.953	.957	.969	.987	6.010	.036	.063	.091	.117	.141
89	.978	.982	.995	6.013	.037	.064	.093	.123	.152	.178
90	6.003	6.007	6.020	.040	.064	.092	.123	.155	.186	.216
91	6.028	6.033	6.046	6.066	6.091	6.121	6.153	6.187	6.221	6.253
92	.053	.058	.071	.092	.118	.149	.183	.219	.255	.291

Contd...

Table A.17 Contd.

Response (%)	Expected probit, Y									
	Y = 5.0-5.9; 51-100% response									
	5.0	5.1	5.2	5.3	5.4	5.5	5.6	5.7	5.8	5.9
93	.078	.083	.097	.118	.146	.178	.213	.251	.290	.329
94	.103	.108	.122	.144	.173	.206	.243	.283	.324	.366
95	.128	.133	.148	.171	.200	.234	.273	.315	.359	.404
96	6.153	6.159	6.174	6.197	6.227	6.263	6.303	6.347	6.393	6.441
97	.178	.184	.199	.223	.254	.291	.333	.379	.428	.479
98	.203	.209	.225	.249	.281	.320	.363	.411	.462	.517
99	.228	.234	.250	.276	.309	.348	.393	.443	.497	.554
100	.253	.259	.276	.302	.336	.376	.423	.475	.531	.592

Response (%)	Expected probit, Y									
	Y = 6.0-6.9; 51-100% response									
	6.0	6.1	6.2	6.3	6.4	6.5	6.6	6.7	6.8	6.9
51	4.631	4.473	4.269	4.006	3.667	3.233	2.677	1.964	1.049	–
52	.672	.519	.321	.064	.734	.310	.767	2.070	.175	0.022
53	.713	.565	.372	.122	.800	.387	.857	.176	.302	.175
54	.755	.611	.424	.181	.867	.464	.947	.283	.429	.327
55	.796	.657	.475	.239	.934	.541	3.037	.389	.555	.480
56	4.837	4.703	4.527	4.297	4.001	3.619	3.127	2.495	1.682	0.632
57	.879	.749	.578	.356	.068	.696	.218	.602	.809	.784
58	.920	.795	.630	.414	.134	.773	.308	.708	.935	.937
59	.961	.841	.681	.472	.201	.850	.398	.814	2.062	1.089
60	5.003	.887	.733	.531	.268	.927	.488	.921	.189	.242
61	5.044	4.932	4.784	4.589	4.335	4.005	3.578	3.027	2.315	1.394
62	.085	.978	.836	.647	.401	.082	.668	.133	.442	.546
63	.127	5.024	.887	.706	.468	.159	.758	.240	.569	.699
64	.168	.070	.939	.764	.535	.236	.849	.346	.695	.851
65	.209	.116	.990	.823	.602	.313	.939	.452	.822	2.004
66	5.251	5.162	5.042	4.881	4.669	4.391	4.029	3.559	2.949	2.156
67	.292	.208	.093	.939	.735	.468	.119	.665	3.075	.308
68	.333	.254	.145	.998	.802	.545	.209	.771	.202	.461
69	.375	.300	.196	5.056	.869	.622	.299	.878	.329	.613
70	.416	.346	.248	.114	.936	.700	.390	.984	.455	.766
71	5.457	5.392	5.299	5.173	5.003	4.777	4.480	4.090	3.582	2.918
72	.499	.437	.351	.231	.069	.854	.570	.197	.709	3.070
73	.540	.483	.402	.289	.136	.931	.660	.303	.835	.223
74	.581	.529	.454	.348	.203	5.008	.750	.409	.962	.375
75	.623	.575	.505	.406	.270	.086	.840	.516	4.089	.528
76	5.664	5.621	5.557	5.464	5.336	5.163	4.930	4.622	4.215	3.680

Contd...

Table A.17 Contd.

Y = 6.0-6.9; 51-100% response

Response (%)	Expected probit, Y									
	6.0	6.1	6.2	6.3	6.4	6.5	6.6	6.7	6.8	6.9
77	.705	.667	.608	.523	.403	.240	5.021	.728	.342	.832
78	.747	.713	.660	.581	.470	.317	.111	.835	.469	.985
79	.788	.759	.711	.639	.537	.394	.201	.941	.595	4.137
80	.829	.805	.763	.698	.604	.472	.291	5.047	.722	.290
81	5.870	5.851	5.814	5.756	5.670	5.549	5.381	5.154	4.849	4.442
82	.912	.896	.866	.815	.737	.626	.471	.260	.975	.594
83	.953	.942	.917	.873	.804	.703	.561	.366	5.102	.747
84	.994	.988	.969	.931	.871	.780	.652	.473	.229	.899
85	6.036	6.034	6.020	.990	.938	.858	.742	.579	.355	5.052
86	6.077	6.080	6.072	6.048	6.004	5.935	5.832	5.685	5.482	5.204
87	.118	.126	.123	.106	.071	6.012	.922	.792	.609	.356
88	.160	.172	.175	.165	.138	.089	6.012	.898	.735	.509
89	.201	.218	.226	.223	.205	.166	.102	6.004	.862	.661
90	.242	.264	.278	.281	.272	.244	.192	.111	.988	.814
91	6.284	6.310	6.329	6.340	6.338	6.321	6.283	6.217	6.115	5.966
92	.325	.355	.381	.398	.405	.398	.373	.323	.242	6.118
93	.366	.401	.432	.456	.472	.475	.463	.430	.368	.271
94	.408	.447	.484	.515	.539	.553	.553	.536	.495	.423
95	.449	.493	.535	.573	.605	.630	.643	.642	.622	.576
96	6.490	6.539	6.587	6.631	6.672	6.707	6.733	6.749	6.748	6.728
97	.532	.585	.638	.690	.739	.784	.824	.855	.875	.880
98	.573	.631	.690	.748	.806	.861	.914	.961	7.002	7.033
99	.614	.677	.741	.807	.873	.939	7.004	7.068	.128	.185
100	.656	.723	.793	.865	.939	7.016	.094	.174	.255	.338

Y = 7.0-7.9; 51-100% response

Response (%)	Expected probit, Y									
	7.0	7.1	7.2	7.3	7.4	7.5	7.6	7.7	7.8	7.9
51	–	–	–	–	–	–	–	–	–	–
52	–	–	–	–	–	–	–	–	–	–
53	–	–	–	–	–	–	–	–	–	–
54	–	–	–	–	–	–	–	–	–	–
55	–	–	–	–	–	–	–	–	–	–
56	–	–	–	–	–	–	–	–	–	–
57	–	–	–	–	–	–	–	–	–	–
58	–	–	–	–	–	–	–	–	–	–
59	–	–	–	–	–	–	–	–	–	–
60	0.013	–	–	–	–	–	–	–	–	–
61	0.198	–	–	–	–	–	–	–	–	–
62	.383	–	–	–	–	–	–	–	–	–

Contd...

Table A.17 Contd.

Response (%)	Expected probit, Y									
Y = 7.0-7.9; 51-100% response										
	7.0	7.1	7.2	7.3	7.4	7.5	7.6	7.7	7.8	7.9
63	.568	–	–	–	–	–	–	–	–	–
64	.753	–	–	–	–	–	–	–	–	–
65	.939	–	–	–	–	–	–	–	–	–
66	1.124	–	–	–	–	–	–	–	–	–
67	.309	0.003	–	–	–	–	–	–	–	–
68	.494	.231	–	–	–	–	–	–	–	–
69	.680	.458	–	–	–	–	–	–	–	–
70	.865	.685	–	–	–	–	–	–	–	–
71	2.050	0.913	–	–	–	–	–	–	–	–
72	.235	1.140	–	–	–	–	–	–	–	–
73	.420	.367	–	–	–	–	–	–	–	–
74	.606	.595	0.263	–	–	–	–	–	–	–
75	.791	.822	.545	–	–	–	–	–	–	–
76	2.976	2.050	0.827	–	–	–	–	–	–	–
77	3.161	.277	1.108	–	–	–	–	–	–	–
78	.347	.504	.390	–	–	–	–	–	–	–
79	.532	.732	.672	0.265	–	–	–	–	–	–
80	.717	.959	.954	.618	–	–	–	–	–	–
81	3.902	3.186	2.236	0.971	–	–	–	–	–	–
82	4.087	.414	.518	1.324	–	–	–	–	–	–
83	.273	.641	.800	.677	0.175	–	–	–	–	–
84	.458	.868	3.082	2.030	.621	–	–	–	–	–
85	.643	4.096	.364	.383	1.068	–	–	–	–	–
86	4.828	4.323	3.645	2.736	1.514	–	–	–	–	–
87	5.014	.551	.927	3.089	.961	0.438	–	–	–	–
88	.199	.778	4.209	.442	2.408	1.008	–	–	–	–
89	.384	5.005	.491	.795	.854	.579	–	–	–	–
90	.569	.233	.773	4.148	3.301	2.149	0.581	–	–	–
91	5.754	5.460	5.055	4.501	3.747	2.720	1.317	–	–	–
92	.940	.687	.337	.854	4.194	3.290	2.054	0.356	–	–
93	6.125	.915	.619	5.207	.640	.861	.790	1.316	–	–
94	.310	6.142	.901	.560	5.087	4.431	3.526	2.275	0.542	–
95	.495	.369	6.182	.914	.533	5.002	4.262	3.235	1.806	–
96	6.681	6.597	6.464	6.267	5.980	5.572	4.998	4.194	3.069	1.493
97	.866	.824	.746	.620	6.426	6.143	5.735	5.154	4.333	3.173
98	7.051	7.051	7.028	.973	.873	.713	6.471	6.114	5.596	4.853
99	.236	.279	.310	7.326	7.319	7.284	7.207	7.073	6.859	6.533
100	.421	.506	.592	.679	.766	.854	.943	8.033	8.123	8.213

Table A.18 Random digits

03991	10461	93716	16894	98953	73231	39528	72484	82474	25593
38555	95554	32886	59780	09958	18065	81616	18711	53342	44276
17546	73704	92052	46215	15917	06253	07586	16120	82641	22820
32643	52861	95819	06831	19640	99413	90767	04235	13574	17200
69572	68777	39510	35905	85244	35159	40188	28193	29593	88627
24122	66591	27699	06494	03152	19121	34414	82157	86887	55087
61196	30231	92962	61773	22109	78508	63439	75363	44989	16822
30532	21704	10274	12202	94205	20380	67049	09070	93399	45547
03788	97599	75867	20717	82037	10268	79495	04146	52162	90286
48228	633379	85783	47619	87481	37220	91704	30552	04737	21031
88618	19161	41290	67312	71857	15957	48545	35247	18619	13674
71299	23853	05870	01119	92784	26340	75122	11724	74627	73707
27954	58909	82444	99005	04921	73701	92904	13141	32392	19763
80863	00514	20247	81759	45197	25332	69902	63742	78464	22501
33564	60780	48460	85558	15191	18782	94972	11598	62095	36787
90899	75754	60833	25983	01291	41349	19152	00023	12302	80783
78038	70267	43529	06318	38384	74761	36024	00867	76378	41605
55986	66485	88722	56736	66164	49431	94458	74284	05041	49807
87539	08823	94813	31900	54155	83436	54158	34243	46978	35482
16818	60311	74457	90561	72848	11834	75051	93029	47665	64382
34677	58300	74910	64345	19325	81549	60365	94653	35075	33949
45305	07521	61318	31855	14413	70951	83799	42402	56623	34442
59747	67277	76503	34513	39663	77544	32960	07405	36409	83232
16520	69676	11654	99893	02181	68161	19322	53845	57620	52606
68652	27376	92852	55866	88448	03584	11220	94747	07399	37408
79375	95220	01159	63267	10622	48391	31751	57260	68980	05339
33521	26665	55823	47641	86225	31704	88492	99382	14454	04504
59589	49067	66821	41575	49767	04037	30934	47744	07481	83828
20554	91409	96277	48257	50816	97616	22888	48893	27499	98748
59404	72059	43947	51680	43852	59693	78212	16993	35902	91386
42614	29297	01918	28316	25163	01889	70014	15021	68971	11403
34994	41374	70071	14736	65251	07629	37239	33295	18477	65622
99385	41600	11133	07586	36815	43625	18637	37509	14707	93997
66497	68646	78138	66559	64397	11692	05327	82162	83745	22567
48509	23929	27482	45476	04515	25624	95096	67946	16930	33361
15470	48355	88651	22596	83761	60873	43253	84145	20368	07126
20094	98977	74843	93413	14387	06345	80854	09279	41196	37480
73788	06533	28597	20405	51321	92246	80088	77074	66919	31678
60530	45128	74022	84617	72472	00008	80890	18002	35352	54131
44372	15486	65741	14014	05466	55306	93128	18464	79982	68416

Contd...

Source: Rao, Mitra, Mathai and Ramamurthy (1975); Formula and Tables for Statistical Work.

Table A.18 Contd.

18611	19241	66083	24653	84609	58232	41849	84547	46850	52326
58319	15997	08355	60860	29735	47762	46352	33049	69248	93460
61199	67940	55121	29281	59076	07936	11087	96294	14013	31792
18627	90872	00911	98936	76355	93779	52701	08337	56303	87315
00441	58997	14060	40619	29549	69616	57275	36898	81304	48585
32624	68691	14845	16672	61958	77100	20857	73156	70284	24326
65961	73488	41839	55382	17267	70943	15633	84924	90415	93614
20288	34060	39685	23309	10061	68829	92694	48297	39904	02115
59362	95938	74416	53166	35208	33374	77613	19019	88152	00080
99782	93478	53152	67433	35663	52972	38688	32486	45134	63545
27767	43584	85301	88977	29490	69714	64015	64874	32444	48277
13025	14338	54066	15243	47724	66733	74108	88222	88570	74015
80217	36292	98525	24335	24432	24896	62880	87873	95160	59221
10875	62004	90391	61105	57411	06368	11748	12102	80580	41867
54127	57326	26629	19087	24472	88779	17944	05600	60478	03343
60311	42824	37301	42678	45990	43242	66067	42792	95043	52680
49739	71484	92003	98086	76668	73209	54244	91030	45547	70818
78626	51594	16453	94614	39014	97066	30945	57589	31732	57260
66692	13986	99837	00582	81232	44987	69170	37403	86995	90307
44071	28091	07362	97703	76447	42537	08345	88975	35841	85771
59820	96163	78851	16499	87064	13075	73035	41207	74699	09310
25704	91035	26313	77463	55387	72681	47431	43905	31048	56699
22304	90314	78438	66276	18396	73538	43277	58874	11466	16082
17710	59621	15292	67139	59526	52113	53856	30743	08670	84741
25852	58905	55018	56374	35824	71708	30540	27886	61732	75454
46780	56487	75211	10271	36633	68424	17374	52003	70707	70214
59849	96169	87195	46092	26787	60939	59202	11973	02902	33250
47670	07654	30342	40277	11049	72049	83012	09832	25571	77628
94304	71803	73465	09819	58869	35220	09504	96412	90193	79568
08105	59987	21437	36786	49226	77837	98524	97831	65704	09514
64281	61826	18555	64937	64654	25843	41145	42820	14924	39650
66847	70495	32350	02985	01755	14750	48968	38603	70312	05682
72461	33230	21529	53424	72877	17334	39283	04149	90850	64618
21032	91050	13058	16218	06554	07850	73950	79552	24781	89683
95362	67011	06651	16136	57216	39618	49856	99326	40902	05069
49712	97380	10404	55452	09971	59481	37006	22186	72682	07385
58275	61764	97586	54716	61459	21647	87417	17198	21443	41803
89514	11788	68224	23417	46376	25366	94746	49580	01176	28838
15472	50669	48139	36732	26825	05511	12459	91314	80582	71944
12120	86124	51247	44302	87112	21476	14713	71181	13177	55292
95294	00556	70481	06905	21785	41101	49386	54480	23604	23554
66986	34099	74474	20740	47458	64809	06312	88940	15995	69321
80620	51790	11436	38072	40405	68032	60942	00307	11897	92674
55411	85667	77535	99892	71209	92061	92329	98932	78284	46347
95083	06783	28102	57816	85561	29671	77936	63574	31384	51924

Contd...

Table A.18 Contd.

90726	57166	98884	08583	95889	57067	38101	77756	11657	13897
68984	83620	89747	98882	92613	89719	39641	69457	91339	22502
36401	16489	18059	51061	67667	60631	84054	40455	99396	63680
92638	40333	67054	16067	24700	71594	47468	03577	57649	63266
21036	82808	77501	97427	76479	68562	43321	31370	28977	23896
13173	33365	41468	85149	49554	17994	91178	10174	29420	90438
86716	38746	94559	37559	49678	53119	98189	81851	29651	84215
92281	02262	41615	70360	64114	58660	96717	54244	10701	41393
12470	56500	50273	93113	41794	86861	39448	93136	25722	08564
01016	00857	41396	80504	90670	08289	58137	17820	22751	36518
34030	60726	25807	24260	71529	78920	47648	13885	70669	93406
50259	46345	06170	97965	88302	98041	11947	56203	19324	20504
73959	76145	60808	54444	74412	81105	69181	96845	38525	11600
46874	37088	80940	44893	10408	36222	14004	23153	69249	05747
60883	52109	19516	90120	46759	71643	62342	07589	08899	05985

Table A.19 The TANH^{-1}(Z) transformation for correlation coefficient

r = 0.00 (0.02) 0.18

r	0	2	4	6	8
.0	.00000	.02000	.04002	.06007	.08017
.1	.10034	.12058	.14093	.16139	.18198

r = 0.200 (0.002) 0.858

r	0	2	4	6	8
.20	.20273	.20482	.20690	.20899	.21108
.21	.21317	.21526	.21736	.21946	.22156
.22	.22366	.22576	.22786	.22997	.23208
.23	.23419	.23630	.23842	.24053	.24265
.24	.24477	.24690	.24902	.25115	.25328
.25	.25541	.25755	.25968	.26182	.26396
.26	.26611	.26825	.27040	.27255	.27471
.27	27686	.27902	.28118	.28335	.28551
.28	.28768	.28985	.29203	.29420	.29638
.29	29857	.30075	.30294	.30513	.30732
.30	.30052	.31172	.31392	.31613	.31833
.31	.32055	.32276	.32498	.32720	.32942
.32	.33165	.33388	.33611	.33835	.34059
.33	.34283	.34507	.34732	.34958	.35182
.34	.35409	.35636	.35862	.36089	.36317
.35	.36544	.36772	.37001	.37230	.37459
.36	.37689	.37919	.38149	.38380	.38611
.37	.38842	.39074	.39307	.39539	.39772
.38	.40006	.40240	.40474	.40709	.40944
.39	.41180	.41416	.41653	.41890	.42127
.40	.42365	.42603	.42842	.43081	.43321
.41	.43561	.43802	.44043	.44285	.44527
.42	.44769	.45012	.45256	.45500	.45745
.43	45990	.46235	.46481	.46728	.46975
.44	.47223	.47471	.47720	.47970	.48220
.45	.48470	.48721	.48973	.49225	.49478
.46	.49731	.49985	.50240	.50495	.50751
.47	.51007	.51264	.51522	.51780	.52039
.48	.52298	.52559	.52819	.53081	.53343
.49	.53606	.53870	.54134	.54399	.54664
.50	.54931	.55198	.55465	.55734	.56603
.51	.56273	.56544	.56815	.57087	.57360
.52	.57634	.57908	.58184	.58460	.58737
.53	.59015	.59293	.59572	.59853	.60134
.54	.60416	.60698	.60982	.61266	.61552
.55	.61838	.62125	.62413	.62702	.62992
.56	.63283	.63575	.63868	.64162	.64457
.57	.64752	.65040	.65347	.65646	.65945

Contd...

Table A.19 Contd.

r = 0.200 (0.002) 0.858					
r	0	2	4	6	8
.58	.66246	.66548	.66851	.67155	.67460
.59	.67767	.68074	.68382	.68692	.69003
.60	.69315	.69628	.69942	.70258	.70574
.61	.70892	.71211	.71532	.71853	.72176
.62	.72501	.72826	.73153	.73481	.73811
.63	.74142	.74474	.74808	.75143	.75479
.64	.75817	.76157	.76498	.76840	.77184
.65	.77530	.77877	.78226	.78576	.78928
.66	.79281	.79637	.79993	.80352	.80712
.67	.81074	.81438	.81804	.82171	.82540
.68	.82911	.83284	.83659	.84036	.84415
.69	.84796	.85178	.85563	.85950	.86339
.70	.86730	.87123	.87519	.87916	.88316
.71	.88718	.89123	.89530	.89939	.90350
.72	.90764	.91181	.91600	.92022	.92446
.73	.92873	.93302	.93734	.94169	.94607
.74	.95048	.95491	.96387	.96387	.96840
.75	.97296	.97754	.98216	.98681	.99150
.76	.99622	1.00097	1.00575	1.01058	1.01543
.77	1.02033	1.02526	1.03023	1.03524	1.04028
.78	1.04537	1.05050	1.05567	1.06088	1.06613
.79	1.07143	1.07677	1.08216	1.08760	1.09308
.80	1.09861	1.10419	1.10982	1.11551	1.12124
.81	1.12703	1.13287	1.13877	1.14473	1.15074
.82	1.15682	1.16295	1.16915	1.17541	1.18174
.83	1.18814	1.19460	1.20113	1.20774	1.21442
.84	1.22117	1.22801	1.23492	1.24191	1.24899
.85	1.25615	1.26340	1.27075	1.27818	1.28571

r = 0.860 (0.001) 0.999										
r	0	1	2	3	4	5	6	7	8	9
.86	1.29334	1.29720	1.30108	1.30498	1.30891	1.31287	1.31686	1.32087	1.32491	1.32898
.87	1.33308	1.33721	1.34137	1.34555	1.34977	1.35403	1.35831	1.36282	1.36697	1.37135
.88	1.37577	1.38022	1.38470	1.38922	1.39378	1.39838	1.40301	1.40765	1.41239	1.41714
.89	1.42193	1.42676	1.43163	1.43654	1.44150	1.44651	1.45156	1.45665	1.46179	1.46698
.90	1.47222	1.47751	1.48285	1.48824	1.49368	1.49918	1.50473	1.51034	1.51601	1.52174
.91	1.52752	1.53337	1.53928	1.54526	1.55130	1.55741	1.56359	1.56984	1.57616	1.58256
.92	1.58903	1.59558	1.60221	1.60892	1.61571	1.62260	1.62957	1.63663	1.64379	1.65104
.93	1.65839	1.66584	1.67340	1.68107	1.68885	1.69674	1.70475	1.71288	1.72114	1.72953
.94	1.73805	1.74671	1.75552	1.76447	1.77358	1.78284	1.79227	1.80188	1.81166	1.82162
.95	1.83178	1.84214	1.85270	1.86340	1.87450	1.88574	1.89723	1.90898	1.92100	1.93331
.96	1.94591	1.95882	1.97207	1.98566	1.99961	2.01395	2.02870	2.04388	2.05952	2.07565
.97	2.09230	2.10950	2.12730	2.14574	2.16486	2.18472	2.20539	2.22692	2.24940	2.27291
.98	2.29756	2.32346	2.35074	2.37958	2.41014	2.44266	2.47741	2.51472	2.55499	2.59875
.99	2.64665	2.69958	2.75873	2.82574	2.90307	2.99448	3.10630	3.25039	3.45338	3.80020

Source: Rao, Mitra, Mathai and Ramamurthy (1975); Formula and Tables for Statistical Work.

Table A.20 Cubes and cube roots, fourth powers, factorials, exponentials and natural logarithms

n	n^3	n^4	$n!$	$\log_e n$	e^{-n}	$e^{-n/100}$	$3\sqrt{n}$
1	1	1	1	0.000000	.367879	.990050	1.0000000
2	8	16	2	0.693147	.135335	.980199	1.2599210
3	27	81	6	1.098612	.049787	.970446	1.4422496
4	64	256	24	1.386204	.018316	.960789	1.5874011
5	125	625	120	1.609438	.673795 (2)	.951229	1.7099759
6	216	1296	720	1.791759	.247875 (2)	.941765	1.8171206
7	343	2401	5040	1.945910	.911882 (3)	.932394	1.9129312
8	512	4096	40320	2.079442	.335463 (3)	.923116	2.0000000
9	729	6561	362880	2.197225	.123410 (3)	.913931	2.0800838
10	1000	10000	0.3628800 (7)	2.302585	.453999 (4)	.904837	2.1544347
11	1331	14641	3.9916800 (7)	2.397895	.167017 (4)	.895834	2.2239801
12	1728	20736	4.7900160 (8)	2.484907	.614421 (5)	.886920	2.2804285
13	2197	28561	0.2270208 (9)	2.564949	.226033 (5)	.878095	2.3513347
14	2744	38416	8.7178291 (10)	2.639057	.831529 (6)	.869358	2.4101423
15	3375	50625	1.3076744 (12)	2.708050	.305902 (6)	.860708	2.4662121
16	4096	65536	2.0922790 (13)	2.772580	.112535 (6)	.852144	2.5198421
17	4913	83521	3.5568743 (14)	2.833213	.413994 (7)	.843665	2.5712816
18	5832	104976	6.4023737 (15)	2.890372	.152300 (7)	.835270	2.6207414
19	6859	130321	1.2164510 (17)	2.944439	.560280 (8)	.826959	2.6684016
20	8000	160000	2.4329020 (18)	2.995732	.206115 (8)	.818731	2.7144176
21	9261	194481	5.1090942 (19)	3.044522	.758256 (9)	.810584	2.7589242
22	10648	234256	1.1240007 (21)	3.091042	.278947 (9)	.802519	2.8020393
23	12167	279841	2.5852017 (22)	3.135494	.102619 (9)	.794534	2.8438670
24	13824	331776	6.2044840 (23)	3.178054	.377513 (10)	.786628	2.8844991
25	15625	390625	1.5511210 (25)	3.218876	.138879 (10)	.778801	2.9240177
26	17576	456976	4.0329146 (26)	3.258097	.510909 (11)	.771052	2.9624961
27	19683	531441	1.0888869 (28)	3.295837	.187953 (11)	.763379	3.0000000
28	21952	614656	3.0488834 (29)	3.332205	.691440 (12)	.755784	3.0365890
29	24389	707281	8.8417620 (30)	3.367296	.254367 (12)	.748264	3.0723168
30	27000	810000	2.6525286 (32)	3.401197	.935762 (13)	.740818	3.1072325
31	29791	923521	8.2228387 (33)	3.433987	.344248 (13)	.733447	3.1413807
32	32768	1048576	2.6313084 (35)	3.465736	.126642 (13)	.726149	3.1748021
33	35937	1185921	8.6833176 (36)	3.496508	.465889 (14)	.718924	3.2075343
34	39304	1336336	2.9523280 (38)	3.526361	.171391 (14)	.711770	3.2396118
35	42875	1500625	1.0333148 (40)	3.555348	.630512 (15)	.704688	3.2710663
36	46656	1679616	3.7199333 (41)	3.583519	.231952 (15)	.697676	3.3019272
37	50653	1874161	1.3763753 (43)	3.610918	.853305 (16)	.690734	3.3322219
38	54872	2085136	5.2302262 (44)	3.637586	.313913 (16)	.683861	3.3619754
39	59319	2313441	2.0397882 (46)	3.663562	.115482 (16)	.677057	3.3912114
40	64000	2560000	8.1591528 (47)	3.688879	.424835 (17)	.670320	3.4199519
41	68921	2825761	3.3452327 (49)	3.713572	.156288 (17)	.663650	3.4482172
42	74088	3111096	1.4050061 (51)	3.737670	.574952 (18)	.657047	3.4760266
43	79507	3418801	6.0415263 (52)	3.761200	.211513 (18)	.650509	3.5033981

Contd...

Source: Rao, Mitra, Mathai and Ramamurthy (1975); Formula and Tables for Statistical Work.

Table A.20 Contd.

n	n^3	n^4	$n!$	$\log_e n$	e^{-n}	$e^{-n/100}$	$3\sqrt{n}$
44	85184	3748096	2.6582716 (54)	3.784190	.778113 (19)	.644036	3.5303483
45	91125	4100625	1.1962222 (56)	3.806662	.286252 (19)	.637628	3.5568933
46	97336	4477456	5.5026222 (57)	3.828641	.105306 (19)	.631284	3.583479
47	103823	4879681	2.5862324 (59)	3.850148	.387400 (20)	.625002	3.6088261
48	110592	5308416	1.2413916 (61)	3.871201	.142516 (20)	.618783	3.634412
49	117649	5764801	6.0828186 (62)	3.891820	.524289 (21)	.612626	3.6593057
50	125000	6250000	3.0414093 (64)	3.912023	.192875 (21)	.606531	3.0840315
51	132651	6765201	1.5511188 (66)	3.931826	.709547 (22)	.600496	3.7084298
52	140608	7311616	8.0658175 (67)	3.951244	.261028 (22)	.594521	3.7325112
53	148877	7890481	4.2748833 (69)	3.970292	.960268 (23)	.588605	3.7562858
54	157464	8503056	2.3084370 (71)	3.988984	.353263 (23)	.582748	3.7797631
55	166375	915025	1.2696403 (73)	4.007333	.129958 (23)	.576950	3.8029525
56	175616	9834496	7.1099859 (74)	4.025352	.478089 (24)	.571209	3.8258624
57	185193	10556001	4.0526920 (76)	4.043051	.175879 (24)	.565525	3.8485011
58	195112	11316496	2.3505613 (78)	4.060443	.647023 (25)	.559898	3.8708766
59	205379	12117361	1.3868312 (80)	4.077537	.238027 (25)	.554327	3.8929964
60	216000	12960000	8.3209871 (81)	4.094345	.875651 (26)	.548812	3.9148676
61	226981	13845841	5.0758021 (83)	4.110874	.322134 (26)	.543351	3.9364972
62	238328	14776336	3.1469973 (85)	4.127134	.118506 (26)	.537944	3.9578916
63	250047	15752961	1.9820083 (87)	4.143135	.435961 (27)	.532592	3.9790572
64	262144	1677216	1.2688693 (89)	4.158883	.160381 (27)	.527292	4.0000000
65	274625	1780625	8.2476506 (90)	4.174387	.590009 (28)	.522046	4.0207258
66	287496	18974736	5.4434494 (92)	4.189655	.217052 (28)	.516851	4.0412400
67	300763	20451121	3.6471111 (94)	4.204693	.798490 (29)	.511709	4.0615481
68	314432	21381376	2.4800355 (96)	4.219508	.293748 (29)	.506617	4.0816551
69	328509	22667121	1.7112245 (98)	4.234107	.108064 (29)	.501576	4.1015659
70	343000	24010000	1.1978572 (100)	4.248495	.397545 (30)	.496585	4.1212853
71	357911	25411681	8.5047859 (101)	4.262680	.146249 (30)	.491644	4.1408177
72	373248	26873856	6.1234458 (103)	4.276666	.538019 (31)	.486752	4.1601676
73	389017	28398241	4.4701155 (105)	4.290459	.197926 (31)	.481909	4.1793392
74	405224	29986576	3.3078854 (107)	4.304065	.728129 (32)	.477114	4.1983365
75	421875	31640625	2.4809141 (109)	4.317488	.267864 (32)	.472367	4.2171633
76	438976	33362176	1.8854947 (111)	4.330733	.985415 (33)	.467666	4.2358236
77	456533	35153041	1.4518309 (113)	4.343805	.362514 (33)	.463013	4.2543209
78	474552	37015056	1.1324281 (115)	4.356709	.133361 (33)	.458406	4.2726587
79	493039	38950081	8.9461821 (116)	4.369448	.490609 (34)	.453845	4.2908404
80	512000	40960000	7.1569457 (118)	4.382027	.180485 (34)	.449329	4.3088694
81	531441	43046721	5.7971260 (120)	4.394449	.663968 (35)	.444858	4.3267487
82	551368	45212176	4.7536433 (122)	4.406719	.244260 (35)	.440432	4.3444815
83	571787	47458321	3.9455240 (124)	4.418841	.898583 (36)	.436049	4.3620707
84	592704	49787136	3.3142401 (126)	4.430817	.330570 (36)	.431711	4.3795191
85	614125	52200625	2.8171041 (128)	4.442651	.121610 (36)	.427415	4.3968297
86	636056	54700816	2.4227095 (130)	4.454347	.447378 (37)	.423162	4.4140050
87	658503	57289761	2.1077573 (132)	4.465908	.164581 (37)	.418952	4.4310478
88	681472	59969536	1.8548264 (134)	4.477337	.605460 (38)	.414783	4.4479602

Contd...

Table A.20 Contd.

n	n^3	n^4	$n!$	$\log_e n$	e^{-n}	$e^{-n/100}$	$3\sqrt{n}$
89	753571	68574961	1.3520015 (140)	4.510860	.301441 (39)	.402524	4.4979414
90	778688	71639296	1.2438414 (142)	4.521789	.110894 (39)	.398519	4.5143574
91	804357	74805201	1.1567725 (144)	4.532599	.407956 (40)	.394554	4.5306549
92	830584	7807496	1.0873662 (146)	4.543295	.150079 (40)	.390628	4.5468359
93	857375	81450625	1.0329978 (148)	4.553877	.552108 (41)	.386741	4.5629026
94	884736	84934656	9.9167793 (149)	4.56438	.203109 (41)	.382893	4.5788570
95	912673	88529281	9.6192760 (151)	4.574711	.747197 (42)	.379083	4.5947009
96	941192	92236816	9.4268904 (153)	4.584967	.274879 (42)	.375311	4.6104363
97	970299	96059601	9.3326215 (155)	4.595120	.101122 (42)	.371577	4.6260650
98	1000000	100000000	9.3326215 (157)	4.605170	.372008 (43)	.367879	4.0415888
99	704969	62742241	1.6507955 (136)	4.488636	.222736 (38)	.410656	4.4647451
100	729000	65610000	1.4857160 (138)	4.499810	.819401 (39)	.406570	4.4814047

The number in brackets following $n!$ is the power of 10 and that following e^{-n} is the power of 1/10 by which the given tabular value must be multiplied.

Table A.21 Common logarithms of numbers

N.	0	1	2	3	4	5	6	7	8	9	D.
100	000000	000434	000868	001301	001734	002166	002598	003029	003461	003891	432
1	4321	4751	5181	5609	6038	6466	6894	7321	7748	8174	428
2	8600	9026	9451	9876	010300	010724	011147	011570	011993	012415	424
3	012837	013259	013680	014100	4521	4940	5360	5779	6197	6616	420
4	7033	7451	7868	8284	8700	9116	9532	9947	020361	020775	416
105	021189	021603	022016	022428	022841	023252	023664	024075	4486	4896	412
6	5306	5715	6125	6533	6942	7350	7757	8164	8571	8978	408
7	9384	9789	030195	030600	031004	031408	031812	032216	032619	033021	404
8	033424	033826	4227	4628	5029	5430	5830	6230	6629	7028	400
9	7426	7825	8223	8620	9017	9414	9811	040207	040602	040998	397
110	041393	041787	842182	042576	042969	043362	043755	044148	044540	044932	393
1	5323	5714	6105	6495	6885	7275	7664	8053	8442	8830	390
2	9218	9606	9993	050380	050766	051153	051538	051924	052309	052694	386
3	053078	053463	053846	4230	4613	4996	5378	5760	6142	6524	383
4	6905	7286	7666	8046	8426	8805	9185	9563	9942	060320	379
115	060698	061075	061452	061829	062206	062582	062958	063333	063709	4083	376
6	4458	4832	5206	5580	5953	6326	6699	7071	7443	7815	373
7	8186	8557	8928	9298	9668	070038	070407	070776	071145	071514	370
8	071882	072250	072617	072985	073352	3718	4085	4451	4816	5182	366
9	5547	5912	6276	6640	7004	7368	7731	8094	8457	8819	363
120	079181	079543	079904	080266	080626	080987	081347	081707	082067	082426	360
1	082785	083144	083503	3861	4219	4576	4934	5291	5647	6004	357
2	6360	6716	7071	7426	7781	8136	8490	8845	9198	9552	355
3	9905	090258	090611	090963	091315	091667	092018	092370	092721	093071	352
4	093422	3772	4122	4471	4820	5169	5518	5866	6215	6562	349
125	6910	7257	7604	7951	8298	8644	8990	9335	9681	100026	346
6	100371	100715	101059	101403	101747	102091	102434	102777	103119	3462	343
7	3804	4146	4487	4828	5169	5510	5851	6191	6531	6871	341
8	7210	7549	7888	8227	8565	8903	9241	9579	9916	110253	338
9	110590	110926	111263	111599	111934	112270	112605	112940	113275	3609	335
130	113943	114277	114611	114944	115278	115611	115943	116276	116608	116940	333
1	7271	7603	7934	8265	8595	8926	9256	9536	9915	120245	330
2	120574	120903	121231	121560	121888	122216	122344	122071	123198	3525	328
3	3852	4178	4504	4830	5156	5481	5806	6131	6556	6781	325
4	7105	7429	7753	8076	8399	8722	9045	9368	9690	130012	323
135	130334	130655	130977	131298	131619	131939	132260	132580	132900	3219	321
6	3539	3858	4177	4496	4814	5133	5451	5769	6086	6403	318
7	6721	7037	7354	7671	7937	8303	8618	8934	9249	9564	316
8	9879	140194	140508	140822	141136	141450	141763	142076	142389	142702	314
9	143015	3327	3639	3951	4263	4574	4885	5196	5507	5818	311

Contd...

Log e = 0.434295; log π = 0.497150; log $\sqrt{\pi}$ = 0.248575.

Source: Rao, Mitra, Mathai and Ramamurthy (1975); Formula and Tables for Statistical Work.

Table A.21 Contd.

N.	0	1	2	3	4	5	6	7	8	9	D.
140	146128	146438	146748	147058	147367	147676	147985	148294	148603	148911	309
1	9219	9527	9835	150142	150449	150756	151063	151370	151676	151982	307
2	152288	152594	152900	3205	3510	3815	4120	4424	4728	5032	305
3	5336	5640	5943	6246	6549	6852	7154	7457	7759	8061	303
4	8362	8664	8965	9266	9567	9868	160168	160469	160769	161068	301
145	161368	161667	161967	162266	162564	162863	3161	3460	3758	4055	299
6	4353	4650	4947	5244	5541	5838	6134	6430	6726	7022	297
7	7317	7613	7908	8203	8497	8792	9086	9380	9674	9968	295
8	170262	170555	170848	171141	171434	171725	172019	172311	172603	172895	293
9	3186	3478	3769	4060	4351	4641	4932	5222	5512	5802	291
150	176091	176381	176670	176959	177248	177536	177825	178113	178401	178689	289
1	8977	9264	9552	9839	180126	180413	180699	180986	181272	181558	287
2	181844	182129	182415	182700	2985	3270	3555	3839	4123	4407	285
3	4691	4975	5259	5542	5825	6108	6391	6674	6956	7239	283
4	7521	7803	8084	8366	8647	8928	9209	9490	9771	190051	281
155	190332	190612	190892	191171	191451	191730	192010	192289	192567	2846	279
6	3125	3403	3681	3959	4237	4514	4792	5069	5346	5623	278
7	5900	6176	6453	6729	7005	7281	7556	7832	8107	8382	276
8	8657	8932	9206	9481	9755	200029	200303	200577	200850	201124	274
9	201397	201670	201943	202216	202488	2761	3033	3305	3577	3848	272
160	204120	204391	204663	204934	205204	205475	205746	206016	206286	206556	271
1	6826	7096	7365	7634	7904	8173	8441	8710	8979	9247	269
2	9515	9783	210051	210319	210586	210853	211121	211388	211654	211921	267
3	212188	212454	2720	2986	3252	3518	3783	4049	4314	4579	266
4	4844	5109	5373	5638	5902	6166	6430	6694	6957	7221	264
165	7484	7747	8010	8273	8536	8798	9060	9323	9585	9846	262
6	220108	220370	220631	220892	221153	221414	221675	221936	222196	222456	261
7	2716	2976	3236	3496	3755	4015	4274	4533	4792	5051	259
8	5309	5568	5826	6084	6342	6600	6858	7115	7372	7630	258
9	7887	8144	8400	8657	8913	9170	9426	9682	9938	230193	256
170	230449	230704	230960	231215	231470	231724	231979	232234	232488	232742	255
1	2996	3250	3504	3757	4011	4264	4517	4770	5023	5276	253
2	5528	5781	6033	6285	6537	6789	7041	7292	7544	7795	252
3	8046	8297	8548	8799	9049	9299	9550	9800	240030	240300	250
4	240549	240799	241048	241297	241546	241795	242044	242293	2541	2790	249
175	3038	3286	3534	3782	4030	4277	4525	4772	5019	5266	248
6	5513	5759	6006	6252	6499	6745	6991	7237	7482	7728	246
7	7973	8219	8464	8709	8954	9198	9443	9687	9932	250176	245
8	250420	250664	250908	251151	251395	251638	251881	252125	252338	2610	243
9	2853	3096	3338	3580	3822	4064	4306	4548	4790	5031	242
180	255273	255514	255755	255996	256237	256477	256718	256958	257198	257439	241
1	7679	7918	8158	8098	8637	8377	9116	9335	9594	9833	239
2	260071	260310	260548	260787	261025	261263	261501	261739	261976	262214	238
3	2451	2688	2925	3162	3399	3636	3373	4109	4346	4582	237
4	4818	5054	5290	5525	5761	5996	6232	6467	6702	6937	235

Contd...

Table A.21 Contd.

N.	0	1	2	3	4	5	6	7	8	9	D.
185	7172	7406	7641	7875	8110	8344	8578	8812	9046	9279	234
6	9513	9746	9980	270213	270446	270679	270912	271144	271377	271609	233
7	271842	272074	272306	2338	2770	3001	3233	3454	3696	3927	232
8	4158	4389	4620	4850	5031	5311	5542	5772	6002	6232	230
9	6462	6692	6921	7151	7380	7609	7838	8067	8296	8525	229
190	278754	278982	279211	279439	279667	279895	280123	280351	280578	280806	228
1	281033	281231	281488	281715	281942	282169	2396	2322	2849	3075	227
2	3301	3527	3753	3979	4205	4431	4336	4882	5107	5332	226
3	5557	5782	6097	6232	6436	6331	6905	7130	7354	7578	225
4	7802	8026	8249	8473	8696	8920	9143	9366	9589	9812	223
195	290035	290257	290480	290702	290925	291147	291369	291591	291813	292034	222
6	2256	2478	2699	2920	3141	3353	3384	3804	4025	4246	221
7	4466	4597	4907	5127	5347	5567	5787	6007	6226	6446	220
8	6665	6834	7104	7323	7542	7731	7979	8198	8416	8635	219
9	8853	9071	8289	9507	9725	9943	300161	300378	300595	300813	218
200	301030	301247	301454	301631	301898	302114	302331	302547	302764	302980	217
1	3196	3412	3528	3044	4039	4275	4491	4706	4921	5136	216
2	5351	5566	5781	5996	6211	6425	6639	6854	7068	7282	215
3	7496	7710	7924	8137	8351	8564	8778	8991	9204	9417	213
4	9630	9843	310056	310268	310481	310693	310906	311118	311330	311542	212
205	311754	311966	2177	2389	2600	2812	3023	3234	3445	3656	211
6	3867	4078	4289	4499	4710	4920	5130	5340	5551	5760	210
7	5970	6180	6390	6399	6809	7018	7227	7436	7646	7854	209
8	8063	8272	8481	8699	8898	9106	9314	9522	9730	9938	208
9	320146	320354	320562	320769	320977	321184	321391	321598	321805	322012	207
210	322219	322426	322633	322839	323046	323252	323458	323665	323871	324077	206
1	4282	4488	4694	4899	5105	5310	5516	5721	5926	6131	205
2	6336	6541	6745	6950	7155	7359	7563	7767	7972	8176	204
3	8380	8583	8787	8991	9194	9398	9601	9805	330008	330211	203
4	330414	330617	330819	331022	331225	331427	331630	331832	2034	2236	202
215	2438	2640	2842	3044	3246	3447	3649	3850	4051	4253	202
6	4454	4655	4856	5057	5257	5458	5658	5859	6059	6260	201
7	6460	6660	6860	7060	7260	7459	7659	7858	8058	8257	200
8	8456	8656	8855	9054	9253	9451	9650	9849	340047	340246	199
9	340444	340642	340841	341039	341237	341435	341632	341830	2028	2225	198
220	342423	342620	342817	343014	343212	343409	343606	343802	343999	344196	197
1	4392	4589	4785	4981	5178	5374	5570	5766	5962	6157	196
2	6353	6549	6744	6939	7135	7330	7525	7720	7915	8110	195
3	8305	8500	8694	8889	9083	9278	9472	9666	9860	350054	194
4	350248	350442	350636	350829	351023	351216	351410	351603	351796	1989	193
225	2183	2375	2568	2761	2954	3147	3339	3532	3724	3916	193
6	4108	4301	4493	4685	4876	5068	5260	5452	5643	5834	192
7	6026	6217	6408	6599	6790	6981	7172	7363	7554	7744	191
8	7935	8125	8316	8506	8696	8886	9076	9266	9456	9646	190
9	9835	360025	360215	360404	360593	360783	360972	361161	361350	361539	189

Contd...

Table A.21 Contd.

N.	0	1	2	3	4	5	6	7	8	9	D.
230	361728	361917	362105	362294	362482	362671	362859	363048	363236	363424	188
1	3612	3800	3988	4176	4363	4551	4739	4926	5113	5301	188
2	5488	5675	5862	6049	6236	6423	6610	6796	6983	7169	187
3	7356	7542	7729	7915	8101	8287	8473	8659	8845	9030	186
4	9216	9401	9587	9772	9958	370143	370328	370513	370698	370883	185
235	371068	371253	371437	371622	371806	1991	2175	2360	2544	2728	184
6	2912	3096	3280	3464	3647	3831	4015	4198	4382	4565	184
7	4748	4932	5115	5298	5481	5664	5846	6029	6212	6394	183
8	6577	6759	6942	7124	7306	7488	7670	7852	8034	8216	182
9	8398	8580	8761	8943	9124	9306	9487	9668	9849	380030	181
240	380211	380392	380573	380754	380934	381115	381296	381476	381656	381837	181
1	2017	2197	2377	2557	2737	2917	3097	3277	3456	3636	180
2	3815	3995	4174	4353	4533	4712	4891	5070	5249	5428	179
3	5606	5785	5964	6142	6321	6499	6677	6856	7034	7212	178
4	7390	7566	7746	7923	8101	8279	8456	8634	8811	8989	178
245	9166	9343	9520	9698	9875	390051	390228	390405	390582	390759	177
6	390935	391112	391288	391464	391641	1817	1993	2169	2345	2521	176
7	2697	2873	3048	3224	3400	3575	3751	3926	4101	4277	176
8	4452	4627	4802	4977	5152	5326	5501	5676	5850	6025	175
9	6199	6374	6548	6722	6896	7071	7245	7419	7592	7766	174
250	397940	398114	398287	398461	398634	398808	398981	399154	399328	399501	173
1	9674	9847	400020	400192	400365	400338	400711	400883	401056	401228	173
2	401401	401573	1745	1917	2089	2261	2433	2605	2777	2949	172
3	3121	3292	3464	3635	3807	3978	4149	4320	4492	4663	171
4	4834	5005	5176	5346	5517	5688	5858	6029	6199	6370	171
255	6540	6710	6881	7051	7221	7391	7561	7731	7901	8070	170
6	8240	8410	8579	8749	8918	9087	9257	9426	9595	9764	169
7	9933	410102	410271	410440	410609	410777	410964	411114	411283	411451	169
8	411620	1788	1956	2124	2293	2461	2629	2796	2964	3132	168
9	3300	3467	3635	3803	3970	4137	4305	4472	4639	4806	167
260	414973	415140	415307	415474	415641	415808	415974	416141	416308	416474	167
1	6641	6807	6973	7139	7306	7472	7638	7804	7970	8135	166
2	8301	8467	8633	8798	8964	9129	9295	9460	9625	9791	165
3	9956	420121	420286	420451	420616	420781	420945	421110	421275	421439	165
4	421604	1768	1933	2097	2261	2426	2590	2754	2918	3082	164
265	3246	3410	3574	3237	3901	4065	4228	4392	4555	4718	164
6	4882	5045	5208	5371	5534	5697	5860	6023	6186	6349	163
7	6511	6674	6836	6999	7161	7324	7486	7648	7811	7973	162
8	8135	8297	8459	8621	8783	8944	9106	9268	9429	9591	162
9	9752	9914	430075	430236	430398	430559	430720	430881	431042	431203	161
270	431364	431525	431685	431846	432007	432167	432328	432488	432649	432809	161
1	2959	3130	3290	3450	3610	3770	3930	4090	4249	4409	160
2	4569	4729	4888	5048	5207	5367	5526	5685	5844	6004	159
3	6163	6322	6481	6640	6799	6957	7116	7275	7433	7592	159
4	7751	7909	8067	8226	8384	8542	8701	8859	9017	9175	158

Contd...

Table A.21 Contd.

N.	0	1	2	3	4	5	6	7	8	9	D.
275	9333	9491	9648	9806	9964	440122	440279	440437	440594	440752	158
6	440909	441066	441224	441381	441538	1695	1852	2009	2166	2323	157
7	2480	2637	2793	2950	3106	3263	3419	3576	3732	3889	157
8	4045	4201	4357	4513	4669	4825	4981	5137	5293	5449	156
9	5604	5760	5915	6071	6226	6382	6537	6692	6848	7003	155
280	447158	447313	447468	447623	447778	447933	448088	448242	448397	443552	155
1	8706	8861	9015	9170	9324	9478	9633	9787	9941	450095	154
2	450249	450403	450557	450711	450865	451018	451172	451326	451479	1633	154
3	1786	1940	2093	2247	2400	2553	2703	2859	3012	3165	153
4	3318	3471	3624	3777	3930	4082	4235	4387	4540	4692	153
285	4845	4997	5150	5302	5454	5606	5758	5910	6062	6214	152
6	6366	6518	6670	6821	6973	7125	7276	7428	7579	7731	152
7	7882	8033	8184	8336	8487	8638	8789	8940	9091	9242	151
8	9392	9543	9694	9345	9995	460146	460296	460447	460397	460748	151
9	460898	461048	461198	461348	461499	1649	1799	1948	2098	2248	150
290	462398	462548	462697	462847	462997	463146	463296	463445	463594	463744	150
1	3893	4042	4191	4340	4490	4639	4788	4936	5335	5234	149
2	5383	5532	5680	5329	5977	6126	6274	6423	6571	6719	149
3	6868	7016	7164	7312	7460	7608	7756	7904	8052	8200	148
4	8347	8495	8643	8790	8938	9085	9233	9380	9527	9675	148
295	9822	9969	470116	470263	470410	470557	470704	470851	470998	471145	147
6	471292	471438	1585	1732	1878	2025	2171	2318	2464	2610	146
7	2756	2903	3049	3195	3341	3487	3633	3779	3925	4071	146
8	4216	4362	4508	4653	4799	4944	5090	5235	5381	5526	146
9	5671	5816	5962	6107	6252	6397	6542	6687	6832	6976	145
300	477121	477266	477411	477555	477700	477844	477989	478133	478278	478422	145
1	8566	8711	8335	8999	9143	9287	9431	9575	9719	9863	144
2	480007	480151	480294	480438	480582	480725	480869	481012	481156	481299	144
3	1443	1586	1729	1872	2016	2159	2302	2445	2588	2731	143
4	2874	3016	3159	3302	3445	3587	3730	3872	4015	4157	143
305	4300	4442	4585	4727	4869	5011	5153	5295	5437	5579	142
6	5721	5863	6005	6147	6239	6420	6572	6714	6355	6997	142
7	7138	7280	7421	7593	7704	7345	7986	8127	8269	8410	141
8	8551	8392	8833	8974	9114	9235	9396	9537	9677	9818	141
9	9958	490099	490239	490380	490520	490661	490801	490941	491081	491222	140
310	491362	491502	491642	491782	491922	492062	492201	492341	492481	492621	140
1	2760	2990	3040	3179	3319	3438	3597	3737	3876	4015	139
2	4155	4294	4433	4572	4711	4330	4939	5128	5267	5406	139
3	5544	5683	5822	5960	6999	6238	6376	6515	6653	6791	139
4	6930	7068	7206	7344	7483	7621	7759	7897	8035	8173	138
315	8311	8448	8586	8724	8862	8999	9137	9275	9412	9550	138
6	9687	9824	9962	500099	500236	500374	500511	500648	500785	500922	137
7	501059	501196	501333	1470	1607	1744	1880	2017	2154	2291	137
8	2427	2564	2700	2837	2973	3109	3246	3382	3518	3655	136
9	3791	3927	4063	4199	4335	4471	4607	4743	4878	5014	136

Contd...

Table A.21 Contd.

N.	0	1	2	3	4	5	6	7	8	9	D.
320	505150	505286	505421	505557	505693	505828	505964	506099	506234	506370	136
1	6505	6640	6776	6911	7046	7181	7316	7451	7586	7721	135
2	7856	7991	8126	8260	8395	8530	8664	8799	8934	9068	135
3	9203	9337	9471	9606	9740	9874	510009	510143	510277	510411	134
4	510545	510679	510813	510947	511081	511215	1349	1482	1616	1750	134
325	1883	2017	2151	2234	2418	2551	2684	2818	2951	3084	133
6	3218	3351	3484	3617	3750	3883	4016	4149	4282	4415	133
7	4548	4681	4813	4946	5079	5211	5344	5476	5609	5741	133
8	5874	6006	5139	6271	6403	6535	6668	6800	6932	7064	132
9	7196	7328	7460	7592	7724	7855	7987	8119	8251	8382	132
330	518514	518646	518777	518909	519040	519171	519303	519434	519566	519697	131
1	9828	9959	520090	520221	520353	520484	520615	520745	520876	521007	131
2	521138	521269	1400	1530	1661	1792	1922	2053	2183	2314	131
3	2444	2575	2705	2835	2966	3096	3226	3356	3486	3616	130
4	3746	3876	4006	4136	4266	4396	4526	4656	4785	4915	130
335	5045	5174	5304	5434	5563	5693	5822	5951	6081	6210	129
6	6339	6469	6598	6727	6856	6985	7114	7243	7372	7501	129
7	7630	7759	7888	8016	8145	8274	8402	8531	8660	8788	129
8	8917	9045	9174	9302	9430	9559	9687	9815	9943	530072	128
9	530200	530328	530456	530581	530712	530840	530968	531086	531223	1351	128
340	531479	531607	531734	531862	531990	532117	532245	532372	532500	532627	128
1	2754	2882	3009	3136	3264	3391	3518	3645	3772	3899	127
2	4026	4153	4280	4407	4534	4661	4787	4914	5041	5167	127
3	5294	5421	5547	5674	5800	5927	6053	6180	6306	6432	126
4	6558	6685	6811	6937	7063	7189	7315	7441	7567	7693	126
345	7819	7945	8071	8197	8322	8448	8574	8699	8825	8951	126
6	9076	9202	9327	9452	9578	9703	9829	9954	540079	540204	125
7	540329	540455	540580	540705	540830	540955	541080	541205	1330	1454	125
8	1579	1704	1829	1953	2078	2203	2327	2452	2576	2701	125
9	2825	2950	3074	3199	3323	3447	3571	3696	3820	3944	124
350	544068	544192	544316	544440	544564	544688	544812	544936	545060	545183	124
1	5307	5431	5555	5678	5802	5925	6049	6172	6296	6419	124
2	6543	6686	6789	6913	7036	7159	7282	7405	7529	7652	123
3	7775	7898	8021	8144	8267	3389	8512	8635	8758	8881	123
4	9003	9126	9249	9371	9494	9616	9739	9861	9984	550106	123
355	550228	550351	550473	550595	550717	550840	550962	551084	551206	1328	122
6	1450	1572	1694	1816	1938	2060	2181	2303	2425	2547	122
7	2668	2790	2911	3033	3155	3276	3398	3519	3640	3762	122
8	3883	4004	4126	4247	4368	4489	4610	4731	4852	4973	121
9	5094	5215	5336	5457	5578	5699	5820	5940	6061	6182	121
360	556303	556423	556544	556664	556785	556905	557026	557146	557267	557387	120
1	7507	7627	7748	7868	7988	8108	8228	8349	8469	8589	120
2	8709	8829	8948	9068	9188	9308	9428	9548	9667	9787	120
3	9907	560026	560146	560265	560385	560504	560624	560743	560863	560982	119
4	561101	1221	1340	1459	1578	1698	1817	1936	2055	2174	119

Contd...

Table A.21 Contd.

N.	0	1	2	3	4	5	6	7	8	9	D.
365	2293	2412	2531	2650	2769	2887	3006	3125	3244	3362	119
6	3481	3600	3718	3837	3955	4074	4192	4311	4429	4548	119
7	4666	4784	4903	5021	5139	5257	5376	5494	5612	5730	118
8	5848	5966	6084	6202	6320	6437	6555	6673	6791	6909	118
9	7026	7144	7262	7379	7497	7614	7732	7849	7967	8084	118
370	568202	568319	568436	568554	568671	568788	568905	569023	569140	569257	117
1	9374	9491	9608	9725	9842	9959	570076	570193	570309	570426	117
2	570543	570660	570776	570893	571010	571126	1243	1359	1476	1592	117
3	1709	1825	1942	2058	2174	2291	2407	2523	2639	2755	116
4	2872	2988	3104	3220	3336	3452	3568	3684	3800	3915	116
375	4031	4147	4263	4379	4494	4610	4726	4841	4957	5072	116
6	5188	5303	5419	5534	5650	5765	5880	5996	6111	6226	115
7	6341	6457	6572	6687	6802	6917	7032	7147	7252	7377	115
8	7492	7607	7722	7836	7951	8066	8181	8295	8410	8525	115
9	8639	8754	8868	8983	9097	9212	9326	9441	9555	9669	114
380	579784	579898	580012	580126	580241	580355	580469	580583	580697	580811	114
1	580925	581039	1153	1267	1381	1495	1608	1722	1836	1950	114
2	2063	2177	2291	2404	2518	2631	2745	2858	2972	3085	114
3	3199	3312	3426	3539	3652	3765	3879	3992	4105	4218	113
4	4331	4444	4557	4670	4783	4896	5009	5122	5235	5348	113
385	5461	5574	5686	5799	5912	6024	6137	6250	6362	6475	113
6	6587	6700	6812	6925	7037	7149	7262	7374	7486	7599	112
7	7711	7823	7935	8047	8160	8272	8384	8496	8608	8720	112
8	8832	8944	9056	9167	9279	9391	9503	9615	9726	9838	112
9	9950	590061	590173	590284	590396	590507	590619	590730	590842	590953	112
390	591065	591176	591287	591399	591510	591621	591732	591843	591955	592066	111
1	2177	2288	2399	2510	2621	2732	2843	2954	3064	3175	111
2	3286	3397	3508	2518	3729	3840	3950	4061	4171	4282	111
3	4393	4503	4614	4724	4834	4945	5055	5165	5276	5380	110
4	5496	5606	5717	5827	5937	6047	6157	6267	6377	6487	110
395	6597	6707	6817	6927	7037	7146	7256	7366	7476	7586	110
6	7695	7805	7914	8024	8134	8243	8353	8462	8572	8681	110
7	8791	8900	9009	9119	9228	9337	9446	9556	9665	9774	109
8	9883	9992	600101	600210	600319	600428	600537	600646	600755	600864	109
9	600973	601082	1191	1299	1408	1517	1625	1734	1843	1951	109
400	602060	602169	602277	602386	602494	602603	602711	602819	602928	603036	108
1	3144	3253	3361	3469	3577	3686	3794	3902	4010	4118	108
2	4226	4334	4442	4550	4658	4766	4874	4932	5089	5197	108
3	5305	6513	5521	5628	5736	5844	5951	6059	6166	6274	108
4	6381	6489	6596	6704	6811	6919	7026	7133	7241	7348	107
405	7455	7562	7669	7777	7884	7991	8098	8205	8312	8419	107
6	8526	8633	8740	8847	8954	9081	9167	9274	9381	9488	107
7	9594	9701	9808	9914	610021	610128	610234	610341	610447	610554	107
8	610660	610767	610873	610979	1086	1192	1298	1405	1511	1617	106
9	1723	1829	1936	2042	2148	2254	2360	2466	2572	2678	106

Contd...

Table A.21 Contd.

N.	0	1	2	3	4	5	6	7	8	9	D.
410	612784	612890	612996	613102	613207	613313	613419	613525	613630	613736	106
1	3842	3947	4053	4159	4264	4370	4475	4581	4686	4792	106
2	4897	5003	5108	5213	5319	5424	5529	5634	5740	5845	105
3	5950	6055	6160	6205	6370	6476	6581	6686	6790	6895	105
4	7000	7105	7210	7315	7420	7525	7629	7734	7839	7943	105
415	8048	8153	8257	8362	8466	8571	8676	8780	8884	8989	105
6	9093	9198	9302	9406	9511	9315	9719	9824	9928	620032	104
7	620136	620240	620344	620448	620352	620656	620760	620864	620968	1072	104
8	1176	1280	1304	1488	1592	1695	1799	1903	2007	2110	104
9	2214	2318	2421	2525	2628	2732	2835	2939	3042	3146	104
420	623249	623353	623456	623559	623663	623766	623869	623973	624076	624179	103
1	4282	4385	4488	4591	5895	4798	4901	5004	5107	5210	103
2	5312	5415	5518	5621	5724	5827	5929	6032	6135	6238	103
3	6340	6443	6546	6648	6751	6853	6956	7058	7161	7263	103
4	7366	7468	7571	7673	7775	7878	7980	8082	8185	8287	102
425	8389	8491	8593	8695	8797	8900	9002	9104	9206	9308	102
6	9410	9512	9613	9715	9817	9919	630021	630123	630224	630326	102
7	630428	630530	630631	630733	630035	630936	1038	1139	1241	1342	102
8	1444	1545	1647	1748	1849	1951	2052	2153	2255	2356	101
9	2457	2559	2660	2761	2862	2963	3064	3165	3266	3367	101
430	633468	633569	633670	633771	633872	633973	634074	634175	634276	634376	101
1	4477	4578	4679	4779	4880	4981	5081	5182	5283	5383	101
2	5484	5584	5695	5785	5886	5986	6087	6187	6287	6388	100
3	6488	6588	6688	6789	6889	6989	7089	7189	7290	7390	100
4	7490	7590	7690	7790	7890	7990	8090	8190	8290	8389	100
435	8489	8589	8689	8789	8888	8988	9088	9188	9287	9387	100
6	9486	9586	9686	9785	9889	9984	640084	640183	640283	640382	99
7	640481	640581	640680	640779	640879	640978	1077	1177	1276	1375	99
8	1474	1573	1672	1771	1871	1970	2069	2168	2267	2366	99
9	2465	2563	2662	2761	2860	2959	3058	3156	3255	3354	99
440	643453	643551	643650	643749	643847	643946	644044	644143	644242	644340	98
1	4439	4537	4636	4734	4832	4931	5029	5127	5226	5324	98
2	5422	5521	5619	5717	5815	5913	6011	6110	6208	6306	98
3	6404	6502	6600	6698	6796	6894	6992	7089	7137	7285	98
4	7383	7481	7579	7676	7774	7872	7969	8067	8165	8262	98
445	8360	8458	8555	8653	8750	8848	8945	9043	9140	9237	97
6	9335	9432	9530	9627	9724	9821	9919	650016	650113	650210	97
7	650308	650405	650502	650599	650696	650793	650890	0987	1084	1181	97
8	1278	1375	1472	1569	1666	1762	1859	1956	2053	2150	97
9	2246	2343	2440	2536	2633	2730	2826	2923	3019	3116	97
450	653213	653309	653405	653502	653598	653695	653791	653888	653984	654080	96
1	4177	4273	4369	4465	4562	4658	4754	4850	4946	5042	96
2	5138	5235	5331	5427	5523	5619	5715	5810	5906	6002	96
3	6098	6194	6290	6386	6482	6577	6673	6769	6864	6960	96
4	7056	7152	7247	7343	7438	7534	7629	7725	7820	7919	96

Contd...

Table A.21 Contd.

N.	0	1	2	3	4	5	6	7	8	9	D.
455	8011	8107	8202	8298	8393	8488	8584	8679	8774	8870	95
6	8965	9060	9155	9250	9346	9441	9536	9631	9726	7821	95
7	9916	660011	660106	660201	660296	660391	660486	660581	660676	660771	95
8	660865	0960	1055	1150	1245	1339	1434	1529	1623	1718	95
9	1813	1907	2002	2096	2191	2286	2380	2475	2569	2663	95
460	662758	662852	662947	663041	663135	663230	663324	663418	663512	663607	94
1	3701	3795	3889	3983	4078	4172	4266	4360	4454	4548	94
2	4642	4736	4830	4924	5018	5112	5206	5299	5393	5487	94
3	5581	5675	5769	5862	5956	6050	6143	6237	6331	6424	94
4	6518	6612	6705	6799	6892	6986	7079	7173	7266	7360	94
465	7453	7546	7640	7733	7826	7920	8013	8106	8199	8293	93
6	8386	8479	9872	8665	8759	8852	8945	9038	9131	9224	93
7	9317	9410	9503	9596	9689	9782	9875	9967	670060	670153	93
8	670246	670339	670431	670524	670617	670710	670802	670895	0988	1080	93
9	1173	1265	1358	1451	1543	1636	1728	1821	1913	2005	93
470	672098	672190	672283	672375	672467	672560	672652	672744	672836	672929	92
1	3021	3113	3205	3297	3390	3482	3574	3666	3758	3850	92
2	3942	4034	4126	4218	4310	4402	4494	4586	4677	4769	92
3	4861	4953	5045	5137	5228	5320	5412	5503	5595	5687	92
4	5778	5870	5962	6053	6145	6236	6328	6419	6511	6602	92
475	6694	6785	6876	6968	7059	7151	7242	7333	7424	7516	91
6	7607	7698	7789	7881	7972	8033	8154	8245	8336	8427	91
7	8518	8609	8700	8791	8882	8973	9064	9155	9246	9337	91
8	9428	9519	9610	9700	9791	9882	9973	680063	680154	680245	91
9	680336	680426	680517	680607	680698	680789	680879	0970	1060	1151	91
480	681241	681332	681422	681513	681603	681693	681784	681874	681964	682055	90
1	2145	2235	2326	2416	2306	2596	2686	2777	2867	2957	90
2	3047	3137	3227	3317	3407	3497	3587	3677	3767	3857	90
3	3947	4037	4127	4217	4307	4396	4486	4576	4666	4756	90
4	4885	4935	5025	5114	5204	5294	5383	5473	5563	5652	90
485	5742	5831	5921	6010	6100	6189	6279	6368	6458	6547	89
6	6636	6726	6815	6904	6994	7033	7172	7261	7351	7440	89
7	7529	7618	7707	7796	7886	7975	8064	8153	8242	8331	89
8	8420	8509	8598	8687	8776	8865	8953	9042	9131	9220	89
9	9309	9398	9486	9575	9664	9753	9841	9930	690019	690107	89
490	690196	690285	690373	690462	690550	690639	690728	690816	690905	690993	89
1	1081	1170	1258	1347	1435	1524	1612	1700	1789	1877	88
2	1965	2053	2142	2230	2318	2406	2494	2383	2671	2759	88
3	2847	2935	3023	3111	3199	3287	3375	3463	3551	3639	88
4	3727	3815	3903	3991	4078	4166	4254	4342	4430	4517	88
495	4605	4693	4781	4868	4956	5044	5131	5219	5307	5394	88
6	5482	5569	5657	5744	5832	5919	6007	6094	6182	6269	87
7	6356	6444	6531	6618	6706	6793	6880	6968	7055	7142	87
8	7229	7317	7404	7491	7578	7665	7752	7839	7926	8014	87
9	8101	8188	8275	8362	8449	8535	8622	8709	8796	8883	87

Contd...

Table A.21 Contd.

N.	0	1	2	3	4	5	6	7	8	9	D.
500	698970	699057	699144	699231	699317	699404	699491	699578	699664	699751	87
1	9338	9924	700011	700098	700184	703271	700358	700444	700331	700617	87
2	700704	700790	0877	0963	1030	1136	1222	1309	1395	1482	86
3	1568	1654	1741	1827	1913	1999	2036	2172	2258	2344	86
4	2431	2517	2603	2689	2775	2861	2947	3033	3119	3205	86
505	3291	3377	3463	3549	3635	3721	3807	3893	3979	4065	86
6	4151	4236	4322	4408	4494	4579	4665	4751	4837	4922	86
7	5008	5094	5179	5265	5350	5436	5522	5607	5693	5778	86
8	5864	5949	6035	6120	6206	6291	6376	6462	6547	6632	85
9	6718	6803	6888	6974	7059	7144	7229	7315	7400	7485	85
510	707570	707655	707740	707826	707911	707996	708081	708166	708251	708336	85
1	8421	8506	8591	8676	8761	8846	8931	9015	9100	9185	85
2	9270	9355	9440	9524	9609	9694	9779	9863	9948	710033	85
3	710117	710202	710287	710371	710456	710540	710625	710710	710794	0879	85
4	0963	1048	1132	1217	1301	1385	1470	1554	1639	1723	84
515	1807	1892	1976	2060	2144	2229	2313	2397	2481	2566	84
6	2650	2734	2818	2902	2986	3070	3154	3238	3323	3407	84
7	3491	3575	3659	3742	3826	3910	3994	4078	4162	4246	84
8	4330	4414	4497	4581	4665	4749	4833	4916	5000	5084	84
9	5167	5251	5335	5418	5502	5586	5669	5753	5836	5920	84
520	716003	716087	716170	716254	716337	716421	716504	716588	716671	716754	83
1	6838	6921	7004	7088	7171	7254	7338	7421	7504	7587	83
2	7671	7754	7837	7920	8003	8086	8169	8253	8336	8419	83
3	8502	8585	8668	8751	8834	8917	9000	9083	9165	9248	83
4	9331	9414	9497	9580	9663	9745	9828	9911	9994	720077	83
525	720159	720242	720325	720407	720490	720573	720655	720738	720821	0903	83
6	0986	1068	1151	1233	1316	1398	1481	1563	1646	1728	82
7	1811	1893	1975	2058	2140	2222	2305	2387	2469	2552	82
8	2634	2716	2798	2881	2963	3045	3127	3209	3291	3374	82
9	3456	3538	3620	3702	3784	3866	3948	4030	4112	4194	82
530	724276	724358	724440	724522	724604	724685	724767	724849	724931	725013	82
1	5095	5176	5258	5340	5422	5503	5585	5667	5748	5830	82
2	5912	5993	6075	6156	6238	6320	6401	6483	6564	6646	82
3	6727	6809	6890	6972	7033	7134	7216	7297	7379	7460	81
4	7541	7623	7704	7785	7866	7948	8029	8110	8191	8273	81
535	8354	8435	8516	8597	8678	8759	8841	8922	9003	9084	81
6	9165	9246	9327	9408	9439	9570	9651	9732	9813	9893	81
7	9974	730035	730136	730217	730298	730378	730459	730340	730621	730702	81
8	730782	0863	0944	1024	1105	1186	1266	1347	1428	1508	81
9	1589	1669	1750	1830	1911	1991	2072	2152	2233	2313	81
540	732394	732474	732555	732635	732715	732796	732876	732956	733037	733117	80
1	3197	3278	3358	3438	3518	3598	3679	3759	3839	3919	80
2	3999	4079	4160	4240	4320	4400	4480	4560	4640	4720	80
3	4800	4880	4960	5040	5120	5200	5279	5359	5439	5519	80
4	5599	5679	5759	5838	5918	5998	6078	6157	6237	6317	80

Contd...

Table A.21 Contd.

N.	0	1	2	3	4	5	6	7	8	9	D.
545	6397	6476	6556	6635	6715	6795	6874	6954	7034	7113	80
6	7193	7272	7332	7431	7511	7390	7670	7749	7829	7908	79
7	7987	8037	8146	8225	8805	8384	8463	8543	8622	8701	79
8	8781	8860	8939	9018	9997	9177	9256	9335	9414	9493	79
9	9572	9651	9731	9810	9889	9968	740047	740126	740205	740284	79
550	740363	740442	740521	740500	740678	740757	740836	740915	740994	741073	79
1	1152	1260	1309	1383	1467	1546	1624	1703	1782	1860	79
2	1939	2018	2095	2175	2254	2332	2411	2439	2568	2647	79
3	2725	2804	2892	2961	3039	3118	3196	3275	3353	3431	78
4	3510	3588	3667	3745	3823	3902	3980	4058	4136	4215	78
555	4293	4371	4449	4528	4606	4684	4762	4840	4919	4997	78
6	5075	5153	5231	5309	5887	5465	5543	5621	5699	5777	78
7	5855	5933	6011	6039	6167	6245	6323	6401	6479	6556	78
8	6634	6712	6790	6868	6945	7023	7101	7179	7256	7334	78
9	7412	7489	7567	7645	7722	7800	7878	7955	8033	8110	78
560	748188	748266	748343	748421	748498	748576	748653	748731	748808	748885	77
1	8963	9040	9118	9195	9272	9330	9427	9504	9582	9659	77
2	9736	9814	9391	9968	756245	750123	750200	750277	750354	750431	77
3	750508	750586	750363	750740	0817	0394	0971	1048	1125	1202	77
4	1279	1356	1433	1510	1587	1664	1741	1818	1895	1972	77
565	2048	2125	2202	2279	2356	2433	2509	2586	2663	2740	77
6	2816	2893	2970	3047	3123	3200	3277	3353	3430	3506	77
7	3583	3660	3736	3813	3839	3966	4042	4119	4195	4272	77
8	4348	4425	4501	4578	4654	4730	4807	4883	4960	5036	76
9	5112	5189	5265	5341	5417	5494	5570	5646	5722	5799	76
570	755875	755951	756027	756103	756180	756256	756332	756408	756484	756560	76
1	6636	6712	6788	6864	6940	7016	7092	7168	7244	7320	76
2	7396	7472	7548	7624	7700	7775	7851	7927	8003	8079	76
3	8155	8230	8306	8382	8458	8533	8609	8685	8761	8836	76
4	8912	8988	9063	9139	9214	9290	9366	9441	9517	9592	76
575	9668	9743	9819	9894	9970	760045	760121	760196	760272	760347	75
6	760422	760498	760573	760649	760724	0799	0875	0950	1025	1101	75
7	1176	1251	1326	1402	1477	1552	1627	1702	1778	1853	75
8	1928	2093	2078	2153	2228	2303	2378	2453	2529	2604	75
9	2679	2754	2829	2904	2978	3053	3128	3203	3278	3353	75
580	763428	763503	763578	763653	763727	763802	763877	763952	764027	764101	75
1	4176	4251	4326	4400	4475	4550	4624	4699	4774	4848	75
2	4923	4998	5072	5147	5221	5296	5370	5445	5520	5594	75
3	5669	5743	5818	5892	5966	6041	6115	6190	6264	6338	74
4	6413	6487	6562	6636	6710	6785	6859	6933	7007	7082	74
585	7156	7230	7304	7379	7453	7527	7601	7675	7749	7823	74
6	7898	7972	8046	8120	8194	8268	8342	8416	8490	8564	74
7	8638	8712	8786	8860	8934	9008	9082	9156	9230	9303	74
8	9377	9451	9525	9599	9673	9746	9820	9894	9968	770042	74
9	770115	770189	770263	770336	770410	770484	770557	770631	770705	0778	74

Contd...

Table A.21 Contd.

N.	0	1	2	3	4	5	6	7	8	9	D.
590	770852	770926	770999	771073	771146	771220	771293	771367	771440	771514	74
1	1587	1661	1734	1808	1881	1955	2028	2102	2175	2248	73
2	2322	2395	2468	2542	2615	2688	2762	2835	2908	2981	73
3	3055	3128	3201	3274	3348	3421	3494	3567	3640	3713	73
4	3786	3860	3933	4006	4079	4152	4225	4298	4371	4444	73
595	4517	4590	4663	4736	4809	4882	4955	5028	5100	5173	73
6	5246	5319	5392	5465	5538	5610	5683	5756	5829	5902	73
7	5974	6047	6120	6193	6265	6338	6411	6483	6556	6629	73
8	6701	6774	6846	6919	6992	7064	7137	7209	7282	7354	73
9	7427	7499	7572	7644	7717	7789	7862	7934	8006	8079	72
600	778151	778224	778296	778368	778441	778513	778585	778658	778730	778802	72
1	8874	8947	9019	9091	9163	9236	9308	9380	9452	9524	72
2	9396	9669	9741	9813	9885	9957	780029	780101	780173	780245	72
3	780317	780389	780461	780533	780605	780677	0749	0821	0893	0965	72
4	1037	1109	1181	1253	1324	1396	1468	1540	1612	1684	72
605	1755	1827	1899	1971	2042	2114	2186	2258	2329	2401	72
6	2473	2544	2616	2688	2759	2831	2902	2974	3046	3117	72
7	3189	3260	3332	3403	3475	3546	3618	3689	3761	3832	71
8	3904	3975	4046	4118	4189	4261	4332	4403	4475	4546	71
9	4617	4689	4760	4831	4902	4974	5045	5116	5187	5259	71
610	785330	785401	785472	785543	785615	785686	785757	785828	785899	785970	71
1	6041	6112	6183	6254	6325	6396	6467	6538	6609	6680	71
2	6751	6822	6893	6964	7035	7106	7177	7248	7319	7390	71
3	7460	7531	7602	7673	7744	7815	7885	7956	8027	8098	71
4	8168	8239	8310	8381	8451	8522	8593	8663	8734	8804	71
615	8875	8946	9016	9087	9157	9228	9299	9369	9440	9510	71
6	9581	9651	9722	9792	9863	9933	790004	790074	790144	790215	70
7	790285	790356	790426	790496	790567	790637	0707	0778	0848	0918	70
8	0988	1059	1129	1199	1269	1340	1410	1480	1550	1620	70
9	1691	1761	1831	1901	1971	2041	2111	2181	2252	2322	70
620	792392	792462	792532	792602	792672	792742	792812	792882	792952	793022	70
1	3092	3162	3231	3301	3371	3441	3511	3581	3651	3721	70
2	3790	3860	3930	4000	4070	4139	4209	4279	4349	4418	70
3	4488	4558	4627	4697	4767	4836	4906	4976	5045	5115	70
4	5185	5254	5324	5393	5463	5532	5602	5672	5741	5811	70
625	5880	5949	6019	6088	6158	6227	6297	6366	6436	6505	69
6	6574	6644	6713	6782	6852	6921	6990	7060	7129	7198	69
7	7268	7337	7406	7475	7545	7614	7683	7752	7821	7890	69
8	7960	8029	8098	8167	8236	8305	8374	8443	8513	8582	69
9	8651	8720	8789	8858	8927	8996	9065	9134	9203	9272	69
630	799341	799409	799478	799547	799616	799685	799754	799823	799892	799961	69
1	800029	800098	800167	800236	800305	800373	800442	800511	800580	800648	69
2	0717	0786	0854	0923	0992	1061	1129	1198	1266	1335	69
3	1404	1472	1541	1609	1678	1747	1815	1884	1952	2021	69
4	2089	2158	2226	2295	2363	2432	2500	2568	2637	2705	68

Contd...

Table A.21 Contd.

N.	0	1	2	3	4	5	6	7	8	9	D.
635	2774	2842	2910	2979	3047	3116	3184	3252	3321	3389	68
6	3457	3525	3594	3662	3730	3798	3867	3935	4003	4071	68
7	4139	4208	4276	4344	4412	4480	4548	4616	4685	4753	68
8	4821	4889	4957	5025	5093	5161	5229	5297	5365	5433	68
9	5501	5569	5637	5705	5773	5841	5908	5976	6044	6112	68
640	806180	806248	806316	806384	806451	806519	806587	806655	806723	806790	68
1	6858	6926	6994	7061	7129	7197	7264	7332	7400	7467	68
2	7535	7603	7670	7738	7806	7873	7941	8008	8076	8143	68
3	8211	8279	8346	8414	8481	8549	8616	8684	8751	8818	67
4	8886	8953	9021	9088	9156	9223	9290	9358	9425	9492	67
645	9560	9627	9694	9762	9829	9896	9964	810031	810098	810165	67
6	810233	810300	810367	810434	810501	810569	810636	0703	0770	0837	67
7	0904	0971	1039	1106	1173	1240	1307	1374	1441	1508	67
8	1575	1642	1709	1776	1843	1910	1977	2044	2111	2178	67
9	2245	2312	2379	2445	2512	2579	2646	2713	2780	2847	67
650	812913	812980	813047	813114	813181	813247	813314	813381	813448	813514	67
1	3581	3648	3714	3781	3848	3914	3981	4048	4114	4181	67
2	4248	4314	4381	4447	4514	4581	4647	4714	4780	4847	67
3	4913	4980	5046	5113	5179	5246	5312	5378	5445	5511	66
4	5578	5644	5711	5777	5843	5910	5976	6042	6109	6175	66
655	6241	6308	6374	6440	6506	6573	6639	6705	6771	6838	66
6	6904	6970	7036	7102	7109	7235	7301	7367	7433	7499	66
7	7565	7631	7698	7764	7880	7896	7962	8028	8094	8160	66
8	4226	8292	8358	8424	8490	8556	8622	8688	8754	8820	66
9	8885	8951	9017	9083	9149	9215	9281	9346	9412	9478	66
660	819544	819610	819676	819741	819807	819873	819939	820004	820070	820136	66
1	820201	820267	820333	820399	820464	820530	820595	0661	0727	0792	66
2	0858	0924	0989	1055	1120	1186	1251	1317	1382	1448	66
3	1514	1579	1645	1710	1775	1841	1905	1972	2037	2103	65
4	2168	2233	2299	2364	2430	2495	2560	2626	2691	2756	65
665	2822	2887	2952	3018	3083	3148	3213	3279	3344	3409	65
6	3474	3539	3605	3670	3735	3800	3865	3930	3996	4061	65
7	4126	4191	4256	4321	4386	4451	4516	4581	4646	4711	65
8	4776	4841	4906	4971	5036	5101	5166	5231	5296	5361	65
9	5426	5491	5556	5621	5686	5751	5815	5880	5945	6010	65
670	826075	826140	826204	826269	826334	826399	826464	826528	826593	826658	65
1	6723	6787	6852	6917	6981	7046	7111	7175	7240	7305	65
2	7369	7434	7499	7563	7628	7692	7767	7821	7886	7951	65
3	8015	8080	8144	8209	8273	8338	8402	8457	8531	8595	64
4	8660	8724	8789	8853	8918	8982	9046	9111	9175	9239	64
675	9304	9368	9432	9497	9561	9625	9690	9754	9818	9882	64
6	9947	830011	830075	830139	830204	830268	830332	830396	830460	830525	64
7	830589	0653	0717	0781	0845	0909	0973	1037	1102	1166	64
8	1230	1294	1358	1422	1486	1550	1614	1678	1742	1806	64
9	1870	1934	1998	2062	2126	2189	2253	2317	2381	2445	64

Contd...

Table A.21 Contd.

N.	0	1	2	3	4	5	6	7	8	9	D.
680	832509	832573	832637	832700	832764	832828	832892	832956	833020	833083	64
1	3147	3211	3275	3338	3402	3466	3530	3593	3657	3721	64
2	3784	3848	3912	3975	4039	4103	4166	4230	4294	4357	64
3	4421	4484	4548	4611	4675	4739	4802	4866	4929	4993	64
4	5056	5120	5183	5247	5310	5373	5437	5500	5564	5627	63
685	5691	5754	5817	5881	5944	6007	6071	6134	6197	6261	63
6	6324	6387	6451	6514	6577	6641	6704	6767	6830	6894	63
7	6957	7020	7083	7146	7210	7273	7336	7399	7462	7525	63
8	7588	7652	7715	7778	7841	7904	7967	8030	8093	8156	63
9	8219	8282	8345	8408	8471	8534	8597	8660	8723	8786	63
690	838849	838912	838975	839038	839101	839164	839227	839289	839352	839415	63
1	9478	9541	9804	9667	9729	9792	9855	9918	9981	840043	63
2	840106	840169	840232	840294	840357	840420	840482	840545	840608	0671	63
3	0733	0796	0859	0921	0984	1046	1109	1172	1234	1297	63
4	1359	1422	1485	1547	1610	1672	1735	1797	1860	1922	63
695	1985	2047	2110	2172	2235	2297	2360	2422	2484	2547	62
6	2609	2672	2734	2796	2859	2921	2933	3046	3108	3170	62
7	3233	3295	3357	3420	3482	3544	3606	3669	3731	3793	62
8	3855	3918	3930	4042	4104	4166	4229	4291	4353	4415	62
9	4477	4539	4601	4664	4726	4788	4850	4912	4974	5036	62
700	845098	845160	845222	845284	845346	845408	845470	845532	845594	845656	62
1	5718	5780	5842	5904	5966	6028	6090	6151	6213	6275	62
2	6337	6399	6461	6523	6585	6646	6708	6770	6832	6894	62
3	6955	7017	7079	7141	7202	7264	7326	7388	7449	7511	62
4	7573	7634	7696	7758	7819	7881	7943	8004	8066	8128	62
705	8189	8251	8312	8374	8435	8497	8559	8620	8682	8743	62
6	8805	8866	8928	8989	9051	9112	9174	9235	9297	9358	61
7	9419	9481	9542	9604	9665	9726	9788	9849	9911	9972	61
8	850033	850095	850156	850217	850279	850340	850401	850462	850524	850585	61
9	0646	0707	0769	0830	0891	0952	1014	1075	1136	1197	61
710	851258	851320	851381	851442	851503	851564	851625	851686	851747	851809	61
1	1870	1931	1992	2033	2114	2175	2236	2297	2358	2419	61
2	2480	2541	2602	2663	2724	2785	2846	2907	2968	3029	61
3	3090	3150	3211	3272	3333	3394	3455	3516	3577	3637	61
4	3698	3759	3820	3881	3941	4002	4063	4124	4185	4245	61
715	4306	4367	4428	4488	4549	4610	4670	4731	4792	4852	61
6	4913	4974	5034	5095	5156	5216	5277	5337	5398	5459	61
7	5519	5580	5640	5701	5761	5822	5882	5943	6003	6064	61
8	6124	6185	6245	6306	6336	6427	6487	6548	6608	6668	60
9	6729	6789	6850	6910	6970	7031	7091	7152	7212	7272	60
720	857332	857393	857453	857513	857574	857634	857694	857755	857815	857875	60
1	7935	7995	8056	8116	8176	8236	8297	8357	8417	8477	60
2	8337	8597	8657	8718	8778	8838	8898	8958	9018	9078	60
3	9138	9198	9258	9318	9379	9439	9499	9559	9619	9679	60
4	9739	9799	9859	9918	9978	860038	860098	860158	860218	860278	60

Contd...

Table A.21 Contd.

N.	0	1	2	3	4	5	6	7	8	9	D.
725	860338	860398	860458	860518	860578	0637	0697	0757	0817	0877	60
6	0937	0996	1036	1116	1176	1236	1295	1355	1415	1475	60
7	1534	1594	1634	1714	1773	1833	1893	1952	2012	2072	60
8	2131	2191	2231	2310	2370	2430	2489	2549	2608	2668	60
9	2728	2787	2847	2906	2966	3025	3085	3144	3204	3263	60
730	863323	863382	863442	863501	863561	863620	863680	363739	863799	863858	59
1	3917	3977	4036	4096	4155	4214	4274	4333	4392	4452	59
2	4511	4570	4630	4689	4748	4808	4867	4926	4985	5045	59
3	5104	5163	5222	5282	5341	5400	5459	5519	5578	5637	59
4	5696	5755	5814	5874	5933	5992	6051	6110	6169	6228	59
735	6287	6346	6405	6465	6524	6583	6642	6701	6760	6819	59
6	6878	6937	6996	7035	7114	7173	7232	7291	7350	7409	59
7	7467	7526	7585	7644	7703	7762	7821	7880	7939	7998	59
8	8056	8115	8174	8233	8292	8350	8409	8468	8527	8586	59
9	8644	8703	8762	8821	8879	8938	8997	9056	9114	9173	59
740	869232	869290	869349	869408	869466	869525	869584	869642	869701	869760	59
1	9818	9877	9935	9994	870033	870111	870170	870228	870287	870345	59
2	870404	870462	870521	870579	0638	0696	0755	0813	0872	0930	58
3	0989	1047	1106	1164	1223	1281	1339	1398	1456	1515	58
4	1573	1631	1690	1748	1806	1865	1923	1981	2040	2098	58
745	2156	2215	2273	2331	2389	2448	2506	2564	2622	2681	58
6	2739	2797	2855	2913	2972	3030	3088	3146	3204	3262	58
7	3321	3379	3437	3495	3553	3611	3669	3727	3785	3844	58
8	3902	3960	4018	4076	4134	4192	4250	4308	4366	4424	58
9	4482	4540	4598	4656	4714	4772	4830	4888	4945	5003	58
750	875061	875119	875177	875235	875293	875351	875409	875466	875524	875582	58
1	5640	5698	5756	5813	5871	5929	5987	6045	6102	6160	58
2	6218	6276	6333	6391	6449	6507	6564	6622	6680	6737	58
3	6795	6853	6910	6968	7026	7083	7141	7199	7256	7314	58
4	7371	7429	7487	7544	7602	7659	7717	7774	7832	7889	58
755	7947	8004	8062	8119	8177	8234	8292	8349	8407	8664	57
6	8522	8579	8637	8694	6752	8809	8866	8924	8981	9039	57
7	9096	9153	9211	9268	9325	9383	9440	9497	9555	9612	57
8	9669	9726	9784	9841	9898	9956	880013	880070	880127	880185	57
9	880242	880299	880356	880413	880471	880528	0585	0642	0699	0756	57
760	880814	88871	880928	880985	881042	881099	881156	881213	881271	881328	57
1	1385	1442	1499	1556	1613	1670	1727	1784	1841	1898	57
2	1955	2012	2069	2126	2183	2240	2297	2354	2411	2468	57
3	2525	2581	2638	2695	2752	2809	2866	2923	2980	3037	57
4	3093	3150	3207	3264	3321	3377	3434	3491	3548	3605	57
765	3661	3718	3775	3832	3888	3945	4002	4059	4115	4172	57
6	4229	4285	4342	4399	4455	4512	4569	4625	4682	4739	57
7	4795	4852	4909	4965	5022	5078	5135	5192	5248	5305	57
8	5361	5418	5474	5531	5587	5644	5700	5757	5813	5870	57
9	5926	5983	6039	6096	6152	6209	6265	6321	6378	6434	56

Contd...

Table A.21 Contd.

N.	0	1	2	3	4	5	6	7	8	9	D.
770	886491	886547	886604	886660	886716	886773	886829	886885	886942	886998	56
1	7054	7111	7167	7223	7280	7336	7392	7449	7505	7561	56
2	7617	7674	7730	7786	7842	7898	7955	8011	8067	8123	56
3	8179	8236	8292	8348	8404	8460	8516	8573	8629	8685	56
4	8741	8797	8853	8909	8965	9021	9077	9134	9190	9246	56
775	9302	9358	9414	9470	9526	9582	9638	9694	9750	9806	56
6	9862	9918	9974	890030	890086	890141	890197	890253	890309	890365	56
7	890421	890477	890333	0589	0645	0700	0756	0812	0868	0924	56
8	0980	1035	1091	1147	1203	1259	1314	1370	1426	1482	56
9	1537	1593	1649	1705	1760	1816	1872	1928	1983	2039	56
780	892095	892150	892206	892262	892317	892373	892429	892484	892540	892595	56
1	2651	2707	2762	2818	2873	2929	2985	3040	3096	3151	56
2	3207	3262	3318	3373	3429	3484	3540	3595	3651	3706	56
3	3762	3817	3873	3928	3984	4039	4094	4150	4205	4261	55
4	4316	4371	4427	4482	4538	4593	4648	4704	4759	4814	55
785	4870	4925	4980	5036	5091	5146	5201	5257	5312	5367	55
6	5423	5478	5533	5588	5644	5699	5754	5809	5864	5920	55
7	5975	6030	6085	6140	6195	6251	6306	6361	6416	6471	55
8	6526	6581	6636	6692	6747	6802	6857	6912	6967	7022	55
9	7077	7132	7187	7242	7297	7352	7407	7462	7517	7572	55
790	897627	897682	897737	897792	897847	897902	897957	898012	898067	898122	55
1	8176	8231	8286	8341	8396	8451	8506	8561	8615	8670	55
2	8725	8780	8835	8890	8944	8999	9054	9109	9164	9218	55
3	9273	9328	9383	9437	9492	9547	9602	9656	9711	9766	55
4	9821	9875	9930	9985	900039	900094	900149	900203	900258	900312	55
795	900367	900422	900476	900531	0586	0640	0695	0749	0804	0859	55
6	0913	0968	1022	1077	1131	1186	1240	1295	1349	1404	55
7	1458	1513	1567	1622	1676	1731	1785	1840	1894	1948	54
8	2003	2057	2112	2166	2221	2275	2329	2384	2438	2492	54
9	2547	2601	2655	2710	2764	2818	2873	2927	2981	3036	54
800	903090	903144	903199	903253	903307	903361	903416	903470	903524	903578	54
1	3633	3687	3741	3795	3849	3904	3958	4012	4066	4120	54
2	4174	4229	4283	4337	4391	4445	4499	4553	4607	4661	54
3	4716	4770	4824	4878	4932	4986	5040	5094	5148	5202	54
4	5256	5310	5364	5418	5472	5526	5580	5634	5688	5742	54
805	5796	5850	5904	5958	6012	6066	6119	6173	6227	6281	54
6	6335	6389	6443	6497	6551	6604	6658	6712	6766	6820	54
7	6874	6927	6981	7035	7089	7143	7196	7250	7304	7358	54
8	7411	7465	7519	7573	7626	7680	7734	7787	7841	7895	54
9	7949	8002	8056	8110	8163	8217	8270	8324	8378	8431	54
810	908485	908539	908592	908646	908699	908753	908807	908860	908914	908967	54
1	9021	9074	9128	9181	9235	9289	9342	9396	9449	9503	54
2	9556	9610	9663	9716	9770	9823	9877	9930	9984	910037	53
3	910091	910144	910197	910251	910304	910358	910411	910464	910518	0571	53
4	0624	0678	0731	0784	0838	0891	0944	0998	1051	1104	53

Contd...

Table A.21 Contd.

N.	0	1	2	3	4	5	6	7	8	9	D.
815	1158	1211	1264	1317	1371	1424	1477	1530	1584	1637	53
6	1690	1743	1797	1850	1903	1956	2009	2063	2116	2169	53
7	2222	2275	2328	2381	2435	2488	2541	2594	2647	2700	53
8	2753	2896	2859	2913	2966	3019	3072	3125	3178	3231	53
9	3284	3337	3390	3443	3496	3549	3602	3655	3708	3761	53
820	913814	913867	913920	913973	914026	914079	914132	914184	914237	914290	53
1	4343	4396	4449	4502	4555	4608	4660	4713	4766	4819	53
2	4872	4925	4977	5030	5083	5136	5189	5241	5294	5347	53
3	5400	5453	5505	5558	5611	5664	5716	5769	5822	5875	53
4	5927	5980	6033	6085	6138	6191	6243	6296	6349	6401	53
825	6454	6507	6559	6612	6664	6717	6770	6822	6875	6927	53
6	6980	7033	7085	7138	7190	7243	7295	7348	7400	7453	53
7	7506	7558	7611	7663	7716	7768	7820	7873	7925	7978	52
8	8030	8083	8135	8188	8240	8293	0345	8397	8450	8502	52
9	8555	8607	8659	8712	8764	8816	8869	8921	8973	9026	52
830	919078	919130	919183	919235	919287	919340	919392	919444	919496	919549	52
1	9601	9653	9706	9758	9810	9862	9914	9967	920019	920071	52
2	920123	920176	920228	920280	920332	920384	920436	920489	0541	0593	52
3	0645	0697	0749	0801	0853	0906	0958	1010	1862	1114	52
4	1166	1218	1270	1322	1374	1426	1478	1530	1582	1634	52
835	1686	1738	1790	1842	1894	1946	1998	2050	2102	2154	52
6	2206	2258	2310	2362	2414	2466	2518	2570	2622	2674	52
7	2725	2777	2829	2881	2933	2985	3037	3089	3140	3192	52
8	3244	3296	3348	3399	3451	3503	3555	3607	3658	3710	52
9	3762	3814	3865	3917	3969	4021	4072	4124	4176	4228	52
840	924279	924331	924383	924434	924486	924538	924589	924641	924693	924744	52
1	4796	4848	4899	4951	5003	5054	5106	5157	5109	5261	52
2	5312	5364	5415	5467	5518	5570	5021	5673	5725	5776	52
3	5828	5879	5931	5982	6034	6085	6137	6188	6240	6291	51
4	6342	6394	6445	6497	6548	6600	6051	6702	6754	6805	51
845	6857	6908	6959	7011	7062	7114	7165	7216	7268	7319	51
6	7370	7422	7473	7524	7576	7627	7678	7730	7781	7832	51
7	7883	7935	7986	8037	8086	8140	8191	8242	8293	8345	51
8	8396	8447	8498	8349	8601	8652	8703	8754	8805	8857	51
9	8908	8959	9010	9061	9112	9163	9215	9266	9317	9368	51
850	929419	929470	929521	929572	929623	929674	929725	929776	929827	929879	51
1	9930	9981	930032	930083	930134	930185	930236	930287	930338	930389	51
2	930440	930491	0542	0592	0643	0694	0745	0796	0847	0898	51
3	0949	1000	1051	1102	1153	1204	1254	1305	1356	1407	51
4	1458	1509	1560	1610	1661	1712	1763	1814	1865	1915	51
855	1966	2017	2068	2118	2169	2220	2271	2322	2372	2423	51
6	2474	2524	2575	2626	2677	2727	2778	2829	2879	2930	51
7	2981	3031	3082	3133	3183	3234	3285	3335	3386	3437	51
8	3487	3538	3589	3639	3690	3740	3791	3841	3892	3943	51
9	3993	4044	4094	4145	4195	4246	4296	4347	4397	4448	51

Contd...

Table A.21 Contd.

N.	0	1	2	3	4	5	6	7	8	9	D.
860	934498	934549	934599	934650	934700	934751	934801	934852	934902	934953	50
1	5003	5054	5104	5154	5205	5255	5306	5356	5406	5457	50
2	5507	5558	5608	5658	5709	5759	5809	5860	5910	5960	50
3	6011	6061	6111	6162	6212	6262	6313	6363	6413	6463	50
4	6514	6564	6614	6665	6715	6765	6815	6865	6916	6966	50
865	7016	7066	7117	7167	7217	7267	7317	7367	7418	7468	50
6	7518	7568	7618	7668	7718	7769	7819	7869	7919	7969	50
7	8019	8069	8119	8169	8219	8269	8320	8370	8420	8470	50
8	8520	8570	8620	8670	8720	8770	8820	8870	8920	8970	50
9	9020	9070	9120	9170	9220	9270	9320	9369	9419	9469	50
870	939519	939569	939619	939669	939719	939769	939819	939869	739918	939968	50
1	940018	940068	940118	940168	940218	940267	940317	940367	940417	940467	50
2	0516	0566	0616	0666	0716	0765	0815	0865	0915	0964	50
3	1014	1064	1114	1163	1213	1263	1313	1362	1412	1462	50
4	1511	1561	1611	1660	1710	1760	1809	1859	1909	1958	50
875	2008	2058	2107	2157	2207	2256	2306	2355	2405	2455	50
6	2504	2554	2603	2653	2702	2752	2801	2851	2901	2950	50
7	3000	3049	3099	3148	3198	3247	3297	3346	3396	3445	49
8	3495	3544	3593	3643	3692	3742	3791	3841	3890	3939	49
9	3989	4038	4088	4137	4186	4236	4285	4335	4384	4433	49
880	944483	944532	944581	944631	944680	944729	944779	944828	944877	944927	49
1	4976	5025	5074	5124	5173	5222	5272	5321	5370	5419	49
2	5469	5518	5567	5616	5665	5715	5764	5813	5862	5912	49
3	5961	6010	6059	6108	6157	6207	6256	6305	6354	6403	49
4	6452	6501	6551	6600	6649	6698	6747	6796	6845	6894	49
885	6943	6992	7041	7090	7140	7189	7238	7287	7336	7385	49
6	7434	7483	7532	7581	7630	7679	7728	7777	7826	7875	49
7	7924	7973	8022	8070	8119	8168	8217	8266	8315	8364	49
8	8413	8462	8511	8560	8609	8657	8706	8755	8804	8853	49
9	8902	8951	8999	9048	9097	9146	9195	9244	9292	9341	49
890	949390	949439	949488	949536	949585	949634	949683	949731	949780	949829	49
1	9878	9926	9975	950024	950073	950121	950170	950219	950267	950316	49
2	950365	950414	950462	0511	0560	0608	0657	0706	0754	0803	49
3	0851	0900	0949	0997	1046	1095	1143	1192	1240	1289	49
4	1338	1386	1435	1483	1532	1580	1629	1677	1726	1775	49
895	1823	1872	1920	1969	2017	2066	2114	2163	2211	2260	48
6	2308	2356	2405	2453	2502	2550	2599	2647	2696	2744	48
7	2792	2841	2889	2938	2986	3034	3083	3131	3180	3228	48
8	3276	3325	3373	3421	3470	3518	3566	3615	3663	3711	48
9	3760	3808	3856	3905	3953	4001	4049	4098	4146	4194	48
900	954243	954291	954339	954387	954435	954484	954532	954580	954628	954677	48
1	4725	4773	4821	4869	4918	4966	5014	5062	5110	5158	48
2	5207	5255	5303	5351	5399	5447	5495	5543	5592	5640	48
3	5688	5736	5784	5832	5880	5928	5976	6024	6072	6120	48
4	6168	6216	6265	6313	6361	6409	6457	6505	6553	6601	48

Contd...

Table A.21 Contd.

N.	0	1	2	3	4	5	6	7	8	9	D.
905	6649	6697	6745	6793	6840	6888	6936	6984	7032	7080	48
6	7128	7176	7224	7272	7320	7368	7416	7464	7512	7559	48
7	7607	7655	7703	7751	7799	7847	7894	7942	7990	8038	48
8	8086	8134	8181	8229	8277	8325	8373	8421	8468	8516	48
9	8564	8612	8659	8707	8755	8803	8850	8898	8946	8994	48
910	959041	959089	959137	959185	959232	959280	959328	959375	959423	959471	48
1	9518	9566	9614	9661	9709	9757	9804	9852	9900	9947	48
2	9995	960042	960090	960138	960185	960233	960280	960328	960376	960423	48
3	960471	0518	0566	0613	0661	0709	0756	0804	0851	0899	48
4	0946	0994	1041	1089	1136	1184	1231	1279	1326	1374	48
915	1421	1469	1516	1563	1611	1658	1706	1753	1801	1848	47
6	1895	1943	1990	2038	2085	2132	2180	2227	2275	2322	47
7	2369	2417	2464	2511	2559	2606	2653	2701	2748	2795	47
8	2843	2890	2937	2985	3032	3079	3126	3174	3221	3268	47
9	3316	3363	3410	3457	3504	3552	3599	3646	3693	3741	47
920	963788	963835	963882	963929	963977	964024	964071	964118	964165	964212	47
1	4260	4307	4354	4401	4448	4495	4542	4590	4637	4684	47
2	4731	4778	4825	4872	4919	4966	5013	5061	5108	5155	47
3	5202	5249	5296	5343	5390	5437	5484	5531	5578	5625	47
4	5672	5719	5766	5813	5860	5907	5954	6001	6048	6095	47
925	6142	6189	6236	6283	6329	6376	6423	6470	6517	6564	47
6	6611	6658	6705	6752	6799	6845	6892	6939	6986	7033	47
7	7080	7127	7173	7220	7267	7314	7361	7408	7454	7501	47
8	7548	7595	7642	7688	7735	7782	7829	7875	7922	7969	47
9	8016	8062	8109	8156	8203	8249	8296	8343	8390	8436	47
930	968483	968530	968576	968623	968670	968716	968763	968810	968856	968903	47
1	8950	8996	9043	9090	9136	9183	9229	9276	9323	9369	47
2	9416	9463	9509	9556	9602	9649	9695	9742	9789	9835	47
3	9882	9928	9975	970021	970068	970114	970161	970207	970254	970300	47
4	970347	970393	970440	0486	0533	0579	0626	0672	0719	0765	46
935	0812	0858	0904	0951	0997	1044	1090	1137	1183	1229	46
6	1276	1322	1369	1415	1461	1508	1554	1601	1647	1693	46
7	1740	1786	1832	1879	1925	1971	2018	2064	2110	2157	46
8	2203	2249	2295	2342	2388	2434	2481	2527	2573	2619	46
9	2666	2712	2758	2804	2851	2897	2943	2989	3035	3082	46
940	973128	973174	973220	973266	973313	973359	973405	973451	973497	973543	46
1	3590	3636	3682	3728	3774	3820	3866	3913	3959	4005	46
2	4051	4097	4143	4189	4235	4281	4327	4374	4420	4466	46
3	4512	4558	4604	4650	4696	4742	4788	4834	4880	4926	46
4	4972	5018	5064	5110	5156	5202	5248	5294	5340	5386	46
945	5432	5478	5524	5570	5616	5662	5707	5753	5799	5845	46
6	5891	5937	5983	6029	6075	6121	6167	6212	6258	6304	46
7	6350	6396	6442	6488	6533	6579	6625	6671	6717	6763	46
8	6808	6854	6900	6946	6992	7037	7083	7129	7175	7220	46
9	7266	7312	7358	7403	7449	7495	7541	7586	7632	7678	46

Contd...

Table A.21 Contd.

N.	0	1	2	3	4	5	6	7	8	9	D.
950	977724	977769	977815	977861	977906	977952	977998	978043	978069	978135	46
1	8181	8226	8272	8317	8363	8409	8454	8500	8546	8591	46
2	8637	8683	8728	8774	8819	8865	8911	8956	9002	9047	46
3	9093	9138	9184	9230	9275	9321	9366	9412	9457	9503	46
4	9548	9594	9639	9685	9730	9776	9821	9867	9912	9958	46
955	980003	980049	980094	980140	980185	980231	980276	980322	980367	980412	45
6	0458	0503	0549	0594	0640	0685	0730	0776	0821	0867	45
7	0912	0957	1003	1048	1093	1139	1184	1229	1275	1320	45
8	1366	1411	1456	1501	1547	1592	1637	1683	1728	1773	45
9	1819	1864	1909	1954	2000	2045	2090	2135	2181	2226	45
960	982271	982316	982362	982407	982452	982497	982543	982588	982633	982678	45
1	2723	2769	2814	2859	2904	2949	2994	3040	3085	3130	45
2	3175	3220	3265	3310	3356	3401	3446	3491	3536	3581	45
3	3626	3671	3716	3762	3807	3852	3897	3942	3987	4032	45
4	4077	4122	4167	4212	4257	4302	4347	4392	4437	4482	45
965	4527	4572	4617	4662	4707	4752	4797	4842	4887	4932	45
6	4977	5022	5067	5112	5157	5202	5247	5292	5337	5382	45
7	5426	5471	5516	5561	5606	5651	5696	5741	5786	5830	45
8	5875	5920	5965	6010	6055	6100	6144	6189	6234	6279	45
9	6324	6369	6413	6458	6503	6548	6593	6637	6682	6727	45
970	986772	986817	986861	986906	986951	986996	987040	987085	987130	987175	45
1	7219	7264	7309	7353	7398	7443	7488	7532	7577	7622	45
2	7666	7711	7756	7800	7845	7890	7934	7979	8024	8068	45
3	8113	8157	8202	8247	8291	8336	8381	8425	8470	8514	45
4	8559	8604	8648	8693	8737	8782	8826	8871	8916	8960	45
975	9005	9049	9094	9138	9183	9227	9272	9316	9361	9405	45
6	9450	9494	9539	9583	9628	9672	9717	9761	9806	9850	44
7	9895	9939	9983	990028	990072	990117	990161	990206	990250	990294	44
8	990339	990383	990428	0472	0516	0561	0605	0650	0694	0738	44
9	0783	0827	0871	0916	0960	1004	1049	1093	1137	1182	44
980	991226	991270	991315	991359	991403	991448	991492	991536	991580	991625	44
1	1669	1713	1758	1802	1846	1890	1935	1979	2023	2067	44
2	2111	2156	2200	2244	2288	2333	2377	2421	2465	2509	44
3	2554	2598	2642	2686	2730	2774	2819	2863	2907	2951	44
4	2995	3039	3083	3127	3172	3216	3260	3304	3348	3392	44
985	3436	3480	3524	3568	3613	3657	3701	3745	3789	3833	44
6	3877	3921	3965	4009	4053	4097	4141	4185	4229	4273	44
7	4317	4361	4405	4449	4493	4537	4581	4625	4669	4713	44
8	4757	4801	4845	4889	4933	4977	5021	5065	5108	5152	44
9	5196	5240	5284	5328	5372	5416	5460	5504	5547	5591	44
990	995635	995679	995723	995767	995811	995854	995898	995942	995986	996030	44
1	6074	6117	6161	6205	6249	6293	6337	6380	6424	6468	44
2	6512	6555	6599	6643	6687	6731	6774	6818	6862	6906	44
3	6949	6993	7037	7080	7124	7168	7212	7255	7299	7343	44
4	7386	7430	7474	7517	7561	7605	7648	7692	7736	7779	44
995	7823	7867	7910	7954	7998	8041	8085	8129	8172	8216	44
6	8259	8303	8347	8390	8434	8477	8521	8564	8608	8652	44
7	8695	8739	8782	8826	8869	8913	8956	9000	9043	9087	44
8	9131	9174	9218	9261	9305	9348	9392	9435	9479	9522	44
9	9565	9609	9652	9696	9739	9783	9826	9870	9913	9957	43

Appendix B: Mathematical Symbols and Expressions

Symbols

$\neq$ not equal to; e.g., $3 \neq 4$

$>$ greater than; e.g., $5 > 2$

$\geq$ greater than or equal to

$<$ less than; e.g., $3 < 7$

$\leq$ less than or equal to

$|\ |$ absolute value; e.g., $|-7| = 7$

$\rightarrow$ tends to or limiting value is: e.g. $\rightarrow 0$ limiting value is 0

$\sim$ is distributed as; e.g. $X \sim B(n, p)$

(a, b) = open interval of values from a to b e.g. (3, 10) is interval of values from 3 to 10, excluding 3 and 10

$[a, b]$ = closed interval of values from a to b, a and b both included e.g. [3, 10] is interval of values from 3 to 10, including both 3 and 10

w.r.to = with respect to

iff = implies if and only if

Notations and Expressions

1. $\sum_{i=1}^{n} i = \frac{n(n+1)}{2}$

2. $\sum_{i=1}^{n} X_i$ = Sum of X_i's for $i = 1$ to $i = n$

   $= X_1 + X_2 + \ldots\ldots\ldots\ldots + X_n$

   = Sum of $X_1, X_2, \ldots\ldots\ldots, X_n$

3. $\prod_{j=1}^{n} X_j$ = Product of X_j's from $j = 1$ to n

   $= (X_1) \cdot (X_2) \cdot \ldots\ldots\ldots\ldots\ldots (X_n)$

   Note that there are n terms in the product in expression 3.

   e.g. $\prod_{i=1}^{5} i = 1.\ 2.\ 3.\ 4.\ 5 = 120$

4. $n!$ = A product of positive integer n by all the positive integers smaller than it, is usually denoted by $n!$(read "n factorial"). Thus

$$n! = n(n-1)(n-2)\ldots 1 = \prod_{j=0}^{n-1}(n-j) \text{ e.g., } 3! = 3(2)1 = 6$$

5. $0! = 1$ and $1! = 1$
6. $\overline{}$ = over bar; used to indicate an arithmetic average or mean e.g. $\bar{X}$ is mean of X.
7. ^ = hat; used to indicate an estimate rather than a true value; most often appear over Greek letters e.g. $\hat{\theta}$ is an estimator of θ

8. $\binom{n}{x} = {}^{n}c_x = \dfrac{n!}{(n-x)!\,x!}$

$\binom{n}{x}$ is read "*combination of n things taking x at a time*" or more briefly as "*n pick x*"; it is also called a binomial coefficient. Define

$$\binom{n}{x} = 0 \text{ if } x < 0 \text{ or } x > n$$

$$\binom{n}{0} = \binom{n}{n} = 1$$

$$\binom{n}{x} = \binom{n}{n-x}$$

9. $\binom{-n}{k} = \dfrac{(-n)(-n-1)\ldots(-n-k+1)}{k!}$

$$= (-1)^k \frac{n(n+1)\ldots\ldots(n+k-1)}{k!}$$

$$= (-1)^k \binom{n+k-1}{k}$$

10. Γx = Pronounced as gamma x = $\int_0^\infty u^{x-1}e^{-u}du$

$= (x-1)\ \Gamma(x-1)$

Remark if $x = n$, an integer, then

$$\Gamma x = (x-1)!$$

11. $f(x)$ = a function of x or an expression involving a single variable x.
12. $\int f(x)\,dx$ = indefinite integral of $f(x)$ with respect to x

$= I(x)$

13. $\int_a^b f(x)\,dx$ = Definite integral of $f(x)$ with respect to x in the interval (a, b)

$= I(b) - I(a)$

14. If $Y = f(x)$

$$\frac{dY}{dX} = \text{differential of } Y \text{ w.r. to } X$$

15. If $Y = f(X, U, V, T.....)$ i.e. Y is a function of several variables X, U, V ....etc.

$\frac{\partial Y}{\partial X}$ = partial differential of Y w.r. to X = $\frac{\partial f(X, U, V, T, ...)}{\partial X}$

$\frac{\partial Y}{\partial X}\Big|_{t=0}$ = Value of partial differential of $Y = f(X, U, V, t)$ w.r. to X at $t = 0$

16. $\mathbf{X}$ = a vector

$\mathbf{X_2} = \begin{pmatrix} X_1 \\ X_2 \end{pmatrix}$ is a 2 component or 2-dimensional vector

$\mathbf{X}_p = \begin{pmatrix} X_1 \\ X_2 \\ \vdots \\ X_p \end{pmatrix}$ is a p-component or p-dimensional vector

17. H_0 = Null hypothesis or hypothesis to be tested e.g. H_0: $\mu = 4$
18. H_1 = Alternative hypothesis. Possible value of hypothesized parameter if H_0 is not accepted. e.g. if above H_0: $\mu = 4$ is not accepted the researcher may choose H_1: $\mu > 4$ or H_1: $\mu < 4$ or H_1: $\mu \neq 4$.
19. α = Probability of type I error = P[Reject H_0/H_0 is true]
20. β = Probability of type II error = P[Accept H_1/H_1 is not true]

Appendix C: Basics of Matrix Algebra

An efficient way of studying many variables simultaneously is through use of vectors and matrices. Whenever the observed random variable is multi-component, in the multivariate techniques described as multivariate analysis, some of the results of matrix algebra are used.
Matrix notations are very useful because:

1. These enable a highly compact presentation. If a separate symbol for each component of a multivariate variable were used, the equations and expressions showing relationships among the components will be enormously bulky.
2. This enables to give simple compact matrix equations easy to understand and work with.
3. Similarity between univariate and multivariate techniques is evident from matrix expressions and it results in easy solution of multivariate problems.

So in this Appendix some elementary definitions and results (without proof) are presented. There are plenty of books on matrix algebra. The reader interested in details and proofs may consult those, such as Hohn (1964), Turnbull (1960), Horst (1963), Searle (1966) etc.

C.1 Definition and Notations

A rectangular array of real or complex numbers is called a matrix. Unless specified the numbers are real. The array of this kind is represented as

$$\mathbf{A} = \begin{bmatrix} a_{11} & a_{12} & \cdots & a_{1j} & \cdots & a_{1n} \\ a_{21} & a_{22} & \cdots & a_{2j} & \cdots & a_{2n} \\ \cdot & \cdot & & \cdot & & \cdot \\ \cdot & \cdot & & \cdot & & \cdot \\ \cdot & \cdot & & \cdot & & \cdot \\ a_{m1} & a_{m2} & \cdots & a_{mj} & \cdots & a_{mn} \end{bmatrix} \tag{C.1}$$

or $$\mathbf{A} = ((a_{ij})),\ i = 1, 2, ..., m;\ j = 1, 2, ..., n$$

The numbers a_{ij} of this array are called elements. Pairs of double bars || || or parentheses (( )) are also used to show the matrices.

The matrix **A** has m rows and n columns and is called a *matrix of order* $m \times n$.

When $m = n$, the matrix **A** is square matrix of order n, and the elements $a_{11}, a_{22}, \ldots, a_{nn}$ constitute the *main diagonal of* **A** and $a_{n1}, a_{n-1, 2}, \ldots, a_{1n}$ constitute its *secondary diagonal.*

If $m = 1$, the array becomes a *row vector* of n components denoted by **a**, given by

$$\mathbf{a} = (a_1\ a_2\ \ldots\ a_n) \tag{C.2}$$

If $n = 1$, the array becomes a *column vector* of m components denoted by **a**, given by

$$\mathbf{a} = \begin{bmatrix} a_1 \\ a_2 \\ \cdot \\ \cdot \\ \cdot \\ a_m \end{bmatrix} \tag{C.3}$$

(a) If $m = n$, and if $a_{ij} = 0$ for $i \neq j$ and $a_{ij} = 1$ for $i = j$, **A** is called an *identity matrix* or *unit matrix* of order n. It is denoted by $\mathbf{I}_n$.
(b) If $a_{ij} = 0$ for all i and j, **A** is called a *zero matrix.*
(c) A matrix obtained from **A** in (C.1) by interchanging its rows and columns is denoted by **A**' or $\mathbf{A}^T$ is of order $n \times m$ is called *transpose* of **A**.
(d) A *square matrix* **A**, is a *symmetric matrix* if $\mathbf{A}^T = \mathbf{A}$.
(e) If for a square matrix **A**, $\mathbf{A} = -\mathbf{A}^T$ holds, **A** is a *skew-symmetric matrix.*
(f) If each main diagonal element of a square matrix of order n is a constant number α and off diagonal elements are all 0, the matrix is called a *scalar matrix* of order n. It can be written as $\alpha\,\mathbf{I}_n$ also.

C.2 Equality of Matrices

Two matrices $((a_{ij}))_{m\times n}$ and $((b_{ij}))_{m\times n}$ are said to be equal iff $a_{ij} = b_{ij}$ for each pair of i and j. In words, two matrices are equal if they are of the same order and the corresponding elements are all equal.

(a) If there are two matrices **A** and **B**, either **A** = **B** or **A** ≠ **B**.
(b) **A** = **A** (c) If **A** = **B** and **B** = **C** then **A** = **C**.

C.3 Linear Combinations of Matrices

(a) **Addition** If $\mathbf{A} = ((a_{ij}))$ and $\mathbf{B} = ((b_{ij}))$ are of the same dimension or order $m \times n$, then the sum **A** + **B** is defined as

$$\mathbf{A} + \mathbf{B} = ((a_{ij} + b_{ij})) \tag{C.4}$$

A + **B** is also of the order $m \times n$.
(b) If **A**, **B** and **C** are conformable for addition

$$\mathbf{A} + \mathbf{B} = \mathbf{B} + \mathbf{A} \text{ and } \mathbf{A} + (\mathbf{B} + \mathbf{C}) = (\mathbf{A} + \mathbf{B}) + \mathbf{C} \tag{C.5}$$

(c) $-\mathbf{A} = ((-a_{ij}))$ is of order $m \times n$.
(d) $\mathbf{A} - \mathbf{B} = ((a_{ij} - b_{ij}))$ is defined if **A** and **B** are of order $m \times n$ and, one can write $\mathbf{A} - \mathbf{B} = \mathbf{A} + (-\mathbf{B})$ is also of order $m \times n$.
(e) If α is a scalar quantity, $\alpha\mathbf{A} = \mathbf{A}\alpha = ((\alpha a_{ij}))$ (C.6)

(f) **Multiplication** of matrices

Let $\mathbf{A} = ((a_{ij}))$ be of order $m \times p$ and $\mathbf{B} = ((b_{ij}))$ be of order $p \times n$ i.e. the number of columns of **A** = number of rows of **B** then **A** is said to be *comformable* to **B** for *multiplication*. If **C** is the result of multiplication of matrices **A** and **B**, one can write, the product

$$\mathbf{AB} = \mathbf{C} = ((c_{ij})) \text{ is matrix of order } m \times n.$$

The $(i\ j)^{th}$ element of **C** $= c_{ij}$ *is found by multiplying the ith row elements of* **A** *with j^{th} column elements of* **B** *and adding the results*

Symbolically

$$\begin{bmatrix} a_{11} & a_{12} & \dots & a_{1p} \\ a_{21} & a_{22} & \dots & a_{2p} \\ \cdot & & & \\ \cdot & & & \\ \cdot & & & \\ a_{i1} & a_{i2} & \dots & a_{ip} \\ \cdot & & & \\ \cdot & & & \\ \cdot & & & \\ a_{m1} & a_{m2} & \dots & a_{mp} \end{bmatrix} \times \begin{bmatrix} b_{11} & b_{12} & \dots & b_{1j} & \dots & b_{1n} \\ b_{21} & b_{22} & \dots & b_{2j} & \dots & b_{2n} \\ \cdot & & & & & \\ \cdot & & & & & \\ \cdot & & & & & \\ b_{p1} & b_{p2} & \dots & b_{pj} & \dots & b_{pn} \end{bmatrix} = \begin{bmatrix} c_{11} & c_{12} & \dots & c_{1j} & \dots & c_{1n} \\ c_{21} & c_{22} & \dots & c_{2j} & \dots & c_{2n} \\ \vdots & \vdots & \vdots & \vdots & \vdots & \vdots \\ c_{i1} & c_{i2} & \dots & c_{ij} & \dots & c_{in} \\ \cdot & & & & & \\ \cdot & & & & & \\ \cdot & & & & & \\ c_{m1} & c_{m2} & \dots & c_{mj} & \dots & c_{mn} \end{bmatrix}$$

where $c_{ij} = a_{i1}b_{1j} + a_{i2}b_{2j} + \dots + a_{ip}b_{pj} = \sum_{k=1}^{p} a_{ik}b_{kj}$

for $i = 1, 2, \dots, m$ and $j = 1, 2, \dots, n$. (C.7)

Properties of Matrix Multiplication

(i) In general **AB** = **BA**. Sometimes **B** may not be conformable to **A** for multiplication. Even if **A** and **B** both are square matrices **AB** and **BA** may not be equal.

(ii) If **A** is of order $m \times n$, **B** of order $n \times p$ and **C** of order $p \times q$, then

$$(\mathbf{AB})\mathbf{C} = \mathbf{A}(\mathbf{BC})$$

(iii) Assuming conformability of matrices

$$(\mathbf{A} + \mathbf{B})\mathbf{C} = \mathbf{AC} + \mathbf{BC}$$

(iv) Assuming conformability of matrices

$$(\mathbf{A} + \mathbf{B})(\mathbf{C} + \mathbf{D}) = \mathbf{AC} + \mathbf{BC} + \mathbf{AD} + \mathbf{BD}$$

(v) $(\mathbf{AB})' = \mathbf{B}'\mathbf{A}'$, if **A** and **B** are conformable.

(vi) **AI** = **A** = **IA** for all **A**.

C.4 Determinant

A square matrix **A** of order n has associated with it a real valued function of the elements of **A** called *determinant of A*. It is denoted by $|\mathbf{A}|$.

The *value of determinant is calculated* using the quantities minors and cofactors to be defined shortly. To understand the principle consider finding of determinant of a 2×2 matrix

$$\mathbf{A} = \begin{pmatrix} a_{11} & a_{12} \\ a_{21} & a_{22} \end{pmatrix}$$

$$|\mathbf{A}| = \begin{vmatrix} a_{11} & a_{12} \\ a_{21} & a_{22} \end{vmatrix} = a_{11}a_{22} - a_{12}a_{21} \tag{C.8}$$

For a 3 × 3 matrix $A = \begin{bmatrix} a_{11} & a_{12} & a_{13} \\ a_{21} & a_{22} & a_{23} \\ a_{31} & a_{32} & a_{33} \end{bmatrix}$, the determinant of **A** viz. |**A**| can not be written easily using the elements of **A**. A method using cofactors can be given for a matrix of any order, after writing |**A**| for the above 3 × 3 matrix.

$$\begin{aligned} |\mathbf{A}| &= a_{11}\begin{vmatrix} a_{22} & a_{23} \\ a_{32} & a_{33} \end{vmatrix} - a_{12}\begin{vmatrix} a_{21} & a_{23} \\ a_{31} & a_{33} \end{vmatrix} + a_{13}\begin{vmatrix} a_{21} & a_{22} \\ a_{31} & a_{32} \end{vmatrix} \\ &= -a_{21}\begin{vmatrix} a_{12} & a_{13} \\ a_{32} & a_{33} \end{vmatrix} + a_{22}\begin{vmatrix} a_{11} & a_{13} \\ a_{31} & a_{33} \end{vmatrix} - a_{23}\begin{vmatrix} a_{11} & a_{12} \\ a_{31} & a_{32} \end{vmatrix} \\ &= a_{11}\begin{vmatrix} a_{22} & a_{23} \\ a_{32} & a_{33} \end{vmatrix} - a_{21}\begin{vmatrix} a_{12} & a_{13} \\ a_{32} & a_{33} \end{vmatrix} + a_{31}\begin{vmatrix} a_{12} & a_{13} \\ a_{22} & a_{23} \end{vmatrix} \end{aligned} \tag{C.9}$$

Some rules are followed in writing the determinant. These are

(i) Choose any row or a column of the matrix.
(1st row is chosen in 1st RHS of (C.9)
2nd row is chosen in 2nd RHS of (C.9)
1st column is chosen in 3rd RHS of (C.9).

(ii) With each element of the chosen row or column multiply the determinant of a 2 × 2 matrix obtained from **A** by deleting the chosen row, and the column in which the element in the chosen row exists. Thus the 1st row is chosen in first expression and determinant multiplying the element a_{12} is obtained by deleting 1st row and 2nd column of **A**. The remaining elements form 2 × 2 matrix, the determinant of which multiplies a_{12}.

(iii) Give *a* + sign to the term if the subscripts of element in the chosen row (or column) add to an even integer. Give *a* – sign if it adds to an odd number. In other words to each term give sign $(-1)^{i+j}$ if i and j are the subscripts of the element of the chosen row or column, which is multiplied with a determinant of 2 × 2 matrix.

For any matrix **A** of order $n \times n$ this approach of determining the |**A**| can be given as:

$$\begin{aligned} |A| &= \sum_{j=1}^{n} a_{ij}(-1)^{i+j} M_{ij} \quad \text{for any choice of row } i \\ &= \sum_{i=1}^{n} a_{ij}(-1)^{i+j} M_{ij} \quad \text{for any choice of column } j \end{aligned} \tag{C.10}$$

where $\mathbf{M}_{ij}$ is determinant of the $(n-1) \times (n-1)$ submatrix of A after deleting i^{th} row and j^{th} column of **A**.

The determinant $\mathbf{M}_{ij}$ is called the *minor* of a_{ij}
$(-1)^{i+j}\, \mathbf{M}_{ij} = \mathbf{A}_{ij}$ is called the *cofactor* of a_{ij}.
Thus one can write that

$$|A| = \sum_{j=1}^{n} a_{ij}\, A_{ij} = \sum_{i=1}^{n} a_{ij} C_{ij} \tag{C.11}$$

where, A_{ij} and C_{ij} are *cofactor* of corresponding a_{ij}.
Numerical examples may be seen by interested reader in any matrix algebra book.

Equation (C.11) holds for any choice of row i or column j. If **A** is of a large dimension, each of the $(n-1) \times (n-1)$ minors are written in terms of minors of order $(n-2) \times (n-2)$. Subsequently each in terms of minor of order $(n-3) \times (n-3)$, and so on upto minors of order 2×2 ultimately. It is quite evident that calculation of determinant is quite lengthy if manual calculation is done. Computer programs exist and are extensively used by practitioners.

Properties of Determinants

(i) If $|\mathbf{A}| = 0$, **A** is called a *singular matrix.*

(ii) If any two parallel lines of **A** are interchanged the determinant of resulting matrix is $|\mathbf{A}|$.

(iii) $|\mathbf{A}| = |\mathbf{A}^T|$.

(iv) If all the elements of any row or a column of **A** are zero, then $|\mathbf{A}| = 0$.

(v) If any two parallel lines of **A** are identical, $|\mathbf{A}| = 0$.

(vi) By adding to any row of **A**, a constant times any other row of **A**, or by addition of a constant times any column of **A** to another column of **A**, does not alter value of $|A|$.
From the last property another method of computing $|\mathbf{A}|$ is to reduce the matrix **A** into a diagonal matrix or a triangular matrix by applying row or column transformations (i.e. adding a constant times one row or column to another row or column). The value of $|\mathbf{A}|$ then can be obtained as the product of the diagonal elements of the reduced matrix.

(vii) Let $\mathbf{A}_1, \mathbf{A}_2, \ldots, \mathbf{A}_k$ be all of order $p \times p$. Then

$$|\mathbf{A}_1\mathbf{A}_2 \ldots \mathbf{A}_k| = |\mathbf{A}_1|\,|\mathbf{A}_2| \ldots |\mathbf{A}_k| \tag{C.12}$$

(viii) If **A** is of order $p \times n$ and **B** is of order $n \times p$, then

$$|\mathbf{I}_p + \mathbf{AB}| = |\mathbf{I}_n + \mathbf{BA}| \tag{C.13}$$

(ix) If $|\mathbf{A}| = 0$, then one or more rows or columns of **A** are linearly dependent on (are equal to the weighted sum of) the remaining rows or columns of the matrix A.

C.5 Inverse of a Matrix

If **A** is a square matrix of order n, such that $|\mathbf{A}| \neq 0$, then there exists a unique matrix **B** (say) of the same order as **A** such that

$$\mathbf{AB} = \mathbf{BA} = \mathbf{I}_n \tag{C.14}$$

The matrix **B** is then called an *inverse* of **A** and is denoted by $\mathbf{A}^{-1}$.

(a) *Inverse of a diagonal matrix is also a diagonal matrix.*
If

$$\mathbf{D} = \begin{bmatrix} d_1 & 0 & \dots & 0 \\ 0 & d_2 & \dots & 0 \\ \cdot & & & \\ \cdot & & & \\ \cdot & & & \\ 0 & 0 & \dots & d_n \end{bmatrix}, \mathbf{D}^{-1} = \begin{bmatrix} 1/d_1 & 0 & \dots & 0 \\ 0 & 1/d_2 & \dots & 0 \\ \cdot & & & \\ \cdot & & & \\ \cdot & 0 & \dots & 1/d_n \end{bmatrix} \quad \text{(C.15)}$$

(b) *Computation of inverse of a square matrix* **A** *of order n.*

Let **A** = $((a_{ij}))$ and C_{ij} be the cofactor of a_{ij}, then the matrix

$$\mathcal{A} = ((C_{ij}))^T \begin{bmatrix} C_{11} & C_{21} & \dots & C_{n1} \\ C_{12} & C_{22} & \dots & C_{n2} \\ \cdot & & & \\ \cdot & & & \\ \cdot & & & \\ C_{1n} & C_{2n} & \dots & C_{nn} \end{bmatrix} \quad \text{(C.16)}$$

A result is that

$$\mathbf{A}\mathcal{A} = \begin{bmatrix} |\mathbf{A}| & 0 & \dots & 0 \\ 0 & |\mathbf{A}| & \dots & 0 \\ \cdot & & & \\ \cdot & & & \\ \cdot & & & \\ \mathbf{0} & 0 & \dots & |\mathbf{A}| \end{bmatrix} \quad \text{(C.17)}$$

So
$$\mathbf{A}\,\mathcal{A} = \mathcal{A}\,\mathbf{A} = |\mathbf{A}|\, I_n \quad \text{(C.18)}$$

Hence if $|\mathbf{A}| \neq 0$ the inverse of **A** can be computed from this relation
Comparing equations (C.14) and (C.18)

$$\mathbf{A}^{-1} = \mathcal{A}/[|\mathbf{A}|] \quad \text{(C.19)}$$

Thus the steps for computing $\mathbf{A}^{-1}$ by this method are:

1. Replace each element a_{ij} of **A** by its cofactor C_{ij}.
2. Divide each element of this matrix of cofactors by $|\mathbf{A}|$.
3. Transpose the resulting matrix and the matrix obtained is the inverse $\mathbf{A}^{-1}$.

(c) Computation of $\mathbf{A}^{-1}$ from (C.19) is mostly quite tedious. So *inverse of a matrix* **A** *can be found by synthetic elimination.* The steps of this method are:

1. Augment the matrix **A** by an identify matrix I_n as:

$$[\mathbf{A} \mid \mathbf{I}_n]$$

2. Now perform linear transformation on rows and columns of the augmented matrix so that **A** is reduced to I_n. The resulting matrix then is of the form $[I_n \mid \mathbf{A}^{-1}]$.

This method is easier to apply if the matrix is not very big. For large matrices it is quite tedious. Computer programs are now available quite easily to do this task.

A note (i) A row transformation may be of two types:

(a) multiplication of each element of a row by a constant k. This may be written as kR_i.

(b) a linear row transformation is to add or subtract from a row a multiple of another row. In symbols $R_i \pm kR_j$

Similarly one may have the column transformations.

(c) kC_i and (d) $C_i \pm kC_j$.

An appropriate value of k is chosen depending on whether the original matrix is to be reduced to a diagonal matrix, a triangular matrix or an identity matrix.

Example C.1: Reduction of a matrix **A** of order 3 into an identity matrix I_3 using row and column transformations is illustrated.

Solution: Let

$$\mathbf{A} = \begin{bmatrix} a_{11} & a_{12} & a_{13} \\ a_{21} & a_{22} & a_{23} \\ a_{31} & a_{32} & a_{33} \end{bmatrix}. \text{ To reduce } \mathbf{A} \text{ into } \mathbf{I}_3$$

Transformations applied	Matrix **A** reduced to
$R_1 \to R_1/a_{11}$ $R_2 \to R_2/a_{21}$ $R_2 \to R_3/a_{31}$	$\begin{bmatrix} 1 & a_{12}/a_{11} & a_{13}/a_{11} \\ 1 & a_{22}/a_{21} & a_{23}/a_{21} \\ 1 & a_{32}/a_{31} & a_{33}/a_{31} \end{bmatrix}$
$R_2 \to R_2 - R_1$ $R_3 \to R_3 - R_1$	$\begin{bmatrix} 1 & a_{12}/a_{11} & a_{13}/a_{11} \\ 0 & a_{22}/a_{21} - a_{12}/a_{11} & a_{23}/a_{21} - a_{13}/a_{11} \\ 0 & a_{32}/a_{31} - a_{12}/a_{11} & a_{33}/a_{31} - a_{13}/a_{11} \end{bmatrix}$
$C_2 \to C_2 - a_{12}/a_{11}C_1$ $C_3 \to C_3 - a_{13}/a_{11}C_1$	$\begin{bmatrix} 1 & 0 & 0 \\ 0 & a_{22}/a_{21} - a_{12}/a_{11} & a_{23}/a_{21} - a_{13}/a_{11} \\ 0 & a_{32}/a_{31} - a_{12}/a_{11} & a_{33}/a_{31} - a_{13}/a_{11} \end{bmatrix} \equiv \begin{bmatrix} 1 & 0 & 0 \\ 0 & a_{22}^0 & a_{23}^0 \\ 0 & a_{32}^0 & a_{33}^0 \end{bmatrix}$
$R_2 \to R_2/a_{22}^0$	$\begin{bmatrix} 1 & 0 & 0 \\ 0 & 1 & a_{23}^0/a_{22}^0 \\ 0 & a_{32}^0 & a_{33}^0 \end{bmatrix}$
$R_3 \to R_3 - a_{32}^0 R_2$	$\begin{bmatrix} 1 & 0 & 0 \\ 0 & 1 & a_{23}^0/a_{22}^0 \\ 0 & 0 & a_{33}^0 - \left(a_{23}^0/a_{22}^0\right)a_{32}^0 \end{bmatrix} = \begin{bmatrix} 1 & 0 & 0 \\ 0 & 1 & a_{23}^1 \\ 0 & 0 & a_{33}^1 \end{bmatrix}$
$R_2 \to R_2 - R_3\, a_{23}^1/a_{33}^1$ $R_3 \to R_3/a_{33}^1$	$\begin{bmatrix} 1 & 0 & 0 \\ 0 & 1 & 0 \\ 0 & 0 & 1 \end{bmatrix}$

Some Properties of Matrix Inverses

(a) If **A** and **B** both of order n, are non-singular, then

$$(\mathbf{AB})^{-1} = \mathbf{B}^{-1}\mathbf{A}^{-1} \tag{C.20}$$

(b) If **A** is non-singular $|\mathbf{A}^{-1}| = |\mathbf{A}|^{-1}$ (C.21)

(c) If **A** is a non-singular matrix of order $(p + n) \times (p + n)$ is partitioned as

$\mathbf{A} = \begin{pmatrix} \mathbf{A}_{11} & \mathbf{A}_{12} \\ \mathbf{A}_{21} & \mathbf{A}_{22} \end{pmatrix}$ where $\mathbf{A}_{11}$ is $p \times p$; $\mathbf{A}_{12}$ is $p \times n$; $\mathbf{A}_{21}$ is $n \times p$ and $\mathbf{A}_{22}$ is $n \times n$.

Then
$$\mathbf{A}^{-1} = \begin{pmatrix} \mathbf{B}_{11} & \mathbf{B}_{12} \\ \mathbf{B}_{21} & \mathbf{B}_{22} \end{pmatrix} \tag{C.22}$$

where $\mathbf{B}_{11} = (\mathbf{A}_{11} - \mathbf{A}_{12}\mathbf{A}_{22}^{-1}\mathbf{A}_{21})^{-1} = \mathbf{A}_{11.2}^{-1}$ is $p \times p$ matrix

$\mathbf{B}_{22} = (\mathbf{A}_{22} - \mathbf{A}_{21}\mathbf{A}_{11}^{-1}\mathbf{A}_{12})^{-1} = \mathbf{A}_{22.1}^{-1}$ is $n \times n$ matrix

$\mathbf{B}_{12} = -\mathbf{A}_{11}\mathbf{A}_{22.1}^{-1}$ is $p \times n$ matrix and

$\mathbf{B}_{21} = -\mathbf{A}_{22}^{-1}\mathbf{A}_{21}\mathbf{A}_{11.2}^{-1}$ is $n \times p$ matrix (C.23)

(d) If $\mathbf{A} = \begin{pmatrix} \mathbf{A}_{11} & \mathbf{O} \\ \mathbf{O} & \mathbf{A}_{22} \end{pmatrix}$, where $\mathbf{A}_{11}$ is $p \times p$ and $\mathbf{A}_{22}$ is $n \times n$ are non-singular then

$$\mathbf{A}^{-1} = \begin{pmatrix} \mathbf{A}_{11}^{-1} & \mathbf{O} \\ \mathbf{O} & \mathbf{A}_{22}^{-1} \end{pmatrix} \tag{C.24}$$

C.6 Rank of a Matrix

Let **A** be a matrix of order $m \times n$ $(m < n)$. The determinant of a submatrix of order r of **A** is often called a determinant of order r, of the matrix **A**.

A matrix is said to be of rank r iff it has at least one non-singular submatrix of order r and has no non-singular submatrix of order more than r.

If **A** is symmetric then the rank of the matrix can be determined by considering the leading principal minor of **A** and to see if determinants of these are non-zero. The largest order r, of the principal minor which has non-zero determinant is the rank of **A**.

The rank of **A** can also be determined (sometimes quite conveniently) by applying various rows and columns transformations to render one or more of its rows or columns to zeros. Total number of non-zero rows or columns after these transformations gives the rank of the matrix **A**.

As an example consider a matrix of order 3 × 4. The rank of this matrix can not exceed 3, since no determinant of larger order than 3 can be formed from this matrix. All the 4 determinants of order 3 should be examined. If at least one is non-zero, rank will be 3. If all four vanish, rank will be less than or equal to 2. If at least one determinant of order 2 is non-zero, the rank will be equal to 2.

$r(\mathbf{A})$ denotes the rank of the matrix **A**.

Some Properties of ranks

1. If **A** and **B** are conformable for multiplication, then

$$r(\mathbf{A}) \le \min\{r(\mathbf{A}), r(\mathbf{B})\} \tag{C.25}$$

2. If **A** and **B** are $p \times n$ matrices

$$r(\mathbf{A} + \mathbf{B}) \leq r(\mathbf{A}) + r(\mathbf{B}) \tag{C.26}$$

3. If **A** is $p \times n$, $r(\mathbf{A}) \leq \min\,(p.\ n)$ (C.27)
4. If **P** is $m \times m$, **Q** is $n \times n$ and both **P** and **Q** are non-singular, then for **A** which is $m \times n$

$$r(\mathbf{PAQ}) = r(\mathbf{A})$$

5. If $\mathbf{D} = \text{diag}\,(\lambda_1, \lambda_2, ..., \lambda_p)$, then

   $r(\mathbf{D})$ = number of non-zero $\lambda_j' s$. (C.28)

C.7 Trace of a Matrix

Sum of the diagonal elements of a square matrix is called its trace

In symbols, if $\mathbf{A} = ((a_{ij})_{n \times n}$

$$\text{tr}\,(\mathbf{A}) = \sum_{j=1}^{n} a_{ii} \tag{C.29}$$

Some Properties of Trace

1. If **A** is $m \times n$ and **B** is $n \times m$, then

$$\text{tr}\,(\mathbf{AB}) = \text{tr}\,(\mathbf{BA}) \tag{C.30}$$

   So if **x** is $p \times 1$ $\text{tr}\,(\mathbf{x}'\mathbf{x}) = \text{tr}\,(\mathbf{xx}')$
2. The trace of a square matrix is equal to the sum of the latent roots (or characteristic roots, to be defined in Section C.9) of the matrix.
3. If **A** and **B** are of same order

$$\text{tr}\,(\mathbf{A} + \mathbf{B}) = \text{tr}\,(\mathbf{A}) + \text{tr}\,(\mathbf{B}) \tag{C.31}$$

4. If α is a scalar number

$$\text{tr}\,(\alpha\mathbf{A}) = \alpha\ \text{tr}\,(\mathbf{A}) \tag{C.32}$$

5. If **S** is a non-singular matrix of order p and **A** is also of order $p \times p$ then

$$\text{tr}\,(\mathbf{S}^{-1}\mathbf{AS}) = \text{tr}\,(\mathbf{A}) \tag{C.33}$$

6. If a row or column transformation is applied on a matrix the resulting matrix is also of the same order and rank.
7. By applying the row and column transformations, the final matrix has one of the following forms

$$\begin{pmatrix} I_r \\ \cdots \\ O \end{pmatrix}, \begin{pmatrix} I_r & O \end{pmatrix}, \begin{bmatrix} I_r & \cdots & O \\ \cdots & & \cdots \\ O & \cdots & O \end{bmatrix} \text{ and } I_r \tag{C.34}$$

   where r is rank of the original matrix.

   The forms given in equation (C.34) are called normal forms.

C.8 Partitioned Matrices

Sometimes it is desirable to partition a matrix into rectangular blocks of elements. These blocks are the *sub matrices* of the original matrix.

In this section it will be seen how one can treat the original matrix as a matrix whose elements are submatrices.

The manner in which a matrix is partitioned is usually indicated by dashed lines and the resulting blocks are the submatrices.

(a) If **A** is an $m \times n$ matrix and the elements $((a_{ij}))$ are partitioned as follows

$$\begin{array}{c} \quad n_1 \text{ cols } n_2 \text{ cols } \ldots \quad n_p \text{ cols} \\ \mathbf{A} = \left[\begin{array}{cccc} \mathbf{A}_{11} & \mathbf{A}_{12} & \cdots & \mathbf{A}_{1p} \\ \mathbf{A}_{21} & \mathbf{A}_{22} & \cdots & \mathbf{A}_{2p} \\ \cdot & & & \\ \cdot & & & \\ \cdot & & & \\ \mathbf{A}_{q1} & \mathbf{A}_{q2} & \cdots & \mathbf{A}_{qp} \end{array}\right] \begin{array}{l} m_1 \text{ rows} \\ m_2 \text{ rows} \\ \cdot \\ \cdot \\ \cdot \\ m_q \text{ rows} \end{array} \end{array} \tag{C.35}$$

The elements in $n_1, n_2, \ldots, n_p$ columns and $m_1, m_2, \ldots, m_q$ rows (such that $\sum_{j=1}^{p} n_j = n$ and $\sum_{i=1}^{q} m_i = m$) form the blocks or submatrices.

These are denoted by $\mathbf{A}_{ij}$, $i = 1, \ldots, q$ and $j = 1, 2, \ldots, p$.

The dashed lines show how the elements of the matrix **A** are partitioned.

If B is conformable for multiplication with *A*, *B* is of order $n \times p$, then **B** also can be partitioned such that the partitions of *B* are also conformable for multiplication. Thus row partitions of **B** should be same as column partitions of **A**, as given below:

$$\begin{array}{c} \quad l_1 \text{ cols } l_2 \text{ cols } \ldots \quad l_k \text{ cols} \\ B = \left[\begin{array}{cccc} \mathbf{B}_{11} & \mathbf{B}_{12} & \cdots & \mathbf{B}_{1k} \\ \mathbf{B}_{21} & \mathbf{B}_{22} & \cdots & \mathbf{B}_{2k} \\ \cdot & & & \\ \cdot & & & \\ \cdot & & & \\ \mathbf{B}_{p1} & \mathbf{B}_{p2} & \cdots & \mathbf{B}_{pk} \end{array}\right] \begin{array}{l} n_1 \text{ rows} \\ n_2 \text{ rows} \\ \cdot \\ \cdot \\ \cdot \\ n_p \text{ rows} \end{array} \end{array} \tag{C.36}$$

The elements in n_i rows ($i = 1, 2, \ldots, p$) and l_j columns, ($j = 1, 2, \ldots, k$) such that $\sum_{j=1}^{k} l_j = q$ are $\mathbf{B}_{ij}$.

Multiplication of **A** and **B** is now possible.

The column partitioning of **A** is same as row partitioning of **B**, so multiplication of **A** and **B** can be done using submatrices $\mathbf{A}_{ij}$ and $\mathbf{B}_{jr}$. The product of **A** and **B** can be written as

$$\mathbf{AB} = \left((C_{ir})\right)_{q \times k} = \sum_{j=1}^{p} \mathbf{A}_{ij}\mathbf{B}_{jr} \quad \text{for } i = 1, 2, \ldots, q \;\; r = 1, 2, \ldots, k \tag{C.37}$$

(b) Partitioning of **A** of order $m \times n$ can also be done *row wise* and that of **B** of order $r \times s$ *column wise*.

Thus

$$\mathbf{A} = \begin{bmatrix} \mathbf{A}_1 \\ \mathbf{A}_2 \\ \cdot \\ \cdot \\ \cdot \\ \mathbf{A}_m \end{bmatrix}, \text{ where } \mathbf{A}_i = [a_{i1} \ldots\ldots a_{in}] \text{ is } i^{\text{th}} \text{ row of } \mathbf{A}. \tag{C.38}$$

and

$$\mathbf{B} = [\mathbf{B}_1 \mathbf{B}_s \ \ldots\ldots \ \mathbf{B}_s] \text{ where } \mathbf{B}_j = \begin{bmatrix} b_{1j} \\ b_{2j} \\ \cdot \\ \cdot \\ \cdot \\ b_{rj} \end{bmatrix} \text{ is the } j^{\text{th}} \text{ column of } \mathbf{B}$$

$$j = 1, 2, \ldots, s \tag{C.39}$$

(c) **A** can also be partitioned as:

$$\mathbf{A} = \begin{bmatrix} \mathbf{A}_1 & 0 & 0 & \ldots & 0 \\ 0 & \mathbf{A}_2 & 0 & \ldots & 0 \\ \cdot & & & & \\ \cdot & & & & \\ \cdot & & & & \\ 0 & 0 & 0 & \ldots & \mathbf{A}_r \end{bmatrix} \text{ where } \mathbf{A}_i \text{ is a submatrix of } \mathbf{A}$$

$$i = 1, \ldots, r \tag{C.40}$$

(d) Suppose **A**: $(p + n) \times (p + n)$ is partitioned as

$$\mathbf{A} = \begin{bmatrix} \mathbf{A}_{11} & \mathbf{A}_{12} \\ \mathbf{A}_{21} & \mathbf{A}_{22} \end{bmatrix} \begin{matrix} p \text{ rows} \\ n \text{ rows} \end{matrix} \tag{C.41}$$

Then if $\mathbf{A}_{11}$ and $\mathbf{A}_{22}$ are non-singular matrices

$$|\mathbf{A}| = |\mathbf{A}_{22}| \,.\, |\mathbf{A}_{11} - \mathbf{A}_{12}\mathbf{A}_{22}^{-1}\mathbf{A}_{21}| = |\mathbf{A}_{22}||\mathbf{A}_{11.2}| \tag{C.42}$$

and

$$|\mathbf{A}| = |\mathbf{A}_{11}| \,.\, |\mathbf{A}_{22} - \mathbf{A}_{21}\mathbf{A}_{11}^{-1}\mathbf{A}_{12}| = |\mathbf{A}_{11}||\mathbf{A}_{22.1}| \tag{C.43}$$

(e) If **A** is partitioned as in (C.41) then

$$\mathbf{A}^T = \begin{pmatrix} \mathbf{A}'_{11} & \mathbf{A}'_{21} \\ \mathbf{A}'_{12} & \mathbf{A}'_{22} \end{pmatrix} \tag{C.44}$$

C.9 Characteristic Roots and Vectors

In many problems of physical and social sciences, a problem arises that for a square matrix **A** of order *n*, it is required to find the scalars λ and the non-zero vectors **X**, which satisfy the following equation

$$\mathbf{AX} = \lambda\mathbf{X} \tag{C.45}$$

This problem is known as the *characteristic value problem.*

Equation (C.45) is equivalent to

$$(\mathbf{A} - \lambda\mathbf{I}_n)\,\mathbf{X} = 0 \tag{C.46}$$

A non-zero solution of **X** exists, iff

$$|\mathbf{A} - \lambda\mathbf{I}_n| = 0 \tag{C.47}$$

This equation is called *characteristic equation* of **A**.

LHS of equation (C.47) gives a polynomial of degree n in λ.

The n roots of the characteristic equation are called the *characteristic roots* (*or characteristic values or latent roots or eigen values*) of the matrix **A**.

If λ_i, $i = 1, 2, ..., n$ is a characteristic root of **A**, the equation (C.46) is satisfied for $\lambda = \lambda_i$ and has a non-zero solution for **X**. Every such solution **X** is called a *characteristic vector* of **A**. A solution $\mathbf{X} = \mathbf{X}_i$ for $\lambda = \lambda_I$ in (C.46) is called *characteristic vector* of **A** *corresponding* to *characteristic root* λ_i.

Characteristic vectors are also called *eigen vectors or latent vectors.*

Some Results

(a) If $\mathbf{A}_{n\times n}$ is a symmetric matrix, all the characteristic roots are real.

(b) If λ_i, $i = 1, 2, ..., n$; are the characteristic roots of $\mathbf{A}_{n\times n}$

$$|\mathbf{A}| = \prod_{i=1}^{n} \lambda_i \tag{C.48}$$

(i) Thus $r(\mathbf{A}) < n$, iff at least one λ_i is zero.

(ii) Rank of a symmetric matrix $\mathbf{A}_{n\times n}$ is the number of non zero characteristic roots of **A**.

(c) Let $\lambda_1, \lambda_2, ..., \lambda_k$ be distinct characteristic roots of a matrix **A** and $\mathbf{X}_1, \mathbf{X}_2, ..., \mathbf{X}_k$ be any non-zero characteristic vectors associated with these roots, respectively. Then $\mathbf{X}_1, \mathbf{X}_2, ..., \mathbf{X}_k$ are linearly independent.

(d) Characteristic roots of a diagonal matrix are the diagonal elements themselves.

(e) If **A** is a symmetric matrix

(i) The eigen vectors $\mathbf{X}_i$ and $\mathbf{X}_j$ corresponding to two eigen values λ_i and $\lambda_j (\lambda_i \neq \lambda_j)$ are orthogonal

i.e.

$$\mathbf{X}_i' \mathbf{X}_j = 0 \tag{C.49}$$

(ii) Let $\mathbf{\Gamma}_{n\times n}$ be the matrix consisting of the normalized eigen vectors of **A** as its columns. Then

$$\mathbf{\Gamma}' \mathbf{A} \mathbf{\Gamma} = \text{diag}\,(\lambda_1, \lambda_2, ..., \lambda_n) \tag{C.50}$$

C.10 Vector Spaces

Some concepts, such as size of an organism, an individual's health status, mental abilities of a school child, prices of household commodities etc. can not be represented by a single number. These have several components and each of these components can be represented by a single number.

For example the prices of 5 household commodities may be represented as $(p_1, p_2, p_3, p_4, p_5)$. Such an ordered set of prices may be represented by **p**, which is a *row vector* of order 1 × 5.

Thus a vector of order $n \times 1$, is an ordered set of n numbers, which is a *column vector* of n components given by

$$\mathbf{X} = \begin{bmatrix} X_1 \\ X_2 \\ \cdot \\ \cdot \\ \cdot \\ X_n \end{bmatrix} \tag{C.51}$$

X_i's are real numbers and are called components or elements of vector **X**.

If **X** is a column vector of n components **X′** is a row vector of order $1 \times n$, written as

$$\mathbf{X}' = [X_1, X_2, ..., X_n] \tag{C.52}$$

Special Kinds of Vectors

(a) *Null vector or zero vector* is a vector with all the components equal to zero. It is denoted by **0** and may be column vector or a row vector.

(b) *Unit Vector.* If all the elements of a row vector or a column vector are 1, it is called a unit vector, and is denoted by **1.** It may be a row vector or a column vector.

(c) Vectors **X** and **Y** are said to be *equal*, if they are of the same order and corresponding elements are equal. Thus $\mathbf{X}_{n\times1}$ and $\mathbf{Y}_{n\times1}$ are equal (denoted by **X** = **Y**), iff $X_i = Y_i$ for $i = 1, 2, ..., n$.

Vector addition is indicated by + sign and is defined for two vectors of the same order and gives a unique vector in the same set of vectors (or vector space).

Thus
$$\mathbf{X} + \mathbf{Y} = \mathbf{Z} \tag{C.53}$$
where $Z_i = X_i + Y_i$; $i = 1, 2, ..., n$.

Negative Vector

If **X** is vector of order $n \times 1$, –**X** is also $n \times 1$ vector consisting of elements $-X_i$, $i = 1, 2, ..., n$

For such vector $\mathbf{X} + (-\mathbf{X}) = \mathbf{0}$ (C.54)

Vector subtraction is indicated by – sign between two vectors of same order.

$$\mathbf{X} - \mathbf{Y} = \mathbf{S} \text{ where } S_i = X_i - Y_i \text{ for } i = 1, 2, ..., n \tag{C.55}$$

Scalar Multiplication of Vector

If k is a real number

$$k\mathbf{X} = \begin{bmatrix} kX_1 \\ kX_2 \\ \cdot \\ \cdot \\ \cdot \\ kX_n \end{bmatrix} \tag{C.56}$$

The following results hold

(i) $\mathbf{X} + \mathbf{Y} = \mathbf{Y} + \mathbf{X}$
(ii) $(\mathbf{X} + \mathbf{Y}) + \mathbf{Z} = \mathbf{X} + (\mathbf{Y} + \mathbf{Z})$
(iii) $\mathbf{X} + \mathbf{0} = \mathbf{X}$
(iv) $k_1(k_2\mathbf{X}) = (k_1k_2)\,\mathbf{X}$, where k_1 and k_2 are two scalar numbers
(v) $k(\mathbf{X} + \mathbf{Y}) = k\mathbf{X} + k\mathbf{Y}$
(vi) $(k_1 + k_2)\,\mathbf{X} = k_1\mathbf{X} + k_2\mathbf{X}$
(vii) $\mathbf{0X} = \mathbf{0}$, $1\mathbf{X} = \mathbf{X}$, $-1\mathbf{X} = -\mathbf{X}$

A collection of vectors (with the associated field of scalars) which satisfy the above results (i) to (vii) and $\mathbf{X} + (-\mathbf{X}) = \mathbf{0}$ is called a linear vector space $\mathcal{V}_F$.

A linear subspace, subspace or linear manifold in a vector space $\mathcal{V}_F$ is a subset of vectors $\mathcal{M}$ closed under addition and scalar multiplication. In symbols

$$\text{if } \mathbf{X} \text{ and } \mathbf{Y} \in \mathcal{M} \text{ then } (c\mathbf{X} + d\mathbf{Y}) \in \mathcal{M} \tag{C.57}$$

for any scalars c and d.

Any such subset $\mathcal{M}$ is itself a vector space.

Let $\mathbf{X}_1, \mathbf{X}_2, ..., \mathbf{X}_k$ are a set S of vectors. *All linear combinations* $\left(= \sum_{i=1}^{k} c_i X_i\right)$ *is a subspace* spanned by S may be denoted as $\mathcal{M}(S)$.

If $\mathbf{A}$ is of order $m \times n$ with real elements, and $\mathbf{X}_{n\times1}$ is a vector of n unknown variables X_i, $i = 1, 2, ..., n$, then the totality of solutions $\mathbf{x}_{n\times1}$ of the m linear equations, written in matrix form as

$$\mathbf{AX} = \mathbf{0} \tag{C.58}$$

constitutes a linear subspace.

Basis of a Vector Space

A set of vectors $\mathbf{X}_1, \mathbf{X}_2, ..., \mathbf{X}_k$ is said to be *linearly dependent* if there exist scalars $C_1, C_2, ..., C_k$ (not all zero), such that

$$\sum_{i=1}^{k} C_i\mathbf{X}_i = \mathbf{0} \tag{C.59}$$

If (C.59) holds only when all C_i's are zero, the set of vectors is *linearly independent*.

Some Results

(i) Null vector is a dependent set.

(ii) Any set of vectors containing a null vector is a dependent set.

(iii) A set of non zero vectors is a dependent set if and only if one of vectors is a linear combination of the others in the set.

A linearly independent subset of vectors in a vector space $\mathcal{V}$, which generates or spans $\mathcal{V}$ is called a *basis* of $\mathcal{V}$.

If one chooses sequentially non-null vectors X_1, X_2, ... in $\mathcal{V}$ such that no $\mathbf{X}_i$ is dependent on its predecessors. In this process suppose after choice of kth vector $\mathbf{X}_k$, there is no linearly independent vector in $\mathcal{V}$. *Then* X_i, X_2, ..., X_k *is a basis of* $\mathcal{V}$ *and k is the dimension of the vector space* $\mathcal{V}$.

Some Results About Basis

(i) Every vector space has a basis

(ii) If $\mathbf{X}_1, \mathbf{X}_2, ..., \mathbf{X}_k$ and $\mathbf{Y}_1, \mathbf{Y}_2, ..., \mathbf{Y}_s$ are two sets of basis vectors of a vector space $\mathcal{V}$, then $s = k$.

(iii) Every vector in $\mathcal{V}$ has a unique representation in terms of a given basis.

Scalar Product of Two Vectors

If **X** and **Y** are two vectors of the same order, their scalar product is a scalar quantity denoted by

$$\mathbf{X}'\mathbf{Y} = \sum_i X_i Y_i \tag{C.60}$$

C.11 Quadratic Forms

Let **X** be $p \times 1$ vector and $\mathbf{A} = ((a_{ij}))$ be a symmetric matrix of order p.
Then the function $Q(\mathbf{X}) = \mathbf{X}'\mathbf{A}\mathbf{X}$ is called a quadratic form in **X**. It can be seen that

$$Q(\mathbf{X}) = \mathbf{X}'\mathbf{A}\mathbf{X} = \sum_{i=1}^{p}\sum_{j=1}^{p} X_i X_j a_{ij} \tag{C.61}$$

For example if $\mathbf{X} = \begin{pmatrix} X_1 \\ X_2 \end{pmatrix}$ and $\mathbf{A} = \begin{pmatrix} a_{11} & a_{12} \\ a_{12} & a_{22} \end{pmatrix}$

$$\mathbf{X}'\mathbf{A}\mathbf{X} = [X_1 X_2]\begin{bmatrix} a_{11} & a_{12} \\ a_{12} & a_{22} \end{bmatrix}\begin{bmatrix} X_1 \\ X_2 \end{bmatrix} = \begin{bmatrix} a_{11}X_1 + a_{12}X_2 & a_{12}X_1 + a_{22}X_2 \end{bmatrix}\begin{bmatrix} X_1 \\ X_2 \end{bmatrix}$$

$$= a_{11}X_1^2 + a_{12}X_2X_1 + a_{12}X_1X_2 + a_{22}X_2^2$$

$$= a_{11}X_1^2 + 2a_{12}X_1X_2 + a_{22}X_2^2$$

(a) If $Q(\mathbf{X}) > 0$ for all $\mathbf{X} \neq \mathbf{0}$, $Q(\mathbf{X})$ is called a *positive definite quadratic form* and the matrix **A** is a *positive definite matrix* and it is often written that $\mathbf{A} > 0$.

(b) If $Q(\mathbf{X}) < 0$ for all $\mathbf{X} \neq \mathbf{0}$, $Q(\mathbf{X})$ is called a *negative definite quadratic form* and the matrix **A** is a *negative definite matrix* and it is written that $\mathbf{A} < 0$.

(c) If $Q(\mathbf{X}) \geq 0$ for all $\mathbf{X} \neq \mathbf{0}$, $Q(\mathbf{X})$ is called a *positive semidefinite quadratic form*, and the matrix **A** is a *positive semidefinite matrix* and it is written as $\mathbf{A} \geq 0$.

(d) If none of these holds, Q is called an *indefinite quadratic form* and **A** is called an *indefinite matrix*.

Some Properties

Some of the properties given by Press (1972) and Dillon and Goldstein (1984) are given below:

1. If **A** > 0, then all the eigen values $\lambda_1, \lambda_2, ..., \lambda_p$ of **A** are positive. If **A** ≥ 0, then $\lambda_i \geq 0$ for i = 1, 2, ..., p and $\lambda_i = 0$ for some i.
2. If **A** > 0, **A** is nonsingular and hence |**A**| > 0.
3. If **A** > 0 then $\mathbf{A}^{-1} > 0$.
4. Let **A** be $n \times n$, **B** be $p \times n$, $p \leq n$ and $r(\mathbf{B}) = s$.

   (a) If **A** > 0, then
   BAB′ > 0 if $s = p$ and **BAB′** ≥ 0 if $s < p$
   (b) If **A** > 0 and **C** is $p \times p$ and nonsingular, then **C′AC** > 0
   (c) If **A** ≥ 0, then **BAB′** ≥ 0

5. If **A** ≥ 0 and **B** > 0 are $p \times p$ matrices then non-zero characteristic roots of $\mathbf{B}^{-1}\mathbf{A}$ are positive.
6. If **A** > 0, there exists at least one matrix **B** for which **A** = **BB′**.
7. *Hadamard's Inequality* (1893): If $\mathbf{A} = ((a_{ij}))_{p \times p} > 0$, $|\mathbf{A}| \leq \prod_{i=1}^{p} a_{ii}$
8. If **A** > 0, there exists a nonsingular matrix **M** such that **MAM′** = **I**
9. Let **A** and **B** be symmetric matrices of same order and $\lambda_j(\mathbf{A})$ and $\lambda_j(\mathbf{B})$ be j^{th} latent roots of **A** and **B**, respectively. Then
   if $\mathbf{A} - \mathbf{B} \geq 0$, $\lambda_j(\mathbf{A}) \geq \lambda_j(\mathbf{B})$ for all j.
   (Proof in Beckenbach and Bellman (1961)).

C.12 Matrix Factorization

The following ways of matrix factorization given by Press (1972) and Anderson (1972) are quite useful in multivariate analysis, so are given below.

(a) If **A** is symmetric of order p, there exists an orthogonal matrix **Γ** such that

$$\mathbf{A} = \mathbf{\Gamma D_\lambda \Gamma'} \tag{C.62}$$

where $\mathbf{D}_\lambda = \text{diag}(\lambda_1, \lambda_2, ..., \lambda_p)$ and λ_j's are latent roots of **A**, some of which may be zero and some may be multiple roots. The columns of **Γ** are an orthonormal set of latent vectors of **A**.

(b) Let **A** > 0, **B** ≥ 0. Then there exists a nonsingular matrix **P**, such that

$$\mathbf{PAP'} = \mathbf{I} \text{ and } \mathbf{PBP'} = \text{diag}(\lambda_1, \lambda_2, ..., \lambda_p) \tag{C.63}$$

where λ_j's are the root of $|\mathbf{B} - \lambda\mathbf{A}| = 0$.

(c) *Vinograd's Theorem* (1950): Let **A** and **B** be of order $p \times n$ such that **AA′** = **BB′**. Then there exists an orthogonal matrix **Δ** of order $n \times n$ satisfying the relation that **A** = **BΔ**.

(d) Let **A** be a $p \times n$ matrix ($p \leq n$), then there exists orthogonal matrix **Γ** of order p and **Δ** of order n, such that

$$\mathbf{A} = \mathbf{\Gamma}(\mathbf{D}_\lambda \mathbf{O})\,\mathbf{\Delta}' \tag{C.64}$$

where $\mathbf{D}_\lambda = \text{diag}(\lambda_1, \lambda_2, ..., \lambda_p)$ and $|\mathbf{AA}' - \lambda^2 \mathbf{I}| = 0$

If $n = p$ and $\mathbf{A} = \mathbf{A}'$ and choosing $\mathbf{\Delta} = \mathbf{\Gamma}$, (C.64) reduces to (C.62)

(e) If $\mathbf{A}$ is $p \times n (p \leq n)$ and (C.64) hold

$$\mathbf{D}_\lambda^* = \text{diag}\left(\lambda_1^*, \lambda_2^*, ..., \lambda_p^*\right) \tag{C.65}$$

where $\lambda_i^* = \dfrac{1}{\lambda_i}$ if $\lambda_i \neq 0$ and $\lambda_i^* = \lambda_i$ otherwise.

The *pseudoinverse* of $\mathbf{A}$ is given by

$$\mathbf{A}^+ = \Delta\left(\mathbf{D}_\lambda^* \mathbf{O}\right)\Gamma' \tag{C.66}$$

(f) If $\mathbf{A}$ and $\mathbf{B}$ are symmetric matrices of order p and $\mathbf{AB} = \mathbf{BA}$ then there exists an orthogonal matrix $\mathbf{P}$ of order p such that

$$\mathbf{P'AP} = \text{diag}(\lambda_{a1}, \lambda_{a2}, ..., \lambda_{ap}) \text{ and}$$
$$\mathbf{P'BP} = \text{diag}(\lambda_{b1}, \lambda_{b2}, ..., \lambda_{bp}) \tag{C.67}$$

where λ_{aj} and λ_{bj}, $j = 1, 2, ..., p$ are the characteristic roots of $\mathbf{A}$ and $\mathbf{B}$, respectively.

(g) Let $\mathbf{A}$ be of order $m \times n$ of rank r, then there is a nonsingular matrix $\mathbf{P}$ of order m and an orthogonal matrix $\mathbf{Q}$ of order n such that

$$\mathbf{A} = \mathbf{P}\begin{pmatrix} \mathbf{I}_r & \mathbf{0} \\ \mathbf{0} & \mathbf{0} \end{pmatrix}\mathbf{Q} \tag{C.68}$$

C.13 Idempotent Matrix

A square symmetric matrix $\mathbf{A}$ of order p is said to be *idempotent* if

$$\mathbf{A}^2 = \mathbf{A} \tag{C.69}$$

Some properties of idempotent matrices as given in Press (1972) and Rao (1972) are

(a) Latent roots of an idempotent matrix $\mathbf{A}$ are 1 or 0.

(b) If $\mathbf{A}$ is idempotent and $\mathbf{A} \geq 0$

$$R(\mathbf{A}) = \text{tr}(\mathbf{A})$$

(c) If $\mathbf{A}$ is idempotent, $(\mathbf{I} - \mathbf{A})$ is also idempotent.

(d) If $\mathbf{A}$ is idempotent and symmetric, then $\mathbf{A} \geq 0$.

(e) If $\mathbf{A}$ is idempotent and $\mathbf{A} \geq 0$, there exists an orthogonal matrix Γ such that

$$\Gamma\mathbf{A}\Gamma' = \begin{pmatrix} \mathbf{I}_r & \mathbf{0} \\ \mathbf{0} & \mathbf{0} \end{pmatrix} \tag{C.70}$$

where $r = r(\mathbf{A})$ is the number of non zero latent roots of $\mathbf{A}$.

C.14 Generalized Inverse of a Matrix

Let $\mathbf{A}$ be a matrix of order $m \times n$ of any rank. A *generalized inverse or g-inverse* of $\mathbf{A}$ is an $n \times m$ matrix $\mathbf{A}^-$, such that $\mathbf{X} = \mathbf{A}^-\mathbf{Y}$ is a solution of the equation $\mathbf{AX} = \mathbf{Y}$, where $\mathbf{Y}$ is a vector belonging to the vector space spanned by **A** (cf. Rao (1973)). Some of the properties of g-inverse are given by Rao (1972), a few of which are

(a) If $\mathbf{A}^-$ is a g-inverse of $\mathbf{A}$, then $\mathbf{AA^-A} = \mathbf{A}$ (C.71)
and if $\mathbf{AA^-A} = \mathbf{A}$, then $\mathbf{A}^-$ is a g-inverse of $\mathbf{A}$.
(b) $\mathbf{A^-A} = \mathbf{H}$ is idempotent and $r(\mathbf{H}) = r(\mathbf{A})$
(c) $r(\mathbf{A}^-) \geq r(\mathbf{A})$

C.15 Matrix Calculus

Let **b** be an m component vector and $f(\mathbf{b})$ is a function of **b**. Then

$$\frac{df(\mathbf{b})}{d\mathbf{b}} = \begin{bmatrix} df(\mathbf{b})/db_1 \\ df(\mathbf{b})/db_2 \\ \cdot \\ \cdot \\ \cdot \\ df(\mathbf{b})/db_m \end{bmatrix} = \begin{bmatrix} df(b_1, b_2, ..., b_m)/db_1 \\ df(b_1, b_2, ..., b_m)/db_2 \\ \cdot \\ \cdot \\ \cdot \\ df(b_1, b_2, ..., b_m)/db_m \end{bmatrix} \tag{C.72}$$

The differentials of some functions of a vector are as given below:

(a) If $f(\mathbf{b})$ is a constant c, for all choices of **b**, then

$$\frac{df(\mathbf{b})}{d\mathbf{b}} = \frac{d(c)}{d\mathbf{b}} = \mathbf{0} \tag{C.73}$$

(b) If $f(\mathbf{b}) = a_1b_1 + a_2b_2 + ... + a_mb_m = \mathbf{a'b} = \mathbf{b'a}$, $\dfrac{df(\mathbf{b})}{db_i} = a_i$

So

$$\frac{df(\mathbf{b})}{d\mathbf{b}} = \frac{d(\mathbf{a'b})}{d\mathbf{b}} = \mathbf{a} = \frac{d(\mathbf{b'a})}{d\mathbf{b}} \tag{C.74}$$

(c) If $f(\mathbf{b}) = \mathbf{b'Ab}$ where **A** is symmetric, then

$$\frac{df(\mathbf{b})}{d\mathbf{b}} = \frac{d(\mathbf{b'Ab})}{d\mathbf{b}} = \frac{d\sum_i \sum_j a_{ij}b_ib_j}{d\mathbf{b}}$$

$$= 2\begin{bmatrix} \mathbf{a}_1'\mathbf{b} \\ \mathbf{a}_2'\mathbf{b} \\ \cdot \\ \cdot \\ \cdot \\ \mathbf{a}_m'\mathbf{b} \end{bmatrix} = 2\begin{bmatrix} \mathbf{b'a}_1 \\ \mathbf{b'a}_2 \\ \cdot \\ \cdot \\ \cdot \\ \mathbf{b'a}_m \end{bmatrix} = 2\mathbf{Ab} \tag{C.75}$$

where $\mathbf{a}_i$ is i^{th} row of **A**.

Reader interested in differential and integration of functions of matrices may consult the book by Press (1972).

Appendix D: Elements of Set Theory

In order to define probability, events should be defined first. Discussion of basic concepts of probability and probability distribution using some ideas and results of set theory becomes very easy and convenient to understand. This subject is also an extensive one but only a few basic rules will be given and explained.

Some Definitions

Set: A set is a collection of objects. Sets are usually denoted by capital letters $A, B, C, \ldots$ etc. To indicate which objects are contained in the set A, the following three methods are used:

(a) List the contents of A, separated by comma and enclosed in a bracket as $A = \{1, 2, 4, 8, 10\}$ describes that set A consists of integers 1, 2, 4, 8 and 10. $A = \{HH, HT, TH, TT\}$ describes the result of toss of two coins, H denoting the head and T the tail.

(b) The set A may be described in words. Thus one can say that A consists of all real numbers between 2 and 4 both inclusive.

(c) A set in (b) can be denoted as $A = \{x \mid 2 \leq x \leq 4\}$ which indicates that A consists of all real numbers between 2 and 4, both inclusive.

Elements

The individual objects making up the collection in the set A are called elements or points or members of A. When a is a member of set A it can be expressed as $a \in A$. When a does not belong to A one can say it by writing $a \notin A$.

Universal Set

It is the set of all the objects under consideration in a given discussion and is denoted by Ω. An element of Ω may be denoted by ω.

Empty or Null Set

If a set has no point or no element, it is called a null set or empty set and is denoted by ϕ. For example a set of real numbers x such that $x^2 + 1 = 0$ is a null set.

Subset

If every element of a set A is also an element of B, A is defined to be a subset of B and it is denoted by $A \subset B$ or $B \supset A$. This is read as 'A is contained in B' or 'B contains A'.

Equivalent or Equal Sets

If $A \subset B$ and $B \subset A$ it means that A and B contain the same members and then A and B are said to be equivalent. This is indicated by writing $A = B$.

Complement of a Set

Is denoted by $\bar{A}$ or A^C. It consists of all the elements not in A but in Ω. Thus $A^C = \Omega - A = \{x \mid x \notin A\}$.

A graphic design, known as Venn diagram can be used to describe combination of sets. In each diagram shaded region is the considered set.

Union

Let A and B be two sets in Ω, then the set of points that belong to either A or B or to both A and B, is called union of A and B and is written as $A \cup B$.

Intersection

Let A and B be two sets in Ω, then the set consisting of all the points belonging to A and B both is defined as intersection of A and B and is written as $A \cap B$ or AB.

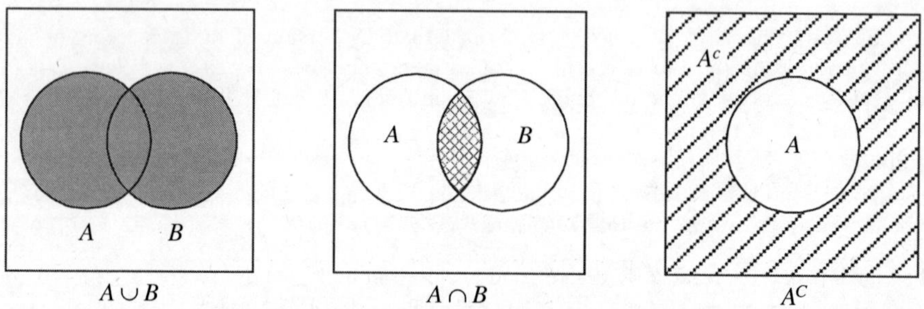

Fig. D.1. $A \cup B$, $A \cap B$ or AB and A^C or $\bar{A}$.

The set operations of union, intersection and complement *satisfy quite a number of laws*, some of which are:

1. *Commutative laws*: $A \cup B = B \cup A$ and $A \cap B = B \cap A$.
2. *Associative laws*: $A \cup (B \cup C) = (A \cup B) \cup C$ and $A \cap (B \cap C) = (B \cap A) \cap C$.
3. *Distributive laws*: $A \cap (B \cup C) = (A \cap B) \cup (A \cap C)$ and $A \cup (B \cap C) = (A \cup B) \cap (A \cup C)$.

There are some identities also involving these operations

1. $(A^C)^C = A$
2. $A\,\Omega = A$; $A \cup \Omega = \Omega$; $A\,\phi = \phi$; and $A \cup \phi = A$
3. $AA^C = \phi$; $A \cup A^C = \Omega$; $A \cap A = A$ and $A \cup A = A$
4. $\overline{(A \cup B)} = \bar{A} \cap \bar{B}$ and $\overline{(A \cap B)} = \bar{A} \cup \bar{B}$. These are also known as *De Morgan's laws*
5. $A - B = A\bar{B}$

References

1. Adams, J.C., Brainerd, W.S., Martin, J.T., Smith, B.T. and Wagener, J.L., 1992, *FORTRAN 90 Handbook*, McGraw Hill, New York.
2. Adams, J.C., Brainerd, W.S., Martin, J.T., Smith, B.T. and Wagener, J.L., 1997, *FORTRAN 95 Handbook*, MIT Press, Cambridge, Massachusetts.
3. Afifi, A.A. and Clark, V., 1984, *Computer Aided Multivariate Analysis*, Belmont, California: Lifetime Learning Publications.
4. Altschul, S.F., 1991, "Amino Acid Substitution Matrices from an Information Theoretic Perspective", *Jour. Mol. Biol.*, 219, pp. 555-565.
5. Altschul, S.F., Gish, W., Miller, W., Meyers, E.W. and Lipman, D.J., 1990, "Basic Local Alignment Search Tool", *Jour. Mol. Biol.*, 215, pp. 403-410.
6. Anderberg, M.R., 1973, *Cluster Analysis for Applications,* Academic Press.
7. Anderson, T.W., 1958, *An Introduction to Multivariate Statistical Analysis,* John Wiley & Sons, New York.
8. Anderson, T.W., 1972, *An Introduction to Multivariate Statistical Analysis*, Wiley Eastern Pvt. Ltd., New Delhi (John Wiley & Sons Inc. NY).
9. Anderson, T.W., 1963, "Asymptotic Theory for Principal Component Analysis", *Annals of Math. Statist.*, 34, No. 2, pp. 122-148.
10. Anderson, T.W. and Rubin, H., 1956, *Statistical Inference in Factor Analysis. Proceedings of the Third Berkeley Symposium on Mathematical Statistics and Probability*, University of California Press Berkeley and LA USA, Vol. V, pp. 111-150.
11. Armstrong, J.S., 1967, "Deviation of Theory by Means of Factor Analysis or Tom Swift and His Electric Factor Analysis Machine", *The Amer. Statistician*, 21, pp. 17-21.
12. Baldi, P. and Brunak, S., 2003, *Bioinformatics, The Machine Learning Approach*, Affiliated East West Press.
13. Bartlett, M.S., 1937a, "Some Examples of Statistical Methods of Research in Agriculture and Applied Biology", *Jour. Roy. Statist. Soc. Suppl.*, 4, pp. 137-170.
14. Bartlett, M.S., 1937b, "Properties of Sufficiency and Statistical Tests", *Proc. Royal Statist. Soc. Ser. A.*, 160, pp. 268-282.
15. Bartlett, M.S., 1954, "A Note on the Multiplying Factors for Various χ^2-approximations", *Jour. Royal Statist. Soc.*, B 16: pp. 296-298.
16. Behrens, W.U., 1929, "Ein Beitrag Zur Fehlerberechnung bei Wenigen Beobachtungen", *Landwirtsch Jahrbucher*, 68, pp. 607-837.
17. Bell, C.G. and Newell, A., 1980, *Computer Structure, Readings and Examples*, 2nd Ed., McGraw Hill, New York.

18. Bennett, S. and Bowers, D., 1976, *An Introduction to Multivariate Techniques for Social and Behavioural Sciences*, The Macmillan Press Ltd., London and Basingstoke.
19. Berkson, J., 1944, Application of the Logistic Function to Bio-assay", *Jour. of Amer. Statist. Assoc.*, 39, pp. 357-365.
20. Bioinformatics Institute of India, Noida, New Delhi, 2003, *Biostatistics,* Bioinfo. Inst. of India, New Delhi.
21. Blackith, R.E. and Reyment, R.A., 1971, *Multivariate Morphometrics*, Academic Press, London.
22. Bliss, C.I., 1935a, "The Calculation of the Dosage-mortality Curve", *Annals of Applied Biology*, 22, pp. 134-167.
23. Bliss, C.I., 1935b, "The Comparison of Dosage Mortality Data", *Annals of Applied Biology*, 22, pp. 307-333.
24. Bliss, C.I., 1938, "The Determination of Dosage – Mortality Curves from Small Numbers", *Quarterly Journal of Pharmacy and Pharmacology*, 11, pp. 192-216.
25. Bliss, C.I., 1940, "Quantitative Aspects of Biological Assay", *Jour. of Amer. Pharmaceutical Assoc.*, 33, pp. 225-245.
26. Bliss, C.I. and Cattell, Mc.K., 1943, Biological Assay, *Annual Review of Physiology*, 5, pp. 479-539.
27. Boneau, C.A., 1960, "The Effects of Violations of Assumptions Underlying the t-test", *Psych. Bull.*, 57, pp. 49-64.
28. Bose, R.C. and Nair, K.R., 1939, Partially Balanced Incomplete Block Design, *Sankhya*, 4, pp. 337-372.
29. Bose, R.C. and Roy, S.N., 1938, The Exact Distribution of the Studentized D^2-statistic, *Sankhya*, 4, pp. 10-38.
30. Box, G.E.P., 1949, "A General Distribution Theory for a Class of Likelihood Ratio Criteria", *Biometrika*, 36, pp. 317-346.
31. Box, G.E.P., 1953, Non Normality and Tests on Variances, *Biometrika*, 40, pp. 318-355.
32. Box, G.E.P. and Cox, D.R., 1964, "An Analysis of Transformations", *Jour. Royal Statist. Soc.*, B 26, pp. 211-243.
33. Box, G.E.P. and Jenkins, G.M., 1978, "*Time Series Analysis, Forecasting and Control*", Revised Ed. Holden Day, San Fransisco, USA.
34. Box, G.E.P. and Pierce, D.A., 1970, "Distribution of Residual Autocorrelations in Autoregressive Integrated Moving Average Time Series Models", *Jour. of Amer. Statist. Assoc.*, 65, pp. 1509-1526.
35. Box, J.F., 1978, *R.A. Fisher, The Life of a Scientist*, Wiley, New York.
36. Breslow, N.E., 2005, *Handbook of Epidemiology*, Springer.
37. Burr, E.H., 1960, The Distribution of Kendall's Score S for a Pair of Tied Rankings, *Biometrika*, 47, pp. 151-171.
38. Burt, C., 1966, The Early History of Multivariate Techniques in Psychological Research, *MBR*, 1, pp. 24-42.
39. Busby, M. and Stultz, R.A., 2000, *Microsoft Office*, BPB Publications, New Delhi.
40. Cain, A.J., 1962, The Evolution of Taxonomic Principles, in *Microbial Classification* (Eds. G.C. Ainsworth and PHA Sneath), Cambridge Univ. Press, pp. 1-13.
41. Chandel, SR.S., 1964, *A Handbook of Agricultural Statistics*, Achal Prakashan Mandir.
42. Chapman, S.J., 1998a, *Introduction to Fortran 90/95 for Scientists and Engineers*, McGraw Hill, Boston.

43. Chapman, S.J., 1998b, *Fortran 90/95 for Scientists and Engineers*, McGraw Hill, Boston.
44. Clatworthy, W.H., 1955, "Partially Balanced Incomplete Blocks Design with Two Associate Classes and Two Treatments Per Block", *Nat. Bur. Standard Jour. Res.*, 54, pp. 174-189.
45. Cochran, W.G., 1947, "Some Consequences When the Assumptions for Analysis of Variance are Not Satisfied", *Biometrics*, 3, pp. 22-38.
46. Cochran, W.G., 1952, "The χ^2-test of Goodness of Fit", *Annals of Math. Statist.*, 23, pp. 315-345.
47. Cochran, W.G., 1954, "Some Methods for Strengthening the Common χ^2-tests", *Biometrika*, 10, pp. 417-451.
48. Cochran, W.G., 1974, *Sampling Techniques*, John Wiley, USA.
49. Cochran, W.G. and Cox, G.M., 1957, 1974, *Experimental Designs*, Wiley Eastern Pvt. Ltd.
50. Cooper, J., 2001, *Teach Yourself Mac OS 85 Using MS DOS 6.22.* Adobe e-Book.
51. Cormack, R.M., 1971, "A Review of Classification (with discussion)", *Jour. of Royal Statist. Soc.*, A 134, pp. 321-367.
52. Cox, D.R., 1958, *Planning of Experiments*, John Wiley & Sons, NY, USA.
53. Croxton, F.E., Cowden, D.J. and Klein, S., 1975, *Applied General Statistics*, 3rd Edition. Prentice Hall of India Pvt. Ltd, New Delhi.
54. Dale, M., 1939, "Biological Standardization", *Analyst*, 64, pp. 554-567.
55. Daniel, W.W., 1987, *Biostatistics: A Foundation for Analysis in Health Sciences*, John Wiley.
56. Das, M.N. and Giri, N.C., 1979, *Design and Analysis of Experiments,* Wiley Eastern Pvt. Ltd.
57. Das, S., 2005, "Gene Finding Using Hidden Markov Model", *Unpublished Project Report Submitted in the Dept. of Computer Science*, BHU, India.
58. Dayhoff, M.O., 1978, "Survey of New Data and Computer Methods of Analysis", *In Atlas of Protein Sequence and Structure vol 5, suppl. 3*, National Biomedical Research Foundation, Georgetown Univ., Washington DC.
59. Dickey, D.A. and Fuller, W.A., 1979, "Distribution of the Estimators of Autoregressive Time Series with a Unit Root", *Jour. of Amer. Statist. Assoc.*, 74, pp. 427-431.
60. Diday, E. and Simon, J.C., 1976, "Cluster Analysis", in *Communication and Cybernetics 10 Digital Pattern Recognition* (Ed. K.S. Fu), Springer Verlag, Berlin, pp. 47-94.
61. Dienes, S.S., 1996, *Microsoft Office Professional: Instant Reference.* BPB Publications, New Delhi.
62. Dillon, W.R. and Goldstein, M., 1984, *Multivariate Analysis, Methods and Applications*, John Wiley & Sons, USA.
63. Dixon, W.J., 1985, *BMDP Statistical Software Manual*, University of California Press.
64. Dixon, W.J. and Massey, F.J., 1969, *Introduction of Statistical Analysis* 3rd Edition, McGraw Hill, Kogakusha.
65. Duncan, D.B., 1951, "A Significance Test for Difference between Ranked Treatments in an Analysis of Variance", *Var. Jour. Sci.*, 2, pp. 171-189.
66. Duncan, D.B., 1955, "Multiple Range and Multiple F-tests", *Biometrics*, 11, pp. 1-42.
67. Duncan, R., 1988, *Advanced MS-DOS Programming*, Microsoft Press.
68. Duncan, R., 1988, *MS-DOS Encyclopedia.* Microsoft Press, Redmond, WA.
69. Dunnett, C.W., 1955, "A Multiple Comparison Procedure for Comparing Several Treatments with Controls", *Jour Amer. Statist. Assoc.*, 50, pp. 1096-1121.
70. Dyson, P. and Coleman, P., 2000, *Windows 2000 Professional: In Record Time*, Sybex Inc.
71. Emmens, C.W., 1940, "The Dose/response Relation for Certain Principles of the Pituitary Glands and the Serum and Urine of Pregnancy", *Jour of Endocrinology*, 2, 194-225.

72. Engeman, R.M., Sugihara, R.T., Pank, L.F. and Dusenberry, W.E., 1994, "A Comparison of Plotless Density Estimators Using Monte Carlo Simulation", *Ecology*, 75(6), pp. 1769-1779.
73. Everitt, B., 1980, *Cluster Analysis, 2nd Ed.* Heinemann Educational Books, London.
74. Ewens, W.J. and Grant, G.R., 2001, 2005, *Statistical Methods in Bioinformatics: An Introduction*, Springer-Verlag, New York.
75. Feller, W., 1966, *An Introduction to Probability Theory and its Applications.* John Wiley, New York.
76. Fieller, E.C., 1940, "The Biological Standardization of Insulin", *Jour. of Royal Statist. Soc.*, Supplement 7, pp. 1-64.
77. Finney, D.J., 1945, "The Microbiological Assay of Vitamins: The Estimate and its Precision", *Quarterly Jour. of Pharmacy and Pharmacology*, 18, pp. 77-82.
78. Finney, D.J., 1947, "The Principles of Biological Assay", *Jour. of the Royal Statist. Soc.*, Supplement 9, pp. 46-91.
79. Finney, D.J., 1947, *Probit Analysis: A Statistical Treatment of the Sigmoid Response Curve (2nd Edition)*: Cambridge University Press, London.
80. Finney, D.J., 1964, *Statistical Method in Biological Assay (2nd Edition)* Charles Griffin & Company Limited, London.
81. Finney, D.J., 1981, *Probit Analysis (3rd Edition),* S. Chand & Company Ltd., New Delhi, India.
82. Fisher, R.A., 1921, "On the Probable Error of a Coefficient Correlation Deduced From a Small Sample", *Metron* 1, Part 4, pp. 3-32.
83. Fisher, R.A., 1940, "An Examination of the Different Possible Solution of a Problem in Incompleted Blocks", *Ann. Eugen.*, 10, pp. 52-75.
84. Fisher, R.A., 1947, *The Design of Experiments.* Oliver and Boyd.
85. Fisher, R.A. and Yates, F., 1950, *Contribution to Mathematical, Statistics*, Wiley, New York.
86. Fisher, R.A. and Yates, F., 1963, *Statistical Tables for Biological, Agricultural and Medical Research, 6th Ed.* Hafner, New York.
87. Fisher, R.A. and Yates, F., 1963, *Statistical Tables for Biological, Agricultural and Medical Research (6th Edition),* Oliver and Boyd, Edinburg.
88. Forthofer, R.N. and Lee, E.S., 2006, *Introduction to Biostatistics: A Guide to Design, Analysis and Discovery*, Academic Press (An imprint of Elsevier).
89. Friedman, M., 1937, "The Use of Ranks to Avoid the Assumption of Normality Implicated in the Analysis of Variance", *Jour. Amer. Statist. Assoc.*, 32, pp. 675-701.
90. Friedman, M., 1940, "A Comparison of Alternate Test of Significance for the Problems of m Rankings", *Ann. Math. Statist.*, 11, pp. 86-92.
91. Frutcher, B., 1954, *Introduction to Factor Analysis,* D Van Nostrand Co. Ltd, New York.
92. Fuller, W.A., 1976, *Introduction to Statistical Time Series*, John Wiley & Sons, New York.
93. Garnett, J.C.M., 1919, "On Certain Independent Factors in Mental Measurement", *Proc. Royal Soc. of London* 96, pp. 91-111.
94. Gautier, R., 1945, "The Health Organization and Biological Standardization", *Bulletin of the Health Organization of the League of Nations* 12, pp. 1-75.
95. Gibbons, J.D., 1971, *Non Parametric Statistical Inference.* McGraw Hill, Kogakusha.
96. Gill, J.L., 1971, "Analysis of Data with Heterogeneous Variances", *Jour. of Dairy Science* 54 No. 3, pp. 369-373.
97. Girshick, M.A., 1936, "Principal Components", *Jour. Amer. Statist. Assoc.*, 31, pp. 519-528.
98. Glimour, J.S.L., 1951, "The Development of Taxonomic Theory Since 1851", *Nature*, 168, pp. 400-402.

99. Goon, A.M., Gupta, M.K. and Dasgupta, B., 1983, *Fundamentals of Statistics Vol 1.* World Press Pvt. Ltd. Kolkata, India.
100. Goon, A.M., Gupta, M.K. and Dasgupta, B., 2002, *Fundamentals of Statistics Vol 2.* World Press Pvt. Ltd. Kolkata, India.
101. Gordon, A.D., 1981, *Classification*, Chapman and Hall, London, New York.
102. Gorsuch, R.L., 1974, *Factor Analysis*, Saunders, Philadelphia.
103. Gottfried, B.S., 1991, *Programming with C, Schaum Outline Series*, Tata McGraw Hill Publishing Co. Ltd., New Delhi.
104. Govindarajulu, Z., 2001, *Statistical Techniques in Bioassay*, Karger.
105. Gower, J.C., 1971, "A General Coefficient of Similarity and Some of its Properties", *Biometrics* 27, pp. 857-874.
106. Gujrati, D.N., 2004, *Basic Econometrics*, Tata McGraw Hill Publication Co. Ltd, New Delhi 110008.
107. Harman, H.H., 1976, *Modern Factor Analysis*, University of Chicago Press, Chicago.
108. Harman, H.H. and Fukuda, Y., 1966, "Resolution of the Heywood Case in the Minres Solution", *Psychometrika* 31, pp. 563-71.
109. Harman, H.H. and Jones, W.H., 1966, "Factor Analysis by Minimizing Residuals (Minres)", *Psychometrika* 31, pp. 351-68.
110. Harris, R.J., 1975, *A Primer of Multivariate Statistics*, Academic Press, Inc. USA.
111. Henikoff, S. and Henikoff J.G. (1992), Amino acid substitution matrices, Proteins Struct. Funct. Genet. 17, pp. 10915-19.
112. Hartigan, J.A., 1975, *Clustering Algorithms*, Wiley, New York.
113. Hartley, P., 1935, "International Biological Standards", *Pharmaceutical Journal* 81, pp. 625-627.
114. Hartley, P., 1945, "International Biological Standards: Prospect and Retrospect", *Proceedings of the Royal Society of Medicine*, 39, pp. 45-58.
115. Henikoff, S. and Henikoff, J.G., 1994a, "Position Based Sequence Weights", *Jour. of Mol. Biol.*, 243, pp. 574-578.
116. Henikoff, S. and Henikoff, J.G., 1994b, "Protein Family Classification Based on Searching a Data Base of Blocks", *Genomics*, 19, pp. 97-104.
117. Hill, A.B., 1979, *A Short Text Book of Medical Statistics*, 10th Edition. ELBS and Hodder and Stoughton, London.
118. Hoel, P.G., 1971, *Introduction to Mathematical Statistics*, 4th Edition, John Wiley & Sons, Inc.
119. Hotelling, H., 1931, "The Generalization of Students' Ratio", *Ann. Math. Statist.*, 2, p. 360.
120. Hotelling, H.,. 1933, "Analysis of a Complex of Statistical Variables into Principal Components", *Jour. Edu. Psych.*, 24, pp. 417-441, 498-520.
121. Hotelling, H., 1935, "The Most Predictable Criterion", *Jour. of Edu. Psych.*, 26, pp. 139-142.
122. Hotelling, H., 1936, "Relations Between Two Sets of Variates", *Biometrika*, 28, pp. 321-377.
123. Hotelling, H., 1951a, "Abraham Wald", *Amer. Statist.*, 5, pp. 18-19.
124. Hotelling, H., 1951b, "The Impact of R.A. Fisher on Statistics", *Jour. Amer. Statist. Assoc.*, 46, pp. 35-46.
125. Hotelling, H., 1953, "New Light on the Correlation Coefficient and its Transform", *Jour. of Royal Statist. Soc.*, B 15, pp. 193-232.
126. Hsu, P.L., 1938, "Notes on Hotelling's Generalized T^2", *Ann. Math. Statist.*, 8, pp. 231-243.
127. Hunt, R. and Shelly, J., 1982, *Computers and Common Sense*, 2nd Edition Prentice Hall of India Pvt. Ltd. New Delhi.

128. Hayne, D.W., 1949, "Two Methods of Estimating Populations from Trapping Records", *Jour. Mammal.*, 30, pp. 99-411.
129. *India*, 1993, Publication Division, Ministry of Information and Broadcasting, Govt. of India, New Delhi.
130. Irwin, J.O., 1937, "Statistical Method Applied to Biological Assays", *Jour. of Royal Statist. Soc.*, Supplement 4, pp. 1-60.
131. Jain, J.L., 2008, *Fundamentals of Biochemistry*, S. Chand & Co., New Delhi.
132. Jain, S. and Aggarwal, B.M., 2000, *Using the Internet in India – A Practical Guide*, Mashbra Industries Pvt. Ltd. (Publication Division), New Delhi.
133. Johnson, N.L. and Kotz, S., 1969, *Discrete Distributions*, John Wiley and Sons, Inc. Singapore.
134. Johnson, N.L. and Kotz, S., 1994, *Continuous Univariate Distributions-1,* John Wiley and Sons, Singapore.
135. Johnson, N.L. and Kotz, S., 1970, 1995, *Continuous Distribution – 2*, John Wiley & Sons, Inc.
136. Johnson, N.L. and Kotz, S., 2000, *Continuous Multivariate Distributions*, John Wiley and Sons, Singapore.
137. Johnson, S.C., 1967, "Hierarchical Clustering Schemes", *Psychometrika*, 32, pp. 241-254.
138. Jones, N.C. and Pevzner, P.A., 2005, *An Introduction to Bioinformatics Algorithms*, Ane Books, New Delhi 110 002.
139. Jöreskog, K.G., 1967, "Some Contribution to Maximum Likelihood Factor Analysis", *Psychometrika* 32, pp. 443-482.
140. Joshi, Y.P., 2003, *An Introduction to FORTRAN 90/95 Syntax and Programming*, Allied Publishers Pvt. Ltd., India.
141. Karian, Z.A. and Dudewicz, W.J., 2001, *Modern Statistical System and GPSS Simulation*, WH Freeman and Co. USA.
142. Kar, S., 2003, *Effect of Environmental Pollution on Physical and Mental Health, Unpublished Ph.D. Thesis*, Submitted at Deptt. of Zoology, BHU, India.
143. Kelley, T.L., 1935, *Essential Traits of Mental Life: Harvard Studies in Education No. 26*. Harvard University Press Cambridge, Mass.
144. Kendall, M.G., 1938, "A New Measure of Rank Correlation", *Biometrika* 30, pp. 81-93.
145. Kendall, M.G. and Babington, S.B., 1939, "The Problem of m Rankings", *Ann. Math. Statist.*, 10, pp. 275-287.
146. Kerninghan, B.W. and Ritchie, D.M., 1991, *The C Programming Language*, 2nd Ed. Prentice Hall Inc., Englewood Cliffs, NJ, USA.
147. Kettell, J.A. and Hart-Davis, G., 2003, *Microsoft Office The complete References*, McGraw Hill.
148. Keul, M., 1952, "The Use of the "Studentized Range" in Connection with Analysis of Variance", *Euphytica* 1, pp. 112-122.
149. King, W.R., 1967, "Structural Analysis and Descriptive Discriminant Functions", *Jour. of Marketing Res.*, 7, No. 2, pp. 39-43.
150. Kochler, K.J. and Larntz, K., 1980, "An Empirical Investigation of Goodness of Fit Statistics for Sparse Multinomials", *Jour. Amer. Statist. Assoc.*, 75, pp. 336-344.
151. Kramer, C.Y., 1956, "Extension of Multiple Range Tests to Group Means with Unequal Numbers of Replications", *Biometrics*, 12, pp. 307- 310.
152. Kramer, M.S., 1988, *Clinical Epidemiology and Biostatistics*, Springer Verlag, Berlin Heidelberg New York, USA.
153. Kruskal, W.H., 1952, "A Non Parametric Test for Several Sample Problem", *Ann. Math. Statist.*, 12, pp. 525-540.

154. Kruskal, W.H. and Wallis, W.A., 1952, "Use of Ranks in One Criterion Analysis of Variance", *Jour. Amer. Statist. Assoc.*, 47, pp. 583-621.
155. Kumar, R., 1986, *Programming with FORTRAN 77*, Tata McGraw Hill Publishing Co. Ltd., New Delhi.
156. Laplace, P.S., 1812, *Theorie, Analytiquedes Probabilites*, Paris: Gauthier Villars.
157. Law, A.M. and Kelton, W.D., 1991, *Simulation, Modelling and Analysis*, McGraw Hill, Inc.
158. Lawley, D.N., 1940, "The Estimation of Factor Loadings by the Method of Maximum Likelihood", *Proc. of Royal Statist. Soc.*, Ed. 60, pp. 64-82.
159. Lawely, D.N. and Maxwell, A.E., 1963, *Factor Analysis as a Statistical Method*, Butterworth, London.
160. Lesk, A.M., 2003, *Introduction to Bioinformatics*, Oxford Univ. Press, New Delhi.
161. Lippman, S.B. and Lajone, J., 2003, *C^{++} Primer*, Pearson Education (Singapore) Pvt. Ltd. Indian Branch, Delhi, 110 092.
162. Ljung, G.M. and Box, G.E.P., 1978, "On a Measure of Lack of Fit in Time Series Models", *Biometrika*, 66, pp. 66-72.
163. Mahalanobis, P.C., 1938, "Professor Ronald Aylmer Fisher", *Sankhya*, 4, pp. 265-272.
164. Makridakis, S., Wheelwright, S.C. and Hyndman, R.J., 1998, *Forecasting Methods and Application* 3rd Ed. John Wiley NY, USA.
165. Mann, H.B. and Whitney, D.R., 1947, "On a Test Whether One of the Two Random Variables is Stochastically Larger Than the Other", *Ann. Math. Statist.*, 18, pp. 50-60.
166. Mather, K., 1951, "R.A. Fisher's Statistical Methods for Research Workers, An Appreciation", *Jour. Amer. Statist. Assoc.*, 46, pp. 51- 54.
167. Mc Mullen, L., 1947, *Foreword, in E.S. Pearson and John Wishart (eds.) "Student's" Collected Papers*, Biometrika Office, University College, London.
168. MCSE, 2007, *Windows 1998*, IT Pass Zone LLC.
169. Menger, K., 1952, "The Formative Years of Abraham Wald and His Work in Geometry", *Ann. Math. Statist.*, 23, pp. 13-20.
170. Meyer, P.L., 1975, *Introductory Probability and Statistical Applications*, Amerind Publication Co. Pvt. Ltd New Delhi, Bombay, Calcutta, New York.
171. Miles, A.A., 1948, "Some Observations on Biological Standards", *Analyst.*, 73, pp. 530-538.
172. Miles, A.A. and Perry, W.L.M., 1950, "Third International Digitalis Standard", *Bulletin of the World Health Organization* 2, pp. 655-672.
173. Milton, J.S. and Tsokos, J.O., 1983, *Statistical Methods in the Biological and Health Sciences*, Mc Graw Hill , Inc.
174. Mohan, R., 2007, *Statistical System of India – Some Reflections*, Inaugural address by the author, Deputy Governor of RBI on Statistics Day and Annual Conference on Financial Statistics, Dept. of Statistical Analysis and Computer Services, Mumbai, 29 June 2007.
175. Mood, A.M., 1950, *An Introduction to Theory of Statistics*, McGraw Hill NY, USA.
176. Mood, A.M., Graybill, F.A. and Boes, D.C., 1974, *Introduction to the Theory of Statistics, 3rd Edn.* McGraw Hill Kogakusha Ltd.
177. Morrison, D.F., 1967, *Multivariate Statistical Methods*, McGraw Hill, San Fransisco, USA.
178. Mount, D.W., 2001, *Bioinformatics*, Cold Spring Harbor Laboratory Press, USA.
179. Mulaik, S.A., 1972, *The Foundations of Factor Analysis,* McGraw Hill, San Francisco.
180. Murthy, M.N., 1961, *Introduction to Sampling Theory, Lecture Notes*, ISI Calcutta.
181. Needleman, S.B. and Wunsch, C.D., 1970, "A General Method Applicable to the Search for Similarities in Amino Acid Sequence of Two Proteins", *Jour. Mol. Biol.*, 48, pp. 443-453.

182. Newman, D., 1939, "The Distribution of Range in Samples from a Normal Population, Expressed in Terms of An Independent Estimate of Standard Deviation", *Biometrika* 31, pp. 20-30.
183. Neyman, J., 1935, "Statistical Problems in Agricultural Experimentation", *Jour. Royal Statist. Soc.*, 11, No. 2: pp. 130-136.
184. Neyman, J. and Pearson, E.S., 1931, "On the Problem of k Samples", *Bull. Acad. Polon. Sci. Lett. Ser.*, A 3: pp. 460-481.
185. O'Leary, 1999, *Microsoft Word 2000*, McGraw Hill.
186. Olmstead, P.S. and Tukey, J.W., 1947, "A Corner Test for Association", *Ann. Math. Statist.*, 18, pp. 495-513.
187. Overall, J.E. and Klett, J., 1972, *Applied Multivariate Analysis*, McGraw Hill Book Co.
188. Owen, D.B., 1962, *Handbook of Statistical Tables*, Addison Wesley, Reading, Massachusetts.
189. Pandey, M., 1996, *Factor Analysis of Disparities in Socio-economic Condition of Weaker Section of Rural Population in Eastern UP.* UGC Sponsored Project Report (Unpublished.)
190. Pattan, M., 1990, *Qualitative Evaluation of Research Methods*, Sage Publication & Newbury Park, California, USA.
191. Pearson, E.S., 1936, "Karl Pearson, An Appreciation of Some Aspects of His Life and Work, Part I-1857-1906", *Biometrika* 28, pp. 193-257.
192. Pearson, E.S., 1938, Karl Pearson, "An Appreciation of Some Aspects of His Life and Work, Part II-1906-1936", *Biometrika* 29, pp. 161-248.
193. Pearson, E.S. and Hartley, H.O., 1942, "The Probability Integral of Range in Samples of n Observations from a Normal Population", *Biometrika* 32, pp. 301-310.
194. Pearson, E.S. and Hartley, H.O., 1943, "Tables of the Probability Integral of Studentized Range", *Biometrika* 33, pp. 89-99.
195. Pearson, K., 1896, "Mathematical Contributions to the Theory of Evolution-III, Regression", *Heredity and Panmixia. Phil. Trans.*, 187A: pp. 253-318.
196. Pearson, K., 1901, "On Lines and Planes of Closest Fit to Systems of Points in Space", *Phil. Mag. Series*, 2, pp. 559-72.
197. Pindyck, R.S. and Rubinfield, D.L., 1998, *Econometric Models and Economic Forecasts 4th Ed.*, McGraw Hill, NY, USA.
198. Plauger, P.J., 1992, *The Standard C Library*, Prentice Hall, Englewood Cliffs, NJ, USA.
199. Podlin, J., 1999, *Teach Yourself MS Excel 2000 Programming*, Sams Publishing.
200. Posten, H.O., Yeh, H.C. and Owen, D.B., 1982, "Robustness of the Two Sample t-test Under Violation of the Homogeneity of Variance Assumption", *Commun. Statist. – Theory and Methods*, 11, pp. 109-126.
201. Prasad, N. and Mookerjee, S., 1985, "Transcription Pattern in Some Indian Ecotypes of *Hydra*", *Indian Jour. Exp. Biol.*, 23, pp. 689-691.
202. Press, S.J., 1972, *Applied Multivariate Analysis*, Holt Rinehart and Winston, Inc., USA.
203. Rabiner, L.R., 1989, "A Tutorial on Hidden Markov Models and Selected Applications in Speech Recognition", *Proc. IEEE*, Vol. 77, pp. 257-286.
204. Rai, S.K., Maurya, N.K. and Singh, K.S., 2006, *Development of Cluster Techniques for Microarray Data Analysis*, Unpublished Project Report Submitted in the Dept. of Computer Science, BHU, India.
205. Raj, D., 1966, "Some Remarks on a Simple Procedure of Sampling Without Replacement", *Jour. Amer. Statist. Assoc.*, 61, pp. 391-397.
206. Rajaraman, V., 1974, *Computer Programming in FORTRAN IV*, Prentice Hall of India Pvt. Ltd., New Delhi.

207. Rajaraman, V., 1982, *Principles of Computer Programming*, Prentice Hall of India Pvt. Ltd., New Delhi.
208. Rajaraman, V., 1997, *Computer Programming in FORTRAN 90 and 95*, Prentice Hall of India Pvt. Ltd.
209. Rajaraman, V., 1995, 2004, *Fundamentals of Computers*, Prentice Hall of India Pvt. Ltd., New Delhi.
210. Randell, B., 1973, *The Origin of Digital Computers – Selected Papers*, Springer Verlag, New York.
211. Rao, C.R., 1955, "Estimation and Tests of Significance in Factor Analysis", *Psychometrica* 20, pp. 93-111.
212. Rao, C.R., 1972, *Linear Statistical Inference and Applications*, John Wiley & Sons, USA.
213. Rao, C.R., 1973, *Linear Statistical Inference and its Applications*, John Wiley & Co., New York.
214. Rastogi, N., Pandey, M. and Mookerjee, S., 1992, "Statistical Analysis of Geographical Variability in Sixteen Ecotypes of *Hydra* from India", *Evol. Biol.*, 6, pp. 195-210.
215. Rastogi, N. and Pandey, M., 1995, "Criteria Necessary for the Separation of Indian Ecotypes of *Hydra* into Ecological Clusters", *Evol. Biol.*, 8 and 9, pp. 191-201.
216. Rastogi, S.C., Menderatta, N. and Rastogi, P., 2004, *Bioinformatics, Concepts, Skills and Applications*, CBS Publishers and Distributors, New Delhi.
217. Rastriya Computer Sakshartha Mission, 2001, *Computer Fundamentals and Microsoft Disk Operating Systems*, MCC Software, Developments and Training Pvt. Ltd.
218. Rastriya Computer Sakshartha Mission, 2002, *MS Window 2000 and MS Office 2000* (Word, Excel, Power Point, and Access), MCC Software, Developments and Training Pvt. Ltd.
219. Rijkoort, P.J., 1952, "A Generalization of Wilcoxon's Test", *Proceedings of Koninkl. Ned. Akad. Wetenschap*, A 55, pp. 394-404.
220. Rohatagi, V.K., 1984, *Statistical Inference*, John Wiley & Sons, Inc.
221. Roscoe, J.T. and Byars, J.A., 1971, "Sample Size Restraints Commonly Imposed on the Use of the Chi-square Statistic", *Jour. of Amer. Statist. Assoc.*, 66, pp. 755-759.
222. Rummel, R.J., 1970, *Applied Factor Analysis,* North Western University Press, Evanston, IL, USA.
223. SAS Institute Inc., 1985a, *SAS Introductory Guide for Personal Computers, Version 6,* Ed. SAS Institute Inc., Cary, NC, USA.
224. SAS Institute Inc., 1985b, *SAS Language Guide for Personal Computers, Version 6*, Ed. SAS Institute Inc., Cary, NC, USA.
225. SAS Institute Inc., 1987a, *SAS/STAT Guide for Personal Computers, Version 6*, Ed. SAS Institute Inc., Cary, NC, USA.
226. SAS Institute Inc., 1987b, *SAS/GRAPH Guide for Personal Computers, Version 6*, Ed. SAS Institute Inc., Cary, NC, USA.
227. Satterthwaite, F.E., 1946, "An Approximate Distribution of Estimates of Variance Components", *Biometrics Bull.*, 2, pp. 110-114.
228. Saxena, P.N., 1950, "A Simplified Method of Estimating Latin Squares by Mac Mahon's Differential Operators, Part II, The 7 x 7 Latin Squares", *Jour. Indian Soc. of Agril. Stat.*, 3, pp. 24-79.
229. Scheffé, H., 1961, *The Analysis of Variance*, John Wiley & Sons, Inc. USA.
230. Scheffé, H., 1970, "Practical Solutions to Behrens' Fisher Problem", *Jour. of Amer. Statist. Assoc.*, 65, pp. 1501-1508.
231. Schwartz, K., 2000, *Quick Reference Guide Microsoft Excel 2000*, Cathehy Veseckey Dale Publishing.

232. Sen, K., Agarwal, M. and Goswami, A.K., 2006, "Distribution Related to Runs in Three State Trials", In *Statistical Advances in Biosciences and Bioinformatics* (Ed. M. Pandey), Allied Publishers, New Delhi.
233. Sharma, P.D., 2005, *Ecology and Environment*, Rastogi Publication, Meerut, India.
234. Singh, B.N., 1989, "Inversion Polymorphism in Indian Populations of *Drosophila Ananassae*", *Hereditus* 118, pp. 133-138.
235. Smith, T.F. and Waterman, M.S., 1981a, "Identification of Common Molecular Subsequences", *Jour. Mol. Biol.*, 147, pp. 195-197.
236. Smith, T.F. and Waterman, M.S., 1981b, "Comparison of Biosequences", *Adv. Appl. Math.*, 2, pp. 482-489.
237. Sneath, P.H.A., 1957, "The Application of Computers to Taxonomy", *Jour. Gen. Microbiology*, 17, pp. 201-226.
238. Sneath, P.H.A. and Sokal, R.R., 1973, *Numerical Taxonomy*, Freeman, San Francisco.
239. Sokal, R.R. and Rohlf, F.J., 1973, *Introduction to Biostatistics*, WH Freeman and Company San Francisco.
240. Sokal, R.R. and Rolhf, F.J., 1995, *Biometry*, Freeman & Co., NY, USA.
241. Southwood, T.R.E. and Henderson, P.A., 2000, *Ecological Methods*, Blackwell Science Ltd. USA.
242. Spearman, C., 1904, "General Intelligence, Objectively Determined and Measured", *Amer. Jour. of Psych.*, 15, pp. 201-293.
243. Spencer, R., 1984, "Cluster Analysis", *Byte September*, pp. 29-130 and 423-426.
244. SPSS Inc., 1999, *SPSS Base 10.0 User's Guide*, SPSS Inc. Chicago, IL, USA.
245. Srivastava, A.B.L., 1958, "Effect of Non-normality on the Power Function of t-test", *Biometrika* 45, pp. 421-429.
246. Steel, R.G.D. and Torrie, J.H., 1928, "On the Use and Interpretation of Certain Test Criteria for Purpose of Statistical Inference Pt. II", *Biometrika* 20, pp. 263-294.
247. Steel, R.G.D. and Torrie, J.H., 1984, *Principles and Procedures of Statistics: A Biomedical Approach*, McGraw Hill International.
248. Stephen, M.A., 1974, "EDF Statistics for Goodness of Fit and Some Comparison", *Jour. Amer. Statist. Assoc.*, 69, pp. 730-737.
249. Stevens, A., 1989, *Teach Yourself DOS 5.0.* Tata McGraw Hill.
250. *'Student'* (WS Gossett), 1908, "The Probable Error of a Mean", *Biometrika* 6, pp. 1-25.
251. *'Student'* (Gossett WS), 1927, "Errors of Routine Analysis", *Biometrika* 19, pp. 151-164.
252. Sukhatme, P.V. and Sukhatme, B.V., 1976, *Sampling Theory of Surveys with Applications*, Iowa State University Press, Ames Iowa, USA and ISAS New Delhi.
253. Sundaram, S.M., 2002, *Some Multivariate Analysis of the Water Quality Data on River Ganga at Varanasi*, Unpublished PhD Thesis, Submitted at Dept. of Zoology, BHU, India.
254. Sundaram, S.M., 2005, *A Text Book on Statistics*. Shruthi Publishers, Chennai, India.
255. Tabachnick, B.G. and Fidel, L.S., 1983, *Using Multivariate Statistics*, Harper and Row, New York.
256. Taunenbaum, A.S., 1984, *Structured Computer Organization* 2nd Ed. Prentice Hall Inc, Englewood Cliffs NJ, USA.
257. Taunenbaum, A.S., 1993, *Computer Networks*, Prentice Hall of India Pvt. Ltd. New Delhi.
258. Thomas, J.A., 1983, "A Quick Method for Estimating Butterfly Numbers During Surveys", *Biol. Conserv.*, 27, pp. 195-211.
259. Thurstone, L.L., 1931, "Multiple Factor Analysis", *Psych. Rev.*, 38, pp. 408-427.
260. Thurstone, L.L., 1947, *Multiple Factor Analysis*, Univ. of Chicago Press, Chicago.
261. Tintner, G., 1952, "Abraham Wald's Contribution to Econometrics", *Ann. Math. Statist.*, 23, pp. 21-28.

262. Tukey, J.W., 1953, *The Problem of Multiple Comparisons*, Department of Statistics, Princeton University, NJ (Unpublished).
263. Uspenskey, J.U., 1937, *Introduction to Mathematical Probability* McGraw Hill Book Company Inc., New York and London.
264. Vigyan Prasar, 1999, "Different Generation of Computers", *Employment News* 2-8 October.
265. Wald, A., 1947, *Sequential Analysis*, Wiley, New York.
266. Wald, A. and Wolfowitz, J., 1940, "On a Test Whether Two Samples are From the Same Population", *Ann. Math. Statist.*, 11, pp. 147-162.
267. Walker, H.M., 1958, "The Contributions of Karl Pearson", *Jour. Amer. Statist. Assoc.*, 53, pp. 11-27.
268. Welch, B.L., 1938, "The Significance of the Difference Between Two Means When the Population Variances are Unequal", *Biometrika* 29, No. 3-4, pp. 350-362.
269. Welch, B.L., 1947, "The Generalization of Student's Problem When Several Population Variances are Involved", *Biometrika*, 34, pp. 28-35.
270. Westhead, D.R., Parish, J.H. and Twyman, R.M., 2003, *Bioinformatics, Instant Notes Series*, Viva Books Pvt. Ltd., New Delhi.
271. Wetherill, G.B., Duncomb, P., Kenward, M., Kollestrom, J., Paul, S.R. and Vowden, B.J., 1986, *Regression Analysis with Application*, Chapmann and Hall, London.
272. Wilcoxon, F., 1945, "Individual Comparisons by Ranking Methods", *Biometrics* 1, pp. 80-83.
273. Wilson, E.B. and Worcester, J., 1943a, "The Determination of LD_{50} and its Sampling Error in Bio-assay", *Proceedings of the National Academy of Sciences* 29, pp. 79-85.
274. Wilson, E.B. and Worcester, J., 1943b, "The Determination of LD_{50} and its Sampling Error in Bio-assay II", *Proceedings of the National Academy of Sciences* 29, pp. 114-120.
275. Wilson, E.B. and Worcester, J., 1943c, "The Determination of ED_{50} and its Sampling Error in Bio-assay III", *Proceedings of the National Academy of Sciences* 29, pp. 257-262.
276. Wolfowitz, J., 1952, "Abraham Wald, 1902-1950", *Ann. Math. Statist.*, 23, pp. 1-13.
277. Wrigley, C. and Neuhaus, J.O., 1952, "A Refactorization of Burt – Pearson Matrix With the ORDVAC Electronic Computer", *British Jour. of Statist. Psycho.*, 65, pp. 105-108.
278. Wrigley, C. and Neuhaus, J.O., 1955, "The Use of an Electronic Computer in Principal Axes Factor Analysis", *Jour. Edu. Psycho.*, 46, pp. 31-41.
289. Yates, F., 1933, "The Principle of Orthogonality Confounding in Replicated Experiments", *Jour. of Agr. Sc.*, 23, pp. 108-145.
280. Yates, F., 1934, "Contingency Tables Involving Small Numbers and χ^2-test", *Jour. Royal Statist. Soc.*, (Suppl.) 1, pp. 217-235.
281. Yates, F., 1936, "Incomplete Randomized Blocks", *Ann. Eugen.*, 7, pp. 120- 140.
282. Yates, F., 1951, "The Influence of Statistical Methods for Research Workers on the Development of the Science of Statistics", *Jour. Amer. Statist. Assoc.*, 46, pp. 19-34.
283. Yule, G.U., 1897a, "On the Significance of Bravais' Formulae for Regression etc. in the Case of Skew Correlation", *Proceedings of Royal Statist. Soc.*, 60, pp. 477-489.
284. Yule, G.U., 1897b, "On the Theory of Correlation", *Jour. Royal. Statist. Soc.*, 60, pp. 812-854.
285. Yule, G.U. and Kendall, M.G., 1958, *An Introduction to the Theory of Statistics*, 9th Ed. Charles Griffin & Co. Ltd., London.
286. Zar, J.H., 1928, "On the Use and Interpretation of Certain Test Criteria for Purposes of Statistical Inference Pt. I", *Biometrika*, 20, pp. 175- 240.
287. Zar, J.H., 2004, *Biostatistical Analysis, 4th Ed.* Pearson Education Singapore Pvt. Ltd.

Referred Websites

http://cplus.about.com/od/introductiontoprogramming/a/cppbeginners.html
http://en.wikipedia.org
www.brij.com
www.census.india.gov.in
http://stats.oecd.org/glossary/detail.asp?ID=125
www.yu/08
www:dsc.vic.gov.au/dse/wcmn203.nsf/linkreview/
www.fao.org/docrep/W3241E
www3.interscience.wiley.com/Journal/
http://www.socialresearchmethods.net/kb/sampnon.php
http://en.wikipedia.org/wiki /snowball_ sampling
http://en.wikipedia.org/wiki/clinical_trial
http://www.bmj.com/epidem/
http://www.childrenmerry.org/stats/definitions
http://www.jerrydallal.com/LHSP/
http://www.lib.stat.cmu.edu/apstat/
http://www.fortran.com/tools.html
http://www.nag.co.uk

Subject Index

Author Index